CW01025161

Transport by Advection and Diffusion: Momentum, Heat, and Mass Transfer

Transport by Advection and Diffusion: Momentum, Heat, and Mass Transfer

Ted D. Bennett
University of California, Santa Barbara

WILEY

Vice President & Publisher *Don Fowley*
Associate Publisher *Daniel Sayre*
Assistant Editor *Alexandra Spicehandler*
Marketing Manager *Christopher Ruel*
Marketing Assistant *Ashley Tomeck*
Senior Product Designer *Tom Kulesa*
Media Specialist *Andre Legaspi*
Senior Production Manager *Janis Soo*
Associate Production Manager *Joel Balbin*
Production Editor *Yee Lyn Song*
Cover Designer *Seng Ping Ngieng*
Cover Illustration *Ted D. Bennett*

This book was set in 9.5/11.5 Palatino by MPS Limited and printed and bound by Courier Kendallville. The cover was printed by Courier Kendallville.

This book is printed on acid free paper ♾.
Founded in 1807, John Wiley & Sons, Inc. has been a valued source of knowledge and understanding for more than 200 years, helping people around the world meet their needs and fulfill their aspirations. Our company is built on a foundation of principles that include responsibility to the communities we serve and where we live and work. In 2008, we launched a Corporate Citizenship Initiative, a global effort to address the environmental, social, economic, and ethical challenges we face in our business. Among the issues we are addressing are carbon impact, paper specifications and procurement, ethical conduct within our business and among our vendors, and community and charitable support. For more information, please visit our website: www.wiley.com/go/citizenship.

Library of Congress Cataloging-in-Publication Data:

Bennett, Ted D., 1965-
Transport by advection and diffusion : momentum, heat, and mass transfer / Ted D. Bennett.
pages cm
Includes bibliographical references and index.
ISBN 978-0-470-63148-5
1. Transport theory. 2. Diffusion processes. I. Title.
TP156.T7B465 2013
530.13′8—dc23

2012016769

Printed in the United States of America

10 9 8 7 6 5 4 3 2 1

To Lei, Annette, and Ava

Preface

This text covers material for an introductory-level graduate course or advanced undergraduate course that draws attention to the intellectual coherence of transport. While not intended to replace specialized treatises that introduce terminology and organization of subject matter for a narrow benefit, this text provides a broad treatment of transport phenomena in the coverage of a wide array of topics. A general framework for transport phenomena is revealed through the development of differential equations that employ transport principles and conservation laws. In application, these governing equations must be solved. Therefore, significant attention is given to the mathematical treatment of these equations, which is a powerful, if not essential, way to build understanding of the associated physics.

The common features of transport phenomena provide the basis for simultaneous development of momentum, heat, and mass transport. This commonality is emphasized throughout the text for maximum pedagogical benefit. For example, the momentum equation is derived from the basic elements of diffusion and advection transport, rather than using the traditional approach that follows from Newton's second law. The essential difference is whether viscous effects in the flow are treated as diffusion of momentum or as stresses imparting momentum on the fluid. Both descriptions are equivalent, but in this text, the transport perspective is advanced first to emphasize an essentially equivalent treatment given to all the transport equations.

This text is organized into relatively short chapters that address concise topics. The first three chapters are devoted to some preliminary subjects: Chapter 1 reviews some fundamental thermodynamics; Chapter 2 introduces basic transport principles; and a cursory overview of index notation is given in Chapter 3. The next two chapters are devoted to developing transport equations from the principles of conservation laws and transport phenomena. In Chapter 4, transport equations are developed to reveal the common advection and diffusion transport terms. However, significant differences between various transport equations are exposed by the addition of source terms, which are considered in Chapter 5. Chapter 6 reviews some elementary aspects of problem formulation and solution requirements associated with differential equations.

As will be seen, in many problems transport is dominated by either diffusion or advection, encouraging the insignificant process to be dropped from the description. In Chapters 7 through 13 of this text, problems in which diffusion describes the main features of transport are considered. Chapters 7 through 11 treat diffusion transport in transient one-dimensional and steady two-dimensional problems. The scope of diffusion transport is extended to moving boundary problems in Chapter 12 and lubrication theory in Chapter 13.

Chapters 14 through 22 of this text look at problems in which advection describes the main features of transport. Chapter 14 and 15 discuss ideal plane flows, which is applied to airfoil problems in Chapter 16. Two other important classes of advection problems are discussed: open-channel flows in Chapters 18 and 19, and high-speed gas dynamics in Chapters 20, 21, and 22.

Chapter 24 is the first chapter devoted to convection transport, in which both diffusion and advection play a comparable role. The topic of convection is carried throughout the remainder of the text. Through Chapter 29 transport is assumed to occur in laminar flows. Chapters 24, 25, and 26 are devoted to boundary layer problems, and

Chapters 27 and 28 are concerned with internal flows. Chapter 29 looks at the significance of nonconstant fluid properties on the solution to transport equations.

Some elementary concepts of turbulence are introduced in Chapters 30 and 32, in the context of the mixing length model. The mixing length model is used to solve the time-averaged transport equations for fully developed internal flows bounded by smooth surfaces in Chapters 31 and 33, and bounded by rough surfaces in Chapter 34. The mixing length model is also used to solve the turbulent boundary layer problem in Chapter 35. Finally, the k-epsilon model of turbulence is discussed in Chapter 36, and applied to fully developed transport in Chapter 37.

Interspersed among the main topics of this text are sections devoted to building the mathematical tools required to solve equations that govern problems of interest. For example, the method of separation of variables provides a systematic approach to solving linear partial differential equations. This topic is developed in Chapters 7 through 9, first in the context of transient diffusion and then steady-state diffusion in multiple spatial directions. In Chapter 14 it is demonstrated that some steady-state irrotational incompressible flows governed by advection also lead to a linear equation that can be solved by the method of separation of variables.

Flows governed by nonlinear advection prove to be among the most difficult transport equations to solve, and MacCormack integration is introduced in Chapter 17 as a numerical recipe to address some of these problems. MacCormack integration is used to solve open channel flow problems in Chapter 18 and 19, and to solve problems in gas dynamics in Chapters 20, 21, and 22.

Some problems can be solved using a similarity variable to transform linear and nonlinear governing equations into more easily solved ordinary differential equations. The similarity solution is introduced in Chapter 10, where it is applied to transient diffusion problems, and is applied to moving boundary problems in Chapter 12. This technique also finds great utility in solving laminar boundary layer convection problems treated in Chapters 24, 25, and 26. Similarity solutions of linear governing equations will give rise to linear ordinary differential equations with nonconstant coefficients that may be solved by the method of power series solutions, which is folded into Chapter 10. However, similarity solutions of nonlinear governing equations will give rise to nonlinear ordinary differential equations for which numerical solutions are required. Chapter 23 discusses fourth-order Runge-Kutta integration of ordinary differential equations that arise in convection transport treated in Chapter 25 and subsequent chapters.

A few numerical tasks in this text will require the use of finite differencing methods. For example, MacCormack integration is developed in Chapter 17 for application to equations describing advection transport. MacCormack integration is used to solve open-channel flows in Chapters 18 and 19, and high-speed gas flows in Chapters 20, 21, and 22. The finite differencing method is also applied to convection equations describing turbulent transport; the boundary layer equations are solved using the mixing length model in Chapter 35, and the equations for fully developed transport are solved using the k-epsilon model in Chapter 37.

Although the text is not developed with the use of commercial computational software in mind, the mathematical attention given to solving transport equations could easily be coupled to such an activity. The material in this text has been developed with the idea that programming languages, using freely available compilers, can be employed for advanced problems in transport where analytical techniques are not feasible. Engaging the mathematical problems fully (whether the approach is analytical or numerical) can demystify the process of establishing solutions and provide an empowering experience.

Carrying out one's own solutions to problems encourages a healthy level of skepticism in the results, and the process of identifying wrong results will teach critical thinking skills. Not contemplating carefully the meaning of results that are accepted at face value is a tendency that the inexperienced can easily fall into with commercial software. Therefore, the pedagogical role of commercial software should be contemplated with the idea that the best method of solving a problem for the first time may be different from the tenth time.

Finally, it is hoped that the students who use this textbook to learn about transport phenomena will have the same experience of discovery as the author had in writing it.

TDB

Organization of Text

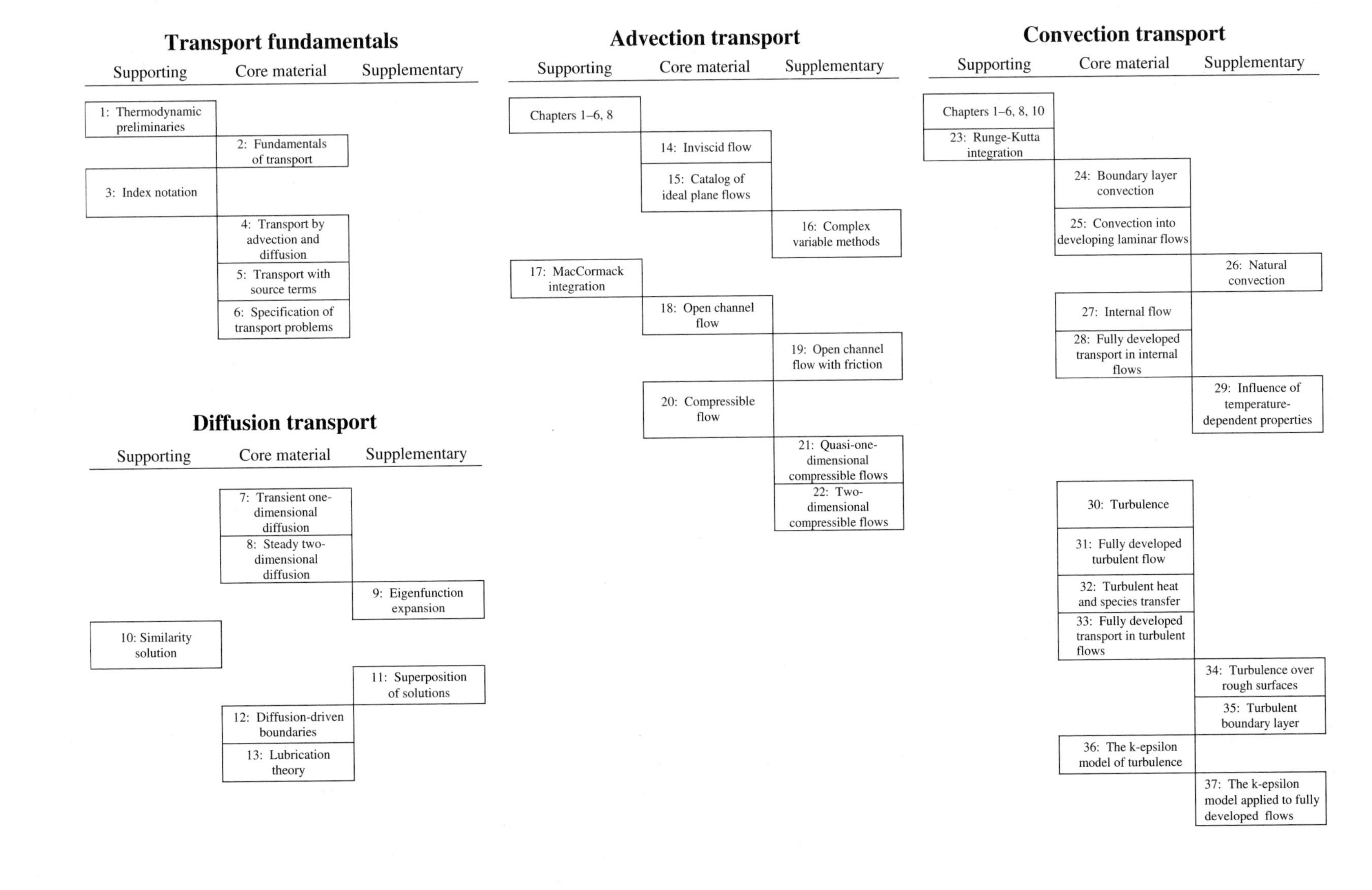

Brief Contents

Chapter 1 Thermodynamic Preliminaries 1

Chapter 2 Fundamentals of Transport 12

Chapter 3 Index Notation 25

Chapter 4 Transport by Advection and Diffusion 36

Chapter 5 Transport with Source Terms 50

Chapter 6 Specification of Transport Problems 66

Chapter 7 Transient One-Dimensional Diffusion 82

Chapter 8 Steady Two-Dimensional Diffusion 103

Chapter 9 Eigenfunction Expansion 119

Chapter 10 Similarity Solution 140

Chapter 11 Superposition of Solutions 159

Chapter 12 Diffusion-Driven Boundaries 172

Chapter 13 Lubrication Theory 188

Chapter 14 Inviscid Flow 206

Chapter 15 Catalog of Ideal Plane Flows 224

Chapter 16 Complex Variable Methods 234

Chapter 17 MacCormack Integration 249

Chapter 18 Open Channel Flow 265

Chapter 19 Open Channel Flow with Friction 284

Chapter 20 Compressible Flow 296

Chapter 21 Quasi-One-Dimensional Compressible Flows 315

Chapter 22 Two-Dimensional Compressible Flows 333

Chapter 23 Runge-Kutta Integration 344

Chapter 24 Boundary Layer Convection 359

Chapter 25 Convection into Developing Laminar Flows 376

Chapter 26 Natural Convection 399

Chapter 27 Internal Flow 412

Chapter 28 Fully Developed Transport in Internal Flows 429

Chapter 29 Influence of Temperature-Dependent Properties 447

Chapter 30 Turbulence 465

Chapter 31 Fully Developed Turbulent Flow 479

Chapter 32 Turbulent Heat and Species Transfer 507

Chapter 33 Fully Developed Transport in Turbulent Flows 517

Chapter 34 Turbulence over Rough Surfaces 545

Chapter 35 Turbulent Boundary Layer 565

Chapter 36 The K-Epsilon Model of Turbulence 581

Chapter 37 The K-Epsilon Model Applied to Fully Developed Flows 589

Appendix A 606

Index 611

Contents

Chapter 1 Thermodynamic Preliminaries 1

1.1 The First and Second Laws of Thermodynamics 1
1.2 Fundamental Equations 2
  1.2.1 The Maxwell Relations 4
  1.2.2 Internal Energy Expressed in Measurable Variables 5
  1.2.3 Enthalpy Expressed in Measurable Variables 6
1.3 Ideal Gas 7
1.4 Constant Density Solid or Liquid 8
1.5 Properties of Mixtures 9
1.6 Summary of Thermodynamic Results 9
1.7 Problems 10

Chapter 2 Fundamentals of Transport 12

2.1 Physics of Advection and Diffusion 12
2.2 Advection Fluxes 14
  2.2.1 Advection Transport in a Binary Mixture 15
  2.2.2 Summary of Advection Transport 16
2.3 Diffusion Fluxes 17
  2.3.1 Heat Diffusion 18
  2.3.2 Momentum Diffusion 18
  2.3.3 Species Diffusion 20
  2.3.4 Summary of Diffusion Laws 21
2.4 Reversible vs. Irreversible Transport 22
2.5 Looking Ahead 23
2.6 Problems 23

Chapter 3 Index Notation 25

3.1 Indices 25
3.2 Representation of Cartesian Differential Equations 26
3.3 Special Operators 27
  3.3.1 Surface Normal Operator 27
  3.3.2 Kronecker Delta Operator 28
  3.3.3 Alternating Unit Tensor Operator 30
  3.3.4 Proof of a Vector Identity 30
3.4 Operators in Non-Cartesian Coordinates 31
3.5 Problems 34

Chapter 4 Transport by Advection and Diffusion 36

4.1 Continuity Equation 37
4.2 Transport of Species 39
  4.2.1 Transport in a Binary Mixture 40
4.3 Transport of Heat 42
4.4 Transport of Momentum 43
4.5 Summary of Transport Equations without Sources 44
4.6 Conservation Statements from a Finite Volume 44
4.7 Eulerian and Lagrangian Coordinates and the Substantial Derivative 46
4.8 Problems 48

Chapter 5 Transport with Source Terms 50

5.1 Continuity Equation 51
5.2 Species Equation 51
5.3 Heat Equation (without Viscous Heating) 52
5.4 Momentum Equation 54
5.5 Kinetic Energy Equation 55
5.6 Heat Equation (with Viscous Heating) 57
5.7 Entropy Generation in Irreversible Flows 58
5.8 Conservation Statements Derived from a Finite Volume 59
  5.8.1 Continuity 59
  5.8.2 Momentum 60
  5.8.3 Total Energy 61
5.9 Leibniz's Theorem 62
5.10 Looking Ahead 63
5.11 Problems 64

Chapter 6 Specification of Transport Problems 66

6.1 Classification of Equations 66
6.2 Boundary Conditions 67
6.3 Elementary Linear Examples 69
  6.3.1 Gravity Driven Flow on an Inclined Surface 69
  6.3.2 Heat Transfer across a Liquid Film 70
  6.3.3 Groundwater Contamination 71

6.4 Nonlinear Example 73
6.4.1 Steady-State Evaporation 73
6.5 Scaling Estimates 75
6.5.1 Scaling of Gravity-Driven Flow on an Inclined Surface 75
6.5.2 Scaling of Groundwater Contamination 76
6.5.3 Scaling Simplification to a Governing Equation 76
6.6 Problems 78

Chapter 7 Transient One-Dimensional Diffusion 82

7.1 Separation of Time and Space Variables 83
7.1.1 Problem with Homogeneous Equation and Boundary Conditions 83
7.1.2 Demonstration of Orthogonality 86
7.1.3 Problem with Nonhomogeneous Equation and Boundary Conditions 87
7.2 Silicon Doping 89
7.3 Plane Wall With Heat Generation 93
7.4 Transient Groundwater Contamination 97
7.5 Problems 101

Chapter 8 Steady Two-Dimensional Diffusion 103

8.1 Separation of Two Spatial Variables 103
8.2 Nonhomogeneous Conditions on Nonadjoining Boundaries 105
8.3 Nonhomogeneous Conditions on Adjoining Boundaries 107
8.3.1 Bar Heat Treatment 108
8.4 Nonhomogeneous Condition in Governing Equation 111
8.4.1 Steady Rectangular Duct Flow 111
8.5 Looking Ahead 115
8.6 Problems 115

Chapter 9 Eigenfunction Expansion 119

9.1 Method of Eigenfunction Expansion 119
9.1.1 Species Transport (Silicon Doping) 120
9.1.2 Heat Transfer (Bar Heat Treatment) 122
9.1.3 Momentum Transport (Duct Flow) 125
9.2 Non-Cartesian Coordinate Systems 127
9.2.1 Cartesian Coordinates 128
9.2.2 Cylindrical Coordinates 128
9.2.3 Spherical Coordinates 130
9.3 Transport in Non-Cartesian Coordinates 130
9.3.1 Pin Fin Cooling 131
9.3.2 Transient Heat Transfer in a Sphere 136
9.4 Problems 139

Chapter 10 Similarity Solution 140

10.1 The Similarity Variable 140
10.2 Laser Heating of a Semi-Infinite Solid 142
10.3 Transient Evaporation 146
10.4 Power Series Solution 148
10.5 Mass Transfer with Time-Dependent Boundary Condition 152
10.6 Problems 157

Chapter 11 Superposition of Solutions 159

11.1 Superposition in Time 159
11.1.1 Duhamel's Theorem 161
11.1.2 Semi-Infinite Fluid Bounded by a Plate Set in Motion 162
11.2 Superposition in Space 164
11.2.1 Product-Superposition of Solutions 165
11.2.2 Method of Images 167
11.3 Problems 169

Chapter 12 Diffusion-Driven Boundaries 172

12.1 Thermal Oxidation 172
12.2 Solidification of an Undercooled Liquid 174
12.3 Solidification of a Binary Alloy from an Undercooled Liquid 178
12.3.1 Heat Transfer 178
12.3.2 Species Transfer 179
12.3.3 Coupling Heat and Species Transport 182
12.4 Melting of a Solid Initially at the Melting Point 183
12.5 Problems 186

Chapter 13 Lubrication Theory 188

13.1 Lubrication Flows Governed by Diffusion 188
13.2 Scaling Arguments for Squeeze Flow 189
13.2.1 Scaling Continuity 190
13.2.2 Scaling Momentum 190
13.3 Squeeze Flow Damping in an Accelerometer Design 191
13.3.1 Scaling Analysis 192
13.3.2 Flow Damping Coefficient 193
13.4 Coating Extrusion 194
13.4.1 Scaling Arguments 195
13.4.2 Final Coating Thickness 195

13.5 Coating Extrusion on a Porous Surface 198
13.6 Reynolds Equation for Lubrication Theory 202
13.7 Problems 203

Chapter 14 Inviscid Flow 206

14.1 The Reynolds Number 207
14.2 Inviscid Momentum Equation 208
14.3 Ideal Plane Flow 209
14.4 Steady Potential Flow through a Box with Staggered Inlet and Exit 210
14.5 Advection of Species through a Box with Staggered Inlet and Exit 215
14.6 Spherical Bubble Dynamics 217
14.6.1 Effect of Viscosity and Surface Tension 219
14.7 Problems 221

Chapter 15 Catalog of Ideal Plane Flows 224

15.1 Superposition of Simple Plane Flows 224
15.2 Potential Flow over an Aircraft Fuselage 225
15.3 Force on a Line Vortex in a Uniform Stream 227
15.4 Flow Circulation 229
15.5 Potential Flow over Wedges 231
15.5.1 Pressure Gradient along a Wedge 231
15.6 Problems 233

Chapter 16 Complex Variable Methods 234

16.1 Brief Review of Complex Numbers 234
16.2 Complex Representation of Potential Flows 235
16.3 The Joukowski Transform 236
16.4 Joukowski Symmetric Airfoils 238
16.5 Joukowski Cambered Airfoils 240
16.6 Heat Transfer between Nonconcentric Cylinders 242
16.7 Transport with Temporally Periodic Conditions 244
16.8 Problems 246

Chapter 17 MacCormack Integration 249

17.1 Flux-Conservative Equations 249
17.2 MacCormack Integration 250
17.2.1 Stability of Numerical Integration 251
17.2.2 Addition of Viscosity for Numerical Stability 252
17.2.3 Numerical Solution to Burgers' Equation 253
17.3 Transient Convection 255
17.3.1 Groundwater Contamination 256
17.4 Steady-State Solution of Coupled Equations 259
17.5 Problems 262

Chapter 18 Open Channel Flow 265

18.1 Analysis of Open Channel Flows 265
18.2 Simple Surface Waves 267
18.3 Depression and Elevation Waves 268
18.4 The Hydraulic Jump 269
18.5 Energy Conservation 271
18.6 Dam-Break Example 273
18.6.1 Analytic Description 274
18.6.2 Numerical Description 276
18.7 Tracer Transport in the Dam-Break Problem 280
18.8 Problems 280

Chapter 19 Open Channel Flow with Friction 284

19.1 The Saint-Venant Equations 284
19.2 The Friction Slope 286
19.3 Flow through a Sluice Gate 287
19.3.1 Numerical Solution to the Spillway Flow 288
19.3.2 Solutions to the Spillway Flow 291
19.4 Problems 293

Chapter 20 Compressible Flow 296

20.1 General Equations of Momentum and Energy Transport 296
20.1.1 Flow Equations Far from Boundaries 297
20.2 Reversible Flows 298
20.3 Sound Waves 299
20.4 Propagation of Expansion and Compression Waves 300
20.5 Shock Wave (Normal to Flow) 302
20.6 Shock Tube Analytic Description 304
20.7 Shock Tube Numerical Description 307
20.8 Shock Tube Problem with Dissimilar Gases 311
20.9 Problems 312

Chapter 21 Quasi-One-Dimensional Compressible Flows 315

21.1 Quasi-One-Dimensional Flow Equations 315
21.1.1 Continuity Equation 316
21.1.2 Momentum Equation 316
21.1.3 Energy Equation 317

21.2 Quasi-One-Dimensional Steady Flow Equations without Friction 318
21.2.1 Isentropic Flows 319
21.2.2 Flow through a Converging-Diverging Nozzle 322
21.3 Numerical Solution to Quasi-One-Dimensional Steady Flow 323
21.3.1 Unsteady Flow Equations without Friction 323
21.3.2 Boundary Conditions 325
21.3.3 Initial Conditions and Convergence 326
21.3.4 Converging-Diverging Nozzle Example 327
21.4 Problems 330

Chapter 22 Two-Dimensional Compressible Flows 333

22.1 Flow through a Diverging Nozzle 333
22.1.1 Boundary Conditions and Initial Condition 338
22.1.2 Illustrative Result 339
22.1.3 Nozzle with a Transient Inlet Temperature 341
22.2 Problems 342

Chapter 23 Runge-Kutta Integration 344

23.1 Fourth-Order Runge-Kutta Integration of First-Order Equations 344
23.2 Runge-Kutta Integration of Higher Order Equations 347
23.3 Numerical Integration of Bubble Dynamics 349
23.4 Numerical Integration with Shooting 351
23.4.1 Bisection Method 353
23.4.2 Newton-Raphson Method 354
23.5 Problems 355

Chapter 24 Boundary Layer Convection 359

24.1 Scanning Laser Heat Treatment 359
24.2 Convection to an Inviscid Flow 363
24.3 Species Transfer to a Vertically Conveyed Liquid Film 369
24.4 Problems 374

Chapter 25 Convection into Developing Laminar Flows 376

25.1 Boundary Layer Flow over a Flat Plate (Blasius Flow) 376
25.2 Species Transfer across the Boundary Layer 383
25.3 Heat Transfer across the Boundary Layer 387
25.4 A Correlation for Forced Heat Convection from a Flat Plate 389
25.5 Transport Analogies 390
25.6 Boundary Layers Developing on a Wedge (Falkner-Skan Flow) 392
25.6.1 Heat and Mass Transfer for Flows over a Wedge 393
25.7 Viscous Heating in the Boundary Layer 394
25.8 Problems 396

Chapter 26 Natural Convection 399

26.1 Buoyancy 399
26.2 Natural Convection from a Vertical Plate 400
26.3 Scaling Natural Convection from a Vertical Plate 401
26.4 Exact Solution to Natural Convection Boundary Layer Equations 404
26.5 Problems 411

Chapter 27 Internal Flow 412

27.1 Entrance Region 412
27.2 Heat Transport in an Internal Flow 414
27.3 Entrance Region of Plug Flow between Plates of Constant Heat Flux 415
27.4 Plug Flow between Plates of Constant Temperature 417
27.5 Fully Developed Transport Profiles 419
27.5.1 Scaling of the Fully Developed Temperature Field 419
27.5.2 Constant Self-Similar Transport Profiles 421
27.6 Fully Developed Heat Transport in Plug Flow between Plates of Constant Heat Flux 421
27.7 Fully Developed Species Transport in Plug Flow Between Surfaces of Constant Concentration 424
27.8 Problems 426

Chapter 28 Fully Developed Transport in Internal Flows 429

28.1 Momentum Transport in a Fully Developed Flow 429
28.2 Heat Transport in a Fully Developed Flow 430
28.2.1 Heat Transport with Isothermal Boundaries 432
28.2.2 Heat Transport with Constant Heat Flux Boundaries 435
28.2.3 Downstream Development of Temperature in a Heat Exchanger 437

28.3 Species Transport in a Fully Developed Flow 441
28.3.1 Contaminant Leaching from a Constant Concentration Pipe Wall 441
28.4 Problems 444

Chapter 29 Influence of Temperature-Dependent Properties 447

29.1 Temperature-Dependent Conductivity in a Solid 447
29.1.1 Solution by Regular Perturbation 448
29.1.2 Numerical Solution by Runge-Kutta Integration 450
29.2 Temperature-Dependent Diffusivity in Internal Convection 451
29.3 Temperature-Dependent Gas Properties in Boundary Layer Flow 457
29.4 Problems 462

Chapter 30 Turbulence 465

30.1 The Transition to Turbulence 466
30.2 Reynolds Decomposition 468
30.3 Decomposition of the Continuity Equation 469
30.4 Decomposition of the Momentum Equation 470
30.5 The Mixing Length Model of Prandtl 471
30.6 Regions in a Wall Boundary Layer 473
30.6.1 The Viscous Sublayer: (Advection) $\ll$ (Diffusion) and $\varepsilon_M \ll \nu$ 473
30.6.2 Inner Region: (Advection) $\ll$ (Diffusion) and $\varepsilon_M \gg \nu$ 474
30.6.3 Outer Region: (Advection) $\sim$ (Diffusion) and $\varepsilon_M \gg \nu$ 475
30.7 Parameters of the Mixing Length Model 476
30.8 Problems 477

Chapter 31 Fully Developed Turbulent Flow 479

31.1 Turbulent Poiseuille Flow Between Smooth Parallel Plates 480
31.2 Turbulent Couette Flow between Smooth Parallel Plates 485
31.3 Turbulent Poiseuille Flow in a Smooth-Wall Pipe 488
31.4 Utility of the Hydraulic Diameter 490
31.5 Turbulent Poiseuille Flow in a Smooth Annular Pipe 490
31.6 Reichardt's Formula for Turbulent Diffusivity 495
31.6.1 Turbulent Poiseuille Flow Between Smooth Parallel Plates 496
31.7 Poiseuille Flow with Blowing between Walls 497
31.8 Problems 504

Chapter 32 Turbulent Heat and Species Transfer 507

32.1 Reynolds Decomposition of the Heat Equation 507
32.2 The Reynolds Analogy 508
32.3 Thermal Profile Near the Wall 510
32.3.1 The Diffusion Sublayer: (Advection) $\ll$ (Diffusion) and $\varepsilon_H \ll \alpha$ 510
32.3.2 Inner Region: (Advection) $\ll$ (Diffusion) and $\varepsilon_H \gg \alpha$ 510
32.3.3 Outer Region: (Advection) $\sim$ (Diffusion) and $\varepsilon_H \gg \alpha$ 512
32.4 Mixing Length Model for Heat Transfer 513
32.5 Mixing Length Model for Species Transfer 514
32.6 Problems 515

Chapter 33 Fully Developed Transport in Turbulent Flows 517

33.1 Chemical Vapor Deposition in Turbulent Tube Flow with Generation 517
33.2 Heat Transfer in a Fully Developed Internal Turbulent Flow 522
33.3 Heat Transfer in a Turbulent Poiseuille Flow between Smooth Parallel Plates 523
33.3.1 Summary of Turbulent Momentum Transport 523
33.3.2 Turbulent Heat Transfer with Constant Temperature Boundary 524
33.3.3 Heat Transfer with Constant Heat Flux Boundary 529
33.4 Fully Developed Transport in a Turbulent Flow of a Binary Mixture 532
33.5 Problems 543

Chapter 34 Turbulence over Rough Surfaces 545

34.1 Turbulence over a Fully Rough Surface 546
34.1.1 Inner Region: (Advection) $\ll$ (Diffusion) and $\varepsilon_M \gg \nu$ 546
34.2 Turbulent Heat and Species Transfer from a Fully Rough Surface 547
34.2.1 Heat Transfer through the Inner Region: (Advection) $\ll$ (Diffusion) and $\varepsilon_H \gg \alpha$ 548

34.3 Application of the Rough Surface Mixing Length Model 549
34.3.1 Species Transfer across Couette Flow 552
34.4 Application of Reichardt's Formula to Rough Surfaces 553
34.4.1 Turbulent Poiseuille Flow between Rough Parallel Plates 554
34.4.2 Turbulent Heat Convection in Flow between Fully Rough Parallel Isothermal Plates 558
34.5 Problems 563

Chapter 35 Turbulent Boundary Layer 565

35.1 Formulation of Transport in Turbulent Boundary Layer 565
35.1.1 Finite Difference Representation of the Momentum Equation 568
35.1.2 Finite Difference Representation of the Continuity Equation 570
35.1.3 Marching Scheme for Numerical Solution 571
35.1.4 Results of Momentum Transport 573
35.2 Formulation of Heat Transport in the Turbulent Boundary Layer 575
35.2.1 Finite Difference Representation of the Heat Equation 576
35.2.2 Results of the Heat Equation 578
35.3 Problems 580

Chapter 36 The K-Epsilon Model of Turbulence 581

36.1 Turbulent Kinetic Energy Equation 581
36.1.1 Inner Region Scaling of the Turbulent Kinetic Energy 584
36.2 Dissipation Equation for Turbulent Kinetic Energy 585
36.3 The Standard K-Epsilon Model 586
36.4 Problems 587

Chapter 37 The K-Epsilon Model Applied to Fully Developed Flows 589

37.1 K-Epsilon Model for Poiseuille Flow between Smooth Parallel Plates 589
37.2 Transition Point between Mixing Length and K-Epsilon Models 591
37.3 Solving the K and E Equations 593
37.4 Solution of the Momentum Equation with the K-Epsilon Model 597
37.5 Turbulent Diffusivity Approaching the Centerline of the Flow 598
37.6 Turbulent Heat Transfer with Constant Temperature Boundary 601
37.7 Problems 604

Appendix A 606

Index 611

Chapter 1

Thermodynamic Preliminaries

1.1 The First and Second Laws of Thermodynamics
1.2 Fundamental Equations
1.3 Ideal Gas
1.4 Constant Density Solid or Liquid
1.5 Properties of Mixtures
1.6 Summary of Thermodynamic Results
1.7 Problems

Transport equations follow from conservation principles. However, conserved quantities are not necessarily those that are most easily measured. Thermodynamics provides a connection between conserved and measurable quantities—for example, how the energy content of a flow relates to variables like temperature, density, and pressure. When a fluid behaves as an ideal gas or as an incompressible liquid, the thermodynamic relations between conserved and measurable quantities take on particularly simple forms. In practice, such cases are broadly applicable. In other more specialized situations, appropriate thermodynamic assumptions about a flow, such as being reversible and isentropic, may significantly simplify the description. For the treatment of compressible flows, a thermodynamic equation of state is required. Therefore, this chapter reviews the important thermodynamic relations needed for subsequent use in derivation and solution of transport equations.

1.1 THE FIRST AND SECOND LAWS OF THERMODYNAMICS

The *first law of thermodynamics* states the principle of energy conservation. Expressions of the first law vary depending on the forms of energy and transport being considered. In the generic terms of internal energy, heat, and work, the first law could be stated for a closed system as:

> "The increase in the internal energy of a system is equal to the amount of energy added to the system by heat, minus the amount of energy lost through work done by the system on its surroundings."

A *closed system* prohibits mass transfer through the surfaces surrounding the system. For *open systems*, the first law statement should be modified to include the energy associated with mass transport. The first law and other conservation principles are used to construct transport equations in Chapters 4 and 5.

The *second law of thermodynamics* prohibits the destruction of entropy. *Entropy* is a thermodynamic measure of disorder. Although the second law of thermodynamics may seem less familiar than the first law, it is manifested in physical behaviors that are ubiquitous. For example, the second law prohibits heat from spontaneously flowing from a lower temperature to a higher temperature. This principle accounts for the direction of transport associated with the diffusion laws developed in Chapter 2.

The second law of thermodynamics stipulates that the entropy of an isolated system remains the same if all processes undertaken by the system are reversible, but increases if some processes are irreversible. For nonisolated systems, heat transfer to and from the surroundings can transfer entropy. The second law of thermodynamics requires that the rise in entropy for the system must satisfy

$$dS \geq \frac{\delta Q}{T}, \tag{1-1}$$

where δQ is the amount of heat transfer to the system, and T is the temperature of the system boundary across which heat is transferred. The reversible change in entropy associated with heat transfer to the system is given by $dS = \delta Q/T$. Any additional rise in entropy is the result of irreversible entropy generation. The second law of thermodynamics prohibits destruction of entropy. In transport phenomena, one utility of the second law lies in the idealization of reversible adiabatic flows, as discussed in Chapter 20 for compressible flows. When a flow is adiabatically reversible, each fluid element traveling with the flow becomes isolated to entropy transport, with the result that $dS = 0$. This imposes a useful constraint by which the thermodynamic properties, or *state variables*, of the fluid are related throughout the flow.

For a more complete discussion of the first and second laws, interested readers should refer to thermodynamics texts, such as references [1] and [2].

1.2 FUNDAMENTAL EQUATIONS

It is possible to think about the thermodynamic state of a system in terms of variables that influence its energy content. For example, the differential change in the internal energy of an *open system* is described by a fundamental equation of thermodynamics:

$$dU = TdS - Pd\forall + \sum_{i}^{n} \mu_i dN_i, \tag{1-2}$$

where U, S, and $\forall$ are the internal energy, entropy, and volume of the system, respectively. T and P are the thermodynamic temperature and pressure. N_i and μ_i are the number and electrochemical potential of the ith species of a system having n constituents. As illustrated in Figure 1-1, one can interpret TdS as the reversible heat transfer to the system, $Pd\forall$ as the expansion work done by the system, and $\Sigma\mu dN$ as the electrochemical potential of species

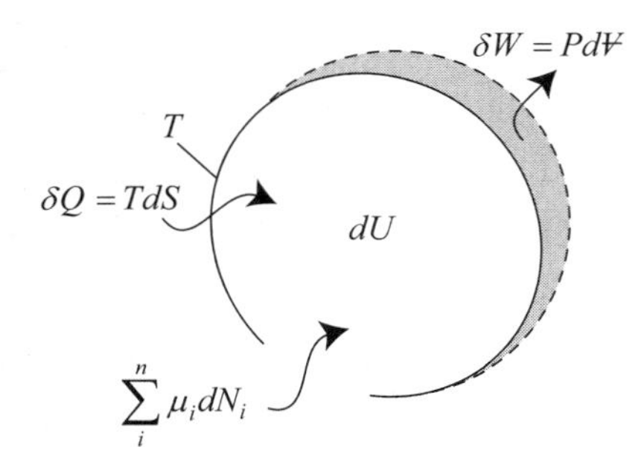

Figure 1-1 Internal energy change in an open system.

transferred to the system. Therefore, this fundamental equation can be understood as a first law statement, saying that the rise in internal energy equals the heat transfer to the system minus the work done by the system plus the chemical potential added by species transfer to the system. As a fundamental equation, it is suitable for simple compressible systems subject to heat and mass transfer. Once the fundamental equation of a particular system is identified, other thermodynamic information can be ascertained from it.

The fundamental equation (1-2) indicates that internal energy is a function of $(2+n)$ variables: $dU(S, \mathcal{V}, N_1 \ldots N_n)$, where n is the number of nonfixed species in the system. This equation provides thermodynamic definitions for temperature, pressure, and chemical potential:

$$T = \left(\frac{\partial U}{\partial S}\right)_{\mathcal{V}, N_1, \ldots, N_n}, \quad P = -\left(\frac{\partial U}{\partial \mathcal{V}}\right)_{S, N_1, \ldots, N_n}, \quad \text{and} \quad \mu_i = \left(\frac{\partial U}{\partial N_i}\right)_{S, \mathcal{V}, N_1, \ldots, N_{i-1}, N_{i+1}, \ldots, N_n} \tag{1-3}$$

For many topics treated in this text, the composition of the system may be considered fixed ($dN_i = 0$). Therefore, for a closed system (one in which no species are transferred across the system boundaries) the fundamental equation becomes Gibbs equation [3]:

$$dU = TdS - Pd\mathcal{V}. \tag{1-4}$$

U, S, and $\mathcal{V}$ are all *extensive* properties because they denote amounts that are dependent on the extent of the system. *Intensive* properties, including T and P, are independent of the extent of the system.

Since T and P are defined by partial derivates of $U(S, \mathcal{V})$, T and P must also be describable as functions of $T(S, \mathcal{V})$ and $P(S, \mathcal{V})$. Such relationships, expressing a relation between independent thermodynamic parameters, are called *equations of state*. The familiar equations of states have combined $T(S, \mathcal{V})$ and $P(S, \mathcal{V})$ to eliminate S, and are expressions of the form $P(\mathcal{V}, T)$. The ideal gas law $P\mathcal{V} = N\hat{R}T$ is the most common example, where N is the number of moles and $\hat{R}$ is the universal gas constant. The carrot is being used to denote quantities reported on a per-unit-mole basis.

Transport equations will be used to describe thermodynamic variables expressed on a per-unit-volume or per-unit-mass basis. Variables that are expressed on a per-unit-mass basis are typically denoted with lowercase letters (e.g., u, s, and $\mathcal{v}$). These intensive variables can also be expressed on a per-unit-volume basis through a product with density $\rho = 1/\mathcal{v}$. For example, if u is the internal energy per unit mass of a substance, then ρu is the internal energy per unit volume. In terms of intensive variables, the ideal gas law becomes $P\mathcal{v} = RT$ or $P = \rho RT$, where the gas constant R is now related to the molecular weight of the gas $\hat{M}$ through the relation $R = \hat{R}/\hat{M}$.

Expressed in terms of intensive variables, the Gibbs equation (1-4) for systems of fixed composition can be written as

$$du = Tds - Pd\mathcal{v}. \tag{1-5}$$

The thermodynamic definitions of pressure and temperature can be written in terms of intensive variables as

$$T = \left(\frac{\partial u}{\partial s}\right)_{\mathcal{v}} \quad \text{and} \quad P = -\left(\frac{\partial u}{\partial \mathcal{v}}\right)_{s}. \tag{1-6}$$

Gibbs equation (1-5) indicates that the specific internal energy of the state is a function of the specific entropy and the specific volume. This observation suggests the thermodynamic state postulate: *The state of a simple compressible system of fixed composition is completely specified by two independent state variables.*

Sometimes it is useful to express the metric for energy in alternate forms. For example, enthalpy is related to internal energy through the definition:

$$h = u + P\boldsymbol{v} \quad \text{or} \quad h = u + P/\rho \,. \tag{1-7}$$

Differentiating the relation between enthalpy and internal energy and eliminating internal energy from the resulting expression with Gibbs equation (1-5) gives a second equivalent form of the fundamental equation for fixed composition systems:

$$dh = Tds + \boldsymbol{v}dP \quad \text{or} \quad dh = Tds + dP/\rho \,. \tag{1-8}$$

Alternate, but equivalent, thermodynamic definitions for temperature and density can be derived from this second fundamental equation,

$$T = \left(\frac{\partial h}{\partial s}\right)_P \quad \text{and} \quad \boldsymbol{v} = \frac{1}{\rho} = \left(\frac{\partial h}{\partial P}\right)_s . \tag{1-9}$$

1.2.1 The Maxwell Relations

Maxwell's thermodynamic relations express equivalence of change between different thermodynamic variables [4]. These relations are extremely useful when the change in one variable, which cannot be measured directly, may be expressed in terms of the change in a second variable that can be directly measured. The Maxwell relations can be derived from the observations that follow.

In accordance with the state postulate, any state variable can be completely specified by two independent state variables. Therefore, let $u = u(x, y)$, $s = s(x, y)$, and $\boldsymbol{v} = \boldsymbol{v}(x, y)$, where x and y are any two independent intensive properties. Based on rules of partial differentiation:

$$du = \left(\frac{\partial u}{\partial x}\right)_y dx + \left(\frac{\partial u}{\partial y}\right)_x dy, \tag{1-10}$$

$$ds = \left(\frac{\partial s}{\partial x}\right)_y dx + \left(\frac{\partial s}{\partial y}\right)_x dy, \tag{1-11}$$

and

$$d\boldsymbol{v} = \left(\frac{\partial \boldsymbol{v}}{\partial x}\right)_y dx + \left(\frac{\partial \boldsymbol{v}}{\partial y}\right)_x dy. \tag{1-12}$$

Substituting these expressions into Gibbs equation (1-5) yields

$$\left(\frac{\partial u}{\partial x}\right)_y dx + \left(\frac{\partial u}{\partial y}\right)_x dy = T\left(\frac{\partial s}{\partial x}\right)_y dx + T\left(\frac{\partial s}{\partial y}\right)_x dy - P\left(\frac{\partial \boldsymbol{v}}{\partial x}\right)_y dx - P\left(\frac{\partial \boldsymbol{v}}{\partial y}\right)_x dy. \tag{1-13}$$

Satisfying Eq. (13) requires that

$$\left(\frac{\partial u}{\partial x}\right)_y = T\left(\frac{\partial s}{\partial x}\right)_y - P\left(\frac{\partial v}{\partial x}\right)_y \tag{1-14}$$

and

$$\left(\frac{\partial u}{\partial y}\right)_x = T\left(\frac{\partial s}{\partial y}\right)_x - P\left(\frac{\partial v}{\partial y}\right)_x. \tag{1-15}$$

Differentiating the above equations by y and x, respectively, yields

$$\left(\frac{\partial^2 u}{\partial y \partial x}\right) = \left(\frac{\partial T}{\partial y}\right)_x \left(\frac{\partial s}{\partial x}\right)_y + T\left(\frac{\partial^2 s}{\partial y \partial x}\right) - \left(\frac{\partial P}{\partial y}\right)_x \left(\frac{\partial v}{\partial x}\right)_y - P\left(\frac{\partial^2 v}{\partial y \partial x}\right) \tag{1-16}$$

and

$$\left(\frac{\partial^2 u}{\partial x \partial y}\right) = \left(\frac{\partial T}{\partial x}\right)_y \left(\frac{\partial s}{\partial y}\right)_x + T\left(\frac{\partial^2 s}{\partial x \partial y}\right) - \left(\frac{\partial P}{\partial x}\right)_y \left(\frac{\partial v}{\partial y}\right)_x - P\left(\frac{\partial^2 v}{\partial x \partial y}\right). \tag{1-17}$$

Because the order of mixed differentiation of an analytic function of two variables is irrelevant, the following terms are equivalent:

$$\left(\frac{\partial^2 u}{\partial x \partial y}\right) = \left(\frac{\partial^2 u}{\partial y \partial x}\right), \quad \left(\frac{\partial^2 s}{\partial x \partial y}\right) = \left(\frac{\partial^2 s}{\partial y \partial x}\right), \quad \text{and} \quad \left(\frac{\partial^2 v}{\partial x \partial y}\right) = \left(\frac{\partial^2 v}{\partial y \partial x}\right). \tag{1-18}$$

Therefore, subtracting Eq. (1-16) from (1-17) results in a general expression that yields many of Maxwell's thermodynamic relations:

$$\left(\frac{\partial T}{\partial x}\right)_y \left(\frac{\partial s}{\partial y}\right)_x - \left(\frac{\partial P}{\partial x}\right)_y \left(\frac{\partial v}{\partial y}\right)_x = \left(\frac{\partial T}{\partial y}\right)_x \left(\frac{\partial s}{\partial x}\right)_y - \left(\frac{\partial P}{\partial y}\right)_x \left(\frac{\partial v}{\partial x}\right)_y. \tag{1-19}$$

Maxwell equations can be realized from this result by letting x and y be different combinations of independent state properties. For example, with $x = T$ and $y = v$, the general expression (1-19) reduces to

$$\left(\frac{\partial s}{\partial v}\right)_T = \left(\frac{\partial P}{\partial T}\right)_v. \tag{1-20}$$

Since entropy is not a quantity that can be measured directly, the Maxwell equation (1-20) establishes a useful relationship between changes in entropy and changes in pressure. Utility for this result will arise in the next section, and other useful Maxwell equations can easily be obtained from the general expression (1-19) with alternate choices of independent state variables (see Problem 1-2).

1.2.2 Internal Energy Expressed in Measurable Variables

The fundamental equation for $du(s, \rho)$, given by Eq. (1-5), has limited experimental utility for characterizing changes in internal energy, since entropy is not a readily measurable quantity. Consequently, it is useful to replace entropy with another thermodynamic variable, like temperature. To express the change in internal energy with respect to changes in T and ρ, one may write:

$$du(T,\rho) = \left(\frac{\partial u}{\partial T}\right)_\rho dT + \left(\frac{\partial u}{\partial \rho}\right)_T d\rho. \tag{1-21}$$

Both $(\partial u/\partial T)_\rho$ and $(\partial u/\partial \rho)_T$ are related to material properties of the substance. In fact, the former is the definition of the first form of specific heat:

$$C_v = \left(\frac{\partial u}{\partial T}\right)_v. \tag{1-22}$$

To establish $(\partial u/\partial \rho)_T$, Gibbs equation (1-5) is evaluated to show

$$\left(\frac{\partial u}{\partial \rho}\right)_T = T\left(\frac{\partial s}{\partial \rho}\right)_T + \frac{P}{\rho^2}. \tag{1-23}$$

To remove the explicit dependency on entropy, the Maxwell relation (1-20) is cast into the form

$$\left(\frac{\partial P}{\partial T}\right)_\rho = \left(\frac{\partial s}{\partial v}\right)_T = -\rho^2\left(\frac{\partial s}{\partial \rho}\right)_T. \tag{1-24}$$

With Eqs. (1-22) through (1-24), Eq. (1-21) may be rewritten as:

$$du(T,\rho) = C_v dT + \frac{1}{\rho^2}\left(P - T\frac{\partial P}{\partial T}\bigg|_\rho\right)d\rho. \tag{1-25}$$

In this form, the change in internal energy is expressed as a function of measurable thermodynamic variables.

1.2.3 Enthalpy Expressed in Measurable Variables

Motivated by the desire to express the fundamental equation (1-8) in terms of readily measurable thermodynamic variables, one can write

$$dh(T,P) = \left(\frac{\partial h}{\partial T}\right)_P dT + \left(\frac{\partial h}{\partial P}\right)_T dP, \tag{1-26}$$

where both $(\partial h/\partial T)_P$ and $(\partial h/\partial P)_T$ are related to material properties of the substance. The former is the definition of the second form of specific heat:

$$C_p = \left(\frac{\partial h}{\partial T}\right)_p. \tag{1-27}$$

With the fundamental equation Eq. (1-8), $(\partial h/\partial P)_T$ can be expressed as

$$\left(\frac{\partial h}{\partial P}\right)_T = T\left(\frac{\partial s}{\partial P}\right)_T + \frac{1}{\rho}. \tag{1-28}$$

Furthermore, with $x = P$ and $y = T$, the general expression for Maxwell's relations (1-19) reduces to

$$\left(\frac{\partial s}{\partial P}\right)_T = -\left(\frac{\partial v}{\partial T}\right)_P = \frac{1}{\rho^2}\left(\frac{\partial \rho}{\partial T}\right)_P. \tag{1-29}$$

Combining Eqs. (1-27) through (1-29) into Eq. (1-26) yields

$$dh(T,P) = C_p dT + \left(\frac{T}{\rho^2}\frac{\partial \rho}{\partial T}\bigg|_P + \frac{1}{\rho}\right)dP. \tag{1-30}$$

Equations (1-25) and (1-30) offer two expressions that describe the energy content of a substance in terms of measurable thermodynamic variables. However, to work with these equations requires an equation of state that relates pressure, temperature, and density. The ideal gas law equation of state is considered first.

1.3 IDEAL GAS

A *gas* is distinct from a liquid by the ability to expand when it is not physically confined and held under pressure. An *ideal gas* is defined by the familiar equation of state:

$$Pv = RT \quad \text{or} \quad P = \rho RT, \tag{1-31}$$

where R is the ideal gas constant. The ideal gas law fulfills the expectation that a function $P(v, T)$ must follow from the fundamental equation of thermodynamics for a fixed-composition simply compressible system, as discussed in Section 1.2. The ideal gas law conforms to the experimentally determined behavior of $P(v, T)$ for most gases in the limit of low pressure. The ideal gas law finds utility in a large number of engineering problems.

For an ideal gas, the change in internal energy expressed by Eq. (1-25) simplifies with the observation that

$$P - T\frac{\partial P}{\partial T}\bigg|_\rho = P - \rho RT = 0. \tag{1-32}$$

Therefore, change in internal energy depends only on the change in temperature of an ideal gas through the relation

$$du = C_v dT. \tag{1-33}$$

For an ideal gas, the change in enthalpy expressed by Eq. (1-30) simplifies with the observation that

$$\left(\rho + T\frac{\partial \rho}{\partial T}\bigg|_P\right) = \rho - \frac{P}{RT} = 0. \tag{1-34}$$

Therefore, a change in enthalpy depends only on the change in temperature of an ideal gas through the relation

$$dh = C_p dT. \tag{1-35}$$

Differentiating the relation between internal energy and enthalpy (1-7) and applying the ideal gas relations Eqs. (1-31) through (1-35), it is straightforward to show that

$$dh = du + RdT \quad \text{or} \quad C_p dT = C_v dT + RdT. \tag{1-36}$$

From this last result, it is evident that for an ideal gas

$$C_p = C_v + R. \tag{1-37}$$

Equations (1-33), (1-35), and (1-37) are useful results for quantifying the energy content of an ideal gas.

1.4 CONSTANT DENSITY SOLID OR LIQUID

A *liquid* is distinguished from a *solid* by the ability to flow. However, both liquids and solids do not have the ability to expand freely due to strong intermolecular attractions. For a constant density solid or liquid, where $dv = 0$, the fundamental equation (1-5) becomes

$$du = Tds. \tag{1-38}$$

Furthermore, the change in internal energy expressed by Eq. (1-25) becomes

$$du = C_v dT. \tag{1-39}$$

Therefore, the internal energy of a constant density substance (solid or liquid) depends only on temperature (as was also the case for an ideal gas).

Equation (1-7) can be differentiated to express a change in enthalpy as

$$dh = du + vdP + Pdv. \tag{1-40}$$

For a constant density substance, the change in enthalpy can be expressed using Eq. (1-39), as

$$dh = C_v dT + vdP. \tag{1-41}$$

However, for a typical solid or liquid, $C_v \gg v$. For example, water at 25°C has $C_v = 4180\ \text{J/kg/K}$ and $v = 0.001\ \text{m}^3/\text{kg}$. Therefore, for modest changes in pressure, the second term can often be neglected, and the change in enthalpy of a solid or liquid becomes

$$dh \approx C_v dT. \tag{1-42}$$

Equation (1-42) can only be reconciled with the definition $(\partial h/\partial T)|_P = C_p$ when

$$C_v \approx C_p = C. \tag{1-43}$$

Therefore, the two forms of specific heat are nearly identical for an incompressible solid or liquid, and the distinguishing subscript is often dropped from notation. Equations (1-39), (1-42), and (1-43) are useful results for quantifying energy content in a constant density solid or liquid.

1.5 PROPERTIES OF MIXTURES

Most matter is a composition of constituent elements. For example, air is composed of oxygen and nitrogen. However, it is not always productive to draw attention to this fact unless changes in composition occur during transport. Thermodynamic properties of mixtures can be expressed in terms of the constituent elements. This is done either on a molar or a mass basis, depending on which is more convenient. Molar densities c_i (concentrations) and mass densities ρ_i both sum to the mixture values

$$\sum_{i=1}^{n} c_i = c \quad \text{and} \quad \sum_{i=1}^{n} \rho_i = \rho, \tag{1-44}$$

where n is the number of constituents in the mixture. In other words, the total molar concentration c of a mixture equals the sum of all the constituent concentrations c_i, and the mass density ρ of a mixture equals the sum of all the partial densities ρ_i.

Often it is convenient to describe mixtures in terms of molar or mass fractions. The molar fraction χ_i is the ratio of the moles of the ith species to the total number of moles in the mixture. Similarly, the mass fraction ω_i is the ratio of the ith species mass to the total mass of the mixture. Therefore, in terms of the total concentration and density of the mixture,

$$\chi_i = c_i/c \quad \text{and} \quad \omega_i = \rho_i/\rho. \tag{1-45}$$

Notice that by definition, the mixture fractions sum to unity:

$$\sum_{i=1}^{n} \chi_i = 1 \quad \text{and} \quad \sum_{i=1}^{n} \omega_i = 1. \tag{1-46}$$

Mixture properties reported on a per-unit-mole basis are calculated from constituent values using molar fractions. For example, the molecular weight $\hat{M}$ of a mixture is the mixture mass on a per-unit-mole basis. Therefore, the molecular weight of a mixture can be calculated from

$$\hat{M} = \sum_{i=1}^{n} \chi_i \hat{M}_i. \tag{1-47}$$

Mixture properties that are reported on a per-unit-mass basis are calculated from constituent values using mass fractions. This includes all the specific properties discussed in this chapter, such as internal energy u, enthalpy h, and entropy s. Since the specific heats and the ideal gas constant of a mixture are quantified on a per-unit-mass basis as well, they can be calculated from the mass fractions as

$$C_{\mathcal{v}} = \sum_{i=1}^{n} \omega_i C_{\mathcal{v},i}, \quad C_p = \sum_{i=1}^{n} \omega_i C_{p,i}, \quad \text{and} \quad R = \sum_{i=1}^{n} \omega_i R_i, \tag{1-48}$$

where $C_{\mathcal{v},i}$ and $C_{p,i}$ are the specific heats and R_i the ideal gas constant of the ith species.

1.6 SUMMARY OF THERMODYNAMIC RESULTS

Table 1-1 summarizes some of the thermodynamic relations for ideal gases and constant density liquids or solids. These results will be used in deriving and manipulating transport equations for heat, mass, and momentum transport in later chapters. It is left as

Table 1-1 Some thermodynamic relations

	Specific energies: (J/kg)		
total: $e = u + v^2/2$	internal: u	kinetic: $v^2/2$	enthalpy: $h = u + P/\rho$
	Ideal gas: $P = \rho RT$		
$du = C_{\not{v}} dT$	$dh = C_p dT$	$C_p - C_{\not{v}} = R$	
$ds = C_{\not{v}} dT/T - R d\rho/\rho$	$ds = C_p dT/T - R dP/P$	$ds = C_{\not{v}} dP/P - C_p d\rho/\rho$	
	Constant density: (solid or liquid)		
$du = CdT$	$dh \approx CdT$	$ds = CdT/T$	$C_p \approx C_{\not{v}} = C$

an exercise (see Problems 1-1 and 1-3) to demonstrate the validity of the expressions shown in Table 1-1 for changes in entropy (ds) for an ideal gas and incompressible liquid.

The entropy relations in Table 1-1 will find utility in the treatment of *isentropic flows* ($ds = 0$), where fluid changes are both adiabatic and reversible. This constrains the way in which thermodynamic properties of the fluid may change in a flow. For example, the ideal gas relations for $ds(T, \rho)$, $ds(T, P)$, and $ds(P, \rho)$ in Table 1-1 may be easily integrated for a *calorically perfect gas,* defined as being when C_p and $C_{\not{v}}$ are constant. In this case, integration, with $ds = 0$, between two states in the flow yields the results

$$\left(\frac{T_2}{T_1}\right) = \left(\frac{\rho_2}{\rho_1}\right)^{\gamma - 1} = \left(\frac{P_2}{P_1}\right)^{\frac{\gamma-1}{\gamma}} \tag{1-49}$$

where

$$\gamma = C_p/C_{\not{v}}. \tag{1-50}$$

Equation (1-49) provides thermodynamic relations between any two points in a reversible adiabatic flow of a calorically perfect ideal gas. Another interesting feature of reversible adiabatic flows is that once $ds = 0$ is enforced, the transport of energy is no longer independent from the transport of momentum, as will be demonstrated in Chapter 20.

1.7 PROBLEMS

1-1 The fundamental equation given by Eq. (1-2) can be rewritten as

$$dS = \frac{1}{T} dU + \frac{P}{T} d\not{V} - \sum_i^n \frac{\mu_i}{T} dN_i.$$

Use this entropy function to provide thermodynamic definitions for temperature, pressure, and chemical potential. Derive expressions for $ds(T, \not{v})$ for an ideal gas and $\rho = const.$ liquid.

1-2 Derive six Maxwell equations using combinations of s, $\not{v}$, T, and P for the independent state properties.

1-3 Starting with

$$ds(T, P) = \left(\frac{\partial s}{\partial T}\right)_P dT + \left(\frac{\partial s}{\partial P}\right)_T dP,$$

demonstrate that for an ideal gas,

$$ds(T,P) = C_p\frac{dT}{T} - R\frac{dP}{P}.$$

1-4 A fluid is depressurized by 5 kPa passing through a valve. Without performing work, the change in enthalpy across the valve is zero. Assuming that the fluid is incompressible water with a density of $\rho = 1000\ \text{kg/m}^3$ and specific heat of $C_v = 4180\ \text{J/(kgK)}$, determine the change in temperature across the valve. Assuming that the fluid is an ideal gas, determine the change in temperature across the valve when the fluid is depressurized by 5 kPa. If gas in a tank is depressurized by 5 kPa from an initial pressure and temperature of 106 kPa and 30°C, respectively, what is the temperature change of the gas remaining in the tank, assuming that $\gamma = C_p/C_v = 1.4$?

1-5 Air is composed of 21% O_2 and 79% N_2. These gases have molecular weights of $M_w(O_2) = 32$ g/mol and $M_w(N_2) = 28$ g/mol, and have ideal gas constants of $R(O_2) = 259.8\ \text{J/(kgK)}$ and $R(N_2) = 296.8\ \text{J/(kgK)}$. Calculate the molecular weight and ideal gas constant mixture properties of air.

REFERENCES

[1] A. Bejan, *Advanced Engineering Thermodynamics*, Third Edition. Hoboken: John Wiley & Sons, 2006.

[2] H. B. Callen, *Thermodynamics and an Introduction to Thermostatistics*, Second Edition. New York: John Wiley & Sons, 1985.

[3] J. W. Gibbs, "Part 1: Graphical Methods in the Thermodynamics of Fluids," *Transactions of the Connecticut Academy*, **2**, 309 (1873).

[4] J. C. Maxwell, *Theory of Heat*. London: Longmans, Green, and Co, 1871; reprinted 2001 (Dover, New York).

Chapter 2

Fundamentals of Transport

2.1 Physics of Advection and Diffusion
2.2 Advection Fluxes
2.3 Diffusion Fluxes
2.4 Reversible vs. Irreversible Transport
2.5 Looking Ahead
2.6 Problems

This chapter introduces the basic notion of two kinds of transport, advection and diffusion. The means to quantify advection and diffusion fluxes is needed for the development of transport equations, undertaken in Chapters 4 and 5. It is demonstrated in this chapter that the physical picture and mathematical treatment of transport by advection and diffusion remains essentially the same, irrespective of the specific fluid property being considered. As a consequence, equations developed later for species, heat, and momentum transport will all share a common form.

2.1 PHYSICS OF ADVECTION AND DIFFUSION

Two basic transport mechanisms are sufficient for describing a large number of transport phenomena. The first is *advection*, which is transport by bulk motion. If matter is moving, then it makes sense that the properties of matter are carried with that motion. The second transport mechanism is *diffusion*, which can take place in the absence of bulk motion. Diffusion is a subtler phenomenon that lacks a macroscopic explanation. For example, although it is understood that heat "flows" from hot to cold, one may not have ever considered why. The answer lies in molecular motion. Even though a gas, for instance, may seem static at the macroscopic level, at room temperature air molecules are moving around with speeds on the order of 460 m/s! However, the air molecules travel only about 65 nm before colliding with other molecules at atmospheric pressure. Diffusion is the transport of properties through these and other molecular-level interactions, which redistributes fluid properties from regions of high content to regions of low content.

Figure 2-1 illustrates the diffusion of momentum in a gas overlying a stationary wall. The "zigzag" course of a molecular trajectory and the straight course of the average bulk flow are shown at two distances from the wall. Averaged over time, the random molecular speed and trajectory becomes the local flow velocity. For the case illustrated, the momentum of the flow ρv_x increases with distance from the wall (ρ is assumed constant). Immediately adjacent to the stationary wall at y_1, flow momentum is lost each time a molecule hits the wall. However, momentum is gained each time a molecule at y_1 is hit by another molecule originating from y_2, a distance further from the wall where the flow velocity is greater. Therefore, as long as the velocity at y_2 is maintained, there will be

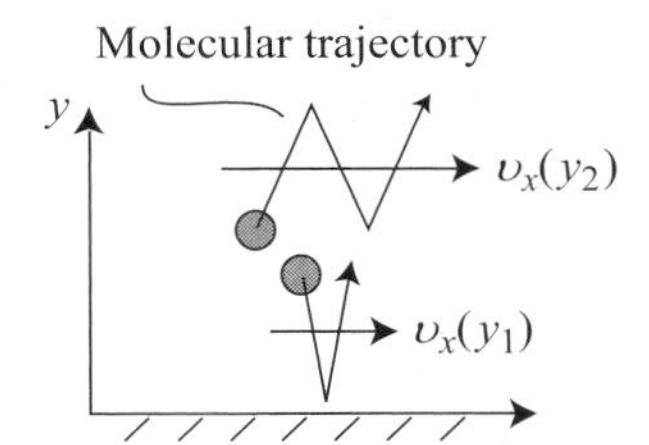

Figure 2-1 Diffusion transport.

a continuous transport of momentum toward the wall through these intermolecular collisions. If the velocity at y_2 is not maintained, the momentum lost to the wall will eventually bring the flow to rest.

The preceding description illustrates how momentum diffusion is a molecular phenomenon that, in the presence of gradients in momentum (velocity), causes momentum to be transported from regions of high momentum content to regions of low momentum content. Heat and mass diffusion are analogous to momentum diffusion. In the presence of a temperature gradient, internal energy (such as the average molecular kinetic energy) is transported through intermolecular collisions from regions of high temperature to regions of low temperature. Similarly, mass diffusion occurs when species of molecules randomly migrate. However, to have any net effect requires the existence of a gradient in species.

Diffusion in a liquid or a solid is conceptually similar to that of a gas, but the nature of "particle" interactions is much more complex. Fortunately, one does not need to consider the molecular processes in a detailed way in order to describe diffusion transport mathematically at a *continuum* level. In continuum theory, the scales of molecular interactions (i.e., the mean free path between particle collisions) are small compared with the important length scales of transport. At the continuum level, diffusion fluxes are simply proportional to the gradient in properties that can be detected macroscopically.

Many problems can be classified as the result of either advection transport or diffusion transport, or a combination of both, as illustrated schematically in Figure 2-2. The transport of isolated heated blocks on a conveyor belt illustrates the features of heat advection in the absence of diffusion. In contrast, a stationary bar subjected to a temperature gradient illustrates heat diffusion in the absence of advection. When advection and diffusion are acting simultaneously, the transport process is called *convection*. Since advection and diffusion are independent, they can complement or oppose each other. Both advection and diffusion transport occur in fluids, but they are not necessarily equally important to all flows. For example, consider a flow adjacent to a stationary surface. Approaching the surface, the fluid velocity and therefore advection transport in

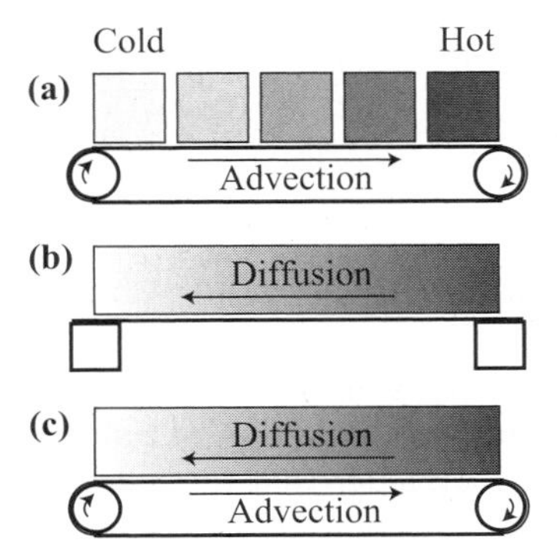

Figure 2-2 Transport of heat by: (a) advection, (b) diffusion, and (c) a combination of both.

the fluid must approach zero. Accordingly, sufficiently close to the surface diffusion becomes the dominant transport mechanism in the flow. However, at distances further from the surface, as the speed of the fluid picks up, advection can become the dominant transport process.

It is important to note that in much of the fluid mechanics literature the term "convection" is used to refer to advection (transport by bulk fluid motion). In this text, the terminology of heat and mass transfer literature is adopted, where convection is the combination of advection and diffusion transport.

2.2 ADVECTION FLUXES

Advection fluxes are simple to quantify. In a unit of time, the volume of fluid crossing a unit area of an imaginary plane depends on the velocity $\vec{v}$ of the fluid, as illustrated in Figure 2-3. The amount of some fluid property X carried across the plane depends on the fluid content of that property on a per-unit-volume basis $(X/\forall)$. The advection flux of X is given by

$$\left.\begin{matrix} \textit{Advection} \\ \textit{flux of } X \end{matrix}\right\} = \frac{X}{\forall}\vec{v} \tag{2-1}$$

where $(X/\forall)\vec{v}$ has dimensions of X per unit time per unit area. Notice that the direction of transport is given by the vector sense of the fluid velocity.

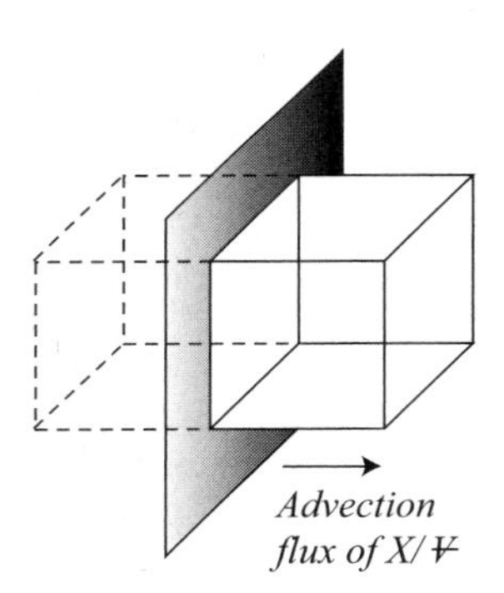

Figure 2-3 Advection flux as carried by bulk fluid motion.

As an example, consider the advection flux of heat. The heat carried per unit mass is the internal energy u. Therefore, on a unit volume basis, $(X/\forall) = \rho u$ and the advection flux of internal energy becomes

$$\text{Internal energy: } (\rho u)\vec{v}. \tag{2-2}$$

As another example, consider the advection flux of momentum. Momentum on a per-unit-volume basis is $(X/\forall) = \rho\vec{v}$. Therefore, the advection flux of momentum becomes

$$\text{Momentum: } (\rho\vec{v})\vec{v}. \tag{2-3}$$

Notice that the vector product $\vec{v}\vec{v}$ is a tensor that contains nine terms (in three-dimensional space). Each term identifies a direction of transport combined with a component of momentum. To illustrate, consider that the advection of the y-component of momentum (associated with the velocity component v_y) can be moved in the x-direction by the v_x velocity component. Therefore, $(\rho v_x)v_y$ would be the advection flux of the

x-component of momentum transported in the y-direction. Keeping track of all the vector senses is facilitated by the use of *index notation* introduced in Chapter 3.

As a final example, consider the advection flux of fluid mass. Since mass carried in a unit volume is simply the fluid density, the advection flux of mass must be

$$\text{Total mass: } \rho\vec{v}. \tag{2-4}$$

In mass transfer of species, the velocity of each constituent in the mixture may potentially be different. For this reason, the "velocity" of a mixture can be somewhat ambiguous. Up to this point, the flow velocity $\vec{v}$ has represented a mass-averaged value. However, to keep track of a fluid on a concentration basis, where c is the total molar concentration of the fluid, the flow velocity should be identified with a molar average value $\vec{v}^*$. In this case, the advection flux, with respect to moles of fluid, is expressed by $c\vec{v}^*$. One typically expects that $\vec{v} = \vec{v}^*$ for homogeneous mixtures. However, when concentration gradients exist in the fluid, diffusion causes the various species in the mixture to propagate at different net velocities (resulting from advection plus diffusion). When this is the case, the mass-averaged and molar-averaged velocities are not the same, $\vec{v} \neq \vec{v}^*$, for species of different molecular weights, as will be demonstrated for a binary mixture in the next section.

2.2.1 Advection Transport in a Binary Mixture

For a fluid mixture, it is easy to imagine a situation in which the velocity of each component is different from the others. When this is the case, mass-averaged and molar-averaged velocities of the flow are different for mixtures of elements with different molecular weights. To illustrate, consider the mass flux carried by advection in a binary fluid, where $\vec{v}_A$ and $\vec{v}_B$ are the independent velocities of the two components A and B, as illustrated in Figure 2-4. Notice that $\vec{v}_A$ and $\vec{v}_B$ are not averaged velocities, for which reference to either a mass or a molar basis is needed. In terms of the constituent velocities, the advection flux of mass is $\rho_A\vec{v}_A + \rho_B\vec{v}_B$, where ρ_A and ρ_B are the partial densities of components A and B, respectively. This is reconciled with the description of the average transport of mass by advection $\rho\vec{v} = (\rho_A + \rho_B)\vec{v}$, if, and only if, the flow velocity $\vec{v}$ is defined by the mass-averaged value such that $(\rho_A + \rho_B)\vec{v} = \rho_A\vec{v}_A + \rho_B\vec{v}_B$ or

$$\vec{v} = \omega_A\vec{v}_A + \omega_B\vec{v}_B \quad \text{(mass-averaged velocity)}. \tag{2-5}$$

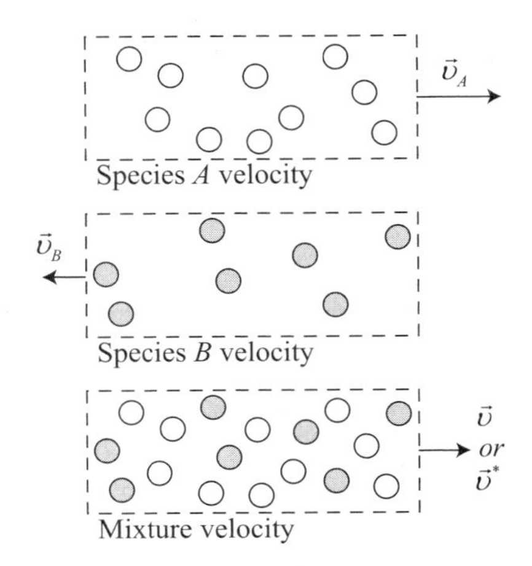

Figure 2-4 Decomposition of a mixture velocity.

Here, $\omega_A = \rho_A/\rho$ and $\omega_B = \rho_B/\rho$ are the mass fractions of the two components of the fluid having a total density $\rho = \rho_A + \rho_B$. When quantifying mass transport, or transport of other quantities on a per-unit-mass basis, the flow velocity should be the mass-averaged value $\vec{v}$. In fluid mechanics most problems are handled with the mass-averaged velocity, since momentum transport is directly related to the mass of the fluid.

The molar flux carried by advection in a binary fluid is $c_A\vec{v}_A + c_B\vec{v}_B$, where again $\vec{v}_A$ and $\vec{v}_B$ are the independent velocities of the two components. This molar flux is reconciled with the description of the average transport of total moles: $c\vec{v}^* = (c_A + c_B)\vec{v}^*$, if, and only if, the flow velocity $\vec{v}^*$ is defined by the molar average such that $(c_A + c_B)\vec{v}^* = c_A\vec{v}_A + c_B\vec{v}_B$ or

$$\vec{v}^* = \chi_A\vec{v}_A + \chi_B\vec{v}_B \quad \text{(molar-averaged velocity)}. \tag{2-6}$$

Here, $\chi_A = c_A/c$ and $\chi_B = c_B/c$ are the molar fractions of the two components in the fluid having a total concentration $c = c_A + c_B$. When quantifying molar transport, or transport of other quantities on a per-unit mole basis, the flow velocity must be the molar averaged value $\vec{v}^*$. Mass transport problems in which the total molar concentration is constant are handled most effectively with a description of the molar fluxes as opposed to the mass fluxes.

2.2.2 Summary of Advection Transport

Table 2-1 summarizes expressions for advection fluxes of different properties. With the exception of species transport, advection fluxes are shown with respect to the mass-averaged velocity. In the column for species transport of Table 2-1, expressions for both the mass flux and the molar flux are shown.

Notice that the mass flux attributed to advection $\rho_A\vec{v}$ is different from the net flux of species A as described by $\rho_A\vec{v}_A$. The former expresses a flux based on the rate with which the mass-average fluid velocity carries the mass content of species A, while the latter expresses a flux based on the actual velocity with which the mass content of species A is moving. The difference, given by

$$j_A = \rho_A(\vec{v}_A - \vec{v}), \tag{2-7}$$

is the flux contribution that comes from diffusion. As expressed by Eq. (2-7), j_A is a mass diffusion flux seen relative to coordinates moving with the mass-averaged velocity $\vec{v}$. In contrast, if diffusion were observed from coordinates moving with the molar-averaged velocity, the flux of moles of species A would appear to be

$$J_A^* = c_A(\vec{v}_A - \vec{v}^*). \tag{2-8}$$

Typical transport problems address the goal of quantifying the combined effect of advection and diffusion. Therefore, defining diffusion as the difference between the net

Table 2-1 Advection fluxes

	Heat	Mass	Momentum	Species A	Kinetic energy
Advection flux:	$(\rho u)\vec{v}$	$(\rho)\vec{v}$	$(\rho\vec{v})\vec{v}$	$(\rho_A)\vec{v}$ or $(c_A)\vec{v}^*$	$\rho(v^2/2)\vec{v}$
Units:	$\left(\frac{J}{m^2 s}\right)$	$\left(\frac{kg}{m^2 s}\right)$	$\left(\frac{kg\ m/s}{m^2 s}\right)$	$\left(\frac{kg}{m^2 s}\right)$ or $\left(\frac{mol.}{m^2 s}\right)$	$\left(\frac{J}{m^2 s}\right)$

transport and the contribution from advection, as is done in Eqs. (2-7) and (2-8), is not entirely satisfactory. Instead, diffusion fluxes need to be quantified by an independent method, as is discussed in the next section.

2.3 DIFFUSION FLUXES

The physical arguments presented in Section 2.1 suggest that a diffusion flux is related to the presence of a gradient in the property experiencing diffusion. The simplest statement consistent with this observation is a linear relation, where the diffusion flux is proportional to the gradient as illustrated in Figure 2-5. In this illustration, the fluid property of interest is given on a per-unit-volume basis by $(X/\forall)$. Therefore, a linear diffusion law should have the form

$$\left.\begin{array}{r}\textit{diffusion}\\ \textit{flux of } X\end{array}\right\} = -(\textit{diffusivity})\vec{\nabla}\left(\frac{X}{\forall}\right). \tag{2-9}$$

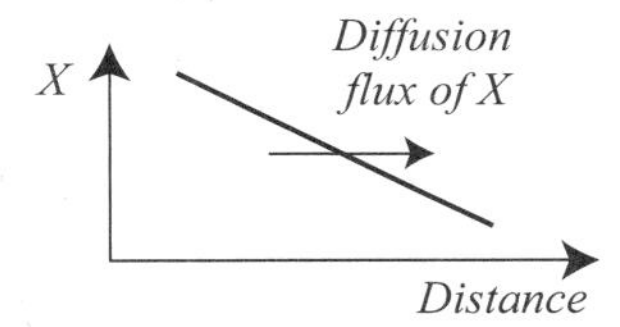

Figure 2-5 Diffusion relative to a gradient.

Notice that the direction of transport is given by the vector sense of the gradient, but is opposite in sign because the diffusion flux occurs from the high content region to the low content region. The proportionality constant between the diffusion flux and the gradient is the *diffusivity*.

Since the physical picture of diffusion presented in Section 2.1 is essentially the same for different properties, one might suspect that the diffusivity constants associated with various diffusion laws are related. Indeed, for the simplest of molecular interactions they are related. A gas with a number density $n = N/\forall$ has a volumetric heat capacity of $nc_{\forall}$ and a volumetric mass of nm, where $c_{\forall}$ and m are the specific heat and mass of individual molecules. Kinetic theory predicts that the mass diffusivity of a gas is $(n)(\lambda\bar{v}/3)$, the heat diffusivity is $(c_{\forall}n)(\lambda\bar{v}/3)$, and the momentum diffusivity is $(mn)(\lambda\bar{v}/3)$, where $\bar{v}$ is the mean molecular speed of the gas and λ is the mean distance between molecular collisions. Therefore, the diffusivities predicted by kinetic theory for mass, heat, and momentum transport in a gas differ only with respect to the coefficients of 1, $c_{\forall}$, and m, respectively.

From the kinetic theory of gasses, the mean molecular speed is found to be $\bar{v} = \sqrt{8k_BT/(\pi m)}$ (where k_B is the Boltzmann constant). Additionally, using the mean distance between molecular collisions given by the hard sphere model $\lambda = 1/(\sqrt{2}\pi nd^2)$ (where d is the molecular diameter), it is seen that kinetic theory predicts that diffusivities should increase with temperature as $\sqrt{T}$ and are independent of the density (n) of the gas. More details of kinetic theory can be found in reference [1].

Shortcomings of simple kinetic theory, however, can be demonstrated with experimental measurements. For real gases, the cohesive forces between atoms can cause diffusivity to increase more rapidly with temperature than simple kinetic theory suggests. Furthermore, in liquids it is found that momentum diffusivity decreases with increased temperature, as the average distance between molecules increases. Therefore, in practice, diffusivities should be considered empirical properties that are dependent on the thermodynamic state.

2.3.1 Heat Diffusion

Consider internal energy as given by ρu on a per-unit-volume basis. To simplify the discussion, consider an incompressible substance, for which the specific heats are indistinguishable $C_v \approx C_p = C$. Adopting the form of a linear diffusion law, the heat transport in the x-direction should be expressed in terms of the gradient of internal energy in the x-direction as

$$q_x = -\alpha \frac{\partial(\rho u)}{\partial x}, \tag{2-10}$$

where α is the thermal diffusivity. However, since internal energy is not easily measured, it is more convenient to express the heat flux in terms of the temperature gradient using the thermodynamic relation for an incompressible fluid, $du = CdT$. Heat diffusion can then be expressed as

$$q_x = -k \frac{\partial T}{\partial x}, \tag{2-11}$$

where $k = \rho C \alpha$ is the thermal conductivity. Heat diffusion in the other directions can be expressed as $q_y = -k\, \partial T/\partial y$ and $q_z = -k\, \partial T/\partial z$. The linear relation between the heat diffusion flux and the temperature gradient is known as *Fourier's law* [2]. In practice, this linear relation is used to define thermal conductivity k without stipulating that density is a constant. In symbolic notation Fourier's law is written as

$$\vec{q} = -k\vec{\nabla} T. \tag{2-12}$$

2.3.2 Momentum Diffusion

Consider diffusion of momentum. As observed while discussing advection in Section 2.2, there are two vector senses to momentum transport: the momentum component and the transport direction. For simplicity, consider a unidirectional constant density flow in which v_x is only a function of y and $v_y = 0$. This is illustrated by Couette flow,* as shown in Figure 2-6, in which ρv_x is the momentum content on a per-unit-volume basis. Again adopting the form of a linear diffusion law, the momentum flux in the y-direction should be given by

$$M_{yx} = -\nu \frac{\partial(\rho v_x)}{\partial y} \quad \text{(unidirectional flow)}, \tag{2-13}$$

where the coefficient ν is the momentum diffusivity. In practice, momentum diffusion is expressed in terms of gradients in velocity rather than gradients in momentum (since velocity is more observable than momentum). Notice that the fluid velocity is the specific

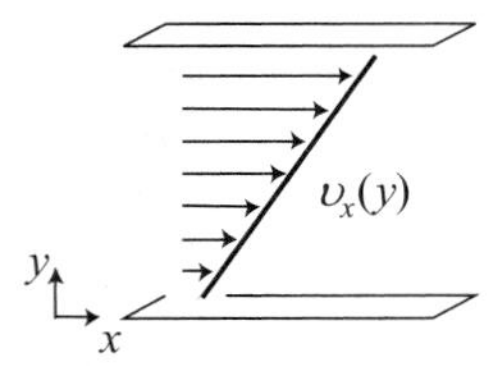

Figure 2-6 Couette flow.

*This type of flow is named in honor of the French physicist Maurice Marie Alfred Couette (1858–1943), who used this geometry of flow to measure viscosity.

momentum per unit mass. Therefore, for the x-component of momentum diffusing in the y-direction, one can write

$$M_{yx} = -\mu \frac{\partial v_x}{\partial y} \quad \text{(unidirectional flow)}, \tag{2-14}$$

where $\mu = \rho\nu$ is the fluid viscosity. In the literature, the momentum diffusivity ν is also referred to as the *kinematic viscosity* and μ is referred to as *dynamic viscosity.* There are nine combinations of momentum diffusion that can be written for the three spatial directions and three momentum components. The subscript on the velocity indicates the momentum component, and the spatial derivative relates to the direction of diffusion transport.

The linear relation between the momentum diffusion flux and the velocity gradient is referred to as *Newton's viscosity law* [3]. Fluids that obey this law are referred to as Newtonian. Newton's observations of this law were made in the context of stresses exerted within the flow, rather than momentum fluxes. However, it is understood that the momentum flux resulting from diffusion creates a corresponding *shear stress* τ (force per unit area) in the flow. Since this stress is a reaction to the momentum flux, there is a difference in sign between the two quantities. For example, $M_{yx} = -\tau_{yx}$, where τ_{yx} is the force in the x-direction acting on a unit surface area normal to the y-direction. By definition, shear stresses exclude any normal stress components associated with pressure. As will be discussed in Chapter 5, pressure can impart momentum to the flow as well, but not through diffusion.

It is not trivial to generalize Newton's viscosity law to a nonunidirectional flow. A property of the law that one might not initially recognize is that the momentum flux tensor M_{ji} must be symmetric, requiring that

$$M_{yx} = -\mu\left(\frac{\partial v_x}{\partial y} + \frac{\partial v_y}{\partial x}\right). \tag{2-15}$$

Interchanging appearances of x with y and v_x with v_y demonstrates that symmetry. For the unidirectional flow with $v_y = 0$, the momentum flux $M_{yx} = -\mu\, \partial v_x/\partial y$ is the same as before. However, symmetry of M_{ji} dictates a momentum flux M_{xy} in the x-direction in addition to the M_{yx} flux in the y-direction, as illustrated in Figure 2-7. If this symmetry did not exist, a fluid element in the flow would experience an unbalanced force moment, giving rise to unbounded angular momentum. Although this is clearly unacceptable, it is not immediately apparent what physical effect is responsible for the required momentum diffusion flux in the streamwise direction of Couette flow, given that there is no momentum gradient in that direction. An explanation is offered by the observation that the streamwise direction diffusion occurs because of the distortion of the unit cell experiencing shear, as illustrated in Figure 2-7. Surfaces of two adjacent unit cells must slide relative to one another to accommodate this distortion. The relative motion of these

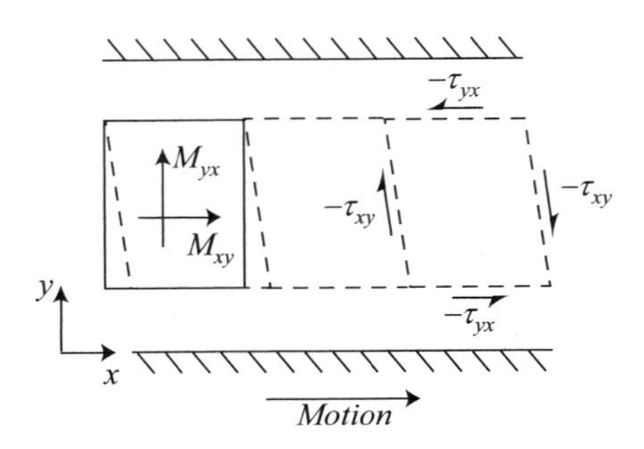

Figure 2-7 Momentum diffusion fluxes and corresponding shear stresses.

adjacent surfaces continuously brings fluid surfaces of different streamwise momentum into contact. Therefore, streamwise momentum must be continuously transported parallel to the streamwise direction to keep the flow unidirectional (i.e., to keep $\partial v_x/\partial x = 0$).

When density is not constant, Newton's viscosity law can be generalized to provide the momentum flux tensor:

$$\vec{\vec{M}} = -\mu[\vec{\nabla}\vec{v} + (\vec{\nabla}\vec{v})^t] + [(2/3)\mu - \kappa](\vec{\nabla} \cdot \vec{v})\vec{\vec{I}}. \quad (2\text{-}16)$$

The notation $()^t$ is used to express the transpose of a matrix, and $\vec{\vec{I}}$ is the identity matrix. The first term in Newton's viscosity law (2-16) is the contribution to the momentum flux tensor from the behavior of a constant density flow. Added to this is the momentum diffusion flux associated with expansion of the flow, as related to the divergence of the velocity field $\vec{\nabla} \cdot \vec{v}$. This term only contributes to the diagonal terms of the momentum flux tensor, as enforced with the identity matrix $\vec{\vec{I}}$. The coefficient to this term includes the dilatational viscosity κ, which is zero for monatomic gases at low densities. Development of Newton's viscosity law from first principles can be found in reference [4].

2.3.3 Species Diffusion

Consider a species A of molar concentration c_A. As discussed in Section 2.2.2, the molar diffusion flux of A can be defined by the difference between total species flux $c_A\vec{v}_A$ (as related to the velocity $\vec{v}_A$ of the species A) and the advective flux $c_A\vec{v}^*$ (as related to the molar-averaged velocity of the flow $\vec{v}^*$). In other words, the molar diffusion flux is given by $c_A(\vec{v}_A - \vec{v}^*)$. Analogously, the mass diffusion flux is given by $\rho_A(\vec{v}_A - \vec{v})$, where ρ_A is the mass density of species A and $\vec{v}$ is the mass-averaged velocity of the flow.

For simplicity, consider diffusion in a binary fluid comprised of species A and B. Based on earlier experience with diffusion laws, the molar diffusion flux of species A might be expressed in terms of the gradient in concentration as $-\text{Đ}^{\chi}_{AB}\vec{\nabla}c_A$ or the gradient in mole fraction as $-c\,\text{Đ}^{\chi}_{AB}\vec{\nabla}\chi_A$, where Đ^{χ}_{AB} is the A species molar diffusivity through B and $\chi_A = c_A/c$ is the molar fraction of A. These two expressions do not lead to equivalent definitions for Đ^{χ}_{AB} unless the total molar concentration $c = c_A + c_B$ is constant. To define the species diffusivity, the latter suggestion is adopted, such that the molar diffusion flux would be written as

$$c_A(\vec{v}_A - \vec{v}^*) = -\,c\text{Đ}^{\chi}_{AB}\vec{\nabla}\chi_A. \quad (2\text{-}17)$$

By analogy, the mass diffusion flux law is written as

$$\rho_A(\vec{v}_A - \vec{v}) = -\rho\,\text{Đ}^{\omega}_{AB}\vec{\nabla}\omega_A, \quad (2\text{-}18)$$

where Đ^{ω}_{AB} is the A species mass diffusivity through B and $\omega_A = \rho_A/\rho$. The wisdom of defining the diffusion laws in this manner is discovered through the observation that they result in species diffusivities Đ^{χ}_{AB} and Đ^{ω}_{AB} that are equivalent. This is demonstrated in Problem 2-3, by showing that

$$\frac{\chi_A(\vec{v}_A - \vec{v}^*)}{\vec{\nabla}\chi_A} = \text{Đ}^{\chi}_{AB} = \text{Đ}^{\omega}_{AB} = \frac{\omega_A(\vec{v}_A - \vec{v})}{\vec{\nabla}\omega_A}. \quad (2\text{-}19)$$

Dropping the notational distinction between molar and mass diffusivities, the diffusion laws for species transport are written:

$$\vec{j}_A^* = c_A(\vec{v}_A - \vec{v}^*) = -c\, Ð_{AB} \vec{\nabla} \chi_A \quad \text{(molar flux)} \tag{2-20}$$

and

$$\vec{j}_A = \rho_A(\vec{v}_A - \vec{v}) = -\rho\, Ð_{AB} \vec{\nabla} \omega_A \quad \text{(mass flux)}. \tag{2-21}$$

The linear relation between the diffusion flux and the molar or mass concentration gradient is known as *Fick's law* [5] of diffusion.

In a binary system, an expression for the species B diffusion flux can also be written. In terms of the mass flux, the diffusion law is

$$\vec{j}_B = \rho_B(\vec{v}_B - \vec{v}) = -\rho Ð_{BA} \vec{\nabla} \omega_B, \tag{2-22}$$

where $Ð_{BA}$ is the B species mass diffusivity through A and $\omega_B = \rho_B/\rho$. It is not immediately apparent how the diffusivity $Ð_{AB}$ of species A may be related to the diffusivity $Ð_{BA}$ of species B. However, by summing the diffusion laws,

$$\vec{j}_A + \vec{j}_B = \rho_A(\vec{v}_A - \vec{v}) + \rho_B(\vec{v}_B - \vec{v}) = -\rho\, Ð_{AB} \vec{\nabla} \omega_A - \rho Ð_{BA} \vec{\nabla} \omega_B, \tag{2-23}$$

it is revealed, after use of the mass fraction relation $\omega_A + \omega_B = 1$, that

$$\begin{aligned} \overbrace{\rho_A \vec{v}_A + \rho_B \vec{v}_B - (\rho_B + \rho_A)\vec{v}}^{=0} &= -\rho\Big(Ð_{AB} \vec{\nabla} \omega_A + Ð_{BA} \vec{\nabla}(1 - \omega_A)\Big) \\ &= -\rho(Ð_{AB} - Ð_{BA}) \vec{\nabla} \omega_A. \end{aligned} \tag{2-24}$$

Since the definition of the mass-averaged velocity (2-5) forces the sum of the diffusion fluxes appearing on the left-hand side of Eq. (2-24) to be zero, it is concluded that the diffusivities for species A and B must be the same:

$$Ð_{AB} = Ð_{BA}. \tag{2-25}$$

It is useful to observe that for the lower limits of species concentrations, Fick's diffusion laws take the form

$$\vec{j}_A = -\rho\, Ð_A \vec{\nabla}(\rho_A/\rho) \approx -\, Ð_A \vec{\nabla} \rho_A \quad (\rho_A \ll \rho) \tag{2-26}$$

$$\vec{j}_A^* = -c\, Ð_A \vec{\nabla}(c_A/c) \approx -\, Ð_A \vec{\nabla} c_A \quad (c_A \ll c) \tag{2-27}$$

since the total concentration (c) or density (ρ) values of the fluid remain virtually unchanged by the presence of small amounts of species A. In this case, there is no distinction between the molar-averaged and mass-averaged velocities of the fluid: $\vec{v}^* \approx \vec{v}$. This dilute species approximation will arise in many problems of interest in later chapters.

2.3.4 Summary of Diffusion Laws

Table 2-2 summarizes the form of the different diffusion laws. Notice that the species $\vec{j}_A$ and heat $\vec{q}$ diffusion laws describe fluxes of scalar quantities, but the momentum $\vec{\vec{M}}$ diffusion law describes a flux of a vector quantity. The vector nature of $\vec{j}_A$ and $\vec{q}$ arises from the

Table 2-2 Diffusion fluxes

Fourier's law (for heat flux)

$$\vec{q} = -k\vec{\nabla}T \ (\rho = const. \text{ or } \rho \neq const.)$$

Fick's law (for species flux)

Mass flux: $\vec{j}_A = -Đ_A\vec{\nabla}\rho_A \ (\rho = const.)$ or $\vec{j}_A = -\rho\, Đ_A\vec{\nabla}\omega_A \ (\rho \neq const.)$

Molar flux: $\vec{J}_A^* = -Đ_A\vec{\nabla}c_A \ (c = const.)$ or $\vec{J}_A^* = -c\, Đ_A\vec{\nabla}\chi_A \ (c \neq const.)$

Newton's viscosity law (for momentum flux) ($\vec{\vec{M}} = -\vec{\vec{\tau}}$)

$$\vec{\vec{M}} = -\mu(\vec{\nabla}\vec{v} + (\vec{\nabla}\vec{v})^t) \ (\rho = const.) \text{ or } \vec{\vec{M}} = -\mu\left[\vec{\nabla}\vec{v} + (\vec{\nabla}\vec{v})^t - \frac{2}{3}(\vec{\nabla}\cdot\vec{v})\vec{\vec{I}}\right] (\rho \neq const.\ \kappa = 0)$$

direction of transport, and the tensor nature of $\vec{\vec{M}}$ arises from the direction of transport in combination with the vector nature of momentum.

The diffusion fluxes given in Table 2-2 are examples of *constitutive laws*. Even though they attest to physical relations, they cannot be considered undisputable in all contexts. Although the linear diffusion relations adequately describe an abundance of common situations, they can prove inadequate when molecular interactions are complex. For example, consider what might happen if long polymeric chains were the agents of diffusion instead of particles. It seems reasonable that these chains would interact over many different-length scales in a way that would defeat the simple linear behavior of interactions that occur with only one characteristic-length scale (such as the mean distance between molecular collisions). Therefore, even though many problems solved in transport use linear diffusion laws satisfactorily, one should be aware that there is no assured truth of linear behavior in every circumstance.

2.4 REVERSIBLE VS. IRREVERSIBLE TRANSPORT

A distinguishing thermodynamic feature between the two kinds of transport discussed in this chapter is that advection transport is reversible while diffusion transport is not. Recall that a process is defined to be reversible when no entropy is created, as discussed in Section 1.1.

To illustrate the irreversibility of heat diffusion, consider a plane wall whose front surface is held at T_1 and back surface is held at $T_2 < T_1$, as illustrated in Figure 2-8. Heat passes through the wall at a constant rate $Q = Aq = -Ak\partial T/\partial x$ by diffusion. A certain amount of entropy can be transported reversibly with heat transfer, as given by $dS = \delta Q/T$. Therefore, the rate of reversible entropy transport across the front surface of the wall is $\dot{S}_1 = Q/T_1$, while the rate of reversible entropy transport across the back surface is $\dot{S}_2 = Q/T_2$. Due to the temperature difference, more entropy leaves the back surface than is introduced across the front surface ($\dot{S}_2 > \dot{S}_1$). Consequently, the transport of heat by diffusion within the wall must be creating entropy at a rate of

$$\dot{S}_{gen} = \dot{S}_2 - \dot{S}_1 = Q(1/T_2 - 1/T_1). \tag{2-28}$$

Heat transfer through the wall without entropy generation requires that $T_1 = T_2$. However, without a temperature gradient, heat diffusion does not occur. A flow of heat from the cold side to the hot side would require entropy destruction, which is prohibited by the second law of thermodynamics. Consequently, one concludes that diffusion transport of heat is irreversible. By extension, other forms of diffusion transport are also irreversible,

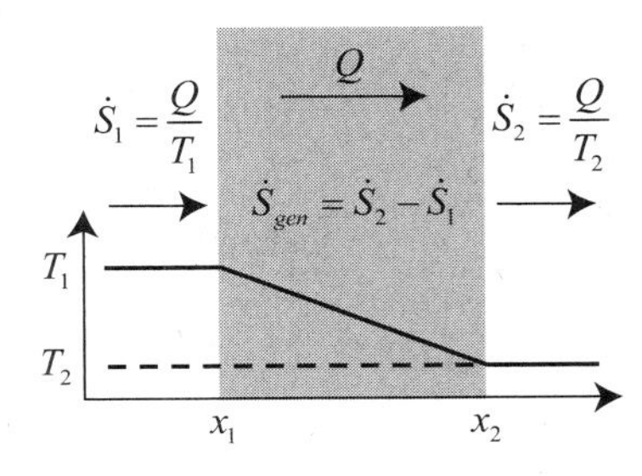

Figure 2-8 Entropy generation during heat transfer through a wall.

as will be demonstrated for species transport in Problem 2.4 and momentum diffusion in Section 5.7.

In contrast, transport by advection is reversible. The advection flux of entropy is given by $(\rho s)\vec{v}$, where s is the entropy of the fluid on a per-unit-mass basis. No increase in disorder occurs within a system through simple translation of fluid elements. Since no entropy generation occurs, the transport of entropy can be reversed by reversal of the flow velocity direction. Once a flow is identified as reversible, thermodynamic equations can be employed as useful constraints when solving transport equations. This approach is discussed in Chapter 20 for *inviscid* compressible flows, since inviscid flows lack all the diffusion terms associated with irreversible transport.

2.5 LOOKING AHEAD

Conservation equations for conserved properties will be formulated in Chapters 4 and 5. These statements will be expressed in a differential form as transport equations ready for application to specific problems. It will be seen that most of the equations of transport share a similar mathematical form. In general, four kinds of terms appear in such equations: one associated with transient change, one associated with advection transport, one associated with diffusion transport, and one or more associated with "sources" or "sinks" of transported properties. However, before developing the transport equations, a brief detour into index notation is taken in Chapter 3. This notation will be an immense aid in keeping straight the meaning of vector and tensor terms used in transport equations.

2.6 PROBLEMS

2-1 In addition to internal energy, fluids carry kinetic energy. Hypothesize expressions for the advection and diffusion fluxes of kinetic energy in a unidirectional flow (such as the Couette flow illustrated, where $v_x(y)$ and $v_y = 0$). In what directions do advection and diffusion transport kinetic energy? How is the diffusion flux of kinetic energy related to the diffusion flux of momentum? Show that the diffusion flux of kinetic energy corresponds dimensionally to the rate of work performed on a unit area of the fluid.

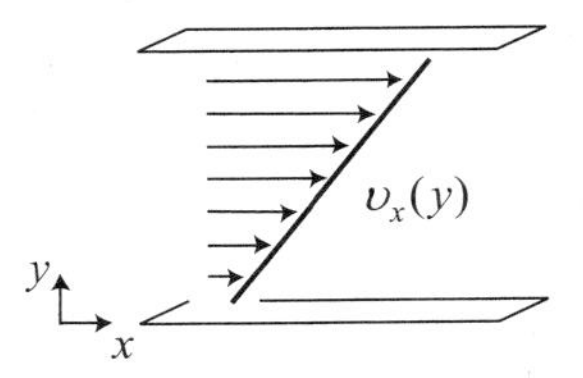

2-2 Express the advection of heat downstream in an incompressible pipe flow in terms of T_m and v_m, where $v_m = \int_A v_z dA/A$ and $T_m = \int_A v_z T dA/(v_m A)$. Balance the change in advected heat with the heat lost by convection to the wall: $q_{conv.} = h(T_m - T_s)$. Find an expression for dT_m/dz. Assuming that h is not a function of z, solve for $T_m(z)$ using $T_m(0) = T_m(z = 0)$ as an initial condition. Express the change in mean temperature as a function of the dimensionless quantities:

$$\mathrm{Nu}_D = \frac{h(2R)}{k} \quad \text{(Nusselt number)}, \qquad \mathrm{Pr} = \frac{\nu}{\alpha} \quad \text{(Prandtl number)}$$

and

$$\mathrm{Re}_D = \frac{v_m(2R)}{\nu} \quad \text{(Reynolds number)}.$$

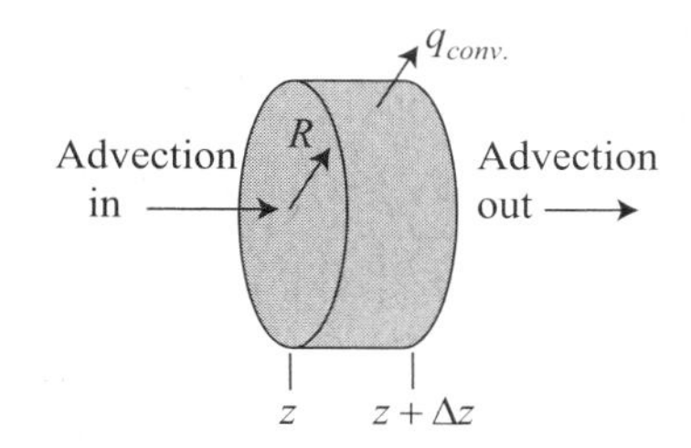

2-3 Show that the diffusion laws $c_A(\vec{v}_A - \vec{v}^*) = -c\, Ð^{\chi}_{AB}\vec{\nabla}\chi_A$ and $\rho_A(\vec{v}_A - \vec{v}) = -\rho\, Ð^{\omega}_{AB}\vec{\nabla}\omega_A$ for a binary system lead to equivalent definitions for the species diffusivities $Ð^{\chi}_{AB}$ and $Ð^{\omega}_{AB}$.

2-4 Consider a container of two ideal gases, at the same temperature and pressure, separated by a partition. When the partition is removed, a concentration gradient is established in the container that drives diffusion transport. Eventually, a homogeneous mixture of the two gases is formed. Demonstrate that the diffusion transport occurring in this process is not reversible.

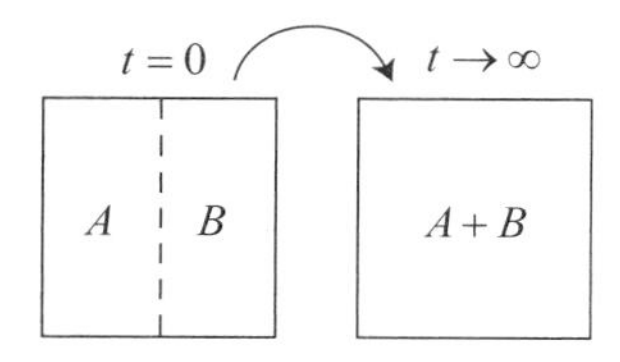

REFERENCES

[1] T. Gombosi, *Gaskinetic Theory*. Cambridge: Cambridge University Press, 1994.
[2] J. B. J. Fourier, *The Analytical Theory of Heat*, translated by Alexander Freeman. Cambridge: Cambridge University Press, 2011.
[3] I. Newton, *Philosophiae Naturalis Principia Mathematica*, First Edition, 1687 (Project Gutenberg EBook, 2009).
[4] G. K. Batchelor, *An Introduction to Fluid Dynamics*. Cambridge: Cambridge University Press, 1967.
[5] A. Fick, "On Liquid Diffusion," *Philosophical Magazine and Journal of Science*, **10**, 31 (1855).

Chapter 3

Index Notation

3.1 Indices
3.2 Representation of Cartesian Differential Equations
3.3 Special Operators
3.4 Operators in Non-Cartesian Coordinates
3.5 Problems

This chapter contains a cursory overview of the use of index notation. Albert Einstein [1] introduced an index-based notation that has become widely used in the physical sciences. The notation provides a simple way to keep track of relations between vector quantities. It also provides a compact form for writing generalized equations that expand into multiple spatial directions. Manipulating conservation equations without such a notation is quite laborious.

3.1 INDICES

Index notation is an alternative to symbolic notation for writing vector and tensor expressions compactly. Index notation, also known as *Cartesian notation*, provides clarity in vector and tensor arithmetic. To illustrate the use of index notation, consider the system of equations that results from the product between a second-order tensor $\vec{\vec{B}}$ and a vector $\vec{C}$, as symbolically expressed by $\vec{A} = \vec{\vec{B}} \cdot \vec{C}$. Following the rules of matrix algebra, the product

$$\begin{pmatrix} a_x \\ a_y \\ a_z \end{pmatrix} = \begin{pmatrix} b_{xx} & b_{xy} & b_{xz} \\ b_{yx} & b_{yy} & b_{yz} \\ b_{zx} & b_{zy} & b_{zz} \end{pmatrix} \begin{pmatrix} c_x \\ c_y \\ c_z \end{pmatrix} \tag{3-1}$$

must satisfy three equations:

$$\begin{aligned} a_x &= b_{xx}c_x + b_{xy}c_y + b_{xz}c_z \\ a_y &= b_{yx}c_x + b_{yy}c_y + b_{yz}c_z\,. \\ a_z &= b_{zx}c_x + b_{zy}c_y + b_{zz}c_z \end{aligned} \tag{3-2}$$

In index notation, multiple equations can be expressed with a *free index*, such that the above system of equations (3-2) can be written as

$$a_i = b_{xi}c_x + b_{yi}c_y + b_{zi}c_z, \quad \text{for} \quad i = x,\ y \text{ or } z. \tag{3-3}$$

The free index, "i," indicates that expressions like (3-3) actually represent several equations, one for each value that the free index can take.

If an index appears twice in a term, it is called a *dummy index* and represents a summation over all possible values of that index. Consequently, with a dummy index, the sum appearing in expression (3-3) can be written more compactly as

$$a_i = b_{ji} c_j. \tag{3-4}$$

In this expression, the j-index is recognized as a dummy index because of its appearance twice in the $b_{ji}c_j$ term. Because j is a dummy index, the term $b_{ji}c_j$ actually represents a sum of terms over each value that the j-index can take. An index can never occur more than twice in the same term. For example, $b_{jj}c_j$ has no clear meaning and is not allowed.

Index notation is often superior to symbolic notation in clarity of presentation. When a confusing operation appears in symbolic notation, it is common for a new operator symbol to be defined to preserve clarity. Consider the symbolic expression $\vec{\nabla} \cdot \vec{\nabla} \vec{a}$, which contains two operators and one variable, all having some vector sense. There is a dot product between two of the vector senses, but which two? In all likelihood the dot product is between the vector senses of the two gradient operators, in which case this expression is more clearly written in symbolic notation as $\nabla^2 \vec{a}$. In contrast, this term would be written as $\partial_j \partial_j a_i$ using index notation. Here there is no confusion over the meaning, since the dummy index j is used to establish which two vector senses are involved in the dot product. As another example, consider the expression $\partial_i a_j \partial_i a_j$, which is a scalar term resulting from a double dot product. In index notation it is clear how each vector sense is paired in the dot products. To approach the same clarity in symbolic notation, a new symbol is required: $\vec{\nabla} \vec{a} : \vec{\nabla} \vec{a}$.

3.2 REPRESENTATION OF CARTESIAN DIFFERENTIAL EQUATIONS

The Navier-Stokes equations,* which will be derived in Chapter 5, enforce conservation of momentum in a fluid. In a form suitable for incompressible flow, with constant viscosity, the Navier-Stokes equations can be written in index notation as

$$\partial_o v_i + v_j \partial_j v_i = \nu\, \partial_j \partial_j v_i - \frac{1}{\rho} \partial_i P + g_i. \tag{3-5}$$

The time derivative is denoted by ∂_o and spatial derivatives are written as ∂_i or ∂_j (where the indices "i" and "j" refer to the spatial coordinates), such that

$$\partial_o(\ ) = \frac{\partial(\ )}{\partial t}, \tag{3-6}$$

and with $i = x, y$ or z:

$$\partial_i(\ ) = \frac{\partial(\ )}{\partial x}, \frac{\partial(\ )}{\partial y} \quad \text{or} \quad \frac{\partial(\ )}{\partial z} \quad \text{(and is the same for } j\text{)}. \tag{3-7}$$

Notice that there is one free index i in the Navier-Stokes equations. Therefore this is an equation that is true for all possible values of the i-index (i.e., x, y, and z). In other words, the Navier-Stokes equations are comprised of three equations, one for each spatial

---

*The equations are named in honor of the French fluid dynamicist Claude-Louis Navier (1785–1836) and the Cambridge mathematician and physicist George Gabriel Stokes (1819–1903).

direction. Notice also that several of the terms have a dummy index j. Each of these terms is an abbreviation for the sum of three terms over all values of the j-index (x, y, and z).

To expand the Navier-Stokes equations, consider the equation corresponding to the x-direction:

$$\partial_o v_x + v_j \partial_j v_x = \nu \partial_j \partial_j v_x - \frac{1}{\rho} \partial_x P + g_x, \tag{3-8}$$

where each occurrence of the i-index has been replaced with "x". Next, expanding each term in which the dummy index j appears, gives

$$\frac{\partial v_x}{\partial t} + v_x \frac{\partial v_x}{\partial x} + v_y \frac{\partial v_x}{\partial y} + v_z \frac{\partial v_x}{\partial z} = \nu \left(\frac{\partial^2 v_x}{\partial x^2} + \frac{\partial^2 v_x}{\partial y^2} + \frac{\partial^2 v_x}{\partial z^2} \right) - \frac{1}{\rho} \frac{\partial P}{\partial x} + g_x, \tag{3-9}$$

where ∂_o has been replaced with the explicit form of the time derivative. This is the x-direction momentum equation, expressed in Cartesian coordinates. The other two equations for the y- and z-components of momentum can be expanded in a similar fashion to obtain the complete set of Navier-Stokes equations.

3.3 SPECIAL OPERATORS

Operators are required to describe some vector and tensor actions. For example, ∂_j is an operator indicating that a derivative is taken with respect to the j-index. Other special operators in index notation are the surface normal n_j, the Kronecker delta δ_{ji}, and alternating unit tensor ε_{ijk}.

3.3.1 Surface Normal Operator

The surface normal operator n_j is evaluated based on the local orientation of the surface being described. Numerically, it evaluates to the dot product between the surface unit normal vector and the unit direction vector specified by the j-index. Figure 3-1 illustrates how the surface normal operator is evaluated over the segments of a particular surface.

The surface normal operator has utility in expressing fluxes across control surfaces. For example, the total heat transfer rate of diffusion across a control surface A would be expressed as

$$Q = \int_A (-k \partial_j T) n_j dA. \tag{3-10}$$

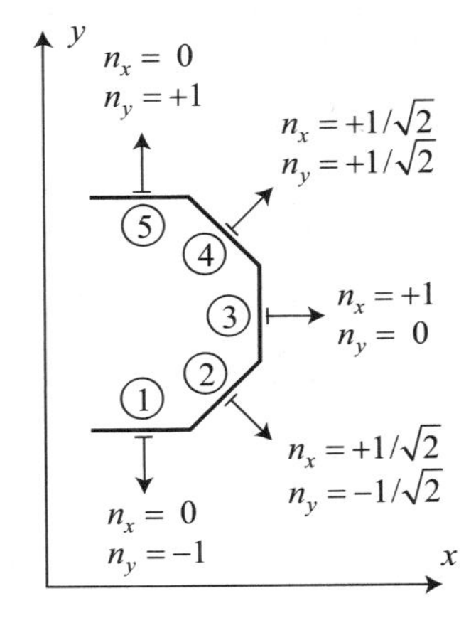

Figure 3-1 Example showing evaluation of the surface normal operator n_i.

Notice that the spatial gradient operator gives the vector sense for the diffusion heat flux $-k\partial_j T$, and the surface normal operator gives the vector sense needed for the area integral with respect to $n_j dA$. For the segmented surface illustrated in Figure 3-1, this expression expands in two-dimensions, with $j = x$ and y, to yield

$$\begin{aligned} Q = & \int_{A_1} \left[-k\partial_x T \overbrace{n_x}^{0} - k\partial_y T \overbrace{n_y}^{-1} \right] dA + \int_{A_2} \left[-k\partial_x T \overbrace{n_x}^{+1/\sqrt{2}} - k\partial_y T \overbrace{n_y}^{-1/\sqrt{2}} \right] dA \\ & + \int_{A_3} \left[-k\partial_x T \overbrace{n_x}^{+1} - k\partial_y T \overbrace{n_y}^{0} \right] dA + \int_{A_4} \left[-k\partial_x T \overbrace{n_x}^{+1/\sqrt{2}} - k\partial_y T \overbrace{n_y}^{+1/\sqrt{2}} \right] dA \\ & + \int_{A_5} \left[-k\partial_x T \overbrace{n_x}^{0} - k\partial_y T \overbrace{n_y}^{+1} \right] dA \end{aligned} \quad (3\text{-}11)$$

The final result simplifies to

$$\begin{aligned} Q = & \int_{A_1} [+k\partial_y T] dA + \int_{A_2} [-k\partial_x T + k\partial_y T] dA/\sqrt{2} + \int_{A_3} [-k\partial_x T] dA \\ & + \int_{A_4} [-k\partial_x T - k\partial_y T] dA/\sqrt{2} + \int_{A_5} [-k\partial_y T] dA. \end{aligned} \quad (3\text{-}12)$$

3.3.2 Kronecker Delta Operator

The *Kronecker delta* operator[†] evaluates to either 0 or 1 by the rule

$$\delta_{ji} = \begin{cases} 1 & \text{if } i = j \\ 0 & \text{if } i \neq j \end{cases} \quad (3\text{-}13)$$

One utility of the Kronecker delta is illustrated in the mathematical description of hydrostatic pressure. The application of hydrostatic pressure to a surface is described by $-P\delta_{ji} n_j$, where P is the pressure magnitude and n_j is the surface normal operator. The operator δ_{ji} ensures that the impose force (whose direction is given by the i-index) can act only in the direction of the surface unit normal (whose direction is given by the j-index). The negative sign $(-P)$ is needed because pressure acts in the opposite direction to that given by the surface unit normal.

To further illustrate, consider the force of pressure acting on the surface illustrated in Figure 3-1. The force is calculated by integrating the effect of pressure over the surface area:

$$F_i = \int_A (-P\delta_{ji}) n_j dA. \quad (3\text{-}14)$$

Suppose the x-direction component of this force is to be evaluated. With $i = x$, the force expression expands in two dimensions with $j = x$ and y, to yield

---

[†]This operator is named in honor of the German mathematician Leopold Kronecker (1823–1891).

$$F_x = \int_A \left[-P \overbrace{\delta_{xx}}^{1} n_x - P \overbrace{\delta_{yx}}^{0} n_y \right] dA = \int_A [-P n_x] dA. \tag{3.15}$$

For the segmented surface illustrated in Figure 3-1, the area integral is evaluated as

$$\begin{aligned} F_x = &\int_{A_1} \left[-P \overbrace{n_x}^{0} \right] dA + \int_{A_2} \left[-P \overbrace{n_x}^{+1/\sqrt{2}} \right] dA \\ &+ \int_{A_3} \left[-P \overbrace{n_x}^{+1} \right] dA + \int_{A_4} \left[-P \overbrace{n_x}^{+1/\sqrt{2}} \right] dA + \int_{A_5} \left[-P \overbrace{n_x}^{0} \right] dA. \end{aligned} \tag{3-16}$$

The final result simplifies to

$$F_x = \int_{A_2} [-P] dA/\sqrt{2} + \int_{A_3} [-P] dA + \int_{A_4} [-P] dA/\sqrt{2}. \tag{3-17}$$

It is noteworthy that the hydrostatic pressure contributes only three components to the state of stress in a fluid, since the force exerted by pressure can only act in parallel (and opposite) to the unit normal direction of a surface. In general, the state of stress in a fluid is described by a *stress tensor* σ_{ji} that has nine components associated with the three possible unit normal directions and the three possible spatial directions of a force. The difference between the total-stress tensor σ_{ji} and contributions coming from the pressure $P\delta_{ji}$ describes the state of the shear stress in the fluid τ_{ji}. Figure 3-2 illustrates the decomposition of the total stress tensor into contributions from shear stress and pressure. In symbolic notation, this relation is described by

$$\vec{\vec{\sigma}} = \vec{\vec{\tau}} - P\vec{\vec{I}} = \begin{pmatrix} \tau_{xx} & \tau_{xy} & \tau_{xz} \\ \tau_{yx} & \tau_{yy} & \tau_{yz} \\ \tau_{zx} & \tau_{zy} & \tau_{zz} \end{pmatrix} - \begin{pmatrix} P & 0 & 0 \\ 0 & P & 0 \\ 0 & 0 & P \end{pmatrix} \tag{3-18}$$

where $\vec{\vec{I}}$ is the identity matrix. In index notation this expression is written as

$$\sigma_{ji} = \tau_{ji} - P\delta_{ji}, \tag{3-19}$$

where the Kronecker delta operator serves the purpose of the identity matrix. As discussed in Section 2.3.2, the shear stress in the fluid is a direct reaction to the momentum fluxes resulting from diffusion—that is $\tau_{ji} = -M_{ji}$.

Unfortunately, no consensus exists as to the ordering of subscripts on tensor terms, where the indices are used to identify both a surface normal direction and a momentum

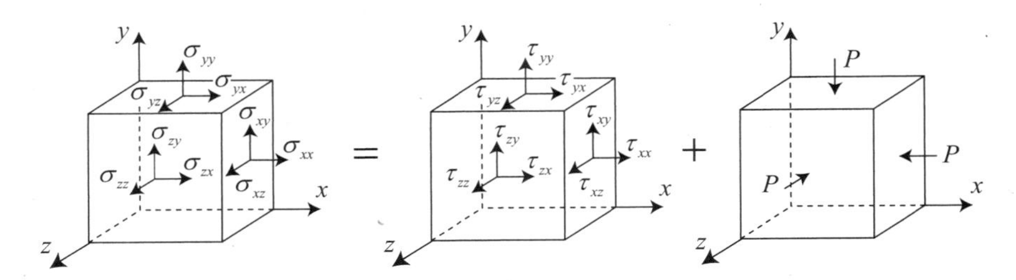

Figure 3-2 Decomposition of the total stress tensor into shear stress and pressure.

flux or stress component. In this text, a convention common to engineering disciplines is followed. In this convention, the first index is the surface unit normal and the second is the momentum flux or stress component:

$$M_{ji} \leftarrow \text{momentum component}, \ j \leftarrow \text{surface normal direction} \qquad \tau_{ji} \leftarrow \text{force component}, \ j \leftarrow \text{surface normal direction} \tag{3-20}$$

3.3.3 Alternating Unit Tensor Operator

The *alternating unit tensor*, also known as the *Levi-Civita epsilon*,[‡] evaluates to either +1, −1, or 0 by the rule:

$$\varepsilon_{ijk} = \begin{cases} +1 & \text{if } i,\ j,\ k \ \text{ in forward order,} \quad \text{such as } x,\ y,\ z;\ \ z,\ x,\ y;\ \ y,\ z,\ x \\ -1 & \text{if } i,\ j,\ k \ \text{ in backward order,} \quad \text{such as } z,\ y,\ x;\ \ x,\ z,\ y;\ \ y,\ x,\ z \\ 0 & \text{otherwise} \end{cases} \tag{3-21}$$

The alternating unit tensor permits the cross product $\vec{a} \times \vec{b}$ to be expressed in index notation as $\varepsilon_{ijk} a_j b_k$. This will be useful when describing the *vorticity* $\vec{\omega}$ of a flow, which is twice the local angular velocity of a fluid element, and is defined mathematically by

$$\vec{\omega} = \vec{\nabla} \times \vec{v} \quad \text{or} \quad \omega_i = \varepsilon_{ijk} \partial_j v_k. \tag{3-22}$$

3.3.4 Proof of a Vector Identity

To illustrate the brute force application of index notation, the expression $\vec{\nabla} \times (\vec{\nabla} \cdot a)$, where a is a scalar, is evaluated to demonstrate the well-known vector identity: $\vec{\nabla} \times (\vec{\nabla} \cdot a) = 0$. In index notation $\vec{\nabla} \times (\vec{\nabla} \cdot a)$ is expressed as $\varepsilon_{ijk} \partial_j \partial_k a$. Noting that this expression has three components, $i = x, y$, and z, one can start by evaluating the case when $i = x$. Expanding $\varepsilon_{xjk} \partial_j \partial_k a$ with respect to the first dummy index j yields

$$\varepsilon_{xjk} \partial_j \partial_k a = \overbrace{\varepsilon_{xxk}}^{=0} \partial_x \partial_k a + \varepsilon_{xyk} \partial_y \partial_k a + \varepsilon_{xzk} \partial_z \partial_k a. \tag{3-23}$$

Notice that the first term in the expansion is zero by the rules of the alternating unit tensor, irrespective of what value the k-index becomes. Now expand the second dummy index k. For the expression $\varepsilon_{xyk} \partial_y \partial_k a$ one could write out three terms, but two of them will be zero. By the rules of the alternating unit tensor, the only nonzero term is when $k = z$. Likewise, for $\varepsilon_{xzk} \partial_z \partial_k a$, the only nonzero term arises when $k = y$. Therefore, expansion of the second dummy index k yields

$$\varepsilon_{xjk} \partial_j \partial_k a = \varepsilon_{xyz} \partial_y \partial_z a + \varepsilon_{xzy} \partial_z \partial_y a. \tag{3-24}$$

Finally, the rules of the alternating unit tensor are applied, $\varepsilon_{xyz} = +1$ and $\varepsilon_{xzy} = -1$, to determine the sign of the remaining terms. Because y and z are independent variables, $\partial_z \partial_y a = \partial_y \partial_z a$. Therefore, it is determined that

$$\varepsilon_{xjk} \partial_j \partial_k a = \partial_y \partial_z a - \partial_z \partial_y a = \partial_y \partial_z a - \partial_y \partial_z a = 0. \tag{3-25}$$

[‡]This operator is named in honor of the Italian mathematician and physicist Tullio Levi-Civita (1873–1941).

It is straightforward to show that for the remaining cases, when $i = y$ and z, the expression $\varepsilon_{ijk}\partial_j\partial_k a$ also evaluates to zero. Therefore one concludes that $\varepsilon_{ijk}\partial_j\partial_k a = 0$.

3.4 OPERATORS IN NON-CARTESIAN COORDINATES

Sometimes problems are formulated more simply in coordinate systems other than Cartesian. Frequently used alternatives are cylindrical and spherical coordinate systems, illustrated in Figure 3-3. Cylindrical and spherical coordinates are orthogonal curvilinear systems. Orthogonally requires that the three directional vectors are mutually perpendicular at any point. In curvilinear systems, variation of a single coordinate can follow a curved path, and differential distances along the path of one directional component may depend on the magnitude of another. For example, a differential distance in the θ-direction is $r\,d\theta$ for both cylindrical and spherical coordinate systems, and a differential distance in the ϕ-direction of spherical coordinates is $r \sin\theta\, d\phi$. For this reason, operators involving spatial derivatives may evaluate differently in different coordinate systems. Table 3-1 defines simple derivatives taken of a scalar quantity s, and the curl and divergence operators acting a vector quantity v_k, for Cartesian, cylindrical, and spherical coordinates.

The operators defined in Table 3-1 can be used to evaluate other operators, such as the *Laplacian*.§ The Laplacian of Φ can be evaluated by substituting $v_i = \partial_i\Phi$ into the expression for $\partial_i v_i$ yielding $\partial_i\partial_i\Phi$. Following these steps, the Laplace operator evaluates in the different coordinate systems to:

$$\text{Cartesian} \quad (i = x, y, z): \quad \partial_i\partial_i\Phi = \frac{\partial^2\Phi}{\partial x^2} + \frac{\partial^2\Phi}{\partial y^2} + \frac{\partial^2\Phi}{\partial z^2} \tag{3-26}$$

$$\text{Cylindrical} \quad (i = r, \theta, z): \quad \partial_i\partial_i\Phi = \frac{1}{r}\frac{\partial}{\partial r}\left(r\frac{\partial\Phi}{\partial r}\right) + \frac{1}{r^2}\frac{\partial^2\Phi}{\partial\theta^2} + \frac{\partial^2\Phi}{\partial z^2} \tag{3-27}$$

$$\text{Spherical} \quad (i = r, \theta, \phi): \quad \partial_i\partial_i\Phi = \frac{1}{r^2}\frac{\partial}{\partial r}\left(r^2\frac{\partial\Phi}{\partial r}\right) + \frac{1}{r^2 \sin\theta}\frac{\partial}{\partial\theta}\left(\sin\theta\frac{\partial\Phi}{\partial\theta}\right) + \frac{1}{r^2 \sin^2\theta}\frac{\partial^2\Phi}{\partial\phi^2}. \tag{3-28}$$

Two important operators omitted from Table 3-1 are $\partial_i v_j$, describing the gradient of a vector quantity, and $\partial_i\partial_i v_j$, describing the divergence of a vector quantity. The first operator

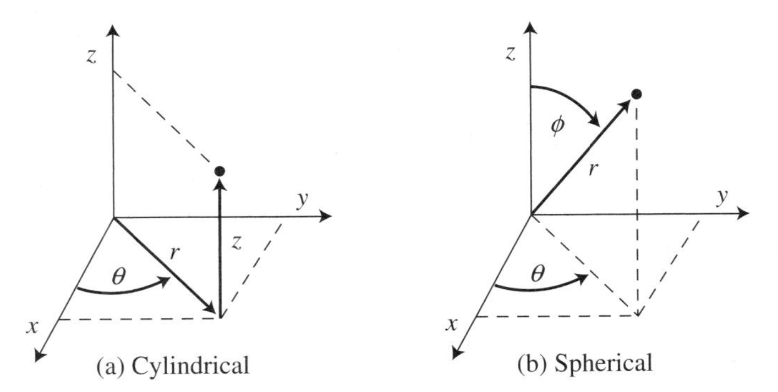

Figure 3-3 Alternative coordinate systems.

§This operator is named in honor of the French mathematician and astronomer Pierre-Simon, Marquis de Laplace (1749–1827).

Table 3-1 Common operators in alternative coordinate systems

(a) *Cartesian*	$i=x$	$i=y$	$i=z$
$\partial_i s =$	$\dfrac{\partial s}{\partial x}$	$\dfrac{\partial s}{\partial y}$	$\dfrac{\partial s}{\partial z}$
$\varepsilon_{ijk}\partial_j v_k =$	$\dfrac{\partial v_z}{\partial y}-\dfrac{\partial v_y}{\partial z}$	$\dfrac{\partial v_x}{\partial z}-\dfrac{\partial v_z}{\partial x}$	$\dfrac{\partial v_y}{\partial x}-\dfrac{\partial v_x}{\partial y}$
$\partial_j v_j =$	$\dfrac{\partial v_x}{\partial x}+\dfrac{\partial v_y}{\partial y}+\dfrac{\partial v_z}{\partial z}$		
(b) *Cylindrical*	$i=r$	$i=\theta$	$i=z$
$\partial_i s =$	$\dfrac{\partial s}{\partial r}$	$\dfrac{1}{r}\dfrac{\partial s}{\partial \theta}$	$\dfrac{\partial s}{\partial z}$
$\varepsilon_{ijk}\partial_j v_k =$	$\dfrac{1}{r}\dfrac{\partial v_z}{\partial \theta}-\dfrac{\partial v_\theta}{\partial z}$	$\dfrac{\partial v_r}{\partial z}-\dfrac{\partial v_z}{\partial r}$	$\dfrac{1}{r}\dfrac{\partial}{\partial r}(r\,v_\theta)-\dfrac{1}{r}\dfrac{\partial v_r}{\partial \theta}$
$\partial_j v_j =$	$\dfrac{1}{r}\dfrac{\partial}{\partial r}(r\,v_r)+\dfrac{1}{r}\dfrac{\partial v_\theta}{\partial \theta}+\dfrac{\partial v_z}{\partial z}$		
(c) *Spherical*	$i=r$	$i=\theta$	$i=\phi$
$\partial_i s =$	$\dfrac{\partial s}{\partial r}$	$\dfrac{1}{r}\dfrac{\partial s}{\partial \theta}$	$\dfrac{1}{r\sin\theta}\dfrac{\partial s}{\partial \phi}$
$\varepsilon_{ijk}\partial_j v_k =$	$\dfrac{1}{r\sin\theta}\dfrac{\partial}{\partial\theta}(\sin\theta\,v_\phi)-\dfrac{1}{r\sin\theta}\dfrac{\partial v_\theta}{\partial\phi}$	$\dfrac{1}{r\sin\theta}\dfrac{\partial v_r}{\partial\phi}-\dfrac{1}{r}\dfrac{\partial}{\partial r}(r v_\phi)$	$\dfrac{1}{r}\dfrac{\partial}{\partial r}(r v_\theta)-\dfrac{1}{r}\dfrac{\partial v_r}{\partial\theta}$
$\partial_j v_j =$	$\dfrac{1}{r^2}\dfrac{\partial}{\partial r}(r^2 v_r)+\dfrac{1}{r\sin\theta}\dfrac{\partial}{\partial\theta}(\sin\theta\,v_\theta)+\dfrac{1}{r\sin\theta}\dfrac{\partial v_\phi}{\partial\phi}$		

describes a tensor, whose components are given in Table 3-2. The second operator describes a vector, whose components are given in Table 3-3. The expressions $\partial_i v_j$ and $\partial_i\partial_i v_j$ cannot be evaluated correctly by treating v_j as a scalar quantity. For this reason, in index notation expressions like $v_j\partial_j T$ and $v_j\partial_j v_r$ evaluate to significantly different expressions in curvilinear systems. For example, in cylindrical coordinates ($j=r,\theta,z$),

$$v_j\partial_j T = v_r\frac{\partial T}{\partial r}+v_\theta\frac{1}{r}\frac{\partial T}{\partial\theta}+v_z\frac{\partial T}{\partial z} \tag{3-29}$$

but

$$v_j\partial_j v_r = v_r\frac{\partial v_r}{\partial r}+v_\theta\left(\frac{1}{r}\frac{\partial v_r}{\partial\theta}-\frac{v_\theta}{r}\right)+v_z\frac{\partial v_r}{\partial z}. \tag{3-30}$$

Notice the appearance of $v_\theta\, v_\theta/r$ in the expression for $v_j\partial_j v_r$ that has no equivalent term in the expression for $v_j\partial_j T$. Similarly, in index notation expressions like $\partial_j\partial_j T$ and $\partial_j\partial_j v_i$ are evaluated differently in curvilinear systems. For example, in cylindrical coordinates ($j=r,\theta,z$):

$$\partial_j\partial_j T = \frac{1}{r}\frac{\partial}{\partial r}\left(r\frac{\partial T}{\partial r}\right)+\frac{1}{r^2}\frac{\partial^2 T}{\partial\theta^2}+\frac{\partial^2 T}{\partial z^2} \tag{3-31}$$

Table 3-2 Components of $\partial_i v_j$ in alternative coordinates

(a) ***Cartesian***

$\partial_j v_i$	$j = x$	$j = y$	$j = z$
$i = x$	$\dfrac{\partial v_x}{\partial x}$	$\dfrac{\partial v_x}{\partial y}$	$\dfrac{\partial v_x}{\partial z}$
$i = y$	$\dfrac{\partial v_y}{\partial x}$	$\dfrac{\partial v_y}{\partial y}$	$\dfrac{\partial v_y}{\partial z}$
$i = z$	$\dfrac{\partial v_z}{\partial x}$	$\dfrac{\partial v_z}{\partial y}$	$\dfrac{\partial v_z}{\partial z}$

(b) ***Cylindrical***

$\partial_j v_i$	$j = r$	$j = \theta$	$j = z$
$i = r$	$\dfrac{\partial v_r}{\partial r}$	$\dfrac{1}{r}\dfrac{\partial v_r}{\partial \theta} - \dfrac{v_\theta}{r}$	$\dfrac{\partial v_r}{\partial z}$
$i = \theta$	$\dfrac{\partial v_\theta}{\partial r}$	$\dfrac{1}{r}\dfrac{\partial v_\theta}{\partial \theta} + \dfrac{v_r}{r}$	$\dfrac{\partial v_\theta}{\partial z}$
$i = z$	$\dfrac{\partial v_z}{\partial r}$	$\dfrac{1}{r}\dfrac{\partial v_z}{\partial \theta}$	$\dfrac{\partial v_z}{\partial z}$

(c) ***Spherical***

$\partial_j v_i$	$j = r$	$j = \theta$	$j = \phi$
$i = r$	$\dfrac{\partial v_r}{\partial r}$	$\dfrac{1}{r}\dfrac{\partial v_r}{\partial \theta} - \dfrac{v_\theta}{r}$	$\dfrac{1}{r \sin\theta}\dfrac{\partial v_r}{\partial \phi} - \dfrac{v_\phi}{r}$
$i = \theta$	$\dfrac{\partial v_\theta}{\partial r}$	$\dfrac{1}{r}\dfrac{\partial v_\theta}{\partial \theta} + \dfrac{v_r}{r}$	$\dfrac{1}{r \sin\theta}\dfrac{\partial v_\theta}{\partial \phi} - \dfrac{v_\phi}{r}\cot\theta$
$i = \phi$	$\dfrac{\partial v_\phi}{\partial r}$	$\dfrac{1}{r}\dfrac{\partial v_\phi}{\partial \theta}$	$\dfrac{1}{r \sin\theta}\dfrac{\partial v_\phi}{\partial \phi} + \dfrac{v_r}{r} + \dfrac{v_\theta}{r}\cot\theta$

Table 3-3 Components of $\partial_j \partial_j v_i$ in alternative coordinates

(a) ***Cartesian***

$$\partial_j \partial_j v_x = \frac{\partial^2 v_x}{\partial^2 x} + \frac{\partial^2 v_x}{\partial^2 y} + \frac{\partial^2 v_x}{\partial^2 z}$$

$$\partial_j \partial_j v_y = \frac{\partial^2 v_y}{\partial^2 x} + \frac{\partial^2 v_y}{\partial^2 y} + \frac{\partial^2 v_y}{\partial^2 z}$$

$$\partial_j \partial_j v_z = \frac{\partial^2 v_z}{\partial^2 x} + \frac{\partial^2 v_z}{\partial^2 y} + \frac{\partial^2 v_z}{\partial^2 z}$$

(b) ***Cylindrical***

$$\partial_j \partial_j v_r = \frac{\partial}{\partial r}\left(\frac{1}{r}\frac{\partial}{\partial r}(r v_r)\right) + \frac{1}{r^2}\frac{\partial^2 v_r}{\partial \theta^2} + \frac{\partial^2 v_r}{\partial z^2} - \frac{2}{r^2}\frac{\partial v_\theta}{\partial \theta}$$

$$\partial_j \partial_j v_\theta = \frac{\partial}{\partial r}\left(\frac{1}{r}\frac{\partial}{\partial r}(r v_\theta)\right) + \frac{1}{r^2}\frac{\partial^2 v_\theta}{\partial \theta^2} + \frac{\partial^2 v_\theta}{\partial z^2} + \frac{2}{r^2}\frac{\partial v_r}{\partial \theta}$$

$$\partial_j \partial_j v_z = \frac{1}{r}\frac{\partial}{\partial r}\left(r\frac{\partial v_z}{\partial r}\right) + \frac{1}{r^2}\frac{\partial^2 v_z}{\partial \theta^2} + \frac{\partial^2 v_z}{\partial z^2}$$

(c) ***Spherical***

$$\partial_j \partial_j v_r = \frac{\partial}{\partial r}\left(\frac{1}{r^2}\frac{\partial (r^2 v_r)}{\partial r}\right) + \frac{1}{r^2 \sin\theta}\frac{\partial}{\partial \theta}\left(\sin\theta \frac{\partial v_r}{\partial \theta}\right) + \frac{1}{r^2 \sin^2\theta}\frac{\partial^2 v_r}{\partial \phi^2} - \frac{2}{r^2 \sin\theta}\left(\frac{\partial(\sin\theta\, v_\theta)}{\partial \theta} + \frac{\partial v_\phi}{\partial \phi}\right)$$

$$\partial_j \partial_j v_\theta = \frac{1}{r^2}\frac{\partial}{\partial r}\left(r^2 \frac{\partial v_\theta}{\partial r}\right) + \frac{1}{r^2}\frac{\partial}{\partial \theta}\left(\frac{1}{\sin\theta}\frac{\partial}{\partial \theta}(\sin\theta\, v_\theta)\right) + \frac{1}{r^2 \sin^2\theta}\frac{\partial^2 v_\theta}{\partial \phi^2} + \frac{2}{r^2}\frac{\partial v_r}{\partial \theta} - \frac{2\cot\theta}{r^2 \sin\theta}\frac{\partial v_\phi}{\partial \phi}$$

$$\partial_j \partial_j v_\phi = \frac{1}{r^2}\frac{\partial}{\partial r}\left(r^2 \frac{\partial v_\phi}{\partial r}\right) + \frac{1}{r^2}\frac{\partial}{\partial \theta}\left(\frac{1}{\sin\theta}\frac{\partial}{\partial \theta}(\sin\theta\, v_\phi)\right) + \frac{1}{r^2 \sin^2\theta}\frac{\partial^2 v_\phi}{\partial \phi^2} + \frac{2}{r^2 \sin\theta}\left(\frac{\partial v_r}{\partial \phi} + \cot\theta \frac{\partial v_\theta}{\partial \phi}\right)$$

but

$$\partial_j \partial_j v_r = \frac{\partial}{\partial r}\left(\frac{1}{r}\frac{\partial}{\partial r}(r v_r)\right) + \frac{1}{r^2}\frac{\partial^2 v_r}{\partial \theta^2} + \frac{\partial^2 v_r}{\partial z^2} - \frac{2}{r^2}\frac{\partial v_\theta}{\partial \theta}. \tag{3-32}$$

Notice the appearance of $(2/r^2)(\partial v_\theta/\partial \theta)$ in the expression for $\partial_j \partial_j v_r$ that has no equivalent term in the expression for $\partial_j \partial_j T$. Notice as well the difference in terms related to the radial derivatives.

As a further example, suppose it is necessary to expand the Navier-Stokes equation for the radial velocity component v_r into spherical coordinates ($j = r, \theta, \phi$):

$$\partial_o v_r + v_j \partial_j v_r = \nu\, \partial_j \partial_j v_r - \frac{1}{\rho}\partial_r P + g_r. \tag{3-33}$$

In spherical coordinates, the operators in the Navier-Stokes equations are evaluated as

$$\partial_o v_r = \frac{\partial v_r}{\partial t}$$

$$v_j \partial_j v_r = v_r(\partial_r v_r) + v_\theta(\partial_\theta v_r) + v_\phi(\partial_\phi v_r)$$

$$= v_r \frac{\partial v_r}{\partial r} + v_\theta\left(\frac{1}{r}\frac{\partial v_r}{\partial \theta} - \frac{v_\theta}{r}\right) + v_\phi\left(\frac{1}{r \sin\theta}\frac{\partial v_r}{\partial \phi} - \frac{v_\phi}{r}\right) \quad \text{(from Table 3-2c)}$$

$$\nu\, \partial_j \partial_j v_r = \nu\left[\frac{\partial}{\partial r}\left(\frac{1}{r^2}\frac{\partial (r^2 v_r)}{\partial r}\right) + \frac{1}{r^2 \sin\theta}\frac{\partial}{\partial \theta}\left(\sin\theta \frac{\partial v_r}{\partial \theta}\right)\right.$$

$$\left. + \frac{1}{r^2 \sin^2\theta}\frac{\partial^2 v_r}{\partial \phi^2} - \frac{2}{r^2 \sin\theta}\left(\frac{\partial(\sin\theta\, v_\theta)}{\partial \theta} + \frac{\partial v_\phi}{\partial \phi}\right)\right] \quad \text{(from Table 3-3c)}$$

$$\frac{-1}{\rho}\partial_r P = \frac{-1}{\rho}\frac{\partial P}{\partial r} \quad \text{(from Table 3-1c).}$$

Assembling these terms into the full radial momentum equation for spherical coordinates yields a quite lengthy expression. Often problems undertaken in cylindrical or spherical coordinates are symmetric, leading to significant simplification of the final equation.

Pipe flow is an important geometry of flow that enjoys symmetry about the axis of the pipe. This requires $\partial v_z/\partial \theta = 0$ and $v_\theta = 0$ (in the absence of ridged body rotation). It is left as an exercise (see Problem 3-10) to show that the Navier-Stokes equation for v_z, in cylindrical coordinates, reduces to

$$\frac{\partial v_z}{\partial t} + v_r \frac{\partial v_z}{\partial r} + v_z \frac{\partial v_z}{\partial z} = \nu\left(\frac{\partial}{\partial r}\left(r \frac{\partial v_z}{\partial r}\right) + \frac{\partial^2 v_z}{\partial z^2}\right) - \frac{1}{\rho}\frac{\partial P}{\partial z} + g_z \tag{3-34}$$

for an axisymmetric flow.

3.5 PROBLEMS

3-1 Verify that $v_j \partial_j v_i = \partial_i(v^2/2) - \varepsilon_{ijk} v_j \omega_k$, where $\omega_k = \varepsilon_{klm}\partial_l v_m$. This will be a useful result for the description of irrotational flows ($\omega_k = 0$).

3-2 Newton's viscosity law for the diffusion flux of momentum in a $\rho \neq const.$ fluid is $M_{ji} = -\mu(\partial_j v_i + \partial_i v_j) + (2/3)\mu \delta_{ij} \partial_k v_k$. For a $\rho = const.$ fluid, $\partial_k v_k = 0$. Show that for a $\rho = const.$ fluid $\partial_j M_{ji} = -\mu \partial_j \partial_j v_i$.

3-3 Which of the following expressions are "sensible" in index notation? Here a, b, c, d, and e are arbitrary quantities.

$$
\begin{array}{ll}
a = b_i c_{ij} d_j & a_i = b_i + c_{ij} d_{ji} e_i \\
a = b_i c_i + d_j & a_\ell = \varepsilon_{ijk} b_j c_k \\
a_i = \delta_{ij} b_i + c_i & a_{ij} = b_{ji} \\
a_k = b_i c_{ki} & a_{ij} = b_i c_j + e_{jk} \\
a_k = b_k c + d_i e_{ik} & a_{k\ell} = b_i c_{ki} d_\ell
\end{array}
$$

3-4 Show by the expansion of terms that $\varepsilon_{ijk} u_i u_j = 0$ and $\varepsilon_{ijk} t_{jm} t_{mk} = 0$, if t_{ij} is symmetric.

3-5 The thermal energy equation can be written in the form

$$\rho \partial_o u + \rho v_i \partial_i u - \partial_i (k \partial_i T) + M_{ji} \partial_j v_i + P \partial_i v_i = 0.$$

Express the thermal energy equation in symbolic notation and in two-dimensional Cartesian coordinates. Do not change the variables of the equation (e.g., substitute other expressions for u or M_{ji}).

3-6 Expand each of the following expressions for three-dimensional space (in x, y, and z):

$$\omega_i = \varepsilon_{ijk} \partial_j v_k$$

$$\partial_i v_i$$

$$\delta_{ij} \delta_{ij}.$$

3-7 Which expressions are not allowed, and why?

$$a_{ij} + c_{ij} = b_{ijk}$$

$$\varepsilon_{ijk} \varepsilon_{pjq} = 2\delta_{ip}$$

$$\beta_{in} H_{ij} K_{jm} - \alpha_{ij} L_{jm}$$

$$u_i u_{ji} = \nu \alpha_{ij} \partial_k \partial_k u_i$$

3-8 $\omega_k = \varepsilon_{klm} \partial_l v_m$ is the vorticity of a flow. Demonstrate that the integral of $\varepsilon_{ijk} v_j \omega_k$ is zero along a streamline by demonstrating that $\varepsilon_{ijk} v_j \omega_k$ is perpendicular to the local velocity v_i. In other words, show that the dot product between the two yields $v_i \varepsilon_{ijk} v_j \omega_k = 0$.

3-9 Demonstrate that $\varepsilon_{kli} \varepsilon_{ijm} \partial_l (\omega_j v_m) = \partial_m (\omega_k v_m) - \partial_j (\omega_j v_k)$. This result is useful when taking the curl of the Navier-Stokes equations to derive the vorticity equation.

3-10 Show that with $v_\theta = 0$, the Navier-Stokes equation for v_z reduces to Eq. (3-34) for axisymmetric flow in cylindrical coordinates.

REFERENCE

[1] A. Einstein, *The Meaning of Relativity.* Princeton: Princeton University Press, 1979.

Chapter 4

Transport by Advection and Diffusion

4.1 Continuity Equation
4.2 Transport of Species
4.3 Transport of Heat
4.4 Transport of Momentum
4.5 Summary of Transport Equations without Sources
4.6 Conservation Statements from a Finite Volume
4.7 Eulerian and Lagrangian Coordinates and the Substantial Derivative
4.8 Problems

This chapter takes a first look at the conservation statements for total mass, species, heat, and momentum transport. Conservation statements will be formulated from the ability to mathematically express advection and diffusion fluxes of these properties, as developed in Sections 2.2 and 2.3. The resulting transport equations will be incomplete in some cases because they will not include potentially important source terms. These source terms are addressed separately in Chapter 5.

All expressions of conservation laws can begin with a descriptive statement that the rate of increase of some conservable property in a control volume must equal the net transport of that property to the volume, as illustrated in Figure 4-1. For a differential volume it will be explicitly shown that the net fluxes into the control volume $(in - out)$ must equal the negative value of the divergence of fluxes: $-\partial_j(fluxes)_j$. Therefore, it will be shown that for a differential volume, conservation statements can be written as

$$\begin{array}{rcl} (increase) & = & (in - out) \\ \downarrow & & \downarrow \\ \text{differential volume: } \partial_o(stored) & = & -\partial_j(fluxes)_j \end{array} \quad . \tag{4-1}$$

After this is demonstrated for the total mass of a fluid by a direct but cumbersome approach in the next section, Eq. (4-1) can serve as a starting point for more complicated transport equations. Equation (4-1) will also be derived starting from a finite control

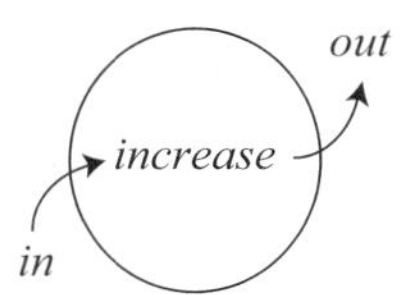

Figure 4-1 Conservation balance.

volume in Section 4.6. This approach warrants particular attention because it avails the integral form of conservation laws, prior to casting them into differential form. In later chapters it will be discovered that discontinuous behavior exhibited by nonlinear transport equations can cause transport properties to be nondifferentiable. The integral form of the conservation laws is an appropriate starting point for analysis of transport across such discontinuities.

4.1 CONTINUITY EQUATION

The conservation statement for the mass of a fluid is the *continuity equation*. The continuity equation considers transport caused by bulk motion of a fluid (advection), without attention to its composition. Because the composition of the fluid is not considered, diffusion is not a contributing factor to transport described by the continuity equation. Justification for this is elaborated on in Section 4.2, where species conservation within a mixture is considered.

Consider a control volume in a steady two-dimensional flow. Steady-state conditions require that the mass flow into the control volume exactly equals the mass flow out. The continuity equation is derived for this situation by considering the differential volume shown in Figure 4-2. The dimension of the volume into the page is unity.

The rate of mass advection across each face of the control volume shown in Figure 4-2 is found from $(\rho v_j)n_j dA$. This expression is the dot product between the advection flux ρv_j and the differential area $n_j dA$. This product is positive if the mass flux is out of the control volume, since the unit normal vector for the differential area points outward from the control volume. There are four faces on the control volume to consider. The left-hand face (1) is at a position x and has an area of $1\Delta y$, with a unit normal in the negative x-direction. The right-hand face (3) is a distance Δx further in the x-direction, with a unit normal in the positive x-direction. The bottom face (2) is at a position y and has an area of $1\Delta x$, with a unit normal in the negative y-direction. The top face (4) is a distance Δy further in the y-direction, with a unit normal in the positive y-direction. The net rate at which mass flows out of the control volume is obtain by adding the contribution of $(\rho v_j)n_j dA$ at each face, as tabulated in Figure 4-2. Since mass is conserved, this sum must be zero—that is, $(out) - (in) = 0$. Therefore,

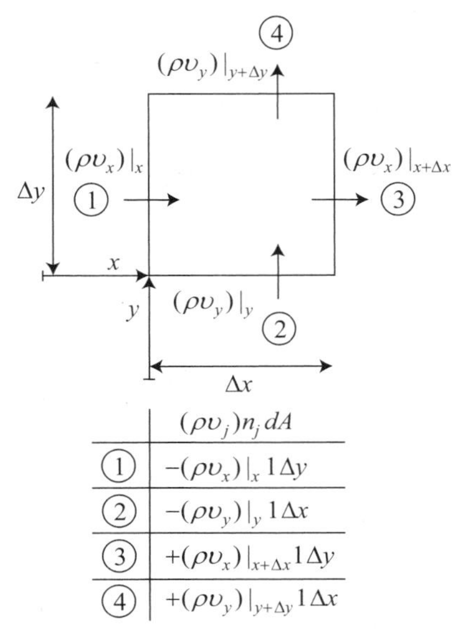

	$(\rho v_j)n_j\,dA$
1	$-(\rho v_x)\|_x\,1\Delta y$
2	$-(\rho v_y)\|_y\,1\Delta x$
3	$+(\rho v_x)\|_{x+\Delta x}\,1\Delta y$
4	$+(\rho v_y)\|_{y+\Delta y}\,1\Delta x$

Figure 4-2 Differential volume analysis of continuity.

$$(\rho v_x)|_{x+\Delta x}\, 1\Delta y + (\rho v_y)|_{y+\Delta y}\, 1\Delta x - (\rho v_x)|_x\, 1\Delta y - (\rho v_y)|_y\, 1\Delta x = 0. \tag{4-2}$$

Dividing the conservation statement by $(1\ \Delta x\ \Delta y)$ and rearranging yields

$$\frac{(\rho v_x)|_{x+\Delta x} - (\rho v_x)|_x}{\Delta x} + \frac{(\rho v_y)|_{y+\Delta y} - (\rho v_y)|_y}{\Delta y} = 0. \tag{4-3}$$

In the limit that $\Delta x, \Delta y \to 0$ (i.e., become differential distances), one obtains

$$\frac{\partial(\rho v_x)}{\partial x} + \frac{\partial(\rho v_y)}{\partial y} = 0. \tag{4-4}$$

This is the differential form of the continuity equation for a steady two-dimensional flow. One can interpret the first term $\partial(\rho v_x)/\partial x$ as the increase in advection mass flux crossing the differential volume in the x-direction. The second term $\partial(\rho v_y)/\partial y$ is the increase in advection mass flux crossing the differential volume in the y-direction. Since their sum must be zero to conserve mass, an increase in the advection flux in one direction must be accompanied by a decrease in the other direction. Collectively one says that the *divergence* of the advection mass flux is zero. In index notion this would be written as

$$\partial_i(\rho v_i) = 0. \tag{4-5}$$

There are situations in which the divergence of the mass advection flux is not zero. If more mass advection out of the control volume occurs than mass advection into the control volume, the divergence would be positive. This could happen only if mass were somehow being created in the control volume, or the amount of mass stored in the control volume was decreasing with time. The first situation is discounted as violating the principle of mass conservation. However, the second situation describes an unsteady flow. One can revise the continuity statement for an unsteady flow to say that the rate of mass storage in the differential volume equals the net rate of mass flow into the volume, or $(increase) = (in - out)$. For the control volume illustrated in Figure 4-2, one writes

$$(1\Delta x\ \Delta y)\frac{\partial \rho}{\partial t} = (\rho v_x)|_x\, 1\Delta y + (\rho v_y)|_y\, 1\Delta x - (\rho v_x)|_{x+\Delta x}\, 1\Delta y - (\rho v_y)|_{y+\Delta y}\, 1\Delta x, \tag{4-6}$$

where $\partial\rho/\partial t$ multiplied by the dimensions of the volume $(1\Delta x\ \Delta y)$ is the rate of increase of mass in the control volume. Note that $(in - out)$ is opposite in sign to $(\rho v_j)n_j dA$ (tabulated in Figure 4-2), which is positive going out. Dividing the new conservation statement by $(1\Delta x\ \Delta y)$, and taking the limit $\Delta x, \Delta y \to 0$ yields

$$\frac{\partial \rho}{\partial t} + \frac{\partial}{\partial x}(\rho v_x) + \frac{\partial}{\partial y}(\rho v_y) = 0. \tag{4-7}$$

This is the differential form of the continuity statement for an unsteady two-dimensional flow. It is interpreted as saying the rate of mass increase in a differential volume (first term) plus the net rate of mass advection out of the differential volume (second two terms) equals zero. It may be generalized with index notation to

$$\partial_o \rho + \partial_j(\rho v_j) = 0. \tag{4-8}$$

The process of deriving the continuity equation and the mathematical interpretation of the resulting differential form is worth reviewing. All conservation statements build on the concepts just introduced and the extension of these concepts to other conservation statements is not complex. Derivation of the continuity equation has demonstrated that the net transport rate to a differential volume is balanced by the negative divergence of fluxes. Therefore, the descriptive conservation statement (*increase*) = (*in* − *out*) may be written mathematically as Eq. (4-1) for a differential volume. In the present consideration of mass conservation, the amount of mass stored in a unit volume is ρ, and the only fluxes transporting mass across the differential volume are advection: ρv_i. Therefore, the mathematical form that the continuity equation takes can be written directly from Eq. (4-1) as

$$\begin{array}{ccc} \frac{\partial}{\partial t}(stored) & = & -\partial_j(fluxes)_j \\ \downarrow & & \downarrow \\ \overbrace{\partial_o(\underbrace{\rho}_{stored})} & = & \overbrace{-\partial_j(\underbrace{\rho v_j}_{advection})} \end{array} \tag{4-9}$$

Following the same logic, the continuity equation can also be readily derived for the total molar concentration of the fluid:

$$\partial_o c + \partial_j(c\, v_j^*) = 0. \tag{4-10}$$

In this statement, v_j^* is the molar-average velocity of the flow, while in the previous continuity equation (4-8), v_j was the mass-averaged value, as discussed in Section 2.2.1.

4.2 TRANSPORT OF SPECIES

A conservation statement for a species in a fluid is constructed next. Unlike the continuity equation, one must consider diffusion transport as well as advection to keep track of the species. It is assumed that no reactions or other sources of the species, call it species "A", are present in the fluid. The mass of species A in a unit volume is ρ_A, which is a fraction of the total density ρ of the fluid. The total density ρ of the fluid is given by the sum of all the partial densities:

$$\rho = \rho_A + \rho_B + \cdots. \tag{4-11}$$

The advection flux for species A is written as $\rho_A v_j$. Consulting Table 2-2 for the general case in which $\rho \neq const.$, the diffusion flux for species A is written as $-\rho Đ_A \vec{\nabla}(\rho_A/\rho)$. Therefore, the conservation statement for species A is written:

$$\begin{array}{ccc} \frac{\partial}{\partial t}(stored) & = & -\partial_j(fluxes)_j \\ \downarrow & & \downarrow \\ \overbrace{\partial_o(\underbrace{\rho_A}_{stored})} & = & \overbrace{-\partial_j(\underbrace{\rho_A v_j}_{advection} + \underbrace{-\rho Đ_A \partial_j(\rho_A/\rho)}_{diffusion})}. \end{array} \tag{4-12}$$

Instead of keeping track of the density of species A, it is often convenient to track the mass fraction $\omega_A = \rho_A/\rho$, where the total density of the fluid ρ is governed by the continuity equation derived in Section 4.1. Introducing this definition yields

$$\partial_o(\rho\omega_A) + \partial_j(\rho\omega_A\, v_j) = \partial_j(\rho Đ_A \partial_j \omega_A). \tag{4-13}$$

The advection term (which has been moved to the left-hand side of the equation) and the transient storage term are expanded to reveal the expression of continuity,

$$\partial_o(\rho\omega_A) + \partial_j(\rho\omega_A\, v_j) = \rho\partial_o\omega_A + \omega_A \underbrace{[\partial_o\rho + \partial_j(\rho v_j)]}_{by\ continuity = 0} + \rho v_j \partial_j \omega_A. \tag{4-14}$$

Therefore, the species conservation statement (4-13) simplifies to

$$\rho\partial_o\omega_A + \rho v_j\partial_j\omega_A = \partial_j(\rho Đ_A \partial_j\omega_A). \tag{4-15}$$

If $\rho Đ_A = const.$, then

$$\partial_o\omega_A + v_j\partial_j\omega_A = Đ_A\partial_j\partial_j\omega_A \quad (\rho Đ_A = const.). \tag{4-16}$$

It is left as an exercise (see Problem 4.2) to show that the species transport equation may be derived in terms of the molar fraction as

$$\partial_o\chi_A + v_j^*\partial_j\chi_A = Đ_A\partial_j\partial_j\chi_A \quad (cĐ_A = const.). \tag{4-17}$$

Notice that in the molar transport equation, advection is defined with the molar-averaged velocity v_j^* instead of the mass-averaged velocity v_j.

4.2.1 Transport in a Binary Mixture

Consider a binary system of species A and B, in which the flux of the B species is zero. This may be accomplished by fixing the coordinate system to the motion of species B. Diffusion drives species A through species B. However, this simultaneously introduces an advection flux proportional to the average speed of the two species.

Suppose a system of unidirectional transport, as illustrated in Figure 4-3, has achieved a steady-state. In terms of the mass content of the flow, the transport equations of both species can be written from Eq. (4-13) as

$$A: \quad 0 = -\partial_x(\rho_A v_x - \rho Đ_{AB}\partial_x\omega_A) \tag{4-18}$$

$$B: \quad 0 = -\partial_x(\rho_B v_x - \rho Đ_{BA}\partial_x\omega_B). \tag{4-19}$$

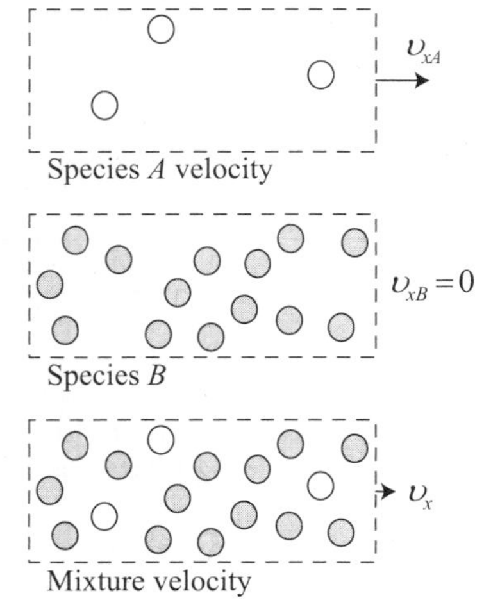

Figure 4-3 Transport of one species through a stationary second species.

In a binary system, the mass diffusivity is the same for both species $Đ_{AB} = Đ_{BA}$, as demonstrated in Section 2.3.3. In general, solving for species transport is complicated by the fact that the mass-averaged flow velocity is a function of the mass fraction content. This makes the transport equations nonlinear.

To establish the mass-averaged flow velocity, the transport equation for species B can be integrated once, yielding

$$\omega_B v_x - Đ_{AB} \partial_x \omega_B = D_1 \quad (= 0), \tag{4-20}$$

where $\omega_B = \rho_B / \rho$. However, since the mass flux of B is zero, the integration constant is $D_1 = 0$. Therefore, transport of B yields the relation $\omega_B v_x = Đ_{AB} \partial_x \omega_B$, which, with $\omega_A + \omega_B = 1$, can be used to determine the flow velocity:

$$v_x = \frac{Đ_{AB} \partial_x \omega_B}{\omega_B} = \frac{-Đ_{AB} \partial_x \omega_A}{1 - \omega_A}. \tag{4-21}$$

This flow is maintained by the production or removal of species A at a boundary, which is known as *Stefan flow*.* The flow velocity can be substituted back into the transport equation for species A for the result

$$\partial_x \left(\rho_A \frac{\partial_x \omega_A}{1 - \omega_A} + \rho \partial_x \omega_A \right) = 0. \tag{4-22}$$

In this form, the transport equation reveals the relative importance of advection and diffusion fluxes. Specifically, the ratio of advection to diffusion transport is given by

$$\frac{\text{advection flux}}{\text{diffusion flux}} = \frac{\rho_A}{\rho(1 - \omega_A)} = \frac{\omega_A}{1 - \omega_A}. \tag{4-23}$$

In the lower limit of species A mass fractions, the advection term becomes negligible compared to the diffusion term. This simplification carries over to situations where the mass flux of B is nonzero as well. In this case, the mixture velocity is simply the velocity of species B. Therefore, in the lower limit of species A mass fractions, the mixture velocity is independent of the mixture content of species A. This limiting case offers a significant simplification in that the fluid velocity becomes independent of species A, making the species transport equation linear. When the concentration of species A is sufficiently dilute, transport of A no longer has a significant impact on the total concentration of the mixture. When this is the case, the total concentration can be assumed constant (so long as the fluid is incompressible), and the species transport equation (4-17) can be written in terms of the dilute species A concentration:

$$\partial_o c_A + v_j \partial_j \, c_A = Đ_A \partial_j \partial_j \, c_A \quad (c, Đ_A = const.). \tag{4-24}$$

In this *dilute species transport* equation for incompressible flow, there is no distinction between the molar-averaged velocity and mass-averaged velocity.

---

*This flow is named after the Slovene physicist, mathematician, and poet Joseph Stefan for his work on calculating evaporation rates.

The continuity equation for the total mixture (4-8) can be derived from the sum of the transient transport equations written for species A and B. Using Eq. (4-13), the transient species equations are written in the form

$$A: \quad \partial_o \rho_A = -\partial_j(\rho_A v_j - \rho Đ_{AB} \partial_j \omega_A) \tag{4-25}$$

$$B: \quad \partial_o \rho_B = -\partial_j(\rho_B v_j - \rho Đ_{BA} \partial_j \omega_B), \tag{4-26}$$

where the mass diffusivity is the same for both species $Đ_{AB} = Đ_{BA}$. Summing these two equations yields

$$\partial_o(\rho_A + \rho_B) + \partial_j\big((\rho_A + \rho_B)v_j\big) = \partial_j\big(\rho Đ_{AB} \partial_j(\omega_A + \omega_B)\big). \tag{4-27}$$

Since $\rho_A + \rho_B = \rho$ and $\partial_j(\omega_A + \omega_B) = \partial_j(1) = 0$, the result of summing the species transport equations over both species yields

$$\partial_o \rho + \partial_j(\rho v_j) = 0, \tag{4-28}$$

which is the same as the continuity equation (4-8). Notice that the effect of diffusion vanishes from the transport description of the mixture density. Similar arguments can be made starting with the molar transport equations for both species, with the result that continuity of the mixture given by Eq. (4-10) can be derived, as shown in Problem 4-4.

4.3 TRANSPORT OF HEAT

The steps to derive the heat equation closely follow those for conservation of species in Section 4.2. The amount of internal energy in a unit volume is ρu. The advection flux of heat is $\rho u v_j$ and the diffusion flux of heat is $-k\partial_j T$ (see Table 2-2). Therefore, the heat equation may be written as

$$\begin{array}{ccc} \dfrac{\partial}{\partial t}(storage) & = & -\partial_j(fluxes)_j \\ \downarrow & & \downarrow \\ \overbrace{\partial_o(\underbrace{\rho u}_{stored})} & = & \overbrace{-\partial_j(\underbrace{\rho u v_j}_{advection} + \underbrace{-k\partial_j T}_{diffusion})} \end{array} \tag{4-29}$$

This formulation ignores the possibility of coupling between thermal and mechanical forms of energy (discussed in Sections 5.3 and 5.6). The advection term in Eq. (4-29) is moved to the left-hand side of the equation and is expanded along with the transient storage term to reveal the expression of continuity, which is dropped.

$$\partial_o(\rho u) + \partial_j(\rho u\, v_j) = \rho \partial_o u + u \underbrace{[\partial_o \rho + \partial_j(\rho v_j)]}_{by\ continuity\ =\ 0} + \rho v_j \partial_j u. \tag{4-30}$$

This simplification puts the heat equation into the form

$$\rho \partial_o u + \rho v_j \partial_j u = \partial_j(k \partial_j T). \tag{4-31}$$

To express internal energy in terms of temperature, one must decide what kind of fluid is being considered. Here, assume that the fluid is incompressible, $\rho = const.$, for which

$du = CdT$. (Section 5.3 treats the possibility of a compressible gas.) Therefore, for an incompressible liquid, the heat equation becomes

$$\rho C \partial_o T + \rho C v_j \partial_j T = \partial_j (k \partial_j T) \quad (\rho = const.). \tag{4-32}$$

Or, assuming that thermal conductivity is a constant,

$$\partial_o T + v_j \partial_j T = \alpha \, \partial_j \partial_j T \quad (\rho, k = const.), \tag{4-33}$$

where $\alpha = k/(\rho C)$ is the thermal diffusivity. It should be noted that, in addition to the assumptions of constant fluid properties, Eq. (4-33) neglects viscous heating (a mechanism by which mechanical energy is transformed into thermal energy), which is discussed in Section 5.6.

4.4 TRANSPORT OF MOMENTUM

Next, consider conservation of momentum, for which the amount of momentum in a unit volume is ρv_i. The i-index allows one to talk simultaneously about any one of the three directional components of momentum. The advection flux of momentum is $\rho v_i v_j$ and the diffusion flux of momentum is M_{ji}. The j-index is used to keep track of the directions of transport. For simplicity, consider the case of an incompressible flow when $\rho = const.$, where $M_{ji} = -\mu(\partial_j v_i + \partial_i v_j)$ (see Table 2-2). In this case, the conservation of momentum statement becomes

$$\begin{array}{ccc} \dfrac{\partial}{\partial t}(storage) & = & -\partial_j (fluxes)_j \\ \downarrow & & \downarrow \\ \overbrace{\partial_o(\underbrace{\rho v_i}_{stored})} & = & \overbrace{-\partial_i(\underbrace{\rho v_i v_j}_{advection} + \underbrace{-\mu(\partial_j v_i + \partial_i v_j)}_{diffusion})} \end{array} \quad (\rho = const.) \tag{4-34}$$

This formulation ignores the possibility that sources of momentum exist in the flow (as discussed in Section 5.4). The order of differentiation of independent variables i and j can be switched, such that $\partial_j(\partial_i v_j) = \partial_i(\partial_j v_j)$. And, since $\partial_j v_j = 0$ for an incompressible fluid, the diffusion term in Eq. (4-34) simplifies to

$$\partial_j[\mu(\partial_j v_i + \partial_i v_j)] = \mu \partial_j(\partial_j v_i). \quad (\rho = const.). \tag{4-35}$$

Additionally, the storage and advection terms in Eq. (4-34) can be expanded to reveal the continuity equation, which is dropped:

$$\partial_o(\rho v_i) + \partial_j(\rho v_i v_j) = \rho \partial_o v_i + \underbrace{v_i \partial_o \rho + v_i \partial_j(\rho v_j)}_{by\ continuity = 0} + \rho v_j \partial_j v_i. \tag{4-36}$$

These simplifications put the momentum equation into the form

$$\rho(\partial_o v_i + v_j \partial_j v_i) = \partial_j(\mu \partial_j v_i) \quad (\rho = const.). \tag{4-37}$$

By dividing through by density, and assuming the dynamic viscosity μ to be a constant, the momentum equation becomes

$$\partial_o v_i + v_j \partial_j v_i = \nu \, \partial_j \partial_j v_i \quad (\rho, \mu = const.). \tag{4-38}$$

The momentum diffusivity $\nu = \mu/\rho$ is also known as the *kinematic viscosity*. In addition to the assumptions of constant fluid properties, the momentum equation (4-38) has limited utility because sources of momentum, such as pressure gradients that often drive a flow, have been omitted. This consideration is addressed in Chapter 5 when source terms to the transport equations are considered.

4.5 SUMMARY OF TRANSPORT EQUATIONS WITHOUT SOURCES

Table 4-1 summarizes the transport equations that have been derived in this chapter. Notice that, with the exception of the continuity equation (4-8), every transport equation has the form

$$\underbrace{\partial_o(\ )}_{storage} + \underbrace{v_j\partial_j(\ )}_{advection} = \underbrace{\partial_j\partial_j(\ )}_{diffusion} . \tag{4-39}$$

Given the redundancy in form, these transport equations can easily be committed to memory. However, when using these equations, one should be mindful that they assume constant fluid properties and that no sources exist. This latter assumption is quite restrictive, and it will be necessary to generalize these transport equations further in Chapter 5 to cover a wider range of situations. The most glaring omission is the effect of pressure on momentum transport. Gradients in pressure act as a source of momentum that can accelerate a flow or offset losses of momentum to bounding surfaces.

Table 4-1 Transport equations in the absence of source terms and with constant fluid properties

	Transport equation
Species	
($\rho Ð_A = const.$)	$\partial_o\omega_A + v_j\partial_j\omega_A = Ð_A\partial_j\partial_j\omega_A$
($cÐ_A = const.$)	$\partial_o\chi_A + v_j^*\partial_j\chi_A = Ð_A\partial_j\partial_j\chi_A$
Heat	
($\rho, k = const.$)	$\partial_o T + v_j\partial_j T = \alpha\, \partial_j\partial_j T$
Momentum	
($\rho, \mu = const.$)	$\partial_o v_i + v_j\partial_j v_i = \nu\, \partial_j\partial_j v_i$

4.6 CONSERVATION STATEMENTS FROM A FINITE VOLUME

In the preceding sections, the conservation equations were derived for a differential volume. This may be generalized to any finite control volume, as illustrated in Figure 4-4, with the statement that

$$\text{finite volume:} \quad \overset{(increase)}{\int_{\forall} \frac{\partial}{\partial t}(stored)d\forall} = \overset{(in - out)}{-\int_A (fluxes)_j n_j dA} \tag{4-40}$$

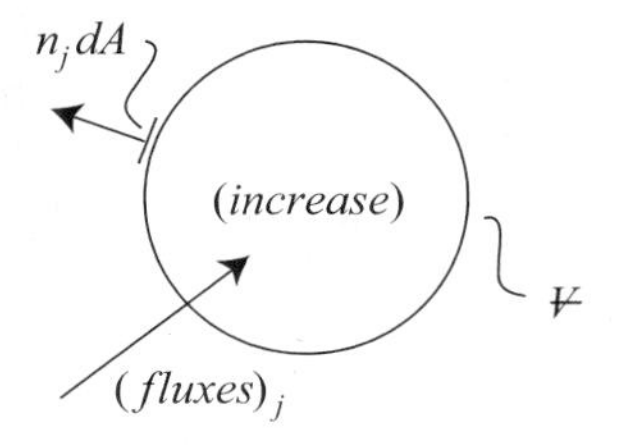

Figure 4-4 Arbitrary control volume.

The left-hand side of Eq. (4-40) evaluates the rate at which the content of a fluid property within the control volume $\forall$ increases with time. When the control volume is stationary, the ordering of the spatial integration and the time differentiation is inconsequential. The rate of content increase is balanced by the right-hand side of Eq. (4-40), which describes fluxes of the fluid property integrated over the surface A that bounds the control volume. Since $(fluxes)_j n_j dA$ are positive *leaving* the control volume (the surface unit normal n_j points outward), the negative sign preceding the area integral in Eq. (4-40) is needed to evaluate fluxes *entering* the control volume.

To cast the integral statement into a differential form, *Gauss's theorem*[†] may be applied. In words, Gauss's theorem says that the outward flux of a quantity through a closed surface area is equal to the volume integral of the divergence of that quantity over the volume inside the surface. Mathematically, Gauss's theorem is expressed as

$$\int_A (\ )_j n_j \, dA = \int_\forall \partial_j (\ )_j d\forall, \tag{4-41}$$

where the absent quantity associated with the parentheses "$(\ )_j$" is a vector (or tensor) function.

To cast the integral statement into a differential form, the sequence of steps outlined in Table 4-2 can be followed. After Gauss's theorem is applied, terms in the conservation

Table 4-2 Steps between the integral and differential forms of a conservation statement

(0)	$(increase)$	$=$	$(in - out)$	
	$\downarrow$		$\downarrow$	Express statement for control volume...
(1)	$\int_\forall \frac{\partial}{\partial t}(stored) d\forall$	$=$	$-\int_A (fluxes)_j n_j \, dA$	
	$\downarrow$		$\downarrow$	Apply Gauss's theorem...
(2)	$\int_\forall \frac{\partial}{\partial t}(stored) d\forall$	$=$	$-\int_\forall \partial_j (fluxes)_j \, d\forall$	
	$\downarrow$		$\downarrow$	Collect terms under common integral...
(3)	$\int_\forall \left[\frac{\partial}{\partial t}(stored) \right.$	$=$	$\left. -\partial_j (fluxes)_j \right] d\forall$	
	$\downarrow$		$\downarrow$	Argue integrand is identically true...
(4)	$\frac{\partial}{\partial t}(stored)$	$=$	$-\partial_j (fluxes)_j$	

[†]Gauss's theorem is named in honor of the German mathematician and scientist Johann Carl Friedrich Gauss (1777–1855).

statement can be collectively written under one volume integral, as shown in step (3) of Table 4-2. For this integral to be satisfied, for any choice of volume, the differential statement appearing as the integrand to the volume integral must be identically true. This last conceptual argument results in the differential form of the conservation equation. This differential statement was the starting point for the conservation equations derived in the earlier sections, which started with consideration of a differential volume rather than the finite volume employed in the current derivation.

In Chapter 5, conservation equations that include important source terms will be derived. Rather than starting from an arbitrary control volume, equations could be derived for a differential volume starting with the statement that

$$\begin{array}{rccccc} & (\mathit{increase}) & = & (\mathit{in}-\mathit{out}) & + & (\mathit{generation}) \\ & \downarrow & & \downarrow & & \downarrow \\ \text{for a differential volume:} & \dfrac{\partial}{\partial t}(\mathit{stored}) & = & -\vec{\nabla}\cdot(\mathit{fluxes}) & + & (\mathit{sources}) \end{array}, \tag{4-42}$$

where the (*sources*) are expressed on a per-unit-volume basis. It is left as an exercise (see Problem 4-3) to show that this differential statement can be derived from the initial consideration of a finite control volume, following the steps outlined in Table 4-2.

4.7 EULERIAN AND LAGRANGIAN COORDINATES AND THE SUBSTANTIAL DERIVATIVE

A flow can be viewed from two important frames of reference. When the observer is stationary, such that a flow sweeps fluid properties past the observer, the frame of reference is *Eulerian*,[‡] as illustrated in Figure 4-5(a). This is the frame of reference used to derive the transport equations listed in Table 4-1. In contrast, the flow can be observed from a frame of reference that is moving with the fluid. This is known as a *Lagrangian description*,[§] as illustrated in Figure 4-5(b). Notice that a particle is labeled in the flow at time t and $t+\Delta t$. The particle remains stationary relative to the Lagrangian coordinates, but is swept past the Eulerian coordinates.

The time rate of change of some quantity evaluated with respect to a fluid element moving with the flow is known as the *substantial derivative* or *material derivative*. The substantial derivative is mathematically defined as

$$\underbrace{\frac{D(\ )}{Dt}}_{\mathit{Lagrangian}} = \underbrace{\partial_o(\ ) + v_j\partial_j(\ )}_{\mathit{Eulerian}}. \tag{4.43}$$

The substantial derivative is interpreted as being the rate of change of some fluid property in the Lagrangian frame of reference, traveling with the flow. In the Eulerian frame of reference, that change is seen to occur transiently, for a stationary position, and spatially, in the progression of the flow field. Notice that the notion of a steady-state flow, for which the term $\partial_o(\ )$ is zero, is an Eulerian concept; a steady-state flow does not imply that a material fluid element experiences no change while moving with the flow.

[‡]The Eulerian perspective is named after the Swiss mathematician and physicist Leonhard Euler (1707–1783).

[§]The Lagrangian perspective is named after the French mathematician Joseph-Louis Lagrange (1736–1813).

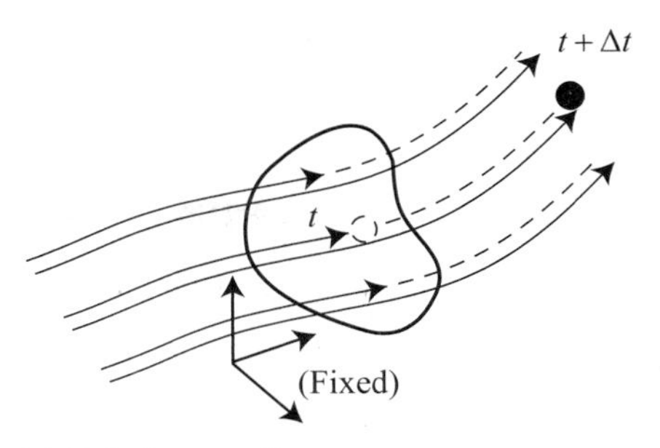

(a) Eulerian coordinates

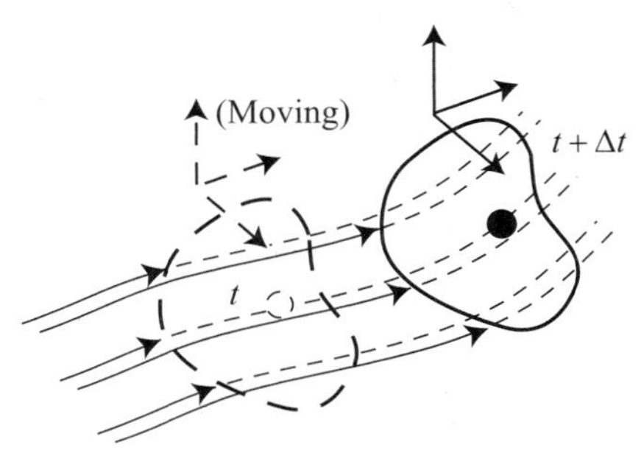

(b) Lagrangian coordinates

Figure 4-5 A material volume viewed from Eulerian coordinates (a) and Lagrangian coordinates (b). A particle in the flow is labeled at time t and time $t+dt$ for both coordinates.

In terms of the substantial derivative, the transport equations (without sources) for a constant property fluid are written as

$$\text{Species:} \qquad \frac{D\omega_A}{Dt} = \text{Đ}_A \partial_j \partial_j \omega_A \quad (\rho \text{Đ}_A = const.) \tag{4-44}$$

$$\text{Heat:} \qquad \frac{DT}{Dt} = \alpha\, \partial_j \partial_j T \qquad (\rho,\, k = const.) \tag{4-45}$$

$$\text{Momentum:} \qquad \frac{Dv_i}{Dt} = \nu\, \partial_j \partial_j v_i \qquad (\rho,\, \mu = const.). \tag{4-46}$$

In this Lagrangian frame of reference, there is an absence of bulk fluid motion that occurs when the observer is moving with the fluid. Therefore, diffusion is the sole transport term for the conservation statements written from the Lagrangian viewpoint.

In terms of the substantial derivative, the continuity equation becomes

$$\frac{D\rho}{Dt} = -\rho \partial_i v_i. \tag{4-47}$$

If density does not change for a control mass moving with the flow, then the continuity equation becomes $\partial_i v_i = 0$. This is the formal definition of an *incompressible flow*, and is strictly a Lagrangian concept that follows from the notion that $D\rho/Dt = 0$. Notice that this definition does not prohibit density from being a function of time and space $\rho(t, x)$. For example, if two immiscible incompressible liquids flow together, such as oil and water, the flow density changes in time and space even though $D\rho/Dt = 0$.

The substantial derivative also has utility in expressing thermodynamic statements that concern fluid properties transported with the flow. For example, the fundamental thermodynamic equation $dh = Tds + dP/\rho$ [Eq. (1-8)] imposes on a flow the condition:

$$\frac{Dh}{Dt} = T\frac{Ds}{Dt} + \frac{1}{\rho}\frac{DP}{Dt}. \tag{4-48}$$

Therefore, if the flow were isentropic $Ds/Dt = 0$, one could utilize the relation

$$\rho\frac{Dh}{Dt} = \frac{DP}{Dt}, \tag{4-49}$$

which will prove useful in the context of the energy equation developed for compressible flows in Chapter 20.

4.8 PROBLEMS

4-1 An incompressible fluid is squeezed between two plates by the motion of the top plate in the z-direction. If the position of the top plate changes vertically as dH/dt, show that the following is true:

$$\partial_o H = -\partial_i G_i, \text{ where } G_i = \int_0^H v_i dz \text{ and } i = x \text{ and } y.$$

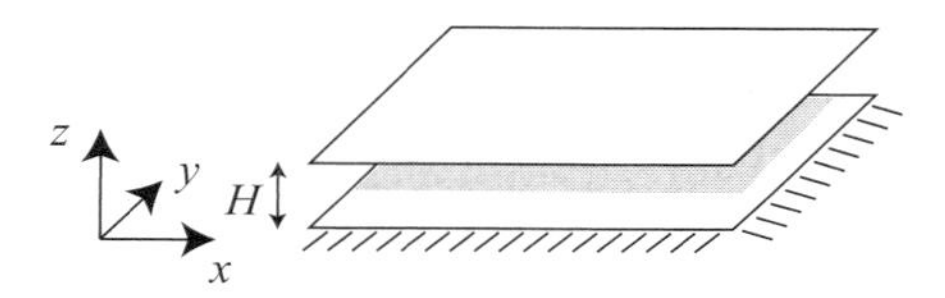

4-2 Derive the species transport equation in terms of the molar fraction to demonstrate the result

$$\partial_o \chi_A + v_j^* \partial_j \chi_A = Đ_A \partial_j \partial_j \chi_A \quad (cĐ_A = const.).$$

4-3 Starting with a finite control volume in which generation occurs, prove that for a differential volume the conservation statement should read:

$$\frac{\partial}{\partial t}(stored) = -\vec{\nabla}\cdot(fluxes) + (sources)$$

where (*sources*) express the generation on a per-unit-volume basis.

4-4 Derive the continuity equation for the total mixture (4-10) from the sum of the molar concentration transport equations written for species A and B.

4-5 Derive the transient energy equation for a control volume in a solid that experiences volumetric generation q_{gen}. Make simplifying assumptions associated with transport in a solid from the outset. Take the internal energy to be a function of temperature and volume $u(T, \mathcal{v})$ and show that the energy equation can be expressed in terms of temperature as if thermal conductivity and density are assumed constant: $\partial_o T = \alpha\partial_i\partial_i T + q_{gen}/(\rho C_{\mathcal{v}})$.

4-6 Consider the transient heat equation derived in Problem 4-5, with a laser as the source of q_{gen} for a one-dimensional problem. The laser intensity will decay into a semi-infinite material as the function $I(x) = I_o \exp(-x/\delta_{opt})$, where x is the distance from the irradiated surface, I_o

is the incident radiant flux, and δ_{opt} is the optical penetration depth for the laser. Use Gauss's theorem to demonstrate that the laser energy absorption per unit volume is given by $q_{gen} = (I_o/\delta_{opt})\exp(-x/\delta_{opt})$.

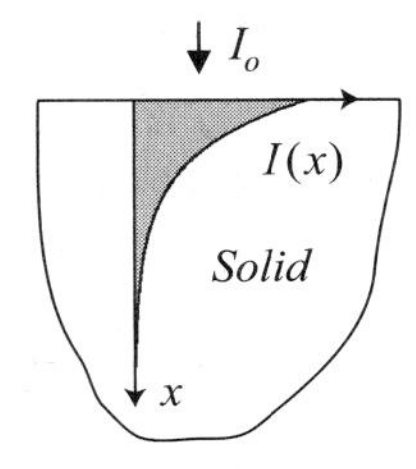

4-7 Explain the different interpretations for $D\rho/Dt = 0$, $\partial\rho/\partial t = 0$, and $\rho = const$. Which of these statements imply that $\partial_i v_i = 0$?

Chapter 5

Transport with Source Terms

5.1 Continuity Equation
5.2 Species Equation
5.3 Heat Equation (without Viscous Heating)
5.4 Momentum Equation
5.5 Kinetic Energy Equation
5.6 Heat Equation (with Viscous Heating)
5.7 Entropy Generation in Irreversible Flows
5.8 Conservation Statements Derived from a Finite Volume
5.9 Leibniz's Theorem
5.10 Looking Ahead
5.11 Problems

This chapter takes a second look at the transport equations for momentum, heat, and species, paying attention to sources of each. A modification of the conservation statement used in Chapter 4 is required to include generation or source terms. The full conservation statement requires that the rate of increase of some conservable property in a control volume equals the net transport of that property to the volume plus the net difference between all sources and sinks of that property within, or acting on, the control volume. For a differential volume, the conservation statement can be written as

$$\begin{array}{lccccc} & (\textit{increase}) & = & (\textit{in} - \textit{out}) & + & (\textit{sources} - \textit{sinks}) \\ & \downarrow & & \downarrow & & \downarrow \\ \text{for a differential volume:} & \dfrac{\partial}{\partial t}(\textit{stored}) & = & -\partial_j(\textit{fluxes})_j & + & (\textit{generation}) \end{array} \tag{5-1}$$

As previously discussed in Section 4.1, when the conservation statement is written for a differential volume, the net fluxes for $(in - out)$ can be expressed as the negative divergence of fluxes $-\vec{\nabla} \cdot (\textit{fluxes})$. The net difference between all sources and sinks accounts for generation, which added to the conservation statement is expressed on a per-unit-volume basis. Careful inclusion of all relevant sources (and sink) terms will help generalize the transport equations considered in Chapter 4. Although some simplifying assumptions will still be made, the transport equations derived in this chapter will be in a form useful for solving a great number of problems.

In this chapter the conservation equations will additionally be derived in integral form for a finite control volume. The procedure will differ slightly from what was introduced in

Chapter 4, by initially analyzing a control volume of fluid that is traveling with the flow. Leibniz's theorem will then be applied to address the change of perspective that occurs when going from a coordinate system traveling with the flow to one that is stationary.

5.1 CONTINUITY EQUATION

The conservation statement for total mass, the continuity equation, was derived in Chapter 4. Since, in general, mass is neither created nor destroyed, no modification to the previous continuity equation is required to account for sources. The continuity equation is restated here for completeness:

$$\partial_o \rho + \partial_j(\rho v_j) = 0. \tag{5-2}$$

In terms of the substantial derivative (see Section 4.7) the continuity equation can be written as

$$\frac{D\rho}{Dt} = -\rho \partial_j v_j. \tag{5-3}$$

In this form, the continuity equation says that the density of a material fluid element moving with the flow changes as a result of the divergence of the velocity field $\partial_j v_j$. For an *incompressible flow*, $D\rho/Dt = 0$ and $\partial_j v_j = 0$. It is noteworthy that the incompressibility of a flow is a hydrodynamic characteristic, while the incompressibility of a fluid is a thermodynamic characteristic. In some situations, a flow may behave as though incompressible even though the fluid itself is a compressible gas.

5.2 SPECIES EQUATION

Although mass is neither created nor destroyed, it can undergo a change in chemical form. A species, call it species "A", can appear or disappear as a consequence of a chemical reaction. Therefore, this source (or depletion) of species mass can be added to the rest of the conservation statement as a reaction rate r_A specified on a per-unit-volume basis. Following the form of Eq. (5-1), the species transport equation is written as

$$\partial_o(\overbrace{\rho_A}^{\text{stored}}) = -\partial_j[\overbrace{\rho_A v_j}^{\text{advection}} + \overbrace{-\rho Đ_A \partial_j(\rho_A/\rho)}^{\text{diffusion}}] + \overbrace{r_A}^{\text{generation}}. \tag{5-4}$$

The nature of the reaction term is problem-specific. For example, it might be the case that a reaction causes a depletion of species A at a rate proportional to the concentration of species A. Or, creation of species A may occur at a rate proportional to some other factor, such as the concentration of a second species B, or some thermodynamic variable such as temperature.

Following the steps performed in Section 4.2, the mass fraction $\omega_A = \rho_A/\rho$ definition is introduced and advection is combined with the storage term on the left-hand side of the equation, such that the appearance of the continuity equation can be removed. The equation that results from these steps is

$$\rho \partial_o \omega_A + \rho v_j \partial_j \omega_A = \partial_j(\rho Đ_A \partial_j \omega_A) + r_A, \tag{5-5}$$

which is identical to Eq. (4-15) except for the appearance of the source term r_A. If $\rho Đ_A = const.$ for the fluid, the species transport equation becomes

$$\partial_o \omega_A + v_j \partial_j \omega_A = Đ_A \partial_j \partial_j \omega_A + r_A/\rho. \qquad (\rho Đ_A = const.) \tag{5-6}$$

In terms of the substantial derivative, the species transport equation is written as

$$\frac{D\omega_A}{Dt} = Đ_A \partial_j \partial_j \omega_A + r_A/\rho. \qquad (\rho Đ_A = const.) \tag{5-7}$$

In this form, the species transport equation says that the concentration of species in a control mass of fluid moving with the flow changes as a result of diffusion transport and a reaction rate.

5.3 HEAT EQUATION (WITHOUT VISCOUS HEATING)

If a flow is compressible, mechanical energy may be converted to heat (internal energy) by compression, and a source term needs to be added to the heat equation developed in Section 4.3. Figure 5-1 illustrates the conversion of work to heat from the perspective of a differential volume. The rate of compression work done on a unit volume of the fluid is $-P\partial_j v_j$. This is the force per unit area (pressure) times the rate of compression (negative divergence of the velocity field). With the addition of this source term, the heat equation becomes

$$\partial_o(\overbrace{\rho u}^{stored}) = -\partial_j(\overbrace{\rho u v_j}^{advection} + \overbrace{-k\partial_j T}^{diffusion}) + \overbrace{-P\partial_j v_j}^{generation}. \tag{5-8}$$

It should be noted that viscous generation of heat is still omitted from this form of the heat equation. Viscous heating tends to be small, except for flows experiencing high rates of shear. This heat generation term will be considered later in Section 5.6, after the mathematical form of this term is discovered from contemplation of the kinetic energy equation in Section 5.5.

Following the steps performed in Section 4.3, the heat equation can be rewritten as

$$\rho\partial_o u + \rho v_j \partial_j u = \partial_j(k\partial_j T) - P\partial_j v_j, \tag{5-9}$$

which is similar to Eq. (4-31), except for the appearance of the compressive source term $-P\partial_j v_j$. Written in terms of the substantial derivative, the heat equation becomes

$$\rho\frac{Du}{Dt} = \partial_j(k\partial_j T) - P\partial_j v_j. \tag{5-10}$$

The continuity equation Eq. (5-2) can be used to show that the compressive source term $(-P\partial_j v_j)$ is related to the rate of change in fluid density traveling with the fluid. Expressed as such, the compression term becomes

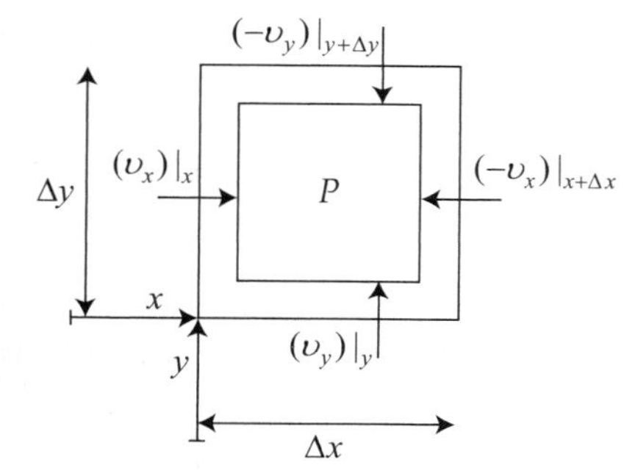

Figure 5-1 Mechanical energy converted to heat by compression.

$$-P\partial_i v_i = \frac{P}{\rho}\frac{D\rho}{Dt} \quad \text{or} \quad -P\partial_i v_i = \frac{DP}{Dt} - \rho\frac{D(P/\rho)}{Dt}. \tag{5-11}$$

The latter form allows P/ρ to be grouped with internal energy in the heat equation:

$$\rho\frac{D}{Dt}(u + P/\rho) = \partial_j(k\partial_j T) + \frac{DP}{Dt}. \tag{5-12}$$

Using the thermodynamic relation between internal energy and enthalpy, $h = u + P/\rho$, the heat equation becomes

$$\rho\frac{Dh}{Dt} = \partial_j(k\partial_j T) + \frac{DP}{Dt}. \tag{5-13}$$

If it is assumed that the compressible fluid behaves as an ideal gas, then changes in enthalpy depend only on temperature, $dh = C_p dT$, and

$$\rho C_p\frac{DT}{Dt} = \partial_j(k\partial_j T) + \frac{DP}{Dt} \quad (dh = C_p dT). \tag{5-14}$$

Furthermore, if the thermal conductivity is constant, the heat equation can be written in the form

$$\frac{DT}{Dt} = \alpha\,\partial_j\partial_j T + \frac{1}{\rho C_p}\frac{DP}{Dt} \quad (dh = C_p dT,\ k = const.). \tag{5-15}$$

Or equivalently,

$$\partial_o T + v_j\partial_j T = \alpha\,\partial_j\partial_j T + \frac{1}{\rho C_p}\frac{DP}{Dt} \quad (dh = C_p dT,\ k = const.). \tag{5-16}$$

For many external flows, dynamic effects are insufficient to cause significant compression of the gas and $DP/Dt \approx 0$. In such cases, the pressure everywhere in the flow is close to the ambient value. Flows in which pressure is constant are called *isobaric*. When the DP/Dt term is dropped, the heat equation appears identical to Eq. (4-33) in Section 4.3. However, there is an important interpretational difference. Equation (4-33) was derived for a constant density liquid ($\rho = const.$) where the sum of the storage and advection of internal energy is written as $\rho C(\partial_o T + v_j\partial_j T)$. However, the expression $\rho C_p(\partial_o T + v_j\partial_j T)$ in the current heat equation accounts also for the rate of work being done on the gas by changes in density at constant pressure; in this case, changes in density are brought about by changes in temperature.

More generally, if a gas does not exhibit ideal behavior, enthalpy will depend on two thermodynamic variables. If enthalpy is expressed as $h(T, P)$, then using the results of Section 1.2.3,

$$\rho\frac{Dh}{Dt} = \rho C_p\frac{DT}{Dt} + \rho\left(\frac{\partial h}{\partial P}\right)_T\frac{DP}{Dt} = \rho C_p\frac{DT}{Dt} + \left(1 + \frac{T}{\rho}\left(\frac{\partial\rho}{\partial T}\right)_P\right)\frac{DP}{Dt}. \tag{5-17}$$

Combining this with the heat equation (5-13) gives

$$\rho C_p\frac{DT}{Dt} = \partial_j(k\partial_j T) - \frac{T}{\rho}\left(\frac{\partial\rho}{\partial T}\right)_P\frac{DP}{Dt}. \tag{5-18}$$

When thermal conductivity is a constant, the heat equation may be rewritten as

$$\partial_o T + v_j \partial_j T = \alpha\, \partial_j \partial_j T + \frac{\beta T}{\rho C_p} \frac{DP}{Dt} \quad (k = const.) \tag{5-19}$$

where

$$\beta = -\frac{1}{\rho} \left(\frac{\partial \rho}{\partial T} \right)_P \tag{5-20}$$

is the thermal expansion coefficient. If the fluid is a constant density liquid, $\beta = 0$, $C_p = C_{\mathrm{v}} = C$, and Eq. (5-19) becomes the same as Eq. (4-33). If the fluid is an ideal gas, $\beta = 1/T$ and Eq. (5-19) becomes the same as Eq. (5-16).

5.4 MOMENTUM EQUATION

Momentum can be created in a flow by any force imbalance acting on the fluid. Two forces that arise frequently in flows are pressure and gravity. Figure 5-2 illustrates the gravitational force acting in the x-direction as well as pressure imbalance in the x-direction acting on the faces of the differential volume. Fluid passing through the differential volume will experience an increase in x-direction momentum related to the sum of $-\partial P/\partial x$, due to the pressure imbalance, and ρg_x, due to the component of gravity acting in the x-direction. This is generalized to describe all three spatial directions by substituting the i-index for "x". Therefore, $-\partial_i P$ plus ρg_i represents the force imbalance in the i-index direction, which appears as a source in the momentum equation:

$$\partial_o(\overbrace{\rho v_i}^{stored}) = -\partial_j[\overbrace{\rho v_i v_j}^{advection} + \overbrace{M_{ji}}^{diffusion}] + \overbrace{-\partial_i P + \rho g_i}^{generation}. \tag{5-21}$$

The diffusion term can usually be evaluated with Newton's viscosity law [1]:

$$M_{ji} = -\mu(\partial_j v_i + \partial_i v_j) + (2/3)\mu \delta_{ji} \partial_k v_k. \tag{5-22}$$

Following the steps performed in Section 4.4, the momentum equation can be simplified for a $\rho = const.$ fluid, to

$$\rho(\partial_o v_i + v_j \partial_j v_i) = \partial_j(\mu \partial_j v_i) - \partial_i P + \rho g_i \quad (\rho = const.). \tag{5-23}$$

This result is similar to Eq. (4-37) except for the appearance of the pressure gradient and body force that are associated with sources of momentum. Dividing through by density and assuming μ to be constant yields

$$\partial_o v_i + v_j \partial_j v_i = \nu\, \partial_j \partial_j v_i - (\partial_i P)/\rho + g_i \quad (\rho, \mu = const.). \tag{5-24}$$

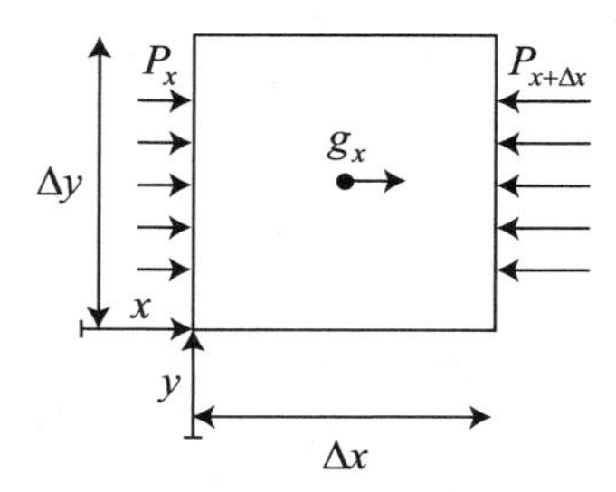

Figure 5-2 Forces acting on a differential fluid volume.

Table 5-1 Summary of transport equations in typical form

Continuity	$\partial_o \rho + \partial_i(\rho v_i) = 0$
	$\overbrace{\partial_o v_i}^{increase} + \overbrace{v_j \partial_j v_i}^{advection} = \overbrace{\nu\, \partial_j \partial_j v_i}^{diffusion} + \overbrace{(-\partial_i P/\rho + g_i)}^{generation}$
Momentum ($\rho, \mu = const.$)	$\partial_o v_i + v_j \partial_j v_i = \nu\, \partial_j \partial_j v_i + (-\partial_i P/\rho + g_i)$
Heat*,** ($k = const.$)	$\partial_o T + v_j \partial_j T = \alpha\, \partial_j \partial_j T + \dfrac{\beta T}{\rho C_p}\dfrac{DP}{Dt}$
Species ($\rho Đ_A = const.$)	$\partial_o \omega_A + v_j \partial_j \omega_A = Đ_A \partial_j \partial_j \omega_A + r_A/\rho$

*Without viscous heat generation.
**$\beta = 1/T$ for an ideal gas and $\beta = 0$ for a constant density liquid.

In terms of the substantial derivative, the momentum equation can be written as

$$\frac{Dv_i}{Dt} = \nu\, \partial_j \partial_j v_i - \frac{1}{\rho}\partial_i P + g_i \quad (\rho, \mu = const.). \tag{5-25}$$

The momentum equations written for a Newtonian fluid are known as the *Navier-Stokes equations.** Equation (5-25) is the incompressible form of the Navier-Stokes equations, for a constant viscosity fluid.

Table 5-1 summarizes the transport equations derived in this chapter. Notice that every transport equation has the form

$$\underbrace{\partial_o(\;)}_{storage} + \underbrace{v_j \partial_j(\;)}_{advection} = \underbrace{\partial_j \partial_j(\;)}_{diffusion} + \underbrace{(\;)}_{generation}. \tag{5-26}$$

It should be remembered that constant fluid properties are assumed for the transport equations summarized in Table 5-1. Later chapters will consider problems in which fluid properties are variable. Although not all conceivable sources have been considered, formulating any of the conservation statements for additional special considerations is not difficult. One noteworthy example is viscous heating, which is absent from the heat equation in Table 5-1, and will be included in Section 5.6.

5.5 KINETIC ENERGY EQUATION

The transport of kinetic energy as a conservation statement alone has little utility, as will be seen shortly. However, this transport equation can be developed to illustrate the presence of viscous dissipation of kinetic energy, which becomes heat. The amount of kinetic energy in a unit volume is quantified by $\rho v^2/2$, where $v^2 = v_i v_i$, and the advection flux of kinetic energy is $\rho(v^2/2)v_j$. Less obviously, the diffusion flux of kinetic energy is given by

$$M_{ji} v_i = j\text{-direction diffusion flux of kinetic energy}, \tag{5-27}$$

*In honor of the French mathematician and physicist Claude-Louis Navier (1785–1836) and Irish mathematician and physicist George Gabriel Stokes (1819–1903).

where M_{ji} is the momentum diffusion flux tensor. This expression can easily be demonstrated for the simple case of a unidirectional flow $v_x(y)$ of a fluid with constant density. In this case, it would be sensible to postulate that the diffusion flux of kinetic energy in the y-direction is given by

$$-\mu\partial(v_x^2/2)/\partial y = -\mu(\partial v_x/\partial y)v_x = -\mu(\partial v_x/\partial y + \overbrace{\partial v_y/\partial x}^{=0})v_x = M_{yx}v_x, \tag{5-28}$$

which demonstrates the claimed relation between the diffusion of kinetic energy and diffusion of momentum. Specifically, the kinetic energy diffusion flux is given by the dot product of the momentum diffusion flux tensor with the velocity vector.

Next, the sources of kinetic energy need to be considered. The force imbalance acting on the differential volume transforms work into kinetic energy at a rate proportional to the fluid velocity. Therefore, the rate of work performed on the fluid by the pressure gradient and gravity is given by

$$(rate\ \ of\ \ work) = v_i(-\partial_i P + \rho g_i). \tag{5-29}$$

Additionally, viscous forces in the fluid will dissipate kinetic energy into heat. However, it is less clear how to quantify this effect. At this point, the viscous dissipation term is temporarily overlooked, and the transport equation for kinetic energy is tentatively written as:

$$\partial_o\overbrace{(\rho v^2/2)}^{stored} = -\ \partial_j[\overbrace{\rho(v^2/2)v_j}^{advection} + \overbrace{M_{ji}v_i}^{diffusion}] + \overbrace{v_i(-\partial_i P + \rho g_i) - \underbrace{(?)}_{viscous\ heat}}^{generation} \tag{5-30}$$

or

$$\partial_o(\rho v^2/2) + \partial_j[\rho(v^2/2)v_j] = -\partial_j[M_{ji}v_i] + v_i(-\partial_i P + \rho g_i) - \underbrace{(?)}_{viscous\ heat}\ . \tag{5-31}$$

Working on the left-hand side of the kinetic energy equation (5-31) yields,

$$\begin{aligned}\partial_o(\rho v^2/2) + \partial_j[\rho(v^2/2)v_j] &= \rho\partial_o(v^2/2) + (v^2/2)\underbrace{[\partial_o\rho + \partial_j(\rho v_j)]}_{by\ continuity\ =\ 0} + \rho v_j\partial_j(v^2/2) \\ &= \rho\frac{D(v^2/2)}{Dt} = \rho v_i\frac{Dv_i}{Dt}.\end{aligned} \tag{5-32}$$

Working on the right-hand side of the kinetic energy equation (5-31), the diffusion term can be expressed as

$$\partial_j(M_{ji}v_i) = v_i\partial_j M_{ji} + M_{ji}\partial_j v_i\ . \tag{5-33}$$

Therefore, the transport equation for kinetic energy becomes

$$\rho v_i\frac{Dv_i}{Dt} = -v_i\partial_j M_{ji} - M_{ji}\partial_j v_i + v_i(-\partial_i P + \rho g_i)\ -\ \underbrace{(?)}_{viscous\ heat}\ . \tag{5-34}$$

or

$$\underbrace{v_i\left(\rho\frac{Dv_i}{Dt}+\partial_j M_{ji}+\partial_i P-\rho g_i\right)}_{by\ momentum\ Eq.=0}=-M_{ji}\partial_j v_i-\underbrace{(?)}_{viscous\ heat}. \tag{5-35}$$

The appearance of the momentum equation in the last form of the kinetic energy equation causes the left-hand side of Eq. (5-35) to evaluate to zero. Since conservation of kinetic energy must be reconciled with the remaining terms, in becomes apparent that the required expression for viscous heating is

$$(viscous\ \ heat)=-M_{ji}\partial_j v_i. \tag{5-36}$$

Armed with this information, the correct initial transport equation for kinetic energy should be written as

$$\partial_o(\overbrace{\rho v^2/2}^{stored})=-\partial_j[\overbrace{\rho(v^2/2)v_j}^{advection}+\overbrace{M_{ji}v_i}^{diffusion}]+\overbrace{v_i(-\partial_i P+\rho g_i)}^{source}-\overbrace{-M_{ji}\partial_j v_i}^{sink}. \tag{5-37}$$

Then, with the loss of kinetic energy to thermal energy accounted for, the transport equation for kinetic energy reduces to

$$v_i\left(\rho\frac{Dv_i}{Dt}+\partial_j M_{ji}+\partial_i P-\rho g_i\right)=0. \tag{5-38}$$

Of course, this is simply a dot product between the momentum equation and the velocity field. Therefore, any flow solution satisfying the momentum equation will also satisfy the kinetic energy equation. Because it is redundant to impose both equations on a solution, the transport equation for kinetic energy is superfluous. However, one can take comfort in the knowledge that kinetic energy is conserved by flows that satisfy the momentum equation.

Another important outcome of the present investigation is the ability to quantify viscous dissipation. Since the viscous dissipation is a sink term ($-M_{ji}\partial_j v_i$) in the kinetic energy equation, it should reappear as a source term in the heat equation when viscous heating is accounted for.

5.6 HEAT EQUATION (WITH VISCOUS HEATING)

For flows with low to moderate rates of shear, viscous generation of heat is often negligible compared with other terms in the heat equation. However, if this source of thermal energy is included in the heat equation, the result for a compressible gas becomes

$$\partial_o(\overbrace{\rho u}^{stored})=-\partial_j(\overbrace{\rho u v_j}^{advection}+\overbrace{-k\partial_j T}^{diffusion})+\overbrace{-P\partial_j v_j+-M_{ji}\partial_j v_i}^{generation}. \tag{5-39}$$

The term ($-M_{ji}\partial_j v_i$), which was subtracted from the kinetic energy equation as a loss of energy, is now added to the heat equation as a source. Following the steps performed in Section 5.3, the heat equation can be written as

$$\rho C_p\partial_o T+\rho C_p v_j\partial_j T=\partial_j(k\partial_j T)+\beta T\frac{DP}{Dt}-M_{ji}\partial_j v_i, \tag{5-40}$$

which is identical to Eq. (5-18) except for the appearance of viscous heating. Dividing both sides of the equation by ρC_p, and taking k to be a constant, yields

$$\partial_o T+v_j\partial_j T=\alpha\,\partial_j\partial_j T+\frac{\beta T}{\rho C_p}\frac{DP}{Dt}-\frac{M_{ji}\partial_j v_i}{\rho C_p}\quad (k=const.), \tag{5-41}$$

where M_{ji} is given by Newton's viscosity law (5-22). The heat equation can also be derived by subtracting the mechanical energy equation from the total energy equation, as done in Problem 5.2. The total energy equation accounts for both kinetic and thermal forms of energy, and is derived later in Section 5.8.3 starting with a finite control volume.

5.7 ENTROPY GENERATION IN IRREVERSIBLE FLOWS

As discussed in Section 2.4, transport is *irreversible* if entropy generation occurs. Such transport is irreversible because the production of entropy in the process would need to become a destruction of entropy in the reversed process—which is not allowed by the second law of thermodynamics. Entropy rise in a flow can be investigated with the heat equation. The heat equation can be expressed in the form (see Problem 5.2):

$$\rho \frac{Du}{Dt} = -\partial_i q_i - P\partial_i v_i - M_{ji}\partial_j v_i, \tag{5-42}$$

where $q_i = -k\partial_i T$ by Fourier's law and M_{ji} is given by Newton's viscosity law (5-22). The fundamental thermodynamic equation $du = Tds + (P/\rho^2)d\rho$ requires a flow to obey the relation

$$\frac{Du}{Dt} = T\frac{Ds}{Dt} + \frac{P}{\rho^2}\frac{D\rho}{Dt}. \tag{5-43}$$

Therefore, the heat equation can be written in terms of entropy as

$$\rho T\frac{Ds}{Dt} + \frac{P}{\rho}\frac{D\rho}{Dt} = -\partial_i q_i - P\partial_i v_i - M_{ji}\partial_j v_i. \tag{5-44}$$

Using continuity in the form $D\rho/Dt = -\rho\partial_i v_i$, the heat equation simplifies to a transport equation for entropy:

$$\rho\frac{Ds}{Dt} = \frac{-\partial_i q_i}{T} + \frac{-M_{ji}\partial_j v_i}{T}. \tag{5-45}$$

Although heat diffusion is irreversible, part of the entropy rise in Eq. (5-45) is reversible. Specifically, the amount of entropy transfer that occurs with heat transfer at constant temperature is reversible. Therefore, for clearer interpretation, Eq. (5-45) can be written as

$$\rho\frac{Ds}{Dt} = \overbrace{-\partial_i\left(\frac{q_i}{T}\right)}^{\text{reversible}} + \underbrace{\overbrace{\frac{-q_i\partial_i T}{T^2}}^{\text{irreversible}} + \overbrace{\frac{-M_{ji}\partial_j v_i}{T}}^{\text{irreversible}}}_{\text{entropy generation}}. \tag{5-46}$$

Notice that the entropy generation related to heat transfer is given by

$$-q_i\frac{\partial_i T}{T^2} = k\frac{\partial_i T\partial_i T}{T^2}, \tag{5-47}$$

with the use of Fourier's law. This quantity is guaranteed to be positive since thermal conductivity is a positive number. Likewise, entropy generation related to viscous heating is guaranteed to be positive by Newton's viscosity law, since the dynamic viscosity is a positive number.

It can be inferred from Eq. (5-46) that flows including any form of diffusion transport are irreversible. Therefore, reversible flows exclude viscous flows because of the appearance of $(-M_{ji}\partial_j v_i)/T$ as a source of entropy generation, and exclude flows with heat diffusion because of the appearance of $(-q_i\partial_i T)/T^2$ as a source of entropy generation. Without diffusion transport, the entropy equation (5-46) simplifies to

$$\frac{Ds}{Dt} = 0 \quad \text{(reversible flow)} \tag{5-48}$$

which is the definition of an *isentropic flow*, and is synonymous to a reversible flow.

5.8 CONSERVATION STATEMENTS DERIVED FROM A FINITE VOLUME

Conservation equations, including source terms, can be derived from a finite control volume, as discussed in Section 4.6. Figure 5-3 illustrates the inclusion of source terms acting on the surface area and body of the control volume. The control volume approach is preferred when an integral expression is the desired result, but also yields the differential form of the conservation equations with the use of Gauss's theorem, as discussed in Section 4.6. Quasi-one-dimensional equations will be derived with the integral approach in later chapters; these equations are differential with respect to one direction only, and finite with respect to others. The control volume approach will also be used to analyze discontinuities in flows, such as hydraulic jumps (Chapter 18) and shock waves (Chapter 20), where the differential form of conservation equations fail.

Table 5-2 summarizes the terms appearing in the conservation equations for continuity, momentum, and total energy. It should be noted that two forms of energy are considered: internal u and kinetic $v^2/2$. Therefore, advection and diffusion fluxes must account for both forms. Additionally, sources of momentum and energy can be introduced to the flow through both surface and volumetric effects. Therefore, source terms include both a surface area integral and a volumetric integral, as shown in the last two columns of Table 5-2. A negative sign precedes the pressure term since pressure acts in an opposing direction to the surface normal vector.

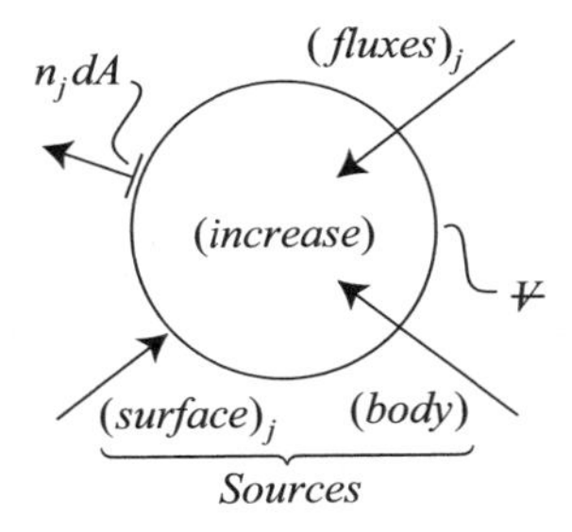

Figure 5-3 Analysis of an arbitrary control volume.

5.8.1 Continuity

The continuity equation is derived in integral form starting with

$$\begin{array}{ccc} (increase) & = & (in - out) \\ \downarrow & & \downarrow \\ \int_{\forall} \frac{\partial}{\partial t}(stored)\,d\forall & = & -\int_A (fluxes)_j n_j\, dA \\ \downarrow & & \downarrow \\ \int_{\forall} \partial_o(\rho)\,d\forall & = & -\int_A (\rho\, v_j) n_j\, dA. \end{array} \tag{5-49}$$

Table 5-2 Terms appearing in the conservation equations for continuity, momentum, and energy

	Stored	*Fluxes*		*Sources*	
		Advection	*Diffusion*	*Surface*	*Body*
Continuity:	ρ	$\rho\, v_j$			
Momentum:	ρv_i	$\rho v_i\, v_j$	M_{ji}	$-P\delta_{ji}$	ρg_i
Energy: (internal + kinetic)	$\rho(u+\frac{v^2}{2})$	$\rho(u+\frac{v^2}{2})v_j$	$q_j + M_{ji}v_i$	$-P\delta_{ji}v_i$	$\rho g_i v_i$
	$\int_{\forall} \partial_o(\;)d\forall$	$\int_A (\;)n_j dA$	$\int_A (\;)n_j dA$	$\int_A (\;)n_j dA$	$\int_{\forall} (\;)d\forall$

$M_{ji} = -\mu(\partial_j v_i + \partial_i v_j) + (2/3)\mu\delta_{ji}\partial_k v_k$ - Newton's law for momentum diffusion
$q_j = -k\partial_j T$ - Fourier's law for heat diffusion

The differential form of the continuity equation can be obtained by applying Gauss's theorem to transform the area integral of the advection flux into a volume integral. Subsequently, the terms are written under a common volume integral:

$$\begin{array}{ccc} \int_{\forall} \partial_o \rho \, d\forall & = & -\int_{\forall} \partial_j(\rho\, v_j) d\forall \\ \downarrow & & \downarrow \\ \int_{\forall} [\partial_o \rho & = & -\partial_j(\rho\, v_j)]d\forall. \end{array} \tag{5-50}$$

From the last step, it is concluded that the differential expression appearing in the integrand must be identically true if the choice of volume integral in the fluid is to be arbitrary. Therefore, the differential form of the continuity equation must be satisfied:

$$\partial_o \rho + \partial_j(\rho v_j) = 0. \tag{5-51}$$

5.8.2 Momentum

The momentum equation is derived from integral form starting with

$$\begin{array}{ccccc} (increase) & = & (in - out) & + & (sources - sinks) \\ \downarrow & & \downarrow & & \downarrow \\ \int_{\forall} \frac{\partial}{\partial t}(stored) d\forall & = & -\int_A (fluxes)_j n_j dA & + & \int_A (surface)_j n_j dA + \int_{\forall} (body) d\forall \\ \downarrow & & \downarrow & & \downarrow \\ \int_{\forall} \partial_o(\rho v_i) d\forall & = & -\int_A (\rho v_i\, v_j + M_{ji}) n_j dA & + & \int_A (-P\delta_{ji}) n_j dA + \int_{\forall} (\rho g_i) d\forall. \end{array} \tag{5-52}$$

The differential form of the momentum equation can be obtained by applying Gauss's theorem, and writing the volume integrals under a common integral:

$$\begin{array}{ccccc} \int_{\forall} \partial_o(\rho v_i) d\forall & = & -\int_{\forall} \partial_j(\rho v_i\, v_j + M_{ji}) d\forall & + & \int_{\forall} \partial_j(-P\delta_{ji}) d\forall + \int_{\forall} (\rho g_i) d\forall \\ \downarrow & & \downarrow & & \downarrow \\ \int_{\forall} [\partial_o(\rho v_i) & = & -\partial_j(\rho v_i\, v_j + M_{ji}) & - & \partial_j(P\delta_{ji}) + \rho g_i] d\forall. \end{array} \tag{5-53}$$

Notice that $\partial_j(P\delta_{ji})$ is nonzero only when $\partial_j(P\delta_{ji}) = \partial_i P$. From the last step, it is concluded that the differential expression appearing in the integrand of the volume integral must be identically true, such that the differential form of the momentum equation is

$$\partial_o(\rho v_i) + \partial_j(\rho v_i v_j) = -\partial_j M_{ji} - \partial_i P + \rho g_i. \tag{5-54}$$

With the use of continuity, the left-hand side can be rewritten such that the momentum equation becomes

$$\rho(\partial_o v_i + v_j \partial_j v_i) = -\partial_j M_{ji} - \partial_i P + \rho g_i. \tag{5-55}$$

5.8.3 Total Energy

Up to this point, the total energy equation has not been derived. Rather, equations for two forms of energy, heat and kinetic, were derived separately. Making use of the notation $e = (u + v^2/2)$, the total energy equation can be derived from integral form, starting with

$$\begin{array}{ccccc}
(\textit{increase}) & = & (\textit{in} - \textit{out}) & + & (\textit{sources} - \textit{sinks}) \\
\downarrow & & \downarrow & & \downarrow \\
\int_{V} \frac{\partial}{\partial t}(\textit{stored})dV & = & -\int_{A} (\textit{fluxes})_j n_j dA & + & \int_{A} (\textit{surface})_j n_j dA + \int_{V} (\textit{body}) dV \\
\downarrow & & \downarrow & & \downarrow \\
\int_{V} \partial_o(\rho e) dV & = & -\int_{A} (\rho e\, v_j + q_j + M_{ji} v_i) n_j dA & + & \int_{A} (-P\delta_{ji} v_i) n_j dA + \int_{V} (\rho g_j v_j) dV.
\end{array} \tag{5-56}$$

The differential form of the total energy equation is obtained by applying Gauss's theorem, and writing the integrals under a common volume integral:

$$\begin{array}{ccccc}
\int_{V} \partial_o(\rho e) dV & = & -\int_{V} \partial_j(\rho e\, v_j + q_j + M_{ji} v_i) dV & + & \int_{V} \partial_j(-P\delta_{ji} v_i) dV + \int_{V} (\rho g_j v_j) dV \\
\downarrow & & \downarrow & & \downarrow \\
\int_{V} \big[\partial_o(\rho e) & = & -\partial_j(\rho e\, v_j + q_j + M_{ji} v_i) & - & \partial_j(P\delta_{ji} v_i) + \rho g_j v_j\big] dV.
\end{array} \tag{5-57}$$

Again, notice that $\partial_j(P\delta_{ji}v_i)$ is nonzero only when $\partial_j(P\delta_{ji}v_i) = \partial_j(Pv_j)$. From the last step, it is concluded that the differential expression appearing in the integrand of the volume integral must be identically true, such that the differential form of the total energy equation is

$$\partial_o(\rho e) + \partial_j(\rho e\, v_j) = -\partial_j q_j - \partial_j(M_{ji} v_i) - \partial_j(P v_j) + \rho g_j v_j. \tag{5-58}$$

With the use of continuity, the left-hand side can be rewritten such that the total energy equation becomes

$$\rho(\partial_o e + v_j \partial_j e) = -\partial_j q_j - \partial_j(M_{ji} v_i) - \partial_j(P v_j) + \rho g_j v_j. \tag{5-59}$$

Most often, it is not convenient to use $e = (u + v^2/2)$ as the dependent variable of a problem. Consequently, the energy equation is typically cast into an alternate form before it is solved, such as illustrated in Chapter 20 for compressible flows. Since conservation of kinetic energy is already covered by the momentum equation, additional heat conservation is often handled more simply by satisfying the heat equation (5-41) than by applying the total energy equation (5-59). When a flow is incompressible, there is no mechanism by which heat can be converted into kinetic energy, and the momentum equation may be solved independent of the heat equation. However, because of advection, the heat equation is always coupled to the solution of the momentum equation when there is fluid flow.

5.9 LEIBNIZ'S THEOREM

In the preceding sections, transport equations were derived in a stationary or *Eulerian frame* of reference. Sometimes conservation statements are initially derived in a Lagrangian frame, and subsequently switched to an Eulerian frame. The *Lagrangian frame* of reference is one moving with the fluid, as discussed in Section 4.7. In this frame of reference, the volume identifying a control mass in the fluid can distort with time. Additionally, diffusion is the only mechanism of transport between the control mass and the surrounding flow. Leibniz's theorem is used to transform the time derivative of a volume integral in the Lagrangian frame of reference into an equivalent expression for the Eulerian frame, where the control volume becomes fixed in time. The *theorem of Leibniz*† states

$$\frac{d}{dt}\int_{\forall(t)}(\ )d\forall = \int_{\forall}\frac{\partial(\ )}{\partial t}d\forall + \int_{A}(\ )v_j n_j dA, \tag{5-60}$$

where the absent quantity inside the parentheses "( )" can be a scalar, vector, or tensor function. The volume $\forall$ associated with the integral on the left-hand side of equation (5-60) has a boundary that moves as a function of time with the fluid velocity v_j in the Lagrangian perspective. However, the volume $\forall$ associated with the integral on the right-hand side of the equation is fixed in the Eulerian perspective, and bounded by a surface area A.

In fluid mechanics, Eq. (5-60) is also referred to as *Reynolds' transport theorem* [2], for his application of this statement to conserved properties of a flow. In this context, the rate at which the content of a fluid property within the control volume increases with time is equated to net sources of that fluid property plus the net effect of diffusion transport through the bounding surface of the volume. Advection is not explicitly considered initially, because the control volume is moving with the fluid. However, advection is revealed through the use of Leibniz's theorem. To illustrate, consider conservation of species applied to an arbitrary volume moving with the fluid:

$$\begin{array}{ccccc}
(increase) & = & (in - out) & + & (sources - sinks)\\
\downarrow & & \downarrow & & \downarrow\\
\dfrac{d}{dt}\displaystyle\int_{\forall(t)}(stored)d\forall & = & -\displaystyle\int_{A(t)}(fluxes)_j n_j dA & + & \displaystyle\int_{\forall(t)}(reaction)d\forall\\
\downarrow & & \downarrow & & \downarrow\\
\dfrac{d}{dt}\displaystyle\int_{\forall(t)}(\rho_A)d\forall & = & -\displaystyle\int_{A(t)}(\underbrace{-\rho D_A\partial_j(\rho_A/\rho)}_{diffusion})n_j dA & + & \displaystyle\int_{\forall(t)}(r_A)d\forall.
\end{array} \tag{5-61}$$

---

†Leibniz's theorem is named after the German mathematician and philosopher Gottfried Wilhelm Leibniz (1646–1716).

Notice that in the Lagrangian frame of reference, diffusion is the only transport mechanism with which to express fluxes of species between the control volume and the flow. Applying Leibniz's theorem produces the conservation statement in an Eulerian frame of reference:

$$\begin{array}{ccccccc} \frac{d}{dt}\int\limits_{\forall(t)}(\rho_A)\,d\forall & & = -\int\limits_{A(t)}(-\rho \mathcal{D}_A \partial_j(\rho_A/\rho))n_j dA & + & \int\limits_{\forall(t)}(r_A)d\forall \\ \downarrow & & \downarrow & & \downarrow \\ \int\limits_{\forall}\frac{\partial(\rho_A)}{\partial t}d\forall + \int\limits_{A}\underbrace{(\rho_A v_j)}_{advection}n_j dA & = & -\int\limits_{A}\underbrace{(-\rho \mathcal{D}_A \partial_j(\rho_A/\rho))}_{diffusion}n_j dA & + & \int\limits_{\forall}(r_A)d\forall. \end{array} \quad (5\text{-}62)$$

Notice now the appearance of the advection term in the Eulerian frame of reference. Next, Gauss's theorem is used to express area integrals as volume integrals:

$$\begin{array}{ccccccc} \int\limits_{\forall}\frac{\partial(\rho_A)}{\partial t}d\forall + \int\limits_{A}(\rho_A v_j)n_j dA & = & -\int\limits_{A}(-\rho \mathcal{D}_A \partial_j(\rho_A/\rho))n_j dA & + & \int\limits_{\forall}(r_A)d\forall \\ \downarrow & & \downarrow & & \downarrow \\ \int\limits_{\forall}\partial_o(\rho_A)d\forall + \int\limits_{\forall}\partial_j(\rho_A v_j)d\forall & = & \int\limits_{\forall}\partial_j(\rho \mathcal{D}_A \partial_j(\rho_A/\rho))d\forall & + & \int\limits_{\forall}(r_A)d\forall \\ \downarrow & & \downarrow & & \downarrow \\ \int\limits_{\forall}[\partial_o(\rho_A) + \partial_j(\rho_A v_j) & = & \partial_j(\rho \mathcal{D}_A \partial_j(\rho_A/\rho)) & + & r_A]d\forall. \end{array} \quad (5\text{-}63)$$

After collecting all the differential terms under a common volume integral, the usual argument is made that the integrand expression is identically true. For the current example, this results in a differential equation that is identical to Eq. (5-4).

5.10 LOOKING AHEAD

In some cases, the transport equations derived in this chapter exhibit significant complexity. However, many practical problems admit substantial simplifications that can be made to these more general equations. It is important to look for ways to simplify the mathematical description of problems without losing important physics. Developing intuition for the physically significant terms in transport equations is an important skill. This will be accomplished by first looking at problems governed by diffusion transport alone in Chapters 7 through 13, followed by problems governed by advection transport alone in Chapters 14 through 22. In Chapter 24 and thereafter, convection problems in which both advection and diffusion are important to transport will be considered. Within these broad classifications of flows, other means by which transport equations can be simplified are identified. This includes the use of quasi-one-dimensional equations, as applied to advection transport in open channel flows (Chapters 18 and 19) and in compressible flows (Chapter 21). Convection transport can be simplified through the use of the boundary layer approximation, as applied in Chapters 24 through 26 to laminar flows and in Chapter 35 to turbulent flow. Convection is also simplified by the state of fully developed transport as discussed in Chapter 27, and applied to laminar flows in Chapter 28 and to turbulent flows in Chapters 31, 33, 34, and 37.

5.11 PROBLEMS

5-1 Vorticity is defined as $\omega_k = \varepsilon_{kli}\partial_l v_i$ ($\vec{\omega} = \vec{\nabla} \times \vec{v}$), and is twice the angular velocity of a fluid element. Derive the transport equation for vorticity by writing a conservation statement for ($\rho\omega_k$) over a control volume. Postulate a constitutive law for the diffusion flux of vorticity. A source term (generation) for vorticity, $\rho\omega_j\partial_j v_k$, is also required to reflect the fact that fluid "stretching" can increase vorticity. (This phenomenon is analogous to an ice skater spinning with her arms held horizontally outward, who can increase her spinning rate by raising her arms over her head). Use index notation throughout the derivation, and express your result in terms of the substantial derivative of vorticity. You may assume that viscosity is a constant. Demonstrate the validity of your equation by taking the curl of the incompressible Navier-Stokes equations, and compare the result with your transport equation for vorticity.

5-2 By taking the dot product of the momentum equation (5-55) with v_i, show that the mechanical energy equation is

$$\rho\frac{D(v^2/2)}{Dt} = -v_i\partial_j M_{ji} - v_i\partial_i P + \rho v_i g_i.$$

By subtracting the mechanical energy equation from the total energy equation (5-59), show that the thermal energy equation is given by

$$\rho\frac{Du}{Dt} = -\partial_i q_i - P\partial_i v_i - M_{ji}\partial_j v_i.$$

Show that this result is the same as Eq. (5-41) derived in Section 5.6. Expand the thermal energy equation in two-dimensional Cartesian coordinates.

5-3 Starting with a control volume in the Lagrangian frame of reference, derive an integral statement for the conservation of momentum. From this integral statement, derive the differential forms for the momentum equation in the Eulerian and Lagrangian frames of reference for a variable density flow.

5-4 In some flows, the pressure field is relatively unaltered from the hydrostatic condition (no flow). Solve the momentum equation for the hydrostatic pressure field, letting $\rho = \rho_\infty$. If at some location in the fluid, the density experiences a small change $\rho = \rho_\infty + \Delta\rho$, explain why this might not affect the pressure field. Evaluate the momentum equation for a local condition where $\rho = \rho_\infty + \Delta\rho$, when $\Delta\rho \ll \rho$. Assume the density change $\Delta\rho$ in the fluid is caused by a temperature change ΔT, as related through the thermal expansion coefficient: $\beta = -(\partial\rho/\partial T)_P/\rho$. For the coordinates shown in the illustration, demonstrate that the x-direction momentum equation is

$$\frac{Dv_x}{Dt} = \nu\, \partial_j\partial_j v_x + \beta g \Delta T.$$

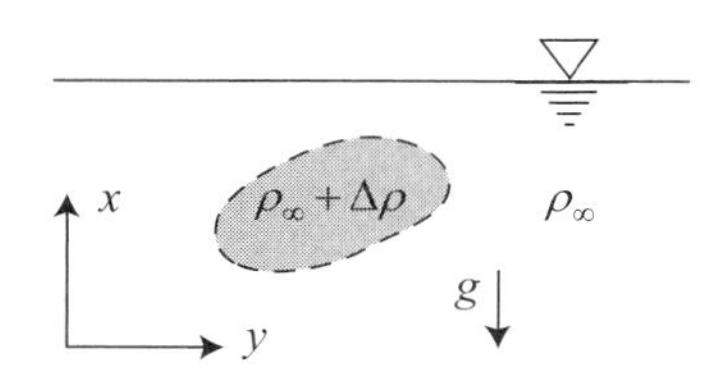

Provide a physical interpretation for each term in the momentum equation.

5-5 Consider a flow governed by the momentum equation derived in Problem 5-4, where the effects of buoyancy drive the flow. Assuming the geometry of the flow is two-dimensional, how many unknown dependent variables are there in the momentum equation? Provide as many additional equations as are required to solve this flow problem. Simplify these equations for a flow that experiences relatively small changes in density. Make any other simplifications to the governing equations that are appropriate, providing some justification.

5-6 Consider a reversible flow that is adiabatic ($q_j = 0$) and isentropic ($Ds/Dt = 0$). Demonstrate that for such a flow, the energy equation is satisfied by the momentum equation.

REFERENCES

[1] G. K. Batchelor, *An Introduction to Fluid Dynamics*. Cambridge: Cambridge University Press, 1967.

[2] O. Reynolds, *On the Sub-Mechanics of the Universe*. Published for the Royal Society of London, Cambridge: Cambridge University Press, 1903.

Chapter 6

Specification of Transport Problems

6.1 Classification of Equations
6.2 Boundary Conditions
6.3 Elementary Linear Examples
6.4 Nonlinear Example
6.5 Scaling Estimates
6.6 Problems

Solutions to physical problems are constrained by the conservation concepts developed in Chapters 4 and 5, which are manifested as transport equations written in differential form. These differential equations specify how a *dependent variable* can change with respect to one or more *independent variables* (describing space and time) without violating a conservation law. It is useful to classify these equations in a number of ways. Additionally it is important to understand how boundary conditions constrain integration of these equations and how estimates of their solution can be obtained.

6.1 CLASSIFICATION OF EQUATIONS

Differential equations describe an unknown function, in terms of its derivatives and independent variables. An equation that is differential with respect to more than one independent variable is a *partial differential equation* (PDE). In contrast, an *ordinary differential equation* (ODE) contains derivatives of the unknown function with respect to only one independent variable. *Nonlinear equations* contain terms that are a product of the dependent variable and itself, or its dependencies, such as derivatives of the dependent variable. For example, consider the incompressible Navier-Stokes equations:

$$\partial_o v_i + v_j \partial_j v_i = \nu\, \partial_j \partial_j v_i - \frac{1}{\rho} \partial_i P + g_i \quad (i = x, y \text{ or } z). \tag{6-1}$$

This is a nonlinear equation for v_i by virtue of the term $v_j \partial_j v_i$. However, the heat equation in the form

$$\partial_o T + v_j \partial_j T = \alpha\, \partial_j \partial_j T \tag{6-2}$$

is a *linear equation* for T, so long as v_j and α is independent of T.

An equation describing a dependent variable, call it ϕ, is *homogeneous* if $\phi = 0$ is a particular solution to that equation. (Note that when ϕ is identically zero, it must also be the case that $\partial_o \phi = 0$ and $\partial_j \phi = 0$.) Therefore, the Navier-Stokes equations (6-1) are *nonhomogeneous* by virtue of the terms $\partial_i P/\rho$ and g_i. However, the heat equation (6-2) is homogeneous, since $T = 0$ is a solution to that equation.

An easy test of whether or not an equation is both linear and homogeneous is to multiply appearances of the dependent variable by an arbitrary constant. If the resulting expression is unaltered from the initial equation, the equation must be linear and homogeneous. To illustrate with the heat equation (6-2), appearances of T are replaced with cT:

$$\partial_o(cT) + v_j\partial_j(cT) = \alpha\, \partial_j\partial_j(cT), \tag{6-3}$$

which is the same as

$$c(\partial_o T + v_j\partial_j T) = c(\alpha\, \partial_j\partial_j T) \quad \rightarrow \quad \partial_o T + v_j\partial_j T = \alpha\, \partial_j\partial_j T. \tag{6-4}$$

Since the resulting expression is the same as the initial equation, the heat equation (6-2) is linear and homogeneous.

6.2 BOUNDARY CONDITIONS

Although a governing equation specifies how a dependent variable can change spatially, and in time, it is impractical to carry out that description throughout all space and all time. It is necessary to prescribe in some way the dependent variable at the edges of a spatial domain and at some point in time. The number of these conditions needed to uniquely identify a solution depends on the governing equation's order and type. The order of an equation refers to the highest order of differentiation of the unknown function with respect to the dependent variable. For example, $\partial_j\partial_j\phi = 0$ is a second-order equation. An equation with a second-order spatial derivative will be integrated twice with respect to that variable. This introduces two integration constants that can only be determined by two independent statements concerning the condition of the dependent variable at one or more boundaries. These are *boundary conditions*. Furthermore, boundary conditions are required for each spatial direction in which integration is performed. The time dependence of most governing equations is first order, requiring that one additional condition be specified for integration in time. This is usually given by the *initial condition*.

Several types of boundary conditions arise frequently in transport problems and are discussed in this section. Establishing the correct boundary conditions can be as critical to a quantitative solution as formulating the correct governing equation. However, the gravity of this is often under-appreciated when the task of selecting boundary conditions can seem deceivingly straightforward. In reality, the largest uncertainty in many real situations will be associated with the idealization of boundary conditions.

It makes physical sense to try to specify the quantity, or the flux, of a physical property at the edges of a spatial domain. Therefore, boundary conditions are used to prescribe the value of the dependent variable or its spatial derivative, as related to a diffusion flux. Attention will be restricted to the following three kinds of boundary conditions that involve the dependent variable ϕ and its derivative $\partial_j\phi$:

$$\text{First kind:} \quad \phi = a \tag{6-5}$$

$$\text{Second kind:} \quad \partial_j\phi\ n_j = b \tag{6-6}$$

$$\text{Third kind:} \quad \partial_j\phi\ n_j + h\ \phi = c \quad (h = const.). \tag{6-7}$$

The simplest mathematical situation that can arise for these three kinds of boundary conditions is when a, b, and c are zero, corresponding to homogeneous boundary conditions. The next simplest scenario is when a, b, and c are nonzero constants. However, in

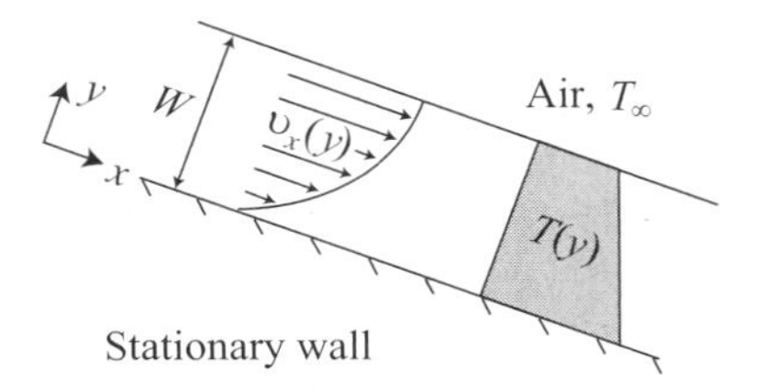

Figure 6-1 Heated film flowing on an inclined surface.

general, a, b, and c could also be functions of space and time. The boundary conditions are nonhomogeneous when a, b, and c are nonzero.

To illustrate the appearance of various kinds of boundary conditions, consider the gravity driven flow of a liquid film over a heated surface, as shown in Figure 6-1. The film is assumed to have a constant thickness a, and the flow is hydrodynamically *fully-developed*, with the properties that $v_y = 0$ and $\partial v_x/\partial x = 0$. (A full discussion of fully-developed transport is undertaken in Chapter 27.)

Boundary conditions imposed on the liquid film constrain the behavior of the functions describing velocity v_x and temperature T in the flow as they approach the boundaries. For example, the *no-slip condition* specifies that the velocity of a fluid in contact with a stationary wall must be zero:

$$v_x(y = 0) = 0 \quad \text{(no-slip condition)}. \tag{6-8}$$

This is a homogeneous boundary condition of the first kind. At the *free surface* of the liquid film, it can be argued that there is no momentum transfer to the overlying air. The free surface condition assumes that the air density is so much less than the liquid film density that it cannot be a significant recipient of a momentum flux. Therefore, without a diffusion flux of momentum crossing the boundary, the free surface boundary condition requires that

$$\partial v_x/\partial y\big|_{y=a} = 0 \quad \text{(free surface condition)}. \tag{6-9}$$

This is a homogeneous boundary condition of the second kind.

If the liquid film is heated by the wall with a specified heat flux q_s, the appropriate boundary condition imposes Fourier's law at the wall:

$$-k\partial T/\partial y\big|_{y=0} = q_s \quad \text{(heat flux boundary condition)}, \tag{6-10}$$

where k is the thermal conductivity of the liquid film. This is a nonhomogeneous boundary condition of the second kind. Notice that diffusion is the only transport mechanism that can carry heat away from the surface of the wall.

One might assume that heat is lost from the liquid film to the overlying air. It is understood that heat transfer is driven by a temperature difference between the top surface of the liquid film $T_s = T(y = a)$ and the ambient air temperature T_∞. It is convenient to define a convection coefficient h for this process that satisfies the relationship

$$q = h(T_s - T_\infty), \tag{6-11}$$

where q is the heat flux driven by the temperature difference $(T_s - T_\infty)$. This relationship is known as *Newton's convection law*. In general, h characterizes the nontrivial transport of heat by the combined effects of advection and diffusion. Therefore, determining h from first principles is one of the subjects of this text. However, when Newton's law of cooling is invoked as a boundary condition, one assumes that the task of determining h has already been accomplished. In this case, to describe the heat loss from the top surface of the liquid film, one could write the boundary condition

$$[-k\,\partial T/\partial y = h(T - T_\infty)]\big|_{y=a} \quad \text{(convective boundary condition).} \tag{6-12}$$

In words, this boundary condition states that the rate of heat transfer to the free surface by conduction (left-hand side of equation) equals the rate of heat lost by convection to the overlying air (right-hand side of equation). This is a nonhomogeneous boundary condition of the third kind.

6.3 ELEMENTARY LINEAR EXAMPLES

The application of boundary conditions to the integration of linear equations is illustrated in this section. The governing equations derived in Chapters 4 and 5 are drawn upon to describe fluid motion, heat, and mass transfer. The elementary problems addressed in this section require integration of linear ordinary differential equations. In contrast, the less straightforward task of integrating a nonlinear ordinary differential equation is illustrated in Section 6.4. A more complete discussion of solutions to differential equations can be found in the many books devoted to the subject, such as references [1] and [2].

6.3.1 Gravity Driven Flow on an Inclined Surface

Consider the problem of determining the velocity distribution of the liquid film illustrated in Figure 6-1. The fluid velocity $v_x(y)$ is governed by the steady-state form of the momentum equation:

$$\overbrace{v_j\partial_j v_x}^{advection} = \overbrace{\nu\partial_j\partial_j v_x}^{diffusion} + \overbrace{g_x}^{source}. \tag{6-13}$$

Because the liquid film is subject to the ambient pressure of the overlying air (which is assumed constant), no pressure gradient exists in the streamwise direction of the flow. Therefore, the pressure gradient term is excluded as a source of momentum in Eq. (6-13).

Using the rules of index notation, the momentum equation is expanded in two-dimensional space ($j = x, y$) for

$$v_x\overbrace{\frac{\partial v_x}{\partial x}}^{=0} + \overbrace{v_y}^{=0}\frac{\partial v_x}{\partial y} = \nu\left(\overbrace{\frac{\partial^2 v_x}{\partial x^2}}^{=0} + \frac{\partial^2 v_x}{\partial y^2}\right) + g_x. \tag{6-14}$$

With the simplifications that result for a fully developed flow ($v_y = 0$, $\partial v_x/\partial x = 0$), the governing equation for v_x becomes

$$\nu\frac{\partial^2 v_x}{\partial y^2} + g_x = 0, \tag{6-15}$$

which is a nonhomogeneous ordinary differential equation. It is interesting to note that the advection terms have completely dropped from the governing equation because of the fully developed hydrodynamic condition. Since Eq. (6-15) is a second-order equation, two boundary conditions are required to specify the two integration constants that will arise. In Section 6.2, the two boundary conditions were identified as

$$y = 0: \quad v_x(y = 0) = 0 \tag{6-16}$$

and

$$y = a: \quad \partial v_x/\partial y\big|_{y=a} = 0. \tag{6-17}$$

Integrating Eq. (6-15) twice yields

$$\nu\, v_x = -\frac{g_x}{2}y^2 + C_1 y + C_2. \tag{6-18}$$

Appling the boundary conditions reveals that

$$y = 0: \quad 0 = -0 + 0 + C_2 \quad \text{or} \quad C_2 = 0; \tag{6-19}$$

$$y = a: \quad \nu \left.\frac{\partial v_x}{\partial y}\right|_{y=a} = -g_x a + C_1 + 0 = 0 \quad \text{or} \quad C_1 = g_x a. \tag{6-20}$$

Therefore, the solution to the velocity field in the liquid film is given by

$$v_x = (g_x/\nu)(a - y/2)y. \tag{6-21}$$

The shear stress imparted by the liquid film flow on the inclined surface can be calculated from the solution:

$$\tau_{yx} = \mu \left.\frac{\partial v_x}{\partial y}\right|_{y=0} = \rho a g_x \tag{6-22}$$

This is simply the component of the liquid film's weight per unit area acting in the x-direction.

6.3.2 Heat Transfer across a Liquid Film

The problem illustrated in Figure 6-1 is revisited to determine the temperature distribution across the film $T(y)$. A situation that occurs some distance down the length of the inclined surface, where no streamwise accumulation of heat in the fluid is experienced, will be described. At this distance, $\partial T/\partial x = 0$, and only heat transfer across the film occurs. This is analogous to the fully-developed hydrodynamic conditions, discussed in Section 6.3.1, where $\partial v_x/\partial x = 0$. The steady-state temperature in the liquid film is governed by the transport equation:

$$\overbrace{v_j \partial_j T}^{\text{advection}} = \overbrace{\alpha\, \partial_j \partial_j T}^{\text{diffusion}}, \tag{6-23}$$

which is a form of the heat equation derived in Chapter 4. This form of the heat equation assumes that the liquid properties do not change significantly with temperature, and that viscous generation of heat is small compared to heat transfer across the film.

Using the rules of index notation, the heat equation is expanded in two-dimensional space ($j = x, y$) for

$$v_x \overbrace{\frac{\partial T}{\partial x}}^{=0} + \overbrace{v_y}^{=0} \frac{\partial T}{\partial y} = \alpha\left(\overbrace{\frac{\partial^2 T}{\partial x^2}}^{=0} + \frac{\partial^2 T}{\partial y^2}\right). \tag{6-24}$$

With the simplifications that result from fully-developed conditions in the liquid film, the governing equation for T becomes

$$\frac{\partial^2 T}{\partial y^2} = 0, \tag{6-25}$$

which is a homogeneous ordinary differential equation. Advection terms have again completely dropped from the governing equation because of the fully-developed conditions. Since Eq. (6-25) is a second-order differential equation, two boundary conditions are required to specify the two integration constants that will arise. In Section 6.2, the two boundary conditions were identified as

$$y = 0: \quad -k\partial T/\partial y|_{y=0} = q_s \tag{6-26}$$

$$y = a: \quad [-k\,\partial T/\partial y = h(T - T_\infty)]_{y=a}. \tag{6-27}$$

Integrating Eq. (6-25) twice yields

$$T = C_1 y + C_2. \tag{6-28}$$

Substituting this expression for T into the two boundary conditions,

$$y = 0: \quad -k\,C_1 = q_s \tag{6-29}$$

$$y = a: \quad -k\,C_1 = h(C_1 a + C_2 - T_\infty), \tag{6-30}$$

permits determination of $C_1 = -q_s/k$ and $C_2 = T_\infty + q_s(1/h + a/k)$. Therefore, the temperature profile in the liquid film becomes

$$T = T_\infty + q_s\left(\frac{1}{h} + \frac{a - y}{k}\right). \tag{6-31}$$

Owing to the fact that advection does not contribute to heat transport across the liquid layer, the same temperature distribution would result even if the liquid film were not moving.

6.3.3 Groundwater Contamination

Consider a situation in which water flows through the ground at a uniform speed U until it reaches a reservoir, as illustrated in Figure 6-2. Bacteria live in the reservoir at a concentration of c_R. It is desired to make a prediction of the steady-state concentration profile of bacteria in the groundwater as a function of distance from the reservoir $c_B(x)$. Assuming that bacteria concentrations are dilute, as discussed in Section 4.2.1, the transport equation governing the steady-state groundwater bacteria concentration is

$$\overbrace{v_j^* \partial_j c_B}^{advection} = \overbrace{Đ_B \partial_j \partial_j c_B}^{diffusion} \tag{6-32}$$

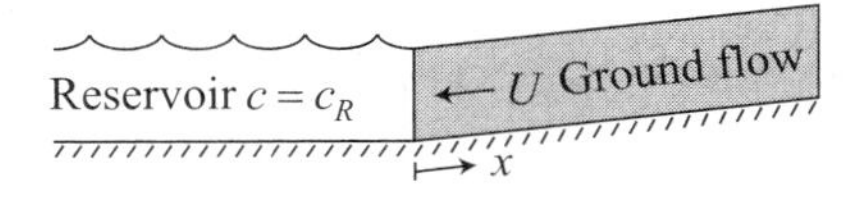

Figure 6-2 Groundwater flow to a reservoir.

Since transport in this problem is one-dimensional, the governing species equation is expanded with $j = x$:

$$-U\frac{\partial c_B}{\partial x} = Đ_B\frac{\partial^2 c_B}{\partial x^2} \tag{6-33}$$

where $v_x^* = -U$. Following from the assumption of a dilute state, the transport of bacteria concentration has a negligible effect on the flow velocity of the groundwater. This assumption makes the governing differential equation linear.

Equation (6-33) is a homogeneous second-order ordinary differential equation requiring two boundary conditions. It is suggested that suitable boundary conditions for this problem are:

$$x = 0: \quad c_B\,(x = 0) = c_R \tag{6-34}$$

$$x \to \infty: \quad c_B\,(x \to \infty) = 0. \tag{6-35}$$

The governing equation (6-33) can be written in a characteristic form:

$$Đ_B\, m^2 + Um = 0, \tag{6-36}$$

revealing the roots $m = 0$ and $-U/Đ_B$ that lead to a solution of the form

$$c_B = C_1 e^0 + C_2 e^{-U\,x/Đ_B}. \tag{6-37}$$

Substituting this expression for c_B into the two boundary conditions yields

$$x = 0: \quad C_1 + C_2 e^0 = c_R, \tag{6-38}$$

$$x \to \infty: \quad C_1 + C_2 e^{-\infty} = 0. \tag{6-39}$$

This dictates that $C_1 = 0$ and $C_2 = c_R$. Therefore, the solution for c_B becomes

$$c_B = c_R\, e^{-U\,x/Đ_B}. \tag{6-40}$$

A concerned citizen may want to know what the steady-state flux of bacteria is from the reservoir into the groundwater. However, the total flux is the sum of contributions from advection and diffusion, which is zero:

$$\begin{aligned} n_B &= (v_x c_B)\big|_{x=0} + \left(-Đ_B\frac{\partial c_B}{\partial x}\right)\bigg|_{x=0} \\ &= \left(-Uc_R\, e^{-U\,x/Đ_B}\right)\big|_{x=0} + \left(-\,Đ_B\frac{-U}{Đ_B}c_R\, e^{-U\,x/Đ_B}\right)\bigg|_{x=0} = 0 \end{aligned} \tag{6-41}$$

This result is required for steady-state conditions, where the bacteria concentration in the groundwater is no longer increasing with time. A better question for the concerned citizen to ask is: "How far into the groundwater has the bacteria traveled?" If the penetration distance δ_B is defined by the distance at which $c_B(x = \delta_B)/c_R = 0.01$ (i.e., where the concentration is 1% of the reservoir value), the upstream penetration distance of bacteria into the groundwater is given by

$$0.01 = e^{-U\delta_B/Đ_B} \quad \text{or} \quad \delta_B = -\ln(0.01)\,Đ_B/U. \tag{6-42}$$

An even better question that our concerned citizen should have asked some time ago is: "If no bacterium is in the groundwater today, how long will it take for the groundwater to become fully contaminated?" This question requires a more difficult transient solution, as will be sought in later chapters. The transient equation of bacteria transport into the groundwater can be integrated by the separation of variables technique, as discussed in Chapter 7, or by numerical integration, as discussed in Chapter 17.

6.4 NONLINEAR EXAMPLE

When a governing equation is nonlinear, integration by analytical means becomes more challenging, or impossible. To arrive at solutions to many nonlinear equations will require numerical techniques, such as fourth-order Runge-Kutta integration introduced in Chapter 23. However, some nonlinear equations do have analytical solutions, as demonstrated by the following problem.

6.4.1 Steady-State Evaporation

Consider water evaporation in the system illustrated in Figure 6-3. Water vapor is transported from the liquid interface at $x = 0$ by diffusion through an air column of length L. Water is treated as one element in a pseudo-binary system in which air is treated as the second element. The water vapor is in equilibrium with the liquid surface at $x = 0$, where the mole fraction of water is χ_{W0}. The water vapor flux is introduced to the dry air stream at $x = L$, where the water vapor mole fraction is zero.

The transport equation for this problem can be formulated in terms of either the mass fraction or mole fraction of water vapor. In both cases, gases that obey the ideal gas law require that

$$\frac{P}{RT} = \rho \quad \text{and} \quad \frac{P}{\hat{R}T} = \frac{N}{\forall} = c. \tag{6-43}$$

If conditions in the gas are isothermal (constant T) and isobaric (constant P), the total density of the mixture ρ is nonconstant because of density's dependency on composition (R changes with composition). In contrast, the total concentration of the mixture c is constant. Therefore, for isothermal and isobaric conditions, the transport equation expressed in terms of mole fractions is more simply integrated than the equation expressed in terms of mass fractions.

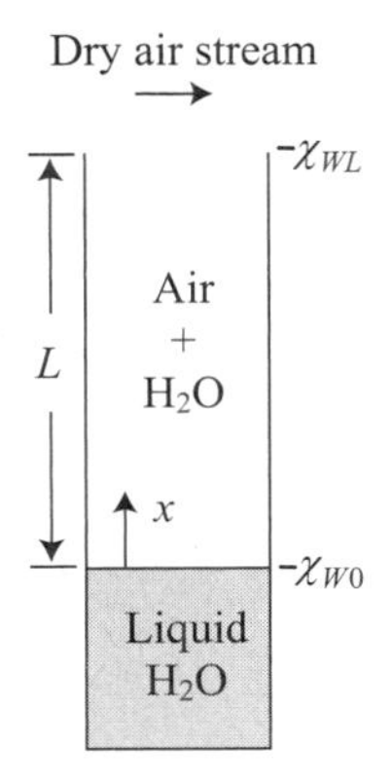

Figure 6-3 Water transport in air column.

For steady-state conditions, the species transport equations can be written in terms of the mole fractions of water and air components in the mixture:

$$\text{transport of water :} \qquad 0 = -\frac{\partial}{\partial x}\left(\overbrace{c_W v_x^*}^{advection} - \overbrace{c\, Ð_{WA}\frac{\partial \chi_W}{\partial x}}^{diffusion} \right). \tag{6-44}$$

$$\text{transport of air :} \qquad 0 = -\frac{\partial}{\partial x}\left(\overbrace{c_A v_x^*}^{advection} - \overbrace{c\, Ð_{WA}\frac{\partial \chi_A}{\partial x}}^{diffusion} \right). \tag{6-45}$$

These equations are written in a form that explicitly expresses the fact that the sum of advection and diffusion transport of either species is constant everywhere in the air column. The transport equation for air (6-45) can be integrated once, yielding

$$c\left(\chi_A v_x^* - Ð_{WA}\frac{\partial \chi_A}{\partial x} \right) = B_1 \quad (= 0). \tag{6-46}$$

However, since the air flux is zero at the interface with the water reservoir, the integration constant must evaluate to $B_1 = 0$. Therefore, the air transport equation (6-46) with $\chi_A = (1 - \chi_W)$ can be used to determine the *Stefan flow* velocity:

$$v_x^* = \frac{-Ð_{WA}}{1-\chi_W}\frac{\partial \chi_W}{\partial x}. \tag{6-47}$$

The flow velocity can be substituted back into the transport equation for water vapor (6-44), yielding

$$\frac{\partial}{\partial x}\left(c_W \frac{Ð_{WA}}{1-\chi_W}\frac{\partial \chi_W}{\partial x} + c\, Ð_{WA}\frac{\partial \chi_W}{\partial x} \right) = 0 \quad \text{or} \quad \frac{d}{dx}\left(\frac{cÐ_{WA}}{1-\chi_W}\frac{d\chi_W}{dx} \right) = 0 \tag{6-48}$$

When the total concentration c and $Ð_{WA}$ are constant, the transport equation can easily be integrated. After integrating once, the transport equation can be expressed in terms of the first integration constant C_1 as

$$\frac{d(1-\chi_W)}{1-\chi_W} = -C_1 dx. \tag{6-49}$$

With the second integration, the solution becomes

$$\ln(1-\chi_W) = -C_1 x + C_2 \quad \text{or} \quad 1 - \chi_W = \exp(-C_1)^x \exp(C_2) = D_1^x D_2, \tag{6-50}$$

after renaming the integration constants. The boundary conditions allow the integration constants to be determined:

$$\chi_W(x=0) = \chi_{W0} \quad \rightarrow \quad 1 - \chi_{W0} = D_2 \tag{6-51}$$

$$\chi_W(x=L) = 0 \quad \rightarrow \quad 1 - 0 = D_1^L D_2, \tag{6-52}$$

after which, the solution for the water molar fraction can be determined:

$$\chi_W(x) = 1 - \ (1-\chi_{W0})^{1-x/L}. \tag{6-53}$$

The molar water vapor flux N_W from the liquid can be determined from the solution by summing the contributions from advection and diffusion:

$$N_W = c_W v_x^* - c\,Đ_{WA}\frac{\partial \chi_W}{\partial x}. \tag{6-54}$$

Using Eq. (6-47) for the molar-averaged velocity, the water vapor flux (6-54) can be expressed at the surface of the liquid, yielding

$$N_W = \left[\frac{-cĐ_{WA}}{1-\chi_W}\frac{\partial \chi_W}{\partial x}\right]_{x=0}. \tag{6-55}$$

Using the solution (6-53) to evaluate the water vapor gradient and the ideal gas law to evaluate the total molar concentration, the flux of water from the liquid surface can be determined:

$$N_W = \frac{-P}{\hat{R}T}\frac{Đ_{WA}}{L}\ln(1-\chi_{W0}). \tag{6-56}$$

6.5 SCALING ESTIMATES

Since governing equations provide the means to establish an exact solution to a well-defined problem, it seems reasonable that one should also be able to estimate a result without undertaking an exact solution by consulting these same equations. This process is called *scaling analysis*, and is an important practice for several reasons. For example, after deriving an exact result, it is wise to question whether the solution is indeed correct. Scaling analysis can provide an order of magnitude prediction of a result, with which the exact solution can be compared. Furthermore, in some instances, an order-of-magnitude knowledge of the solution may be sufficient to make an engineering decision. If that is the case, one may choose to conclude analysis with simple scaling arguments, particularly when faced with deriving an exact solution to a problem that is particularly difficult. Finally, an exact result might be described by an equation in which one or more of the terms are sufficiently "small" that they can be neglected without significant impact on the solution. Such small terms are identified through scaling analysis. Such simplifications to governing equations will be undertaken with the lubrication approximation introduced in Chapter 13 and the boundary layer approximation introduced in Chapter 24.

6.5.1 Scaling of Gravity-Driven Flow on an Inclined Surface

To demonstrate scaling analysis, consider the problem of determining the shear stress imparted by the film flow on an inclined heated surface, as discussed in Section 6.3.1 and illustrated with Figure 6-1. Besides the physical properties of the fluid, the important scales of this problem are the gravity component g_x and the film thickness a. Another important but unknown scale described by the solution is the free surface fluid velocity $v_x(y=a) = U$. With the scales $y \sim a$ and $v_x \sim U$ the governing equation for the flow, Eq. (6-15), can be scaled as

$$\underbrace{\nu\frac{\partial^2 v_x}{\partial y^2}}_{\displaystyle\downarrow\atop\displaystyle \nu\frac{U}{a^2}} \;\begin{matrix}=\\ \\ \sim\end{matrix}\; \underbrace{-g_x}_{\displaystyle\downarrow\atop\displaystyle g_x}. \tag{6-57}$$

Notice that the sign of terms used in the scaling arguments is irrelevant to the objective. Scaling of the governing equation dictates that the free surface velocity will scale as

$$U \sim \frac{g_x a^2}{\nu}. \tag{6-58}$$

This result can be used to determine the scale of the shear stress on the inclined surface:

$$\tau_{yx} = \mu \left.\frac{\partial v_x}{\partial y}\right|_{y=0} \sim \frac{\mu U}{a} = \frac{\mu}{a}\frac{g_x a^2}{\nu} = \rho g_x a. \tag{6-59}$$

The fact that the scaling result $\tau_{yx} \sim \rho g_x a$ is identical to the exact solution given by Eq. (6-22) is fortuitous. Normally one can only expect the scaling result to agree with the exact solution to within an order of magnitude, as illustrated in the next example.

6.5.2 Scaling of Groundwater Contamination

Consider the problem of determining the upstream groundwater penetration distance of bacteria from a reservoir, as discussed in Section 6.3.3 and illustrated with Figure 6-2. The important scales of this problem are the groundwater flow speed U, the reservoir bacteria concentration c_R, and the physical properties of the fluid. Notice that there is no geometric scale imposed on this transport problem. Therefore, the penetration distance δ_B is the only physically significant scale for upstream distances, and its determination is an objective of the solution. Scaling $v_x \sim U$, $c_B \sim c_R$, and $x \sim \delta_B$, the governing equation (6-33) for bacteria transport into the groundwater can be scaled as

$$\underbrace{-U\frac{\partial c_B}{\partial x}}_{\downarrow \atop U\frac{c_R}{\delta_B}} = \underbrace{Đ_B\frac{\partial^2 c_B}{\partial x^2}}_{\downarrow \atop Đ_B\frac{c_R}{\delta_B^2}} \quad \text{with} \quad U\frac{c_R}{\delta_B} \sim Đ_B\frac{c_R}{\delta_B^2}. \tag{6-60}$$

Therefore, the governing equation dictates that the upstream bacteria penetration distance should scale as

$$\delta_B \sim \frac{Đ_B}{U}. \tag{6-61}$$

The exact solution, as given by Eq. (6-42), is $\delta_B = 4.6\ Đ_B/U$. Therefore, the scaling estimate of the bacteria penetration distance is seen to provide the same order of magnitude result as the exact solution.

6.5.3 Scaling Simplification to a Governing Equation

To illustrate how scaling arguments can be useful in identifying small terms in a governing equation, consider the problem illustrated by Figure 6-4, showing the transient laser heating of a semi-infinite solid. It is desired to determine the temperature rise in the solid in response to a short laser pulse of duration τ_p. Heat is introduced to the solid over a surface area $L \times L$.

Since advection transport does not occur in a solid, transient heat transfer into the semi-infinite body is governed by the equation

$$\overbrace{\partial_o T}^{\text{increase}} = \overbrace{\alpha\, \partial_j \partial_j T}^{\text{diffusion}}. \tag{6-62}$$

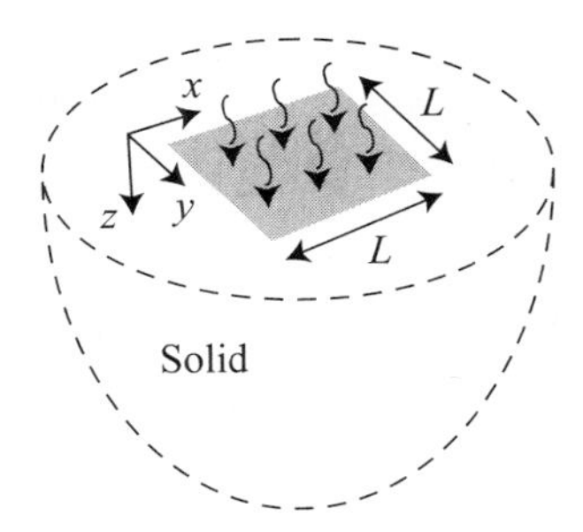

Figure 6-4 Laser-heated patch on a semi-infinite solid.

To perform scaling analysis, it is recognized that the temperature solution is related to a number of characteristic scales. There is an imposed timescale associated with the laser pulse τ_p. Also, there is an imposed geometric length scale associated with the heated patch L. There is a characteristic length scale for heat diffusion into the solid, which is the thermal penetration depth δ_T. The thermal penetration depth is not an imposed scale on the problem, but relates to the solution. Also related to the solution is a characteristic temperature rise ΔT.

To perform scaling, the heat equation is expanded into three dimensions by the rules of index notation, with the result

$$\frac{\partial T}{\partial t} = \alpha\left(\frac{\partial^2 T}{\partial x^2} + \frac{\partial^2 T}{\partial y^2} + \frac{\partial^2 T}{\partial z^2}\right). \tag{6-63}$$

Scaling of the heat equation can be performed to determine the conditions required for $\delta_T < L$, and establish an appropriately simplified heat equation for this problem. Scaling is performed by replacing variables in the heat equation with the characteristic scales:

$$t \sim \tau_p, \quad x \sim L, \quad y \sim L, \quad \text{and} \quad z \sim \delta_T. \tag{6-64}$$

Applied to the heat equation, scaling gives

$$\begin{array}{ccccccc} \underbrace{\frac{1}{\alpha}\frac{\partial T}{\partial t}} & = & \underbrace{\frac{\partial^2 T}{\partial x^2}} & + & \underbrace{\frac{\partial^2 T}{\partial y^2}} & + & \underbrace{\frac{\partial^2 T}{\partial z^2}} \\ \downarrow & & \downarrow & & \downarrow & & \downarrow \\ \frac{\Delta T}{\alpha \tau_p} & \sim & \frac{\Delta T}{L^2} & \text{``}+\text{''} & \frac{\Delta T}{L^2} & \text{``}+\text{''} & \frac{\Delta T}{\delta_T^2} \end{array}. \tag{6-65}$$

The relative importance of heat spreading in the x- and y-directions can be compared to heat penetration in the z-direction by forming a ratio of the related diffusion terms:

$$\frac{\textit{heat spreading}}{\textit{heat penetration}} \sim \frac{\Delta T / L^2}{\Delta T / \delta_T^2} = \left(\frac{\delta_T}{L}\right)^2. \tag{6-66}$$

Therefore, so long as $(\delta_T/L)^2 \ll 1$, the effect of heat spreading can be ignored relative to penetration, and the governing equation simplifies to

$$\frac{\partial T}{\partial t} = \alpha \frac{\partial^2 T}{\partial z^2}. \tag{6-67}$$

However, since the thermal penetration depth is not imposed on the problem, it remains to be seen how the imposed scales will relate to the assertion that $(\delta_T/L)^2 \ll 1$. After heat spreading is neglected, the remaining terms in the heat equation dictate that

$$\underbrace{\frac{1}{\alpha}\frac{\partial T}{\partial t}}_{\downarrow} = \underbrace{\frac{\partial^2 T}{\partial z^2}}_{\downarrow} \qquad \frac{\Delta T}{\alpha \tau_p} \sim \frac{\Delta T}{\delta_T^2} \quad \text{or} \quad \delta_T \sim \sqrt{\alpha \tau_p}\,. \tag{6-68}$$

Therefore, to ensure that $(\delta_T/L)^2 \ll 1$ requires

$$(\sqrt{\alpha\tau_p}/L)^2 \ll 1 \quad \text{or} \quad \tau_p \ll L^2/\alpha. \tag{6-69}$$

So long as the heat delivered by the laser pulse is short compared with the time scale L^2/α, the heat transfer problem can be satisfactorily described with the transient one-dimensional heat equation (6-67), as opposed to the transient three-dimensional heat equation (6-63). This simplifies the solution procedure tremendously. Even though the transient one-dimensional heat equation is still a partial differential equation, a relatively simple analytic solution to this problem is found in Chapter 11.

6.6 PROBLEMS

6-1 Consider laser heating of a plane wall of thickness L. The laser intensity I_o (which is uniform over the surface area) will decay in the material of the wall as $I_o \exp(-x/\delta_{opt})$, where x is the distance from the irradiated surface and δ_{opt} is the optical penetration depth. The temperature distribution in the wall is governed by the equation:

$$\partial_o T = \alpha\, \partial_i \partial_i T + q_{gen}/(\rho C)$$

where $q_{gen} = (I_o/\delta_{opt})\exp(-x/\delta_{opt})$. Solve for the steady-state ($\partial T/\partial t = 0$) temperature distribution in the wall. Assume that no heat spreading occurs, $\partial T/\partial y = \partial T/\partial z = 0$, and that heat transfer can be considered one-dimensional, in the direction of x. The front surface of the wall can be taken as adiabatic (with no heat lost) and the back surface is held at a constant temperature, $T(x = L) = T_b$. Present your solution in terms of the nondimensional parameters:

$$\theta = \frac{T - T_b}{I_o L/k}, \quad \eta = \frac{x}{L}, \quad \Lambda = \frac{\delta_{opt}}{L}.$$

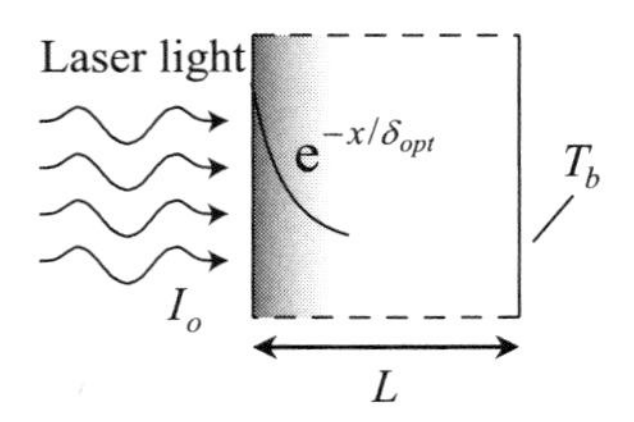

6-2 A solid rod is fed into a furnace with a speed V, as shown. The rod passes through a vacuum and has a low radiation emissivity, such that negligible heat is lost to the environment. Determine the steady-state distance (δ_T) over which the rod is at an elevated temperature by solving the heat equation for the boundary conditions: $T(x = 0) = T_m$ and $T(x \to \infty) = T_\infty$.

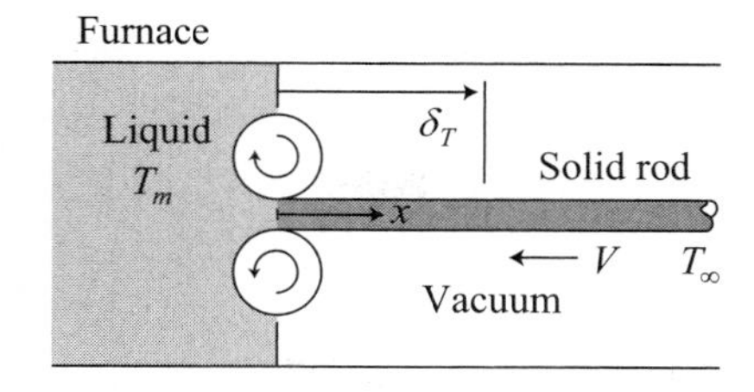

6-3 In the absence of body-forces, the mechanical energy equation is given by

$$\rho \frac{D}{Dt}\left(\frac{v_i v_i}{2}\right) = -v_i \partial_j M_{ji} - v_i \partial_i P,$$

where $M_{ji} = -\mu(\partial_j v_i + \partial_i v_j)$. Simplify the mechanical energy equation for a two-dimensional steady, incompressible, pressure-driven flow through a slot. You may assume that $-dP/dx = const.$ and $dv_x/dx = 0$. Solve the mechanical energy equation.

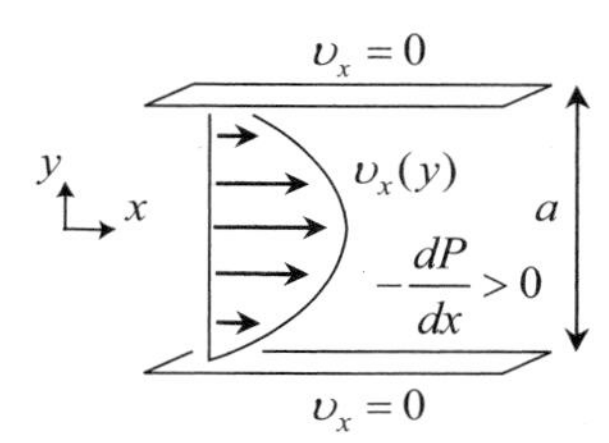

6-4 Establish the steady-state temperature distribution of air moving between two parallel plates, including the effect of viscous dissipation. The flow is driven by motion of the top plate. The velocity distribution between the plates is $v_x = U(y/h)$ and $v_y = 0$. The energy equation for an ideal gas is given by

$$\rho C_p \frac{DT}{Dt} = \partial_j(k \partial_j T) + \frac{DP}{Dt} - M_{ji} \partial_j v_i,$$

where $M_{ji} = -\mu(\partial_j v_i + \partial_i v_j)$. Express the energy equation as simply as possible in Cartesian coordinates. Solve the energy equation for the temperature distribution between the plates, noting any approximations made.

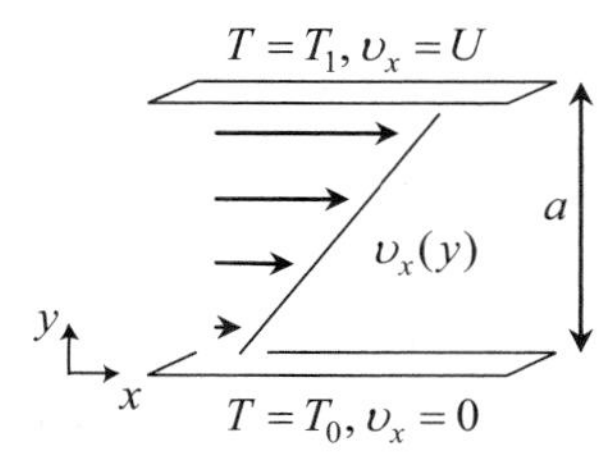

6-5 A large solid plate of thickness t_s overlies a liquid film of thickness t_ℓ on an inclined surface, as shown. Both the plate and the liquid are free to move down the inclined wall under the force of gravity. Find the terminal speed of the plate's descent. For what condition might the weight of the liquid be ignored? For what condition might the weight of the plate be ignored?

6-6 A constant pressure gradient dP/dx forces fluid in the positive x-direction through a long narrow duct of height a. The walls of the duct are porous, permitting a uniform cross flow of $v_y = v_o$. Flows of this type can be used to concentrate large molecules in the fluid near the upper wall. Determine the governing equation and boundary conditions required to determine $v_x(y)$. Solve for $v_x(y)$. Suppose the flow has a dilute concentration of some molecular species "A". The top wall of the duct has a concentration $c_A = c_2$, and the bottom wall a concentration $c_A = c_1$. Determine the governing equation and boundary conditions required to determine the concentration profile $c_A(y)$. Solve for $c_A(y)$.

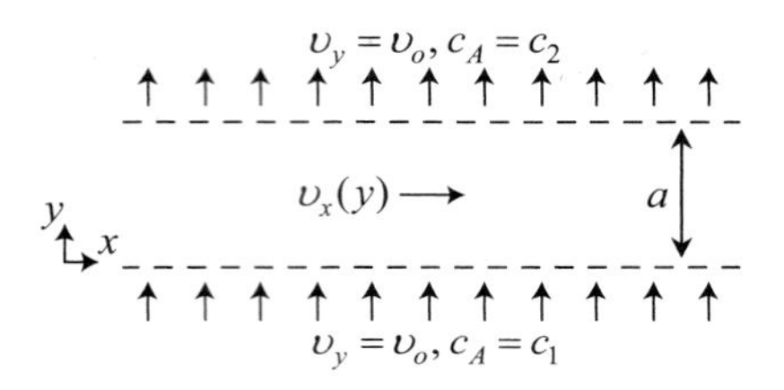

6-7 Consider a bar being heat treated by a laser, as shown. The bar moves with a speed U past the laser. The top surface of the bar is irradiated with negligible optical penetration. Demonstrate the requirements on the imposed scales of this problem that lead to the simplified form of the heat equation:

$$U\frac{\partial T}{\partial x} = \alpha\frac{\partial^2 T}{\partial y^2}.$$

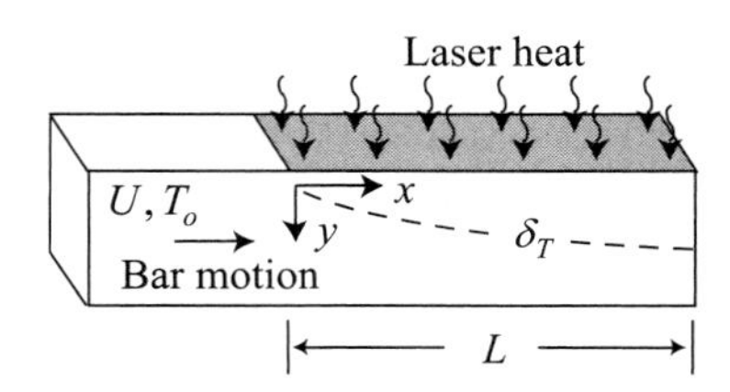

For the heat equation to simplify to this form, what requirement is imposed on the dimensionless group UL/α?

6-8 Consider the evaporation tube illustrated in Figure 6-3. Suppose the air is replaced by an inert gas having exactly twice the molecular weight of water. For this case, derive the governing differential equation for the mass fraction distribution of water in the evaporation tube. Demonstrate that the water mass fraction distribution in the tube is given by

$$\omega_W(x) = \frac{2}{1 + \left(\dfrac{2}{1+\omega_{W0}} - 1\right)^{1-x/L}} - 1,$$

and demonstrate the equivalence of this result with the molar fraction distribution given by Eq. (6-53).

REFERENCES

[1] P. Blanchard, R. L. Devaney, and G. R. Hall, *Differential Equations*, Third Edition. Belmont: Thompson Brooks/Cole, 2006.

[2] D. Zwillinger, *Handbook of Differential Equations*, Third Edition. Boston: Academic Press, 1997.

Chapter 7

Transient One-Dimensional Diffusion

7.1 Separation of Time and Space Variables
7.2 Silicon Doping
7.3 Plane Wall With Heat Generation
7.4 Transient Groundwater Contamination
7.5 Problems

The elementary problems considered in Chapter 6 were governed by simple, ordinary differential equations. However, more complicated problems can involve two or more independent variables. For example, a problem might require two spatial variables or one temporal and one spatial variable. Therefore, strategies are needed for integrating partial differential equations. Separation of variables is well suited for solving linear partial differential equations for problems confined to geometrically simple domains. In this chapter, separation of variables is discussed in the context of transient one-dimensional diffusion problems of heat, mass, and momentum transport. However, separation of variables is not entirely limited to diffusion transport, as will be illustrated in Section 7.4. Additionally, steady-state ideal planar flows, discussed in Chapter 14, describe a circumstance in which advection is governed by a linear partial differential equation of two spatial variables that can be solved by separation of variables.

Separation of variables employs a strategy of transforming a single partial differential equation into two ordinary differential equations that may be integrated independently. The approach for any given problem is not unique. However, the technique is based on a few strategies that are applicable to a wide variety of situations. The essential idea of separation of variables is that a solution to the problem for $\phi(x, t \; or \; y)$ can be sought in the form $\phi = X(x)Y(t \; or \; y)$. When this form of a solution is applied to the governing equation, and expressions involving x are separated from expressions involving t or y, two ordinary differential equations for X and Y result. This allows the solution to a partial differential equation to be expressed in terms of elementary solutions of two ordinary differential equations. The main complication that arises involves nonhomogeneous conditions that are encountered. This complication is unavoidable because every problem of interest has at least one nonhomogeneous condition relating to a source of heat, mass, or momentum.

Unfortunately, the nonhomogeneous condition typically cannot be described by the functional form of the elementary functions describing X or Y. This incompatibility would appear to prohibit a solution from being expressed in terms of X and Y. However, the separation of variables technique utilizes elementary functions that can be used to "build" the required nonhomogeneous condition by superposition. This process of superposition requires the problem to be linear, and the elementary functions used to build the nonhomogeneous condition should exhibit orthogonality.

In this chapter, strategies for solving partial differential equations involving one temporal and one spatial variable are developed. The method is extendable to problems having more than one spatial variable. Additionally, some transient problems involving multiple spatial variables are easily handled by a simple product-superposition of solutions of one-dimensional problems, as illustrated in Chapter 11. Steady-state equations involving two spatial variables alone are solved by separation of variables in Chapter 8. In Chapter 9, a variant of the separation of variables technique called eigenfunction expansion is developed.

7.1 SEPARATION OF TIME AND SPACE VARIABLES

To develop the separation of variables technique for partial differential equations involving one temporal and one spatial variable, consider a governing equation with the characteristic form of one-dimensional transient diffusion:

$$\frac{\partial \phi}{\partial t} = \frac{\partial^2 \phi}{\partial x^2}. \tag{7-1}$$

This equation is linear and homogeneous. With appropriate diffusion coefficients, this equation can describe any number of problems in heat, mass, or momentum transfer. To solve this equation, two boundary conditions and one initial condition must be specified by the problem statement. This leads to many permutations of problems that can be solved by the transient diffusion equation. As a first illustration of the separation of variables technique, a problem with homogeneous boundary conditions and nonhomogeneous initial condition will be solved in Section 7.1.1. Subsequently, a problem with nonhomogeneous boundary conditions and initial condition will be solved for a nonhomogeneous governing equation in Section 7.1.3. This will serve to outline the solution technique that will be applied to a few specific problems in Sections 7.2 through 7.4. An extensive treatment of separation of variables can be found in books such as references [1] and [2].

7.1.1 Problem with Homogeneous Equation and Boundary Conditions

Figure 7-1 illustrates an "ideal" problem in which both the boundary conditions and governing equation are homogeneous. For the ideal problem, the nonhomogeneous term appears as an initial condition. The specific forms of the boundary conditions and initial condition are unidentified as of yet. As will be demonstrated momentarily, the spatial part of the transient heat equation can be described by elementary trigonometric functions. This problem is ideal for separation of variables because the homogeneous boundary conditions allow the nonhomogeneous initial condition to be satisfied by superposition of the trigonometric functions. This is facilitated by the orthogonality of the trigonometric functions for the typical boundary conditions discussed in Chapter 6.

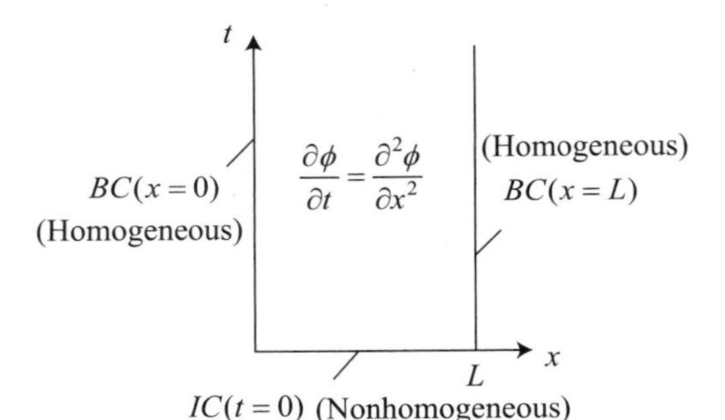

Figure 7-1 Ideal problem having homogeneous boundary conditions (BC) and nonhomogeneous initial condition (IC).

To illustrate the procedure of separation of variables, consider a specific problem prescribed by the boundary conditions:

$$\text{at } x = 0: \qquad \phi(x = 0) = 0 \tag{7-2}$$

$$\text{at } x = L: \qquad \partial\phi/\partial x\big|_{x=L} = 0 \tag{7-3}$$

with the initial condition:

$$\text{at } t = 0: \qquad \phi(t = 0) = A. \tag{7-4}$$

A solution is assumed to have the form $\phi = X(x)Y(t)$, which is substituted back into the governing equation for

$$\frac{\partial(XY)}{\partial t} = \frac{\partial^2(XY)}{\partial x^2}. \tag{7-5}$$

Separating variables, such that functions of t and functions of x fall on opposite sides of the equation, yields

$$\frac{1}{Y}\frac{\partial Y}{\partial t} = \frac{1}{X}\frac{\partial^2 X}{\partial x^2} \qquad (= -\lambda_n^2). \tag{7-6}$$

Since the left-hand side of Eq. (7-6) is some function of t, and the right-hand side is some function of x, the only way to assure equality of these two functions is to force them both to equal the same constant. The separation constant, $-\lambda_n^2$, is squared so that it will not appear as $\sqrt{\lambda_n}$ later in the solution, and is negative to give a solution for X in terms of trigonometric functions (as will be demonstrated shortly). The index "n" on the separation constant will be used later to identify all the different discrete values that the separation constant could be. In terms of the separation constant, two ordinary differential equations for $Y(t)$ and $X(x)$ emerge from (7-6) that may be integrated directly:

$$\begin{array}{ccc} \dfrac{dY}{dt} + \lambda_n^2 Y = 0 & & \dfrac{d^2X}{dx^2} + \lambda_n^2 X = 0 \\ \downarrow & \text{and} & \downarrow \\ Y(t) = C_1 e^{-\lambda_n^2 t} & & X(x) = C_2\cos(\lambda_n x) + C_3\sin(\lambda_n x) \end{array} \tag{7-7}$$

Notice that the homogeneous boundary conditions on the original problem for ϕ can be satisfied by imposing them on the problem for $X(x)$ alone. Specifically, if $X(0) = 0$, then $\phi(x = 0) = X(0)Y(t) = 0$ irrespective of what $Y(t)$ is. Furthermore, if $dX/dx|_{x=L} = 0$, then $(d\phi/\partial x)|_{x=L} = (dX/dx|_{x=L})Y(t) = 0$, also irrespective of what $Y(t)$ is. However, assigning $Y(0) = A$, in an attempt to satisfy the initial condition, does not satisfy the required condition that $\phi(t = 0) = X(x)Y(0) = A$. Therefore, any nonhomogeneous conditions in the problem for ϕ must be addressed only after the solution for $\phi = X(x)Y(t)$ is reconstructed.

For the example at hand, both homogeneous boundary conditions are addressed in the problem for $X(x)$:

$$\text{at } x = 0: \quad \phi(x = 0) = 0 \quad \rightarrow \quad X(0) = C_2 \overbrace{\cos(0)}^{=1} + C_3 \overbrace{\sin(0)}^{=0} = 0 \tag{7-8}$$

$$\text{at } x = L: \quad \partial\phi/\partial x|_{x=L} = 0 \quad \rightarrow \quad dX/dx|_{x=L} = -C_2\lambda_n \sin(\lambda_n L) + C_3\lambda_n\cos(\lambda_n L) = 0. \tag{7-9}$$

The $x = 0$ boundary condition dictates that $C_2 = 0$. Additionally, the second boundary condition at $x = L$ is satisfied when $dX/dx|_{x=L} = C_3\lambda_n \cos(\lambda_n L) = 0$. However, choosing C_3 to be zero would make $X(x)$ identically zero everywhere. Alternatively, the second boundary condition may be satisfied by restricting the possible choices of the separation constant λ_n. Specifically, the second boundary condition on X can be satisfied when $\cos(\lambda_n L) = 0$, such that

$$\lambda_n L = \frac{\pi}{2}, \frac{3\pi}{2}, \frac{5\pi}{2}, \cdots \quad \text{or} \quad \lambda_n = \frac{(2n+1)\pi}{2L} \quad \textit{for } n = 0, 1, 2, \ldots. \tag{7-10}$$

Notice that there is an infinite, but discrete, set of possible choices for the separation constant. This is a consequence of the periodic nature of the trigonometric function.

Since the nonhomogeneous initial condition cannot be satisfied with $Y(t)$ alone, the solution for $\phi = X(x)Y(t)$ is reconstructed:

$$\phi = C_n \sin(\lambda_n x) e^{-\lambda_n^2 t} \tag{7-11}$$

The two integration constants C_1 and C_3 have been combined into C_n. The "n" subscript acknowledges that the final integration constant is dependent on the specific value of the separation constant λ_n.

Figure 7-2 illustrates what the solution (so far) would look like at $t = 0$ for a few of the possible choices of λ_n. Notice in all cases that Eq. (7-11) satisfies the governing equation and the spatial boundary conditions. However, it is apparent from Figure 7-2 that the initial condition $\phi = A$ cannot be satisfied with any individual choice of λ_n.

Since the governing equation and boundary conditions are linear and homogeneous, the sum of any number of solutions will also be a solution. Therefore, a solution can be sought that is a summation of Eq. (7-11) over all possible values of the separation constant:

$$\phi = \sum_{n=0}^{\infty} C_n \sin(\lambda_n x) e^{-\lambda_n^2 t}. \tag{7-12}$$

Each term in the series must be weighted by a judicious value of C_n in order to satisfy the initial condition. Therefore, the final task is to determine the appropriate values of C_n. Starting with a statement of the initial condition $\phi(t = 0) = A$:

$$A = \sum_{n=0}^{\infty} C_n \sin(\lambda_n x) \tag{7-13}$$

both sides are multiplied by $\sin(\lambda_m x)$ and integrated with respect to x from 0 to L:

$$\int_0^L \sin(\lambda_m x) A dx = \int_0^L \sum_{n=0}^{\infty} C_n \sin(\lambda_m x) \sin(\lambda_n x) dx. \tag{7-14}$$

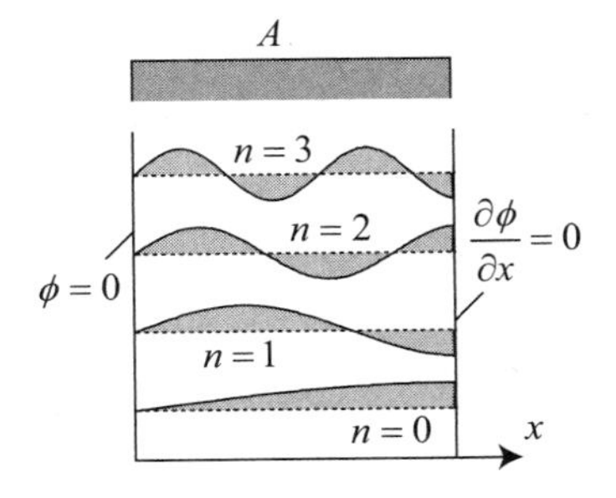

Figure 7-2 $\text{Sin}(\lambda_n x)$ for a few choices of λ_n.

Notice that each term in the series on the right-hand side of the equation can be integrated independently. However,

$$\int_0^L \sin(\lambda_m x)\sin(\lambda_n x)dx = \left[\frac{\sin\{(\lambda_m - \lambda_n)x\}}{2(\lambda_m - \lambda_n)} - \frac{\sin\{(\lambda_m + \lambda_n)x\}}{2(\lambda_m - \lambda_n)}\right]_0^L = \begin{cases} 0 & if\ m \neq n \\ L/2 & if\ m = n \end{cases}. \quad (7\text{-}15)$$

That this integral is nonzero (L/2 in this case) only when $\lambda_m = \lambda_n$ reflects an *orthogonality property* of the $\sin(\lambda_n x)$ function, for the set of λ_n. Therefore, only one nonzero term exists on the right-hand side of Eq. (7-14). Accordingly, one is left with an expression that can be evaluated for values of C_n:

$$\int_0^L \sin(\lambda_n x)Adx = C_n\frac{L}{2} \quad \text{or} \quad C_n = \frac{2A[1-\cos(\lambda_n L)]}{\lambda_n L} = \frac{2A}{\lambda_n L} \quad (7\text{-}16)$$

With this result for C_n substituted into Eq. (7-12), the final solution can be presented as

$$\phi = \sum_{n=0}^{\infty} \frac{2A}{\lambda_n L}\sin(\lambda_n x)e^{-\lambda_n^2 t}, \quad (7\text{-}17)$$

where

$$\lambda_n = \frac{(2n+1)\pi}{2L}.$$

7.1.2 Demonstration of Orthogonality

A critical step in the separation-of-variables solution is the ability to determine the correctly weighted coefficients for each term in the series, such that the nonhomogeneous condition is satisfied by the solution. This is made possible by the orthogonality property of $X_n(x)$ for the associated set of separation constants. In general, the function $X_n(x)$ having this characteristic is called an *eigenfunction*, and the corresponding separation constants λ_n are called *eigenvalues.*

The orthogonality of the eigenfunction can be demonstrated for an appropriate set of eigenvalues. To illustrate, consider the problem of the previous section, where $X_n(x) = \sin(\lambda_n x)$ was the eigenfunction that satisfies the equation

$$\frac{d^2X_n}{dx^2} + \lambda_n^2 X_n = 0. \quad (7\text{-}18)$$

Furthermore, the *eigencondition* for the problem requires the eigenvalues to satisfy the condition $dX_n/dx|_{x=L} = \lambda_n \cos(\lambda_n L) = 0$. To demonstrate orthogonality, the equation describing the eigenfunction is written for two independent indices, *m* and *n*:

$$\frac{d^2X_m}{dx^2} + \lambda_m^2 X_m = 0 \quad \text{and} \quad \frac{d^2X_n}{dx^2} + \lambda_n^2 X_n = 0. \quad (7\text{-}19)$$

Multiplying the equation for X_m by X_n and multiplying the equation for X_n by X_m, the two resulting equations can be subtracted for the result

$$X_n\frac{d^2X_m}{dx^2} - X_m\frac{d^2X_n}{dx^2} + (\lambda_m^2 - \lambda_n^2)X_m X_n = 0. \quad (7\text{-}20)$$

However, since

$$X_n \frac{d^2 X_m}{dx^2} - X_m \frac{d^2 X_n}{dx^2} = \frac{d}{dx}\left(X_n \frac{dX_m}{dx} - X_m \frac{dX_n}{dx}\right), \tag{7-21}$$

Eq (7-20) can be integrated once for the result:

$$\left(X_n \frac{dX_m}{dx} - X_m \frac{dX_n}{dx}\right)\Bigg|_0^L = \left(\lambda_n^2 - \lambda_m^2\right) \int_0^L X_m X_n dx. \tag{7-22}$$

Observe that the combination of the eigenfunction and eigencondition force the left-hand side of Eq. (7-22) to zero since

$$X_m(x = 0) = X_n(x = 0) = 0 \quad \text{and} \quad \frac{dX_m}{dx}\bigg|_{x=L} = \frac{dX_n}{dx}\bigg|_{x=L} = 0. \tag{7-23}$$

Therefore, Eq. (7-22) demonstrates that

$$\int_0^L X_m X_n dx = 0, \quad \text{when } m \neq n, \tag{7-24}$$

which proves the orthogonality of the eigenfunction for the given set of eigenvalues.

It should be reemphasized that the orthogonality of a function requires an appropriate set of eigenvalues, which are related to the boundary conditions of the problem. When the eigenvalues satisfy homogeneous boundary conditions of the first, second, or third kind (see Section 6.2), the eigenfunction will exhibit orthogonality, as can be demonstrated following the procedure outlined above.

When $m = n$, the nonzero value of the integral:

$$\int_0^L X_n X_n dx \neq 0 \tag{7-25}$$

will need to be evaluated in the separation-of-variables procedure. When $X_n = \sin(\lambda_n x)$ or $X_n = \cos(\lambda_n x)$, it is noted that

$$\int_0^L \sin(\lambda_n x)\sin(\lambda_n x) dx = \left[\frac{x}{2} - \frac{\sin(2\lambda_n x)}{4\lambda_n}\right]_0^L \tag{7-26}$$

and

$$\int_0^L \cos(\lambda_n x)\cos(\lambda_n x) dx = \left[\frac{x}{2} + \frac{\sin(2\lambda_n x)}{4\lambda_n}\right]_0^L. \tag{7-27}$$

7.1.3 Problem with Nonhomogeneous Equation and Boundary Conditions

Unfortunately, the physical problem to be solved may often dictate a different mathematical description than the ideal form of the previous section. Figure 7-3 illustrates a more general situation in which both the boundary conditions and governing equation are nonhomogeneous. (For most problems, the situation will be simpler than what is illustrated here.)

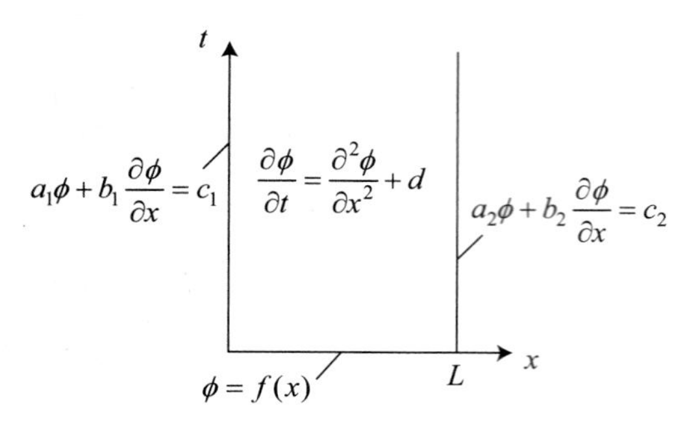

Figure 7-3 Example of a nonhomogeneous problem.

For transient problems, there is a useful strategy for transforming the separation-of-variables task into a problem having the ideal form discussed in Section 7.1.3. The strategy is to decompose the problem for ϕ into the steady-state solution $\phi_{ss} = \phi(t \to \infty)$ plus another problem θ yet to be prescribed. By letting $\phi = \phi_{ss} + \theta$, the problem for ϕ can be decomposed into two secondary problems that linearly combine to give back the original problem. Let the governing equations for the two problems be

$$\begin{array}{rcll} + \quad 0 & = & \partial^2\phi_{ss}/\partial x^2 & + d \\ \partial\theta/\partial t & = & \partial^2\theta/\partial x^2 & \\ \hline \partial(\phi_{ss}+\theta)/\partial t & = & \partial^2(\phi_{ss}+\theta)/\partial x^2 & + d \end{array}, \tag{7-28}$$

which sum back to the original governing equation for ϕ. Notice that the governing equation for ϕ_{ss} is the same as for ϕ with $\partial\phi/\partial t = 0$. Furthermore, notice that the governing equation for θ is homogeneous since the nonhomogeneous term in the equation for ϕ goes into the problem for ϕ_{ss}.

Let the boundary conditions for the ϕ_{ss} and θ problems be

$$\begin{array}{ll} \underline{\text{At } x=0} & \underline{\text{At } x=L} \\ \begin{array}{rcl} a_1\phi_{ss} + b_1\partial\phi_{ss}/\partial x & = & c_1 \\ + \quad a_1\theta + b_1\partial\theta/\partial x & = & 0 \\ \hline a_1(\phi_{ss}+\theta) + b_1\partial(\phi_{ss}+\theta)/\partial x & = & c_1 \end{array} & \begin{array}{rcl} a_2\phi_{ss} + b_2\partial\phi_{ss}/\partial x & = & c_2 \\ + \quad a_2\theta + b_2\partial\theta/\partial x & = & 0 \\ \hline a_2(\phi_{ss}+\theta) + b_2\partial(\phi_{ss}+\theta)/\partial x & = & c_2 \end{array} \end{array},$$

which sum back to the original boundary conditions imposed on ϕ. Notice that the boundary conditions for ϕ_{ss} are the same as for ϕ. Furthermore, notice that the boundary conditions for θ are of the same kind as the boundary conditions for ϕ. However, for the θ problem, all of the boundary conditions are homogeneous, while for the ϕ problem none of the boundary conditions are homogeneous.

The initial condition requires that $\phi_{ss} + \theta(x, t = 0) = \phi(x, t = 0) = f(x)$. Therefore, for the θ problem one requires the initial condition:

$$t = 0: \quad \theta(x, t = 0) = f(x) - \phi_{ss}. \tag{7-29}$$

Determining the solution to the steady-state problem for ϕ_{ss} should be straightforward, since ϕ_{ss} is governed by an ordinary differential equation in terms of the spatial variable x. Once ϕ_{ss} is known, the solution for the problem for θ, which is a partial differential equation, may be obtained by the separation of variables technique outlined in Section 7.1.3. Finally, the solution to the original problem is reconstructed from $\phi = \phi_{ss} + \theta$.

7.2 SILICON DOPING

A silicon wafer can be doped with a donor species using a "spin-on" glass. The spin-on glass has a concentration of phosphorous and is applied to the silicon wafer as a thin film. The fluid dynamic process by which the glass is applied to the wafer is discussed in Chapter 13 (see Problem 13-1). The silicon wafer, with an overcoat of the spin-on glass, is put into a furnace. At high temperature, the phosphorous will diffuse from the glass into the silicon. A description of the transient diffusion of phosphorous into the silicon is desired. Because the silicon is solid, advection cannot contribute to species transport, so long as the concentration of phosphorous is low, as discussed in Section 4.2.1. By geometric considerations, the diffusion process can be considered one-dimensional. Therefore, the governing equation for species transfer reduces to

$$\partial c_P/\partial t = Đ_p \partial^2 c_P/\partial x^2, \tag{7-30}$$

where c_P is the concentration of phosphorous in the silicon.

Since the spatial derivative in the governing equation is second-order, two boundary conditions must be specified. It is assumed that the donor glass is able to keep the surface of the silicon at a constant phosphorous concentration P_o. Additionally, for the problem illustrated in Figure 7-4, there is a diffusion barrier in the silicon at some distance L from the surface. Therefore, the boundary conditions are:

$$\text{at } x = 0: \quad c_P(x = 0) = P_o \tag{7-31}$$

$$\text{at } x = L: \quad \partial c_P/\partial x|_{x=L} = 0. \tag{7-32}$$

Diffusion will not start until the coated silicon is put into a high-temperature oven, at $t = 0$. Therefore, the initial condition is

$$\text{at } t = 0: \quad c_P(t = 0) = 0. \tag{7-33}$$

This problem has a homogeneous governing equation and initial condition, and one nonhomogeneous boundary condition. To bring the nonhomogeneous boundary condition to the initial condition, by the method discussed in Section 7.1.3, one can utilize the steady-state solution that is approached as $t \to \infty$. Assume that c_{ss} is the steady-state concentration distribution, and let $c_p = c_{ss} + \theta$ to decompose the problem for c_p into two secondary problems. The secondary problems must linearly combine to give back the original problem. Let the governing equations for the two problems be

$$\begin{array}{rccl} & 0 & = & Đ_p \ \partial^2 c_{ss}/\partial x^2 \\ + & \partial\theta/\partial t & = & Đ_p \ \partial^2 \theta/\partial x^2 \\ \hline & \multicolumn{3}{c}{\partial(c_{ss} + \theta)/\partial t = Đ_p \ \partial^2 (c_{ss} + \theta)/\partial x^2} \end{array} \tag{7-34}$$

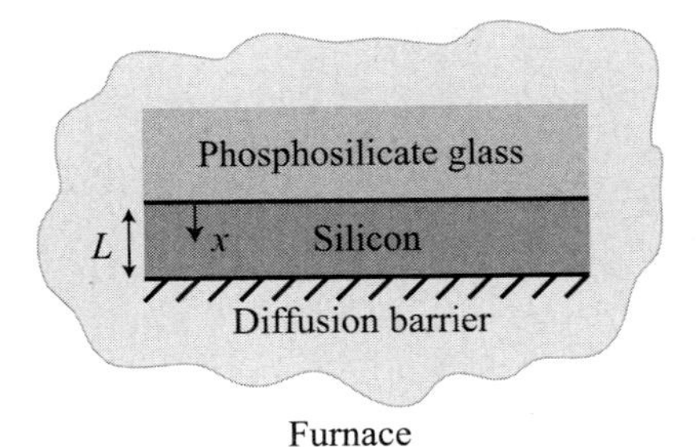

Figure 7-4 Doping of a silicon layer backed by a diffusion barrier.

which sum back to the original governing equation.

Let the boundary conditions for the c_{ss} and θ problems be

$$\begin{array}{lrcl} \underline{\text{At } x = 0} & & & \\ & c_{ss} & = & P_o \\ + & \theta & = & 0 \\ \hline & c_{ss} + \theta & = & P_o \end{array} \qquad \begin{array}{lrcl} \underline{\text{At } x = L} & & & \\ & \partial c_{ss}/\partial x & = & 0 \\ + & \partial \theta/\partial x & = & 0 \\ \hline & \partial(c_{ss} + \theta)/\partial x & = & 0 \end{array}$$

which sum back to the original boundary conditions. Notice that for the θ problem all the boundary conditions are homogeneous.

The initial condition requires that $c_{ss} + \theta(x, t = 0) = c_p(x, t = 0) = 0$. Therefore, for the θ problem, the initial condition is required to be

$$\text{at } t = 0: \quad \theta(x, t = 0) = -c_{ss}. \tag{7-35}$$

Starting with the steady-state problem, governed by

$$0 = Đ_p \ \partial^2 c_{ss}/\partial x^2, \tag{7-36}$$

and subject to the boundary conditions

$$\text{at } x = 0: \quad c_{ss}(x = 0) = P_o \tag{7-37}$$

$$\text{at } x = L: \quad \partial c_{ss}/\partial x\big|_{x=L} = 0, \tag{7-38}$$

it is straightforward to establish that the solution is

$$c_{ss} = P_o. \tag{7-39}$$

The remaining problem for θ is summarized in Figure 7-5. Applying separation of variables, $\theta = X(x)Y(t)$ is substituted into the governing equation for θ, and the variables are separated for

$$\frac{1}{Đ_p}\frac{1}{Y}\frac{\partial Y}{\partial t} = \frac{1}{X}\frac{\partial^2 X}{\partial x^2} \qquad (= -\lambda_n^2) \tag{7-40}$$

Two ordinary differential equations for $Y(t)$ and $X(x)$ may now be constructed in terms of the separation constant λ_n. These equations are integrated directly:

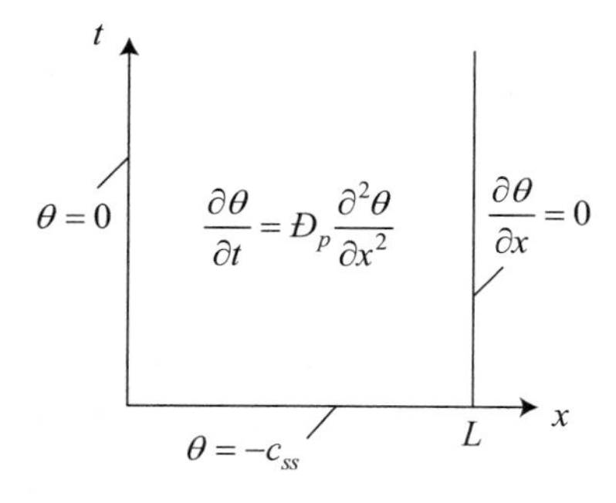

Figure 7-5 Silicon layer transient subproblem.

$$\frac{dY}{dt} + Đ_p\,\lambda_n^2 Y = 0 \quad \text{and} \quad \frac{d^2X}{dx^2} + \lambda_n^2 X = 0$$
$$\downarrow \qquad\qquad\qquad \downarrow \tag{7-41}$$
$$Y(t) = C_1 e^{-Đ_p \lambda_n^2 t} \qquad X(x) = C_2\cos(\lambda_n x) + C_3\sin(\lambda_n x)$$

Applying the homogeneous boundary conditions to the problem for $X(x)$:

$$\text{at } x = 0 : \theta(x = 0) = 0 \quad \rightarrow \quad X(0) = C_2\cos(0) + C_3\sin(0) = 0 \tag{7-42}$$

$$\text{at } x = L : \partial\theta/\partial x|_{x=L} = 0 \rightarrow dX/dx|_{x=L} = -C_2\lambda_n\sin(\lambda_n L) + C_3\lambda_n\cos(\lambda_n L) = 0 \tag{7-43}$$

One finds from the $x = 0$ boundary condition that $C_2 = 0$, which, combined with the $x = L$ boundary condition, requires that the choices of λ_n must satisfy $\cos(\lambda_n L) = 0$. Therefore, it is established that

$$\lambda_n = \frac{(2n+1)\pi}{2L} \quad \textit{for } n = 0, 1, 2, \ldots. \tag{7-44}$$

With $X = C_3\sin(\lambda_n x)$, the solution for $\theta = X(x)Y(t)$ is reconstructed:

$$\theta = C_n\sin(\lambda_n x)\, e^{-Đ_p\,\lambda_n^2 t}. \tag{7-45}$$

The two remaining integration constants C_1 and C_3 have been combined into C_n. Since the equation being solved is linear, the sum of any number of solutions to the governing equation will also be a solution. Therefore, one can propose that

$$\theta = \sum_{n=0}^{\infty} C_n\sin(\lambda_n x)\, e^{-Đ_p\,\lambda_n^2 t}. \tag{7-46}$$

The solution to the problem at hand is constructed by judiciously picking values of C_n to satisfy the initial condition $\theta(t = 0) = -P_o$, such that

$$-P_o = \sum_{n=0}^{\infty} C_n\sin(\lambda_n x). \tag{7-47}$$

To determine C_n, both sides are multiplied by $\sin(\lambda_m x)$ and integrated with respect to x from 0 to L:

$$\int_0^L \sin(\lambda_m x)(-P_o)dx = \int_0^L \sum_{n=0}^{\infty} C_n\sin(\lambda_m x)\sin(\lambda_n x)dx. \tag{7-48}$$

The only nonzero term on the right-hand side of this equation is when $m = n$, and therefore one is left with

$$C_n = \frac{2}{L}\int_0^L \sin(\lambda_n x)(-P_o)dx = \frac{-2P_o[1 - \cos(\lambda_n L)]}{\lambda_n L} = \frac{-2P_o}{\lambda_n L}, \tag{7-49}$$

Code 7-1 Silicon layer doping calculation

```
#include <stdio.h>
#include <math.h>

double Cp_Po(double eta,double tau)
{
   int n;
   double l,soln=0.;
   for (n=0;n<100;++n) {
      l=(2*n+1)*M_PI/2.;
      soln+=( sin(l*eta)/(2*n+1) )*exp(-l*l*tau);
   }
   return ( 1.-(4./M_PI)*soln );
}

int main(void)
{
   double tau,eta;
   FILE *fp=fopen("out.dat","w");
   for (tau=0.001;tau<=10.;tau*=10.) {
      for (eta=.0;eta<=1.;eta+=0.02)
      fprintf(fp,"%e %e\n",eta,Cp_Po(eta,tau));
      fprintf(fp,"\n");
   }
   fclose(fp);
   return 0;
}
```

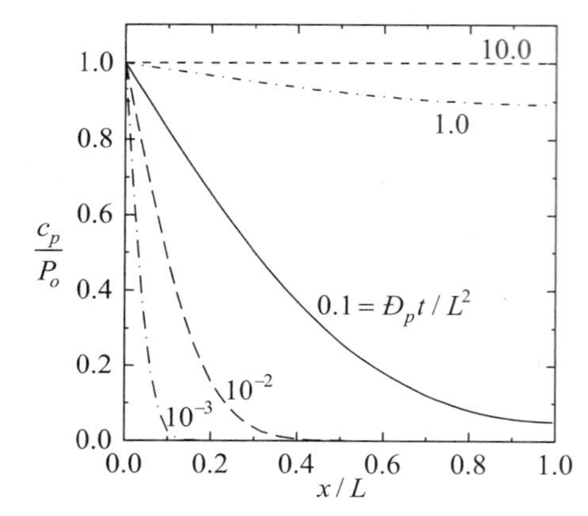

Figure 7-6 Time evolution of phosphorous concentration in silicon layer.

where the eigencondition $\cos(\lambda_n L) = 0$ was used to simplify this expression. Applying this result to Eq. (7-46), and the relation $c_p = c_{ss} + \theta$, the final solution can presented in the form

$$c_p = P_o\left[1 - \frac{4}{\pi}\sum_{n=0}^{\infty}\frac{\sin\left(\frac{(2n+1)\pi}{2}\frac{x}{L}\right)}{(2n+1)}\exp\left(-\left(\frac{(2n+1)\pi}{2}\right)^2\frac{Đ_p t}{L^2}\right)\right]. \tag{7-50}$$

It is impossible to assess by inspection the nature of a function described with a series solution. Therefore, series solutions should always be plotted to assure physical correctness. Although many mathematical packages are adept for such a task, it is suggested that a simple code, such as illustrated in Code 7-1, be written to create a data file that can be plotted. More involved programming tasks will be important for solving problems in later chapters. Therefore, it is advisable to hone one's programming skills on relatively simple tasks now.

Figure 7-6 plots the spatial and temporal solution for phosphorous diffusion into the silicon. Observance of the initial condition and boundary conditions can be verified in the solution. Additionally, the solution yields the time required for the phosphorous concentration in the silicon to rise to the glass value as approximately $\sim 10L^2/Đ_p$. This result is useful for planning the furnace time required to fully dope the silicon wafer.

7.3 PLANE WALL WITH HEAT GENERATION

Consider the solid plane wall illustrated in Figure 7-7 that is initially at a uniform temperature of T_∞. On one side, the wall is bounded by a fluid, also at T_∞. At time zero, two things happen. First, the front surface of the wall (at $x = 0$) is brought into contact with a constant temperature T_b. Second, a chemical reaction in the wall starts, resulting in volumetric heat generation q_{gen}. As the temperature in the wall rises, heat is convected away from the surface of the wall in contact with the fluid (at $x = L$). The convection coefficient for this process is h.

The temperature distribution in the wall, as a function of time, is governed by the transient heat equation with the source term $q_{gen}/\rho C$. Since the problem has only one spatial dimension, the transient heat equation becomes

$$\partial T/\partial t = \alpha\ \partial^2 T/\partial x^2 + q_{gen}/\rho C. \tag{7-51}$$

This is a second-order nonhomogeneous partial differential equation requiring two boundary conditions and an initial condition. For boundary conditions, it is required that

$$\text{at } x = 0: \quad T = T_b \tag{7-52}$$

$$\text{at } x = L: \quad -k\ \partial T/\partial x = h(T - T_\infty). \tag{7-53}$$

The initial condition requires that

$$\text{at } t = 0: \quad T = T_\infty. \tag{7-54}$$

This problem has a nonhomogeneous governing equation, two nonhomogeneous boundary conditions, and a nonhomogeneous initial condition. Following the method discussed in Section 7.1.3 the problem is decomposed such that $T = T_{ss} + \theta$, where T_{ss} is the steady-state temperature solution. The secondary problems for T_{ss} and θ must linearly combine to give back the original problem. The governing equation for T_{ss} is the same as for T with $\partial T/\partial t = 0$. Therefore, the governing equations for the two secondary problems are

$$\begin{array}{rcll} 0 & = & \alpha\partial^2 T_{ss}/\partial x^2 & +\ q_{gen}/\rho C \\ +\ \partial\theta/\partial t & = & \alpha\partial^2\theta/\partial x^2 & \\ \hline \partial(T_{ss}+\theta)/\partial t & = & \alpha\partial^2(T_{ss}+\theta)/\partial x^2 & +\ q_{gen}/\rho C \end{array}, \tag{7-55}$$

which sum back to the original governing equation for T. Notice that the governing equation for θ is homogeneous since the nonhomogeneous term in the equation for T goes into the problem for T_{ss}.

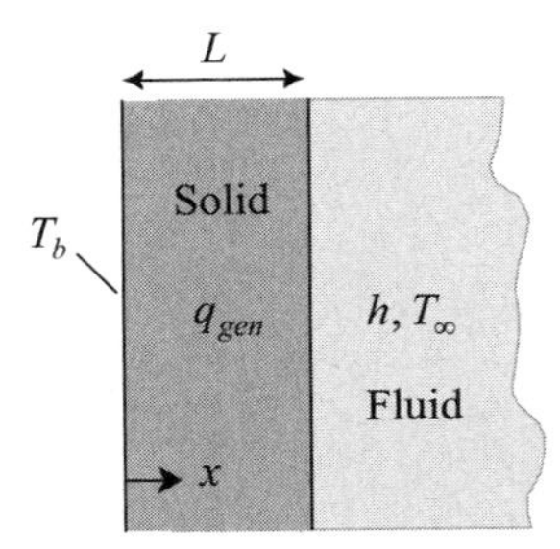

Figure 7-7 A plane wall with heat generation.

Let the boundary conditions for the T_{ss} and θ problems be

$$
\begin{array}{lrcl}
\underline{\text{At } x=0} & & & \\
 & T_{ss} & = & T_b \\
+ & \theta & = & 0 \\
\hline
 & T_{ss}+\theta & = & T_b
\end{array}
\qquad\qquad
\begin{array}{lrcl}
\underline{\text{At } x=L} & & & \\
 & -k\,\partial T_{ss}/\partial x & = & h(T_{ss}-T_\infty) \\
+ & -k\,\partial \theta/\partial x & = & h\,\theta \\
\hline
 & -k\partial(T_{ss}+\theta)/\partial x & = & h(T_{ss}+\theta-T_\infty)
\end{array}
$$

which sum back to the original boundary conditions. Notice that the boundary conditions for T_{ss} are the same as for the original problem for T. Notice furthermore that the boundary conditions for θ are of the same kind as the boundary conditions for T. However, for the θ problem, all the boundary conditions are homogeneous, while for the T problem none of the boundary conditions were homogeneous.

The initial condition requires that $T_{ss}+\theta(x,t=0)=T(x,t=0)=T_\infty$. Therefore, for the θ problem one requires the initial condition:

$$\text{at } t=0: \quad \theta(x,t=0)=T_\infty - T_{ss}. \tag{7-56}$$

Starting with the steady-state problem, the governing equation

$$0 = k\ \partial^2 T_{ss}/\partial x^2 + q_{gen} \tag{7-57}$$

is subject to these boundary conditions:

$$\text{at } x=0: \quad T_{ss}=T_b \tag{7-58}$$

$$\text{at } x=L: \quad -k\ \partial T_{ss}/\partial x = h(T_{ss}-T_\infty) \tag{7-59}$$

The steady-state solution is determined to be:

$$T_{ss}=T_b-\frac{Bi(T_b-T_\infty)}{1+Bi}\frac{x}{L}+\frac{L^2 q_{gen}}{2k}\left(\frac{2+Bi}{1+Bi}\frac{x}{L}-\left(\frac{x}{L}\right)^2\right), \tag{7-60}$$

where

$$Bi = h\,L/k.$$

The remaining problem for θ is summarized in Figure 7-8. Substituting $\theta = X(x)Y(t)$ into the governing equation for θ, and separating the two variables, one finds:

$$\frac{1}{\alpha}\frac{1}{Y}\frac{\partial Y}{\partial t}=\frac{1}{X}\frac{\partial^2 X}{\partial x^2} \qquad (=-\lambda_n^2). \tag{7-61}$$

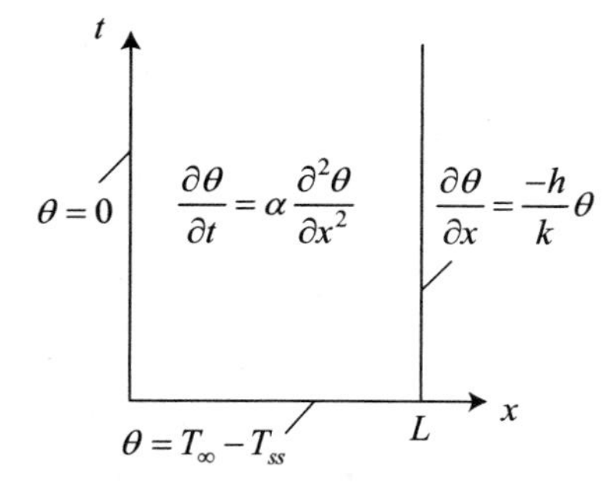

Figure 7-8 Plane wall transient subproblem.

Two ordinary differential equations for $Y(t)$ and $X(x)$ are constructed in terms of the separation constant λ_n. These equations are integrated directly:

$$\begin{array}{ccc} \dfrac{dY}{dt} + \alpha\,\lambda_n^2 Y = 0 & & \dfrac{d^2X}{dx^2} + \lambda_n^2 X = 0 \\ & \text{and} & \\ \downarrow & & \downarrow \\ Y(t) = C_1 e^{-\alpha\lambda_n^2 t} & & X(x) = C_2\cos(\lambda_n x) + C_3\sin(\lambda_n x). \end{array} \quad (7\text{-}62)$$

The homogeneous boundary conditions for θ are applied to the problem for $X(x)$:

$$\text{at } x = 0: \quad C_2\cos(0) + C_3\sin(0) = 0 \quad (7\text{-}63)$$

$$\text{at } x = L: \quad -C_2\lambda_n\sin(\lambda_n L) + C_3\lambda_n\cos(\lambda_n L) = \frac{-h}{k}\left(C_2\cos(\lambda_n L) + C_3\sin(\lambda_n L)\right) \quad (7\text{-}64)$$

From the boundary conditions, one finds that $C_2 = 0$, and that the choices for λ_n must satisfy the following:

$$\lambda_n L\cos(\lambda_n L) = -Bi\sin(\lambda_n L), \quad (7\text{-}65)$$

where $Bi = h\,L/k$. Unlike the previous example, a simple relation for the admissible separation constants is not revealed by Eq. (7-65). Instead, this eigencondition must be numerically solved to determine its roots.

With $X = C_3\sin(\lambda_n x)$, the solution for $\theta = X(x)Y(t)$ can be reconstructed with the two remaining integration constants C_1 and C_3 combined into C_n. To satisfy the initial condition, superposition of solutions with all the possible separation constants is required:

$$\theta = \sum_{n=1}^{\infty} C_n\sin(\lambda_n x)\,e^{-\alpha\,\lambda_n^2 t}. \quad (7\text{-}66)$$

The appropriate values of C_n are determined from the initial condition statement:

$$\theta(x, t = 0) = T_\infty - T_{ss} = \sum_{n=1}^{\infty} C_n\sin(\lambda_n x). \quad (7\text{-}67)$$

To determine the values of C_n, both sides of this last statement are multiplied by $\sin(\lambda_m x)$ and integrated with respect to x from 0 to L:

$$\int_0^L \sin(\lambda_m x)(T_\infty - T_{ss}(x))dx = \int_0^L \sum_{n=0}^{\infty} C_n\sin(\lambda_m x)\sin(\lambda_n x)dx \quad (7\text{-}68)$$

By orthogonality, the only nonzero term on the right-hand side of this equation is when $m = n$, and therefore one is left with

$$\int_0^L \sin(\lambda_n x)(T_\infty - T_{ss}(x))dx = C_n\int_0^L \sin(\lambda_n x)\sin(\lambda_n x)dx = C_n\left[\frac{x}{2} - \frac{\sin(2\lambda_n x)}{4\lambda_n}\right]_0^L. \quad (7\text{-}69)$$

Substituting Eq. (7-60) for $T_{ss}(x)$, the left-hand side of Eq. (7-69) integrates to

$$\int_0^L \sin(\lambda_n x)(T_\infty - T_{ss}(x))dx = \frac{-1}{\lambda_n}\left(T_b - T_\infty + \frac{L^2 q_{gen}}{k(\lambda_n L)^2}(1 - \cos(\lambda_n L))\right). \quad (7\text{-}70)$$

Therefore, the initial condition (7-69) requires that

$$\frac{-1}{\lambda_n}\left(T_b - T_\infty + \frac{L^2 q_{gen}}{k(\lambda_n L)^2}(1 - \cos(\lambda_n L))\right) = C_n\left[\frac{x}{2} - \frac{\sin(2\lambda_n x)}{4\lambda_n}\right]_0^L \quad (7\text{-}71)$$

or

$$C_n = -4\frac{T_b - T_\infty + (L^2 q_{gen}/k)(1 - \cos(\lambda_n L))/(\lambda_n L)^2}{2\lambda_n L - \sin(2\lambda_n L)}. \quad (7\text{-}72)$$

With this result for the integration constants applied to Eq. (7-66), the final solution is constructed with the steady-state solution (7-60) from the original transformation $T = T_{ss} + \theta$, such that

$$T = T_b - \frac{Bi(T_b - T_\infty)}{1 + Bi}\frac{x}{L} + \frac{L^2 q_{gen}}{2k}\left(\frac{2 + Bi}{1 + Bi}\frac{x}{L} - \left(\frac{x}{L}\right)^2\right) - 4\sum_{n=1}^{\infty}\frac{T_b - T_\infty + \frac{L^2 q_{gen}}{k(\lambda_n L)^2}(1 - \cos(\lambda_n L))}{2\lambda_n L - \sin(2\lambda_n L)}\sin(\lambda_n x)\, e^{-\alpha\lambda_n^2 t}. \quad (7\text{-}73)$$

In this solution, the eigenvalues λ_n (separation constants) must satisfy the eigen-condition $Bi\tan(\lambda_n L) + \lambda_n L = 0$ and $Bi = h\,L/k$. The solution to the temperature in the plane wall is plotted in a dimensionless form in Figure 7-9 for the conditions $L^2 q_{gen}/k = 1$ and $Bi = 1$. As expected, the temperature in the wall rises with time until a steady-state is reached. Notice that the steady-state temperature distribution has some curvature that reflects the presence of the volumetric heat source.

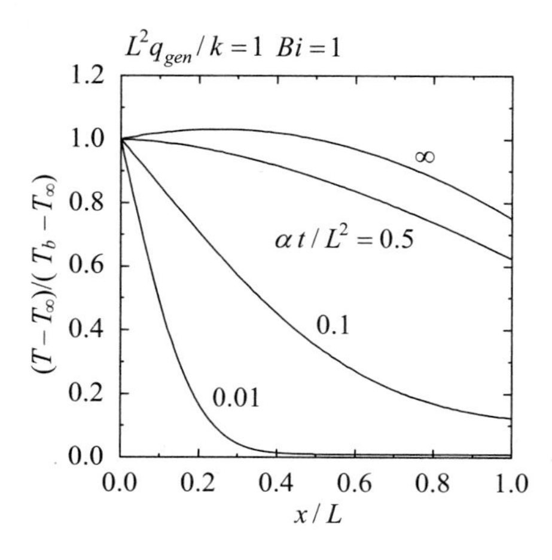

Figure 7-9 Time evolution of temperature in plane wall with generation.

7.4 TRANSIENT GROUNDWATER CONTAMINATION

Although most of the utility of separation of variables lies in solving the equations of diffusion transport, the technique is not strictly limited to such problems. To illustrate, reconsider the problem discussed in Section 6.3.3, in which water flows through the ground at a uniform speed U until it reaches a reservoir, as illustrated in Figure 7-10. Bacteria live in the reservoir at a concentration of c_R and can diffuse upstream into the groundwater. The dilute bacteria concentration in the groundwater is governed by the transient convection transport equation:

$$\frac{\partial c_B}{\partial t} - U\frac{\partial c_B}{\partial x} = Ð_B \frac{\partial^2 c_B}{\partial x^2}, \tag{7-74}$$

where $v_x = -U$. The same boundary conditions are used as for the steady-state problem considered in Section 6.3.3. However, now suppose that the groundwater is initially uncontaminated. Therefore, the conditions to be imposed on bacteria transport in the groundwater are as follows:

$$t = 0: \quad c_B\ (t = 0, x) = 0 \tag{7-75}$$

$$x = 0: \quad c_B\ (x = 0) = c_R \tag{7-76}$$

$$x \to \infty: \quad c_B\ (x \to \infty) = 0 \tag{7-77}$$

To simplify the presentation of this problem, the following dimensionless parameters are defined:

$$\eta = Ux/Ð_B, \quad \tau = U^2 t/Ð_B \text{ and } \theta = c_B/c_R \tag{7-78}$$

The problem statement can now be nondimensionalized, such that the governing convection equation becomes

$$\frac{\partial \theta}{\partial \tau} - \frac{\partial \theta}{\partial \eta} = \frac{\partial^2 \theta}{\partial \eta^2}, \tag{7-79}$$

and conditions imposed on the solution are

$$\theta(\tau = 0, \eta) = 0, \quad \theta(\eta = 0) = 1 \text{ and } \theta(\eta \to \infty) = 0. \tag{7-80}$$

The solution to the convection equation (7-79) lacks spatial functions that have the desired orthogonality for separation of variables. However, this setback can be overcome by assuming a solution of the form $\theta(\tau, \eta) = A(\tau, \eta)\exp(-\eta/2)$, where the requirements on the function $A(\tau, \eta)$ are yet to be determined. When this assumed form of a solution is substituted into the original governing equation, boundary conditions, and initial condition, it can be determined that the function $A(\tau, \eta)$ must satisfy the partial differential equation

$$\frac{\partial A}{\partial \tau} + \frac{A}{4} = \frac{\partial^2 A}{\partial \eta^2} \tag{7-81}$$

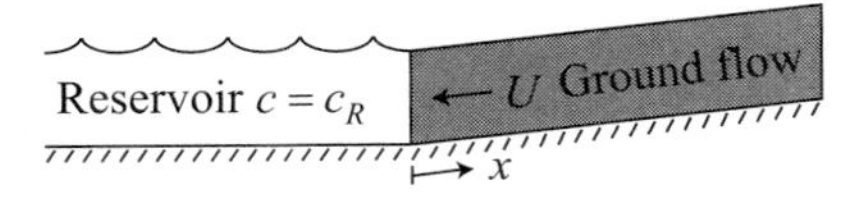

Figure 7-10 Groundwater flow to a reservoir.

and the conditions

$$A(\tau = 0, \eta) = 0, \;\; A(\eta = 0) = 1 \;\text{ and }\; A(\eta \to \infty) = \mathit{finite}. \tag{7-82}$$

Notice that in the $\eta \to \infty$ limit, the requirement that $A(\tau, \eta)$ remains finite is sufficient to ensure that the boundary condition

$$\theta(\tau, \eta \to \infty) = A(\tau, \eta \to \infty)\exp(-\infty/2) = 0 \tag{7-83}$$

is satisfied.

The key factor motivating the transformation to this new problem for $A(\tau, \eta)$ is that the spatial part of equation (7-81) is described by trigonometric functions with the required orthogonality for separation of variables. In order to move the nonhomogeneous conditions to the initial condition of the problem, by the method discussed in Section 7.1.3, $A(\tau, \eta)$ is decomposed in to the steady-state solution $A_{ss}(\eta)$ plus a transient problem $B(\tau, \eta)$:

$$A(\tau, \eta) = A_{ss}(\eta) + B(\tau, \eta). \tag{7-84}$$

The steady-state problem for $A_{ss}(\eta)$ is governed by

$$\frac{\partial^2 A_{ss}}{\partial \eta^2} - \frac{A_{ss}}{4} = 0 \;\text{ with }\; A_{ss}(\eta = 0) = 1 \;\text{ and }\; A_{ss}(\eta \to \infty) = \mathit{finite}, \tag{7-85}$$

and has the solution

$$A_{ss} = \exp(-\eta/2). \tag{7-86}$$

The remaining problem for $B(\tau, \eta)$ is governed by

$$\frac{\partial B}{\partial \tau} + \frac{B}{4} = \frac{\partial^2 B}{\partial \eta^2} \tag{7-87}$$

and is subject to the conditions

$$B(\tau = 0, \eta) = -A_{ss} = -\exp(-\eta/2), \tag{7-88}$$

$$B(\eta = 0) = 0 \;\text{ and }\; B(\eta \to \infty) = \mathit{finite}. \tag{7-89}$$

It can be verified that the problems $A_{ss}(\eta) + B(\tau, \eta)$ sum back to the original problem for $A(\tau, \eta)$.

Assuming the solution to Eq. (7-87) can take the form $B(\eta, \tau) = X(\eta)Y(\tau)$, the governing equation is separated:

$$\frac{1}{Y}\frac{\partial Y}{\partial \tau} + \frac{1}{4} = \frac{1}{X}\frac{\partial^2 X}{\partial \eta^2} \qquad (= -\lambda_n^2). \tag{7-90}$$

Two ordinary differential equations for $Y(\tau)$ and $X(\eta)$ are constructed in terms of the separation constant and integrated:

$$\begin{array}{ccc} \dfrac{dY}{d\tau} + \left(\dfrac{1}{4} + \lambda_n^2\right)Y = 0 & & \dfrac{d^2X}{d\eta^2} + \lambda_n^2 X = 0 \\ \downarrow & \text{and} & \downarrow \\ Y(\tau) = \exp\left[-\left(\dfrac{1}{4} + \lambda_n^2\right)\tau\right] & & X(\eta) = C_2\cos(\lambda_n \eta) + C_3\sin(\lambda_n \eta). \end{array} \tag{7-91}$$

In a typical separation of variables problem, the spatial domain is finite $0 \leq \eta \leq \eta_L$ and only discrete values of the separation constant can satisfy the boundary condition at $\eta = \eta_L$. When the domain is unbounded, as in the current problem, the separation constant is no longer restricted to discrete values. However, for numerical evaluation, it is convenient to approach a solution by assuming instead of $B(\eta \to \infty) = \textit{finite}$, that $B(\eta = \eta_\infty) = 0$, where η_∞ is a sufficiently large number such that $B(\eta = \eta_\infty) \approx B(\eta \to \infty)$. By allowing η_∞ to be finite, the separation constant will be restricted to discrete values, and the solution will have the usual form of an infinite series. To determine the admissible values of the separation constant, the homogeneous boundary conditions are applied to the problem for $X(\eta)$, revealing that

$$\text{at } \eta = 0: \quad B(\eta = 0) = 0 \quad \to \quad X(0) = C_2 \cos(0) + C_3 \sin(0) = 0 \tag{7-92}$$

$$\text{at } \eta = \eta_\infty: \quad B(\eta = \eta_\infty) = 0 \quad \to \quad X(\eta_\infty) = C_2 \cos(\lambda_n \eta_\infty) + C_3 \sin(\lambda_n \eta_\infty) = 0. \tag{7-93}$$

From the boundary conditions, it is revealed that $C_2 = 0$ and the separation constant must satisfy the condition $\sin(\lambda_n \eta_\infty) = 0$. Therefore, the separation constant becomes any of the discrete values expressed by

$$\lambda_n = \frac{n\pi}{\eta_\infty} \quad \textit{for } n = 1, 2, \ldots. \tag{7-94}$$

However, notice that as $\eta_\infty \to \infty$, the distance between the discrete values of the separation constant goes to zero.

With $X = C_3 \sin(\lambda_n \eta)$, the solution for $B(\eta, \tau) = X(\eta)Y(\tau)$ is reconstructed and summed over all possible values of the separation constant:

$$B(\tau, \eta) = \sum_{n=1}^{\infty} C_n \sin(\lambda_n \eta) \exp\left[-\left(\frac{1}{4} + \lambda_n^2\right)\tau\right]. \tag{7-95}$$

The two remaining integration constants C_1 and C_3 have been combined into C_n. A solution for $B(\tau, \eta)$ that satisfies the initial condition requires an appropriate evaluation of the remaining coefficients C_n. Starting with the initial condition:

$$\text{at } t = 0: \quad -\exp(-\eta/2) = \sum_{n=1}^{\infty} C_n \sin(\lambda_n \eta), \tag{7-96}$$

both sides are multiplied by $\sin(\lambda_m \eta)$ and integrated with respect to η from 0 to η_∞:

$$\int_0^{\eta_\infty} -\exp(-\eta/2)\sin(\lambda_m \eta) d\eta = \int_0^{\eta_\infty} \sum_{n=1}^{\infty} C_n \sin(\lambda_m \eta)\sin(\lambda_n \eta) d\eta. \tag{7-97}$$

Exploiting the orthogonality of $\sin(\lambda_n \eta)$ for the discrete values of λ_n given by Eq. (7-94), the only nonzero term on the right-hand side of this equation is when $m = n$. Therefore, the series coefficients evaluate to

$$C_n = \frac{2}{\eta_\infty} \int_0^{\eta_\infty} -\exp(-\eta/2) \sin(\lambda_n \eta) d\eta = \frac{-8\lambda_n}{(1 + 4\lambda_n^2)\eta_\infty}, \tag{7-98}$$

and the solution for $B(\tau, \eta)$ becomes

$$B(\tau, \eta) = \sum_{n=1}^{\infty} \frac{-8\lambda_n \sin(\lambda_n \eta)}{(1 + 4\lambda_n^2)\eta_\infty} \exp\left[-\left(\frac{1}{4} + \lambda_n^2\right)\tau\right]. \tag{7-99}$$

Working backwards through the transformations made in this problem, the original problem has a solution described by

$$\theta(\tau,\eta) = \{A_{ss}(\eta) + B(\tau,\eta)\}\exp(-\eta/2). \tag{7-100}$$

With the results of Eq. (7-86) and Eq. (7-99), the final solution becomes

$$\theta(\tau,\eta) = \exp(-\eta) - \sum_{n=1}^{\infty}\frac{8\lambda_n\sin(\lambda_n\eta)}{(1+4\lambda_n^2)\eta_\infty}\exp\left[-\left(\frac{1}{4}+\lambda_n^2\right)\tau - \frac{\eta}{2}\right] \text{ with } \lambda_n = \frac{n\pi}{\eta_\infty}. \tag{7-101}$$

Note that with $\Delta n = 1$, the series term in Eq. (7-101) can be expressed equivalently as

$$\sum_{n=1}^{\infty}\frac{8\lambda_n\sin(\lambda_n\eta)}{\pi(1+4\lambda_n^2)}\exp\left[-\left(\frac{1}{4}+\lambda_n^2\right)\tau - \frac{\eta}{2}\right]\left(\frac{\Delta n\pi}{\eta_\infty}\right)$$
$$= \sum_{n=1}^{\infty}\frac{8\lambda_n\sin(\lambda_n\eta)}{\pi(1+4\lambda_n^2)}\exp\left[-\left(\frac{1}{4}+\lambda_n^2\right)\tau - \frac{\eta}{2}\right]\Delta\lambda. \tag{7-102}$$

In the limit that $\Delta\lambda \to 0$ (i.e., $\eta_\infty \to \infty$) the summation of terms can be replaced by an integral:

$$\lim_{\Delta\lambda\to 0}\sum_{n=1}^{\infty}\frac{8\lambda_n\sin(\lambda_n\eta)}{\pi(1+4\lambda_n^2)}\exp\left[-\left(\frac{1}{4}+\lambda_n^2\right)\tau - \frac{\eta}{2}\right]\Delta\lambda$$
$$= \int_0^{\infty}\frac{8\lambda\sin(\lambda\eta)}{\pi(1+4\lambda^2)}\exp\left[-\left(\frac{1}{4}+\lambda^2\right)\tau - \frac{\eta}{2}\right]d\lambda. \tag{7-103}$$

Therefore, with the summation replaced by a continuous integral with respect to a variable describing the separation constant, the solution (7-101) becomes

$$\theta(\tau,\eta) = \exp(-\eta) - \int_0^{\infty}\frac{8\lambda\sin(\lambda\eta)}{\pi(1+4\lambda^2)}\exp\left[-\left(\frac{1}{4}+\lambda^2\right)\tau - \frac{\eta}{2}\right]d\lambda. \tag{7-104}$$

Figure 7-11 plots the solution (7-104) to show the rise in groundwater bacteria concentration as a function of dimensionless time and dimensionless distance from the reservoir. The steady-state bacteria concentration profile is achieved when $tU^2/Đ_B \approx 10$, answering the final question that was posed in Section 6.3.3.

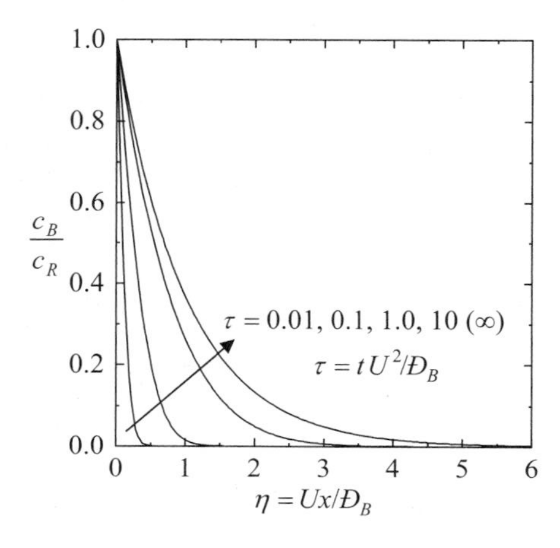

Figure 7-11 Time evolution of bacteria concentration in groundwater.

7.5 PROBLEMS

7-1 A gap between parallel plates of width W is filled with a stationary fluid. At $t = 0$, the top plate is set into motion with a velocity U parallel to the fluid. Find the transient expression for the fluid velocity between the plates $v_x(t, y)$. Nondimensionalize your solution using $\tau = \nu t/W^2$, $\eta = y/W$, and $u = v_x/U$. Plot the development of the flow starting from rest using a few nonequispaced time steps: $\tau = 0.001, 0.01, 0.1$, and 1.0 .

7-2 A plane wall of thickness L is initially at a temperature T_o. At $t = 0$, the front surface is suddenly elevated to a temperature T_1, and heat transfer to the wall occurs. Heat arriving to the back surface of the wall is convected away, such that the back boundary condition is given by $[-k\partial T/\partial x = h(T - T_o)]_{x=L}$, where h is the convection coefficient for heat transfer. Determine the wall temperature as a function of time and space. Plot your solution for some representative numbers.

7-3 Consider the laser heating of a plane wall governed by the transient heat equation:

$$\partial_o T = \alpha\, \partial_i \partial_i T + q_{gen}/(\rho C), \quad \text{where} \quad q_{gen} = (I_o/\delta_{opt})\exp(-x/\delta_{opt}),$$

where δ_{opt} is the optical penetration depth of laser light through the front surface of the wall. In Problem 6-1 the steady-state temperature distribution $T_{ss}(x)$ in the wall was found:

$$\frac{T_{ss}(x) - T_b}{I_o L/k} = 1 - \frac{x}{L} + \frac{\delta_{opt}}{L}\left[\exp\left(\frac{-L}{\delta_{opt}}\right) - \exp\left(\frac{-x}{\delta_{opt}}\right)\right]$$

where L is the wall thickness and T_b is the back surface temperature of the wall. Consider now a transient problem in which the initial wall temperature is T_b. Conduction of heat from the front (laser-heated surface) is still zero and the back surface is still held at a constant temperature T_b. Find the transient solution by assuming that $T = T_{ss} + T^*$. Substitute this expression into the governing equations, boundary conditions, and initial conditions to prescribe the problem for T^*. Solve for T^*. Express your solution for $T = T_{ss} + T^*$ in terms of the dimensionless variables:

$$\theta = \frac{T - T_b}{I_o L/k}, \quad \eta = \frac{x}{L}, \quad \tau = \frac{\alpha t}{L^2}, \quad \text{and} \quad \Lambda = \frac{\delta_{opt}}{L}.$$

For the optical penetration depth of $\Lambda = 0.1$, plot the spatial distribution of temperature θ as a function of time for $\tau = 0.01, 0.1, 1.0$, and 10.0.

7-4 A gap between parallel plates of width W is filled with a fluid. The fluid experiences a pressure gradient dP/dx exerted by a pump. Find the transient expression for the fluid velocity between the plates $v_x(t, y)$ after the pump is turned on. Nondimensionalize your solution using $\tau = \nu t/W^2$, $\eta = y/W$, and $u = \mu v_x/(-W^2 dP/dx)$. Plot the development of the flow starting from rest using a few nonequispaced time steps: $\tau = 0.1, 1.0$, and 10.

7-5 Solve the mass transfer problem mathematically depicted in the illustration. Present the solution for the normalized dimensionless concentration θ in terms of the variables:

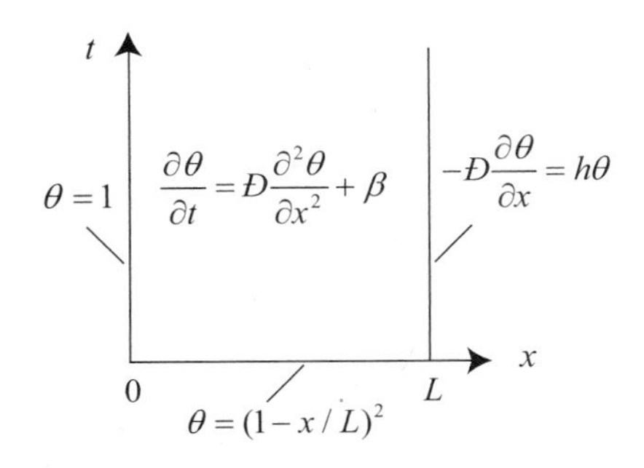

$$S = \frac{\beta L^2}{Đ}, \; Bi_m = \frac{hL}{Đ}, \eta = \frac{x}{L}, \text{ and } \tau = \frac{Đt}{L^2}$$

Considering the case where $S = 1$ and $Bi_m = 1$, plot the solution for τ =0.001, 0.01, 0.1, 1.0, and 1.0 .

REFERENCES

[1] R. Haberman, *Applied Partial Differential Equations (with Fourier Series and Boundary Value Problems)*, Fourth Edition. Upper Saddle River, NJ: Prentice-Hall, 2006.

[2] A. D. Polyanin, *Handbook of Linear Partial Differential Equations for Engineers and Scientists*. Boca Raton, FL: Chapman & Hall/CRC Press, 2002.

Chapter 8

Steady Two-Dimensional Diffusion

8.1 Separation of Two Spatial Variables

8.2 Nonhomogeneous Conditions on Nonadjoining Boundaries

8.3 Nonhomogeneous Conditions on Adjoining Boundaries

8.4 Nonhomogeneous Condition in Governing Equation

8.5 Looking Ahead

8.6 Problems

As seen in Chapter 7 with transient one-dimensional problems, separation of variables employs a strategy of transforming a single partial differential equation into two ordinary differential equations before integration. Separation of variables applied to steady two-dimensional problems, as discussed in this chapter, shares many of the same strategies as applied to transient one-dimensional problems. However, a few of the nuances are sufficiently different to warrant additional treatment of the subject. A more extensive discussion of separation of variables can be found in books such as references [1] and [2].

8.1 SEPARATION OF TWO SPATIAL VARIABLES

Consider a governing equation that is characteristic of steady-state two-dimensional diffusion transport:

$$\frac{\partial^2 \phi}{\partial x^2} + \frac{\partial^2 \phi}{\partial y^2} = 0. \tag{8-1}$$

This equation could describe any number of problems in heat, mass, or momentum transfer. To solve this equation, four boundary conditions must be specified, leading to many permutations of problems. The first step to solving this equation by separation of variables is to assume a solution exists with the form $\phi = X(x)Y(y)$. When this expression is substituted into Eq. (8-1), the resulting equation may be separated such that the left-hand side is some function of y and the right-hand side is some function of x:

$$\frac{-1}{Y}\frac{\partial^2 Y}{\partial y^2} = \frac{1}{X}\frac{\partial^2 X}{\partial x^2} \qquad (= \pm\lambda_n^2). \tag{8-2}$$

The only way to assure equality of these two functions is to force both functions to equal the same constant $\pm\lambda_n^2$. Two sets of ordinary differential equations result from the two possible signs that could be assigned to the separation constant. With $+\lambda_n^2$,

$$\begin{array}{ccc} \dfrac{d^2X}{dx^2} + \lambda_n^2 X = 0 & & \dfrac{d^2Y}{dy^2} - \lambda_n^2 Y = 0 \\ \downarrow & \text{and} & \downarrow \\ X = C_1 \cos(\lambda_n x) + C_2 \sin(\lambda_n x) & & Y = C_3 \cosh(\lambda_n y) + C_4 \sinh(\lambda_n y) \end{array} \quad . \tag{8-3}$$

Or, for the other possible sign, $-\lambda_n^2$,

$$\begin{array}{ccc} \dfrac{d^2X}{dx^2} - \lambda_n^2 X = 0 & & \dfrac{d^2Y}{dy^2} + \lambda_n^2 Y = 0 \\ \downarrow & \text{and} & \downarrow \\ X = C_1 \cosh(\lambda_n x) + C_2 \sinh(\lambda_n x) & & Y = C_3 \cos(\lambda_n y) + C_4 \sin(\lambda_n y) \end{array} \quad . \tag{8-4}$$

Deciding which sign to give the separation constant will depend on the location of nonhomogeneous boundary conditions in the problem. In either case, the recombined solution $\phi = X(x)Y(y)$ will satisfy the governing equation.

One expects to find at least one nonhomogeneous condition in every problem statement. Unfortunately, nonhomogeneous conditions seldom come with the spatial distribution described by the elementary functions cos(), sin(), cosh(), and sinh() that are derived from the separation of variables technique. However, because the governing equation (8-1) is homogeneous and linear, superposition can be employed to "build" up a solution to the nonhomogeneous boundary conditions with the trigonometric functions, as was done for the initial condition of the transient problems treated in Chapter 7. To do this, however, it is necessary to choose the sign of the separation constant such that the cos() and sin() functions appear along the nonhomogeneous boundary or boundaries.

To illustrate, consider the nonhomogeneous boundary condition shown at two different locations in Figure 8-1. In the first case, the nonhomogeneous boundary condition falls on a surface that is parallel to the y-direction. Therefore, the sign of the separation constant is chosen such that the solution for Y contains the trigonometric functions, $Y = C_3 \cos(\lambda_n y) + C_4 \sin(\lambda_n y)$. In contrast, if the nonhomogeneous boundary condition appears on a surface parallel to the x-direction, as shown in the second case, the sign of the separation constant is chosen such that the solution for X contains the trigonometric functions, $X = C_1 \cos(\lambda_n x) + C_2 \sin(\lambda_n x)$. In this manner the sign of the separation constant is chosen to allow superposition of the trigonometric functions to build a solution that satisfies the nonhomogeneous boundary condition.

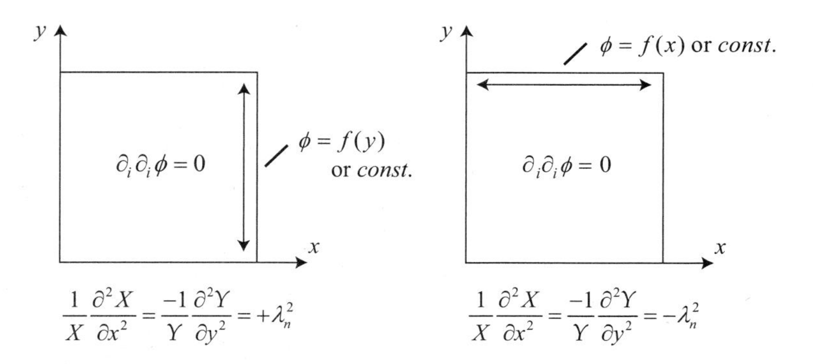

Figure 8-1 Illustration of separation constant sign choice.

When solving two-dimensional problems by separation of variables, it is useful to be aware of alternate forms of the solution to the equation:

$$\frac{d^2Y}{dy^2} - \lambda_n^2 Y = 0$$
$$\downarrow$$
$$\begin{aligned} & Y(y) = C_1 \exp(\lambda_n y) - C_2 \exp(-\lambda_n y) \\ or \quad & Y(y) = C_3 \cosh(\lambda_n y) + C_4 \sinh(\lambda_n y) \\ or \quad & Y(y) = C_5 \cosh(\lambda_n (W - y)) + C_6 \sinh(\lambda_n (W - y)). \end{aligned} \tag{8-5}$$

The preferred form of the solution for $Y(y)$ is dictated by the boundary conditions imposed on a problem. For example, if the solution requires $Y(0) = 0$, the second form of the solution for $Y(y)$ with $C_3 = 0$ most simply satisfies this boundary condition. However, if the solution requires $Y(W) = 0$, the third form of the solution for $Y(y)$ with $C_5 = 0$ most easily satisfies that boundary condition.

8.2 NONHOMOGENEOUS CONDITIONS ON NONADJOINING BOUNDARIES

The simplest problems to solve by separation of variables contain only one nonhomogeneous boundary condition, or two nonhomogeneous boundary conditions on opposing (nonadjoining) sides of the domain. For example, consider the problem illustrated in Figure 8-2. To solve by separation of variables, a solution of the form $\phi = X(x)Y(y)$ is applied to the governing equation and the variables separated to yield

$$\frac{-1}{Y}\frac{\partial^2 Y}{\partial y^2} = \frac{1}{X}\frac{\partial^2 X}{\partial x^2} \qquad (= -\lambda_n^2). \tag{8-6}$$

The sign of the separation constant was selected to allow the two ordinary differential equations for $X(x)$ and $Y(y)$ to emerge:

$$\begin{array}{ccc} \dfrac{d^2X}{dx^2} + \lambda_n^2 X = 0 & & \dfrac{d^2Y}{dy^2} - \lambda_n^2 Y = 0 \\ \downarrow & \text{and} & \downarrow \\ X(x) = C_3 \cos(\lambda_n x) + C_4 \sin(\lambda_n x) & & Y(y) = C_1 \cosh(\lambda_n y) + C_2 \sinh(\lambda_n y) \end{array}. \tag{8-7}$$

Specifically, the sign of the separation constant was chosen such that the solution for $X(x)$ is in terms of the trigonometric functions cos() and sin(). This will permit the nonhomogeneous

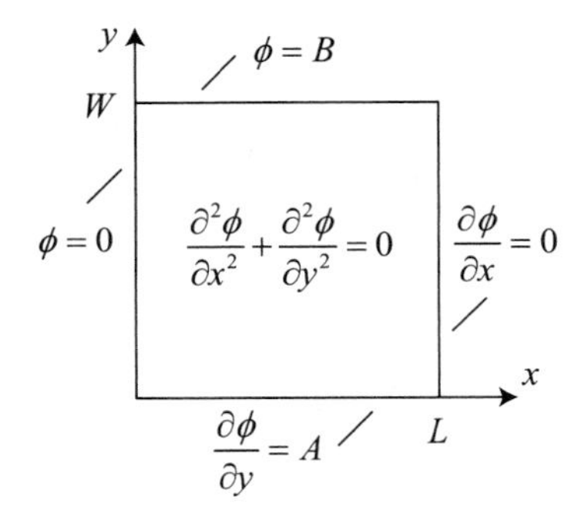

Figure 8-2 Problem with nonhomogeneous conditions on nonadjoining boundaries.

boundary conditions at $y = 0$ and $y = W$ to be satisfied later. Both homogeneous boundary conditions at $x = 0$ and $x = L$ can be applied to the solution for $X(x)$:

$$\text{at} \quad x = 0: \quad \phi(x = 0) = 0 \quad \rightarrow \quad X(0) = C_3 \cos(0) + C_4 \sin(0) = 0 \tag{8-8}$$

$$\text{at} \quad x = L: \quad \partial\phi/\partial x|_{x=L} = 0 \quad \rightarrow \quad dX/dx|_{x=L} = -C_3\lambda_n \sin(\lambda_n L) + C_4\lambda_n \cos(\lambda_n L) = 0. \tag{8-9}$$

The first boundary condition at $x = 0$ requires that $C_3 = 0$. The second boundary condition is satisfied when $dX/dx|_{x=L} = C_4\lambda_n \cos(\lambda_n L) = 0$. This requires that the separation constant be

$$\lambda_n = \frac{(n + 1/2)\pi}{L} \quad \textit{for } n = 0, 1, 2, \ldots. \tag{8-10}$$

Notice that the homogeneous boundary conditions are now satisfied by the solution $\phi = X(x)Y(y)$, since $\phi(x = 0) = X(0)Y(y) = 0$ and $\partial\phi/\partial x|_{x=L} = dX/dx|_{x=L}Y(y) = 0$. However, since neither of the remaining boundary conditions is homogeneous, they cannot be applied to the solution for $Y(y)$ alone. These boundary conditions can only be addressed after the solution for ϕ is reconstructed:

$$\phi = X(x)Y(y) = (C_1C_4 \cosh(\lambda_n y) + C_2C_4 \sinh(\lambda_n y)) \sin(\lambda_n x). \tag{8-11}$$

Notice that for any single choice of λ_n, the solution will not satisfy the remaining boundary conditions. However, since the equation being solved is linear and homogeneous, the sum of all the proposed solutions to the governing equation will also be a solution:

$$\phi = \sum_{n=0}^{\infty} (C_n \cosh(\lambda_n y) + D_n \sinh(\lambda_n y)) \sin(\lambda_n x). \tag{8-12}$$

The integration constants have been renamed as $C_n = C_1C_4$ and $D_n = C_2C_4$, where the index "n" serves as a reminder that the values of these constants are dependent on the specific choice of λ_n. The final task is to determine appropriate values of C_n and D_n that will build a solution satisfying the remaining nonhomogeneous boundary conditions. Starting with the boundary condition at $y = 0$:

$$A = \left.\frac{\partial\phi}{\partial y}\right|_{y=0} = \sum_{n=0}^{\infty} (C_n\lambda_n \sinh(0) + D_n\lambda_n \cosh(0)) \sin(\lambda_n x). \tag{8-13}$$

Both sides of this equation are multiplied by $\sin(\lambda_m x)$ and integrated with respect to x from 0 to L:

$$\int_0^L A \sin(\lambda_m x) dx = \int_0^L \sum_{n=0}^{\infty} (D_n\lambda_n) \sin(\lambda_n x) \sin(\lambda_m x) dx. \tag{8-14}$$

Because of orthogonality, the only nonzero term on the right-hand side of this equation is when $m = n$. Therefore, as long as A is not a function of x, integration yields

$$\frac{A}{\lambda_n}(1 - \cos(\lambda_n L)) = D_n\lambda_n \frac{L}{2} \quad \text{or} \quad D_n = \frac{2A}{L\lambda_n^2}(1 - \cos(\lambda_n L)). \tag{8-15}$$

Next, the boundary condition at $y = W$ requires that

$$B = \phi(y = W) = \sum_{n=0}^{\infty} (C_n \cosh(\lambda_n W) + D_n \sinh(\lambda_n W)) \sin(\lambda_n x). \tag{8-16}$$

Again, both sides of this equation are multiplied by $\sin(\lambda_m x)$ and integrated with respect to x from 0 to L:

$$\int_0^L B \sin(\lambda_m x) dx = \int_0^L \sum_{n=0}^{\infty} (C_n \cosh(\lambda_n W) + D_n \sinh(\lambda_n W)) \sin(\lambda_n x) \sin(\lambda_m x) dx. \tag{8-17}$$

Once more, the only nonzero term on the right-hand side of this equation is when $m = n$. Therefore, as long as B is not a function of x, then integration yields

$$\frac{B}{\lambda_n}(1 - \cos(\lambda_n L)) = (C_n \cosh(\lambda_n W) + D_n \sinh(\lambda_n W))\frac{L}{2}. \tag{8-18}$$

Making use of Eq. (8-15) for D_n, the value of C_n can be determined:

$$C_n = \frac{2}{L\lambda_n}\left[B - \frac{A}{\lambda_n}\sinh(\lambda_n W)\right]\frac{1 - \cos(\lambda_n L)}{\cosh(\lambda_n W)}. \tag{8-19}$$

Applying the results for the two integration constants to Eq. (8-12), the final solution can be written in the form

$$\phi = \sum_{n=0}^{\infty} 2\frac{1 - \cos(\lambda_n L)}{L\lambda_n}\left[\left(B - \frac{A}{\lambda_n}\sinh(\lambda_n W)\right)\frac{\cosh(\lambda_n y)}{\cosh(\lambda_n W)} + \frac{A}{\lambda_n}\sinh(\lambda_n y)\right]\sin(\lambda_n x). \tag{8-20}$$

8.3 NONHOMOGENEOUS CONDITIONS ON ADJOINING BOUNDARIES

Problems that have nonhomogeneous conditions on adjoining boundaries (sharing a corner) should be broken into two problems, each having nonhomogeneous conditions only on opposing boundaries. When superimposed, these two problems must satisfy the conditions of the original problem. Consider, for example, the problem illustrated in Figure 8-3.

The original problem for ϕ has two nonhomogeneous boundary conditions, corresponding to $\phi = 1$ at $x = 1$ and $\partial\phi/\partial y = f(x)$ at $y = 0$. The problem is decomposed with

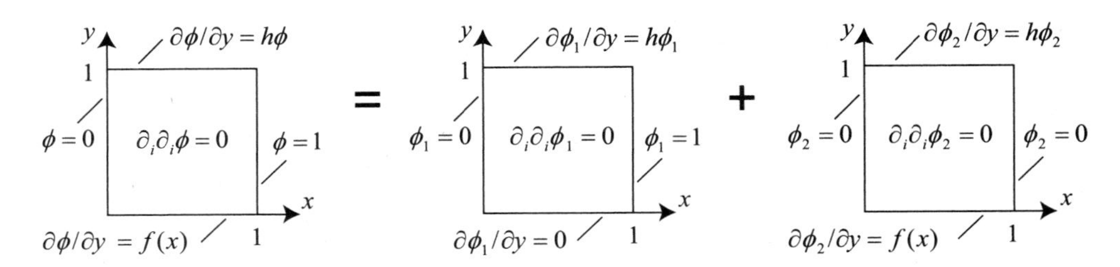

Figure 8-3 Superposition of problems with nonhomogeneous conditions on nonad joining boundaries.

$\phi = \phi_1 + \phi_2$, where the problems defining ϕ_1 and ϕ_2 are shown in Figure 8-3. Notice that both of the new problems have only one nonhomogeneous boundary condition. Furthermore, the superposition of these two solutions will satisfy the original problem. Superposition of the governing equations yields

$$\begin{array}{rrcrl} & \partial^2\theta_1/\partial x^2 & + & \partial^2\theta_1/\partial y^2 & = 0 \\ + & \partial^2\theta_2/\partial x^2 & + & \partial^2\theta_2/\partial y^2 & = 0 \\ \hline & \partial^2(\theta_1+\theta_2)/\partial x^2 & + & \partial^2(\theta_1+\theta_2)/\partial y^2 & = 0 \end{array} \tag{8-21}$$

and superposition of the boundary conditions yields

$$\begin{array}{llcl} \underline{\text{At } x=0} & & & \\ & \phi_1 & = & 0 \\ + & \phi_2 & = & 0 \\ \hline & (\phi_1+\phi_2) & = & 0 \end{array} \qquad \begin{array}{llcl} \underline{\text{At } x=1} & & & \\ & \phi_1 & = & 1 \\ + & \phi_2 & = & 0 \\ \hline & (\phi_1+\phi_2) & = & 1 \end{array}$$

$$\begin{array}{llcl} \underline{\text{At } y=0} & & & \\ & \partial\phi_1/\partial y & = & 0 \\ + & \partial\phi_2/\partial y & = & f(x) \\ \hline & \partial(\phi_1+\phi_2)/\partial y & = & f(x) \end{array} \qquad \begin{array}{llcl} \underline{\text{At } y=1} & & & \\ & \partial\phi_1/\partial y - h\phi_1 & = & 0 \\ + & \partial\phi_2/\partial y - h\phi_2 & = & 0 \\ \hline & \partial(\phi_1+\phi_2)/\partial y - h(\phi_1+\phi_2) & = & 0 \end{array}$$

The problems defined for ϕ_1 and ϕ_2 can be solved by separation of variables using the steps discussed in Section 8.2.

8.3.1 Bar Heat Treatment

Consider a bar of rectangular cross section undergoing a heat treatment. As illustrated in Figure 8-4, the boundary temperature is specified on three sides, while the fourth side of the bar is exposed to a constant heat flux. Heat transfer is governed by the steady diffusion equation:

$$\alpha\left(\frac{\partial^2 T}{\partial x^2} + \frac{\partial^2 T}{\partial y^2}\right) = 0. \tag{8-22}$$

All four boundary conditions are nonhomogeneous. For convenience, the temperature relative to T_o is determined. Therefore, in terms of $\theta = T - T_o$, the problem becomes

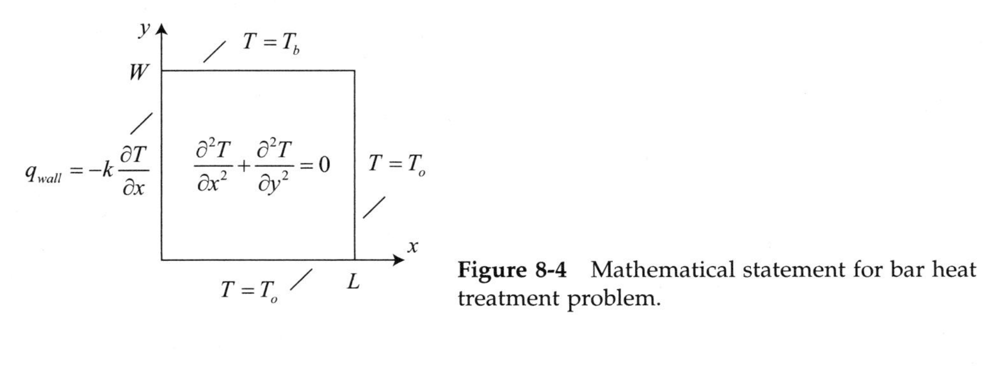

Figure 8-4 Mathematical statement for bar heat treatment problem.

$$\frac{\partial^2\theta}{\partial x^2}+\frac{\partial^2\theta}{\partial y^2}=0. \tag{8-23}$$

subject to the boundary conditions

$$\partial\theta/\partial x|_{x=0}=-q_{wall}/k \qquad \text{and} \qquad \theta(x=L)=0 \tag{8-24}$$

$$\theta(y=0)=0 \qquad \text{and} \qquad \theta(y=W)=T_b-T_o=\theta_b \tag{8-25}$$

The problem for θ has two nonhomogeneous conditions on adjoining boundaries. Therefore, to apply separation of variables, the problem is decomposed by letting $\theta=\theta_1+\theta_2$, such that

$$\begin{array}{rcccl} & \partial^2\theta_1/\partial x^2 & + & \partial^2\theta_1/\partial y^2 & =0 \\ + & \partial^2\theta_2/\partial x^2 & + & \partial^2\theta_2/\partial y^2 & =0 \\ \hline & \partial^2(\theta_1+\theta_2)/\partial x^2 & + & \partial^2(\theta_1+\theta_2)/\partial y^2 & =0 \end{array} \tag{8-26}$$

with

$$\begin{array}{ll} \underline{\text{At } x=0} & \underline{\text{At } x=L} \\ \begin{array}{rrcl} & \partial\theta_1/\partial x & = & -q_{wall}/k \\ + & \partial\theta_2/\partial x & = & 0 \\ \hline & \partial(\theta_1+\theta_2)/\partial x & = & -q_{wall}/k \end{array} & \begin{array}{rrcl} & \theta_1 & = & 0 \\ + & \theta_2 & = & 0 \\ \hline & (\theta_1+\theta_2) & = & 0 \end{array} \\ \underline{\text{At } y=0} & \underline{\text{At } y=W} \\ \begin{array}{rrcl} & \theta_1 & = & 0 \\ + & \theta_2 & = & 0 \\ \hline & (\theta_1+\theta_2) & = & 0 \end{array} & \begin{array}{rrcl} & \theta_1 & = & 0 \\ + & \theta_2 & = & \theta_b \\ \hline & (\theta_1+\theta_2) & = & \theta_b \end{array} \end{array}$$

Now both problems for θ_1 and θ_2 have only one nonhomogeneous boundary condition. Each can be solved by the standard procedure for separation of variables. For θ_1, one finds that

$$\theta_1=\sum_{n=1}^{\infty}C_n\sin(\lambda_n y)\sinh(\lambda_n(L-x)) \quad \lambda_n=n\pi/W \quad n=1,2,\ldots \tag{8-27}$$

satisfies all the homogeneous conditions: $\theta_1(x=L)=0$, $\theta_1(y=0)=0$, and $\theta_1(y=W)=0$. Therefore, the final nonhomogeneous boundary condition $\partial\theta_1/\partial x|_{x=0}=-q_{wall}/k$ must be satisfied by

$$\frac{-q_{wall}}{k}=-\sum_{n=0}^{\infty}C_n\sin(\lambda_n y)\,\lambda_n\cosh(\lambda_n L). \tag{8-28}$$

To determine the correct values of C_n, the orthogonality of the $\sin(\lambda_n y)$ function is exploited:

$$\int_0^W \frac{q_{wall}}{k}\sin(\lambda_m y)dy = \int_0^W \sum_{n=0}^{\infty} C_n \sin(\lambda_n y)\,\lambda_n \cosh(\lambda_n L)\sin(\lambda_m y)dy \tag{8-29}$$

or

$$\frac{q_{wall}}{k\lambda_n}(1-\cos(n\pi)) = C_n\lambda_n\cosh(\lambda_n L)\frac{L}{2}. \tag{8-30}$$

With C_n established from the last result, the solution to the first problem is written:

$$\theta_1 = \sum_{n=1}^{\infty} \frac{2q_{wall}}{kL\lambda_n^2}\left(\frac{1-(-1)^n}{\cosh(\lambda_n L)}\right)\sin(\lambda_n y)\sinh(\lambda_n(L-x)) \tag{8-31}$$

where $\cos(n\pi)$ has been replaced by $(-1)^n$.

For θ_2 one finds that

$$\theta_2 = \sum_{n=1}^{\infty} D_n \cos(\eta_n x)\sinh(\eta_n y) \quad \eta_n = (n-1/2)\pi/L \quad n = 1, 2, \ldots \tag{8-32}$$

satisfies all the homogeneous conditions: $\partial\theta_2/\partial x|_{x=0} = 0$, $\theta_2(x = L) = 0$, and $\theta_2(y = 0) = 0$. At $y = W$, $\theta_2(y = W) = \theta_b$ must be satisfied by

$$\theta_b = \sum_{n=1}^{\infty} D_n \cos(\eta_n x)\sinh(\eta_n W). \tag{8-33}$$

To determine the correct values of D_n, the orthogonality of the $\cos(\eta_n x)$ function is exploited:

$$\int_0^L \theta_b \cos(\eta_m x)dx = \int_0^L \sum_{n=0}^{\infty} D_n \cos(\eta_n x)\sinh(\eta_n W)\cos(\eta_m x)dx. \tag{8-34}$$

Or,

$$\frac{\theta_b}{\eta_n}\sin((2n-1)\pi/2) = D_n \sinh(\eta_n W)\frac{W}{2}. \tag{8-35}$$

With D_n established from the last result, the solution to the second problem is written:

$$\theta_2 = \sum_{n=1}^{\infty} \frac{2\theta_b}{\eta_n W}\frac{(-1)^{n-1}}{\sinh(\eta_n W)}\cos(\eta_n x)\sinh(\eta_n y) \tag{8-36}$$

where $\sin((2n-1)\pi/2)$ has been replaced by $(-1)^{n-1}$.

Finally, the reconstructed solution for $\theta = \theta_1 + \theta_2$ can be written:

$$\theta = \sum_{n=1}^{\infty}\left(\frac{2q_{wall}}{kL\lambda_n^2}\left(\frac{1-(-1)^n}{\cosh(\lambda_n L)}\right)\sin(\lambda_n y)\sinh(\lambda_n(L-x)) + \frac{2\theta_b}{\eta_n W}\frac{(-1)^{n-1}}{\sinh(\eta_n W)}\cos(\eta_n x)\sinh(\eta_n y)\right) \tag{8-37}$$

with

$$\lambda_n = n\pi/W \quad \text{and} \quad \eta_n = (n-1/2)\pi/L. \tag{8-38}$$

8.4 NONHOMOGENEOUS CONDITION IN GOVERNING EQUATION

Problems with nonhomogeneous governing equations can be cumbersome to solve using the standard method of separation of variables. To proceed with this technique it is necessary to transform the original problem to one where the governing equation is homogeneous. This was done in a relatively simple way in Chapter 7 for transient problems by introducing the steady-state solution $\phi_{ss}(x)$, such that $\phi(t,x) = \phi_{ss}(x) + \theta(t,x)$. After splitting the problem, all of the nonhomogeneous conditions, except the initial condition, are associated with the problem for $\phi_{ss}(x)$ that is governed by an ordinary differential equation. Separation of variables is then applied to the problem for $\theta(t,x)$ that has a homogeneous governing equation and homogeneous boundary conditions.

When faced with a nonhomogeneous governing equation for a steady-state problem of two spatial dimensions, one can look for a transformation where $\phi(x,y) = \psi(x \text{ or } y) + \theta(x,y)$. The problem is broken down so that the partial differential equation for $\theta(x,y)$ is homogeneous and the ordinary differential equation for $\psi(x \text{ or } y)$ contains the nonhomogeneous term appearing in the original governing equation.

8.4.1 Steady Rectangular Duct Flow

Consider a duct of rectangular cross section through which a steady *Poiseuille flow** is pumped by a constant pressure gradient, as illustrated in Figure 8-5. The walls of the rectangular duct are separated by the distances $2L$ and $2W$. Because of the symmetry of the problem, a flow description for the upper right quadrant of the duct is sufficient.

Momentum lost to the walls is balanced by the constant pressure gradient down the length of the duct. Since diffusion is the only mechanism for transport of momentum to the walls, the momentum equation governing this problem takes the form

$$\nu\left(\frac{\partial^2 \upsilon_z}{\partial x^2} + \frac{\partial^2 \upsilon_z}{\partial y^2}\right) - \frac{1}{\rho}\frac{\partial P}{\partial z} = 0 \tag{8-39}$$

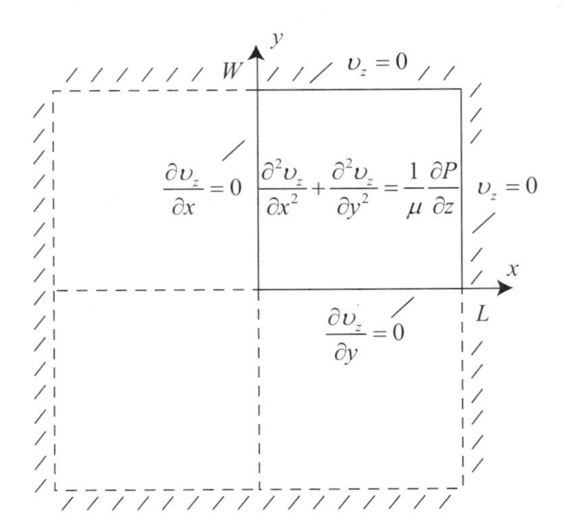

Figure 8-5 Mathematical statement for Poiseuille flow through a rectangular channel.

*Internal flows driven by a pressure gradient are called Poiseuille flow in honor of the French physician Jean Louis Marie Poiseuille (1797–1869) for his experimental work in determining the relation between the flow rate and pressure drop in pipe flow.

and is subject to the boundary conditions

$$\partial v_z/\partial x|_{x=0} = 0 \quad \text{and} \quad v_z(x = L) = 0 \tag{8-40}$$

$$\partial v_z/\partial y|_{y=0} = 0 \quad \text{and} \quad v_z(y = W) = 0. \tag{8-41}$$

The governing equation can be expressed more simply as

$$\frac{\partial^2 v_z}{\partial x^2} + \frac{\partial^2 v_z}{\partial y^2} = B, \quad \text{were} \quad B = (\partial P/\partial z)/\mu. \tag{8-42}$$

Letting $v_z(x, y) = \psi(x) + \theta(x, y)$, the original problem is subdivided:

$$\begin{array}{lccccc} & \partial^2\psi/\partial x^2 & + & 0 & = B \\ + & \partial^2\theta/\partial x^2 & + & \partial^2\theta/\partial y^2 & = 0 \\ \hline & \partial^2(\psi+\theta)/\partial x^2 & + & \partial^2(\psi+\theta)/\partial y^2 & = B \end{array} \tag{8-43}$$

with

$$\begin{array}{lccc} \text{At } x = 0 & & & \\ & \partial\psi/\partial x & = & 0 \\ + & \partial\theta/\partial x & = & 0 \\ \hline & \partial(\psi+\theta)/\partial x & = & 0 \end{array} \qquad \begin{array}{lccc} \text{At } x = L & & & \\ & \psi & = & 0 \\ + & \theta & = & 0 \\ \hline & (\psi+\theta) & = & 0 \end{array}$$

$$\begin{array}{lccc} \text{At } y = 0 & & & \\ & \partial\psi/\partial y & = & 0 \\ + & \partial\theta/\partial y & = & 0 \\ \hline & \partial(\psi+\theta)/\partial y & = & 0 \end{array} \qquad \begin{array}{lccc} \text{At } y = W & & & \\ & \psi & = & \psi \\ + & \theta & = & -\psi \\ \hline & (\psi+\theta) & = & 0 \end{array}$$

Notice that the nonhomogeneous term in the original governing equation is carried over into the governing equation for ψ. Notice also that the only boundary condition in the original problem that cannot be satisfied with the function $\psi(x)$ is at $y = W$. Along this boundary, $\psi(x)$ will not be identically equal to zero. Therefore, the problem for $\theta(x, y)$ is required to have the boundary condition $\theta(x, y = W) = -\psi(x)$, such that sum $\theta + \psi$ evaluates to zero at $y = W$.

The problem for ψ is defined by the ordinary differential equation:

$$\frac{d^2\psi}{dx^2} = B, \tag{8-44}$$

subject to the boundary conditions

$$d\psi/dx|_{x=0} = 0 \quad \text{and} \quad \psi(x = L) = 0. \tag{8-45}$$

It is straightforward to show that ψ has the solution

$$\psi(x) = B(x^2 - L^2)/2. \tag{8-46}$$

The remaining problem for θ is defined by

$$\frac{\partial^2\theta}{\partial x^2}+\frac{\partial^2\theta}{\partial y^2}=0, \tag{8-47}$$

subject to the boundary conditions

$$\partial\theta/\partial x|_{x=0}=0 \quad \text{and} \quad \theta(x=L)=0 \tag{8-48}$$

$$\partial\theta/\partial y\big|_{y=0}=0 \quad \text{and} \quad \theta(y=W)=-\psi. \tag{8-49}$$

One finds by standard separation of variables that

$$\theta=\sum_{n=0}^{\infty}C_n\cos(\lambda_n x)\cosh(\lambda_n y) \quad \text{with} \quad \lambda_n=(n+1/2)\pi/L \quad n=0,1,2,\ldots \tag{8-50}$$

satisfies all the homogeneous conditions imposed on θ at $x=0$, $x=L$, and $y=0$. Only the boundary condition at $y=W$, where $\theta(y=W)=-\psi$, remains to be satisfied by

$$-\psi=\sum_{n=0}^{\infty}C_n\cos(\lambda_n x)\cosh(\lambda_n W). \tag{8-51}$$

The orthogonality property of cos $(\lambda_n x)$ is used to determine the appropriate values of C_n:

$$\int_0^L-\psi\cos(\lambda_m x)dx=\int_0^L\sum_{n=0}^{\infty}C_n\cos(\lambda_n x)\cosh(\lambda_n W)\cos(\lambda_m x)dx. \tag{8-52}$$

Or,

$$\int_0^L(-B(x^2-L^2)/2)\cos(\lambda_n x)dx=C_n\cosh(\lambda_n W)\frac{L}{2}. \tag{8-53}$$

Integrating twice, by parts, the left-hand side of the last result yields:

$$\begin{aligned}&\int_0^L(-B(x^2-L^2)/2)\cos(\lambda_n x)dx\\&=B\left[\left(\frac{L^2-x^2}{2}\right)\frac{\sin(\lambda_n x)}{\lambda_n}-x\frac{\cos(\lambda_n x)}{\lambda_n^2}+\frac{\sin(\lambda_n x)}{\lambda_n^3}\right]\Bigg|_0^L=B\frac{\sin(\lambda_n L)}{\lambda_n^3}.\end{aligned} \tag{8-54}$$

Therefore,

$$C_n=\frac{2B}{L\lambda_n^3}\frac{\sin(\lambda_n L)}{\cosh(\lambda_n W)}=\frac{2B}{L\lambda_n^3}\frac{(-1)^n}{\cosh(\lambda_n W)} \tag{8-55}$$

and

$$\theta=\sum_{n=0}^{\infty}\frac{2B}{L\lambda_n^3}(-1)^n\frac{\cos(\lambda_n x)\cosh(\lambda_n y)}{\cosh(\lambda_n W)}. \tag{8-56}$$

Code 8-1 Steady Channel Flow Solution

```
#include <stdio.h>
#include <math.h>

double vz(double x,double y)
{
   int n;
   double l,val,soln=0.;
   for (n=0;n<100;++n) {
      l=(n+1/2.)*M_PI;
      val=cosh(l*y)/cosh(l);
      val*=2.*cos(l*x)*pow(-1.,n)/l/l/l;
      soln+=val;
   }
   return ( (1.-x*x)/2. - soln );
}
int main(void)
{
   double x,y;
   FILE *fp=fopen("out.dat","w");
   for (x=0.;x<=1.;x+=.05)
      for (y=0.;y<=1.;y+=.05)
         fprintf(fp,"%e %e %e\n",x,y,vz(x,y));
   fclose(fp);
   return 0;
}
```

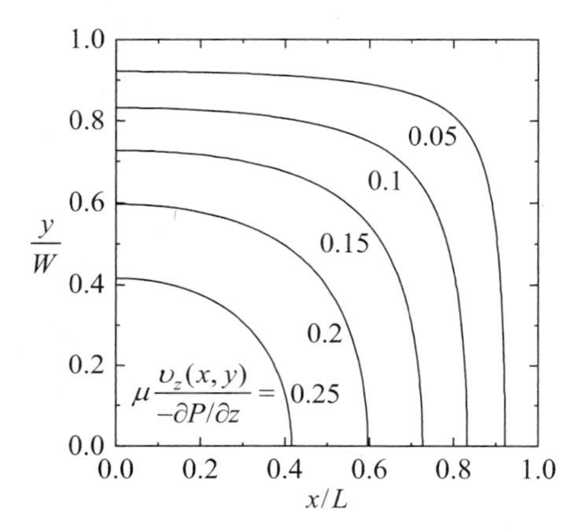

Figure 8-6 Velocity contours for Poiseuille flow through a rectangular channel.

Reconstructing the solution for $v_z(x, y) = \psi(x) + \theta(x, y)$ yields

$$v_z = \frac{1}{\mu}\left(\frac{-\partial P}{\partial z}\right)\left[\frac{L^2 - x^2}{2} - \sum_{n=0}^{\infty} \frac{2}{L\lambda_n^3}(-1)^n \frac{\cos(\lambda_n x)\cosh(\lambda_n y)}{\cosh(\lambda_n W)}\right] \tag{8-57}$$

with

$$\lambda_n = (n + 1/2)\pi/L \quad n = 0, 1, 2, \ldots .$$

This solution is plotted in Figure 8-6, using Code 8-1 to evaluate the series solution. Using the solution (8-57), the total volume flow rate Q_{flow} can be calculated from

$$Q_{flow} = 4\int_0^L\int_0^W v_z dy dx = 4WL\frac{L^2}{3\mu}\left(\frac{-\partial P}{\partial z}\right)\left[1 - \frac{192L}{\pi^5 W}\sum_{n=0}^{\infty}\frac{\tan(\lambda_n W)}{(2n+1)^5}\right] \tag{8-58}$$

This result indicates that the volume flow rate is related to the pressure drop ΔP over a duct of length L and cross-sectional area $A = 4WL$ by

$$\left[\frac{Q_{flow}}{A^2/\mu}\frac{L}{\Delta P} = \frac{1}{12}\frac{L}{W}\left(1 - \frac{192L}{\pi^5 W}\sum_{n=0}^{\infty}\frac{\tan(\lambda_n W)}{(2n+1)^5}\right)\right]_{\substack{duct\\flow}} . \tag{8-59}$$

In contrast, Eduard Hagenbach derived the result for Poiseuille pipe flow [3]:

$$\left[\frac{Q_{flow}}{A^2/\mu}\frac{L}{\Delta P} = \frac{1}{8\pi}\right]_{\substack{pipe\\flow}}. \tag{8-60}$$

8.5 LOOKING AHEAD

Solving nonhomogeneous differential equations of two spatial variables with the approach outlined in Section 8.4 is somewhat cumbersome. The method of eigenfunction expansion, discussed in Chapter 9, offers a more succinct method for solving nonhomogeneous partial differential equations. Although it is based on the same principles as separation of variables, the eigenfunction expansion method has the advantage of being able to handle multiple nonhomogeneous conditions on adjoining boundaries without breaking the problem into two parts. Also treated in Chapter 9 is the solution of partial differential equations in non-Cartesian coordinates. Although the principles of separation of variables and the eigenfunction method remain the same whether in Cartesian or non-Cartesian coordinates, the ordinary differential equations that arise in non-Cartesian coordinates have solutions in terms of less familiar functions.

8.6 PROBLEMS

8-1 Three sides of a long bar are embedded in a matrix material that contains a dilute contaminate species "A". The embedded sides of the bar experience a constant species concentration $c_A = B$. However, the top surface of the bar is unexposed to the matrix material and is washed by a fluid bath (in which $c_A = 0$). The fluid bath provides a convective boundary condition for species transport from the bar. Determine the relative concentration profile ($\phi = c_A/B$) in the bar for steady-state conditions. For the case when $W = H$, evaluate the gross average mass transfer coefficient $\overline{h}_{gross}$ for species transport from the matrix material, through the bar, to the fluid. The gross average mass transfer coefficient is defined in terms of the species flux from the top surface of the bar by

$$\overline{h}_{gross}(B - 0)\ W = \int_0^W J_A^*(y = H)dx.$$

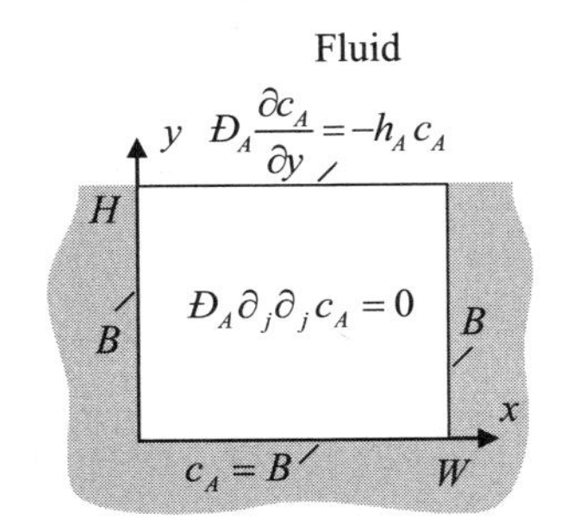

Present the result for $\overline{h}_{gross}$ in terms of a Sherwood number defined by

$$Sh = \frac{\overline{h}_{gross} W}{Đ_A} = \frac{1}{Đ_A B}\int_0^W \left(-Đ_A \frac{\partial c_A}{\partial y}\bigg|_{y=H}\right)dx.$$

8-2 For the illustrated heat transfer problem, find a solution by superposition of two problems $T_1(x,y)$ and $T_2(x,y)$, each having one nonhomogeneous boundary condition. Fully specify these two problems mathematically. Sketch the solution contours for $T_1(x,y)$ and $T_2(x,y)$.

Sketch the solution contours for the original problem $T(x,y) = T_1(x,y) + T_2(x,y)$. Determine the solution for $T(x,y)$.

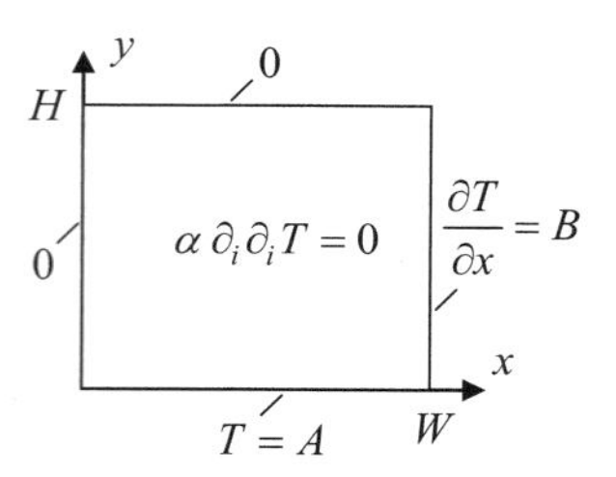

8-3 Consider a flow confined to a long open channel. The width of the channel is W and the thickness of the flow is H. The floor of the channel is sloped such that a component of gravity acts in the direction of the flow: $g_z = g \sin\theta$. Additionally, one of the vertical walls moves with velocity $v_z = -U$. The speed of the moving wall is maintained such that there is no net flow in the z-direction. Decompose the mathematical problem for $v_z(x,y)$ into two sub-problems $v_z(x,y) = v_1(x,y) + v_2(x,y)$, each having only one nonhomogeneous condition. Fully specify these two problems mathematically. Sketch the solution contours for $v_1(x,y)$ and $v_2(x,y)$. Solve this problem for $v_z(x,y) = v_1(x,y) + v_2(x,y)$. Sketch the solution contours for $v_z(x,y)$. Determine the relation between U and $\sin\theta$ that gives rise to no net flow when $W = H$.

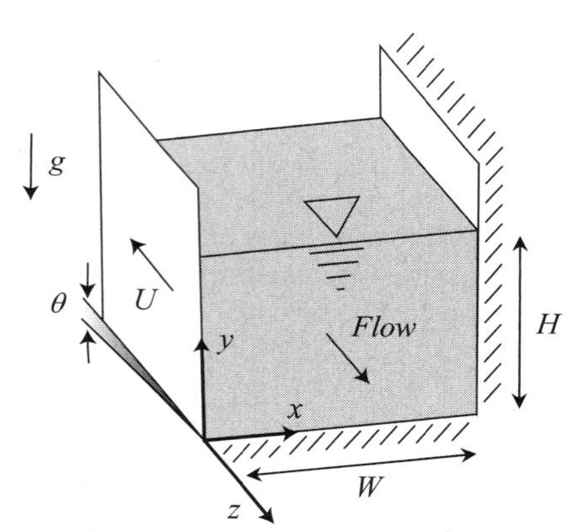

8-4 Consider a duct of rectangular cross section ($W \times H$). Determine the average coefficient of friction c_f for steady Poiseuille flow through the duct defined by

$$c_f = \frac{2\tau_m}{\rho v_m^2}$$

where τ_m and v_m are the average wall shear stress and mean velocity of the flow, respectively. Demonstrate that the duct flow solution for c_f has the form $c_{f,duct} = const./\mathrm{Re}_D$, where

$$\mathrm{Re}_D = \frac{v_m D}{\nu},$$

and the hydraulic diameter is

$$D = \frac{4 \times Area}{Perimeter} = \frac{4\,WH}{2W + 2H}.$$

Determine the constant in the result for $c_{f,duct}$ when $W/H = 4$. Compare the duct flow result with the pipe flow result: $c_{f,pipe} = 16/\mathrm{Re}_D$.

8-5 Consider the composite bar shown. The bottom half of the bar $(0 \le y \le H/2)$ is comprised of a material having the thermal conductivity k_1. The top half of the bar $(H/2 \le y \le H)$ is comprised of a material having the thermal conductivity k_2. Using the dimensionless temperature variable:

$$\theta = \frac{T - A}{B - A},$$

show that the temperature solution in the bar can be expressed by a separation of variables solution with the form:

$$\theta_1(0 \le y \le H/2) = \sum_{n=1}^{\infty} C_n \sin(\lambda_n x) \sinh(\lambda_n y)$$

$$\theta_2(H/2 \le y \le H) = \sum_{n=1}^{\infty} \sin(\lambda_n x)(D_n \cosh(\lambda_n(H/2 - y)) + E_n \sinh(\lambda_n(H/2 - y)))$$

where

$$\lambda_n = n\pi/W \quad n = 1, 2, \ldots .$$

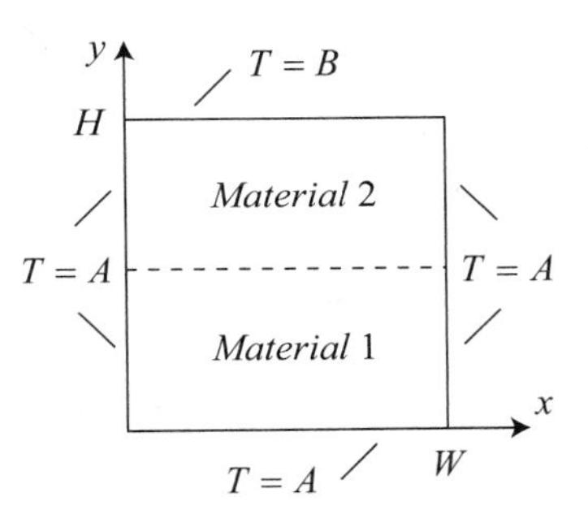

By imposing the interface conditions

$$\theta_1(y = H/2) = \theta_2(y = H/2) \quad \text{and} \quad -k_1 \partial\theta_1/\partial y|_{y=H/2} = -k_2 \partial\theta_2/\partial y|_{y=H/2},$$

determine the required relationships between C_n, D_n, and E_n. Complete the solution to demonstrate that the temperature field in the bar is

$$\theta_1\left(0 \le y \le \frac{H}{2}\right) = \sum_{n=1}^{\infty} \frac{(1 - (-1)^n) \sin\left(n\pi \frac{x}{W}\right) \sinh\left(n\pi \frac{y}{W}\right)}{\frac{n\pi}{2}\left(1 + \frac{k_1}{k_2}\right) \sinh\left(\frac{n\pi}{2}\frac{H}{W}\right) \cosh\left(\frac{n\pi}{2}\frac{H}{W}\right)}$$

$$\theta_2\left(\frac{H}{2} \le y \le H\right) = \sum_{n=1}^{\infty} \frac{(1-(-1)^n)\sin\left(n\pi\frac{x}{W}\right)}{\frac{n\pi}{2}\left(1+\frac{k_1}{k_2}\right)} \left(\frac{\cosh\left(n\pi\frac{H}{W}\left(\frac{1}{2}-\frac{y}{H}\right)\right)}{\cosh\left(\frac{n\pi}{2}\frac{H}{W}\right)} - \frac{k_1}{k_2} \frac{\sinh\left(n\pi\frac{H}{W}\left(\frac{1}{2}-\frac{y}{H}\right)\right)}{\sinh\left(\frac{n\pi}{2}\frac{H}{W}\right)} \right)$$

REFERENCES

[1] R. Haberman, *Applied Partial Differential Equations (with Fourier Series and Boundary Value Problems)*, Fourth Edition. Upper Saddle River, NJ: Prentice-Hall, 2006.
[2] A. D. Polyanin, *Handbook of Linear Partial Differential Equations for Engineers and Scientists.* Boca Raton, FL: Chapman & Hall/CRC Press, 2002.
[3] S. P. Sutera and R. Skalak, "The History of Poiseuille's Law," *Annual Review of Fluid Mechanics*, **25**, 1 (1993).

Chapter 9

Eigenfunction Expansion

9.1 Method of Eigenfunction Expansion

9.2 Non-Cartesian Coordinate Systems

9.3 Transport in Non-Cartesian Coordinates

9.4 Problems

The method of eigenfunction expansion is an extension of the separation-of-variables method discussed in Chapters 7 and 8. To understand eigenfunction expansion, it is useful to first summarize the separation-of-variables approach, so that the connection between the two methods will be apparent.

The separation-of-variables method assumes a solution of the form $\phi = X(x)Y(t\ or\ y)$. ϕ is the dependent variable governed by the partial differential equation to be solved. When the assumed solution is substituted into the governing equation, and the variables separated, two ordinary differential equations emerge that are linked by a separation constant. For the sake of discussion, suppose all nonhomogeneous conditions are imposed on boundaries associated with the t or y variables. The problem for $X(x)$, therefore, has homogeneous boundary conditions. The solution for $X(x)$ is expressed in terms of trigonometric functions and satisfies both boundary conditions in part by requiring discrete choices of the separation constant λ_n. The solution to the problem for $X(x)$ is referred to as the *eigenfunction*. The discrete choices of the separation constant are referred to as the *eigenvalues*. Since the solution for $X(x)$ is expressed in terms of trigonometric functions, a series representation of the nonhomogeneous boundary condition(s) can be built by superposition of all the solutions corresponding to each value of λ_n. However, this requires determining the correct "weighting" of each superimposed solution, which is made possible by the orthogonality property of $X(x)$ for the associated set of eigenvalues.

9.1 METHOD OF EIGENFUNCTION EXPANSION

The method of eigenfunction expansion starts with an assumed form for the solution:

$$\phi = \sum_{n=0}^{\infty} Y_n(t\ or\ y)X_n(x). \tag{9-1}$$

The choice of eigenfunction $X_n(x)$ and corresponding eigenvalues λ_n is decided by the problem for $X(x)$ using the standard separation-of-variables approach described in Chapters 7 and 8. However, this determination is carried out for an entirely homogeneous problem modeled after the original nonhomogeneous problem. For example, if the governing equation has a nonhomogeneous term, it is dropped in the model problem. Additionally, if the original problem has nonhomogeneous boundary conditions, the model problem has boundary conditions of the same type, but that are homogeneous.

The second part of eigenfunction expansion is to determine $Y_n(t \text{ or } y)$ such that the original problem with all the nonhomogeneous conditions is satisfied. It must be understood that Y_n is not the same as the Y arising in standard separation of variables. The required form of Y_n is dictated by an ordinary differential equation that is found by multiplying the governing equation for ϕ by the eigenfunction $X_m(x)$ and integrating with respect to x. For a problem confined to a domain $0 \leq x \leq L$, this last step requires evaluation of the expression

$$\int_0^L (PDE \text{ for } \phi) X_m(x)\ dx = 0. \tag{9-2}$$

The nature of integration in this step varies in detail from problem to problem, and therefore is best illustrated by an example. However, the outcome is that a nonhomogeneous ordinary differential equation for $Y_n(t \text{ or } y)$ is produced. Some of the nonhomogeneous conditions of the original problem (there can be more than one) are incorporated into this equation for $Y_n(t \text{ or } y)$. After Y_n is determined, the final solution for ϕ is constructed from Eq. (9-1), and any remaining nonhomogeneous conditions are subsequently dealt with. One major advantage of the eigenfunction expansion method is the ability to handle multiple nonhomogeneous conditions in a problem more directly than with standard separation of variables.

For newcomers to this method, this overview of the eigenfunction expansion method will initially be difficult to comprehend completely. However, the following examples will illustrate the various elements of the method that are highlighted in this introduction. Additional coverage of the eigenfunction expansion method can be found in books such as references [1] and [2].

9.1.1 Species Transport (Silicon Doping)

Reconsider the silicon doping example of Section 7.2. Recall that in that example, species transfer occurs from a donor material, phosphosilicate glass, to a thin silicon layer that is backed by a diffusion barrier. Species transfer by diffusion starts when the system is put into a high-temperature oven, at $t = 0$. The governing equation is

$$\partial c_P/\partial t = Ð_p \partial^2 c_P/\partial x^2, \tag{9-3}$$

and is subject to the boundary conditions and initial condition for this problem:

$$\text{at} \quad x = 0: \quad c_P(x = 0) = P_o \tag{9-4}$$

$$\text{at} \quad x = L: \quad \partial c_P/\partial x|_{x=L} = 0 \tag{9-5}$$

$$\text{at} \quad t = 0: \quad c_P(t = 0) = 0. \tag{9-6}$$

$c_P(t, x)$ is the concentration of the donor species, phosphorous, and P_o is the concentration of phosphorous imposed on the front surface of the thin silicon layer. The method of eigenfunction expansion starts with an assumed form for the solution:

$$c_p = \sum_{n=0}^{\infty} Y_n(t) \overbrace{\sin(\lambda_n x)}^{X_n(x)} \tag{9-7}$$

with

$$\lambda_n = \frac{(2n+1)\pi}{2L} \quad \textit{for } n = 0, 1, 2, \ldots. \tag{9-8}$$

Notice that the homogeneous version of this problem has the boundary conditions

$$c_P(x=0) = 0 \quad \text{and} \quad \partial c_P/\partial x|_{x=L} = 0 \quad \text{(homogeneous)}. \tag{9-9}$$

These homogeneous boundary conditions are satisfied by the proposed form of $X(x) = \sin(\lambda_n x)$, since

$$X_n(0) = \sin(0) = 0 \quad \text{and} \quad dX_n/dx|_{x=L} = \lambda_n \cos(\lambda_n L) = 0. \tag{9-10}$$

Next, $Y_n(t)$ is determined such that the original problem with all the nonhomogeneous conditions can be satisfied. The required form of $Y_n(t)$ is found by multiplying the governing equation by the eigenfunction $\sin(\lambda_m x)$ and integrating over the domain $0 \le x \le L$:

$$\int_0^L \left(\frac{1}{Đ_p} \frac{\partial c_P}{\partial t} - \frac{\partial^2 c_P}{\partial x^2} \right) \sin(\lambda_m x) dx = 0. \tag{9-11}$$

Breaking up the task of integration, the first term in Eq. (9-11) evaluates to

$$\int_0^L \frac{1}{Đ_p} \frac{\partial c_P}{\partial t} \sin(\lambda_m x) dx = \frac{1}{Đ_p} \int_0^L \left[\sum_{n=0}^{\infty} \frac{dY_n}{dt} \sin(\lambda_n x) \sin(\lambda_m x) \right] dx = \frac{1}{Đ_p} \frac{dY_m}{dt} \frac{L}{2}. \tag{9-12}$$

Notice that the assumed solution for c_P, given by Eq. (9-7), was applied to this expression prior to integration.

The second term in Eq. (9-11) requires integration by parts to be performed twice for the result:

$$\int_0^L \frac{\partial^2 c_P}{\partial x^2} \sin(\lambda_m x) dx = \left(\frac{\partial c_P}{\partial x} \sin(\lambda_m x) - c_P \lambda_m \cos(\lambda_m x) \right) \Big|_0^L - \int_0^L c_P \, \lambda_m^2 \sin(\lambda_m x) dx. \tag{9-13}$$

Since $\partial c_P/\partial x|_{x=L} = 0$, $\sin(0) = 0$, and $\cos(\lambda_m L) = 0$, this result simplifies to

$$\begin{aligned} \int_0^L \frac{\partial^2 c_P}{\partial x^2} \sin(\lambda_m x) dx &= c_P(x=0) \lambda_m \cos(0) - \int_0^L c_P \, \lambda_m^2 \sin(\lambda_m x) dx \\ &= P_o \lambda_m - \int_0^L \left[\sum_{n=0}^{\infty} Y_n(t) \sin(\lambda_n x) \right] \lambda_m^2 \sin(\lambda_m x) dx \end{aligned} \tag{9-14}$$

or

$$\int_0^L \frac{\partial^2 c_P}{\partial x^2} \sin(\lambda_m x) dx = P_o \lambda_m - Y_m(t) \lambda_m^2 \frac{L}{2}. \tag{9-15}$$

Notice that the nonhomogeneous boundary condition $c_P(x=0) = P_o$ is applied to this result, and the assumed solution for c_P, given by Eq. (9-7), is inserted into the final integration with respect to x.

Applying the integration results (9-12) and (9-15) to Eq. (9-11) yields

$$\frac{1}{Ð_p}\frac{dY_n}{dt}\frac{L}{2} - P_o\lambda_m + Y_m(t)\lambda_m^2\frac{L}{2} = 0 \tag{9-16}$$

or

$$\frac{dY_n}{dt} + Ð_p\lambda_n^2 Y_n = \frac{2P_o D_p \lambda_n}{L}. \tag{9-17}$$

The method of eigenfunction expansion has presented the required form of $Y_n(t)$ in terms of an ordinary differential equation. The initial condition $c_P(t=0) = 0$ can only be satisfied if $Y_n(0) = 0$. Therefore, Eq. (9-17) is solved using this initial condition for the result

$$Y_n = \frac{2P_o}{L\lambda_n}\left(1 - e^{-Ð_p\lambda_n^2 t}\right). \tag{9-18}$$

Reconstructing the solution for c_P from Eq. (9-7) yields

$$c_p = \sum_{n=0}^{\infty}\frac{2P_o}{L\lambda_n}\left(1 - e^{-Ð_p\lambda_n^2 t}\right)\sin(\lambda_n x). \tag{9-19}$$

This is equivalent to the solution obtained by traditional separation of variables in Section 7.2. The only disadvantage to the solution expressed by Eq. (9-19) is that since $\sin(0) = 0$, one might incorrectly conclude that this solution does not satisfy the boundary condition $c_P(x=0) = P_o$. In fact, the solution approaches the correct boundary value $c_P(x \to 0) = P_o$, even though a discontinuity at $x = 0$ exists in the series solution. Furthermore, since

$$\sum_{n=0}^{\infty}\frac{2}{L\lambda_n}\sin(\lambda_n x) = 1, \tag{9-20}$$

the solution for c_P can be rewritten as

$$c_p = P_o - \sum_{n=0}^{\infty}\frac{2P_o}{\lambda_n L}\sin(\lambda_n x)\, e^{-Ð_p\lambda_n^2 t}, \tag{9-21}$$

which is the same result as obtained in Section 7.2. In this form, it is clear that $c_P(x=0) = P_o$. As demonstrated here, the method of eigenfunction expansion does not always give the "best" analytical form of the solution. For small x, the solution given by Eq. (9-19) requires a large number of terms in the series be evaluated before a reasonably accurate result can be obtained.

9.1.2 Heat Transfer (Bar Heat Treatment)

Reconsider the heat treatment example of Section 8.3.1. Recall that in this example a bar of rectangular cross section is subjected to both temperature and heat flux boundary conditions. Mathematically, the problem is governed by the diffusion equation:

$$\frac{\partial^2\theta}{\partial x^2} + \frac{\partial^2\theta}{\partial y^2} = 0, \tag{9-22}$$

where the relative temperature is defined by $\theta = T - T_o$. The solution to this equation is subject to the boundary conditions

$$\partial\theta/\partial x|_{x=0} = -q_{wall}/k \quad \text{and} \quad \theta(x = L) = 0 \tag{9-23}$$

$$\theta(y = 0) = 0 \quad \text{and} \quad \theta(y = W) = T_b - T_a = \theta_b. \tag{9-24}$$

Separation of variables on the homogeneous version of this problem suggests

$$X_n(x) = \cos(\lambda_n x) \tag{9-25}$$

with

$$\lambda_n = \frac{(2n+1)\pi}{2L} \quad \textit{for } n = 0, 1, 2, \ldots. \tag{9-26}$$

Notice that the homogeneous version of this problem has the boundary conditions

$$\partial\theta/\partial x|_{x=0} = 0 \quad \text{and} \quad \theta(x = L) = 0 \quad \text{(homogeneous)}. \tag{9-27}$$

These homogeneous boundary conditions are satisfied by the proposed form of $X(x) = \cos(\lambda_n x)$ since

$$dX_n/dx|_{x=0} = \lambda_n \sin(0) = 0 \quad \text{and} \quad X_n(x = L) = \cos(\lambda_n L) = 0. \tag{9-28}$$

Therefore, the method of eigenfunction expansion starts with the assumed form for the solution:

$$\theta = \sum_{n=0}^{\infty} Y_n(y) \overbrace{\cos(\lambda_n x)}^{X_n(x)}. \tag{9-29}$$

Next, $Y_n(y)$ is determined such that the original problem, with all the nonhomogeneous conditions, is satisfied. The required form of $Y_n(y)$ is found by multiplying the governing equation by the eigenfunction $\cos(\lambda_m x)$ and integrating over the domain $0 \le x \le L$:

$$\int_0^L \left(\frac{\partial^2\theta}{\partial x^2} + \frac{\partial^2\theta}{\partial y^2}\right)\cos(\lambda_m x)dx = 0. \tag{9-30}$$

Integration by parts performed twice on the first term in the governing equation (9-30) yields

$$\int_0^L \frac{\partial^2\theta}{\partial x^2}\cos(\lambda_m x)dx = \left(\frac{\partial\theta}{\partial x}\cos(\lambda_m x) + \theta\lambda_m \sin(\lambda_m x)\right)\Bigg|_0^L - \int_0^L \theta\, \lambda_m^2 \cos(\lambda_m x)dx. \tag{9-31}$$

Because $\theta(x = L) = 0$, $\cos(\lambda_m L) = 0$, and $\sin(0) = 0$, this result simplifies to

$$\begin{aligned}\int_0^L \frac{\partial^2\theta}{\partial x^2}\cos(\lambda_m x)dx &= -\frac{\partial\theta}{\partial x}\bigg|_{x=0} - \int_0^L \theta\, \lambda_m^2 \cos(\lambda_m x)dx \\ &= -\frac{\partial\theta}{\partial x}\bigg|_{x=0} - \int_0^L \left[\sum_{n=0}^{\infty} Y_n(y)\cos(\lambda_n x)\right] \lambda_m^2 \cos(\lambda_m x)dx\end{aligned} \tag{9-32}$$

or

$$\int_0^L \frac{\partial^2\theta}{\partial x^2}\cos(\lambda_m x)dx = \frac{q_{wall}}{k} - Y_n\lambda_n^2\frac{L}{2}. \tag{9-33}$$

Integration of the second term in the governing equation (9-30) yields

$$\int_0^L \frac{\partial^2\theta}{\partial y^2}\cos(\lambda_m x)dx = \int_0^L \left[\sum_{n=0}^{\infty}\frac{d^2Y_n}{dy^2}\cos(\lambda_n x)\right]\cos(\lambda_m x)dx = \frac{d^2Y_n}{dy^2}\frac{L}{2}. \tag{9-34}$$

Applying the integration results (9-33) and (9-34) to Eq. (9-30) yields

$$\frac{q_{wall}}{k} - Y_n\lambda_n^2\frac{L}{2} + \frac{d^2Y_n}{dy^2}\frac{L}{2} = 0 \tag{9-35}$$

or

$$\frac{d^2Y_n}{dy^2} - \lambda_n^2 Y_n = -\frac{2q_{wall}}{k\,L}. \tag{9-36}$$

The ordinary differential equation for Y_n may be integrated:

$$Y_n = C_1\,\sinh[\lambda_n(W-y)] + C_2\,\sinh(\lambda_n y) + \frac{2q_{wall}}{k\,L\lambda_n^2}. \tag{9-37}$$

Two boundary conditions remain to be imposed on the problem, one at $y = 0$ and the other at $y = W$. Because the boundary condition $\theta(y = 0) = 0$ is homogeneous, satisfying this condition can be accomplished by forcing $Y_n(0) = 0$:

$$Y_n(0) = C_1\,\sinh(\lambda_n W) + 0 + \frac{2q_{wall}}{k\,L\lambda_n^2} = 0, \tag{9-38}$$

such that

$$C_1 = \frac{-2q_{wall}}{k\,L\lambda_n^2\,\sinh(\lambda_n W)}. \tag{9-39}$$

Therefore,

$$Y_n = C_2\,\sinh(\lambda_n y) + \frac{2q_{wall}}{k\,L\lambda_n^2}\left(1 - \frac{\sinh[\lambda_n(W-y)]}{\sinh(\lambda_n W)}\right). \tag{9-40}$$

In contrast, the last remaining boundary condition $\theta(y = W) = \theta_b$ is nonhomogeneous. This condition cannot be successfully imposed on the function $Y_n(y)$ alone. Therefore, the final nonhomogeneous boundary condition must be applied to the reconstructed solution for θ:

$$\theta = \sum_{n=0}^{\infty}\left[C_n\sinh(\lambda_n y) + \frac{2q_{wall}}{k\,L\lambda_n^2}\left(1 - \frac{\sinh[\lambda_n(W-y)]}{\sinh(\lambda_n W)}\right)\right]\cos(\lambda_n x). \tag{9-41}$$

The final integration constant C_2 has been renamed as C_n, since its value can be different for each eigenvalue λ_n. The appropriate value of C_n for each eigenvalue needs to be determined so that, when all the terms in the series are summed, the final boundary condition $\theta(y = W) = \theta_b$ is satisfied. Starting with the statement for the final boundary condition:

$$\text{at} \quad y = W: \quad \theta_b = \sum_{n=0}^{\infty} \left[C_n \sinh(\lambda_n W) + \frac{2q_{wall}}{k\,L\lambda_n^2} \right] \cos(\lambda_n x). \tag{9-42}$$

both sides are multiplied by $\cos(\lambda_m x)$ and integrated in x from 0 to L:

$$\int_0^L \theta_b \cos(\lambda_m x) dx = \int_0^L \sum_{n=0}^{\infty} \left[C_n \sinh(\lambda_n W) + \frac{2q_{wall}}{k\,L\lambda_n^2} \right] \cos(\lambda_n x) \cos(\lambda_m x) dx \tag{9-43}$$

or

$$\frac{\theta_b}{\lambda_n} \sin(\lambda_m L) = \left[C_n \sinh(\lambda_n W) + \frac{2q_{wall}}{k\,L\lambda_n^2} \right] \frac{L}{2}. \tag{9-44}$$

From the result of this integration, the series solution coefficients can be determined:

$$C_n = \frac{2\theta_b}{L\lambda_n} \frac{\sin(\lambda_m L)}{\sinh(\lambda_n W)} - \frac{2q_{wall}}{k\,L\lambda_n^2} \frac{1}{\sinh(\lambda_n W)}. \tag{9-45}$$

With this result, the final solution for the temperature field in the bar becomes

$$\theta = \sum_{n=0}^{\infty} \frac{2}{L\lambda_n} \left[\frac{\theta_b \sin(\lambda_m L) \sinh(\lambda_n y)}{\sinh(\lambda_n W)} + \frac{q_{wall}}{k\lambda_n} \left(1 - \frac{\sinh(\lambda_n y) + \sinh[\lambda_n (W - y)]}{\sinh(\lambda_n W)} \right) \right] \cos(\lambda_n x), \tag{9-46}$$

where

$$\lambda_n = \frac{(2n+1)\pi}{2L} \quad \textit{for } n = 0, 1, 2, \ldots. \tag{9-47}$$

This is equivalent to the result obtained in Section 8.3.1 by superposition of two subproblems solved by standard separation of variables. This solution illustrates that the eigenfunction expansion method can address multiple nonhomogeneous terms more succinctly than standard separation of variables. However, the tradeoff is that the eigenfunction expansion method does not always produce the best behaved series, as the last example in Section 9.1.1 illustrated.

9.1.3 Momentum Transport (Duct Flow)

Reconsider the fluid flow through a rectangular duct described in Section 8.4.1. Recall that the flow velocity in the cross section of the duct is governed by

$$\frac{\partial^2 v_z}{\partial x^2} + \frac{\partial^2 v_z}{\partial y^2} = B, \quad \text{where} \quad B = (\partial P/\partial z)/\mu. \tag{9-48}$$

The boundary conditions for the quarter domain of the duct are given by

$$\partial v_z/\partial x|_{x=0} = 0 \quad \text{and} \quad v_z(x = L) = 0 \tag{9-49}$$

$$\partial v_z/\partial y|_{y=0} = 0 \quad \text{and} \quad v_z(y = W) = 0 \tag{9-50}$$

Separation of variables for the homogeneous version of this problem yields

$$X_n(x) = \cos(\lambda_n x) \quad \text{with} \quad \lambda_n = (n + 1/2)\pi/L \quad n = 0, 1, 2, \ldots. \tag{9-51}$$

Notice that the homogeneous boundary conditions for v_z at $x = 0$ and $x = L$ are satisfied by $X_n(x) = \cos(\lambda_n x)$, since

$$dX_n/dx|_{x=0} = \lambda_n \sin(0) = 0 \quad \text{and} \quad X_n(x = L) = \cos(\lambda_n L) = 0. \tag{9-52}$$

The method of eigenfunction expansion starts with an assumed form for the solution:

$$v_z = \sum_{n=0}^{\infty} Y_n(y) \overbrace{\cos(\lambda_n x)}^{X_n(x)}. \tag{9-53}$$

The required form of $Y_n(y)$ is found by multiplying the governing equation by the eigenfunction $\cos(\lambda_m x)$ and integrating over the domain $0 \le x \le L$:

$$\int_0^L \left(\frac{\partial^2 v_z}{\partial x^2} + \frac{\partial^2 v_z}{\partial y^2} - B \right) \cos(\lambda_m x) dx = 0. \tag{9-54}$$

Integration by parts performed twice on the first term in the governing equation (9-54) yields

$$\int_0^L \frac{\partial^2 v_z}{\partial x^2} \cos(\lambda_m x) dx = \left(\frac{\partial v_z}{\partial x} \cos(\lambda_m x) + v_z \lambda_m \sin(\lambda_m x) \right) \Bigg|_0^L - \int_0^L v_z \, \lambda_m^2 \cos(\lambda_m x) dx. \tag{9-55}$$

Since $v_z(x = L) = 0$, $\cos(\lambda_m L) = 0$, $\sin(0) = 0$, and $\partial v_z/\partial x|_{x=0} = 0$, this simplifies to

$$\begin{aligned} \int_0^L \frac{\partial^2 v_z}{\partial x^2} \cos(\lambda_m x) dx &= -\int_0^L v_z \, \lambda_m^2 \cos(\lambda_m x) dx \\ &= -\int_0^L \left[\sum_{n=0}^{\infty} Y_n(y) \cos(\lambda_n x) \right] \lambda_m^2 \cos(\lambda_m x) dx \end{aligned} \tag{9-56}$$

or

$$\int_0^L \frac{\partial^2 v_z}{\partial x^2} \cos(\lambda_m x) dx = -Y_n \lambda_n^2 \frac{L}{2}. \tag{9-57}$$

Integrating the second term in Eq. (9-54) yields

$$\int_0^L \frac{\partial^2 v_z}{\partial y^2} \cos(\lambda_m x) dx = \int_0^L \left[\sum_{n=0}^{\infty} \frac{d^2 Y_n}{dy^2} \cos(\lambda_n x) \right] \cos(\lambda_m x) dx = \frac{d^2 Y_n}{dy^2} \frac{L}{2}. \tag{9-58}$$

Integrating the third term in Eq. (9-54) yields

$$\int_0^L B \cos(\lambda_m x) dx = B \frac{\sin(\lambda_m x)}{\lambda_m} \Bigg|_0^L = B \frac{\sin(\lambda_m L)}{\lambda_m}. \tag{9-59}$$

Applying the integration results (9-57), (9-58), and (9-59) to Eq. (9-54) yields an ordinary differential equation for $Y_n(y)$:

$$-Y_n\lambda_n^2\frac{L}{2}+\frac{d^2Y_n}{dy^2}\frac{L}{2}-B\frac{\sin(\lambda_n L)}{\lambda_n}=0 \tag{9-60}$$

or

$$\frac{d^2Y_n}{dy^2}-\lambda_n^2Y_n=2B\frac{(-1)^n}{L\lambda_n}. \tag{9-61}$$

The ordinary differential equation for Y_n is integrated:

$$Y_n=C_1\cosh(\lambda_n y)+C_2\sinh(\lambda_n y)-\frac{2B(-1)^n}{L\lambda_n^3}. \tag{9-62}$$

To satisfy the remaining two homogeneous boundary conditions, $\partial v_z/\partial y\big|_{y=0}=0$ and $v_z(y=W)=0$, the solution for Y_n should satisfy

$$dY_n/dy\big|_{y=0}=0 \quad \text{and} \quad Y_n(y=W)=0. \tag{9-63}$$

The first boundary condition $dY_n/dy\big|_{y=0}=0$ requires that $C_2=0$, and the second boundary condition requires that

$$Y_n(y=W)=C_1\cosh(\lambda_n W)-\frac{2B(-1)^n}{L\lambda_n^3}=0 \tag{9-64}$$

or

$$C_1=\frac{2B(-1)^n}{L\lambda_n^3\cosh(\lambda_n W)}. \tag{9-65}$$

Therefore, the solution for $Y_n(y)$ becomes

$$Y_n=-\frac{2B(-1)^n}{L\lambda_n^3}+\frac{2B(-1)^n\cosh(\lambda_n y)}{L\lambda_n^3\cosh(\lambda_n W)}. \tag{9-66}$$

Using this result for $Y_n(y)$ to construct the solution for v_z from Eq. (9-53) yields

$$v_z=\frac{2}{\mu}\left(\frac{-\partial P}{\partial z}\right)\sum_{n=0}^{\infty}\frac{(-1)^n}{L\lambda_n^3}\left[1-\frac{\cosh(\lambda_n y)}{\cosh(\lambda_n W)}\right]\cos(\lambda_n x) \quad \text{with} \quad \lambda_n=\frac{(n+1/2)\pi}{L}. \tag{9-67}$$

This is equivalent to the result obtained in Section 8.4.1. This solution illustrates that the eigenfunction expansion method can handle nonhomogeneous governing equations more directly than the standard separation-of-variables approach used in Section 8.4.1.

9.2 NON-CARTESIAN COORDINATE SYSTEMS

The method of eigenfunction expansion, as well as the standard separation of variables discussed in Chapters 7 and 8, can be applied to problems in non-Cartesian coordinates. To better understand the application of these methods to non-Cartesian coordinates, it is useful to highlight some common features of these methods applied to Cartesian coordinates.

9.2.1 Cartesian Coordinates

In Cartesian coordinates, separation of variables leads to ordinary differential equations for a function $X(x)$ with the forms:

$$\frac{d^2X}{dx^2} + \lambda_n^2 X = 0 \;\rightarrow\; X = C_1\cos(\lambda_n x) + C_2\sin(\lambda_n x) \tag{9-68}$$

$$\frac{d^2X}{dx^2} - \lambda_n^2 X = 0 \;\rightarrow\; X = C_3\cosh(\lambda_n x) + C_4\sinh(\lambda_n x) \tag{9-69}$$

Solutions to these ordinary differential equations are expressed in terms of trigonometric and hyperbolic functions. For problems in Cartesian coordinates, the trigonometric functions $\cos(\lambda_n x)$ and $\sin(\lambda_n x)$ are the eigenfunctions and the separation constants λ_n are the eigenvalues. Boundary conditions on a given problem dictate what discrete values of λ_n are suitable for the solution. Typical boundary conditions for diffusion-transport problems, imposed on the domain $0 \le x \le L$, give rise to separation constants for which the trigonometric functions $\cos(\lambda_n x)$ and $\sin(\lambda_n x)$ are orthogonal:

$$\int_0^L \cos(\lambda_n x)\cos(\lambda_m x)dx = 0 \quad \text{and} \quad \int_0^L \sin(\lambda_n x)\sin(\lambda_m x)dx = 0 \quad \text{for} \quad n \neq m. \tag{9-70}$$

When $n = m$,

$$\int_0^L \cos^2(\lambda_n x)dx = \left[\frac{x}{2} + \frac{\sin(2\lambda_n x)}{4\lambda_n}\right]_0^L \tag{9-71}$$

and

$$\int_0^L \sin^2(\lambda_n x)dx = \left[\frac{x}{2} - \frac{\sin(2\lambda_n x)}{4\lambda_n}\right]_0^L. \tag{9-72}$$

9.2.2 Cylindrical Coordinates

Separation of variables in cylindrical coordinates leads to ordinary differential equations for a function $R_\nu(r)$ with the forms

$$r\frac{d}{dr}\left(r\frac{dR_\nu}{dr}\right) + \left[(\lambda_n r)^2 - \nu^2\right]R_\nu = 0 \quad \rightarrow \quad R_\nu = C_1 J_\nu(\lambda_n r) + C_2 Y_\nu(\lambda_n r) \tag{9-73}$$

$$r\frac{d}{dr}\left(r\frac{dR_\nu}{dr}\right) - \left[(\lambda_n r)^2 + \nu^2\right]R_\nu = 0 \quad \rightarrow \quad R_\nu = C_3 I_\nu(\lambda_n r) + C_4 K_\nu(\lambda_n r). \tag{9-74}$$

Solutions to these ordinary differential equations are expressed in terms of Bessel functions $J_\nu(\lambda_n r)$ and $Y_\nu(\lambda_n r)$, and modified Bessel functions $I_\nu(\lambda_n r)$ and $K_\nu(\lambda_n r)$, which are illustrated in Figure 9-1. The integer subscript "ν" indicates the order of the functions. As illustrated in Figure 9-1, the Bessel functions have the limiting behaviors $J_\nu(\infty) = Y_\nu(\infty) = K_\nu(\infty) = 0$, while $I_\nu(\infty) = +\infty$, and $J_0(0) = I_0(0) = 1$, $J_1(0) = I_1(0) = 0$, $Y_\nu(0) = -\infty$, and $K_\nu(0) = +\infty$. Some useful properties of the Bessel functions and modified Bessel functions are given in Table 9-1.

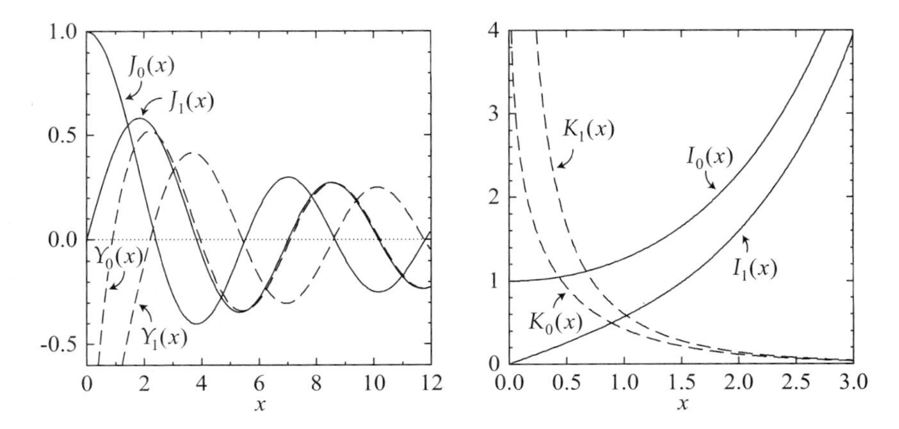

Figure 9-1 Bessel functions $J_\nu(x)$ and $Y_\nu(x)$, and modified Bessel functions $I_\nu(x)$ and $K_\nu(x)$.

Table 9-1 Properties of the Bessel functions and modified Bessel functions

(a) Derivatives

$$\frac{d}{dz}[z^{+\nu}W_\nu(\beta z)] = \begin{cases} +\beta z^{+\nu}W_{\nu-1}(\beta z) & for \quad W = J, Y, I \\ -\beta z^{+\nu}W_{\nu-1}(\beta z) & for \quad W = K \end{cases}$$

$$\frac{d}{dz}[z^{-\nu}W_\nu(\beta z)] = \begin{cases} -\beta z^{-\nu}W_{\nu+1}(\beta z) & for \quad W = J, Y, K \\ +\beta z^{-\nu}W_{\nu+1}(\beta z) & for \quad W = I \end{cases}$$

(b) Integration*

$$\int z^{+\nu}W_{\nu-1}(\beta z)dz = +\frac{z^{+\nu}}{\beta}W_\nu(\beta z) \quad for \quad W = J, Y, I$$

$$\int z^{-\nu}W_{\nu+1}(\beta z)dz = -\frac{z^{-\nu}}{\beta}W_\nu(\beta z) \quad for \quad W = J, Y, K$$

$$\int zW_\nu^2(\beta z)dz = \frac{z^2}{2}\left[W_\nu'^2(\beta z) + \left(1 - \left(\frac{\nu}{\beta z}\right)^2\right)W_\nu^2(\beta z)\right] \quad for \quad W = J, Y$$

(c) Recurrence relations* for $W = J, Y, I$

$$W_{\nu-1}(z) + W_{\nu+1}(z) = \frac{2\nu}{z}W_\nu(z) \qquad W_{\nu-1}(z) - \frac{\nu}{z}W_\nu(z) = W_\nu(z)$$

$$W_{\nu-1}(z) - W_{\nu+1}(z) = 2W_\nu'(z) \qquad -W_{\nu+1}(z) + \frac{\nu}{z}W_\nu(z) = W_\nu'(z)$$

*W' is used to denote dW/dz

For problems in cylindrical coordinates, the Bessel functions $J_\nu(\lambda_n r)$ and $Y_\nu(\lambda_n r)$ are the eigenfunctions. Like their trigonometric counterparts, the Bessel functions $J_\nu(\lambda_n r)$ and $Y_\nu(\lambda_n r)$ are oscillatory. Typical boundary conditions for diffusion-transport problems, imposed on the domain $r_i \leq r \leq r_o$, give rise to separation constants for which the Bessel functions $J_\nu(\lambda_n r)$ and $Y_\nu(\lambda_n r)$ are orthogonal. The orthogonality of Bessel functions obeys the relations:

$$\int_{r_i}^{r_o} J_\nu(\lambda_n r)J_\nu(\lambda_m r)\, r\, dr = 0 \quad \text{and} \quad \int_{r_i}^{r_o} Y_\nu(\lambda_n r)Y_\nu(\lambda_m r)\, r\, dr = 0 \quad \text{for} \quad n \neq m. \tag{9-75}$$

When $n = m$,

$$\int_{r_i}^{r_o} J_\nu^2(\lambda_n r)\, r\, dr = \left[\frac{r^2}{2}\left(J_\nu^2(\lambda_n r) - J_{\nu-1}(\lambda_n r)J_{\nu+1}(\lambda_n r)\right)\right]\Bigg|_{r_i}^{r_o} \tag{9-76}$$

$$\int_{r_i}^{r_o} Y_\nu^2(\lambda_n r)\, r\, dr = \left[\frac{r^2}{2}\left(Y_\nu^2(\lambda_n r) - Y_{\nu-1}(\lambda_n r)Y_{\nu+1}(\lambda_n r)\right)\right]\Bigg|_{r_i}^{r_o}. \tag{9-77}$$

Notice that the integrands in the orthogonality relations are weighted by the distance r.

9.2.3 Spherical Coordinates

Separation of variables in spherical coordinates leads to ordinary differential equations for a function $R_\nu(r)$ with the forms

$$\frac{d}{dr}\left(r^2\frac{dR_\nu}{dr}\right) + \left[(\lambda_n r)^2 - \nu(\nu+1)\right]R_\nu = 0 \quad \rightarrow \quad R_\nu = r^{-1/2}\left[C_1 J_{\nu+1/2}(\lambda_n r) + C_2 Y_{\nu+1/2}(\lambda_n r)\right] \tag{9-78}$$

$$\frac{d}{dr}\left(r^2\frac{dR_\nu}{dr}\right) - \left[(\lambda_n r)^2 + \nu(\nu+1)\right]R_\nu = 0 \quad \rightarrow \quad R_\nu = r^{-1/2}\left[C_3 I_{\nu+1/2}(\lambda_n r) + C_4 K_{\nu+1/2}(\lambda_n r)\right]. \tag{9-79}$$

Solutions to these ordinary differential equations can be expressed in terms of Bessel functions and modified Bessel functions of order $(\nu + 1/2)$. Some solutions can also be expressed in terms of trigonometric and hyperbolic functions. For example, when $\nu = 0$:

$$\frac{d}{dr}\left(r^2\frac{dR_0}{dr}\right) + (\lambda_n r)^2 R_0 = 0 \quad \rightarrow \quad R_0 = D_1\frac{\sin(\lambda_n r)}{r} + D_2\frac{\cos(\lambda_n r)}{r} \tag{9-80}$$

$$\frac{d}{dr}\left(r^2\frac{dR_0}{dr}\right) - (\lambda_n r)^2 R_0 = 0 \quad \rightarrow \quad R_0 = D_3\frac{\sinh(\lambda_n r)}{r} + D_4\frac{\cosh(\lambda_n r)}{r}. \tag{9-81}$$

Typical boundary conditions for diffusion-transport problems, imposed on the domain $r_i \le r \le r_o$, give rise to separation constants for which the solution functions are orthogonal. For $\nu = 0$, orthogonality of the solution functions obeys the relations

$$\int_{r_i}^{r_o} \frac{\cos(\lambda_n r)}{r}\frac{\cos(\lambda_m r)}{r}\, r^2 dr = 0 \quad \text{and} \quad \int_{r_i}^{r_o} \frac{\sin(\lambda_n r)}{r}\frac{\sin(\lambda_m r)}{r}\, r^2 dr = 0 \quad \text{for} \quad n \neq m. \tag{9-82}$$

Notice that the integrands are weighted by the distance r^2, and that the resulting integrals become identical to the orthogonality relations in Cartesian coordinates.

9.3 TRANSPORT IN NON-CARTESIAN COORDINATES

Evaluation of the Laplace operator is of particular importance to solving diffusion equations. As discussed in Section 3.4, for cylindrical and spherical coordinates the Laplace operator evaluates to:

$$\text{Cylindrical} \quad (i = r, \theta, z): \quad \partial_i \partial_i \Phi = \frac{1}{r}\frac{\partial}{\partial r}\left(r\frac{\partial \Phi}{\partial r}\right) + \frac{1}{r^2}\frac{\partial^2 \Phi}{\partial \theta^2} + \frac{\partial^2 \Phi}{\partial z^2} \tag{9-83}$$

$$\text{Spherical} \quad (i = r, \theta, \phi): \quad \partial_i \partial_i \Phi = \frac{1}{r^2}\frac{\partial}{\partial r}\left(r^2\frac{\partial \Phi}{\partial r}\right) + \frac{1}{r^2 \sin\theta}\frac{\partial}{\partial \theta}\left(\sin\theta\frac{\partial \Phi}{\partial \theta}\right) + \frac{1}{r^2 \sin^2\theta}\frac{\partial^2 \Phi}{\partial \phi^2} \tag{9-84}$$

Analysis of steady-state and transient diffusion problems in non-Cartesian coordinates is illustrated in the following subsections.

9.3.1 Pin Fin Cooling

As a first example, consider the cylindrical pin fin shown in Figure 9-2. The pin is used to enhance heat transfer to a fluid between two walls. The walls are at an elevated temperature T_w and are separated by a distance $2L$. The fluid flowing over the pin has a free stream temperature of T_∞, and convection of heat to the fluid is characterized by the convection coefficient h. The heat transfer rate from the surface of the pin is given by

$$\left[-k\frac{\partial T}{\partial r} = h(T - T_\infty)\right]_{r=r_o}. \tag{9-85}$$

For cylindrical coordinates, the steady-state heat diffusion equation evaluates to

$$\begin{array}{ccc} \partial_o T = & & \alpha\, \partial_j \partial_j T \\ \downarrow & & \downarrow \\ 0 & = & \alpha\left[\dfrac{1}{r}\dfrac{\partial}{\partial r}\left(r\dfrac{\partial T}{\partial r}\right) + \overbrace{\dfrac{1}{r^2}\dfrac{\partial^2 T}{\partial \theta^2}}^{=0} + \dfrac{\partial^2 T}{\partial z^2}\right]. \end{array} \tag{9-86}$$

Defining a new temperature variable,

$$\phi = \frac{T - T_\infty}{T_w - T_\infty} \tag{9-87}$$

the heat equation becomes

$$\frac{1}{r}\frac{\partial}{\partial r}\left(r\frac{\partial \phi}{\partial r}\right) + \frac{\partial^2 \phi}{\partial z^2} = 0. \tag{9-88}$$

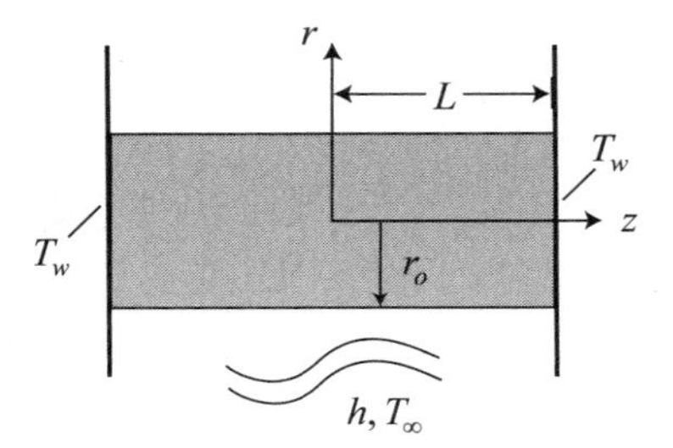

Figure 9-2 Cooling pin fin.

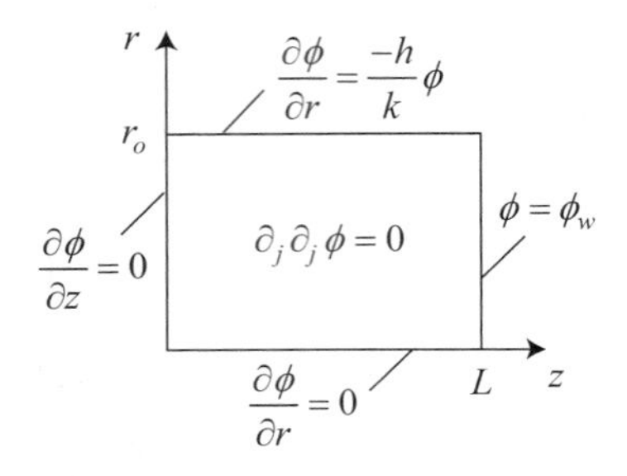

Figure 9-3 Mathematical statement for pin fin cooling.

The mathematical problem for ϕ is summarized in Figure 9-3. Symmetry of the pin boundary conditions require that

$$\text{mid-pin:} \quad \partial\phi/\partial z|_{z=0} = 0 \tag{9-89}$$

and

$$\text{centerline:} \quad \partial\phi/\partial r|_{r=0} = 0. \tag{9-90}$$

The remaining nonadiabatic surfaces have these boundary conditions:

$$\text{surface:} \quad \left[\frac{\partial \phi}{\partial r} = \frac{-h}{k}\phi\right]_{r=r_o} \tag{9.91}$$

$$\text{wall:} \quad \phi(z = L) = 1. \tag{9-92}$$

The temperature field in the pin fin is sought by the method of eigenfunction expansion, starting with an assumed form for the solution:

$$\phi = \sum_{n=0}^{\infty} Z_n(z) \overbrace{J_0(\lambda_n r)}^{R_n(r)}. \tag{9-93}$$

Notice that the eigenfunction choice $R_n(r) = J_0(\lambda_n r)$ satisfies the homogeneous boundary condition at the centerline of the pin, since $dR/dr|_{r=0} = 0$. The eigenvalues λ_n are chosen to satisfy the homogeneous boundary condition at the pin surface, such that

$$\left[\frac{dR}{dr} = \frac{-h}{k}R\right]_{r=r_o} \quad \text{or} \quad -C_1 J_1(\lambda r_o)\lambda = -(h/k)C_1 J_0(\lambda r_o). \tag{9-94}$$

Therefore, λ_n are the roots of the equation:

$$(\lambda_n r_o) J_1(\lambda_n r_o)/J_0(\lambda_n r_o) = (hr_o/k) = Bi \tag{9-95}$$

The eigenvalues defined by the roots of Eq. (9-95) must be numerically determined, and are dependent on a dimensionless group (hr_o/k) known as the *Biot number.* The Biot number arises in problems where the relative resistance to conduction of heat through a solid is in contrast to the resistance of convection of heat into a fluid. The length scale important to the definition of the Biot number is problem specific. In the current problem the Biot number is defined to be

$$Bi = \frac{hr_o}{k} = \frac{r_o/k}{1/h} \begin{matrix} \leftarrow \textit{conductive resistance} \\ \leftarrow \textit{covective resistance} \end{matrix} \tag{9-96}$$

The required form of $Z_n(z)$ needed for the temperature solution (9-93) is found by multiplying the heat equation (9-88) by $J_0(\lambda_m r)\, r$ and integrating with respect to r over the radius of the pin:

$$\int_0^{r_o} \left(\frac{1}{r}\frac{\partial}{\partial r}\left(r\frac{\partial \phi}{\partial r} \right) + \frac{\partial^2 \phi}{\partial z^2} \right) J_0(\lambda_m r)\, r\, dr = 0 \tag{9-97}$$

Integrating twice by parts, the first term in the governing equation yields

$$\int_0^{r_o} \frac{\partial}{\partial r}\left(r\frac{\partial \phi}{\partial r} \right) J_0(\lambda_m r)\, dr = \left(\frac{\partial \phi}{\partial r} J_0(\lambda_m r)\, r + \phi\, \lambda_m J_1(\lambda_m r)\, r \right)\Bigg|_0^{r_o} - \int_0^{r_o} \phi\, \lambda_m^2 J_0(\lambda_m r)\, r\, dr. \tag{9-98}$$

The differential relations

$$\frac{d}{dr}[J_0(\lambda_m r)] = -\lambda_m J_1(\lambda_m r) \quad \text{and} \quad \frac{d}{dr}[J_1(\lambda_m r)\, r] = \lambda_m\, J_0(\lambda_m r)\, r \tag{9-99}$$

needed for integration by parts can be determined using Table 9-1. With the eigenvalue relation (9-95) and the pin surface boundary condition (9-91) it is determined that

$$\left(\frac{\partial \phi}{\partial r} J_0(\lambda_m r)\, r + \phi\, \lambda_m J_1(\lambda_m r)\, r \right)\Bigg|_0^{r_o} = \left(\frac{\partial \phi}{\partial r} + \frac{h}{k}\phi \right)\Bigg|_{r=r_o} J_0(\lambda_m r_o)\, r_o = 0. \tag{9-100}$$

Consequently, Eq. (9-98) simplifies to

$$\begin{aligned} \int_0^{r_o} \frac{\partial}{\partial r}\left(r\frac{\partial \phi}{\partial r} \right) J_0(\lambda_m r)\, dr &= -\int_0^{r_o} \phi\, \lambda_m^2 J_0(\lambda_m r)\, r\, dr \\ &= -\int_0^{r_o} \left[\sum_{n=0}^{\infty} Z_n(z)\, J_0(\lambda_n r) \right] \lambda_m^2 J_0(\lambda_m r)\, r\, dr \end{aligned} \tag{9-101}$$

Because of orthogonality, only one term in the sum appearing in Eq. (9-101) integrates to a nonzero value. Using Table 9-1, that final integral is evaluated with

$$\int_0^{r_o} J_0^2(\lambda_n r)\, (\lambda_n r)\, d(\lambda_n r) = \frac{(\lambda_n r_o)^2}{2}\left[J_0^2(\lambda_n r_o) + J_1^2(\lambda_n r_o) \right] \tag{9-102}$$

Therefore, Eq. (9-101) simplifies to

$$\int_0^{r_o} \frac{\partial}{\partial r}\left(r\frac{\partial \phi}{\partial r} \right) J_0(\lambda_m r)\, dr = -Z_n(z)\frac{(\lambda_n r_o)^2}{2}\left[J_0^2(\lambda_n r_o) + J_1^2(\lambda_n r_o) \right]. \tag{9-103}$$

Integrating the second term in equation (9-97) yields

$$\begin{aligned} \int_0^{r_o} \frac{\partial^2 \phi}{\partial z^2} J_0(\lambda_m r)\, r\, dr &= \int_0^{r_o} \left[\sum_{n=0}^{\infty} \frac{d^2 Z_n}{dz^2} J_0(\lambda_n r) \right] J_0(\lambda_m r)\, r\, dr \\ &= \frac{d^2 Z_n}{dz^2}\frac{r_o^2}{2}\left[J_0^2(\lambda_n r_o) + J_1^2(\lambda_n r_o) \right]. \end{aligned}$$

The integration results given by Eq. (9-103) and Eq. (9-104) are applied to the governing equation (9-97) for the result

$$\left(-\lambda_n^2 Z_n + \frac{d^2 Z_n}{dz^2}\right)\frac{r_o^2}{2}\left[J_0^2(\lambda_n r_o) + J_1^2(\lambda_n r_o)\right] = 0 \tag{9-105}$$

or

$$\frac{d^2 Z_n}{dz^2} - \lambda_n^2 Z_n = 0. \tag{9-106}$$

This ordinary differential equation for Z_n may be integrated:

$$Z_n = C_1 \cosh(\lambda_n z) + C_2 \sinh(\lambda_n z). \tag{9-107}$$

To satisfy the remaining homogeneous boundary condition $\partial\phi/\partial z|_{z=0} = 0$, the solution for Z_n should satisfy

$$dZ_n/dx|_{z=0} = \lambda_n C_1 \sinh(0) + \lambda_n C_2 \cosh(0) = 0. \tag{9-108}$$

This requires $C_2 = 0$. The remaining nonhomogeneous boundary condition at $z = L$ must be applied to the reconstructed solution for ϕ:

$$\phi = \sum_{n=0}^{\infty} C_n \cosh(\lambda_n z) J_0(\lambda_n r). \tag{9-109}$$

The final integration constant C_1 has been renamed as C_n. The appropriate value of C_n for each eigenvalue λ_n is determined so that when all the terms in the series are summed, the final boundary condition $\phi(z = L) = 1$ is satisfied. Starting with the statement for the boundary condition

$$\text{at } z = L: \quad 1 = \sum_{n=0}^{\infty} C_n \cosh(\lambda_n L) J_0(\lambda_n r), \tag{9-110}$$

both sides are multiplied by $J_o(\lambda_m r)\,(\lambda_m r)\,\lambda_m$ and integrated from 0 to r_o:

$$\int_0^{r_o} J_0(\lambda_m r)\,(\lambda_m r)\,d(\lambda_m r) = \sum_{n=1}^{\infty} C_n \int_0^{r_o} \cosh(\lambda_n L)\,J_0(\lambda_n r)\,J_0(\lambda_m r)\,(\lambda_m r)\,d(\lambda_m r). \tag{9-111}$$

Exploiting orthogonality of the eigenfunction, only one term in the series survives integration, such that

$$\int_0^{r_o} J_0(\lambda_m r)\,(\lambda_m r)\,d(\lambda_m r) = C_n \cosh(\lambda_n L) \int_0^{r_o} J_0^2(\lambda_n r)\,(\lambda_n r)\,d(\lambda_n r). \tag{9-112}$$

Evaluating the remaining two integrals yields

$$(\lambda_n r_o) J_1(\lambda_n r_o) = C_n \cosh(\lambda_n L)\frac{(\lambda_n r_o)^2}{2}\left[J_0^2(\lambda_n r_o) + J_1^2(\lambda_n r_o)\right]. \tag{9-113}$$

Code 9-1 Pin fin coooling

```
#include <stdio.h>
#include <math.h>
#include <gsl/gsl_sf_bessel.h>

inline double fnc(double x,double Bi) {
   return x*gsl_sf_bessel_J1(x)
        -Bi*gsl_sf_bessel_J0(x);
}

double eigenvalue(int n,double Bi) {
   double x,dx=1.;
   double val,hi,lo=0.;
   for (x=.1e-6; x<1.e3; lo=hi,x+=dx) {
      hi=fnc(x,Bi);
      if (hi*lo<0.&&--n<0) break;
   }
   if (hi>0) {hi=x; lo=x-dx;}
   else      {lo=x; hi=x-dx;}
   do { // bisection to find root value
      x=(lo+hi)/2.;
      val=fnc(x,Bi);
      if (val<0.) lo=x;
      else        hi=x;
   } while ( fabs(hi-lo)>1.e-6 );
   return x;
}

double T(double r,double z,double Bi)
{
   double l,val,soln=0.;
   for (n=0;n<100;++n) {
      l=eigenvalue(n,Bi);
      val=cosh(l*z)/cosh(l);
      val*=2.*Bi*gsl_sf_bessel_J0(l*r)
            /gsl_sf_bessel_J0(l)/(l*l+Bi*Bi);
      soln+=val;
   }
   return soln;
}

int main(void)
{
   int n;
   double r,z,Bi=1.;

   FILE *fp=fopen("out.dat","w");
   for (z=0.;z<1.01;z+=.05)
      for (r=0;r<1.01;r+=.05)
         fprintf(fp,"%e %e %e\n",z,r,T(r,z,Bi));
   fclose(fp);
   return 0;
}
```

With the eigenvalue condition given by Eq. (9-95), the expression for the series coefficients becomes

$$C_n = \frac{2Bi}{\cosh(\lambda_n L) J_0(\lambda_n r_o)\left[(\lambda_n r_o)^2 + Bi^2\right]}, \tag{9-114}$$

where $Bi = (hr_o/k)$. With this result, the solution (9-109) can be written in its final form:

$$\frac{T(r,z) - T_\infty}{T_w - T_\infty} = \sum_{n=1}^{\infty} \frac{2Bi\, J_0(\lambda_n r)\cosh(\lambda_n z)}{\cosh(\lambda_n L) J_0(\lambda_n r_o)\left[(\lambda_n r_o)^2 + Bi^2\right]}. \tag{9-115}$$

Code 9-1 evaluates Eq. (9-115) for the temperature field in the pin. Isotherms for a cross section of the pin are shown in Figure 9-4.

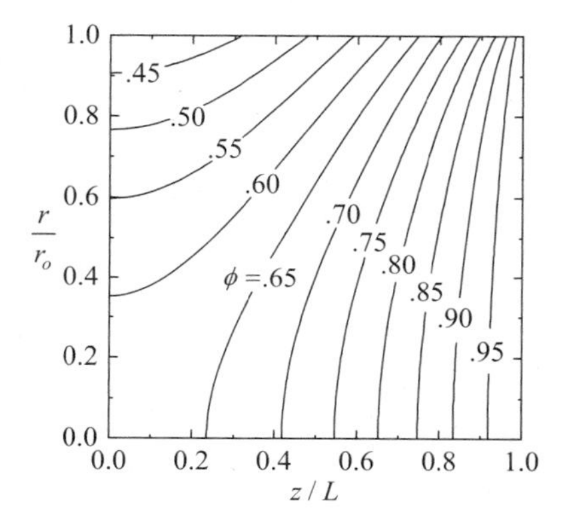

Figure 9-4 Isotherms in cooling pin.

The adiabatic boundaries at $z = 0$ and $r = 0$ are easily identified, as well as the constant wall temperature at $z = L$.

9.3.2 Transient Heat Transfer in a Sphere

Consider a solid sphere quenched in a fluid bath, as illustrated in Figure 9-5. The initial temperature of the sphere is $T = T_o$. For $t > 0$, the surface of the sphere equals the boiling temperature T_B of the fluid. Therefore, heat transfer into the sphere is subject to the following conditions:

$$\text{at} \quad t = 0 \quad T = T_o \tag{9-116}$$

$$\begin{aligned} \text{for} \quad t > 0 \quad & T(r = r_o) = T_B \\ & dT/dr|_{r=0} = 0 \end{aligned} \tag{9-117}$$

For spherical coordinates, the transient heat diffusion equation evaluates to

$$\begin{array}{ccc} \partial_o T = & & \alpha\, \partial_j \partial_j T \\ \downarrow & & \downarrow \\ \dfrac{\partial T}{\partial t} & = & \alpha \left[\dfrac{1}{r^2} \dfrac{\partial}{\partial r} \left(r^2 \dfrac{\partial T}{\partial r} \right) + \dfrac{1}{r^2 \sin\theta} \dfrac{\partial}{\partial \theta} \left(\sin\theta \overbrace{\dfrac{\partial T}{\partial \theta}}^{=0} \right) + \dfrac{1}{r^2 \sin^2\theta} \overbrace{\dfrac{\partial^2 T}{\partial \phi^2}}^{=0} \right]. \end{array} \tag{9-118}$$

Defining the new variables,

$$\Phi = \frac{T - T_B}{T_o - T_B}, \quad \eta = \frac{r}{r_o} \quad \text{and} \quad \tau = \frac{\alpha t}{r_o^2}, \tag{9-119}$$

the heat equations becomes

$$\frac{\partial \Phi}{\partial \tau} = \frac{1}{\eta^2} \frac{\partial}{\partial \eta} \left(\eta^2 \frac{\partial \Phi}{\partial \eta} \right) \tag{9-120}$$

The mathematical problem for Φ is summarized in Figure 9-6. To illustrate application of standard separation of variables to a problem in non-Cartesian coordinates, this approach is applied here.

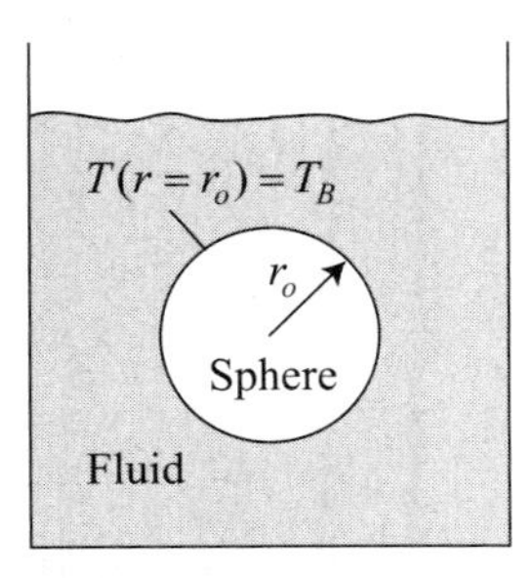

Figure 9-5 Solid sphere quenched in fluid bath.

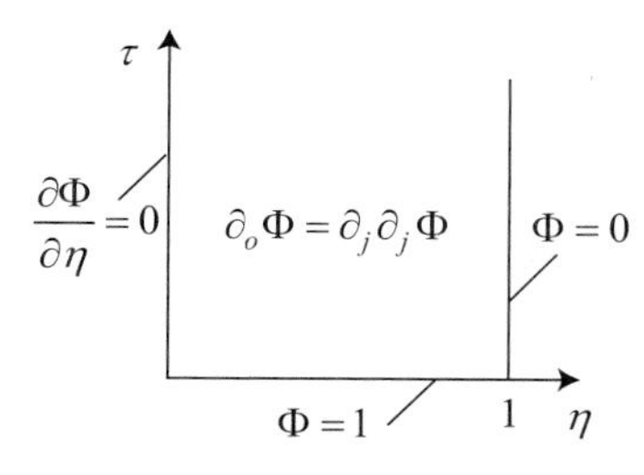

Figure 9-6 Problem statement for heat transfer to sphere.

Assume that a solution exists having the form

$$\Phi = R(\eta)Y(\tau). \tag{9-121}$$

Substituting this into the governing equation (9-120) and separating variables yields

$$\frac{1}{Y}\frac{\partial Y}{\partial \tau} = \frac{1}{R}\frac{1}{\eta^2}\frac{\partial}{\partial \eta}\left(\eta^2 \frac{\partial R}{\partial \eta}\right) = -\lambda_n^2 \tag{9-122}$$

where $-\lambda_n^2$ is the separation constant. Separation of variables leads to two ordinary differential equations:

$$\begin{array}{cc} \dfrac{dY}{d\tau} + \lambda_n^2 Y = 0 & \dfrac{d}{d\eta}\left(\eta^2 \dfrac{dR}{d\eta}\right) + (\lambda_n \eta)^2 R = 0 \\ \downarrow & \downarrow \\ Y = C_1 \exp(-\lambda_n^2 \tau) & R = C_2 \dfrac{\cos(\lambda_n \eta)}{\eta} + C_3 \dfrac{\sin(\lambda_n \eta)}{\eta} \end{array} \tag{9-123}$$

For R to remain finite as $\eta \to 0$, it is required that $C_2 = 0$. Furthermore, the value of the separation constant must be restricted to satisfy the homogeneous surface boundary condition:

$$\text{at} \quad \eta = 1: \quad R = 0 = C_3 \frac{\sin(\lambda_n \cdot 1)}{1}, \quad \text{so} \quad \lambda_n = n\pi \quad \text{with} \quad n = 1, 2, 3, \ldots. \tag{9-124}$$

Notice that the gradient of R at the center of the sphere is given by:

$$\frac{dR}{d\eta} = C_3 \frac{\lambda_n \eta \cos(\lambda_n \eta) - \sin(\lambda_n \eta)}{\eta^2} \tag{9-125}$$

That $dR/d\eta|_{\eta=0} = 0$ can be demonstrated by L'Hôpital's rule:

$$\left.\frac{dR}{d\eta}\right|_{\eta=0} = C_3 \lim_{\eta \to 0} \frac{\dfrac{d}{d\eta}(\lambda_n \eta \cos(\lambda_n \eta) - \sin(\lambda_n \eta))}{d(\eta^2)/d\eta} = -\lambda_n C_3 \lim_{\eta \to 0} \frac{\lambda_n \sin(\lambda_n \eta)}{2} = 0 \tag{9-126}$$

Therefore, the solution for $\Phi = R(\eta)Y(\tau)$ takes the form $C_n\{\sin(\lambda_n \eta)/\eta\}\exp(-\lambda_n^2 \tau)$, where the two integration constants C_1 and C_3 have been combined into C_n. To satisfy the nonhomogeneous initial condition, a superposition of all possible solutions, corresponding to each λ_n, is required. Therefore, the proposed solution takes the form

$$\Phi = \sum_{n=1}^{\infty} C_n \frac{\sin(\lambda_n \eta)}{\eta} \exp(-\lambda_n^2 \tau). \tag{9-127}$$

This solution is required to satisfy the initial condition

$$\text{at } t = 0: \quad 1 = \sum_{n=1}^{\infty} C_n \frac{\sin(\lambda_n \eta)}{\eta}. \tag{9-128}$$

Making use of orthogonality, the integration constant C_n can be determined from

$$\int_0^1 1 \frac{\sin(\lambda_m \eta)}{\eta} \eta^2 d\eta = \int_0^1 \sum_{n=1}^{\infty} C_n \frac{\sin(\lambda_n \eta)}{\eta} \frac{\sin(\lambda_m \eta)}{\eta} \eta^2 d\eta \tag{9-129}$$

or

$$-\cos(\lambda_n)/\lambda_n = C_n \int_0^1 \sin^2(\lambda_n \eta)\, d\eta \quad (\textit{for } n = m). \tag{9-130}$$

Therefore,

$$C_n = -(2/\lambda_n)\cos(\lambda_n) \quad \text{or} \quad C_n = -(2/\lambda_n)(-1)^n. \tag{9-131}$$

Finally, the solution (9-127) for Φ can be expressed as

$$\Phi = \frac{T - T_B}{T_o - T_B} = \sum_{n=1}^{\infty} \frac{-2(-1)^n}{\lambda_n} \frac{\sin(\lambda_n \eta)}{\eta} \exp(-\lambda_n^2 \tau), \tag{9-132}$$

where

$$\lambda_n = n\pi. \tag{9-133}$$

This solution is plotted in Figure 9-7. Heat is seen diffusing outward toward the constant temperature boundary condition $\Phi(t, r = r_o) = 0$. The sphere approaches a uniform temperature as time becomes large, $\Phi(t \to \infty, r) \to 0$. The solution reveals that the center temperature $\Phi(r = 0)$ drops 99% from its initial value when $\tau = \alpha\, t/r_o^2 = 0.54$.

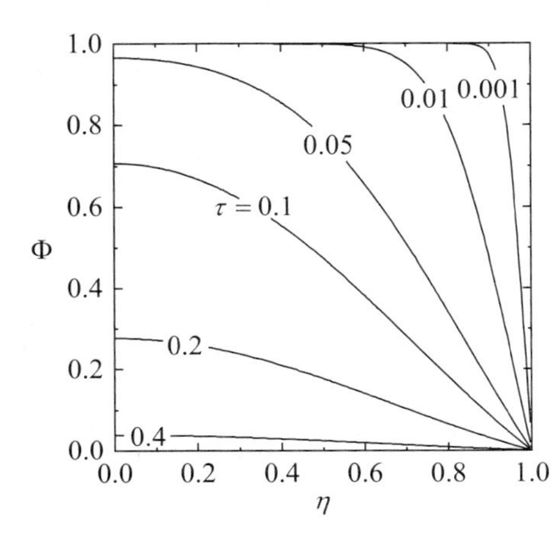

Figure 9-7 Time evolution of temperature during sphere quenching.

9.4 PROBLEMS

9-1 Solve the steady-state dilute species diffusion problem illustrated using the eigenvalue expansion method. Sketch what the solution should look like.

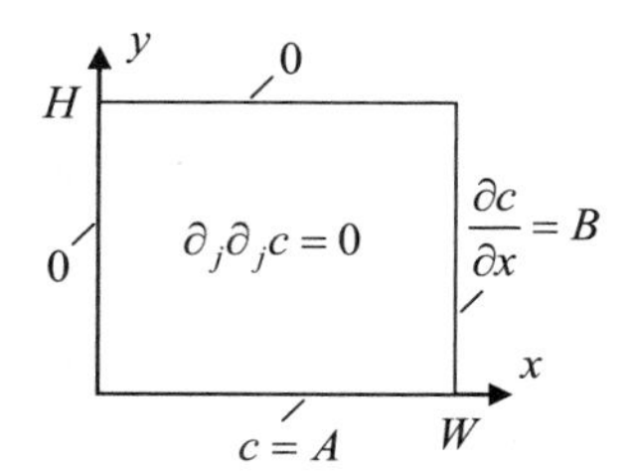

9-2 Solve the illustrated flow problem by the eigenvalue expansion method. Sketch what the solution should look like.

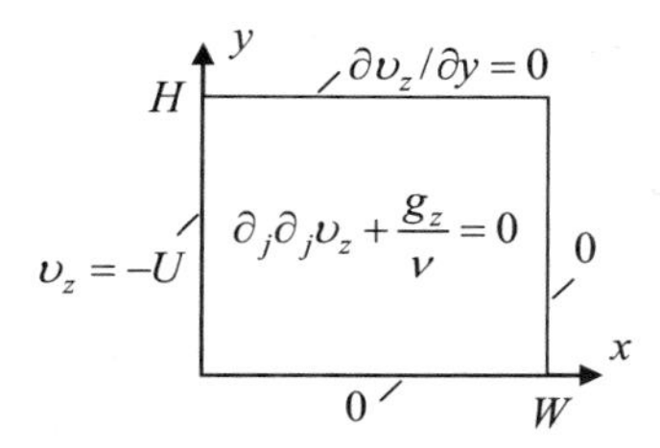

9-3 A horizontal pipe of radius r_o is filled with an initially stationary fluid. After a pump is turned on at $t = 0$, the fluid experiences an axial pressure gradient $-dP/dz$. Find the transient expression for the fluid velocity in the pipe $v_z(t, r)$ for $t > 0$, using the eigenvalue expansion method.

9-4 Solve Problem 9-3 by standard separation of variables.

9-5 A horizontal pipe is filled with an initially stationary fluid in a weightless environment. At $t = 0$, the pipe starts rotating about its axis with constant angular velocity ω. Using standard separation of variables, find the transient expression for the fluid velocity in the pipe $v_\theta(t, r)$ for $t > 0$.

9-6 Solve the sphere quenching problem of Section 9.3.2 for a convective surface boundary condition. Assuming a bath temperature of T_∞, the new conditions to be imposed on the problem are:

$$\begin{array}{lll} at & t = 0 & T = T_o \\ for & t > 0 & [-k\, dT/dr = h(T - T_\infty)]_{r=r_o} \\ & & dT/dr|_{r=0} = 0 \end{array}$$

Determine the temperature of the sphere as a function of time. At what dimensionless time ($\tau = \alpha t / r_o^2$) does the center temperature drop 99% from the initial value?

REFERENCES

[1] G. Cain and G. H. Meyer, *Separation of Variables for Partial Differential Equations; An Eigenfunction Approach*. Boca Raton, FL: Chapman & Hall/CRC, 2006.

[2] A. D. Polyanin, *Handbook of Linear Partial Differential Equations for Engineers and Scientists*. Boca Raton, FL: Chapman & Hall/CRC Press, 2002.

Chapter 10

Similarity Solution

10.1 The Similarity Variable
10.2 Laser Heating of a Semi-infinite Solid
10.3 Transient Evaporation
10.4 Power Series Solution
10.5 Mass Transfer with Time-Dependent Boundary Condition
10.6 Problems

Separation of variables, as discussed in Chapters 7 through 9, employs one strategy of simplifying the task of integrating a partial differential equation (PDE) by representing its solution in terms of ordinary differential equations (ODEs). A *similarity solution*, which is the topic of this chapter, also replaces the task of integrating a PDE with that of integrating an ODE. However, the original PDE is transformed into a single ODE using a *similarity variable* that is related to the original dependent variables of the PDE. Sometimes a similarity solution can only be found when the dependent variable also undergoes a particular scaling. Although the applicability of the similarity solution technique is not as general as separation of variables, a large number of important problems can be solved by this method. The requirements of the technique are that, first, a similarity variable can be found that transforms the governing PDE into an ODE, and, second, the conditions imposed on the original PDE problem (i.e., initial and/or boundary conditions) "collapse" in a way that can be imposed on the new problem described with respect to the similarity variable. The required collapse in conditions most frequently happens for transport problems involving a semi-infinite domain, as will be illustrated for a number of problems in this chapter.

10.1 THE SIMILARITY VARIABLE

Physical situations that admit a similarity solution generally have a single boundary that is important to the transport problem. Similarity solutions employ a similarity variable that scales distances from that boundary by some function of the second independent variable of the problem. The simplest approach to finding a suitable similarity variable is through dimensional analysis of the governing PDE. To illustrate the technique, consider a simple transient diffusion equation for heat transfer in one spatial direction:

$$\frac{\partial T}{\partial t} = \alpha \frac{\partial^2 T}{\partial x^2}. \tag{10-1}$$

The governing equation suggests that a dimensional relation exists between time and space:

$$\frac{1}{\alpha\, t} \sim \frac{1}{x^2}. \tag{10-2}$$

Therefore, a similarity variable that scales the spatial variable by the temporal variable can be proposed

$$\eta = \frac{x}{\sqrt{\alpha\, t}}. \tag{10-3}$$

To transform the governing equation (10-1) into an ODE requires expressing all partial derivatives with derivatives involving the new similarity variable. To this end, the temporal partial derivative can be transformed with

$$\frac{\partial(\,)}{\partial t} = \left.\frac{\partial \eta}{\partial t}\right|_x \frac{\partial(\,)}{\partial \eta} = \frac{-\eta}{2t}\frac{\partial(\,)}{\partial \eta}. \tag{10-4}$$

The spatial partial derivative can be transformed with

$$\frac{\partial(\,)}{\partial x} = \left.\frac{\partial \eta}{\partial x}\right|_t \frac{\partial(\,)}{\partial \eta} = \frac{\eta}{x}\frac{\partial(\,)}{\partial \eta}, \tag{10-5}$$

and the second spatial partial derivative can be transformed with

$$\frac{\partial^2(\,)}{\partial x^2} = \frac{\partial}{\partial x}\left(\frac{\eta}{x}\frac{\partial(\,)}{\partial \eta}\right) = \frac{\eta}{x}\frac{\partial}{\partial x}\left(\frac{\partial(\,)}{\partial \eta}\right) = \left(\frac{\eta}{x}\right)^2 \frac{\partial^2(\,)}{\partial \eta^2}. \tag{10-6}$$

Notice that η/x is not a function of x, and may be brought outside the derivative with respect to x. Applying the above transformations to the governing equation gives

$$\overbrace{\frac{-\eta}{2t}\frac{\partial T}{\partial \eta}}^{\partial T/\partial t} = \overbrace{\alpha\left(\frac{\eta}{x}\right)^2 \frac{\partial^2 T}{\partial \eta^2}}^{\alpha\,\partial^2 T/\partial x^2}. \tag{10-7}$$

The remaining appearances of x and t are grouped into expressions of η for the result

$$\frac{\partial^2 T}{\partial \eta^2} + \frac{\eta}{2}\frac{\partial T}{\partial \eta} = 0. \tag{10-8}$$

Because all occurrences of x and t have disappeared, the remaining equation for T is an ODE in terms of η. It is noteworthy that the transformed governing equation is still second-order, but now only involves one independent variable. Therefore, there are a reduced number of conditions that can be imposed on Eq. (10-8) relative to the original governing equation (10-1).

It should be observed that Eq. (10-3) defining η is not the only possibility. For example, introducing an additional constant "A" into the definition simply changes the coefficients of the transformed governing equation in a trivial way. With

$$\eta = \frac{x}{\sqrt{A\alpha t}}, \tag{10-9}$$

Eq. (10-1) transforms to

$$\frac{\partial^2 T}{\partial \eta^2} + A\frac{\eta}{2}\frac{\partial T}{\partial \eta} = 0. \tag{10-10}$$

Scaling of Eq. (10-1) also suggests other potential choices of similarity variables that can successfully transform the governing PDE into an ODE. However, choices such as $\eta = \alpha t/x^2$ and $\eta = x^2/\alpha t$ are ill advised when x is a measure of distance from the boundary of importance. The similarity variable $\eta = \alpha t/x^2$ pushes the important part of the solution $(x \to 0)$ to $\eta \to \infty$, which is clearly not desirable. Furthermore, the similarity variable $\eta = x^2/\alpha t$ leads to a transformation of the temperature gradient, where

$$\frac{\partial T}{\partial x} = \left.\frac{\partial \eta}{\partial x}\right|_t \frac{\partial T}{\partial \eta} = \frac{2x}{\alpha t}\frac{\partial T}{\partial \eta} \qquad \left(\text{with} \quad \eta = \frac{x^2}{\alpha t}\right). \tag{10-11}$$

Since the solution will likely require $\partial T/\partial x|_{x=0}$ be finite for all $t > 0$, this last proposed similarity variable has the negative influence of forcing $\partial T/\partial \eta|_{\eta=0} \to \infty$ as $x \to 0$. Therefore, the practical choices for the similarity variable are more limited than what is implied by scaling arguments alone. *As a rule of thumb, the similarity variable should be chosen to scale linearly with respect to the variable measuring distance from the imposed boundary condition on the problem.*

Although a suitable similarity variable can transform the governing PDE into an ODE, success of the method also relies on the ability to address all original conditions imposed on the problem. A similarity solution cannot be attained if the boundary conditions are dependent on the original independent variables in a way that cannot be fully transformed with the similarity variable. Additionally, since there is a reduction in the number of independent variables, one anticipates that the number of conditions that can be imposed on the ODE is reduced from the original PDE. Therefore, a successful similarity solution requires that the conditions imposed on the original problem collapse to an appropriately fewer number when expressed in terms of the similarity variable. These issues are further illustrated by the example discussed in the next section.

10.2 LASER HEATING OF A SEMI-INFINITE SOLID

Consider a semi-infinite solid whose surface is heated by a laser of absorbed intensity I_o, as illustrated in Figure 10-1. The diameter of the beam is large compared with the thermal penetration depth of heating, such that heat transfer can be appropriately described by the transient one-dimensional heat-diffusion equation, as discussed in Section 6.5.3. Furthermore, the optical penetration depth of the laser into the solid is negligible, allowing the heat input to be described with a heat flux boundary condition. An important characteristic of the semi-infinite solid is that heat diffusion never encounters a back boundary. Therefore,

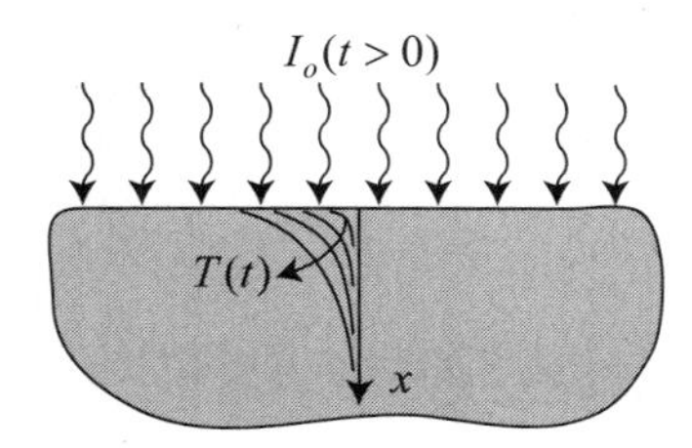

Figure 10-1 Laser heating of a semi-infinite solid.

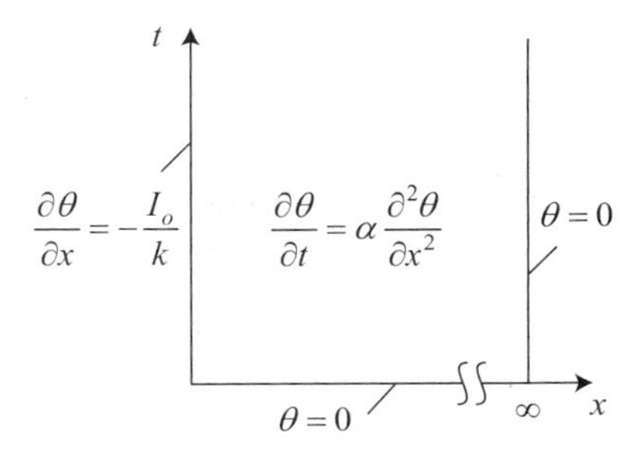

Figure 10-2 Mathematical statement for the laser heating of a semi-infinite solid.

the associated boundary condition is that $T(x \to \infty) = T_o$, where T_o is the initial temperature of the solid before heating.

The mathematical description of this problem is governed by the transient one-dimensional heat diffusion equation

$$\partial_o T = \alpha(\partial^2 T/\partial x^2), \tag{10-12}$$

subject to the initial condition and boundary conditions:

$$T(t=0) = T_o, \quad \partial T/\partial x|_{x=0} = -I_o/k, \quad \text{and} \quad T(x \to \infty) = T_o. \tag{10-13}$$

Letting $\theta = T - T_o$, the mathematical description becomes that illustrated in Figure 10-2.

A similarity solution to this problem is sought by letting

$$\eta = \frac{x}{\sqrt{4\alpha t}}. \tag{10-14}$$

The transformed governing equation becomes

$$\frac{\partial \theta}{\partial t} = \alpha \frac{\partial^2 \theta}{\partial x^2} \quad \rightarrow \quad \frac{\partial^2 \theta}{\partial \eta^2} + 2\eta \frac{\partial \theta}{\partial \eta} = 0. \tag{10-15}$$

Transforming the initial condition and boundary conditions yields

$$\begin{aligned} x = 0 \quad &: \quad \eta = 0 \quad : \quad I_o = -\frac{k\eta}{x} \cdot \frac{d\theta}{d\eta}\bigg|_{\eta=0} \\ x \to \infty \quad &: \quad \eta \to \infty \quad : \quad \theta = 0 \\ t = 0 \quad &: \quad \eta \to \infty \quad : \quad \theta = 0. \end{aligned} \tag{10-16}$$

Notice that the boundary condition at $x = 0$ did not transform into an expression dependent on η alone. This is because the heat flux boundary condition cannot be expressed independently of the original spatial variable x. Therefore, a similarity solution does not exist for this problem, as stated. However, the original problem can be transformed further to make it amenable to a similarity solution. Let a new independent variable for this problem be the diffusion heat flux:

$$q = -k\frac{\partial \theta}{\partial x}, \text{ such that } \theta = \int_x^\theta (q/k)dx. \tag{10-17}$$

The governing equation is transformed by first differentiating both sides with respect to x and introducing a factor of $-k$. Since the order of time and space differentiation can be switched, the governing equation becomes

$$\frac{\partial}{\partial t}\left[-k\frac{\partial \theta}{\partial x}\right] = \alpha\frac{\partial^2}{\partial x^2}\left[-k\frac{\partial \theta}{\partial x}\right] \tag{10-18}$$

or

$$\frac{\partial q}{\partial t} = \alpha\frac{\partial^2 q}{\partial x^2}. \tag{10-19}$$

The initial and boundary conditions of the original problem are transformed into expressions related to the new dependent variable q, with the results summarized in Figure 10-3.

A similarity solution is sought for this new problem by again letting

$$\eta = \frac{x}{\sqrt{4\alpha t}}. \tag{10-20}$$

The transformed governing equation is

$$\frac{\partial q}{\partial t} = \alpha\frac{\partial^2 q}{\partial x^2} \quad \rightarrow \quad \frac{\partial^2 q}{\partial \eta^2} + 2\eta\frac{\partial q}{\partial \eta} = 0 \tag{10-21}$$

as before. However, the transformed initial condition and boundary conditions now become

$$\underbrace{\begin{matrix} q(t=0)=0 \\ q(x\rightarrow\infty)=0 \\ q(x=0)=I_0 \end{matrix}}_{\text{3 original conditions}} \qquad \underbrace{\begin{matrix} \left.\begin{matrix} q(\eta\rightarrow\infty)=0 \\ q(\eta\rightarrow\infty)=0 \end{matrix}\right\}\textit{same} \\ q(\eta=0)=I_0 \end{matrix}}_{\text{2 final conditions}} \tag{10-22}$$

Notice that the initial condition and boundary conditions are free from any dependence on x or t, and that two of the conditions imposed on the original problem have

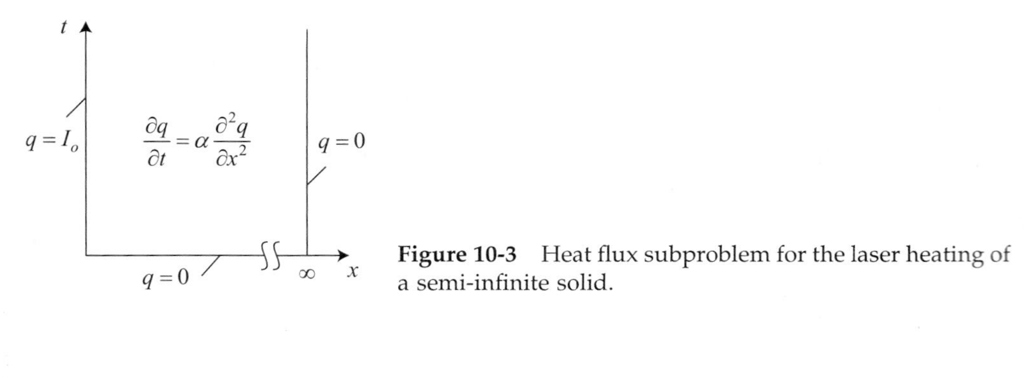

Figure 10-3 Heat flux subproblem for the laser heating of a semi-infinite solid.

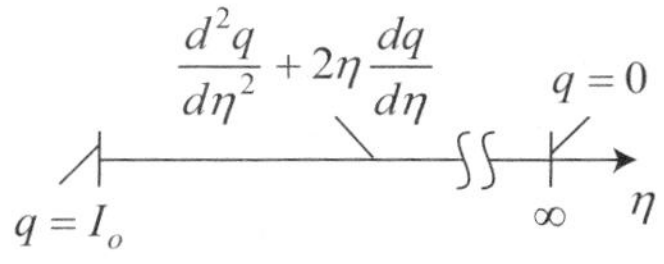

$\frac{d^2q}{d\eta^2}+2\eta\frac{dq}{d\eta}$ $q=0$

$q=I_o$ ∞ η

Figure 10-4 Subproblem using the similarity variable for the laser heating of a semi-infinite solid.

collapsed to one condition on the problem in η. Since the original problem was second-order in x and first order in t, it required three imposed conditions: an initial condition and two boundary conditions. However, the ODE governing the transformed problem is second-order, which admits only two independent conditions. Therefore, the number of conditions imposed on the original problem must be reduced by one to successfully find a similarity solution.

The problem expressed in terms of the similarity variable is illustrated in Figure 10-4. The only task that remains is to integrate the ODE and satisfy the two boundary conditions. Integration is facilitated by observing that the ODE can be rewritten in the form

$$-2\eta d\eta = \frac{d(dq/d\eta)}{(dq/d\eta)}, \tag{10-23}$$

such that integrating once yields

$$dq/d\eta = C_1 \exp(-\eta^2). \tag{10-24}$$

After integrating a second time, the solution to the governing equation can be expressed as

$$q = C_1(\sqrt{\pi}/2)\text{erf}(\eta) + C_2, \tag{10-25}$$

which is expressed in terms of the *error function*:

$$\text{erf}(\eta) = \frac{2}{\sqrt{\pi}}\int_0^{\eta} e^{-\eta^2} d\eta. \tag{10-26}$$

Applying the boundary conditions to determine the integration constants, the solution for q is found to be

$$q = I_o \text{erfc}\left(\frac{x}{\sqrt{4\alpha t}}\right), \tag{10-27}$$

which is expressed in terms of the complementary error function:

$$\text{erfc}(\eta) = 1 - \text{erf}(\eta). \tag{10-28}$$

Returning to the original dependent variable using the transformation given by Eq. (10-17), the solution for $\theta = T - T_o$ can be expressed in terms of the original dependent variables x and t as

$$\frac{T - T_o}{I_o/k} = \int_x^{\infty} \text{erfc}\left(\frac{x}{\sqrt{4\alpha t}}\right)dx = \left[\sqrt{\frac{4\alpha t}{\pi}}\exp\left(-\frac{x^2}{4\alpha t}\right) - x\,\text{erfc}\left(\frac{x}{\sqrt{4\alpha t}}\right)\right]. \tag{10-29}$$

10.3 TRANSIENT EVAPORATION

As further illustration of the similarity solution technique, consider the problem of water evaporating into a column of dry air, as discussed in Section 6.4.1. A transient version of this one-dimensional problem is considered here for the case when the length of the air column shown in Figure 10-5 becomes large, $L \to \infty$. For isobaric and isothermal conditions the total molar concentration everywhere in the column is constant. Therefore, the continuity equation for the total molar concentration of the fluid simplifies to the requirement that

$$\partial_o c + \partial_j (c v_j^*) = 0 \to \partial_x v_x^* = 0 \to v_x^* = const. \tag{10-30}$$

As demonstrated in Section 6.4.1, water evaporation introduces a molar-averaged Stefan flow velocity that can be evaluated at the interface with the liquid water:

$$v_x^* = \frac{-Đ_{WA}}{1-\chi_{W0}} \frac{\partial \chi_W}{\partial x}\bigg|_{x=0}, \tag{10-31}$$

where χ_{W0} is the mole fraction of water vapor at the interface with the liquid. Therefore, assuming that the diffusivity of water in air $Đ_{WA}$ is a constant, the transient molar concentration of water vapor is governed by the transport equation:

$$\frac{\partial \chi_W}{\partial t} + \overbrace{\left(\frac{-Đ_{WA}}{1-\chi_{W0}} \frac{\partial \chi_W}{\partial x}\bigg|_{x=0}\right)}^{v_x^*} \frac{\partial \chi_W}{\partial x} = Đ_{WA} \frac{\partial^2 \chi_W}{\partial x^2}. \tag{10-32}$$

The transport of water vapor is subject to conditions

$$\chi_W(x=0) = \chi_{W0}, \quad \chi_W(x \to \infty) = 0, \quad \text{and} \quad \chi_W(t=0) = 0. \tag{10-33}$$

Unlike most transient convection problems, the magnitude of the flow velocity scales with the diffusion flux. As a result, a simple scaling relation between the independent variables of the problem emerges from the governing equation (10-32). This scaling reveals that

$$\underbrace{\frac{1}{Đ_{WA}} \frac{\partial \chi_W}{\partial t}}_{\downarrow \; \frac{1}{Đ_{WA}\, t}} = \underbrace{\left(\frac{1}{1-\chi_{W0}} \frac{\partial \chi_W}{\partial x}\bigg|_{x=0}\right) \frac{\partial \chi_W}{\partial x}}_{\downarrow \; \frac{1}{x^2}} + \underbrace{\frac{\partial^2 \chi_W}{\partial x^2}}_{\downarrow \; \frac{1}{x^2}} \tag{10-34}$$

$$\frac{1}{Đ_{WA}\, t} \sim \frac{1}{x^2} \;\; \text{"+"} \;\; \frac{1}{x^2}$$

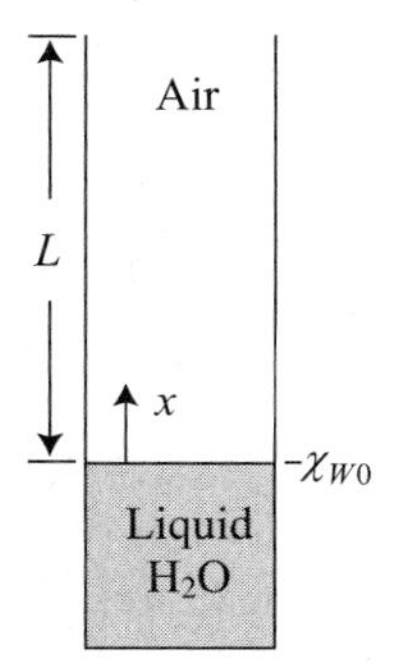

Figure 10-5 Transient water transport through a long air column.

or

$$\frac{1}{Đ_{WA}\, t} \sim \frac{1}{x^2}. \tag{10-35}$$

This suggests attempting a similarity solution to the problem [1]. The similarity variable is defined as

$$\eta = \frac{x}{\sqrt{4Đ_{WA} t}}, \tag{10-36}$$

such that

$$\frac{\partial(\,)}{\partial t} = \frac{-\eta}{2t}\frac{\partial(\,)}{\partial \eta}, \quad \frac{\partial(\,)}{\partial x} = \frac{1}{\sqrt{4Đ_{WA} t}}\frac{\partial(\,)}{\partial \eta}, \quad \text{and} \quad \frac{\partial^2(\,)}{\partial x^2} = \frac{1}{4Đ_{WA} t}\frac{\partial^2(\,)}{\partial \eta^2}. \tag{10-37}$$

Defining the dependent variable for the problem to be

$$\phi = \frac{\chi_W}{\chi_{W0}}, \tag{10-38}$$

the governing equation transforms into the ordinary differential equation

$$\frac{d^2\phi}{d\eta^2} + 2\left(\eta + \frac{\chi_{W0}/2}{1-\chi_{W0}}\left.\frac{d\phi}{d\eta}\right|_{\eta=0}\right)\frac{d\phi}{d\eta} = 0. \tag{10-39}$$

Conditions imposed on the original problem can be transformed into the new problem variables:

$$\underbrace{\begin{array}{c} \chi_W(t=0)=0 \\ \chi_W(x\to\infty)=0 \\ \chi_W(x=0)=\chi_{W0} \end{array}}_{3\ \textit{original conditions}} \qquad \underbrace{\left.\begin{array}{c} \phi(\eta\to\infty)=0 \\ \phi(\eta\to\infty)=0 \end{array}\right\}\textit{same} \\ \phi(\eta=0)=1}_{2\ \textit{final conditions}}, \tag{10-40}$$

where the necessary reduction in the number of conditions is observed. The governing equation (10-39) can be solved by rearranging and integrating once

$$-2\left(\eta - \frac{-\chi_{W0}/2}{1-\chi_{W0}}\left.\frac{d\phi}{d\eta}\right|_{\eta=0}\right)d\eta = \frac{d(d\phi/d\eta)}{(d\phi/d\eta)} \quad \text{or} \quad \frac{d\phi}{d\eta} = C_1 \exp\left[-(\eta+\lambda)^2\right], \tag{10-41}$$

where

$$\lambda = \frac{\chi_{W0}/2}{1-\chi_{W0}}\left.\frac{d\phi}{d\eta}\right|_{\eta=0} \tag{10-42}$$

is a constant related to the solution. After integrating a second time, the solution to (10-39) becomes

$$\phi = C_1(\sqrt{\pi}/2)\,\mathrm{erf}(\eta+\lambda) + C_2. \tag{10-43}$$

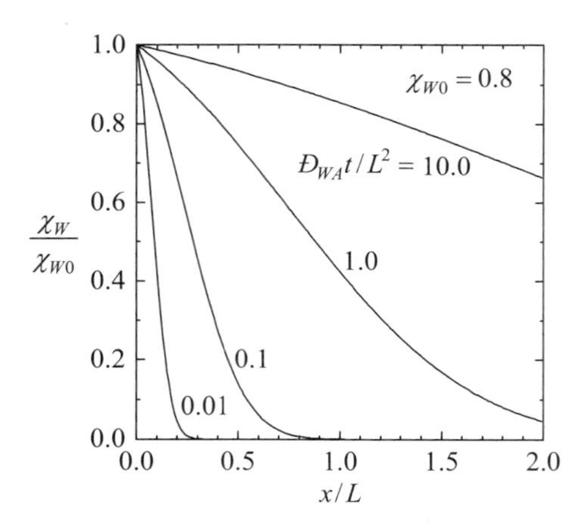

Figure 10-6 Time evolution of the water mole fraction in an air column.

Applying the solution to the boundary conditions (10-40) yields

$$\eta = 0: \qquad 1 = C_1(\sqrt{\pi}/2)\,\mathrm{erf}(\lambda) + C_2 \tag{10-44}$$

$$\eta \to \infty: \qquad 0 = C_1(\sqrt{\pi}/2) + C_2 \tag{10-45}$$

From these equations, the integration constants are determined, such that the solution for ϕ can be written:

$$\phi = \frac{\mathrm{erfc}(\eta + \lambda)}{\mathrm{erfc}(\lambda)}, \quad \text{where} \quad \eta = \frac{x}{\sqrt{4Đ_{WA}t}}. \tag{10-46}$$

Unfortunately, the unknown constant λ appearing in the solution is dependent on the solution itself through Eq. (10-42). However, by substituting the solution (10-46) into Eq. (10-42), a transcendental equation for the unknown constant λ can be derived:

$$\sqrt{\pi}\lambda\, e^{\lambda^2}\,\mathrm{erfc}(\lambda) + \frac{\chi_{W0}}{1 - \chi_{W0}} = 0. \tag{10-47}$$

Therefore, the transient solution to this problem is found by first solving Eq. (10-47) for λ. Then the solution for the transient molar concentration of water in the air column is described by Eq. (10-46). Figure 10-6 illustrates the solution when the water mole fraction at the air–water interface is $\chi_W(x = 0) = \chi_{W0} = 0.8$. Since no length scale is imposed on the problem, L used in the nondimensionalization of space and time in Figure 10-6 is arbitrary.

10.4 POWER SERIES SOLUTION

In the preceding sections, it was fortunate that a closed form solution existed to the ordinary differential equations that arose. However, what would one do for the preceding problems if the erf( ) and erfc( ) functions did not exist in the math libraries of one's computer? Or, what would one do if faced by some other ordinary differential equation for which functions describing the solution have not yet been identified? Generally, the similarity variable transformation yields an ordinary differential equation with non-constant coefficients. In this section, the method of *power series solutions* is demonstrated as an effective approach to solving linear ordinary differential equations with variable coefficients.

To illustrate, consider the mathematical problem arrived at in Section 10.2, where

$$\partial^2 q/\partial\eta^2 + 2\eta(\partial q/\partial\eta) = 0, \quad \text{subject to} \quad q(0) = I_o \quad \text{and} \quad q(\infty) = 0. \tag{10-48}$$

The method of power series solutions starts with the assumption that the desired solution can be found with the form

$$q = \sum_{n=0}^{\infty} a_n \eta^n. \tag{10-49}$$

Substituting the assumed solution into the governing equation yields

$$\sum_{n=2}^{\infty} n(n-1)a_n\eta^{n-2} + 2\eta\sum_{n=1}^{\infty} na_n\eta^{n-1} = 0. \tag{10-50}$$

Notice that the first series starts at $n = 2$ because the preceding terms are all zero. Likewise, the second series starts at $n = 1$. By expanding the power series form of the governing equation, it can be reorganized to have the form

$$(\cdots)\eta^0 + (\cdots)\eta^1 + (\cdots)\eta^2 + \cdots = 0. \tag{10-51}$$

For this equation to be satisfied at every position η, each coefficient of η^n in the series must be identically zero. Therefore, Eq. (10-50) is written in a form where the coefficients to η^n are easily determined:

$$\sum_{n=0}^{\infty} (n+2)(n+1)a_{n+2}\eta^n + 2\sum_{n=1}^{\infty} na_n\eta^n = 0. \tag{10-52}$$

This was accomplished by changing the indexing scheme of the first term in Eq. (10-50) by replacing appearances of n with $n+2$. Each expression of this equation now contains coefficients to η^n. Notice that since the second series does not start until $n = 1$, it does not contribute to the first term $(\cdots)\eta^0$ in the expanded form of the equation. However, when $n \geq 1$, both series contribute to each coefficient of η^n. Fully expanded, the power series form of the governing equation is expressed by

$$\begin{aligned}(2a_2)\eta^0 + (6a_3 + 2a_1)\eta^1 + (12a_4 + 4a_2)\eta^2 + \cdots \\ + ((n+2)(n+1)a_{n+2} + 2na_n)\eta^n + \cdots = 0.\end{aligned} \tag{10-53}$$

By inspection of the coefficients, it can be seen that the governing equation is satisfied if $a_2 = 0$ and $(n+2)(n+1)a_{n+2} + 2na_n = 0$ for $n \geq 1$. The last condition expresses a recursion relation between values of a_n's: specifically, that

$$a_{n+2} = \frac{-2na_n}{(n+2)(n+1)} \quad \text{for} \quad n \geq 1 \tag{10-54}$$

or, equivalently, that

$$a_n = \frac{-2(n-2)a_{n-2}}{n(n-1)} \quad \text{for} \quad n \geq 3. \tag{10-55}$$

Notice that a_n will be a multiple of a_1 for odd values of n and a multiple of a_2 for even values of n. However, since $a_2 = 0$, all even values of a_n are zero. Therefore, the series solution for q becomes

$$q = a_0 + \left(a_1\eta^1 + a_1 \overbrace{\frac{-1}{3}}^{a_3}\eta^3 + a_1 \overbrace{\frac{1}{10}}^{a_5}\eta^5 + \cdots + \overbrace{\frac{-2(n-2)a_{n-2}}{n(n-1)}}^{a_n}\eta^n + \cdots \right). \tag{10-56}$$

Notice that a_0 and a_1 are arbitrary, and represent the two integration constants expected from integrating a second-order differential equation. Factoring out a_1, and changing the indexing scheme to $i = 1, 2, \ldots$ yields

$$q = a_0 + a_1 \left(\overbrace{1}^{c_1}\ \eta^1 + \overbrace{\frac{-1}{3}}^{c_2}\eta^3 + \overbrace{\frac{1}{10}}^{c_3}\ \eta^5 + \cdots + \overbrace{\frac{-(2i-3)c_{i-1}}{(2i-1)(i-1)}}^{c_i}\eta^{2i-1} + \cdots \right). \tag{10-57}$$

Or

$$q = a_0 + a_1\Phi(\eta), \tag{10-58a}$$

where

$$\Phi(\eta) = c_1\eta^1 + c_2\eta^3 + \sum_{i=3}^{\infty} c_i\eta^{2i-1} \tag{10-58b}$$

and

$$c_1 = 1, \ c_2 = -1/3 \quad \text{and} \quad c_i = \frac{-(2i-3)c_{i-1}}{(2i-1)(i-1)}. \tag{10-58c}$$

Figure 10-7 plots the function $\Phi(\eta)$. Notice that as $\eta \rightarrow \infty$, this function approaches a constant value of $\Phi(\infty) = 0.8862$.

At $\eta = 0$, the solution (10-58) evaluates to $q(0) = a_0$. However, for the current problem $q(0) = I_o$, and therefore $a_0 = I_o$. The second boundary condition $q(\infty) = 0$ requires that $0 = I_o + a_1\Phi(\infty)$. Therefore, $a_1 = -I_o/\Phi(\infty)$, and the final solution for q becomes

$$q/I_o = 1 - \frac{1}{\Phi(\infty)}\left(c_1\eta^1 + c_2\eta^3 + \sum_{i=3}^{\infty} c_i\eta^{2i-1} \right). \tag{10-59}$$

Returning to the original dependent variable $\theta = T - T_o$, the solution expressed in terms of x and t becomes

$$\begin{aligned} \theta = \int_x^{\infty} (q/k)dx = \int_x^{\infty} \frac{I_o}{k} \Bigg[1 - \frac{1}{\Phi(\infty)} \Bigg(c_1\left(\frac{x}{\sqrt{4\alpha t}}\right)^1 + c_2\left(\frac{x}{\sqrt{4\alpha t}}\right)^3 \\ + \sum_{i=3}^{\infty} c_i \left(\frac{x}{\sqrt{4\alpha t}}\right)^{2i-1} \Bigg) \Bigg] dx. \end{aligned} \tag{10-60}$$

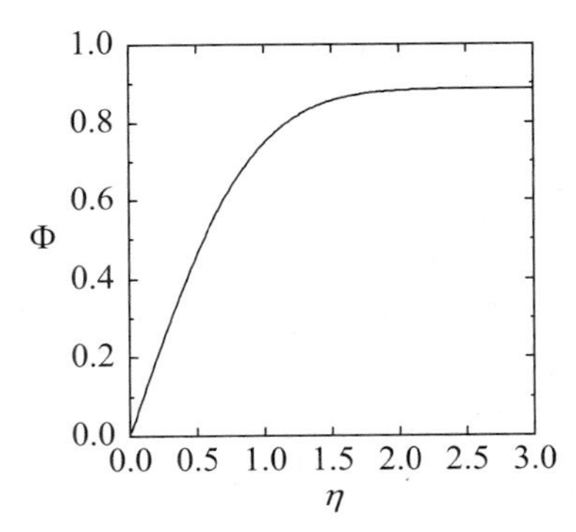

Figure 10-7 Plot of the series function $\Phi(\eta)$, defined by Eq. (10-58b).

After integrating, the result becomes

$$\frac{T - T_o}{LI_o/k} = 2\frac{\sqrt{\alpha t}}{L}\left[\frac{x}{\sqrt{4\alpha t}} - \frac{1}{\Phi(\infty)}\left(\frac{c_1}{2}\left(\frac{x}{\sqrt{4\alpha t}}\right)^2 + \frac{c_2}{4}\left(\frac{x}{\sqrt{4\alpha t}}\right)^4 + \sum_{i=3}^{\infty}\frac{c_i}{2i}\left(\frac{x}{\sqrt{4\alpha t}}\right)^{2i}\right)\right]_x^{\infty} \tag{10-61}$$

where an arbitrary length scale L has been introduced to form the dimensionless group $\sqrt{\alpha t}/L$ for time. The final solution can be expressed as

$$\frac{T - T_o}{LI_o/k} = \frac{\sqrt{4\alpha t}}{L}\left[\Psi(\infty) - \Psi\left(\frac{x}{\sqrt{4\alpha t}}\right)\right] \tag{10-62a}$$

where

$$\Psi\left(\frac{x}{\sqrt{4\alpha t}}\right) = \frac{x}{\sqrt{4\alpha t}} - \frac{1}{\Phi(\infty)}\left(\frac{c_1}{2}\left(\frac{x}{\sqrt{4\alpha t}}\right)^2 + \frac{c_2}{4}\left(\frac{x}{\sqrt{4\alpha t}}\right)^4 + \sum_{i=3}^{\infty}\frac{c_i}{2i}\left(\frac{x}{\sqrt{4\alpha t}}\right)^{2i}\right) \tag{10-62b}$$

with

$$c_1 = 1,\ \ c_2 = -1/3,\ \ c_i = \frac{-(2i-3)c_{i-1}}{(2i-1)(i-1)} \tag{10-62c}$$

and

$$\Phi(\infty) = 0.8862 \quad \text{and} \quad \Psi(\infty) = 0.5642. \tag{10-62d}$$

Equation (10-62) is equivalent to the solution given by Eq. (10-29) in Section 10-2. Equation (10-62) is plotted in Figure 10-8. Notice that the surface temperature rises continuously and can be evaluated from

$$T_s - T_o = \frac{I_o}{k}\Psi(\infty)\sqrt{4\alpha t} = 1.1284\frac{I_o}{k}\sqrt{\alpha t}. \tag{10-63}$$

A thermal penetration depth δ_T can be defined as the distance from the surface at which the diffusion heat flux falls 99% from I_o. The thermal penetration depth is a

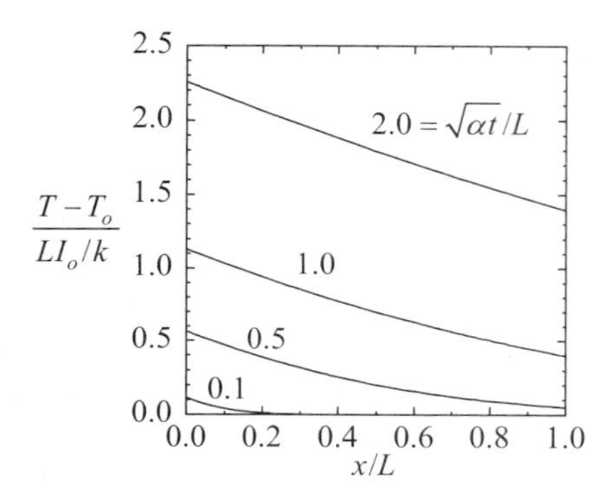

Figure 10-8 Time evolution of temperature in the laser heated semi-infinite solid.

function of time and can be determined from $\eta_{99} = \delta_T/\sqrt{4\alpha t}$, where $\Phi(\eta_{99})/\Phi(\infty) = 0.99$. Numerically, it is determined that η_{99} =1.8214, such that

$$\delta_T = \eta_{99}\sqrt{4\alpha t} = 3.643\sqrt{\alpha t}. \tag{10-64}$$

10.5 MASS TRANSFER WITH TIME-DEPENDENT BOUNDARY CONDITION

Consider transient mass diffusion into a semi-infinite solid. The concentration of some species "A" at the surface is specified as a function of time: $c_A(t, x=0) = c_o(t)$. It is assumed that c_A is sufficiently small such that transport of A has a negligibly small influence on the molar-averaged flow velocity in the solid. The mathematical problem is illustrated in Figure 10-9. It is desired to find the functional form of $c_o(t)$ that could admit a similarity solution to this problem. Since the governing equation has the same form as the heat equation in Section 10.2, it is reasonable to propose the same form of the similarity variable, but expressed in terms of the species diffusivity:

$$\eta = \frac{x}{\sqrt{4D_A t}}. \tag{10-65}$$

If a dimensionless solution is sought for a dependent variable defined by

$$\phi = c(t,x)/c_o(t), \tag{10-66}$$

the problem will have conditions that are all independent of x and t:

$$\begin{array}{ll} c_A(t=0)=0 & \left.\begin{array}{l}\phi(\eta\to\infty)=0\\ \phi(\eta\to\infty)=0\end{array}\right\}\textit{same} \\ c_A(x\to\infty)=0 & \\ \underbrace{c_A(x=0)=c_o(t)}_{\textit{3 original conditions}} & \underbrace{\phi(\eta=0)=1}_{\textit{2 final conditions}} \end{array}, \tag{10-67}$$

Notice that the number of conditions imposed on the problem is reduced from three (in terms of x and t) to two (in terms of η), as is necessary for a similarity solution.

For a similarity solution to exist, the functional form of $c_o(t)$ is dictated by the governing equation. When all derivatives in the governing equation are expressed in terms of η, it will be seen that only certain functional forms of $c_o(t)$ can remove all appearances of t from the transformed equation.

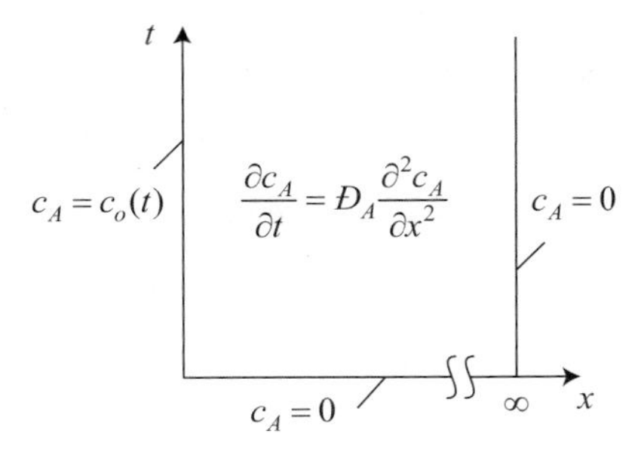

Figure 10-9 Mathematical statement for mass diffusion into a semi-infinite solid with a time-dependent boundary condition.

Substituting $c(t,x) = \phi\, c_o(t)$ into the governing equation yields

$$\phi \frac{\partial c_o}{\partial t} + c_o \frac{\partial \phi}{\partial t} = Đ_A c_o \frac{\partial^2 \phi}{\partial x^2}. \tag{10-68}$$

For the present choice of similarity variable,

$$\frac{\partial (\,)}{\partial t} = \left(\frac{-\eta}{2t}\right)\frac{\partial (\,)}{\partial \eta} \quad \text{and} \quad \frac{\partial^2 (\,)}{\partial x^2} = \left(\frac{\eta}{x}\right)^2 \frac{\partial^2 (\,)}{\partial \eta^2}. \tag{10-69}$$

Therefore, expressing all the derivatives in terms of the similarity variable, the governing equation becomes

$$\phi\left(\frac{-\eta}{2t}\right)\frac{\partial c_o}{\partial \eta} + c_o\left(\frac{-\eta}{2t}\right)\frac{\partial \phi}{\partial \eta} = Đ_A c_o \left(\frac{\eta}{x}\right)^2 \frac{\partial^2 \phi}{\partial \eta^2}. \tag{10-70}$$

Or, multiplying through by $2t$, the remaining appearances of x and t can be expressed in terms of η. Therefore, the governing equation for ϕ becomes

$$c_o \frac{\partial^2 \phi}{\partial \eta^2} + 2\eta\left(\phi \frac{\partial c_o}{\partial \eta} + c_o \frac{\partial \phi}{\partial \eta}\right) = 0. \tag{10-71}$$

However, the governing equation is still not a function of η alone because c_o is a function of time. However, by inspection of Eq. (10-71), it is seen that if

$$\frac{\partial c_o}{\partial \eta} = c_o f(\eta), \tag{10-72}$$

then the governing equation (10-71) becomes

$$\frac{\partial^2 \phi}{\partial \eta^2} + 2\eta\left(f(\eta)\phi + \frac{\partial \phi}{\partial \eta}\right) = 0, \tag{10-73}$$

which is an ordinary differential equation governing ϕ. However, at this point the function $f(\eta)$ is not specified.

The constraint given by Eq. (10-72) for c_o must be changed back into a differential equation with respect to time in order to determine what functional forms of time c_o may have. Since

$$\frac{\partial c_o}{\partial \eta} = \left(\frac{\partial t}{\partial \eta}\right)\frac{\partial c_o}{\partial t} = \left(\frac{-2t}{\eta}\right)\frac{\partial c_o}{\partial t}, \tag{10-74}$$

the constraint equation (10-72) for c_o becomes (after separating variables)

$$\left(\frac{t}{c_o}\right)\frac{\partial c_o}{\partial t} = \frac{-\eta f(\eta)}{2}. \tag{10-75}$$

The only way the left-hand side will equal the right-hand side is if both equal a constant: $(t/c_o)\partial c_o/\partial t = \gamma = -\eta f(\eta)/2$. This requires that $f(\eta) = -2\gamma/\eta$, where γ is an arbitrary constant. Therefore, c_o must satisfy

$$t\frac{\partial c_o}{\partial t} = c_o \gamma, \tag{10-76}$$

which upon integration dictates that

$$c_o = bt^{\gamma}, \tag{10-77}$$

where b and γ are both arbitrary constants. Using the restriction that $f(\eta) = -2\gamma/\eta$ in the governing equation for ϕ yields

$$\frac{\partial^2 \phi}{\partial \eta^2} + 2\eta\frac{\partial \phi}{\partial \eta} - 4\gamma\phi = 0. \tag{10-78}$$

Therefore, the species concentration at the surface of the semi-infinite solid must have the functional form $c_o(t) = bt^{\gamma}$ for a similarity solution to this problem to exist. The resulting concentration distribution through the solid in time is given by $c(x,t) = bt^{\gamma}\phi$, where ϕ must satisfy the governing equation (10-78), subject to the boundary conditions

$$\phi(0) = 1 \quad \text{and} \quad \phi(\infty) = 0. \tag{10-79}$$

A solution for ϕ can be sought using the method of power series, in which it is assumed that

$$\phi = \sum_{n=0}^{\infty} a_n \eta^n. \tag{10-80}$$

Substituting the assumed solution into the governing equation yields

$$\sum_{n=2}^{\infty} n(n-1)a_n\eta^{n-2} + 2\sum_{n=1}^{\infty} na_n\eta^n - 4\gamma\sum_{n=0}^{\infty} a_n\eta^n = 0. \tag{10-81}$$

The power series form of the governing equation is rewritten so that the coefficients to η^n are easily determined:

$$\sum_{n=0}^{\infty} (n+2)(n+1)a_{n+2}\eta^n + 2\sum_{n=1}^{\infty} na_n\eta^n - 4\gamma\sum_{n=0}^{\infty} a_n\eta^n = 0. \tag{10-82}$$

Notice that the second series does not contribute to the first term $(\cdots)\eta^0$ in the expanded form of this equation. However, when $n \geq 1$, all three series contribute to each coefficient of η^n. Therefore, the fully expanded power series form of the governing equation becomes

$$\begin{aligned}(2a_2 - 4\gamma a_0)\eta^0 + (6a_3 + (2-4\gamma)a_1)\eta^1 + \cdots \\ + ((n+2)(n+1)a_{n+2} + (2n-4\gamma)a_n)\eta^n + \cdots = 0.\end{aligned} \tag{10-83}$$

By inspection of the coefficients, it can be seen that the governing equation is satisfied if $a_2 = 2\gamma a_0$ and $(n+2)(n+1)a_{n+2} + (2n-4\gamma)a_n = 0$ for $n \geq 1$. The last condition expresses a recursion relation between values of a_n's. Specifically,

$$a_{n+2} = \frac{-(2n-4\gamma)}{(n+2)(n+1)}a_n \quad \text{for} \quad n \geq 1. \tag{10-84}$$

Or, equivalently,

$$a_n = \frac{-2(n-2-2\gamma)}{n(n-1)}a_{n-2} \quad \text{for} \quad n \geq 3. \tag{10-85}$$

Notice that a_n will be a multiple of a_0 for even values of n and a multiple of a_1 for odd values of n. Therefore, the series solution for ϕ becomes

$$\phi = a_0\eta^0 + a_1\eta^1 + \overbrace{a_0 2\gamma}^{a_2}\eta^2 + \underbrace{a_1\frac{-(1-2\gamma)}{3}}_{a_3}\eta^3 + \cdots + \overbrace{\frac{-2(n-2-2\gamma)}{n(n-1)}a_{n-2}}^{a_n}\eta^n + \cdots \tag{10-86}$$

Notice that a_0 and a_1 are arbitrary. Factoring out a_0 and a_1 and changing the indexing scheme to $i = 1, 2, \ldots$ yields

$$\phi = a_0\Phi(\eta) + a_1\Psi(\eta), \tag{10-87a}$$

where

$$\Phi(\eta) = c_0\eta^0 + c_1\eta^2 + \sum_{i=2}^{\infty} c_i\eta^{2i} \tag{10-87b}$$

with

$$c_0 = 1, \ c_1 = 2\gamma \tag{10-87c}$$

and

$$c_i = \frac{-2(i-1-\gamma)}{i(2i-1)}c_{i-i}. \tag{10-87d}$$

And

$$\Psi(\eta) = d_0\eta^1 + \sum_{i=1}^{\infty} d_i\eta^{2i+1} \tag{10-87e}$$

with

$$d_0 = 1 \quad \text{and} \quad d_i = \frac{-(2i-1-2\gamma)}{i(2i+1)}d_{i-i}. \tag{10-87f}$$

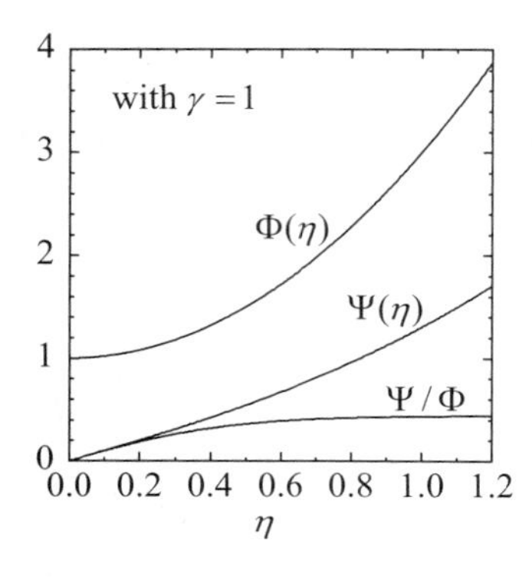

Figure 10-10 Plot of the series functions $\Phi(\eta)$ and $\Psi(\eta)$ defined by Eqs. (10-87b) and (10-87c), with $\gamma = 1$.

Figure 10-10 plots the functions $\Phi(\eta)$, $\Psi(\eta)$, and Ψ/Φ. Notice that $\Phi(0) = 1$, $\Psi(0) = 0$, and

$$(\Psi/\Phi)\big|_{\eta\to\infty} = \Gamma(\gamma), \tag{10-88}$$

where $\Gamma(\gamma)$ is a constant that depends on the value of γ. At $\eta = 0$, the solution for ϕ evaluates to $\phi(0) = a_0$. Since $\phi(0) = 1$, therefore $a_0 = 1$. The second boundary condition $\phi(\infty) = 0$ requires that $0 = \Phi(\infty) + a_1\Psi(\infty)$. Therefore, $a_1 = -(\Phi/\Psi)\big|_{\eta\to\infty} = -1/\Gamma(\gamma)$, and the final solution for ϕ becomes

$$\phi = \Phi(\eta) - \Psi(\eta)/\Gamma(\gamma). \tag{10-89}$$

The solution (10-87) for ϕ is plotted in Figure 10-11 for the special case when $\gamma = 1$ and $\Gamma(1) = 0.4431$.

The flux of species into the semi-infinite solid is expressed by Fick's law:

$$\begin{aligned} J_A^* &= -Đ_A \frac{\partial c_A}{\partial x}\bigg|_{x=0} = -Đ_A c_o(t) \frac{\partial \phi}{\partial x}\bigg|_{x=0} \\ &= -Đ_A b t^{\gamma} \frac{\partial \eta}{\partial x}\frac{\partial \phi}{\partial \eta}\bigg|_{\eta=0} = -\frac{b}{2}\, Đ_A^{1/2}\, t^{\gamma-1/2}\phi'(0). \end{aligned} \tag{10-90}$$

Since $\Phi'(0) = 0$ and $\Psi'(0) = d_0$, the derivative of the solution (10-89) becomes $\phi'(0) = -1/\Gamma(\gamma)$, and

$$J_A^* = \frac{b}{2}\, Đ_A^{1/2}\, \frac{t^{\gamma-1/2}}{\Gamma(\gamma)} \tag{10-91}$$

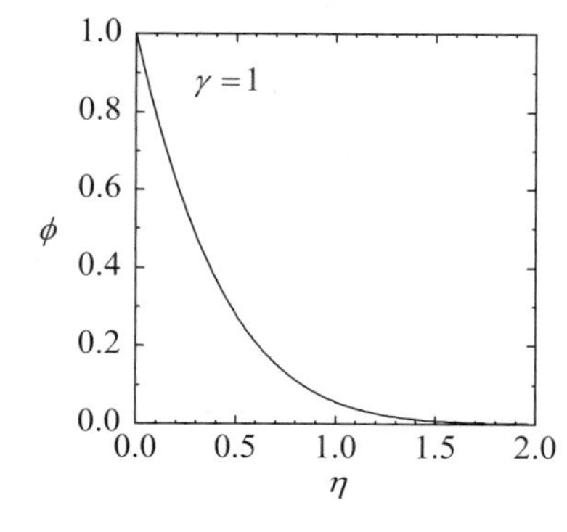

Figure 10-11 Plot of ϕ defined by Eq. (10-66), with $\gamma = 1$.

where

$$\Gamma(\gamma) = (\Psi/\Phi)\Big|_{\eta\to\infty} \tag{10-92}$$

must be evaluated for each case of γ.

10.6 PROBLEMS

10-1 Consider the diffusion problem shown. Attempt a similarity solution and show where this approach fails. Is there any period in time when a similarity solution exists for this problem?

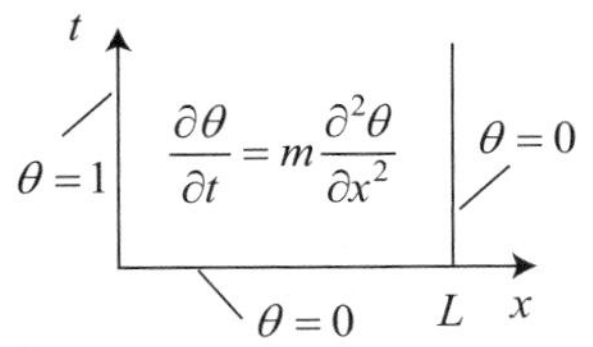

t

$\theta = 1$ $\quad \dfrac{\partial\theta}{\partial t} = m\dfrac{\partial^2\theta}{\partial x^2}$ $\quad \theta = 0$

$\theta = 0$ $\quad L \quad x$

10-2 Consider the problem of laser heating a semi-infinite solid, as discussed in Section 10.2. However, now consider a time-varying laser intensity $I_o(t)$. For what function of time $I_o(t)$ can a similarity solution be found for the temperature field, such that $\theta(t,x) \to \theta(\eta)$. Determine η, and solve for $\theta(\eta)$. Plot the solution for $\theta(t,x)$. What other problem is $\theta(t,x)$ a solution to?

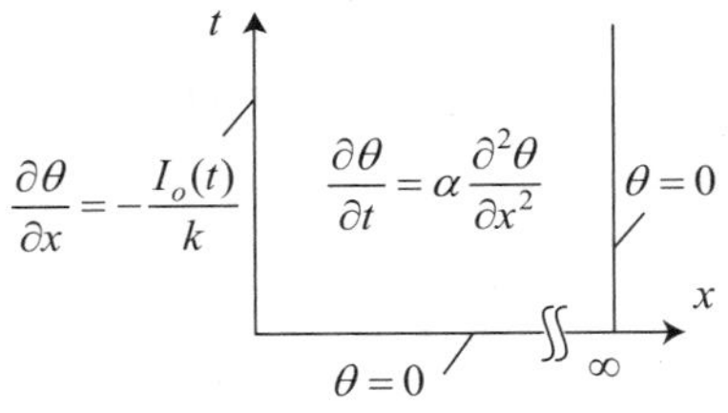

t

$\dfrac{\partial\theta}{\partial x} = -\dfrac{I_o(t)}{k}$ $\quad \dfrac{\partial\theta}{\partial t} = \alpha\dfrac{\partial^2\theta}{\partial x^2}$ $\quad \theta = 0$

x

$\theta = 0$ $\quad \infty$

10-3 A semi-infinite fluid is bounded by a large plate of area A. The massless plate is put into motion by a constant force F at $t = 0$, as illustrated. Suppose a solution to the flow exists, having the form

$$v_x(t,y) = \phi(t,y)\frac{F}{A}\frac{\sqrt{\nu t}}{\mu}.$$

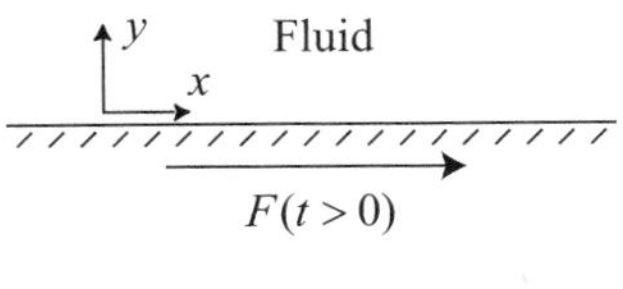

y Fluid

x

$F(t > 0)$

Determine the problem statement for $\phi(t,y)$. Is there a similarity solution to the problem for $\phi(t,y)$? Determine the solution for $\phi(t,y)$, and sketch the solution for $v_x(t,y)$.

10-4 To dope silicon with phosphorous, a spin-on glass with a concentration of phosphorous can be deposited on the surface of a silicon wafer and baked at high temperatures. Develop the governing equations and boundary conditions for the phosphorous transfer between the glass and the silicon materials, assuming that the phosphorous concentration in the glass is initially

$c(t, x \leq 0) = c_o$, while the concentration in the silicon is initially $c(t, x \geq 0) = 0$. The phosphorous diffusivity values for glass and silicon are different. Show that a similarity solution to this problem can be found so long as the glass layer and silicon wafer are "sufficiently" thick. What limitation on time scale does this assumption impose? Solve this problem for the phosphorous concentration profiles in the glass and silicon materials. Find a relation for the phosphorous concentration at the interface between the glass and silicon.

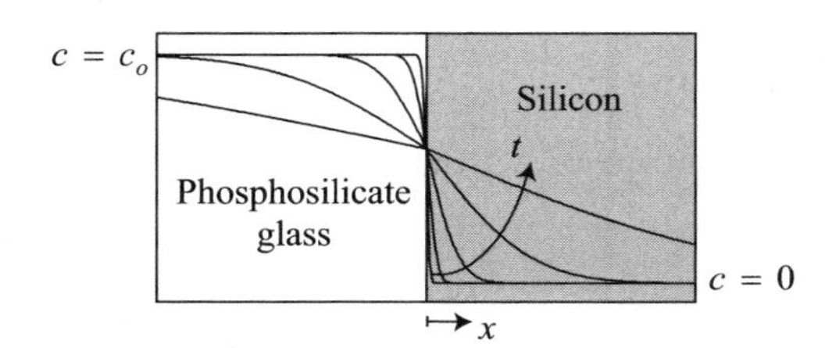

10-5 Consider the equation

$$\left(\frac{\partial^2 f}{\partial x \partial y} + \frac{1}{2x}\frac{\partial f}{\partial y}\right)\frac{\partial f}{\partial y} = \sqrt{\frac{\nu}{xU}}\frac{\partial^3 f}{\partial y^3} + \left(\frac{\partial f}{\partial x} + \frac{1}{2x}f\right)\frac{\partial^2 f}{\partial y^2},$$

where $f(x,y)$ is a dimensionless function, and U and ν are constants with dimensions of velocity and diffusivity, respectively. Propose a similarity variable η to transform this partial differential equation for $f(x,y)$ into an ordinary differential equation for $f(\eta)$. Find the ordinary differential equation for $f(\eta)$. This final equation will arise in the solution of a viscous flow over a flat plate developed in Chapter 25.

10-6 Consider the equation

$$y\frac{\partial \theta}{\partial z} = \beta\frac{\partial^2 \theta}{\partial y^2},$$

subject to the boundary conditions

$$\theta(z=0) = 0, \quad \theta(y=0) = 1, \quad \text{and} \quad \theta(y \to \infty) = 0.$$

Show that a similarity variable can be found to transform the governing equation into

$$\frac{\partial^2 \theta}{\partial \eta^2} + 3\eta^2\frac{\partial \theta}{\partial \eta} = 0.$$

Solve this equation and find an expression for the gradient $\partial\theta/\partial y$ as a function of z along the surface $y = 0$.

REFERENCE

[1] J. H. Arnold, "Studies in Diffusion: III Unsteady-State Vaporization and Adsorption," *Transactions of the American Institute of Chemical Engineers*, **40**, 361 (1944).

Chapter 11

Superposition of Solutions

11.1 Superposition in Time
11.2 Superposition in Space
11.3 Problems

Superposition is a powerful technique by which solutions to complex problems are constructed from relatively simpler problems. Superposition was employed previously in the separation-of-variables techniques developed in Chapters 7 through 9 to build nonhomogeneous initial or boundary conditions. Additionally, superposition was employed in Chapter 8 to decompose linear problems containing multiple nonhomogeneous conditions into simpler problems, each having only one nonhomogeneous condition. The solutions to these simpler problems were superimposed to yield the solution to the original complex problem.

The principle of superposition is extended in this chapter to solve problems having complex temporal conditions, through the use of Duhamel's theorem, and to utilize simple solutions to construct spatial boundaries of more complex problems, by product-superposition and the method of images.

11.1 SUPERPOSITION IN TIME

Consider an extension of the problem addressed in Section 10.2 in which a semi-infinite solid is heated with a laser intensity I_o. Now suppose that the laser is turned off at some time τ_p, as shown in Figure 11-1(b). For $t \leq \tau_p$, the solution is the same as the situation addressed in Section 10.2, where the laser is never turned off.

After the laser is off, it can be assumed that the surface of the semi-infinite solid is adiabatic. A solution for $t > \tau_p$ can be found most simply by superposition of two related

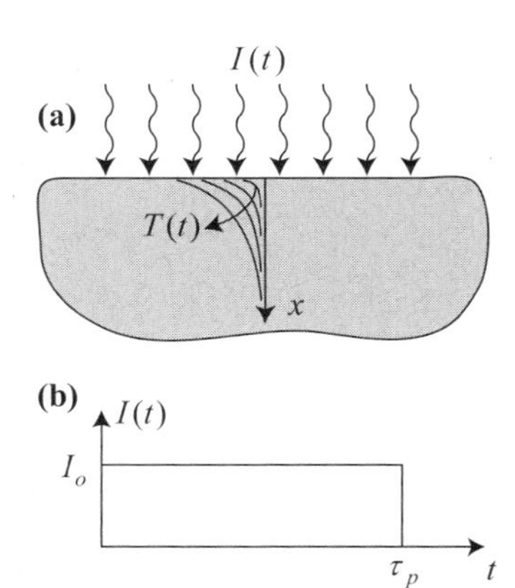

Figure 11-1 Heating of a semi-infinite solid with a finite laser pulse duration.

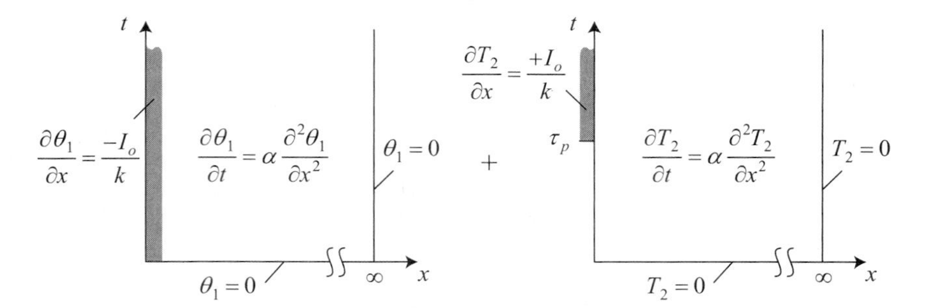

Figure 11-2 Superposition of two problems with different surface boundary conditions.

problems θ_1 and T_2, as shown in Figure 11-2. Superposition is possible because of the linear nature of the problem. As in Section 10.2, $\theta_1 = T_1 - T_o$ is the temperature rise relative to the initial temperature T_o of the solid.

Notice that the second solution T_2 is not "turned on" until $t > \tau_p$, and has a heat flux boundary condition that is opposite in sign to the problem for θ_1. Therefore, when the two solutions are superimposed for $t > \tau_p$, the result is an adiabatic surface condition. From the analysis of Section 10.2, the solution for the first problem θ_1 is known:

$$\theta_1 = \frac{I_o}{k}\left[\sqrt{\frac{4\alpha t}{\pi}}\exp\left(-\frac{x^2}{4\alpha t}\right) - x \cdot erfc\left(\frac{x}{\sqrt{4\alpha t}}\right)\right]. \tag{11-1}$$

The solution for the second problem $T_2(t > \tau_p)$ is obtained by an inspection of Eq. (11-1) for θ_1. By changing the sign on I_o and offsetting time by τ_p, the result for θ_1 is cast into a solution for

$$T_2 = \frac{-I_o}{k}\left[\sqrt{\frac{4\alpha(t-\tau_p)}{\pi}}\exp\left(-\frac{x^2}{4\alpha(t-\tau_p)}\right) - x \cdot erfc\left(\frac{x}{\sqrt{4\alpha(t-\tau_p)}}\right)\right]. \tag{11-2}$$

Finally, the solution to the original problem for $\theta = T - T_o$ is determined from superposition of the solutions for θ_1 and T_2:

$$\theta = \begin{cases} \theta_1 & t \leq \tau_p \\ \theta_1 + T_2 & t > \tau_p \end{cases}. \tag{11-3}$$

To illustrate the solution, τ_p is selected such that $\sqrt{\alpha\tau_p}/L = 2$. Then the solution during heating is the same as illustrated previously in Figure 10-8. For times $t > \tau_p$, the solution during cooling is illustrated in Figure 11-3.

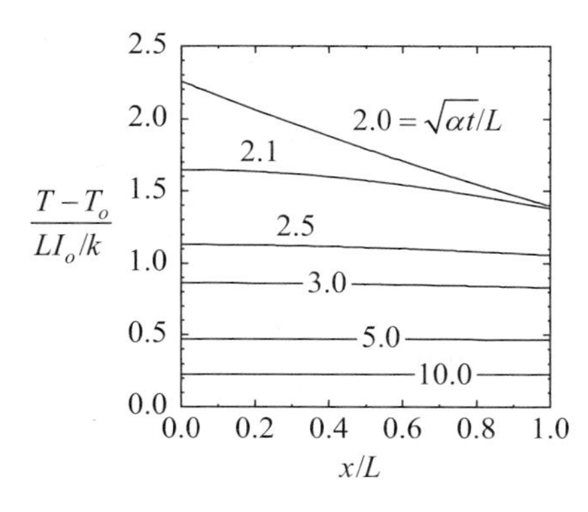

Figure 11-3 Temperature profiles in the semi-infinite solid for time greater than τ_p.

11.1.1 Duhamel's Theorem

Duhamel's theorem makes use of the principle of superposition to build temporal solutions to problems with arbitrary time-varying conditions. To illustrate, consider the last example of laser heating of a semi-infinite solid, but let the laser intensity grow linearly in time from zero. A series of discrete steps can approximate the temporal increase in laser intensity as shown in Figure 11-4.

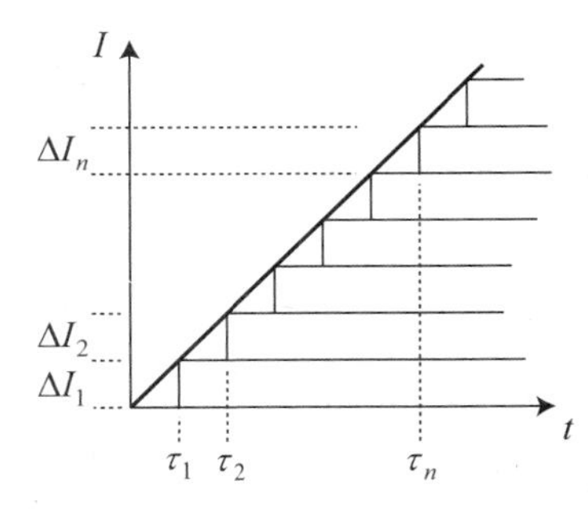

Figure 11-4 A linear function represented by a series of unit steps.

For a unit step in laser intensity ($\Delta I_n = 1$), the "unit-response" of the temperature field $S(x,t) = T(x,t) - T_o$ is known from the analysis in Section 10.2. By equating $I_o = 1$ in Eq. (10-29) and offsetting time by τ_n, the unit-response to laser heating is expressed by

$$S(x, t-\tau_n) = \frac{1}{k}\left[\sqrt{\frac{4\alpha(t-\tau_n)}{\pi}}\exp\left(-\frac{x^2}{4\alpha(t-\tau_n)}\right) - x\cdot erfc\left(\frac{x}{\sqrt{4\alpha(t-\tau_n)}}\right)\right]. \tag{11-4}$$

Therefore, the temperature field resulting from a single step in laser intensity ΔI_1, occurring at time $t = \tau_1$, is given by $\Delta I_1 S(x, t-\tau_1)$. By extension, a solution to the temperature field that develops from a series of laser intensity steps can be built by superposition:

$$\theta = \sum_n \Delta I_n S(x, t-\tau_n), \tag{11-5}$$

where ΔI_n is each incremental change in laser intensity occurring at time τ_n.

Suppose an extension to this superposition technique is desired for a continuously changing laser intensity $I(t)$. Multiplying and dividing the expression given in Eq. (11-5) by $\Delta\tau$, one obtains

$$\theta = \sum_n \frac{\Delta I_n}{\Delta\tau} S(x, t-\tau_n)\Delta\tau. \tag{11-6}$$

In the limit that $\Delta\tau \to 0$, Eq. (11-6) becomes

$$\theta = \int_0^t \frac{dI}{d\tau} S(x, t-\tau)d\tau. \tag{11-7}$$

If I is not continuous, as illustrated in Figure 11-5, then the solution for θ can be expressed as

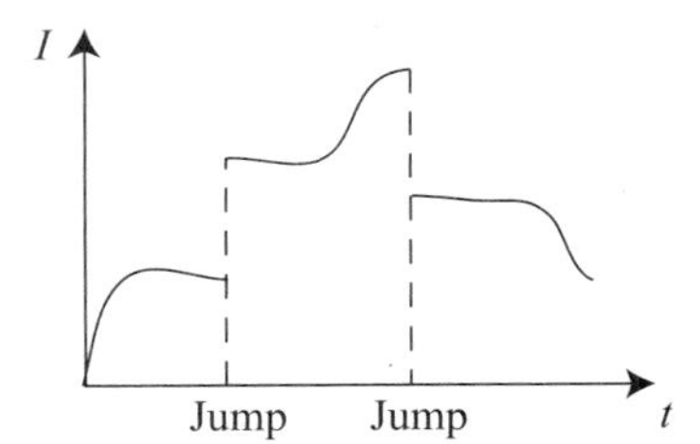

Figure 11-5 A discontinuous function.

$$\theta = \int_0^t \frac{dI}{d\tau} S(x, t-\tau) d\tau + \underbrace{\sum_{i=1}^{N} \Delta I_i S(x, t-\tau_i)}_{for\ N\ jumps} . \tag{11-8}$$

The ability to construct a temporal solution with Eq. (11-8) using the unit-response function is known as *Duhamel's theorem*.* Further application of Duhamel's theorem is illustrated in the next section.

11.1.2 Semi-Infinite Fluid Bounded by a Plate Set in Motion

Consider the problem of a semi-infinite fluid, initially at rest, that is bounded by a plate set into motion at $t = 0$, as shown in Figure 11-6. The motion of the plate is entirely in the x-direction, and does not displace the fluid in the y-direction. The fluid motion is governed by the x-direction momentum equation. However, since $v_y = 0$ and $\partial v_x / \partial x = 0$ for the problem at hand, the advection term in the momentum equation is zero. Furthermore, diffusion of momentum occurs only in the y-direction and there are no pressure gradients or body forces acting in the x-direction. Consequently, the momentum equation for this problem simplifies to

$$\frac{\partial v_x}{\partial t} = \nu \frac{\partial^2 v_x}{\partial y^2} , \tag{11-9}$$

which is subject to the conditions

$$v_x(y, t=0) = 0,\ v_x(y=0, t) = U(t), \quad \text{and} \quad v_x(y \to \infty, t) = 0 . \tag{11-10}$$

The complexity of the surface boundary condition suggests application of Duhamel's theorem. The flow solution is expressed with Duhamel's theorem using the unit-response solution $S(y, t)$ and including one discontinuous jump at $t = t_1$ of $-U_1$:

$$v_x(y, t) = \int_0^t \frac{dU}{d\tau} S(y, t-\tau) d\tau \overbrace{- U_1 S(y, t-t_1)}^{t>t_1} . \tag{11-11}$$

(a)
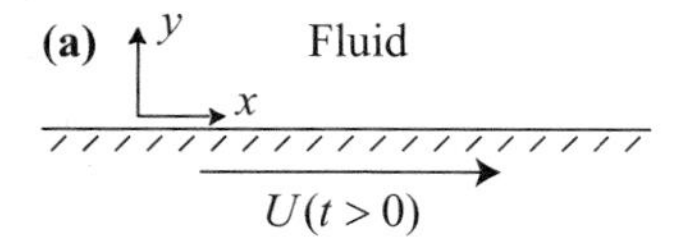

(b)
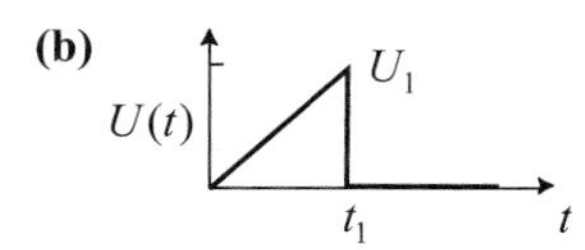

Figure 11-6 A semi-infinite fluid bounded by a plate in motion.

*This theorem is named in honor of the French mathematician Jean-Marie Constant Duhamel (1797–1872).

$S(y,t)$ is the solution to the unit problem for which the original surface boundary condition is replaced with the unit step function $v_x(y=0)=1$. Therefore, the unit-response solution must satisfy

$$\frac{\partial S}{\partial t}=\nu\frac{\partial^2 S}{\partial y^2}, \tag{11-12}$$

subject to

$$S(y,t=0)=0,\quad S(y=0,t)=1,\quad \text{and}\quad S(y\to\infty,t)=0. \tag{11-13}$$

This problem can be solved by the similarity solution method. Let $\eta=y/\sqrt{4\nu t}$, such that

$$\frac{\partial S}{\partial t}=\frac{-\eta}{2t}\frac{\partial S}{\partial \eta}\quad\text{and}\quad\frac{\partial^2 S}{\partial y^2}=\left(\frac{\eta}{y}\right)^2\frac{\partial^2 S}{\partial \eta^2}. \tag{11-14}$$

The governing equation is transformed to the similarity variable, yielding

$$\frac{\partial^2 S}{\partial \eta^2}+2\eta\frac{\partial S}{\partial \eta}=0. \tag{11-15}$$

The initial and boundary conditions become

$$\begin{array}{rclcl} y=0 & : & \eta=0 & : & S=1 \\ y\to\infty & : & \eta\to\infty & : & S=0 \\ t=0 & : & \eta\to\infty & : & S=0 \end{array}\Bigg\}\ \text{two conditions collapse.} \tag{11-16}$$

A solution was found to a mathematically equivalent problem in Section 10.2 for a diffusion heat flux q. Drawing upon that experience, the solution for S is recognized to be

$$S=erfc\left(\frac{y}{\sqrt{4\nu t}}\right). \tag{11-17}$$

Using the unit-response solution, Eq. (11-11) can be evaluated. Notice that for $t\leq t_1$: $\partial U/\partial t=U_1/t_1$, and that for $t>t_1$: $\partial U/\partial t=0$. Therefore Eq. (11-11) evaluates to

$$v_x(y,t)=\begin{cases}\displaystyle\int_0^t (U_1/t_1)S(y,t-\tau)d\tau & t\leq t_1\\ \displaystyle\int_0^{t_1}(U_1/t_1)S(y,t-\tau)d\tau+\int_{t_1}^t 0\,d\tau-U_1S(y,t-t_1) & t>t_1.\end{cases} \tag{11-18}$$

Let $t^*=t/t_1$ and $y^*=y/\sqrt{4\nu t_1}$. Then the solution for the flow can be expressed in a dimensionless form:

$$\frac{v_x(y,t)}{U_1}=\begin{cases}\displaystyle\int_0^{t^*} S(y^*,t^*-\tau)d\tau^* & t^*\leq 1\\ \displaystyle\int_0^{1} S(y^*,t^*-\tau^*)d\tau^*-S(y^*,t^*-1) & t^*>1.\end{cases} \tag{11-19}$$

The integral of the step function solution evaluates to

$$\int_0^{t^*} S(y^*, t^* - \tau^*) d\tau^* = \int_0^{t^*} erfc\left(\frac{y^*}{\sqrt{t^* - \tau^*}}\right) d\tau^*$$

$$= \left[\begin{array}{l} 2\exp\left(\frac{-(y^*)^2}{t^* - \tau^*}\right)\sqrt{\frac{t^* - \tau^*}{\pi}} y^* + \tau^* \\ + \left(2(y^*)^2 + t^* - \tau^*\right) erf\left(\frac{y^*}{\sqrt{t^* - \tau^*}}\right) \end{array} \right]_0^{t^*} \tag{11-20}$$

such that the solution for the flow becomes

$$\frac{v_x(y^*, t^* \le 1)}{U_1} = \left(2(y^*)^2 + t^*\right) erfc\left(\frac{y^*}{\sqrt{t^*}}\right) - 2y^* \sqrt{\frac{t^*}{\pi}} \exp\left(\frac{-(y^*)^2}{t^*}\right) \tag{11-21}$$

and

$$\begin{aligned} \frac{v_x(y^*, t^* > 1)}{U_1} = \frac{2y^*}{\sqrt{\pi}} &\left[\left(\sqrt{t^* - 1}\right) \exp\left(\frac{-(y^*)^2}{t^* - 1}\right) - \sqrt{t^*} \exp\left(\frac{-(y^*)^2}{t^*}\right) \right] \\ &+ \left(2(y^*)^2 + t^*\right) erfc\left(\frac{y^*}{\sqrt{t^*}}\right) - \left(2(y^*)^2 + t^* - 1\right) erfc\left(\frac{y^*}{\sqrt{t^* - 1}}\right) \end{aligned} \tag{11-22}$$

The flow solution is plotted in Figure 11-7.

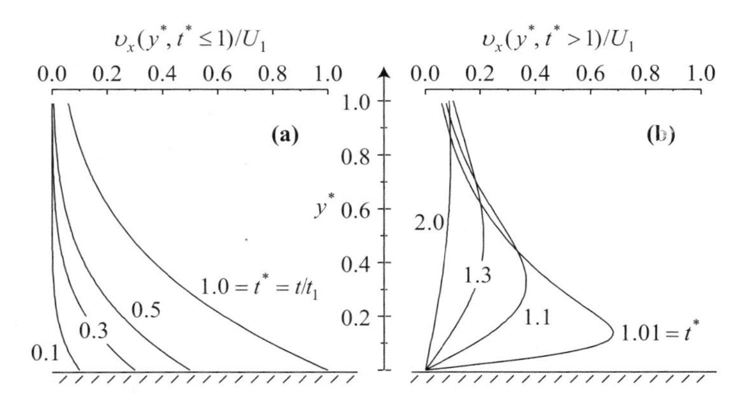

Figure 11-7 Velocity profiles in semi-infinite fluid bounded by a plate in motion.

11.2 SUPERPOSITION IN SPACE

Two methods of superposition in space are discussed in this section. The first, product-superposition of solutions, formulates a solution to a multidimensional problem through the product of two or more lower dimensional solutions. The second, method of images, utilizes superposition of simple solutions over an infinite domain to create domain boundaries.

11.2.1 Product-Superposition of Solutions

Transient problems of more than one spatial dimension can sometimes be solved by the simple *product-superposition of solutions* of one-dimensional problems. For this technique to work, the governing equation and boundary conditions of the problem should be homogeneous and the function describing the initial condition must be expressible by a product of functions, each dependent on only one spatial variable, such as $F(r,z) = g(r)h(z)$.

To illustrate the method, consider a disk, initially at a uniform temperature T_o, being quenched in a liquid bath, as shown in Figure 11-8. The disk has a radius of r_o and a thickness of $2L$. Heat loss to the surrounding fluid at T_∞ is described by a constant convection coefficient h. Defining the temperature variable:

$$\phi = \frac{T - T_\infty}{T_o - T_\infty}, \tag{11-23}$$

heat transfer in the disk is governed by the transient diffusion equation,

$$\begin{array}{ccc} \partial_o \phi & = & \alpha\, \partial_j \partial_j \phi \\ \downarrow & & \downarrow \quad = 0 \\ \dfrac{\partial \phi}{\partial t} & = \alpha & \left[\dfrac{1}{r}\dfrac{\partial}{\partial r}\left(r \dfrac{\partial \phi}{\partial r} \right) + \overbrace{\dfrac{1}{r^2}\dfrac{\partial^2 \phi}{\partial \theta^2}} + \dfrac{\partial^2 \phi}{\partial z^2} \right] \end{array} \tag{11-24}$$

and is subject to the boundary conditions

$$\left[\frac{\partial \phi}{\partial r} = 0 \right]_{r=0}, \quad \left[-k\frac{\partial \phi}{\partial r} = h\phi \right]_{r=r_o}, \tag{11-25}$$

$$\left[\frac{\partial \phi}{\partial z} = 0 \right]_{z=0}, \quad \left[-k\frac{\partial \phi}{\partial z} = h\phi \right]_{z=L}, \tag{11-26}$$

and the initial condition

$$\phi(t=0) = 1. \tag{11-27}$$

A solution is sought with the form $\phi = \phi_1(t,r)\phi_2(t,z)$. Upon substitution $\phi = \phi_1\phi_2$ into the governing equation, it is observed that

$$\phi_2 \left(\frac{\partial \phi_1}{\partial t} - \frac{\alpha}{r}\frac{\partial}{\partial r}\left(r\frac{\partial \phi_1}{\partial r} \right) \right) + \phi_1 \left(\frac{\partial \phi_2}{\partial t} - \alpha \frac{\partial^2 \phi_2}{\partial z^2} \right) = 0. \tag{11-28}$$

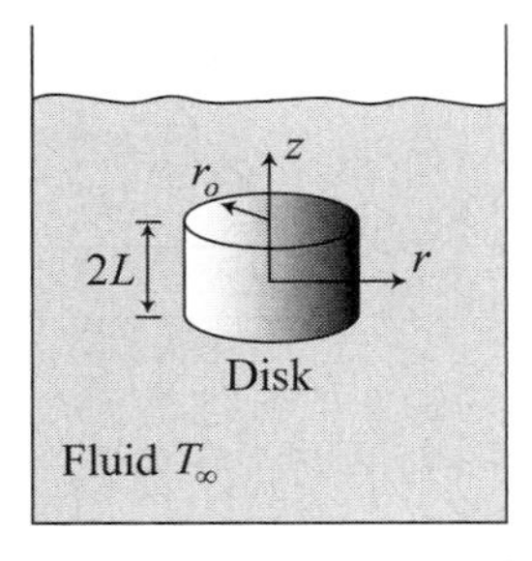

Figure 11-8 A disk quenched in a fluid bath.

Furthermore, upon substitution $\phi = \phi_1\phi_2$ into the boundary conditions, it is observed that

$$\left[\phi_2\left(\frac{\partial \phi_1}{\partial r} = 0\right)\right]_{r=0}, \quad \left[\phi_2\left(-k\frac{\partial \phi_1}{\partial r} = h\phi_1\right)\right]_{r=r_o}, \tag{11-29}$$

$$\left[\phi_1\left(\frac{\partial \phi_2}{\partial z} = 0\right)\right]_{z=0}, \quad \text{and} \quad \left[\phi_1\left(-k\frac{\partial \phi_2}{\partial z} = h\phi_2\right)\right]_{z=L}. \tag{11-30}$$

Finally, upon substitution $\phi = \phi_1\phi_2$ into the initial condition, it is observed that

$$\phi_1(t=0)\phi_2(t=0) = 1. \tag{11-31}$$

Inspection of the governing equation, boundary conditions, and initial condition, described in terms of $\phi_1(t,r)$ and $\phi_2(t,z)$, reveals that a solution can be found if the functions $\phi_1(t,r)$ and $\phi_2(t,z)$ satisfy the problems defined by

$$\frac{\partial \phi_1}{\partial t} = \frac{\alpha}{r}\frac{\partial}{\partial r}\left(r\frac{\partial \phi_1}{\partial r}\right), \quad \text{with} \quad \left[\frac{\partial \phi_1}{\partial r} = 0\right]_{r=0}, \quad \left[-k\frac{\partial \phi_1}{\partial r} = h\phi_1\right]_{r=r_o}, \quad \text{and} \quad \phi_1(t=0) = 1 \tag{11-32}$$

$$\frac{\partial \phi_2}{\partial t} = \alpha\frac{\partial^2 \phi_2}{\partial z^2}, \quad \text{with} \quad \left[\frac{\partial \phi_2}{\partial z} = 0\right]_{z=0}, \quad \left[-k\frac{\partial \phi_2}{\partial z} = h\phi_2\right]_{z=L}, \quad \text{and} \quad \phi_2(t=0) = 1. \tag{11-33}$$

These problems are simpler transient one-dimensional problems describing heat transfer from a cylinder and heat transfer from a slab, as illustrated in Figure 11-9. Standard separation of variables can be performed to demonstrate that

$$\phi_1 = 2\sum_{n=1}^{\infty} \frac{J_0(\lambda_n r)J_1(\lambda_n r_o)}{J_0^2(\lambda_n r_o) + J_1^2(\lambda_n r_o)} \frac{\exp(-\lambda_n^2 \alpha t)}{\lambda_n r_o}, \tag{11-34}$$

where λ_n are the roots of

$$\lambda_n r_o \frac{J_1(\lambda_n r_o)}{J_0(\lambda_n r_o)} = \frac{h r_o}{k}, \tag{11-35}$$

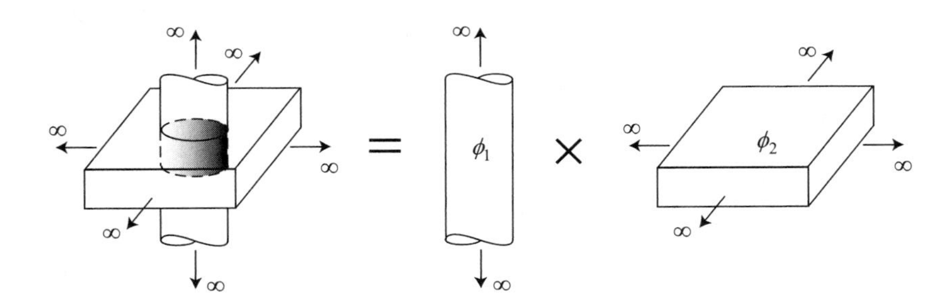

Figure 11-9 Product-superposition of solutions of two one-dimensional problems.

and

$$\phi_2 = 2\sum_{n=1}^{\infty} \frac{\sin(\gamma_n L)\cos(\gamma_n x)}{\gamma_n L + \sin(\gamma_n L)\cos(\gamma_n L)} \exp(-\gamma_n^2 \alpha t), \tag{11-36}$$

where γ_n are the roots of

$$\gamma_n L \tan(\gamma_n L) = \frac{hL}{k}. \tag{11-37}$$

Therefore, the solution to the temperature field in the disk is constructed from

$$\phi = \phi_1 \phi_2 = 4\left(\sum_{n=1}^{\infty} \frac{J_0(\lambda_n r) J_1(\lambda_n r_o)\exp(-\lambda_n^2 \alpha t)}{\lambda_n r_o [J_0^2(\lambda_n r_o) + J_1^2(\lambda_n r_o)]}\right)\left(\sum_{n=1}^{\infty} \frac{\sin(\gamma_n L)\cos(\gamma_n x)\exp(-\gamma_n^2 \alpha t)}{\gamma_n L + \sin(\gamma_n L)\cos(\gamma_n L)}\right). \tag{11-38}$$

11.2.2 Method of Images

The principal utility of the *method of images* is to replicate boundary conditions on planar surfaces by the superposition of related solutions. For example, the method of images can be used to create an adiabatic surface in a heat transfer problem or an impenetrable boundary in a flow problem. In this section, the method is illustrated for a viscous flow and later, in Chapter 15, the method will be applied to inviscid flows.

Consider the situation of an initially stationary fluid bounded by a surface that is set into constant motion U for $t \geq 0$, as illustrated in Figure 11-10(a). If the fluid were semi-infinite, the velocity solution $\upsilon_1(y, t)$ could be found using the similarity variable approach:

$$\upsilon_1(x, t) = U \cdot erfc\left(\frac{y}{\sqrt{4\nu t}}\right). \tag{11-39}$$

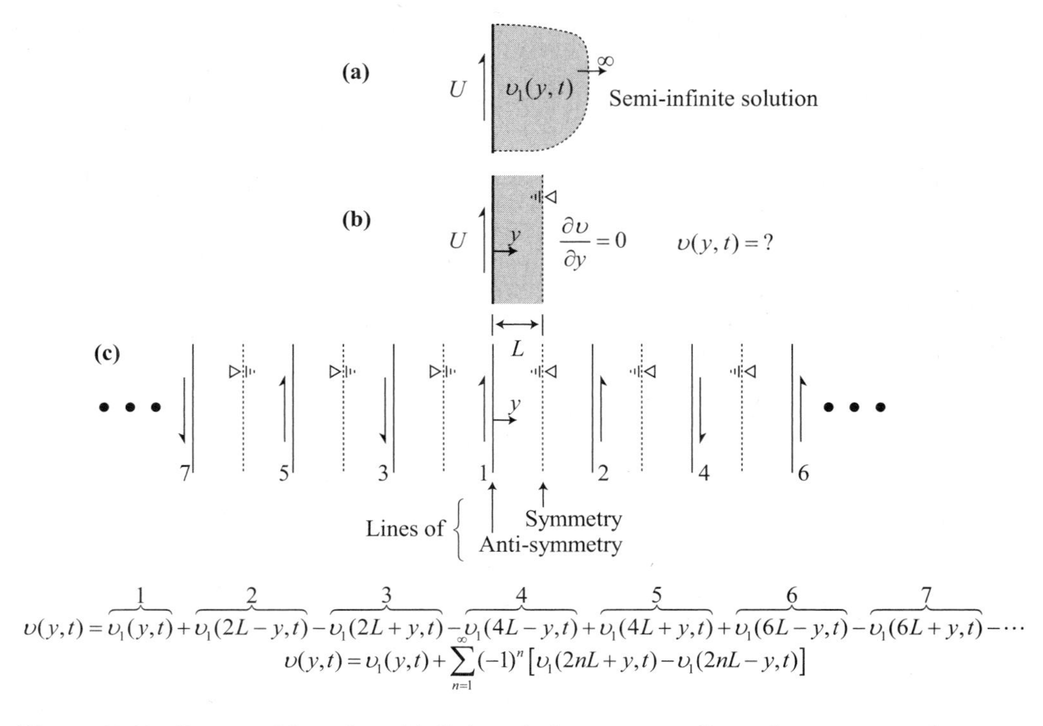

Figure 11-10 Superposition of semi-infinite solutions to create lines of symmetry and anti-symmetry.

However, if the fluid were bounded by a second surface, direct application of the similarity approach would fail. The separation-of-variables techniques could be used to find the velocity solution in the gap between the two surfaces. Alternatively, the semi-infinite fluid solution $v_1(y,t)$ could be employed to build the desired velocity solution in the gap by superposition.

To illustrate, suppose that at a distance L from the moving surface there is a free surface, through which no momentum transfer occurs, as depicted in Figure 11-10(b). To create the solution to this flow, superposition of related solutions is performed in Figure 11-10(c). The surface moving with positive speed U is located at 1 in Figure 11-10(c). If a second surface, also moving with a positive speed U, is placed at position 2, the superposition of these two problems creates a plane of symmetry midway between 1 and 2. Along this plane of symmetry $\partial v/\partial y = 0$, which is the desired condition for a free surface. Unfortunately, momentum from surface 2 eventually diffuses back to position 1, at which time the speed of 1 would begin to exceed U due to the momentum contribution from 2. To prevent this departure from the desired solution, a third surface with negative speed $-U$ is placed at position 3. Momentum transfer from surfaces 2 and 3 will cancel at 1, such that the speed of 1 remains U for all time; surface 1 is a plane of anti-symmetry in the solution. Unfortunately, with the addition of surface 3, the plane of symmetry between surfaces 1 and 2 (at the free surface) is broken. To restore symmetry at the free surface, surface 4 with negative speed $-U$ is added. However, now anti-symmetry in the solution is broken at surface 1, and surface 5 is added. Although it is impossible to simultaneously have anti-symmetry with respect to surface 1 and symmetry midway between surfaces 1 and 2, the process of adding surfaces at ever greater distances has a diminishing effect on the flow between surfaces 1 and 2. Therefore, with a sufficiently large but finite number of surfaces added to the original problem, the desired solution for the flow in the region $0 \leq y \leq L$ can be obtained.

The superposition process of semi-infinite fluid solution $v_1(y,t)$ is illustrated in Figure 11-10(c). Notice that each time the solution $v_1(y,t)$ is applied, the spatial variable y must be offset to the location of the plate that is being added. For example, plate 4 is added to the solution as $-v_1(4L - y, t)$, where the negative sign indicates that this plate is moving downward. The series solution for the flow is given by

$$v(y,t) = v_1(y,t) + \sum_{n=1}^{\infty} (-1)^n [v_1(2nL + y, t) - v_1(2nL - y, t)]. \tag{11-40}$$

The series solution is evaluated in Code 11-1 and plotted in Figure 11-11 for logarithmic steps in time. (For simplicity, the fluid diffusivity in Eq. (11-39) was taken to be $\nu = 1/4$.) Initially, only fluid in close proximity to the moving wall at $y = 0$ is in motion. Momentum diffusion causes the region of fluid motion to grow until it reaches the free surface. The

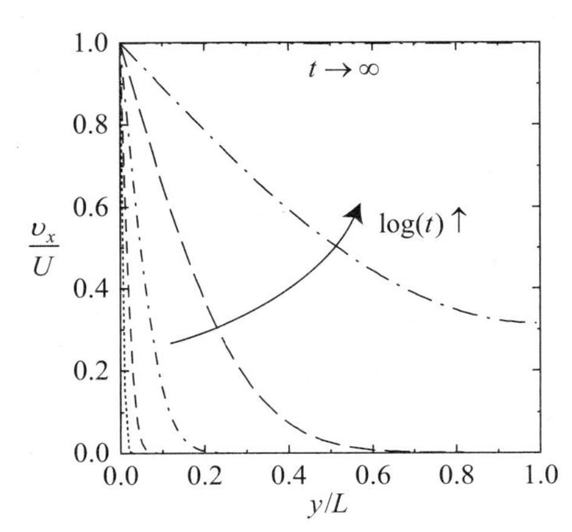

Figure 11-11 Velocity time evolution of fluid layer depicted in Figure 11-10(b).

Code 11-1 Fluid layer set in motion

```
#include <stdio.h>
#include <math.h>
int main ()
{
   int n;
   FILE *fp;
   double y,t,logt,vx;
   fp=fopen("vx.dat","w");
   for (logt=-4.;logt<=5.;logt+=1.) {
      t=pow(10.,logt);
      for (y=0.;y<=1.;y+=.01) {
         vx=erfc(y/sqrt(t));
         for (n=1;n<=1000;++n)
            vx+=pow(-1.,n)
               *(erfc((2.*n+y)/sqrt(t))
               -erfc((2.*n-y)/sqrt(t)));
         fprintf(fp,"%e %e\n",y,vx);
      }
      fprintf(fp,"\n");
   }
   fclose(fp);
   return 0;
}
```

velocity of the free surface increases until eventually the entire fluid layer $0 \leq y \leq L$ reaches a uniform speed U. It can be observed in Figure 11-11 that the free surface condition $\partial v/\partial y = 0$ is preserved throughout this process.

The use of pre-existing solutions is an attractive element of the current approach. However, since each term in the series has a spreading influence that continues to grow with time, the total number of terms in the series required for convergence will always increase with time. Therefore, although the limiting behavior expected of the solution as $t \rightarrow \infty$ can be observed in Figure 11-11 after a finite time and with a finite number of terms in the series, were time integration allowed to continue, the limiting behavior of $v(y, t \rightarrow \infty) = 1$ would eventually be lost without simultaneously increasing the number of terms in the series.

11.3 PROBLEMS

11-1 For a time-dependent laser intensity $I(t)$ incident on a semi-infinite solid, show that the solution for the surface temperature rise $(x = 0)$ as a function of time is

$$\theta_s(t) = \int_0^t \left\{ \frac{2(1-R)}{k} \sqrt{\frac{\alpha(t-\tau)}{\pi}} \frac{d}{d\tau}[I(\tau)] \right\} d\tau.$$

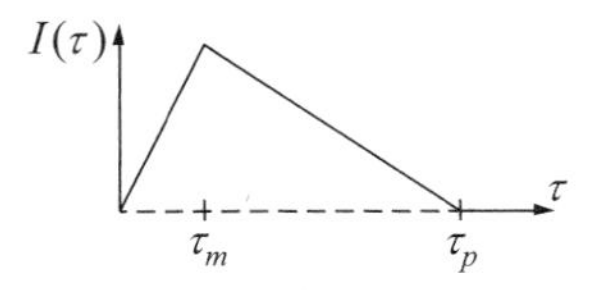

Note that the surface has reflectivity, such that the absorbed intensity is $(1 - R)I(t)$, and all energy absorption is at the surface (not over an optical penetration depth). Consider the situation where the laser pulse coming out of the laser is not a rectangular pulse, but is triangular, and show that for time $\tau_m < t < \tau_p$ that the surface temperature rise is given by

$$\theta_s(\tau_m < t < \tau_p) = \frac{8\phi(1-R)}{3k\tau_m\tau_p} \sqrt{\frac{\alpha}{\pi}} \left(t^{3/2} - \frac{\tau_p(t-\tau_m)^{3/2}}{\tau_p - \tau_m} \right).$$

ϕ is the laser fluence (total energy of the laser pulse per unit area). Finally, show that the maximum surface temperature rise is given by

$$\theta_{s,\max} = (T_{\max} - T_o) = \frac{8\phi(1-R)}{3k} \sqrt{\frac{\alpha/\pi}{2\tau_p - \tau_m}}.$$

11-2 The temperature distribution in a plane wall of thickness L that is being heat-treated with a laser having an optical penetration depth δ_{opt} (significantly less than L) was found in Problem (7-3). Upon nondimensionalizing that solution, one found that for a constant laser flux,

$$\theta(\eta, \tau) = (1 - \eta) + \Lambda\left[e^{-1/\Lambda} - e^{-\eta/\Lambda}\right] - \sum_{n=0}^{\infty} \frac{2(1 + \Lambda\lambda_n^* \exp(-1/\Lambda)(-1)^n)}{\lambda_n^{*2}(1 + \Lambda^2\lambda_n^{*2})} \cos(\lambda_n^* \eta) \exp(-\lambda_n^{*2} \tau)$$

where

$$\theta = \frac{T - T_b}{I_o L/k}, \quad \eta = x/L, \quad \Lambda = \delta_{opt}/L, \quad \tau = (\alpha/L^2)t, \quad \text{and} \quad \lambda_n^* = \lambda_n L = (n + 1/2)\pi.$$

Find an expression for the temperature field when the laser flux increases linearly with time: $I_o = B(t/t_B)$, B, and t_B are constants. For times $t_B < t \leq 2t_B$, the laser intensity decreases linearly until zero, and remains zero for $t > 2t_B$. In what time interval, $t < t_B$, $t_B < t \leq 2t_B$, or $t > 2t_B$, does the maximum temperature in the material occur? Derive a solution for the surface temperature and plot your result for the time interval $0 \leq t \leq 5t_B$.

11-3 Consider a solid sphere submerged in a well-stirred fluid bath. Initially, the fluid and sphere are isothermal at a temperature T_o. However, at $t = 0$ the fluid temperature begins to rise linearly in time, such that $T_B - T_o = ct$. Determine the temperature distribution in the sphere as a function of time, assuming that the surface of the sphere is given by the bath temperature T_B for all time. A related problem in which a sphere is quenched in a liquid bath was solved in Section 9.3.2.

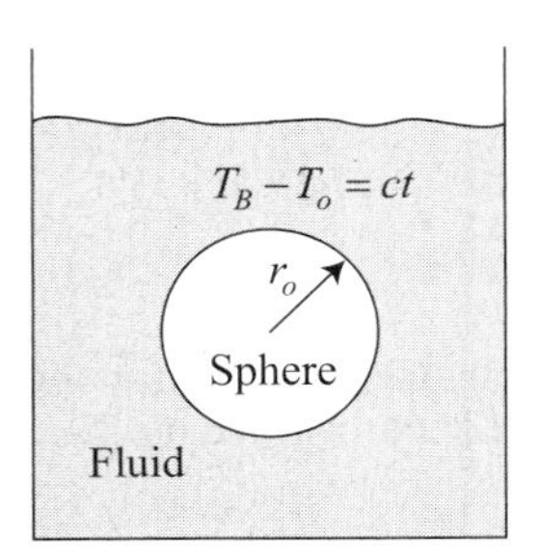

11-4 Plates of a material are heat-treated in the press illustrated below. Initially, the plate is at the ambient temperature T_o. At $t = 0$ the plate is pressed from above with a hot block at T_H. Because the thermal conductivity of the hot block is high, the interface temperature between the plate and the blocks can be assumed T_H. After some time τ_H, the plate is quenched by a cold block (also of high thermal conductivity) at temperatures of T_o.

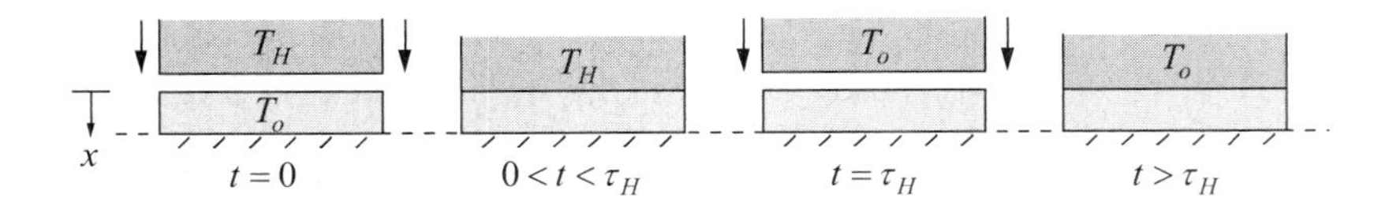

The thickness of the plate is w, and the plate backing is adiabatic. Through separation of variables, the transient temperature ($\theta \equiv T - T_o$) in the plate for $0 < t \leq \tau_H$ is found to be

$$\theta = \theta_H \left[1 - (4/\pi) \sum_{n=0}^{\infty} \frac{\sin(\lambda_n x)}{2n + 1} \exp(-\alpha\lambda_n^2 t)\right], \quad \text{where} \quad \lambda_n = (n + 0.5)\pi/w$$

and $\theta_H \equiv T_H - T_o$. Find the solution for $t > \tau_H$ using the principle of superposition.

11-5 A brick initially at a uniform temperature T_o is quenched in a liquid bath. The brick has dimensions of $2L \times 2H \times 2W$. Heat loss to the surrounding fluid at T_∞ is described by a constant convection coefficient h. Defining the temperature variable

$$\phi = \frac{T - T_\infty}{T_o - T_\infty},$$

derive a solution for the transient temperature distribution in the brick, $\phi(t, x, y, z)$.

11-6 Find the velocity in a fluid layer of thickness L. The fluid is backed by a stationary wall, and the front surface is subject to a constant speed $U(t > 0)$. Assume that the fluid layer is initially stationary, $v(t = 0, x) = 0$. Solve the problem for $v(t, x)$ by superposition, using the related solution $v_1(t, x)$ for a semi-infinite fluid. Plot the solution and demonstrate the correct limiting behavior for $t \rightarrow \infty$.

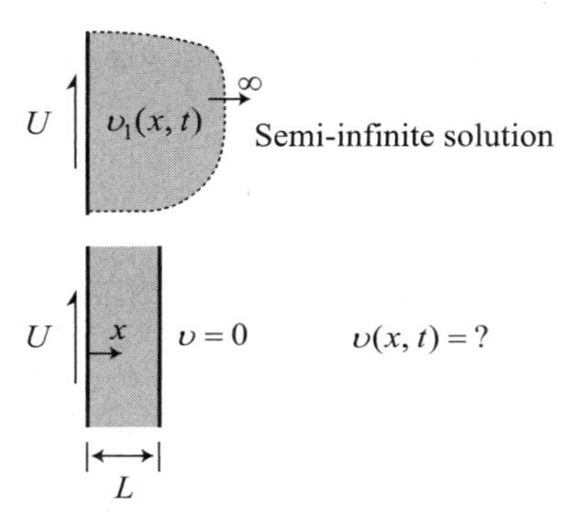

11-7 Find the temperature field in a plane wall of thickness L. The wall has an adiabatic back, and the front surface is subject to a constant heat flux $I_o(t > 0)$. Assume that the relative temperature of the plane wall is initially $\theta(t = 0, x) = 0$. Solve the problem for $\theta(t, x)$ by superposition, using the related solution $\theta_1(t, x)$ for a semi-infinite wall.

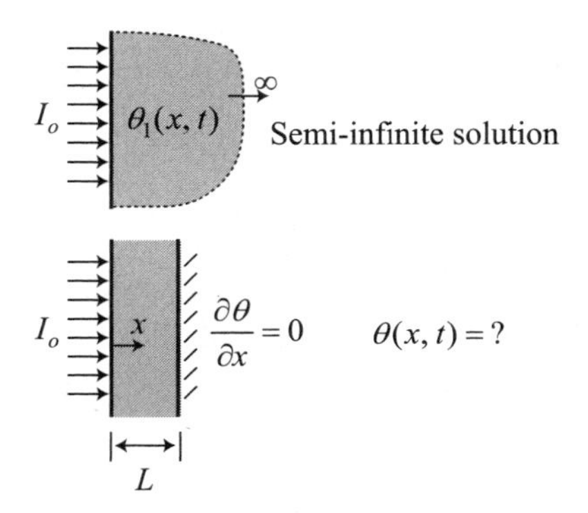

Chapter 12

Diffusion-Driven Boundaries

12.1 Thermal Oxidation

12.2 Solidification of an Undercooled Liquid

12.3 Solidification of a Binary Alloy from an Undercooled Liquid

12.4 Melting of a Solid Initially at the Melting Point

12.5 Problems

There are many examples of dynamic and kinetic processes where interfaces move in response to transport. Examples pertaining to advection transport include hydraulic jumps in open channel flows, treated in Chapter 18, and shockwaves in gas dynamics, treated in Chapter 20. Interface motion that arises in response to diffusion is related to the rate at which transported quantities are consumed (or produced) at the interface. Important processes that can occur at an interface include chemical reactions and phase change.

12.1 THERMAL OXIDATION

A simple but important illustration of a moving boundary problem is thermal oxidation. Silicon dioxide layers (SiO_2) can be grown on silicon wafers in order to form electrically insulating barriers in integrated circuits. Dry thermal oxidation of silicon occurs with a reaction between silicon (Si) and oxygen (O_2):

$$O_2 + Si \rightarrow SiO_2. \tag{12-1}$$

This reaction occurs at the Si-SiO_2 interface, as shown in Figure 12-1. Oxidation of Si is a *first-order reaction* because the reaction rate is proportional to the O_2 concentration. Consequently, the SiO_2 layer thickness $S(t)$ increases at a rate proportional to the concentration of O_2 at the interface:

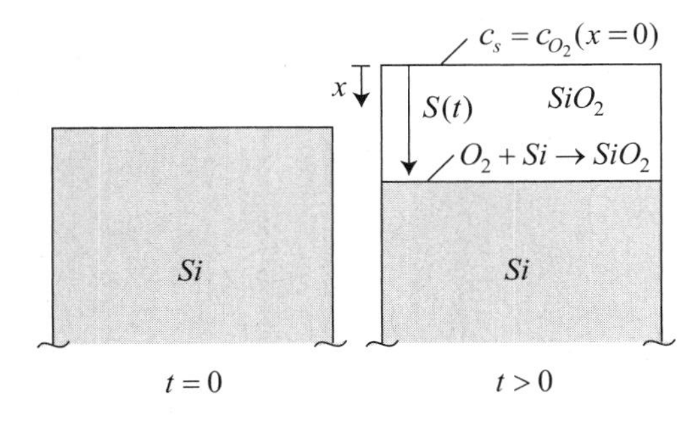

Figure 12-1 Thermal oxidation of silicon.

$$\frac{1}{\hat{v}}\frac{S(t)}{dt} = \kappa\, c_{O_2}(x = S). \tag{12-2}$$

where $\hat{v}$ is the molar volume (m^3/mol) of SiO_2, and κ is the first-order reaction rate coefficient. About 46% of the SiO_2 layer thickness $S(t)$ represents a reduction of the silicon wafer due to the reaction stoichiometry and the difference in densities between Si and SiO_2.

The flux of O_2 diffusing to the interface is balanced by the rate per unit area at which O_2 is being consumed by the reaction:

$$\left[-Đ_{O_2}\frac{dc_{O_2}}{dx} = \kappa c_{O_2}\right]_{x=S(t)}. \tag{12-3}$$

As the thickness of the SiO_2 layer increases, transport of O_2 through the layer is reduced and the reaction rate at the interface decreases accordingly. The transport of O_2 is governed by the diffusion equation:

$$0 \leq x < S(t): \quad \frac{\partial c_{O_2}}{\partial t} = Đ_{O_2}\frac{\partial^2 c_{O_2}}{\partial x^2} \quad (\approx 0). \tag{12-4}$$

It is reasonable to adopt a quasi-steady-state description of the diffusion process ($\partial c_{O_2}/\partial t \approx 0$), if the characteristic time scale for O_2 diffusion through the SiO_2 layer is much less than the overall time scale of the oxidation process. With this simplification, and using the boundary conditions

$$c_s = c_{O_2}(x = 0) \quad \text{and} \quad c_{O_2}(x = S), \tag{12-5}$$

the linear concentration of O_2 through the SiO_2 layer is easily determined:

$$c_{O_2}(x, t) = c_s - \left(\frac{c_s - c_{O_2}(x = S)}{S(t)}\right)x. \tag{12-6}$$

Applying this solution (12-6) to the interface condition (12-3) allows the interface concentration of O_2 to be evaluated as a function of the SiO_2 layer thickness:

$$c_{O_2}(x = S) = \frac{c_s}{1 + \kappa S/Đ_{O_2}}. \tag{12-7}$$

Applying this result to Eq. (12-2) yields

$$\frac{S(t)}{dt} = \frac{\hat{v}\,\kappa c_s}{1 + \kappa S/Đ_{O_2}}. \tag{12-8}$$

Integrating for the SiO_2 layer thickness, with the initial condition $S(t = 0) = 0$, yields the result

$$t = \frac{S}{\hat{v}\,\kappa c_s} + \frac{S^2}{2\hat{v}\,Đ_{O_2}\,c_s}, \tag{12-9}$$

which is the *Deal-Grove model* of thermal oxidation [1]. From this result, it is revealed that initially the SiO_2 layer thickness increases linearly with time

$$S \sim (\hat{v}\,\kappa c_s)\,t \quad \text{(reaction-limited)}, \tag{12-10}$$

while the growth rate is limited by the interface reaction kinetics, and eventually scales with the square root of time

$$S \sim \sqrt{2\hat{\nu} D_{O_2} c_S t} \quad \text{(diffusion-limited)} \tag{12-11}$$

when the growth rate is limited by the O_2 diffusion rate through the SiO_2 layer.

12.2 SOLIDIFICATION OF AN UNDERCOOLED LIQUID

The Stefan problem [2] is associated with the task of tracking a phase-change boundary within a transient temperature field. For example, consider solidification of a pure undercooled liquid having an initial liquid temperature T_o that is below the melting point value T_m, as illustrated in Figure 12-2. The surface at $x = 0$ is adiabatic, meaning that there is no heat transfer from this surface. The instantaneous velocity of the solidification front is related to the rate with which the latent heat Δh_{ls} released by solidification is transported away from the liquid–solid interface. Therefore, with the location of the front denoted by $S(t)$, the propagation speed of solidification into the liquid is governed by the energy balance:

$$\overbrace{\rho_l \Delta h_{ls} \frac{dS}{dt}}^{\text{heat produced}} = \left. \left(\overbrace{k_s \frac{\partial T_s}{\partial x} - k_l \frac{\partial T_l}{\partial x}}^{\text{heat removed}} \right) \right|_{x=S(t)}. \tag{12-12}$$

The subscripts "s" and "l" are used to denote values in the solid and liquid phases, respectively. This equation is known as the *Stefan condition*.* In the current analysis, it is assumed for simplicity that the position of the interface $S(t)$ is not influenced by changes in density associated with either phase change or sensible heating.

Heat transfer in the liquid and solid phases is governed by the transient one-dimensional heat equations:

$$\overbrace{0 \le x < S(t): \quad \frac{\partial T_s}{\partial t} = \alpha_s \frac{\partial^2 T_s}{\partial x^2}}^{\text{solid}} \quad \text{and} \quad \overbrace{S(t) \le x: \quad \frac{\partial T_l}{\partial t} = \alpha_l \frac{\partial^2 T_l}{\partial x^2}}^{\text{liquid}} \tag{12-13}$$

and subject to the conditions:

$$S(t=0) = 0, \quad T_l(t=0) = T_o < T_m, \quad \partial T_s/\partial x|_{x=0} = 0, \quad \text{and} \quad T_l(x \to \infty) = T_o. \tag{12-14}$$

For undercooled conditions, the latent heat release causes the temperature of the liquid adjacent to the solid to increase, resulting in a negative temperature gradient into

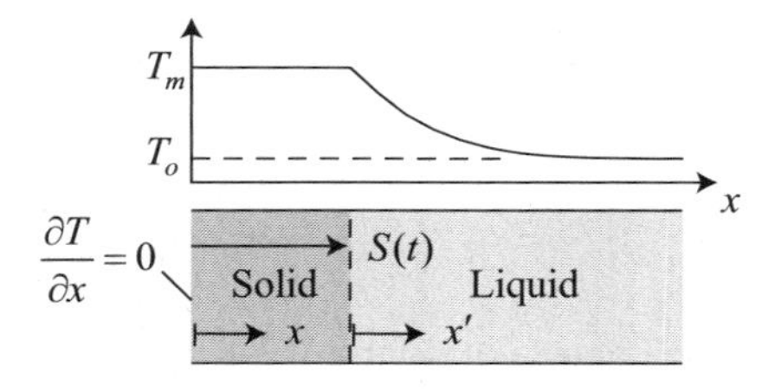

Figure 12-2 Solidification of an undercooled liquid.

*This interface condition is named after the Slovene physicist Jožef Stefan, who introduced the general class of such problems around 1890 in relation to problems of ice formation.

the liquid phase. Since heat transfer through the solid is prohibited by the adiabatic boundary imposed at $x = 0$, the temperature rise approaching the solidification front from the liquid side must come into equilibrium with the solid temperature. This condition leads to the solid phase temperature solution:

$$0 \leq x < S(t): \quad T_s = T_m. \tag{12-15}$$

The solid region grows as the latent heat released to the undercooled liquid is transported away by diffusion. The temperature solution in the liquid can be analyzed from coordinates attached to the liquid–solid interface. The velocity of this interface into the liquid is regulated by the Stefan condition (12-12), with the simplifying factor that the temperature gradient in the solid is zero. Therefore, the interface velocity is

$$\frac{dS}{dt} = -\alpha_l \frac{C_l}{\Delta h_{ls}} \frac{\partial T_l}{\partial x}\bigg|_{x=S}. \tag{12-16}$$

Analysis of the temperature distribution in the liquid is most easily conducted relative to coordinates moving with the solidification front. Letting $x' = x - S(t)$ to denote distances into the liquid, the heat equation becomes

$$x' \geq 0: \quad \frac{\partial T_l}{\partial t} + \overbrace{\alpha_l \frac{C_l}{\Delta h_{ls}} \frac{\partial T_l}{\partial x'}\bigg|_{x'=0}}^{-dS/dt} \frac{\partial T_l}{\partial x'} = \alpha_l \frac{\partial^2 T_l}{\partial x'^2}, \tag{12-17}$$

which includes the advective term associated with the motion of the coordinates. Heat transfer in the liquid is now subject to the conditions:

$$T_l(t = 0) = T_o < T_m, \quad T_l(x' = 0) = T_m, \quad \text{and} \quad T_l(x' \to \infty) = T_o. \tag{12-18}$$

Notice that the magnitude of the flow velocity (12-16) is proportional to the heat diffusion flux. As a result, a simple scaling relation emerges between the independent variables of the governing equation (12-17). This scaling reveals that

$$\underbrace{\frac{1}{\alpha_l}\frac{\partial T_l}{\partial t}}_{\displaystyle \downarrow \atop \displaystyle \frac{1}{\alpha_l t}} = \underbrace{\left(\frac{-C_l}{\Delta h_{ls}} \frac{\partial T_l}{\partial x'}\bigg|_{x'=0}\right) \frac{\partial T_l}{\partial x'}}_{\displaystyle \downarrow \atop \displaystyle \frac{1}{x'^2}} + \underbrace{\frac{\partial^2 T_l}{\partial x'^2}}_{\displaystyle \downarrow \atop \displaystyle \frac{1}{x'^2}} \tag{12-19}$$

with the lower row reading $\frac{1}{\alpha_l t} \sim \frac{1}{x'^2}$ "+" $\frac{1}{x'^2}$

or

$$\frac{1}{\alpha_l t} \sim \frac{1}{x'^2}. \tag{12-20}$$

This scaling suggests attempting a similarity solution to this problem [3], with the proposed similarity variable:

$$\eta = \frac{x'}{\sqrt{4\alpha_l t}}. \tag{12-21}$$

The governing equation may be transformed by representing the derivatives in terms of the similarity variable as

$$\frac{\partial(\,)}{\partial t}=\frac{-\eta}{2t}\frac{\partial(\,)}{\partial\eta},\quad \frac{\partial(\,)}{\partial x'}=\frac{1}{\sqrt{4\alpha_l t}}\frac{\partial(\,)}{\partial\eta},\quad \text{and}\quad \frac{\partial^2(\,)}{\partial x'^2}=\frac{1}{4\alpha_l t}\frac{\partial^2(\,)}{\partial\eta^2}. \tag{12-22}$$

Furthermore, defining a dimensionless temperature:

$$\theta=\frac{T_l-T_o}{\Delta h_{ls}/C_l}, \tag{12-23}$$

the governing equation transforms into the ordinary differential equation

$$\frac{\partial^2\theta}{\partial\eta^2}+2\left(\eta-\frac{1}{2}\frac{\partial\theta}{\partial\eta}\bigg|_{\eta=0}\right)\frac{\partial\theta}{\partial\eta}=0. \tag{12-24}$$

Conditions imposed on the original problem are also transformable to the new problem variables:

$$\underbrace{\begin{array}{c} T_1(t=0)=T_o \\ T_1(x'\to\infty)=T_o \\ T_1(x'=0)=T_m \end{array}}_{3\ original\ conditions} \qquad \underbrace{\begin{array}{c} \left.\begin{array}{c}\theta(\eta\to\infty)=0 \\ \theta(\eta\to\infty)=0\end{array}\right\} same \\ \theta(\eta=0)=\theta_u \end{array}}_{2\ final\ conditions}, \tag{12-25}$$

where the necessary reduction in the number of conditions is observed. The degree of undercooling of the initial liquid state is now described by the dimensionless temperature:

$$\theta_u=\frac{T_m-T_o}{\Delta h_{ls}/C_l}. \tag{12-26}$$

The governing equation (12-24) can be solved by rearranging and integrating once:

$$-2\left(\eta-\frac{1}{2}\frac{d\theta}{d\eta}\bigg|_{\eta=0}\right)d\eta=\frac{d(d\theta/d\eta)}{(d\theta/d\eta)}\quad \text{or}\quad \frac{d\theta}{d\eta}=C_1\exp\left[-(\eta+\lambda)^2\right], \tag{12-27}$$

where

$$\lambda=-\frac{1}{2}\frac{d\theta}{d\eta}\bigg|_{\eta=0} \tag{12-28}$$

is a constant related to the solution. Inspection of Eq. (12-16) reveals that λ is also related to the speed of the solidification front. After a second integration, the solution to the heat equation takes the form

$$\theta=C_1(\sqrt{\pi}/2)\,\mathrm{erf}(\eta+\lambda)+C_2. \tag{12-29}$$

Applying the boundary conditions (12-25) to the solution requires

$$\eta=0:\quad \theta_u=C_1(\sqrt{\pi}/2)\,\mathrm{erf}(\lambda)+C_2 \tag{12-30}$$

$$\eta\to\infty:\quad 0=C_1(\sqrt{\pi}/2)\,+C_2. \tag{12-31}$$

After determining the integration constants, C_1 and C_2, the solution for θ can be expressed in the form

$$\theta = \frac{\theta_u \operatorname{erfc}(\eta + \lambda)}{\operatorname{erfc}(\lambda)}. \tag{12-32}$$

Applying this solution to Eq. (12-28) yields a transcendental equation for the solidification constant λ:

$$\sqrt{\pi}\lambda\, e^{\lambda^2} \operatorname{erfc}(\lambda) = \theta_u. \tag{12-33}$$

Therefore, after solving Eq. (12-33) for λ, the transient temperature in the liquid is described by Eq. (12-32), which, combined with (12-33), can be rewritten in the form

$$\theta = \sqrt{\pi}\lambda\, e^{\lambda^2} \operatorname{erfc}(\eta + \lambda), \quad \text{where} \quad \eta = x'/\sqrt{4\alpha_l t}. \tag{12-34}$$

The propagation distance of the interface into the liquid can be determined by integrating Eq. (12-16) for the result:

$$S(t) = -\sqrt{\alpha_l t}\, \frac{\partial \theta}{\partial \eta}\bigg|_{\eta=0} \quad \text{or} \quad S(t) = \lambda\sqrt{4\alpha_l t}. \tag{12-35}$$

With the interface position known, the temperature solution (12-32) can be expressed in terms of the original position variable $x = S(t) + x'$ for the result:

$$x \geq S(t): \quad T_l = T_o + \frac{(T_m - T_o)\operatorname{erfc}\left(x/\sqrt{4\alpha_l t}\right)}{\operatorname{erfc}(\lambda)}. \tag{12-36}$$

The transient temperature field during solidification is plotted in Figure 12-3 for the case where $\theta_u = 1/2$. An arbitrary length scale L is introduced in this figure to report the spatial and temporal variables in a dimensionless form. As the temperature field penetrates further into the liquid, the temperature gradient at the solidification front becomes less steep. This reduces the rate at which heat is transferred away from the liquid–solid interface, causing the speed of the solidification front to slow.

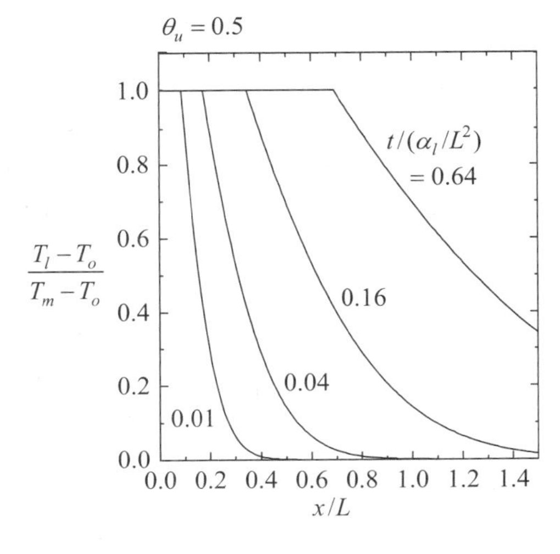

Figure 12-3 Time evolution of temperature during the solidification of an undercooled liquid.

12.3 SOLIDIFICATION OF A BINARY ALLOY FROM AN UNDERCOOLED LIQUID

Directional solidification can be used as a purification technique. Since most impurities are more soluble in the liquid phase, impurities will be rejected from the solid phase during solidification. As a specific example, consider a eutectic binary alloy. The phase diagram, illustrated in Figure 12-4, exhibits unlimited liquid state miscibility of two components A and B, but low or negligible solid state miscibility. Therefore, much of the phase diagram at low temperatures is dominated by a two-phase field of two different solid structures, one that is highly enriched in component A (the α-phase) and one that is highly enriched in component B (the β-phase). Only low concentrations of the solute B are miscible in the solid α-phase of the solvent A. More information about phase diagrams and phase transformations can be found in reference [4].

Suppose a process of solidification occurs in an undercooled liquid, similar to the situation discussed in Section 12.2. However, now the liquid is considered to be a binary alloy composed of a solvent A and a low concentration of a solute B. The alloy has the phase diagram shown in Figure 12-4. A solid forms in the alloy at an interface temperature T_i, below the melting temperature T_{mA} of the pure solvent A. The suppression in the melting point is caused by the solute concentration in the liquid phase $c_{B,l}$. As dictated by the phase diagram, the suppressed melting point temperature of the interface is given by

$$T_i = T_{mA} + m\, c_{B,l} \tag{12-37}$$

where $m < 0$ is the slope of the liquidus line. As shown in Figure 12-4, the solute concentration in the liquid phase $c_{B,l}$ differs from the equilibrium concentration in the solid phase $c_{B,s}$. Therefore, a *partition coefficient* can be defined as the ratio of these two concentrations:

$$K = \frac{c_{B,s}}{c_{B,l}}. \tag{12-38}$$

The solid that emerges from the solidification process has a reduced concentration of solute given by $c_{B,s} = K\, c_{B,l}$. Consequently, solute must be rejected into the liquid phase ahead of the solidification front. However, the resulting rise in liquid solute concentration will further lower the equilibrium temperature at which solid forms. This slows the solidification front associated with heat transfer. Therefore, solidification in an alloy system involves solving the coupled equations for heat and species transport [5].

12.3.1 Heat Transfer

The heat transfer description of solidification from an undercooled state, derived in Section 12.2, can be adapted to the binary alloy considered here by adding to the melting temperature T_{mA} of the pure solvent A the reduction caused by the liquid concentration of

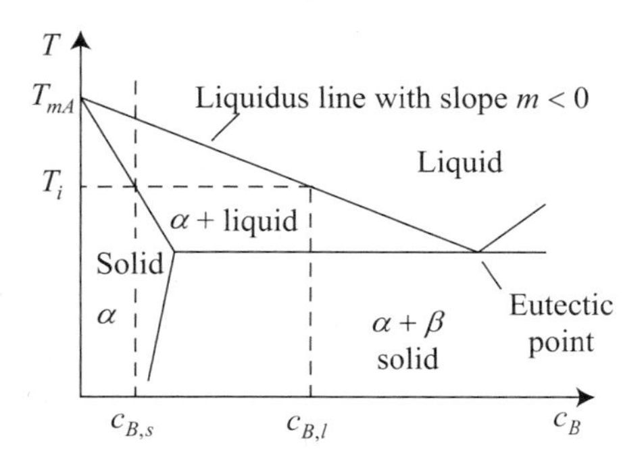

Figure 12-4 Binary eutectic phase diagram.

the solute at the solidification front. For notational simplicity, c_i will denote the liquid phase concentration of the solute B at the interface with the solid ($c_i = c_{B,l}$) and c_o will denote the initial solute concentration in the liquid, far in advance of the solidification front. Therefore, the degree of undercooling of the initial liquid alloy temperature T_o, relative to the interface solidification temperature T_i given by Eq. (12-37), can then be expressed as

$$\frac{T_i - T_o}{\Delta h_{ls}/C_l} = \frac{T_{mA} + mc_{B,l} - T_o}{\Delta h_{ls}/C_l} = \frac{T_{mA} - T_o}{\Delta h_{ls}/C_l} - \frac{-m\, c_o}{\Delta h_{ls}/C_l}\frac{c_i}{c_o} = \theta_{uA} - \theta_{co}\phi_i. \tag{12-39}$$

The dimensionless undercooling of the pure solvent A is defined by

$$\theta_{uA} = \frac{T_{mA} - T_o}{\Delta h_{ls}/C_l}, \tag{12-40}$$

and the reduction in dimensionless undercooling due to the presence of solute B at the interface is given by $\theta_{co}\phi_i$, where

$$\theta_{co} = \frac{-m\, c_o}{\Delta h_{ls}/C_l} \quad \text{and} \quad \phi_i = \frac{c_i}{c_o}. \tag{12-41}$$

The dimensionless temperature θ_{co} is a measure of how much the initial concentration of solute will reduce the undercooling of the liquid state, and ϕ_i is the fraction of solute concentration at the interface relative to the initial value.

Using Eq. (12-39) for the undercooling of a binary alloy, the temperature solution (12-32) derived in Section 12.2 for a pure system can be recast for the alloy system as

$$\theta = \frac{(\theta_{uA} - \theta_{co}\phi_i)\mathrm{erfc}(\eta + \lambda)}{\mathrm{erfc}(\lambda)}, \quad \text{where} \quad \eta = x'/\sqrt{4\alpha_l t} \tag{12-42}$$

and $x' = x - S(t)$ denotes distances from the solidification front into the liquid. The solidification constant λ is still defined by Eq. (12-28). However, using the undercooling of the binary alloy, the transcendental equation (12-33) derived in Section 12.2 for a pure system can be recast for the alloy system as

$$\sqrt{\pi}\lambda\, e^{\lambda^2}\mathrm{erfc}(\lambda) = \theta_{uA} - \theta_{co}\phi_i. \tag{12-43}$$

12.3.2 Species Transfer

Solute transport is driven by propagation of the solidification front when the solute partition coefficient between the liquid and solid phases, defined by Eq. (12-17), is not unity. As new solid forms, solute is rejected into the liquid phase and transported by diffusion. This process is shown schematically in Figure 12-5. The solute transport in the liquid and the solid phases is governed by the transient one-dimensional equations:

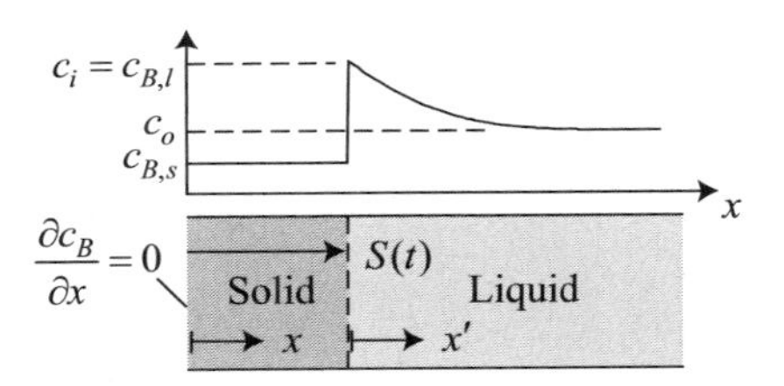

Figure 12-5 Solute concentration profile during solidification of an undercooled liquid.

$$\overbrace{0 \le x < S(t): \quad \frac{\partial c_B}{\partial t} = \mathcal{D}_s \frac{\partial^2 c_B}{\partial x^2}}^{solid} \quad \text{and} \quad \overbrace{x \ge S(t): \quad \frac{\partial c_B}{\partial t} = \mathcal{D}_l \frac{\partial^2 c_B}{\partial x^2}}^{liquid} \tag{12-44}$$

and subject to the conditions

$$S(t=0)=0, \quad c_B(t=0)=c_o, \quad \partial c_B/\partial x\big|_{x=0}=0, \quad \text{and} \quad c_B(x\rightarrow\infty)=c_o. \tag{12-45}$$

In this problem the initial liquid solute concentration is c_o, and the liquid–solid interface is initially at $x = 0$. Solute rejection by the solidification front results in a negative concentration gradient in the liquid, as illustrated in Figure 12-5. Since solute transfer through the solid is prohibited by the boundary condition at $x = 0$, the solute concentration rise in the liquid must establish an equilibrium with the reduced solid concentration specified by the partition coefficient (12-38). This last consideration leads to the solid phase solution:

$$0 \le x < S(t): \quad c_B = K\, c_i, \tag{12-46}$$

where c_i has yet to be determined. As the solid region grows, the solute concentration in the liquid can be analyzed from a coordinate system attached to the liquid–solid interface. Again letting $x' = x - S(t)$ denote distances into the liquid from the solidification front, the species transport equation in the liquid becomes

$$\frac{\partial c_B}{\partial t} + \overbrace{\alpha_l \frac{C_l}{\Delta h_{ls}} \frac{\partial T_l}{\partial x'}\bigg|_{x'=0}}^{-dS/dt} \frac{\partial c_B}{\partial x'} = \mathcal{D}_l \frac{\partial^2 c_B}{\partial x'^2}, \tag{12-47}$$

which includes the advective term associated with the motion of the solidification front. Species transport in the liquid is subject to the conditions

$$c_B(t=0)=c_o, \quad c_B(x'=0)=c_i, \quad \text{and} \quad c_B(x'\rightarrow\infty)=c_o, \tag{12-48}$$

where the concentration of solute at the interface c_i remains an unknown part of the solution.

A similarity solution to the species transport equation is sought, with the same similarity variable definition (12-21) used to solve the heat transfer equation. With the same dimensionless temperature definition (12-23), and defining a dimensionless concentration variable:

$$\phi = c_B/c_o, \tag{12-49}$$

the species equation (12-47) becomes

$$\frac{\partial^2 \phi}{\partial \eta^2} + 2\frac{\alpha_l}{\mathcal{D}_l}\left(\eta - \frac{1}{2}\frac{\partial \theta}{\partial \eta}\bigg|_{\eta=0}\right)\frac{\partial \phi}{\partial \eta} = 0 \tag{12-50}$$

or

$$\frac{\partial^2 \phi}{\partial \eta^2} + 2\, Le(\eta + \lambda)\frac{\partial \phi}{\partial \eta} = 0. \tag{12-51}$$

The final form of the species equation makes use of the definition (12-28) for the solidification constant λ, and the *Lewis number*† defined by

$$Le = \alpha_l / \mathcal{D}_l. \tag{12-52}$$

Conditions imposed on the original problem are transformed to the new problem variables:

$$\underbrace{\begin{aligned} c_B(t=0) &= c_o \\ c_B(x' \to \infty) &= c_o \\ c_B(x'=0) &= c_i \end{aligned}}_{\text{3 original conditions}} \qquad \underbrace{\begin{aligned} &\left.\begin{aligned}\phi(\eta \to \infty) &= 1 \\ \phi(\eta \to \infty) &= 1\end{aligned}\right\} \text{same} \\ &\phi(\eta = 0) = \phi_i \end{aligned}}_{\text{2 final conditions}}, \tag{12-53}$$

where $\phi_i = c_i / c_o$.

The solution to Eq. (12-51) is obtained through steps similar to those used in arriving at the temperature solution (12-32), with the result

$$\phi = 1 + (\phi_i - 1)\frac{\operatorname{erfc}\left((\eta + \lambda)\sqrt{Le}\right)}{\operatorname{erfc}(\lambda\sqrt{Le})}. \tag{12-54}$$

However, this solution cannot be evaluated without knowledge of the solidification constant λ and the solute concentration at the interface ϕ_i.

To establish the liquid side interface concentration requires a species balance at the solidification front. The rate at which solute is rejected by solidification is dictated by the propagation speed of the solidification front and the jump in solute concentration across the liquid–solid interface dictated by the phase diagram. The excess solute rejected by solidification is carried away from the front by diffusion. Therefore, with the location of the solidification front denoted by $S(t)$, the species balance at the interface becomes

$$\overbrace{(c_{B,l} - c_{B,s})\frac{dS}{dt}}^{\text{solute rejected}} = \overbrace{\left(\mathcal{D}_s \frac{\partial c_{B,s}}{\partial x} - \mathcal{D}_l \frac{\partial c_{B,l}}{\partial x}\right)}^{\text{solute removed}}\Bigg|_{x=S(t)}. \tag{12-55}$$

Using the solidification speed given by Eq. (12-16), and the simplifying factor that the concentration gradients in the solid are zero, the species interface balance (12-55) becomes

$$-(c_i - K\,c_i)\alpha_l \frac{C_l}{\Delta h_{ls}} \frac{\partial T_l}{\partial x'}\bigg|_{x'=0} = -\mathcal{D}_l \frac{\partial c_B}{\partial x'}\bigg|_{x'=0}. \tag{12-56}$$

Concentrations on the liquid and solid side of the interface are denoted by $c_{B,l} = c_i$ and $c_{B,s} = K\,c_i$, respectively, and $x' = x - S(t)$ is the distance into the liquid from the solidification front. Using the definitions (12-49), (12-23), and (12-21), for ϕ, θ, and η, respectively, the species balance at the solidification front can be expressed in terms of the solution variables:

$$(1-K)\frac{c_i}{c_o}\frac{\alpha_l}{\mathcal{D}_l}\frac{\partial \theta}{\partial \eta}\bigg|_{\eta=0} = \frac{\partial \phi}{\partial \eta}\bigg|_{\eta=0} \quad \text{or} \quad -2(1-K)\phi_i\, Le\, \lambda = \frac{\partial \phi}{\partial \eta}\bigg|_{\eta=0}. \tag{12-57}$$

---

†The ratio of heat diffusivity to species diffusivity is named in honor of Warren K. Lewis (1882–1975), who was the first head of the Chemical Engineering Department at the Massachusetts Institute of Technology.

Using the solute solution (12-54) to evaluate the concentration gradient at the solidification front in the last expression yields

$$\left(1-\sqrt{\pi Le}(1-K)\lambda\, e^{\lambda^2 Le}\operatorname{erfc}(\lambda\sqrt{Le})\right)\phi_i = 1. \tag{12-58}$$

This transcendental equation expresses the relation between the solidification constant λ and the liquid solute concentration ϕ_i at the solidification front.

12.3.3 Coupling Heat and Species Transport

Combining the transcendental equations for heat transfer (12-43) and species transfer (12-58) allows the liquid solute concentration ϕ_i at the interface to be eliminated from the final equation describing the solidification constant:

$$\theta_{co}+\left(1-\sqrt{\pi Le}(1-K)\lambda\, e^{\lambda^2 Le}\operatorname{erfc}(\lambda\sqrt{Le})\right)\left(\sqrt{\pi}\lambda\, e^{\lambda^2}\operatorname{erfc}(\lambda)-\theta_{uA}\right)=0. \tag{12-59}$$

The transcendental equation (12-59) for the solidification constant λ is a function of the undercooling of the solvent material θ_{uA}, the reduction in undercooling caused by the initial solute concentration θ_{co}, the partition coefficient K, and the relative diffusivity of heat and species transfer given by the Lewis number Le. The physically significant root to the transcendental equation requires $\lambda > 0$.

Combining Eqs. (12-42) and (12-43), the temperature solution in the liquid alloy becomes

$$\theta = \sqrt{\pi}\lambda\, e^{\lambda^2}\operatorname{erfc}(\eta+\lambda). \tag{12-60}$$

However, the similarity variable can be replaced by the original independent variables of the problem using

$$\eta = \frac{x'}{\sqrt{4\alpha_l t}} = \frac{x-S(t)}{\sqrt{4\alpha_l t}} = \frac{x-\lambda\sqrt{4\alpha_l t}}{\sqrt{4\alpha_l t}} = \frac{x}{\sqrt{4\alpha_l t}}-\lambda. \tag{12-61}$$

Then the temperature solution in the liquid becomes

$$x \ge S(t):\quad \frac{T_l-T_o}{\Delta h_{ls}/C_l} = \sqrt{\pi}\lambda\, e^{\lambda^2}\operatorname{erfc}\left(\frac{x}{\sqrt{4\alpha_l t}}\right), \tag{12-62}$$

where λ is determined from Eq. (12-59).

When the interface concentration found from (12-58) is substituted into Eq. (12-54), and the original dependent variables are reintroduced to the solution with Eq. (12-61), the result for the solute concentration in the liquid becomes

$$x \ge S(t):\quad \frac{c_B}{c_o} = 1+\left(\frac{1}{1-\sqrt{\pi Le}(1-K)\lambda\, e^{\lambda^2 Le}\operatorname{erfc}(\lambda\sqrt{Le})}-1\right)\frac{\operatorname{erfc}(x\sqrt{Le}/\sqrt{4\alpha_l t})}{\operatorname{erfc}(\lambda\sqrt{Le})} \tag{12-63}$$

where $S(t)=\lambda\sqrt{4\alpha_l t}$ and the solidification constant λ is established with Eq. (12-59).

Figure 12-6 shows the progression of the solute concentration profile in the material, for the case where the liquid is initially undercooled from the solute melting temperature by $\theta_{uA}=0.6$, undercooling is reduced by the initial solute concentration by $\theta_{co}=0.1$, the partition coefficient is $K=0.1$, and the Lewis number is $Le=1.0$. Notice that the solute

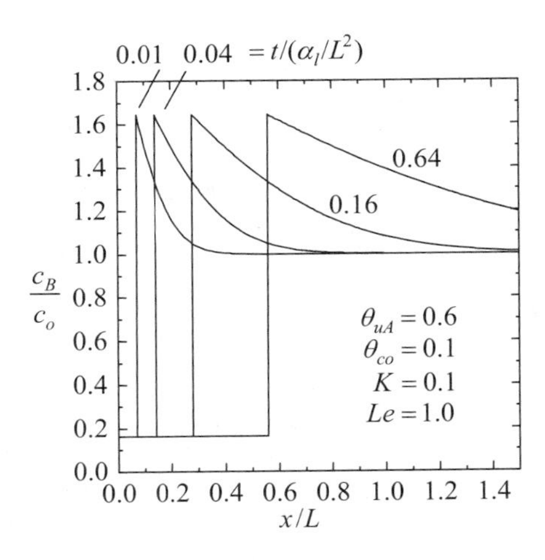

Figure 12-6 Time evolution of solute concentration during the solidification of an undercooled binary liquid.

concentration in the solid is significantly smaller than the initial value in the liquid phase. This consequence of the partition coefficient is exploited in materials processing to purify semiconductors and other materials.

12.4 MELTING OF A SOLID INITIALLY AT THE MELTING POINT

Consider a semi-infinite solid that is initially at the melting temperature. Melting of the solid occurs when the surface is elevated above the melting point, as shown in Figure 12-7. Many aspects of this problem are similar to the solidification problem of Section 12.2. In that problem, the temperature field evolving in front of the solidification front was determined, and a similarity solution was sought for the semi-infinite region $S \leq x < \infty$. However, in the current problem, the temperature field evolves behind the melting front. This requires a solution for the finite region $0 \leq x \leq S(t)$. In general, a solution that evolves between two boundaries does not admit a similarity solution. However, the current situation is an exception by virtue of the fact that the distance $S(t)$ is not an independently imposed length scale. Instead, $S(t)$ is a consequence of the heat transfer equation and the Stefan condition (13-4). The similarity variable required to solve the heat equation is already known. Therefore, to establish the existence of a similarity solution, it is necessary to demonstrate that the Stefan condition at the moving boundary is also satisfied with the same similarity variable.

The instantaneous velocity of the melting front is related to the rate with which the latent heat Δh_{ls} required for melting is transported to the liquid–solid interface. This leads to the Stefan condition:

$$\overbrace{\rho_l \Delta h_{ls} \frac{dS}{dt}}^{\text{heat absorbed}} = \overbrace{\left(\underbrace{k_s \frac{\partial T_s}{\partial x}}_{=0} - k_l \frac{\partial T_l}{\partial x} \right)}^{\text{heat supplied}} \Bigg|_{x=S(t)} . \tag{12-64}$$

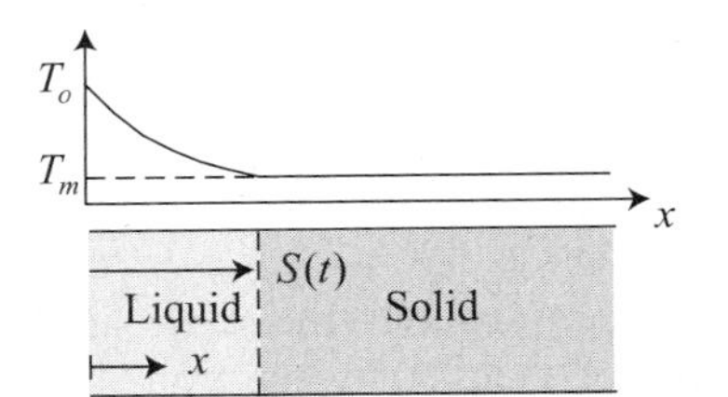

Figure 12-7 Melting of a solid initially at the melting point.

Since the solid is uniformly at the melting point temperature, heat transfer to the melting front from the solid does not occur. Therefore, the interface velocity, as related to the temperature field in the liquid, is given by

$$\frac{dS}{dt} = -\alpha_l \frac{C_l}{\Delta h_{ls}} \frac{\partial T_l}{\partial x}\bigg|_{x=S}. \tag{12-65}$$

For a similarity solution to exist, the condition (12-65) must be expressed in terms of the similarity variable:

$$\eta = \frac{x}{\sqrt{4\alpha_l t}}. \tag{12-66}$$

Notice that there is no benefit in placing the coordinate system on the moving liquid–solid interface, as was done in Section 12-2. Therefore, the spatial variable x remains a measure of distance from the surface of the semi-infinite body. Eliminating appearances of x from the Stefan condition (12-65) yields

$$\frac{dS}{dt} = -\alpha_l \frac{C_l}{\Delta h_{ls}} \frac{\partial \eta}{\partial x} \frac{\partial T_l}{\partial \eta}\bigg|_{\eta=\lambda} = -\alpha_l \frac{C_l}{\Delta h_{ls}} \frac{1}{\sqrt{4\alpha_l t}} \frac{\partial T_l}{\partial \eta}\bigg|_{\eta=\lambda} \tag{12-67}$$

or

$$\sqrt{\frac{4t}{\alpha_l}}\frac{dS}{dt} = \frac{-C_l}{\Delta h_{ls}} \frac{\partial T_l}{\partial \eta}\bigg|_{\eta=\lambda} \quad (= \; B). \tag{12-68}$$

Notice that the location of the melting front, with respect to the similarity variable, has been denoted by $\eta = \lambda$. Since the left-hand side of Eq. (12-68) is some function of the time variable alone, and the right-hand side is some function of the similarity variable alone, this equation can be satisfied only if both sides of the equation equal a separation constant, "B". (This argument also implies that λ must be a constant.) Therefore, the interface position can be determined with respect to the separation constant B as

$$\sqrt{\frac{4t}{\alpha_l}}\frac{dS}{dt} = B \quad \text{or} \quad S = B\sqrt{\alpha_l t}. \tag{12-69}$$

However, since λ is the value of the similarity variable at the interface, it is necessary that

$$\lambda = \frac{S}{\sqrt{4\alpha_l t}} = \frac{B\sqrt{\alpha_l t}}{\sqrt{4\alpha_l t}} = \frac{B}{2}. \tag{12-70}$$

Therefore, with respect to the constant λ, the interface location is required to satisfy

$$S = \lambda\sqrt{4\alpha_l t}, \tag{12-71}$$

and from Eq. (12-68) λ is related to the Stefan condition by

$$\frac{-C_l}{\Delta h_{ls}} \frac{\partial T_l}{\partial \eta}\bigg|_{\eta=\lambda} = 2\lambda. \tag{12-72}$$

With the Stefan condition (12-72) expressed in terms of the similarity variable, as the sole independent variable, the viability of a similarity solution is affirmed.

The heat equation to be solved in the liquid is

$$0 \le x \le S(t): \quad \frac{\partial T_l}{\partial t} = \alpha_l \frac{\partial^2 T_l}{\partial x^2} \tag{12-73}$$

and is subject to the boundary conditions

$$T(x=0) = T_o \quad \text{and} \quad T(x=S) = T_m. \tag{12-74}$$

Notice that no initial condition is required because in the limit that $t \to 0$, the liquid domain also vanishes.

Defining a dimensionless temperature:

$$\vartheta = \frac{T_l - T_m}{T_o - T_m}, \tag{12-75}$$

the governing heat equation (12-73) can be transformed into an ordinary differential equation with respect to the similarity variable (12-66). Therefore, the heat equation in the liquid becomes

$$\frac{\partial^2 \vartheta}{\partial \eta^2} + 2\eta \frac{\partial \vartheta}{\partial \eta} = 0. \tag{12-76}$$

The boundary conditions imposed on the original problem (12-74) are also transformed into the new problem variables:

$$\vartheta(\eta = 0) = 1 \quad \text{and} \quad \vartheta(\eta = \lambda) = 0, \tag{12-77}$$

where the constant λ is an unknown part of the solution. The governing equation (12-76) can be integrated twice for the result:

$$\vartheta = C_1(\sqrt{\pi}/2)\text{erf}(\eta) + C_2. \tag{12-78}$$

Applying this solution to the boundary conditions (12-77) requires

$$\eta = 0: \quad 1 = 0 + C_2 \qquad \text{or} \quad C_2 = 1 \tag{12-79}$$

$$\eta = \lambda: \quad 0 = C_1(\sqrt{\pi}/2)\,\text{erf}(\lambda) + 1 \quad \text{or} \quad C_1 = \frac{-2}{\sqrt{\pi}\text{erf}(\lambda)}. \tag{12-80}$$

Consequently, the solution to the temperature field in the liquid is found to be

$$\vartheta = 1 - \frac{\text{erf}(\eta)}{\text{erf}(\lambda)} \tag{12-81}$$

To determine the unknown constant λ, the solution (12-81) is applied to the Stefan condition (12-72) for the result:

$$2\lambda = \frac{-C_l}{\Delta h_{ls}}(T_o - T_m)\frac{\partial \vartheta}{\partial \eta}\bigg|_{\eta=\lambda} = \frac{C_l}{\Delta h_{ls}}(T_o - T_m)\frac{2e^{-\lambda^2}}{\sqrt{\pi}\text{erf}(\lambda)} \tag{12-82}$$

or

$$\sqrt{\pi}\lambda e^{\lambda^2}\operatorname{erf}(\lambda) = \frac{C_l(T_o - T_m)}{\Delta h_{ls}}. \tag{12-83}$$

The transcendental equation (12-83) is used to determine the constant λ required by the solution (12-81). With knowledge of λ, the transient temperature profile in the semi-infinite body can be expressed in the form

$$0 \le x \le \lambda\sqrt{4\alpha_l t}: \quad T_l = T_o + (T_m - T_o)\operatorname{erf}\left(x/\sqrt{4\alpha_l t}\right)/\operatorname{erf}(\lambda) \tag{12-84}$$

and

$$x > \lambda\sqrt{4Ð_D t}: \quad T_s = T_m. \tag{12-85}$$

The problems illustrated by melting in this section, and solidification in Section 12.2, do not encompass the more general situation that arises when a nontrivial solution evolves on both sides of the moving interface. In this more general situation, the Stefan condition does not simplify for constant conditions on one side of the interface or the other. However, armed with the ability to solve the transport equation on both sides of the interface, the more general situation only requires the additional task of addressing the more complicated Stefan condition, as is explored in Problem 12-3 for heat transfer, and Problem 12-4 for species transfer.

12.5 PROBLEMS

12-1 A novel method for long-term drug administration utilizes a capsule impregnated with a drug at a concentration c_o exceeding the solubility limit of the capsule material. When the capsule is implanted into a patient, the drug is released into solution at the solubility limit c_s. The drug release initiates a depletion layer $S(t)$ that grows in the capsule material, as shown in the illustration. It is assumed that drug transport is by molecular diffusion through the capsule material $x \ge 0$, with a diffusivity $Ð_D$. The drug is released into the patient's tissue $(x = 0)$ at a concentration approaching zero. Determine the transient drug concentration profile $c_D(t, x)$ in the capsule material. Show that the instantaneous drug release rate per unit area of the capsule surface is given by

$$-J_D^* = \lambda \exp(\lambda^2)(c_o - c_s)\sqrt{Ð_D/t},$$

where $\lambda > 0$ is the root to the equation $\sqrt{\pi}\lambda \exp(\lambda^2) erf(\lambda) = c_s/(c_o - c_s)$.

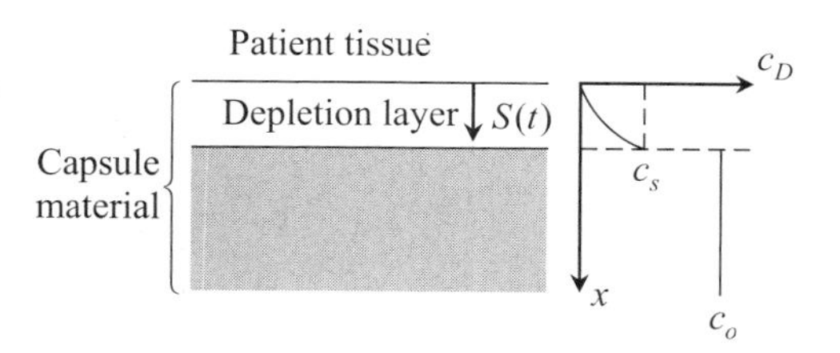

12-2 The spike in drug release rate at $t = 0$, and the subsequent decay, can be undesirable for the drug delivery capsule discussed in Problem 12-1. Consider a capsule impregnated such that the drug concentration above the solubility limit of the capsule material varies linearly with distance from the surface:

$$c_o(x) = c_s + b\,x$$

Show that the Stefan condition, with the drug concentration c_D as the dependent variable, cannot be satisfied using the similarity variable:

$$\eta = x/\sqrt{4Đ_D t}.$$

However, demonstrate that if the dependent variable is scaled as

$$\phi = \frac{c_D - c_s}{\sqrt{4t}},$$

a similarity solution can be found with a constant drug release rate boundary condition:

$$-J_D^* = Đ_D \frac{\partial c_D}{\partial x}\bigg|_{x=0}.$$

Show that the drug concentration profile in the depletion region of the capsule is

$$c_D = c_s + (-J_D^*)\sqrt{\frac{4t}{Đ_D}}\left(\eta - \lambda \frac{\exp(-\eta^2) + \sqrt{\pi}\,\eta\,\mathrm{erf}(\eta)}{\exp(-\lambda^2) + \sqrt{\pi}\,\lambda\,\mathrm{erf}(\lambda)}\right), \quad \text{where} \quad \eta = \frac{x}{\sqrt{4Đ_D t}}$$

and λ is determined from the roots of

$$\frac{1}{\lambda^2} - \frac{\mathrm{erf}(\lambda)/\lambda}{\exp(-\lambda^2) + \sqrt{\pi}\,\lambda\,\mathrm{erf}(\lambda)} = \frac{2bĐ_D}{-J_D^*}.$$

12-3 Suppose a semi-infinite liquid is at an initial temperature T_∞ that is above the melting point. Solidification of the liquid occurs when the surface temperature at $x = 0$ is reduced to a temperature T_o that is below the melting temperature T_m of the body. Determine the transient temperature $T(x, t)$ in the semi-infinite body during solidification.

12-4 Diffusion of a dilute species "A" from a semi-infinite body occurs when the surface concentration c_o is brought to a value less than the initial value c_∞ of the body. Suppose that the diffusivity of the species has a discontinuous dependency on concentration, such that:

$$\text{for} \quad c_A < c_x: \quad Đ_A = Đ_1,$$

and

$$\text{for} \quad c_A > c_x: \quad Đ_A = Đ_2.$$

Determine the transient concentration $c_A(t, x)$ in the semi-infinite body.

REFERENCES

[1] B. E. Deal and A. S. Grove, "General Relationship for the Thermal Oxidation of Silicon," *Journal of Applied Physics*, **36**, 3770 (1965).

[2] A. M. Meirmanov, *The Stefan Problem*, De Gruyter Expositions in Mathematics 3. Berlin/New-York: Walter de Gruyter, 1992.

[3] H. Carslaw and J. Jaeger, *Conduction of Heat in Solids*, Second Edition. Oxford, UK: Clarendon Press, 1959.

[4] M. Hillert, *Phase Equilibria,Phase Diagrams and Phase Transformations—Their Thermodynamic Basis*, Second Edition. Cambridge, UK: Cambridge University Press, 2007.

[5] V. R. Voller, "A Similarity Solution for Solidification of an Under-Cooled Binary Alloy," *International Journal of Heat and Mass Transfer*, **49**, 1981 (2006).

Chapter 13

Lubrication Theory

13.1 Lubrication Flows Governed by Diffusion
13.2 Scaling Arguments for Squeeze Flow
13.3 Squeeze Flow Damping in an Accelerometer Design
13.4 Coating Extrusion
13.5 Coating Extrusion on a Porous Surface
13.6 Reynolds Equation for Lubrication Theory
13.7 Problems

In earlier chapters, flows that were governed by diffusion arose when advection terms in the transport equations were identically zero. This situation occurs in transport without fluid motion, or when fluid motion is perfectly orthogonal to the transport direction of interest. In this chapter, flows are considered where advection terms in the transport equations are small enough to neglect, but are nonzero. The smallness of advection will need to be established by scaling arguments. It will be shown that flows between moving surfaces in close proximity are governed by diffusion, as arising in the use of lubricants.

13.1 LUBRICATION FLOWS GOVERNED BY DIFFUSION

Fluids used as lubricants generally flow in the plane of a gap between two confining surfaces. For an incompressible flow, the continuity equation dictates that the divergence of the velocity field in the gap is zero ($\partial_j v_j = 0$). This can be expressed in Cartesian coordinates and integrated with respect to the distance across the gap H:

$$\int_0^H \left(\frac{\partial v_x}{\partial x} + \frac{\partial v_y}{\partial y} + \frac{\partial v_z}{\partial z} \right) dz = 0 \quad \text{or} \quad \int_0^H \left(\frac{\partial v_x}{\partial x} + \frac{\partial v_y}{\partial y} \right) dz = -v_z\Big|_{z=0}^{z=H} = -\frac{\partial H}{\partial t}. \tag{13-1}$$

In Eq. (13.1), v_z is evaluated as illustrated in Figure 13-1, assuming that vertical motion of the top surface is permitted, $v_z(z = H) = \partial H/\partial t$, but that the lower surface is vertically fixed, $v_z(z = 0) = 0$. To change the order of differentiation and integration in Eq. (13-1) requires the use of Leibniz's theorem:

$$\partial_i \int_0^H (?)dz = \int_0^H \partial_i (?)dz + [(?)\partial_i H]\Big|_{z=0}^{z=H} \quad (i = x \text{ or } y). \tag{13-2}$$

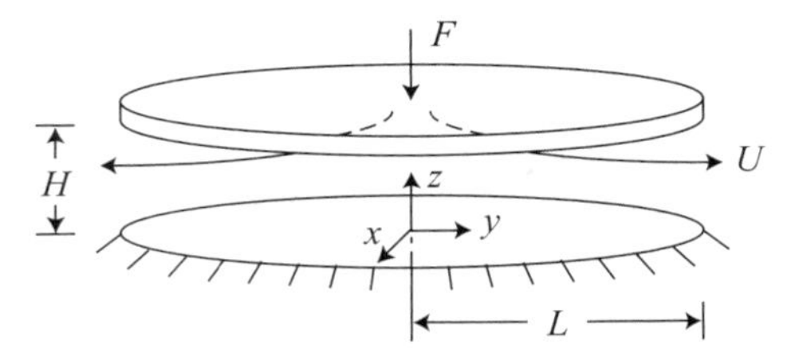

Figure 13-1 Scaling illustration for squeeze flow.

Leibniz's theorem was previously discussed in Section 5.9 in the context of a change in order between time differentiation and spatial integration. Using Leibniz's theorem (13-2) on the integral of the velocity field $v_i(z)$ across the gap yields

$$\partial_i \int_0^H v_i dz = \int_0^H \partial_i v_i dz + \overbrace{[v_i \partial_i H]\big|_{z=0}^{z=H}}^{=0}. \tag{13-3}$$

For squeeze flow, the last term evaluates to zero since $v_i(z=0)=0$ and $v_i(z=H)=0$, for either $i=x$ or y. With this result, the continuity statement given by Eq. (13-1) becomes

$$\text{Continuity:} \qquad -\partial_o H = \partial_i \int_0^H v_i \, dz \quad (i = x \text{ and } y). \tag{13-4}$$

This form of the continuity equation relates the descent of the top surface to the divergence of the integrated velocity distribution across the gap, for an incompressible flow.

In lubrication theory, the transient and advection terms (i.e., the inertial terms) in the momentum equation are "small." To demonstrate this formally requires the use of scaling arguments, as undertaken in the next section. With inertial terms omitted, the momentum equation simplifies to the form

$$\text{Momentum:} \qquad \overbrace{0}^{\partial_o v_i + v_j \partial_j v_x} \approx \nu \partial_j \partial_j v_i - \partial_i P/\rho + g_i. \tag{13-5}$$

In the absence of inertial terms, it is straightforward to solve the momentum equation for the velocity profile across the gap. However, the result is dependent on the local pressure gradient between the plates. Therefore, the velocity profile can be substituted into the continuity equation (14-3), resulting in a differential equation that describes the lateral pressure field between the plates. Once this last equation is solved for the pressure distribution, the description of the velocity profile between the plates is complete.

13.2 SCALING ARGUMENTS FOR SQUEEZE FLOW

Consider a flow between two plates, as illustrated in Figure 13-1, where motion of the top plate towards the bottom plate causes the fluid in the gap to be squeezed outward. Scaling arguments will demonstrate when it is appropriate to assume that the transient and advection terms in the momentum equation are negligible relative to the diffusion term. To perform this analysis it is recognized that the flow solution is related to a number of characteristic scales. For the squeeze flow illustrated, the geometric scales are

$$x,\ y \sim L \quad \text{and} \quad z \sim H, \tag{13-6}$$

and the velocities scales are

$$v_x, v_y \sim U \quad \text{and} \quad v_z \sim \dot{H}. \tag{13-7}$$

The rate of descent of the top plate $\dot{H}$ and velocity scale U are not necessarily imposed on the problem. As illustrated in Figure 13-1, the imposed scale could be a force F that would ultimately be related to $\dot{H}$ and U. The influence of F in the momentum equation is directly associated with the fluid pressure. Therefore, pressure scales as

$$P \sim \frac{F}{L^2}. \tag{13-8}$$

13.2.1 Scaling Continuity

Continuity allows the two velocity scales $\dot{H}$ and U to be related. Starting with the continuity equation in the form of Eq. (13-4), where the index i takes on values of x and y, the scaling of the continuity equation yields

$$\dot{H} \sim v_z \sim \frac{U}{L}H. \tag{13-9}$$

13.2.2 Scaling Momentum

Next, the momentum equation is scaled for the squeeze flow problem. Two momentum equations, corresponding to the x- and y-directions, can be scaled in an equivalent way because $v_x \leftrightarrow v_y$ and $x \leftrightarrow y$ have equivalent scales. Here, the x-direction momentum equation is scaled using the scales identified by Eqs. (13-6) through (13-9), for the result

$$\begin{gathered}
\frac{\partial v_x}{\partial t} + v_x\frac{\partial v_x}{\partial x} + v_y\frac{\partial v_x}{\partial y} + v_z\frac{\partial v_x}{\partial z} = \nu\left(\frac{\partial^2 v_x}{\partial x^2} + \frac{\partial^2 v_x}{\partial y^2} + \frac{\partial^2 v_x}{\partial z^2}\right) - \frac{1}{\rho}\frac{\partial P}{\partial x} \\
\downarrow \\
\frac{U}{L/U}\text{''}+\text{''}U\frac{U}{L}\text{''}+\text{''}U\frac{U}{L}\text{''}+\text{''}\frac{UH}{L}\frac{U}{H} \sim \nu\left(\frac{U}{L^2}\text{''}+\text{''}\frac{U}{L^2}\text{''}+\text{''}\frac{U}{H^2}\right)\text{''}+\text{''}\frac{1}{\rho}\frac{F}{L^3}
\end{gathered} \tag{13-10}$$

Note that $t \sim L/U$ is used in the scaling of $\partial v_x/\partial t$. Several interesting results emerge from this scaling. First, all of the advection terms scale as U^2/L. Additionally, two of the diffusion terms scale as $\nu U/L^2$, while the remaining diffusion term scales as $\nu U/H^2$. Eliminating redundant terms, the scaled momentum equation becomes

$$\overbrace{\frac{U^2}{L}}^{\substack{\textit{intertial forces} \\ \textit{(transient + advection)}}} \sim \overbrace{\nu\left(\frac{U}{L^2} + \frac{U}{H^2}\right)}^{\substack{\textit{viscous forces} \\ \textit{(diffusion)}}}\text{''}+\text{''}\overbrace{\frac{1}{\rho}\frac{F}{L^3}}^{\textit{source}}. \tag{13-11}$$

The relative importance of the inertial terms, associated with advection and transient momentum change, can now be compared with the viscous terms associated with diffusion of momentum:

$$\frac{\textit{inertial terms}}{\textit{viscous terms}} \sim \frac{U^2/L}{\nu(U/L^2 + U/H^2)} = \frac{UL/\nu}{1 + (L/H)^2}. \tag{13-12}$$

The ratio of inertial terms to viscous terms reveals two dimensionless numbers relevant to this scaling. The first is the *Reynolds number* [1]:

$$\text{Re} = \frac{UL}{\nu}. \tag{13-13}$$

The Reynolds number is named after the fluid dynamicist Osborne Reynolds (1842–1912). This number is often said to express the ratio of inertial forces to viscous forces. However, in the context of lubrication theory, one can see from Eq. (13-12) that with a sufficiently small gap $H \ll L$, viscous terms (diffusion) may become large compared with inertial terms (advection) regardless of the magnitude of the Reynolds number. When $H \ll L$, the condition required for diffusion to dominate transport becomes

$$(UH/\nu)(H/L) \ll 1 \quad \text{(for lubrication theory)}. \tag{13-14}$$

This is the product of the Reynolds number (now based on the gap height H) and the aspect ratio of the gap height to flow distance (H/L).

13.3 SQUEEZE FLOW DAMPING IN AN ACCELEROMETER DESIGN

Squeeze flow can be important to the design of a silicon surface micro-machined accelerometer. A simplified accelerometer structure is illustrated in Figure 13-2. Notice the large flat proof mass suspended over a surface. Vertical motion of the proof mass is damped by the viscous forces arising from squeeze flow. Therefore, determining the damping coefficient caused by the flow is an important design element in characterizing the dynamic performance of the accelerometer.

The simplest dynamic model for an accelerometer proof mass is as a driven damped spring oscillator:

$$m\ddot{z} + c\dot{z} + kz = ma_{body}. \tag{13-15}$$

The variables describing the proof mass deflection z are the mass m, damping coefficient c, and spring constant k, as illustrated in Figure 13-3. When a body to which the

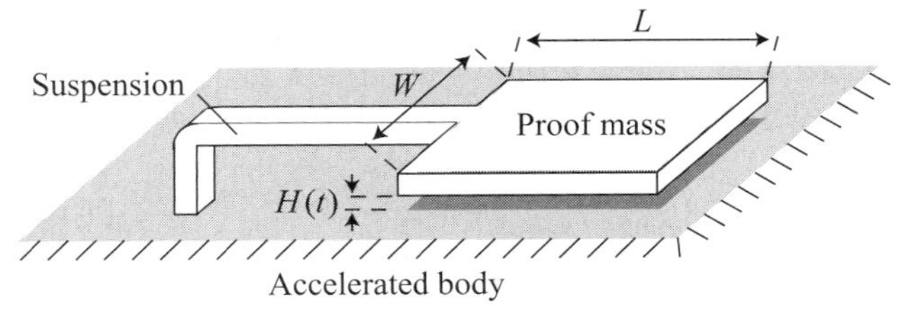

Figure 13-2 Accelerometer geometry.

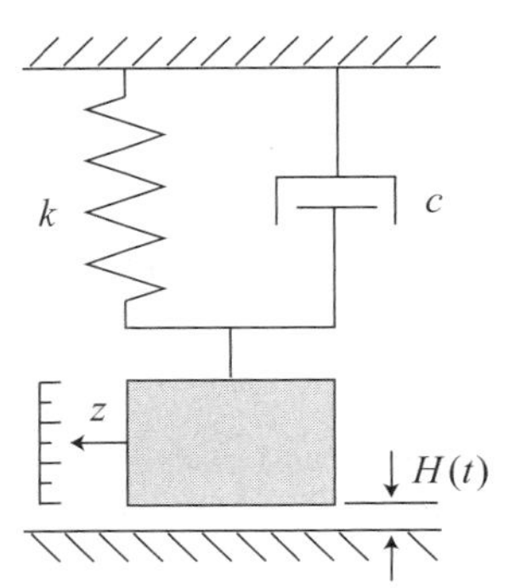

Figure 13-3 Schematic of a damped spring oscillator.

accelerometer is attached accelerates, the inertial force deflects the suspension of the proof mass. Therefore, the displacement of the proof mass is a measure of the acceleration of the body.

It will be assumed that vertical dimensions of the accelerometer design have 1–10 micrometer scales while lateral scales are ~ 100 times greater. To calculate the damping force acting on the proof mass, one must determine the fluid pressure distribution in the gap between the proof mass and underlying surface. The fluid pressure distribution in the gap will be a function of the geometry of the proof mass, the gap height H, and the rate at which the fluid is being squeezed, dH/dt.

13.3.1 Scaling Analysis

Proceeding with the assumption that advection is small compared with diffusion, for a geometry where $H \ll L$, the scaled momentum equation, Eq. (13-11), becomes

$$\frac{\nu U}{H^2} \sim \frac{1}{\rho} \frac{F}{L^3}. \tag{13-16}$$

Suppose an accelerometer is being designed to measure accelerations up to $a_{body} \sim 100\ \text{m/s}^2$ and the overlying proof mass has a thickness h. The force exerted on the flow should scale as

$$F \sim a_{body} (\rho h L^2)_{\substack{proof \\ mass}}. \tag{13-17}$$

Suppose further that the design calls for $H/L \sim 0.01$ and the damping fluid is a gas ($\nu \sim 10^{-5} \text{m}^2/\text{s}$) confined to a gap of $H \sim 10^{-6}\text{m}$. Using Eq. (13-16), the Reynolds number for the flow can be estimated from

$$\frac{UH}{\nu} \sim \frac{a_{body}}{\nu^2 \rho} \frac{H^3}{L^3} (\rho h L^2)_{\substack{proof \\ mass}}. \tag{13-18}$$

Therefore, an order-of-magnitude estimate of the relative importance of inertial terms (advection) compared to viscous terms (diffusion) can be made (for $H/L \ll 1$) using

$$\frac{\textit{inertial terms}}{\textit{viscous terms}} \sim \frac{UH/\nu}{L/H} \sim \frac{a_{body} H^3}{\nu^2} \frac{H^2}{L^2} \frac{(\rho h)_{\substack{proof \\ mass}}}{(\rho H)_{fluid}}. \tag{13-19}$$

Assuming that the thickness of the proof mass has a similar scale to the gap dimension ($h \sim H$), and the weight ratio of the solid proof mass (silicon) to the fluid (gas) is of the order ~ 1000, one estimates that

$$\frac{\textit{inertial terms}}{\textit{viscous terms}} \sim \frac{(100)(10^{-6})^3}{(10^{-5})^2} (0.01)^2 (1000) = 10^{-7}. \tag{13-20}$$

This calculation shows that the inertial terms associated with advection and transient change are entirely negligible in the momentum equation and may be disregarded for the purpose of solving the flow field.

13.3.2 Flow Damping Coefficient

The response of the fluid to a force F exerted by the proof mass permits motion of the proof mass with some velocity $-\dot{H}$. By definition, the damping coefficient for this motion is $c = F/(-\dot{H})$. The relation between F (transmitted to the flow by pressure) and $\dot{H}$ is prescribed by the fluid motion, as governed by the momentum and continuity equations. The momentum equation can be simplified by disregarding the inertial terms. Additionally, the body force g_i can be neglected so long as the weight of the fluid is small compared to the weight of the proof mass. With these simplifications, the momentum equation takes the form

$$\underbrace{\partial_o v_i + v_j \partial_j v_i}_{\approx 0} = \nu \partial_j \partial_j v_i - \partial_i P/\rho + \underbrace{g_i}_{=0} . \tag{13-21}$$

The diffusion term contains the dummy index "j" representing a summation of three terms. However, through dimensional arguments, one expects the terms $\partial_x \partial_x v_i$ and $\partial_y \partial_y v_i$ (which scale as U/L^2) to be small compared with $\partial_z \partial_z v_i$ (which scales as U/H^2). Therefore, the momentum equation becomes

$$\partial_z \partial_z v_i = \frac{1}{\mu} \partial_i P \quad (\text{for } i = x \text{ and } y), \tag{13-22}$$

and is subject to the boundary conditions

$$v_i(z = 0) = 0 \quad \text{and} \quad v_i(z = H) = 0. \tag{13-23}$$

Using the boundary conditions, the momentum equation can be integrated for the result

$$v_i = \frac{\partial_i P}{2\mu}(z^2 - Hz). \tag{13-24}$$

Since the momentum equation describes two dependent variables, v_i and P, a second equation, continuity, is needed to close this problem. Substituting the momentum equation result for v_i into the continuity equation, expressed in the form of Eq. (13-4), yields a governing equation for the pressure distribution underneath the proof mass

$$\partial_i \partial_i P + \beta = 0, \quad \text{where} \quad \beta = 12\mu(-\dot{H})/H^3 . \tag{13-25}$$

Since the flow beneath a rectangular proof mass has two axes of symmetry, one can choose to solve the problem for one-quarter of the domain. The mathematical statement describing this problem is summarized in Figure 13-4, where a new variable $P^* = P - P_o$ is a measure of the pressure rise above the ambient condition.

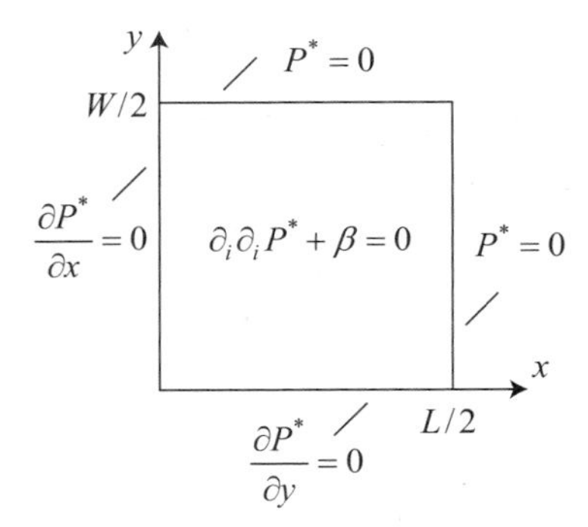

Figure 13-4 Mathematical statement for the squeeze flow problem.

The problem illustrated in Figure 13-4 has been solved previously for an analogous problem, both by separation of variables in Section 8.4.1 and by eigenfunction expansion in Section 9.1.3. Using the separation-of-variables solution, the pressure distribution is expressed by

$$P^* = (\beta/2)\left[(L/2)^2 - x^2\right] + \sum_{n=0}^{\infty} \frac{-4\beta(-1)^n \cos(\lambda_n x)\cosh(\lambda_n y)}{L\lambda_n^3 \cosh(\lambda_n W/2)}, \tag{13-26}$$

where $\lambda_n = (2n+1)\pi/L$. The damping force is obtained by integrating the pressure field over the area of the proof mass

$$F = 4\int_0^{W/2}\int_0^{L/2} P^* dxdy, \tag{13-27}$$

which evaluates to

$$F = 2\beta(LW)^2\phi(L, W) = \left(\frac{24\mu(-\dot{H})}{H^3}\right)(LW)^2\phi(L, W) \tag{13-28}$$

where

$$\phi(L/W) = \left[\frac{L/W}{24} - \sum_{n=0}^{\infty} \frac{8(L/W)^2}{(2n+1)^5\pi^5} \tanh\left((2n+1)\frac{\pi}{2}\frac{W}{L}\right)\right]. \tag{13-29}$$

Finally, the result for the damping coefficient can be determined from

$$c = \frac{F}{-\dot{H}} = \frac{24\mu}{H^3}(LW)^2\phi(L/W). \tag{13-30}$$

The damping coefficient is proportional to the area of the proof mass squared and inversely proportional to the gap dimension cubed. Additionally, the damping coefficient is proportional to the function (13-29) describing the influence of the proof mass length scale ratio L/W.

13.4 COATING EXTRUSION

Consider a method of coating in which a thin coating is extruded onto a plate in a process where the underlying plate is pulled through a die, as illustrated in Figure 13-5. The dimension of the die in the y-direction is assumed to be large. The goal is to determine the final thickness of the coating H_∞. This thickness does not equal the initial value H_o of the coating exiting the die because of the evolving velocity distribution between $x = L$ and $x \rightarrow \infty$. The velocity v_x at the free surface of the coating must be zero initially at the edge of the die ($x = L$) and go to U as $x \rightarrow \infty$. Since the average fluid velocity of the coating increases in this process, the thickness must decrease in order to conserve mass. To quantify the final thickness of the coating requires establishing the mass flow rate under the die.

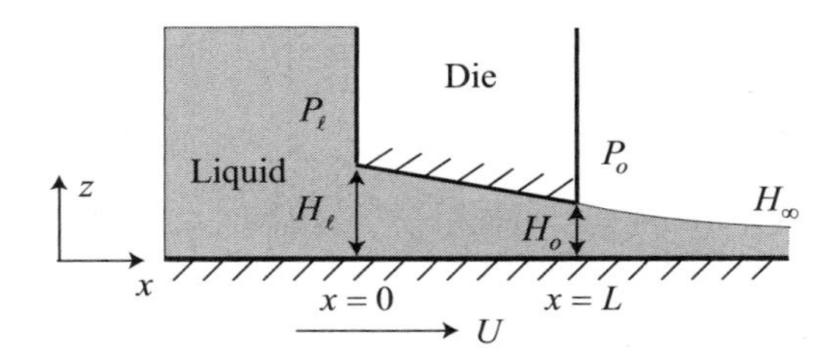

Figure 13-5 Extrusion coating on a flat plate.

13.4.1 Scaling Arguments

Imposed on this problem are the geometric scales, velocity scale, and pressure scale:

$$x \sim L, \ z \sim H, \ v_x \sim U, \quad \text{and} \quad P \sim \Delta P. \tag{13-31}$$

Scaling the continuity equation for an incompressible flow dictates

$$\frac{\partial v_x}{\partial x} + \frac{\partial v_z}{\partial z} = 0 \quad \rightarrow \quad \frac{U}{L} \sim \frac{v_z}{H}. \tag{13-32}$$

Therefore, the vertical velocity component scales as

$$v_z \sim \frac{H}{L} U. \tag{13-33}$$

The x-direction momentum equation can be scaled for the problem at hand:

$$\begin{array}{ccccccc} v_x \dfrac{\partial v_x}{\partial x} & + & v_z \dfrac{\partial v_x}{\partial z} & = & \nu \left(\dfrac{\partial^2 v_x}{\partial x^2} + \dfrac{\partial^2 v_x}{\partial z^2} \right) & - & \dfrac{1}{\rho} \dfrac{\partial P}{\partial x} \\ & & & \downarrow & & & \\ U \dfrac{U}{L} & \text{“}+\text{”} & \dfrac{UH}{L} \dfrac{U}{H} & \sim & \nu \left(\dfrac{U}{L^2} \text{“}+\text{”} \dfrac{U}{H^2} \right) & \text{“}+\text{”} & \dfrac{1}{\rho} \dfrac{\Delta P}{L}. \end{array} \tag{13-34}$$

Eliminating redundant scaling terms, the scaled momentum equation becomes

$$\overbrace{\frac{U^2}{L}}^{\substack{\textit{inertial forces} \\ \textit{(advection)}}} \sim \overbrace{\nu \left(\frac{U}{L^2} + \frac{U}{H^2} \right)}^{\substack{\textit{viscous forces} \\ \textit{(diffusion)}}} \text{“}+\text{”} \overbrace{\frac{1}{\rho} \frac{\Delta P}{L}}^{\textit{source}}. \tag{13-35}$$

Therefore, if $H \ll L$, the condition required for diffusion to dominate transport is that $(UH/\nu)(H/L) \ll 1$, which is the same provision that arose in the squeeze flow problem.

13.4.2 Final Coating Thickness

The momentum equation is written retaining only the diffusion and pressure terms. Expanding into a two-dimensional coordinate system gives

$$0 = \nu \left(\underbrace{\frac{\partial^2 v_x}{\partial x^2}}_{\approx 0} + \frac{\partial^2 v_x}{\partial z^2} \right) - \frac{1}{\rho} \frac{\partial P}{\partial x}. \tag{13-36}$$

Since $H \ll L$, on scaling grounds it is known that $\partial^2 v_x/\partial x^2 \ll \partial^2 v_x/\partial z^2$. Therefore, the governing equation for the coating flow becomes

$$\frac{\partial^2 v_x}{\partial z^2} - \frac{1}{\mu}\frac{\partial P}{\partial x} = 0. \tag{13-37}$$

Since the pressure does not vary with distance z across the gap, $\partial P/\partial x$ is not a function of z. Consequently, the momentum equation can be integrated with respect to z using the boundary conditions

$$v_x(z=0) = U \quad \text{and} \quad v_x(z=H) = 0, \tag{13-38}$$

where the gap dimension H is a function of x. The resulting solution for v_x is

$$v_x = \frac{1}{2\mu}\frac{dP}{dx}(z^2 - Hz) + U\left(1 - \frac{z}{H}\right). \tag{13-39}$$

The volume flux under the die (per unit width of the flow) can be calculated from

$$Q = \int_0^H v_x dz = \frac{-1}{12\mu}\frac{dP}{dx}H^3 + \frac{UH}{2}, \tag{13-40}$$

or

$$\frac{dP}{dx} = 6\mu\left(\frac{U}{H^2} - \frac{2Q}{H^3}\right). \tag{13-41}$$

Therefore, in terms of Q, the velocity solution can be expressed as

$$\frac{v_x(x,z)}{U} = \left(1 - \frac{z}{H}\right)\left[1 + 3\left(\frac{2Q}{UH} - 1\right)\frac{z}{H}\right], \tag{13-42}$$

and has a dependency on x following from the fact that the gap dimension $H(x)$ is a function of downstream distance.

To complete the solution, the continuity equation is invoked, requiring that

$$\int_0^H \left(\frac{\partial v_x}{\partial x} + \frac{\partial v_z}{\partial z}\right) dz = 0 \quad \text{or} \quad \int_0^H \frac{\partial v_x}{\partial x} dz = -v_z\Big|_{z=0}^{z=H}. \tag{13-43}$$

Using Leibniz's theorem (13-2) requires

$$\int_0^H \frac{\partial v_x}{\partial x} dz = \frac{\partial}{\partial x}\int_0^H v_x dz - \underbrace{\left[v_x \frac{\partial H}{\partial x}\right]\Big|_{z=0}^{z=H}}_{=0} = \frac{\partial}{\partial x}\int_0^H v_x dz. \tag{13-44}$$

Therefore, the continuity requirement (13-43) becomes

$$\frac{\partial}{\partial x}\int_0^H v_x dz = -v_z\Big|_{z=0}^{z=H}. \tag{13-45}$$

Since the flow does not penetrate the boundaries, continuity requires

$$\frac{\partial}{\partial x}\int_0^H v_x dz = \frac{\partial Q}{\partial x} = 0. \tag{13-46}$$

Therefore, differentiating expression (13-40) for Q, yields a differential equation governing the pressure distribution in the liquid under the die:

$$\frac{d}{dx}\left(\frac{1}{\mu}\frac{dP}{dx}H^3\right) = 6U\frac{dH}{dx}. \tag{13-47}$$

Since H is solely a function of x, it is straightforward to change the independent variable to H. By using the fact that H is a linear function of distance, the equation for pressure becomes

$$\frac{d}{dH}\left(\frac{dP}{dH}H^3\right) = \frac{-6\mu UL}{H_\ell - H_o}. \tag{13-48}$$

The pressure equation is integrated once for

$$\frac{dP}{dH} = \mu UL\left(AH^{-3} - \frac{6H^{-2}}{H_\ell - H_o}\right). \tag{13-49}$$

By integrating Eq. (13-49) across the die, it is found that the integration constant A must satisfy

$$\int_{H_\ell}^{H_o} dP = P_o - P_\ell = \mu UL\left(-\frac{AH^{-2}}{2} + \frac{6H^{-1}}{H_\ell - H_o}\right)\Bigg|_{H_\ell}^{H_o} \tag{13-50}$$

or

$$A = \frac{-12H_oH_\ell}{H_o^2 - H_\ell^2}\left(1 + \frac{P_\ell - P_o}{6\mu UL}H_oH_\ell\right). \tag{13-51}$$

Applying this result to Eq. (13-49), the pressure gradient under the die can be found:

$$\frac{dP}{dx} = \frac{dH}{dx}\frac{dP}{dH} = \frac{6\mu U}{H^3}\left\{H - \frac{2H_oH_\ell}{H_o + H_\ell}\left(1 + \frac{P_\ell - P_o}{6\mu UL}H_oH_\ell\right)\right\}. \tag{13-52}$$

Combining this result with expression (13-40) for Q allows the volume flow rate to be determined:

$$Q = UH_o\frac{H_\ell}{H_\ell + H_o}\left(1 + \frac{\Pi}{6}\frac{H_\ell}{H_o}\right), \tag{13-53}$$

where a dimensionless pressure is defined by:

$$\Pi = \frac{P_\ell - P_o}{\mu UL/H_o^2}. \tag{13-54}$$

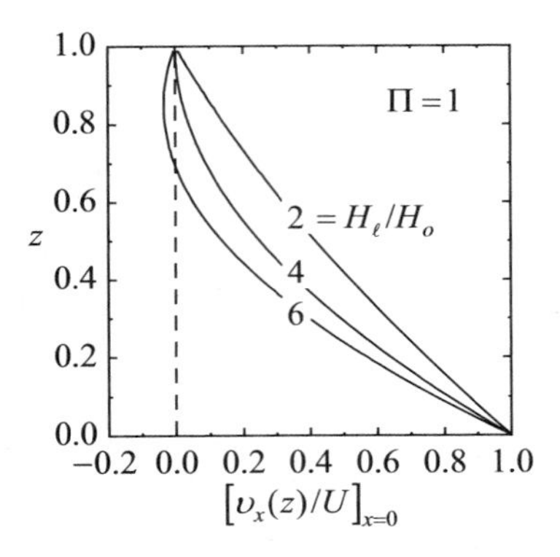

Figure 13-6 Velocity distribution at the die entrance.

Since mass conservation requires that $Q = UH_\infty$, the final coating thickness can be determined from (13-53), which results in the relation

$$\frac{H_\infty}{H_o} = \frac{H_\ell}{H_\ell + H_o}\left(1 + \frac{\Pi}{6}\frac{H_\ell}{H_o}\right). \tag{13-55}$$

With this result, the extrusion process can be designed to achieve the desired coating thickness H_∞.

It is instructive to plot the velocity distribution across the entrance of the die. Substituting Eq. (13-53) for the volume flux under the die into Eq. (13-42) for the velocity distribution, and evaluating the result at the inlet of the die where $H = H_\ell$, the result becomes

$$\left.\frac{v_x(z)}{U}\right|_{x=0} = \left(1 - \frac{z}{H_\ell}\right)\left[1 + 3\left(\frac{1}{H_\ell/H_o + 1}\left(2 + \frac{\Pi}{3}\frac{H_\ell}{H_o}\right) - 1\right)\frac{z}{H_\ell}\right]. \tag{13-56}$$

From the velocity distribution at the entrance of the die (13-56) it is seen that when H_ℓ/H_o becomes large, a portion of the flow field for v_x is forced to become negative. This corresponds to partial recirculation of the flow under the die. The condition for recirculation can be determined from Eq. (13-56) and is given by

$$\frac{1}{H_\ell/H_o + 1}\left(3 + \frac{\Pi}{2}\frac{H_\ell}{H_o}\right) < 1. \tag{13-57}$$

Figure 13-6 shows the velocity distribution across the entrance of the die for a pressure drop of $\Pi = 1$ and several ratios of H_ℓ/H_o. It is seen that when $\Pi = 1$, $H_\ell/H_o = 4$ is the largest tilt ratio of the die possible without recirculation occurring.

13.5 COATING EXTRUSION ON A POROUS SURFACE

Suppose the coating process discussed in Section 13.4 was performed on a surface that had a porous top layer of thickness D, as illustrated in Figure 13-7. Liquid penetrating the porous layer moves with a flux described by Darcy's law. Darcy's law* [2] is a phenomenological equation that assumes the viscous resistance to flow through a porous

---

*Named after the French engineer Henry Philibert Gaspard Darcy (1803–1858).

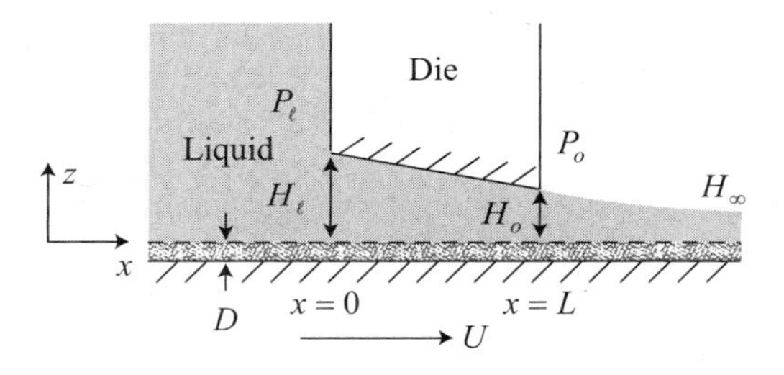

Figure 13-7 Extrusion coating on a porous surface.

medium has a linear dependence. For an isotropic porous medium Darcy's law takes the form

$$v_i^p = -\frac{\kappa}{\mu}(\partial_i P^p - \rho g_i), \tag{13-58}$$

where v_i^p is the volumetric fluid flux (or filtration velocity) and κ is the permeability of the medium. The superscript "p" is used to distinguish flow conditions in the porous medium from the overlying liquid layer. Darcy's law states that the volumetric flux of fluid is proportional to the pressure gradient in the porous medium minus the hydrostatic contribution of gravity.

The incompressible form of the continuity equation may be written for the filtration velocity (in two-dimensions) and integrated over the porous layer:

$$\int_{-D}^{0} \left(\frac{\partial v_x^p}{\partial x} + \frac{\partial v_z^p}{\partial z} \right) dz = 0. \tag{13-59}$$

This is used to express the volumetric flux of fluid leaving the surface of the porous layer:

$$-\int_{-D}^{0} \frac{\partial v_x^p}{\partial x} dz = v_z^p \Big|_{z=-D}^{z=0} = v_z^p(z = 0). \tag{13-60}$$

Using Darcy's law (13-58), the volumetric flux of fluid moving lengthwise in the porous layer can be expressed as

$$v_x^p = -\frac{\kappa}{\mu}\frac{\partial P^p}{\partial x}. \tag{13-61}$$

Substituting this result for v_x^p into (13-60) yields an expression for the volumetric flux of fluid leaving the surface of the porous layer in terms of the pressure field:

$$v_z^p(z = 0) = \frac{\kappa}{\mu} \int_{-D}^{0} \frac{\partial^2 P^p}{\partial x^2} dz. \tag{13-62}$$

If the porous layer is thin, the volumetric flux of fluid leaving the surface of the porous layer can be approximated by [3]:

$$v_z^p(z = 0) \approx \frac{\kappa}{\mu}\frac{\partial^2 P}{\partial x^2} D, \tag{13-63}$$

where $P^p \approx P$ is the pressure in the liquid layer overlying the porous layer.

So long as the no-slip condition, $v_x(z=0)$, is still applicable at the flow boundary with the porous layer, the velocity distribution across the liquid layer will be the same as that derived for the coating process of a nonporous surface, given by Eq. (13-39). However, continuity of the flow in the liquid layer now requires Eq. (13-45) to evaluate to

$$\frac{\partial}{\partial x}\int_0^H v_x dz = -v_z\Big|_{z=0}^{z=H} = v_z(z=0). \tag{13-64}$$

In other words, the lengthwise change in the volumetric flux of the liquid layer is balance by fluid transport out of the porous layer. Using Eq. (13-40) to evaluate the volume flux in the liquid layer, Eq. (13-64) becomes

$$v_z(z=0) = \frac{\partial}{\partial x}\int_0^H v_x dz = \frac{\partial}{\partial x}\left(\frac{-1}{12\mu}\frac{dP}{dx}H^3 + \frac{UH}{2}\right). \tag{13-65}$$

However, continuity across the interface between the porous layer and the liquid layer requires that

$$v_z^p(z=0) = v_z(z=0). \tag{13-66}$$

Therefore, equating Eq. (13-63) and Eq. (13-65) yields

$$\frac{\kappa}{\mu}\frac{\partial^2 P}{\partial x^2}D = \frac{\partial}{\partial x}\left(\frac{-1}{12\mu}\frac{dP}{dx}H^3 + \frac{UH}{2}\right) \tag{13-67}$$

or

$$\frac{d}{dx}\left\{(H^3 + 12\kappa D)\frac{dP}{dx}\right\} = 6\mu U\frac{dH}{dx}. \tag{13-68}$$

Since H is solely a function of x, it is straightforward to change the independent variable to H. By using the fact that H is a linear function of distance, the equation for pressure distribution under the die becomes

$$\frac{d}{dH}\left\{(H^3 + 12\kappa D)\frac{dP}{dH}\right\} = \frac{-6\mu UL}{H_\ell - H_o}. \tag{13-69}$$

Integrating once yields

$$\frac{dP}{dH} = \mu UL\frac{\dfrac{6H}{H_o - H_\ell} + A}{H^3 + 12\kappa D}, \tag{13-70}$$

where the dimensionless integration constant A is evaluated by integrating Eq. (13-70) over the length of the die:

$$P_o - P_\ell = \int_{H_\ell}^{H_o} \mu UL\frac{\dfrac{6H}{H_o - H_\ell} + A}{H^3 + 12\kappa D}dH. \tag{13-71}$$

Introducing the dimensionless variables

$$h = \frac{H}{H_o}, \quad h_\ell = \frac{H_\ell}{H_o}, \quad \Pi = \frac{P_\ell - P_o}{\mu UL/H_o^2}, \quad \text{and} \quad \psi = \frac{\kappa D}{H_o^3}, \tag{13-72}$$

the condition (13-71) imposed on the integration constant A becomes

$$\Pi = \int_1^{h_\ell} \frac{\dfrac{6h}{1-h_\ell} + A}{h^3 + 12\psi} dh. \tag{13-73}$$

Or, solving for A yields

$$A = \left(\int_1^{h_\ell} \frac{6h/(h_\ell - 1)}{h^3 + 12\psi} dh + \Pi \right) \Bigg/ \int_1^{h_\ell} \frac{dh}{h^3 + 12\psi}. \tag{13-74}$$

Once A is determined, (a numerical recipe for this is introduced in Chapter 23) the liquid volume flux under the die can be calculated:

$$Q = \overbrace{UD - \frac{\kappa D}{\mu}\frac{dP}{dx}}^{\text{porous layer}} + \overbrace{-\frac{H^3}{12\mu}\frac{dP}{dx} + \frac{UH}{2}}^{\text{liquid layer}} \tag{13-75}$$

Using Eq. (13-70) to evaluate the pressure gradient with respect to the downstream distance,

$$\frac{dP}{dx} = \mu U \frac{6H + A(H_o - H_\ell)}{H^3 + 12\kappa D}, \tag{13-76}$$

The liquid volume flux becomes

$$Q = -U\left(\kappa D + \frac{H^3}{12}\right)\frac{6H + A(H_o - H_\ell)}{H^3 + 12\kappa D} + UD + \frac{UH}{2}. \tag{13-77}$$

Introducing the dimensionless variables (13-72), the total volume flux under the die becomes:

$$Q = \frac{UH_o}{2} h\left(1 + \frac{2}{h}\left(\psi + \frac{h^3}{12}\right)\frac{A(h_\ell - 1) - 6h}{h^3 + 12\psi}\right) + UD. \tag{13-78}$$

Since mass conservation requires that the volume flux exiting the die equals the volume flux far downstream of the die, $Q(x = L) = U(D + H_\infty)$, the final coating thickness produced by the process can be determined from

$$\frac{H_\infty}{H_o} = \frac{1}{2}\left(1 + 2\left(\psi + \frac{1}{12}\right)\frac{A(H_\ell/H_o - 1) - 6}{1 + 12\psi}\right). \tag{13-79}$$

The constant A needed to evaluate Eq. (13-79) is determined from Eq. (13-74), and is dependent on the pressure difference across the die Π, the permeability of the porous layer ψ, and the tilt ratio of the die h_ℓ.

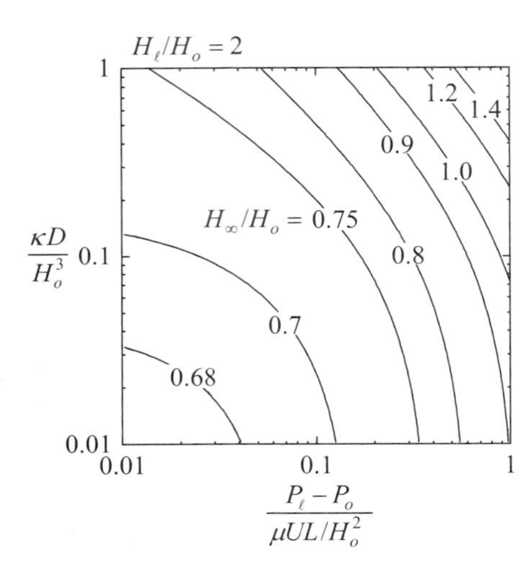

Figure 13-8 Effect of porous layer permeability and die pressure differential on coating thickness for the case when $H_l/H_o = 2$.

The solution for the coating thickness (13-79) is investigated in Figure 13-8 for the case when the tilt ratio of the die is $H_\ell/H_o = 2$. The results show that with greater pressure difference across the die and greater permeability of the porous layer, the coating thickness increases.

13.6 REYNOLDS EQUATION FOR LUBRICATION THEORY

Consider a more general scenario, depicted in Figure 13-9, of a flow driven by the motion of two bounding surfaces. The lower surface moves in the x-direction with velocity U, similar to the coating extrusion example of Section 13.4. However, the top surface can simultaneously move in the z-direction, similar to the squeeze flow problem treated in Section 13.3. The gap between the bottom and top surfaces is considered to be a general function of downstream distance $H(x)$. Additionally, if the top surface has a finite width (into the page), the flow can be squeezed past these edges, leading to a flow and pressure gradient in the y-direction (perpendicular to the plane of the page).

Analysis of the flow illustrated in Figure 13-9 follows steps similar to the previous sections of this chapter. Making use of the lubrication approximations, the momentum equation for v_i is written in the form

$$0 = \nu \frac{\partial^2 v_i}{\partial z^2} - \frac{1}{\rho}\partial_i P \quad (i = x \text{ or } y), \tag{13-80}$$

where inertial terms $(\partial_o v_i + v_j \partial_j v_i)$ have been dropped, and streamwise diffusion (in the x- and y-directions) is neglected relative to the transverse component of diffusion

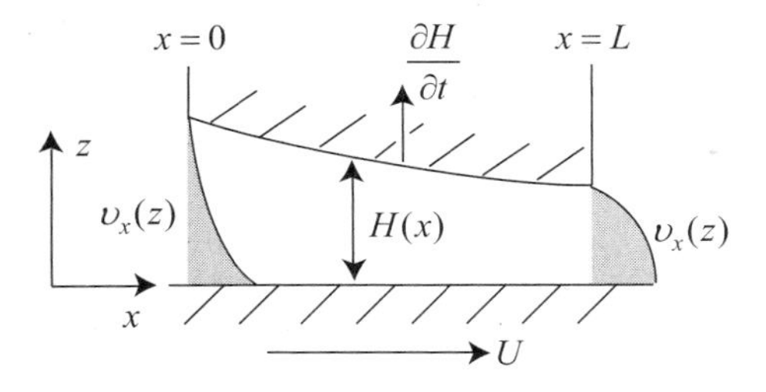

Figure 13-9 Illustration of flow for Reynolds equation.

($\nu\, \partial^2 v_i/\partial z^2$). The momentum equation is integrated with respect to z using the boundary conditions

$$\text{at} : z = 0, \quad v_x = U \qquad \text{and} \quad v_y = v_z = 0 \tag{13-81}$$

$$\text{at} : z = H, \quad v_x = v_y = 0 \quad \text{and} \quad v_z = \partial H/\partial t \tag{13-82}$$

for the results

$$v_x = \frac{1}{2\mu}\frac{dP}{dx}(z^2 - Hz) + U\left(1 - \frac{z}{H}\right) \tag{13-83}$$

$$v_y = \frac{1}{2\mu}\frac{dP}{dy}(z^2 - Hz). \tag{13-84}$$

Substituting the solutions for $v_i(z)$ into the incompressible continuity statement for the flow in the gap given, by Eq. (13-4), yields

$$\frac{1}{\mu}\frac{\partial}{\partial x}\left(H^3\frac{dP}{dx}\right) + \frac{1}{\mu}\frac{\partial}{\partial y}\left(H^3\frac{dP}{dy}\right) = 6U\frac{\partial H}{\partial x} + 12\frac{\partial H}{\partial t}. \tag{13-85}$$

Equation (13-85) describes the pressure distribution in the gap for the flow illustrated in Figure 13-9, and is known as the *Reynolds equation* [4]. When the gap spacing H is only a function of time, Eq. (13-85) simplifies to the governing equation for a squeeze flow. When the gap spacing H is only a function of x (and no pressure gradient across the width of the gap exists, $dP/dy = 0$), Eq. (13-85) simplifies to the governing equation for the coating extrusion example.

13.7 PROBLEMS

13-1 Spin coaters create a thin liquid film by spinning the surface on which the liquid lies. After short times, the fluid spreads on the surface with an angular velocity $v_\theta = \omega r$. By considering the flow at some radius far from the centerline, develop the condition required for diffusion to dominate transport in the momentum equation written for cylindrical coordinates. Make appropriate simplifications to the momentum equation when this condition is satisfied. Obtain an expression for the film thickness as a function of time. Show that for a large amount of time the film thickness changes as

$$H \approx \sqrt{\frac{\nu}{4\omega^2 t}}.$$

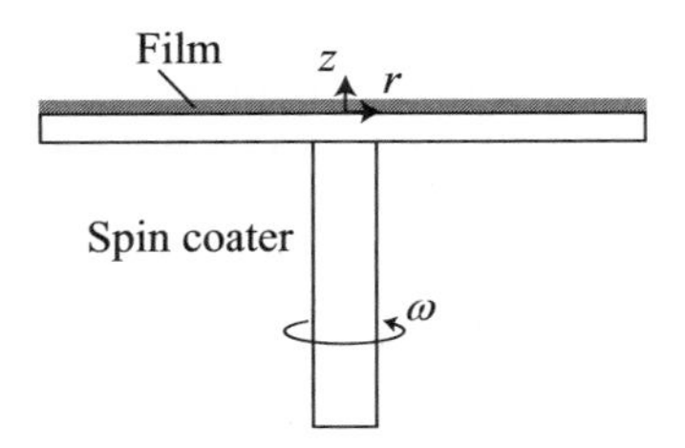

What will happen if the initial film is not uniform in thickness? Will the spatial variation in thinning rates favor the production of a uniform film?

13-2 Consider a bearing consisting of a stepped surface under which a flat surface is moving with speed U, as shown in the figure. Determine the load-bearing force F (per unit width) as a function of the geometry and speed of the bearing top surface. It may be assumed that $H_o \ll L$.

$L/2$ F $L/2$

$2H_o$ H_o

U

13-3 Consider the journal bearing shown. The journal is rotating at an angular speed ω. The displacement of the journal center from the bearing center is the eccentricity, a, as shown. The difference in radii between the bearing and the journal is the clearance, $\varepsilon = R_B - R$. When the gap between the journal and bearing is small, the lubrication thickness varies to a good approximation as $h(x) = h(R\theta) = \varepsilon + a\cos(\theta)$. Find the governing equation for the pressure distribution around the journal. Show that the solution is given by

$$\frac{P(\theta) - P(0)}{\mu\omega(R/\varepsilon)^2} = 6\int_0^\theta \left(\frac{1 + (a/\varepsilon)\cos\theta + A}{[1 + (a/\varepsilon)\cos\theta]^3}\right) d\theta.$$

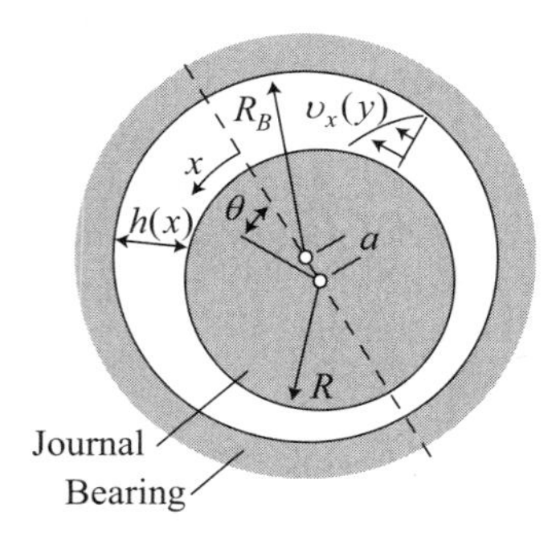

To specify the integration constant A, a final condition must be imposed on the solution. With $P(2\pi) = P(0)$, Sommerfeld [5] found the analytic solution:

$$\frac{P(\theta) - P(0)}{\mu\omega(R/\varepsilon)^2} = \frac{6(a/\varepsilon)\sin\theta\,[2 + (a/\varepsilon)\cos\theta]}{[2 + (a/\varepsilon)^2][1 + (a/\varepsilon)\cos\theta]^2}.$$

This solution, and others specified by different boundary conditions, can be investigated numerically, as developed in Chapter 23 (see Problems 23-5 and 23-6).

13-4 Consider a perforated tabletop of thickness D through which air is blown. The blowing creates an air bearing for a long bar of width L placed on the table, as illustrated. Assume the underside of the tabletop is pressurized to P_o, and that the air flow through the tabletop is incompressible and can be modeled with Darcy's law. Determine the lifting force of the air bearing as a function of H. Evaluate the limiting dependency of the lifting force on H as $H \to 0$ and as $H \to \infty$.

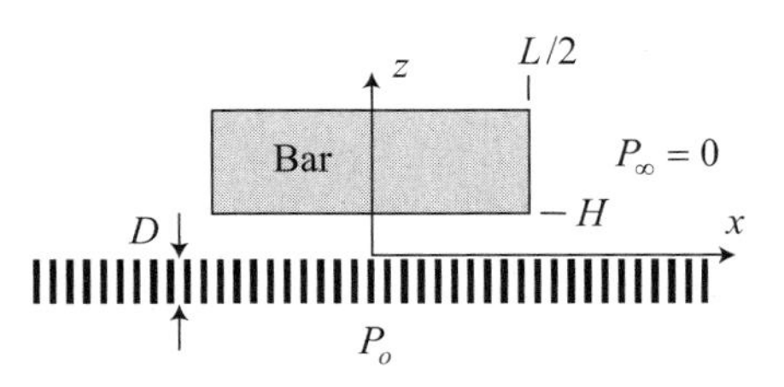

13-5 Consider air blowing through a perforated tabletop, as analyzed in Problem 13-4. Now suppose that the object being supported by the air bearing is a cube with sides of length L. Determine the lifting force of the air bearing as a function of H.

13-6 Consider air blowing through a perforated tabletop, as analyzed in Problem 13.4. Now suppose that the object being supported by the air bearing is a disk of radius r_o. Determine the lifting force of the air bearing as a function of H. Evaluate the limiting dependency of the lifting force on H as $H \to 0$ and as $H \to \infty$.

13-7 Consider the coating of a surface having a porous layer, as discussed in Section 13.5. However, now suppose the die geometry has a simple step-down in height that occurs mid-distance ($x = L/2$) across the die. Reanalyze the problem to solve for the coating thickness H_∞/H_o as a function of the dimensionless variables:

$$h_\ell = \frac{H_\ell}{H_o}, \quad \Pi = \frac{P_\ell - P_o}{\mu UL/H_o^2}, \quad \text{and} \quad \psi = \frac{\kappa D}{H_o^3}.$$

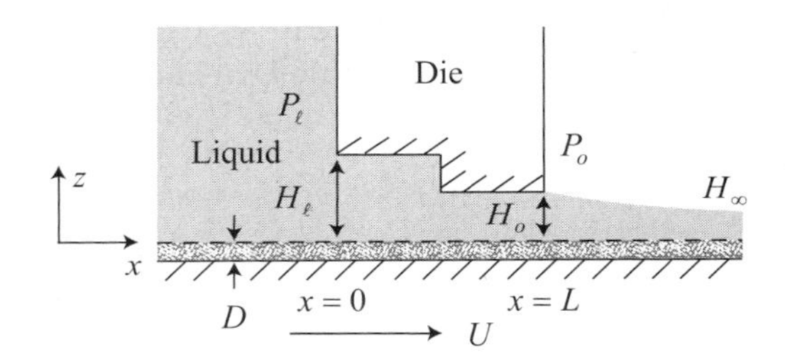

Determine whether the sloped or stepped die geometry gives a thicker coating when $h_\ell = 2$, $\Pi = 0.3$, and $\psi = 0.1$. (For these conditions, Figure 13.8 shows that the sloped die gives a coating thickness of $H_\infty/H_o = 0.8$).

REFERENCES

[1] O. Reynolds, "An Experimental Investigation of the Circumstances which Determine whether the Motion of Water Shall Be Direct or Sinuous, and of the Law of Resistance in Parallel Channels," *Philosophical Transactions of the Royal Society of London*, **174**, 935 (1883).

[2] H. Darcy, *Les Fontaines Publiques de la Ville de Dijon.* Paris, France: Dalmont, 1856.

[3] V. T. Morgan and A. Cameron, "Mechanism of Lubrication in Porous Metal Bearing," in the *Proceedings of the Conference on Lubrication and Wear.* London: Institution of Mechanical Engineers, 151 (1957).

[4] O. Reynolds, "On the Theory of Lubrication and Its Application to Mr. Beauchamp Tower's Experiments, Including an Experimental Determination of the Viscosity of Olive Oil," *Philosophical Transactions of the Royal Society of London*, Pt. 1, **177**, 157 (1886).

[5] A. Sommerfeld, "Zur Hydrodynamischen Theorie der Schmiermittelreibung," Zeitschrift für Mathematik und Physik, **50**, 97 (1904).

Chapter 14

Inviscid Flow

14.1 The Reynolds Number
14.2 Inviscid Momentum Equation
14.3 Ideal Plane Flow
14.4 Steady Potential Flow through a Box with Staggered Inlet and Exit
14.5 Advection of Species through a Box with Staggered Inlet and Exit
14.6 Spherical Bubble Dynamics
14.7 Problems

The no-slip condition between a flow and a solid surface causes a *boundary layer* to form in the region of the flow nearest to the surface. In this region neither diffusion nor advection transport can be ignored. However, outside of the boundary layer the flow may be largely uninfluenced by the effects of the fluid viscosity. In such a case, the flow Figure 14-1 is said to be *inviscid*.

The presence of surfaces is still felt by inviscid flows through the transmission of pressure. The compressibility of a fluid to the pressure field may become a significant effect when flow speeds are comparable to the speed of sound. However, treatment of this is deferred until the discussion of gas dynamics in Chapter 20. In the present chapter, incompressible inviscid flows are investigated, which can include gas flows of moderate speed.

When advection is dominant, transport equations will have the form

$$\frac{\partial(?)}{\partial t} + v_j \partial_j(?) = 0 + (sources) \quad \text{or} \quad \frac{D(?)}{Dt} = 0 + (sources). \tag{14-1}$$

The latter form emphasizes that, in the absence of diffusion, the fluid content of energy, species, momentum, or any other property of the fluid will not change for a material element traveling with the flow unless some source (or sink) is present. Therefore, in

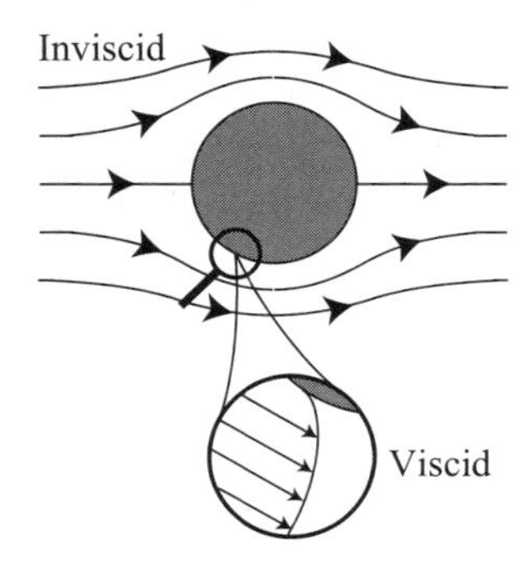

Figure 14-1 Viscid versus inviscid regions of a flow.

describing advection transport of any property, the primary task is to determine the motion of the fluid flow. However, when momentum (v_i) is the quantity of interest, the transport equations become nonlinear. This makes their solution quite difficult in general. Chapter 17 introduces a numerical approach to solving the nonlinear transient advection equation, which is applied to problems in open channel flows (Chapters 18 and 19) and compressible flows (Chapters 20 through 22). Most problems addressed in those chapters are simplified by considering flows that are "directed" to follow a particular path. In such a situation, the challenge is not in describing where the flow goes, but in establishing the state of the flow.

If advection is allowed to transport fluid in more than one spatial direction, the path of the flow becomes a part of the desired solution. Sections 14.3 through 14.5 of this chapter will concentrate on a simplification to this problem that arises when steady advection transport is known to be irrotational, in addition to incompressible. In this case, problems can be described in terms of a stream function, whose solution over the domain of the fluid is governed by an equation identical in form to steady-state diffusion. Since this equation is linear, it is amenable to separation of variables as a solution procedure when a finite domain with well-posed boundary conditions is identified. The weakness of this approach is that, while an irrotational state can be imposed on a solution, there is often no assurance that this necessarily yields the most physically sound description of the flow.

14.1 THE REYNOLDS NUMBER

To say that advection is dominant in regions of the flow that are far from surfaces requires a clearer meaning of "far." To obtain this meaning, the advection and diffusion terms in the momentum equation are nondimensionalized using a characteristic velocity scale U and a characteristic length scale L. Defining the dimensionless variables,

$$v_i^* = v_i/U \quad \text{and} \quad x_i^* = x_i/L, \tag{14-2}$$

the steady-state momentum equation can be transformed

$$\begin{array}{ccccc} v_j\partial_j v_i & = & \nu\partial_j\partial_j v_i & + & (\cdots) \\ \downarrow & & \downarrow & & \\ \dfrac{U^2}{L} v_j^*\partial_j^* v_i^* & = & \dfrac{\nu U}{L^2}\partial_j^*\partial_j^* v_i^* & + & (\cdots) \end{array} \tag{14-3}$$

where ∂_j^* is a derivative with respect to the dimensionless spatial variable x_j^*. Now the dimensionless momentum equation can be written in the form

$$\mathrm{Re}_L v_j^*\partial_j^* v_i^* = \partial_j^*\partial_j^* v_i^* + \quad (\cdots), \tag{14-4}$$

where the Reynolds number is defined by

$$\mathrm{Re}_L = \frac{UL}{\nu}. \tag{14-5}$$

By design, the dimensionless terms $v_j^*\partial_j^* v_i^*$ and $\partial_j^*\partial_j^* v_i^*$ should be of order one. Therefore, it becomes apparent that the meaning of "far" is embodied in the length scale L associated with the Reynolds number Re_L. When Re_L is large, for a corresponding distance L from the surface, the advection term in the momentum equation will be more important than diffusion. However, when there is more than one characteristic length scale in a flow, the significance of the Reynolds number becomes less clear. This was evident in lubrication theory, studied in Chapter 13, where there is a streamwise length scale and a transverse length scale. Even though lubrication flows can have a large Reynolds number, based on the streamwise length scale, diffusion transport dominates the flow characteristics. Similarly, hydrodynamic boundary layers also have two important length scales. Although boundary layers have large Reynolds numbers based on a streamwise length scale, the advection and diffusion terms have the same order of magnitude, as will be shown in Chapter 25.

14.2 INVISCID MOMENTUM EQUATION

When advection dominates transport in a flow, the diffusion term is dropped from the momentum equations:

$$\partial_o v_i + v_j \partial_j v_i = -\partial_i P/\rho + g_i \qquad \text{(inviscid flow)}. \tag{14-6}$$

Furthermore, the advection term can be written (see Problem 3-1) as

$$v_j \partial_j v_i = \partial_i (v_k v_k/2) - \varepsilon_{ijk} v_j \omega_k \tag{14-7}$$

where $\omega_k = \varepsilon_{klm} \partial_l v_m$ is the *vorticity*. Vorticity is a measure of rotation in the flow. Typically, rotation is introduced to a flow through viscous effects. Therefore, flows governed by advection (inviscid flows) have a tendency to be irrotational ($\omega_k = 0$), allowing the advection term to be written more simply as $v_j \partial_j v_i = \partial_i (v_k v_k/2)$. Irrotational flow is also called *potential flow*. For an inviscid and irrotational flow, the momentum equations become

$$\partial_o v_i + \partial_i (v^2/2) = -\partial_i P/\rho + g_i \qquad \text{(inviscid and irrotational)}. \tag{14-8}$$

A body force g_i (such as gravity) can be represented as a potential field through $-g\partial_i h$, where h is a measure of distance along the direction of $-\vec{g}$. The scalar g is the magnitude of $\vec{g}$. To illustrate, consider the situation shown in Figure 14-2, where the distance in the positive x-direction coincides with h. It is seen that

$$\text{for } i \to x, \quad g_x = -g\partial_x(h) = -g\underbrace{(\partial h/\partial x)}_{=1} = -g \tag{14-9}$$

and

$$\text{for } i \to y, \quad g_y = -g\partial_y(h) = -g\underbrace{(\partial h/\partial x)}_{=0} = 0. \tag{14-10}$$

Therefore,

$$g_i = -g\partial_i h. \tag{14-11}$$

Furthermore, a velocity potential function Φ can be defined such that

$$v_i = \partial_i \Phi. \tag{14-12}$$

In terms of the velocity potential function and the gravity potential function, the momentum equations become

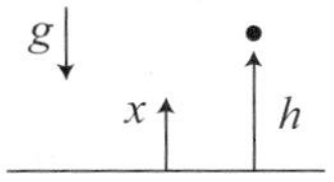

Figure 14-2 Gravity potential.

$$\partial_o \partial_i \Phi + \partial_i (v^2/2) = -\partial_i (P/\rho + gh) \quad \text{(incompressible, inviscid, and irrotational)}, \qquad (14\text{-}13)$$

where ρ and g are taken to be constant (and are brought inside the spatial derivatives). A flow that is both inviscid and incompressible is called an *ideal flow*. Since time and space are independent variables, the order of differentiation $\partial_o \partial_i$ can be reversed. This permits all the terms in the momentum equation to be collected inside a common spatial derivative, such that

$$\partial_i (\partial_o \Phi + v^2/2 + P/\rho + gh) = 0. \qquad (14\text{-}14)$$

Upon integration, the final form of the momentum equation becomes *Bernoulli's equation* [1]:

$$\partial_o \Phi + v^2/2 + P/\rho + gh = C(t). \qquad (14\text{-}15)$$

Since integration was performed with respect to the spatial variables, the integration "constant" in Eq. (14-15) may be a function of time. It should be noted that even if the flow is rotational, Eq. (14-15) will be true along a streamline in the flow since the integral of $\varepsilon_{ijk} v_j \omega_k$ in the advection term (14-7) must be zero along this path. A *streamline* traces the path of a flow, with the attributes of being everywhere tangent to the local fluid velocity. The integral of $\varepsilon_{ijk} v_j \omega_k$ along a streamline can be shown to be zero by demonstrating that $\varepsilon_{ijk} v_j \omega_k$ is perpendicular to the local velocity v_i—in other words, by showing that $v_i \varepsilon_{ijk} v_j \varepsilon_{klm} \partial_l v_m = 0$ (see Problem 3-8).

14.3 IDEAL PLANE FLOW

Once an incompressible ($\rho = const.$) flow is also irrotational, the velocity solution becomes dictated by the flow *kinematic equations*:

$$\varepsilon_{klm} \partial_l v_m = 0 \quad \text{(irrotational)} \qquad (14\text{-}16)$$

and

$$\partial_j v_j = 0 \quad \text{(continuity)}. \qquad (14\text{-}17)$$

Dynamic information, provided by the momentum equation, is unnecessary for integration of the flow kinematic equations (although the constant of integration may still be related to dynamic constraints). In Cartesian coordinates, the kinematic equations for a two-dimensional flow are

$$\frac{\partial v_y}{\partial x} - \frac{\partial v_x}{\partial y} = 0 \quad \text{(irrotational)} \qquad (14\text{-}18)$$

and

$$\frac{\partial v_x}{\partial x} + \frac{\partial v_y}{\partial y} = 0 \quad \text{(continuity)}. \qquad (14\text{-}19)$$

It is clear that these two equations alone may be integrated for the two unknown velocity components. The process of finding a solution to the kinematic equations for planar flows is simplified by the use of a *stream function* $\psi(x, y)$, which is defined by the relations

$$v_x = \frac{\partial \psi}{\partial y} \quad \text{and} \quad v_y = -\frac{\partial \psi}{\partial x}. \qquad (14\text{-}20)$$

One sees that this judicious definition of the stream function automatically satisfies the continuity equation (14-19), since

$$\frac{\partial}{\partial x}\left(\frac{\partial \psi}{\partial y}\right) + \frac{\partial}{\partial y}\left(-\frac{\partial \psi}{\partial x}\right) = 0. \tag{14-21}$$

Therefore, the remaining task is to satisfy the statement that the flow is irrotational. Substituting the stream function definitions into Eq. (14-18) yields:

$$\frac{\partial}{\partial x}\left(-\frac{\partial \psi}{\partial x}\right) - \frac{\partial}{\partial y}\left(\frac{\partial \psi}{\partial y}\right) = 0. \tag{14-22}$$

In other words, the governing equation for an inviscid irrotational two-dimensional flow is simply the Laplace equation:

$$\partial_j \partial_j \psi = 0, \quad j = x, \ y. \tag{14-23}$$

Notice that the Laplace equation has the same mathematical form as steady-state diffusion, even though diffusion is absent from the flow being described. Any function satisfying Eq. (14-23) describes an irrotational incompressible planar flow. In Chapter 15, a number of well-known stream functions that describe relatively simple flow patterns are cataloged. Those flows are unbounded with respect to part of the domain. In the next section, the Laplace equation is solved in a systematic way for a planar flow in a bounded domain using the eigenfunction expansion method (Chapter 9).

It is important to remember that when a solution is derived from the assumption that the flow field is irrotational, there may be a more plausible solution that describes a rotational flow (contrary to the initial assumption). Therefore, care is warranted when labeling a flow as irrotational, since this may lead to a physically implausible solution, as will become evident in the next section.

14.4 STEADY POTENTIAL FLOW THROUGH A BOX WITH STAGGERED INLET AND EXIT

Consider a problem in which a flow moves through a two-dimensional box as shown in Figure 14-3. The flow enters and exits the box with a uniform speed of U. The path of the flow between the inlet and exit is desired. If the problem is characterized by a sufficiently large Reynolds number Re_L (L and W are of the same scale), then diffusion near the walls can be neglected. Proceeding with the assumption that the flow is irrotational and incompressible, the streamlines of the flow are governed by Eq. (14-23).

In potential flow, surfaces, while impenetrable to the flow, do not impose the no-slip condition of a viscous flow; the flow slides along surfaces inviscidly. Of course, in reality a thin boundary layer region adjacent to the wall exists in which viscous effects are important. However, it is assumed that the thickness of the boundary layer is small

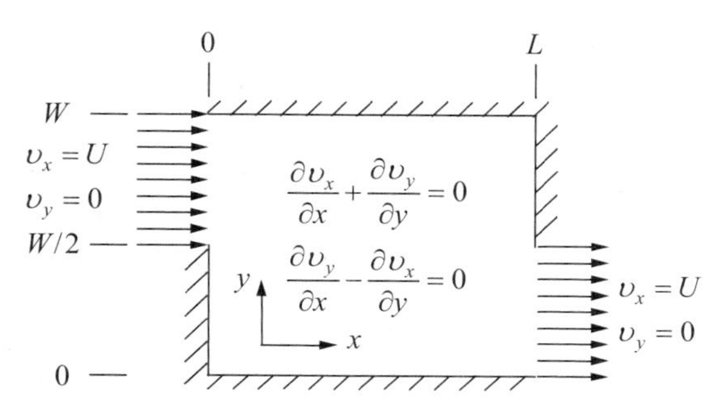

Figure 14-3 Potential flow through a box.

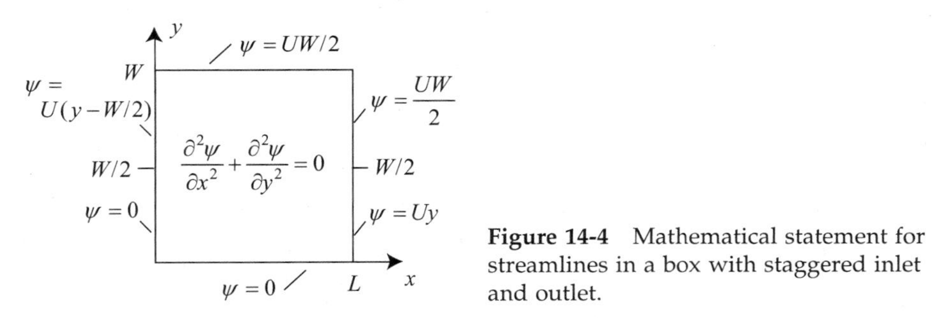

Figure 14-4 Mathematical statement for streamlines in a box with staggered inlet and outlet.

enough to be ignored compared to the other scales of the problem. Because the normal component of the velocity field is zero along any surface, the stream function is a constant along all surfaces. Consider the bottom surface of the box as a simple illustration. Since $v_y = (-\partial\psi/\partial x) = 0$, integrating with respect to x yields $\psi = const.$ along the bottom surface. Notice that there is an arbitrary constant associated with the stream function solution because the velocity field is related to the derivative of the stream function. This permits the value of one streamline in the solution to be arbitrarily set.

Over the box inlet region, shown in Figure 14-3, $v_x = \partial\psi/\partial y = U$. Therefore, integration yields $\psi(x = 0, y > W/2) = Uy + C_1$. If the streamline defining the front and bottom surfaces of the box is chosen to be zero, then $C_1 = -UW/2$. Over the exit region $v_x = \partial\psi/\partial y = U$ and $\psi(x = L, y < W/2) = Uy + C_2$. To satisfy the value of the stream function at the bottom surface of the box, $C_2 = 0$. Since the stream function must be continuous, the streamline defining the top and back surfaces of the box must correspond to $\psi(y = W) = \psi(x = L, y > W/2) = UW/2$.

Having established the boundary conditions for ψ, the mathematical problem is summarized in Figure 14-4. The solution for ψ can be determined by separation of variables (Chapter 8) or eigenfunction expansion (Chapter 9). Here the method of eigenfunction expansion is used, starting with the assumed form of the solution:

$$\psi = \sum_{n=1}^{\infty} X_n(\lambda_n x) Y_n(\lambda_n y) \tag{14-24}$$

where

$$Y_n(\lambda_n y) = \sin(\lambda_n y) \tag{14-25}$$

and

$$\lambda_n = n\pi/W \quad \textit{for } n = 1, 2, \ldots. \tag{14-26}$$

Notice that the homogeneous version of the boundary conditions of the problem for $Y_n(y)$ are satisfied, since $Y_n(y = 0) = \sin(0) = 0$ and $Y_n(y = W) = \sin(\lambda_n W) = 0$. The function $X_n(x)$ is determined such that the original problem, with all the nonhomogeneous conditions, is satisfied. $X_n(x)$ is found by multiplying the governing equation by the eigenfunction $\sin(\lambda_m y)$ and integrating with respect to y:

$$\int_0^W \left(\frac{\partial^2 \psi}{\partial x^2} + \frac{\partial^2 \psi}{\partial y^2} \right) \sin(\lambda_m y) dy = 0. \tag{14-27}$$

Integrating the first term in the governing equation yields

$$\int_0^W \frac{\partial^2 \psi}{\partial x^2} \sin(\lambda_m y) dy = \int_0^W \left[\sum_{n=0}^{\infty} \frac{\partial^2 X_n}{\partial x^2} \sin(\lambda_n y) \right] \sin(\lambda_m y) dy = \frac{\partial^2 X_n}{\partial x^2} \frac{W}{2}. \tag{14-28}$$

Integrating the second term in the governing equation by parts twice yields

$$\int_0^W \frac{\partial^2 \psi}{\partial y^2} \sin(\lambda_m y) dy = \left(\frac{\partial \psi}{\partial y} \sin(\lambda_m y) - \psi \lambda_m \cos(\lambda_m y) \right) \Big|_0^W - \int_0^W \psi \lambda_m^2 \sin(\lambda_m y) dy. \tag{14-29}$$

Since $\sin(\lambda_m W) = 0$, $\psi(y = W) = UW/2$, $\sin(0) = 0$, and $\psi(y = 0) = 0$, this simplifies to

$$\int_0^W \frac{\partial^2 \psi}{\partial y^2} \sin(\lambda_m y) dy = - \frac{UW}{2} \lambda_m \cos(\lambda_m W) - \int_0^W \psi \lambda_m^2 \sin(\lambda_m y) dy \tag{14-30}$$

or

$$\int_0^W \frac{\partial^2 \psi}{\partial y^2} \sin(\lambda_m y) dy = -\frac{UW}{2} \lambda_m \cos(\lambda_m W) - \int_0^W \left[\sum_{n=0}^{\infty} X_n(x) \sin(\lambda_n y) \right] \lambda_m^2 \sin(\lambda_m y) dy$$

$$= -\frac{UW}{2} \lambda_m (-1)^n - X_n(x) \lambda_m^2 \frac{W}{2}. \tag{14-31}$$

The integration results are applied to the governing equation (14-27) to yield

$$\frac{\partial^2 X_n}{\partial x^2} - \lambda_m^2 X_n(x) = U \lambda_m (-1)^n. \tag{14-32}$$

The ordinary differential equation for X_n is integrated:

$$X_n = C_n \sinh[\lambda_n (L - x)] + D_n \sinh(\lambda_n x) - U(-1)^n / \lambda_m. \tag{14-33}$$

Reconstructing the solution for ψ yields

$$\psi = \sum_{n=1}^{\infty} [C_n \sinh[\lambda_n (L - x)] + D_n \sinh(\lambda_n x) - U(-1)^n / \lambda_n] \sin(\lambda_n y). \tag{14-34}$$

Next, the appropriate values of C_n and D_n for each eigenvalue λ_n are determined such that the final nonhomogeneous boundary conditions are satisfied. Starting with

$$\text{at } x = 0: \quad \psi(x = 0) = \sum_{n=1}^{\infty} [C_n \sinh(\lambda_n L) - U(-1)^n / \lambda_n] \sin(\lambda_n y). \tag{14-35}$$

The orthogonality principle is used to determine C_n:

$$\int_0^W \psi(x=0)\sin(\lambda_m y)dy = \int_0^W \sum_{n=1}^{\infty}[C_n \sinh(\lambda_n L) - U(-1)^n/\lambda_n]\sin(\lambda_n y)\sin(\lambda_m y)dy$$
$$= [C_m \sinh(\lambda_m L) - U(-1)^n/\lambda_m]\frac{W}{2}. \tag{14-36}$$

The left hand side integrates to

$$\int_0^W \psi(x=0)\sin(\lambda_m y)dy = \int_{W/2}^W Uy\sin(\lambda_m y)dy - \int_{W/2}^W \frac{UW}{2}\sin(\lambda_m y)dy$$
$$= U\left[\frac{\sin(\lambda_m y)}{\lambda_m^2} + \left(\frac{W}{2} - y\right)\frac{\cos(\lambda_m y)}{\lambda_m}\right]\Bigg|_{W/2}^{W} = -U\left[\frac{\sin(n\pi/2)}{\lambda_m^2} + \frac{W}{2}\frac{(-1)^n}{\lambda_m}\right]. \tag{14-37}$$

Therefore,

$$C_n = \frac{-2U}{W\lambda_n^2}\frac{\sin(n\pi/2)}{\sinh(\lambda_n L)}. \tag{14-38}$$

Next, to determine D_n, the other boundary condition is used:

$$\text{at } x = L: \quad \psi(x=L) = \sum_{n=1}^{\infty}[D_n\sinh(\lambda_n L) - U(-1)^n/\lambda_n]\sin(\lambda_n y). \tag{14-39}$$

Again, using the orthogonality principle:

$$\int_0^W \psi(x=L)\sin(\lambda_m y)dy = \int_0^W \sum_{n=1}^{\infty}[D_n \sinh(\lambda_n L) - U(-1)^n/\lambda_n]\sin(\lambda_n y)\sin(\lambda_m y)dy$$
$$= [D_m \sinh(\lambda_m L) - U(-1)^m/\lambda_m]\frac{W}{2}. \tag{14-40}$$

The left hand side integrates to

$$\int_0^W \psi(x=L)\sin(\lambda_m y)dy = \int_0^{W/2} Uy\sin(\lambda_m y)dy + \int_{W/2}^W \frac{UW}{2}\sin(\lambda_m y)dy$$
$$= U\left[\frac{\sin(\lambda_m y)}{\lambda_m^2} - y\frac{\cos(\lambda_m y)}{\lambda_m}\right]_0^{W/2} - \frac{UW}{2}\left[\frac{\cos(\lambda_m y)}{\lambda_m}\right]_{W/2}^{W} \tag{14-41}$$
$$= U\left[\frac{\sin(n\pi/2)}{\lambda_m^2} - \frac{W}{2}\frac{(-1)^n}{\lambda_m}\right].$$

Therefore,

$$D_n = \frac{2U}{W\lambda_n^2} \frac{\sin(n\pi/2)}{\sinh(\lambda_n L)}, \tag{14-42}$$

and the solution becomes

$$\psi = \sum_{n=1}^{\infty} \frac{2U}{W\lambda_n^2} \left[\frac{\sin(n\pi/2)}{\sinh(\lambda_n L)} (\sinh(\lambda_n x) - \sinh[\lambda_n(L-x)]) - \frac{W\lambda_n(-1)^n}{2} \right] \sin(\lambda_n y) \tag{14-43}$$

or

$$\psi = \sum_{n=1}^{\infty} \frac{2U}{W\lambda_n^2} \left[\frac{\sin(n\pi/2)}{\sinh(\lambda_n L)} (\sinh(\lambda_n x) - \sinh[\lambda_n(L-x)]) \right] \sin(\lambda_n y) + Uy/2. \tag{14-44}$$

The second form of the solution results from the fact that

$$\sum_{n=1}^{\infty} (-1)^n \sin(\lambda_n y)/\lambda_n = -y/2. \tag{14-45}$$

This is more than a cosmetic substitution because the series obtained by differentiating the stream function for $v_x = \partial\psi/\partial y$ converges in the second form but does not converge in the first form.

Figure 14-5 plots the solution for the stream function. Although the results may "please the eye," there are aspects of the solution that lack realism. Recall that the notion that the flow is irrotational has also been imposed on the solution. For the problem at hand, the solution resulting from this approach is unlikely to be observed in a real flow for the following reason: It is implausible that the flow can "sweep out" the lower left-hand corner of the box as shown in Figure 14-5. A more realistic expectation would be for the flow entering the box to detach at the lower edge of the entrance in a way that leaves recirculating fluid trapped in the corner. This recirculation region cannot be revealed in the present solution because its presence would violate the starting premise of an irrotational flow. Although the main flow through the box could still be irrotational, it is no longer bounded in a simple way by the walls of the box alone, due to the presence of the recirculation region.

Given a stream function solution, with corresponding velocity field, the momentum equation specifies a pressure field that must accompany the flow. Using Eq. (14-44) for ψ, the velocity field everywhere in the box can be calculated using the definitions given by Eq. (14-20), for the result

$$v_x = \frac{\partial\psi}{\partial y} = \sum_{n=1}^{\infty} \frac{2U}{W\lambda_n} \left[\frac{\sin(n\pi/2)}{\sinh(\lambda_n L)} (\sinh(\lambda_n x) - \sinh[\lambda_n(L-x)]) \right] \cos(\lambda_n y) + U/2 \tag{14-46}$$

$$v_y = -\frac{\partial\psi}{\partial x} = -\sum_{n=1}^{\infty} \frac{2U}{W\lambda_n} \left[\frac{\sin(n\pi/2)}{\sinh(\lambda_n L)} (\cosh(\lambda_n x) + \cosh[\lambda_n(L-x)]) \right] \sin(\lambda_n y). \tag{14-47}$$

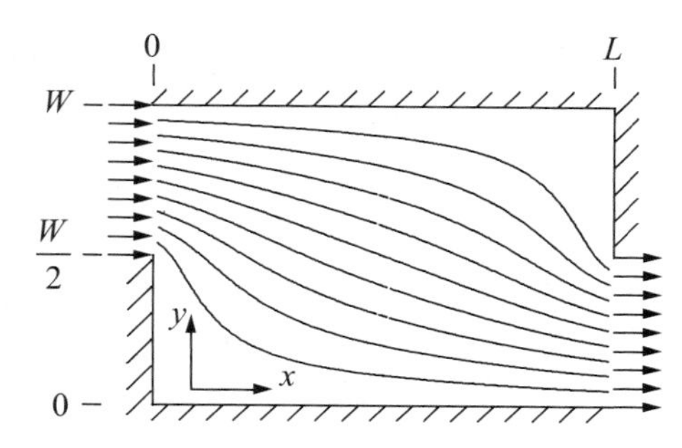

Figure 14-5 Streamlines for potential flow through a box with staggered inlet and outlet.

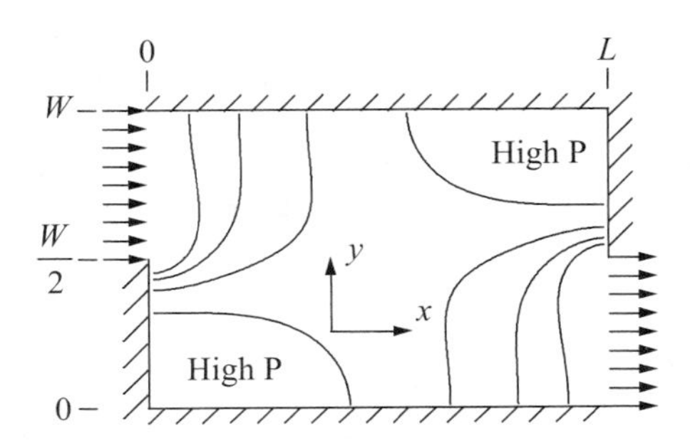

Figure 14-6 Pressure contours for flow through a box with staggered inlet and outlet.

Once the velocity field is known, the steady-state Bernoulli's equation can be evaluated to determine the pressure field. For the present problem, the influence of gravity is neglected, such that one can write

$$v^2/2 + P/\rho + \overbrace{gh}^{neglect} = const. = U^2/2 + P_o/\rho, \tag{14-48}$$

where P_o is a reference pressure corresponding to the inlet (and exit) flow conditions. Therefore, the relative change in pressure depends only on the relative change in kinetic energy:

$$(P - P_o)/\rho = (U^2 - v^2)/2. \tag{14-49}$$

The relative change in pressure is plotted in Figure 14-6. Over most of the interior of the box, the pressure is greater than P_o because the flow speed is reduced from the inlet value. The highest pressures occur in the corners of the box where the flow is slowest, as is evidenced in Figure 14-5 by the large spacing between streamlines.

14.5 ADVECTION OF SPECIES THROUGH A BOX WITH STAGGERED INLET AND EXIT

Suppose a concentration distribution $c(x = 0) = g(y)$ exists across the inlet of the box described in the preceding section, and a steady-state concentration distribution throughout the box is desired. In the absence of diffusion and chemical reactions, the steady-state transport equation for species advection is described by

$$\frac{Dc}{Dt} = 0 \quad \text{or} \quad \underbrace{\frac{\partial c}{\partial t}}_{=0} + v_x \frac{\partial c}{\partial x} + v_y \frac{\partial c}{\partial y} = 0. \tag{14-50}$$

The mathematical problem is summarized in Figure 14-7. Notice that the species equation is first order. Consequently, only one set of boundary conditions (such as inlet conditions) are needed to specify the solution. Since v_x and v_y are complicated functions of x and y, direct integration of the governing equations is not practical. However, as previously noted, the species concentration of a fluid element traveling with the flow is constant in the absence of diffusion and sources. In other words, the species concentration is constant along a streamline for the steady-state problem.

Since a relation between concentration and streamlines can be established, the solution for the concentration distribution in the box can be obtained without formally

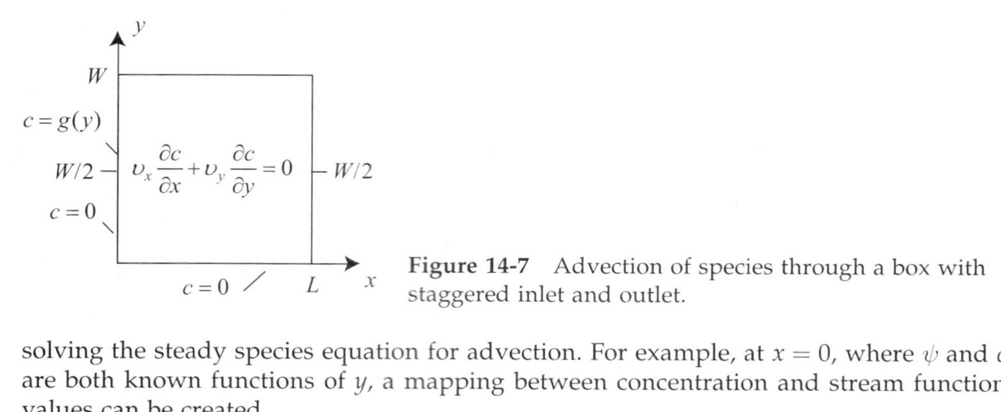

Figure 14-7 Advection of species through a box with staggered inlet and outlet.

solving the steady species equation for advection. For example, at $x = 0$, where ψ and c are both known functions of y, a mapping between concentration and stream function values can be created.

Consider the problem illustrated in Figures 14-4 and 14-7. Suppose the concentration distribution across the inlet of the box is given by

$$g(y) = c_o \sin[\pi(2y/W - 1)]. \tag{14-51}$$

At $x = 0$ and $y \geq W/2$ the stream function boundary condition is $\psi = U(y - W/2)$, or equivalently:

$$y = \psi/U + W/2 \quad \text{(at the inlet)}. \tag{14-52}$$

Therefore, a mapping between the concentration and the stream function value can be created at the inlet: $c = c_o \sin[2\pi\psi/(UW)]$. Since this mapping between concentration and streamlines remains fixed throughout the flow, the concentration anywhere in the box can be found from

$$c(x, y) = c_o \sin\left(\frac{2\pi\psi(x, y)}{UW}\right), \tag{14-53}$$

where $\psi(x, y)$ is evaluated from Eq. (14-44).

For any real fluid in which concentration gradients exists, diffusion will occur. Therefore, the solution presented by Eq. (14-53) is only applicable when the rate of diffusion is negligible compared to the rate of advection. To establish the requirements for this, the species transport equation is scaled in an analogous fashion to the scaling of the momentum equation in Section 14.1. Defining the dimensionless variables,

$$c^* = c/c_o \quad \text{and} \quad x_i^* = x_i/L, \tag{14-54}$$

the steady-state species transport equation is transformed:

$$\begin{aligned} v_j \partial_j c \quad &= \quad Đ \partial_j \partial_j c \quad + (\cdots) \\ \downarrow \quad & \qquad\quad \downarrow \\ \frac{Uc_o}{L} v_j^* \partial_j^* c^* &= \frac{Đc_o}{L^2} \partial_j^* \partial_j^* c^* + (\cdots) \end{aligned} \tag{14-55}$$

where again ∂_j^* is a derivative with respect to the dimensionless spatial variable x_j^*. Now the dimensionless species transport equation can be written in the form

$$\text{Pe}_L \, v_j^* \partial_j^* c^* = \partial_j^* \partial_j^* c^* + (\cdots), \tag{14-56}$$

where a Péclet number* for mass transfer is defined by

$$\mathrm{Pe}_L = \frac{UL}{Ɖ}. \tag{14-57}$$

As long as the scales of L and W are comparable, diffusion transport can be neglected when the Péclet number is large.

It should be noted that the Péclet number is used for both heat and mass transfer, with the only distinction being that the diffusivity term is α for heat transfer and $Ɖ$ for mass transfer. Notice the similarity of the Péclet number to the Reynolds number. Both contrast characteristic magnitudes of advection and diffusion, through the use of a length scale. However, the Reynolds number addresses momentum transport while the Péclet number is used for heat or mass transport.

14.6 SPHERICAL BUBBLE DYNAMICS

Consider a spherical bubble of radius R, surrounded by an infinite domain of a fluid, as illustrated in Figure 14-8. Far from the bubble, the fluid is at a temperature T_∞ and pressure P_∞. Growth or collapse of the bubble is accompanied by radial flow that is inviscid and irrotational. As discussed in Section 14.3, such flows are kinematically constrained. However, the condition that the flow is irrotational is already assured by the radial symmetry of the problem ($v_\theta = v_\phi = 0$). Therefore, only continuity remains as a kinematic constraint to be imposed on v_r. The continuity equation for an incompressible flow is given by

$$\partial_j v_j = 0 \quad \text{or} \quad \partial_j \partial_j \Phi = 0. \tag{14-58}$$

The second form of the continuity equation makes use of the velocity potential definition (14-12).

In spherical coordinates, the continuity equation (14-58) evaluates to

$$\begin{array}{c} \partial_j \partial_j \Phi \qquad\qquad\qquad\qquad = 0 \\ \downarrow \\ \dfrac{1}{r^2}\dfrac{\partial}{\partial r}\left(r^2 \dfrac{\partial \Phi}{\partial r}\right) + \dfrac{1}{r^2 \sin\theta}\dfrac{\partial}{\partial \theta}\left(\sin\theta \overbrace{\dfrac{\partial \Phi}{\partial \theta}}^{=0}\right) + \dfrac{1}{r^2 \sin^2\theta}\overbrace{\dfrac{\partial^2 \Phi}{\partial \phi^2}}^{=0} = 0. \end{array} \tag{14.59}$$

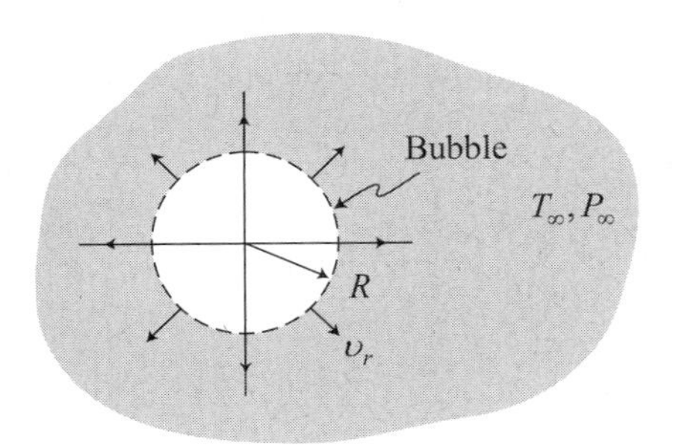

Figure 14-8 Illustration for spherical bubble dynamics.

*The Péclet number is named after the French physicist Jean Claude Eugène Péclet (1793–1857).

Integrating continuity yields

$$r^2 \frac{\partial \Phi}{\partial r} = C_1. \tag{14-60}$$

However, at the surface of the bubble $r = R$

$$v_r = \left.\frac{\partial \Phi}{\partial r}\right|_{r=R} = \dot{R}, \tag{14-61}$$

where $\dot{R}$ is the radial velocity of the bubble expansion. Therefore, the first integration constant evaluates to

$$R^2 \left.\frac{\partial \Phi}{\partial r}\right|_{r=R} = R^2 \dot{R} = C_1 \tag{14-62}$$

such that Eq. (14-60) becomes

$$\frac{\partial \Phi}{\partial r} = \frac{R^2 \dot{R}}{r^2} \quad (= v_r). \tag{14-63}$$

Integrating the continuity equation a second time yields

$$\Phi = \frac{-R^2 \dot{R}}{r} + C_2. \tag{14-64}$$

The second integration constant is (arbitrarily) assigned to be $C_2 = 0$, such that $\Phi(r \to \infty) = 0$.

The continuity equation prescribes the velocity field everywhere in the flow around the bubble, as long as $\dot{R}$ is specified. To develop an equation describing the bubble dynamics ($\dot{R}$) requires consultation of the momentum equation.

As was established in Section 14.2, the momentum equation for an inviscid, irrotational, and incompressible flow integrates to Bernoulli's equation (14-15). Since all radial positions along a streamline must equate to the same time-dependent integration constant $C(t)$, Bernoulli's equation (14-15) dictates that

$$\left[\partial_o \Phi + v^2/2 + P/\rho\right]_{r \to \infty} = \left[\partial_o \Phi + v^2/2 + P/\rho\right]_r, \tag{14-65}$$

In the far field $P(r \to \infty) = P_\infty$, $v_r(r \to \infty) = 0$, and $\Phi(r \to \infty) = 0$. Therefore, using Eq. (14-63) to evaluate v_r and Eq. (14-64) to evaluate Φ, the momentum equation (14-65) dictates that

$$\left[\frac{P_\infty}{\rho}\right] = \left[\frac{\partial}{\partial t}\left(\frac{-R^2 \dot{R}}{r}\right) + \frac{1}{2}\left(\frac{R^2 \dot{R}}{r^2}\right)^2 + \frac{P}{\rho}\right]_r, \tag{14-66}$$

or

$$\frac{P - P_\infty}{\rho} = \frac{R\dot{R}^2 + R^2 \ddot{R}}{r} - \frac{1}{2}\left(\frac{R^2 \dot{R}}{r^2}\right)^2. \tag{14-67}$$

This result describes the pressure field surrounding the bubble. If the pressure field is evaluated at the *outside* surface of the bubble where $r = R$, the result yields *Rayleigh's equation* [2]:

$$R\ddot{R} + \frac{3}{2}\dot{R}^2 = \frac{P(r = R) - P_\infty}{\rho}. \tag{14-68}$$

In the simplest scenario, where boundary effects such as surface tension can be ignored, the pressure inside the bubble P_B is equal to the pressure outside the surface of the bubble $P(r = R) = P_B$.

To illustrate the utility of Rayleigh's equation, suppose that a vapor bubble is created in a superheated liquid at $T_\infty > T_{sat}(P_\infty)$, where $T_{sat}(P_\infty)$ is the equilibrium saturation (boiling point) temperature that corresponds to the surrounding pressure P_∞. The pressure inside the bubble equals the liquid vapor pressure corresponding to the temperature of the surrounding fluid $P_B = P_{sat}(T_\infty) > P_\infty$. In the superheated state, heat transfer through the liquid is not required to drive vapor production in the bubble. The temperature of the vapor inside the bubble is $T_B = T_\infty$. Therefore, with $P(r = R) = P_B = P_{sat}(T_\infty)$, Rayleigh's equation describing the rate of bubble expansion becomes

$$R\ddot{R} + \frac{3}{2}\dot{R}^2 = \frac{P_{sat}(T_\infty) - P_\infty}{\rho}. \tag{14-69}$$

If the bubble grows from zero size, $R(t = 0) = 0$, then the first term in the governing equation remains finite only if $\ddot{R}(t = 0) = 0$. However, from this initial condition, acceleration of the bubbles radius must remain zero for all time $\ddot{R}(t > 0) = 0$, and the bubble growth is a simple linear function of time:

$$R(t) = \left(\frac{2}{3}\frac{P_{sat}(T_\infty) - P_\infty}{\rho}\right)^{1/2} t. \tag{14-70}$$

From other initial conditions, with $R(t = 0) \neq 0$, integration of Eq. (14-69) is nontrivial. Therefore, integration of Rayleigh's equation will be revisited in Chapter 23 with a numerical approach, to address situations that are more complicated.

14.6.1 Effect of Viscosity and Surface Tension

Although there are no viscous terms in the momentum equation, a viscous effect can be felt at the boundary of a bubble. Consider the bubble surface, as illustrated in Figure 14-9, from a frame of reference moving with the boundary. Although the radial flow is zero at the surface of the bubble (with respect to coordinates moving with the surface) the gradient in the radial flow is not. Consequently, there exists a radial diffusion flux of radial

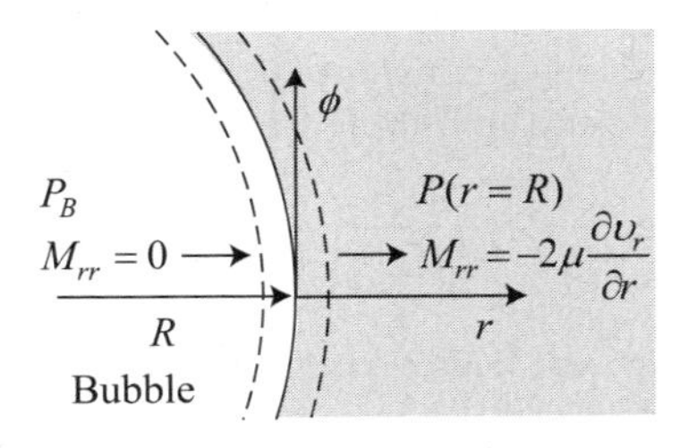

Figure 14-9 Viscous effect at bubble surface. Coordinates are attached to moving surface.

momentum between the bubble surface and the surrounding fluid. However, because the flow is confined to only one side of this boundary, there is a diffusion flux imbalance across the surface of the bubble that must be balanced by a pressure difference $\Delta P_\mu = P_B - P(r = R)$. Using Newton's viscosity law for an incompressible flow, $M_{ji} = -\mu(\partial_j v_i + \partial_i v_j)$, the radial momentum flux can be evaluated:

$$M_{rr} = -2\mu\, \partial_r v_r = -2\mu \frac{\partial v}{\partial r}, \tag{14-71}$$

where the expression for $\partial_r v_r$ is found in Table 3-2 for spherical coordinates. Since the radial velocity gradient at the surface of the bubble evaluates to

$$\left.\frac{\partial v_r}{\partial r}\right|_{r=R} = \left[\frac{\partial}{\partial r}\frac{R^2 \dot{R}}{r^2}\right]_{r=R} = -\frac{2\dot{R}}{R}, \tag{14-72}$$

the pressure imbalance due to viscous effects is

$$\Delta P_\mu = P_B - P(r = R) = -2\mu \left.\frac{\partial v_r}{\partial r}\right|_{r=R} = 4\mu \frac{\dot{R}}{R}. \tag{14-73}$$

The effect of surface tension could also be included at the boundary of the bubble in a more general treatment of the dynamics. Surface tension (alone) gives rise to the pressure imbalance:

$$\left(\Delta P_\sigma = P_B - P(r = R) = \frac{2\sigma}{R}\right)_{spherical}, \tag{14-74}$$

which is the *Young-Laplace pressure*† for a spherical surface [3,4]. Combining the effects of viscosity and surface tension, the pressure drop across the surface of the bubble becomes:

$$\overbrace{P_B - P(r = R)}^{\Delta P_\mu + \Delta P_\sigma} = 4\mu \frac{\dot{R}}{R} + \frac{2\sigma}{R}. \tag{14-75}$$

This can be substituted into Eq. (14-68) for the result

$$R\ddot{R} + \frac{3}{2}\dot{R}^2 + 4\nu\frac{\dot{R}}{R} + \frac{2\sigma}{\rho R} = \frac{P_B - P_\infty}{\rho}, \tag{14-76}$$

which is generally referred to as the *Rayleigh-Plesset* equation‡ [5].

In general, the interior of the bubble may contain both vapor (from the surrounding liquid) and noncondensable gas. Consequently, the total bubble pressure P_B is the sum of the vapor partial pressure p^v and the noncondensable gas partial pressure p^g. The vapor partial pressure p^v in the bubble is a function of temperature. However, since the amount of noncondensable gas in the bubble is a constant (as long as the fluid cannot absorb the gas) the partial pressure of the gas in the bubble can be described by the ideal gas law.

---

†Named after the English scientist Thomas Young (1773–1829) and the French mathematician and astronomer Pierre-Simon Laplace (1749–1827).

‡Named after the English physicist Lord John William Strutt Rayleigh (1842–1919) and the American physicist Milton Spinoza Plesset (1908–1991).

With respect to a reference state denoted by a subscript "o", the partial pressure of the gas p^g can be expressed as

$$\frac{p^g \forall^g}{T^g} = \left(\frac{p^g \forall^g}{T^g}\right)_o \quad \text{or} \quad p^g = p_o^g \left(\frac{R_o}{R}\right)^3 \frac{T^g}{T_o^g}, \tag{14-77}$$

where $\forall = 4\pi R^3/3$ is the volume of the bubble. If the state of the gas in the bubble changes either isothermally ($dT = 0$) or isentropically ($ds = 0$), the partial pressure p^g can be expressed as

$$p^g = p_o^g \left(\frac{R_o}{R}\right)^3 \quad (\text{if } dT = 0) \quad \text{or} \quad p^g = p_o^g \left(\frac{R_o}{R}\right)^{3\gamma} \quad (\text{if } ds = 0). \tag{14-78}$$

With $P(r = R) = P_B = p^v + p^g$, the Rayleigh-Plesset equation (14-76) becomes

$$R\ddot{R} + \frac{3}{2}\dot{R}^2 + 4\nu\frac{\dot{R}}{R} + \frac{2\sigma}{\rho R} = \frac{p^v - P_\infty}{\rho} + \frac{p_o^g}{\rho}\left(\frac{R_o}{R}\right)^{3k}, \tag{14-79}$$

where $k = 1$ for an isothermal process and $k = \gamma$ for an isentropic process. Both cases belong to the more general class of *polytropic processes* where $P\forall^k$ is held constant. Integration of Eq. (14-79) will be investigated numerically in Chapter 23.

14.7 PROBLEMS

14-1 Consider the plane flow illustrated. Determine an expression for $v_y(x, y)$ satisfying incompressible continuity. Determine what values of m yield an ideal plane flow. Determine the pressure distributions for all admissible values of m yielding ideal plane flow.

y

$v_x = C x^m$

x

$v_y = 0$

14-2 Solve analytically for the stream function in the box shown. Assume that the flow is ideal plane flow. Plot the streamlines through the box.

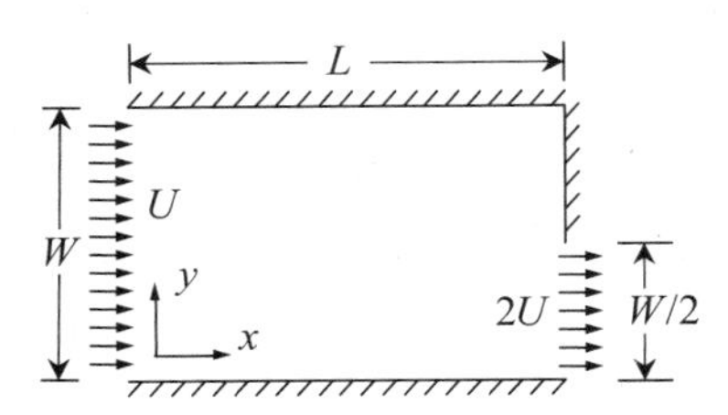

14-3 Solve analytically for the velocity field $v_x(x, y)$ and $v_y(x, y)$ in the box shown. Assume that the flow is ideal plane flow. What is the net force acting on the back wall at $x = L$?

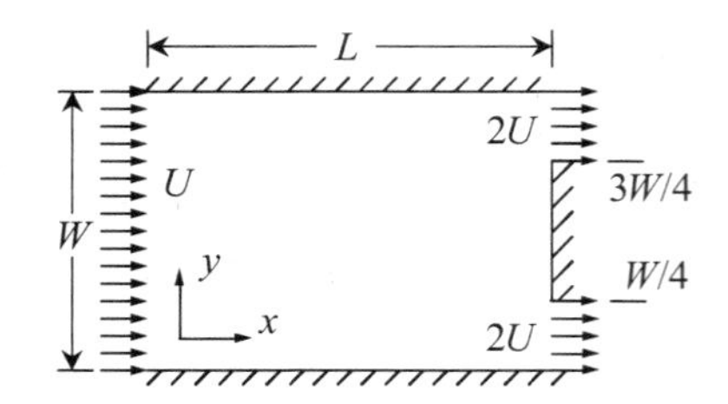

14-4 Consider the ideal potential flow illustrated in Problem 14-3. Suppose that the fluid entering the box has a temperature distribution given by

$$T(y, x=0) = \begin{cases} T_o & 2W/3 < y < W \\ T_1 > T_o & W/3 \leq y \leq 2W/3 \\ T_o & 0 < y < W/3 \end{cases}.$$

Derive the temperature distribution in the box. What is the temperature of the fluid in contact with the wall obstructing the exit?

14-5 Consider the growth of a long cylindrical bubble of radius R. Demonstrate that continuity requires

$$v_r(t, r) = \frac{R}{r}\frac{dR}{dt}.$$

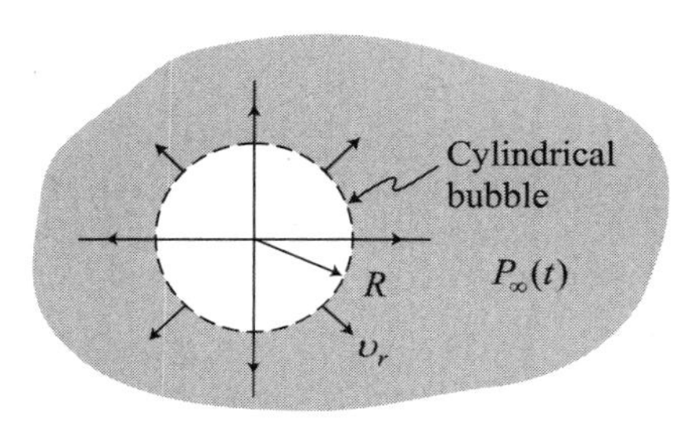

Apply this result to the radial momentum equation. After integrating the momentum equation once, show that for finite $P(r \to \infty) = P_\infty(t)$ the cylindrical bubble growth must obey

$$\frac{d}{dt}\left(R\frac{dR}{dt}\right) = 0.$$

Demonstrate that the fluid pressure outside the surface of the bubble (at $r = R$) is given by

$$P = P_\infty(t) - \frac{\rho}{2}\left(\frac{R_o \dot{R}_o}{r}\right)^2.$$

Including viscous effects and the Young-Laplace pressure for a cylindrical interface,

$$(\Delta P_\sigma = P_B - P(r = R) = \sigma/R)_{cylindrical},$$

demonstrate that the pressure requirement $P_B - P_\infty$ to achieve cylindrical bubble growth is given by

$$P_B(t) - P_\infty(t) = \frac{\sigma}{R_o\sqrt{1 + 2(\dot{R}_o/R_o)t}} - \frac{\rho}{2}\frac{\dot{R}_o^2}{1 + 2(\dot{R}_o/R_o)t}\left(1 - \frac{4\nu}{R_o \dot{R}_o}\right).$$

REFERENCES

[1] D. Bernoulli, *Hydrodynamica, Sive De Viribus et Motibus Fluidorum Commentarii*. Strasbourg, France: J. R. Dulsecker, 1738.

[2] L. Rayleigh, "On the Pressure Developed in a Liquid During Collapse of a Spherical Cavity." *Philosophical Magazine*, **34**, 94 (1917).

[3] P. S. de Laplace, *Sur L'action Capillaire in Supplément au Traité de Mécanique Céleste*, T. 4, Livre X. Paris, France: Gauthier-Villars, 1805.

[4] T. Young, "An Essay on the Cohesion of Fluids."*Philosophical Transactions of the Royal Society of London*, **95**, 65 (1805).

[5] M. S. Plesset, *"The Dynamics of Cavitation Bubbles."* *Journal of Applied Mechanics* (ASME), **16**, 228 (1949).

Chapter 15

Catalog of Ideal Plane Flows

15.1 Superposition of Simple Plane Flows

15.2 Potential Flow over an Aircraft Fuselage

15.3 Force on a Line Vortex in a Uniform Stream

15.4 Flow Circulation

15.5 Potential Flow over Wedges

15.6 Problems

There are a number of relatively simple flow patterns for which the stream functions are well known. These potential flows have characteristically few surfaces that shape the flow, and are otherwise unbounded. The description of these flows finds utility in a limited number of particularly simple situations. However, the power of superposition, as discussed in Chapter 11, significantly expands the utility of the catalog of plane flows that follows.

15.1 SUPERPOSITION OF SIMPLE PLANE FLOWS

The main building blocks required to build a catalog of plane flows are the uniform flow, the line source/sink, and the line vortex, as shown in the first three entries of Table 15-1. The remaining entries in the table are derived by superposition of two or more of these flows. Most of the flows shown in Table 15-1 use the stream function expressed in cylindrical coordinates.

When a source is placed in front of a uniform flow, the flow is forced to separate around the volume of fluid introduced by the source. The simplest scenario of this is the *Rankine half-body*, composed of superposition of a uniform flow and a source, which is useful in describing the flow over the leading edge of a blunt body. The cross-sectional size of the body is controlled with the magnitude of the flow source. More shaping of the blunt body surface requires additional superposition of distributed sources.

Sources alone, superimposed on a uniform flow, generate a body that is unbounded in the downstream direction. To generate a finite-sized object in a uniform flow requires a net strength of sources to be balanced by sinks of equal net strength. The simplest scenario for this is the superposition of a single source and a single sink of equal magnitude, as illustrated in Table 15-1. The dimensions of the resulting (full) *Rankine body* are controlled by the strength and the separation distance of source and sink. If the separation of the equal-strength source and sink goes to zero, as the strength of both goes to infinity (to maintain a finite effect), the flow field becomes that of a doublet. A single doublet superimposed on a uniform flow gives the pattern of a flow over a cylinder, as shown in Table 15-1. The radius of the cylinder is prescribed by the strength of the doublet. If a line vortex is superimposed on the doublet and the uniform flow, the cylinder in the flow appears to be rotating. When superposition of flow patterns has symmetry, the line of

Table 15-1 Catalog of plane flows

Cartesian:	*Cylindrical*:
$\upsilon_x = \dfrac{\partial \psi}{\partial y}$ $\quad \upsilon_y = -\dfrac{\partial \psi}{\partial x}$	$\upsilon_r = \dfrac{1}{r}\dfrac{\partial \psi}{\partial \theta}$ $\quad \upsilon_\theta = -\dfrac{\partial \psi}{\partial r}$

Flow	Stream function
uniform	$\psi = U_x y - U_y x$
vortex	$\psi = \dfrac{-m}{2\pi} \ln r$
source/sink	$\psi = \dfrac{\pm m}{2\pi}\theta$
doublet	$\psi = \dfrac{-m}{2\pi}\dfrac{\sin\theta}{r}$
	uniform + source
	uniform + source + sink
	source + source
	uniform + doublet
	uniform + doublet + vortex

symmetry becomes a plane boundary to the flow. A simple illustration of this, shown in Table 15-1, is for the superposition of two sources. This principle will be useful in creating boundaries on many flow fields, and is called *method of images* (see Section 11.2.2).

15.2 POTENTIAL FLOW OVER AN AIRCRAFT FUSELAGE

The potential airflow around the front end of an aircraft fuselage can be modeled using two sources of strength, m_1 and m_2, superimposed on a uniform flow, as shown in Figure 15-1. Although three-dimensional point sources could be used for a truly three-dimensional fuselage, two-dimensional line sources will be used to describe plane flow. The profile of the fuselage is characterized by the location of three points on the surface (A, B, C) and the maximum downstream dimension H of the fuselage cross section, as shown in Figure 15-1. The first source is at the origin and the second source is in the plane connecting points B and C. The three points on the fuselage are located at $A = (-0.22, 0)$, $B = (0.75, 0.9)$, and $C = (0.75, \ -0.7)$, and the downstream fuselage dimension is $H = 2.6$. Suitable units for the dimensions may be assumed.

The stream function described by superposition of uniform flow and two line sources is given by

$$\psi = \overbrace{Uy}^{uniform} + \overbrace{\frac{m_1\theta_1}{2\pi}}^{source\ 1} + \overbrace{\frac{m_2\theta_2}{2\pi}}^{source\ 2} + \overbrace{C}^{const.} = U\left(y + \frac{m_1'\theta_1}{2\pi} + \frac{m_2'\theta_2}{2\pi} + C'\right). \tag{15-1}$$

Three equations evaluating the stream function at known positions along the fuselage can be written for A, B, and C. Choosing the stream function to be zero on the surface of the fuselage yields

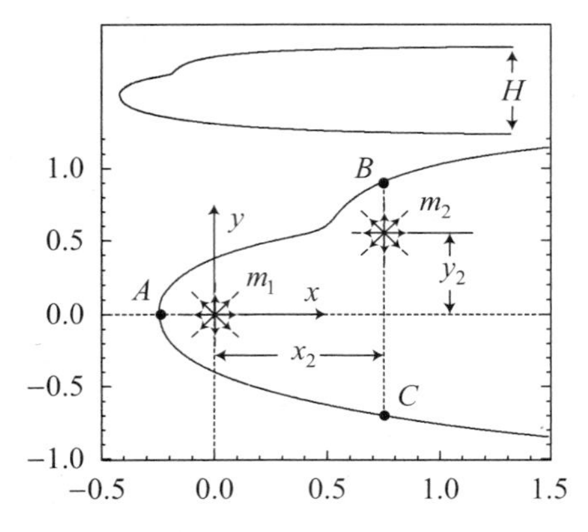

Figure 15-1 Schematic of an aircraft fuselage.

$$\text{Point } A: \quad 0 = U\left[0 + \frac{m_1'}{2} + \frac{m_2'}{2\pi}\left(\pi + \tan^{-1}\frac{y_2}{x_2 - x_A}\right) + C'\right] \tag{15-2}$$

$$\text{Point } B: \quad 0 = U\left[y_B + \frac{m_1'}{2\pi}\tan^{-1}\frac{y_B}{x_2} + \frac{m_2'}{4} + C'\right] \tag{15-3}$$

$$\text{Point } C: \quad 0 = U\left[y_C + \frac{m_1'}{2\pi}\left(2\pi + \tan^{-1}\frac{y_C}{x_2}\right) + \frac{3m_2'}{4} + C'\right] \tag{15-4}$$

There are four unknown variables y_2, m_1', m_2', and C' in these three equations. Therefore, a fourth equation is sought using the known downstream fuselage dimension H. Along the top surface of the fuselage, θ_1 and θ_2 approach 0 as $x \to \infty$. Along the bottom surface of the fuselage, θ_1 and θ_2 approach 2π as $x \to \infty$. Subtracting expressions for the stream function evaluated at the top and bottom surfaces of the fuselage to eliminate C', and introducing H as the distance between the top and bottom surfaces as $x \to \infty$, yields the fourth equation needed to solve the problem:

$$0 = H - m_1' - m_2'. \tag{15-5}$$

Equations (15-2) through (15-5) may be solved for the unknowns:

$$y_2 = 0.55665, \quad m_1' = 1.2445, \quad m_2' = 1.3555, \quad \text{and} \quad C' = -1.4124,$$

yielding the solution for the stream function:

$$\psi = U\left(y + \frac{1.2445}{2\pi}\theta_1 + \frac{1.3555}{2\pi}\theta_2 - 1.4124\right). \tag{15-6}$$

Code 15-1 is used to evaluate the solution. Based on the global position given in Cartesian coordinates (x, y), the program uses a subroutine to find θ_1 and θ_2 for the coordinates centered at the position of each source. The solution for the streamlines around the aircraft fuselage is plotted in Figure 15-2. At the nose of the aircraft, the spacing between streamlines is greatest. This corresponds to a region of the flow with the lowest velocity and highest pressure. One streamline is seen to terminate at the nose of the aircraft, corresponding to the location of a *stagnation point*. A stagnation point is a location in the flow where the fluid comes to rest.

Code 15-1 Flow past an aircraft fuselage

```
#include <stdio.h>
#include <math.h>

void find_polar(double x,double y,
                double *r,double *theta)
{
   *r=sqrt(x*x+y*y);
   if (x>0) {
         if (y>0) *theta=asin(y/(*r));
         else *theta=2.*M_PI+asin(y/(*r));
   } else if (x<0) {
        *theta=M_PI-asin(y/(*r));
   } else *theta=0.;
}

int main()
{
   FILE *fp;
   double m1=1.2445, m2=1.3555, C=-1.4124;
   double x2=0.75, y2=.55665;
   double x,y,Y,r,theta;

   fp=fopen("out.dat","w");
   for (x=-0.5;x<1.5;x+=.01) {
         for (y=-1.5;y<=1.5;y+=.01) {
               Y=y+C;
               find_polar(x,y,&r,&theta);
               Y+=m1*theta/2./M_PI;
               find_polar(x-x2,y-y2,&r,&theta);
               Y+=m2*theta/2./M_PI;
               fprintf(fp,"%e %e %e\n",x,y,Y);
         }
   }
   fclose(fp);
   return 0;
}
```

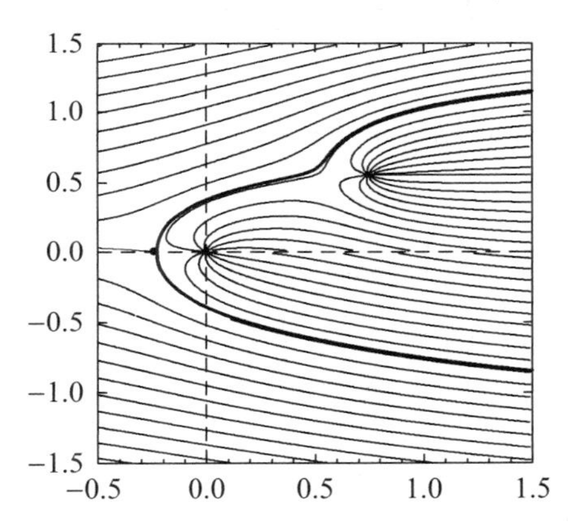

Figure 15-2 Streamlines around an aircraft fuselage.

15.3 FORCE ON A LINE VORTEX IN A UNIFORM STREAM

The flow describing a line vortex in a uniform stream is created by superposition of the two elemental stream functions:

$$\psi = \underbrace{U_x y}_{uniform} + \underbrace{\frac{-m}{2\pi}\ln r}_{vortex} = U_x r \sin\theta - \frac{m}{2\pi}\ln r. \tag{15-7}$$

The streamlines for the flow are illustrated in Figure 15-3. The velocity components of the flow are given by

$$\begin{aligned} v_x &= \frac{\partial\psi}{\partial y} = U_x - \frac{m}{2\pi r}\sin\theta, \\ v_y &= -\frac{\partial\psi}{\partial x} = \frac{m}{2\pi r}\cos\theta. \end{aligned} \tag{15-8}$$

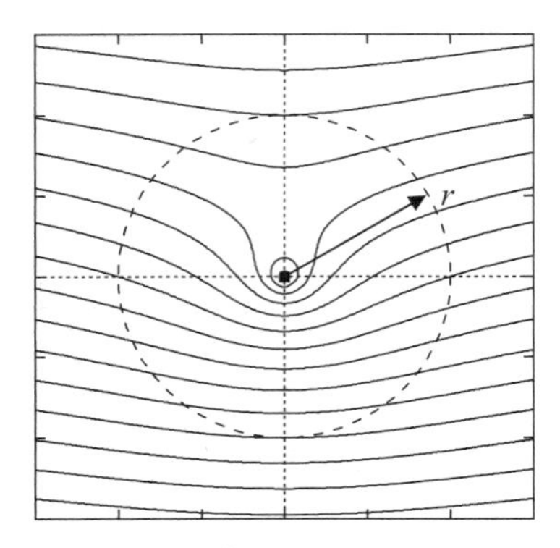

Figure 15-3 Vortex in a uniform flow.

Because the fluid is accelerated around the origin of the vortex, a force is exerted on the flow. The force between the vortex and the flow is not directly evaluated because of the mathematical singularity at the origin of the vortex. To evaluate this force indirectly, a control surface at a distance r from the center of the vortex is analyzed. For a steady flow, momentum conservation is written in integral form for a cylindrical control surface a distance r from the origin:

$$\begin{array}{cccc} (increase) = & (in - out) + & (sources) & \\ \downarrow & \downarrow & \downarrow & \\ 0 & = -\int_{2\pi} v_i \rho v_j n_j r d\theta + & \int_{2\pi} -(P - P_\infty)\delta_{ji} n_j r d\theta & + F_i. \end{array} \tag{15-9}$$

F_i appearing with the source terms is an external force (per unit length) applied to the flow. F_i can be determined by evaluating the rate of momentum change advected through the control surface and the net pressure imbalance on the control surface. The local advection flux through the control surface is proportional to the dot product between the flow velocity and the surface unit normal, and is given by the radial velocity component:

$$v_j n_j = v_r = \frac{1}{r}\frac{\partial \psi}{\partial \theta} = U_x \cos\theta. \tag{15-10}$$

To determine $(P - P_\infty)$ from Bernoulli's equation (14-4) requires evaluating $v_j v_j$:

$$\begin{aligned} v_j v_j = v_x^2 + v_y^2 &= \left(U_x - \frac{m}{2\pi r}\sin\theta\right)^2 + \left(\frac{m}{2\pi r}\cos\theta\right)^2 \\ &= U_x^2 - \frac{m}{\pi r}U_x \sin\theta + \left(\frac{m}{2\pi r}\right)^2. \end{aligned} \tag{15-11}$$

Using this result, Bernoulli's equation, neglecting the influence of gravity, may be evaluated for pressure:

$$\frac{P_\infty - P}{\rho} = \frac{v_j v_j - U_x^2}{2} = \frac{1}{2}\left(\frac{m}{2\pi r}\right)^2 - \frac{m}{2\pi r}U_x \sin\theta. \tag{15-12}$$

With Eqs. (15-10) and (15-12), the integral form of momentum conservation, Eq. (15-9), can be expressed for the x-direction as

$$
\begin{aligned}
0 &= -\int_{2\pi} v_x \rho v_j n_j r d\theta + \int_{2\pi} -(P - P_\infty)\delta_{jx} n_j r d\theta + F_x \\
&\downarrow \qquad\qquad \downarrow \qquad\qquad\qquad\qquad \downarrow \\
0 &= \underbrace{-\int_{2\pi} v_x v_r r d\theta}_{=0} + \underbrace{\int_{2\pi} \left(\frac{P_\infty - P}{\rho}\right)(\cos\theta) r d\theta}_{=0} + \frac{F_x}{\rho}.
\end{aligned}
\tag{15-13}
$$

The result $F_x = 0$ indicates that the flow does not impose drag on the line vortex.

However, the integral conservation statement for momentum expressed in the y-direction is

$$
\begin{aligned}
0 &= -\int_{2\pi} v_y \rho v_j n_j r d\theta + \int_{2\pi} -(P - P_\infty)\delta_{jy} n_j r d\theta + F_y \\
&\downarrow \qquad\qquad \downarrow \qquad\qquad\qquad\qquad \downarrow \\
0 &= -\int_{2\pi} v_y v_r r d\theta + \int_{2\pi} \left(\frac{P_\infty - P}{\rho}\right)(\sin\theta) r d\theta + \frac{F_y}{\rho}.
\end{aligned}
\tag{15-14}
$$

Using Eqs. (15-8) and (15-10) again to evaluate the advection flux integral and Eq. (15-12) to evaluate the pressure integral yields

$$
\begin{aligned}
0 &= -\int_{2\pi} \left(\frac{m}{2\pi r}\cos\theta\right)(U_x \cos\theta) r d\theta + \int_{2\pi} \left(\frac{1}{2}\left(\frac{m}{2\pi r}\right)^2 - \frac{m}{2\pi r} U_x \sin\theta\right)(\sin\theta) r d\theta + \frac{F_y}{\rho} \\
&\downarrow \qquad\qquad \downarrow \qquad\qquad\qquad\qquad \downarrow \qquad\qquad \downarrow \\
0 &= \qquad -\frac{m}{2} U_x \qquad + \qquad -\frac{m}{2} U_x \qquad \frac{F_y}{\rho}.
\end{aligned}
$$

Therefore, $F_y = m\rho U_x$, and the lift force on the line vortex caused by the flow is

$$
Lift = -F_y = (-m)\rho U_x. \tag{15-15}
$$

The negative lift would be reversed by changing the sign of the line vortex strength.

15.4 FLOW CIRCULATION

Identical to a vortex in a uniform flow, as discussed in the previous section, Problem 15-4 demonstrates that the lift force on a rotating cylinder is also given by $Lift = (-m)\rho U_x$, where m is the strength of the vortex used to construct the flow. This exercise suggests (correctly) that the lifting force is not an intrinsic function of the geometry of the object, but rather is a function of the imposed *circulation* of the flow around the object. Circulation can be defined as the closed counterclockwise integral,

$$
\Gamma = \oint_S v_i t_i dS, \tag{15-16}
$$

where t_i is the unit tangent vector to the local integration path segment dS, as shown in Figure 15-4. For a line vortex, the circulation relates simply to the strength coefficient:

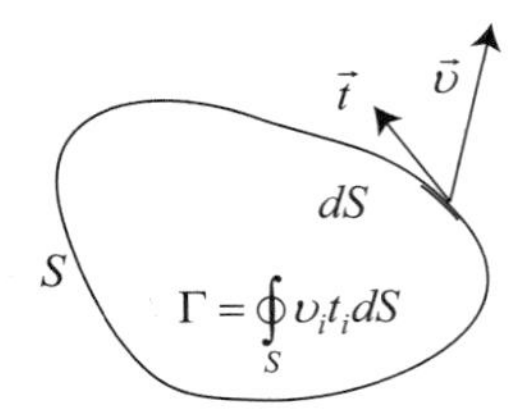

Figure 15-4 Integration path used to calculate circulation.

$$\Gamma = \oint_S v_i t_i dS = \int_0^{2\pi} v_\theta r d\theta = \int_0^{2\pi} -\frac{\partial \psi}{\partial r} r d\theta = \int_0^{2\pi} \frac{m}{2\pi} d\theta = m. \tag{15-17}$$

This result is independent of the integration distance r from the center of the vortex (and a noncircular integration path would yield the same result). Therefore, the lift is related to the circulation of the flow around an object by

$$Lift = \rho U(-\Gamma), \tag{15-18}$$

and is perpendicular to the free stream flow U as shown in Figure 15-5. Equation (15-18) is known as the *Kutta-Joukowski theorem*.*

Lift

U $-\Gamma$

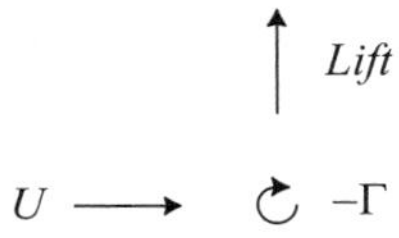

Figure 15-5 Orientation of lift relative to U and the circulation.

Quite complicated flows can be created by the composite superposition of a number of the elemental stream functions in Table 15-1, as illustrated by Figure 15-6. A material surface is created in the flow by a number of distributed sources and sinks, whose strengths sum to zero; the doublet is the simplest example. The lift on that material surface is related to the sum of the strengths of vortices enclosed by the material surface:

$$Lift = -\rho U \sum (\text{vortex strengths}). \tag{15-19}$$

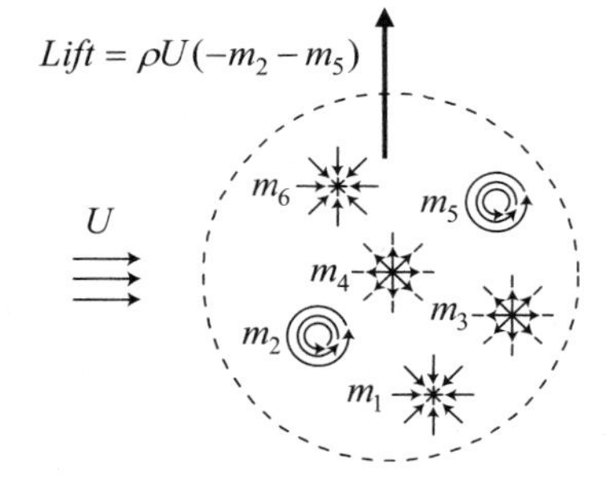

Figure 15-6 Lift on a composite flow.

*The Kutta-Joukowski theorem is named in honor of the Russian scientist Nikolai Yegorovich Zhukovsky (1847–1921) and the German mathematician Martin Wilhelm Kutta (1867–1944).

15.5 POTENTIAL FLOW OVER WEDGES

Another useful stream function, to add to the catalog of plane potential flows, describes various geometries of flows over wedge-like shapes. All wedge flows are derived from the same function,

$$\psi = Ar^n \sin(n\theta), \tag{15-20}$$

but use different values of the constant "n." Table 15-2 illustrates the stream function and a few of the characteristic flows that can be generated.

The wedge flow catalog is useful for evaluating the pressure distribution along a flat surface that accelerates a flow. Additionally, the pressure gradient along a surface is essential information for studying the thin viscous boundary layer region adjacent to the surface. Solving the boundary layer problem, as will be done in Chapter 25, will permit the viscous forces acting on a surface to be determined in addition to the pressure forces.

15.5.1 Pressure Gradient along a Wedge

Consider the wedge flow illustrated by the case $n = 1.2$, as shown in the lower right panel of Table 15-2. Since the viscous region of the boundary layer that forms is very close to the surface of the wedge, it is convenient to assume that the inviscid flow around the wedge is unaffected by the presence of this thin boundary layer. To solve the viscous boundary layer problem, as will be done later in Section 25.6, the pressure gradient $-(\partial P/\partial x)/\rho$ along the surface of the wedge shown in Figure 15-7 will need to be evaluated. Inviscid flow theory is used to determine this pressure gradient.

The momentum equation for an inviscid, irrotational, and incompressible flow is Bernoulli's equation, as developed in Section 14.2. Written for a steady-state flow, Bernoulli's equation is

Table 15-2 Plane wedge flows

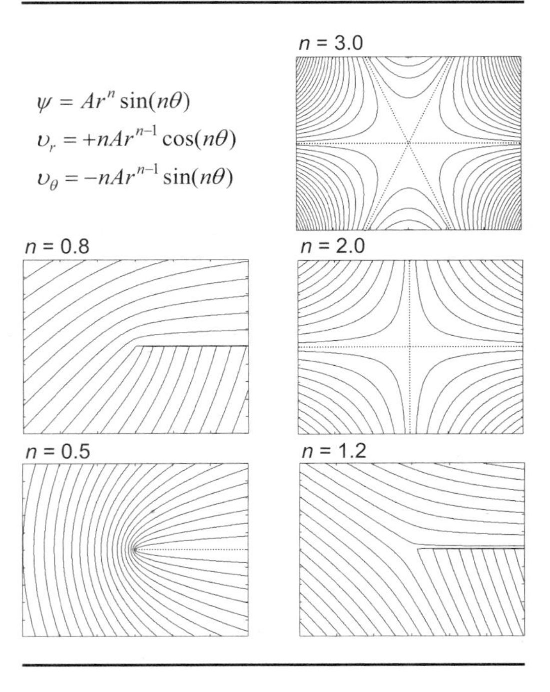

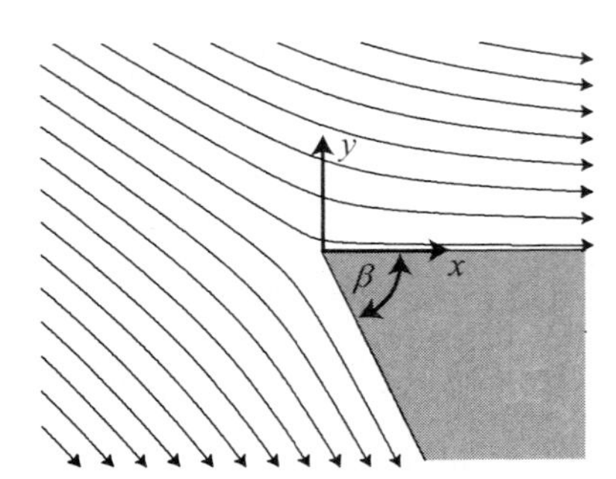

Figure 15-7 Flow over a wedge.

$$\frac{P}{\rho} + \frac{v_e^2}{2} + \underbrace{gh}_{=0} = C. \tag{15-21}$$

Let v_e denote the velocity of the inviscid flow along the surface of the wedge, which is also the velocity of the flow *external* to the viscous boundary layer analyzed in Section 25.6. Differentiating Bernoulli's equation reveals

$$-\frac{1}{\rho}\frac{dP}{dx} = v_e \frac{dv_e}{dx}. \tag{15-22}$$

The flow streamlines over the wedge are given by

$$\psi = Ar^n \sin(n\theta), \tag{15-23}$$

where the wedge angle illustrated in Figure 15-7 is given by

$$\beta = 2\pi \frac{n-1}{n}. \tag{15-24}$$

The velocity field is determined from the stream function definition for cylindrical coordinates:

$$v_r = \frac{1}{r}\frac{\partial\psi}{\partial\theta} = Ar^{n-1}n \cdot \cos(n\theta) \quad \text{and} \quad v_\theta = -\frac{\partial\psi}{\partial r} = -Anr^{n-1}\sin(n\theta). \tag{15-25}$$

Along the surface of the wedge the flow is entirely radial, requiring $v_\theta = 0$ such that $\sin(n\theta) = 0$. However, this equivalently requires that $\cos(n\theta) = 1$. Therefore, v_e is simply v_r along the surface of the wedge, and is given by $v_e = Anx^{n-1}$. Letting $m = n - 1$, the result for v_e may be written as

$$v_e = d\,x^m, \quad \text{where} \quad m = \frac{\beta/\pi}{2 - \beta/\pi} \tag{15-26}$$

and d is a constant. Therefore, potential flow theory reveals that the velocity of the flow along the surface of the wedge grows as a power function of distance from the leading edge. Furthermore, the result for the pressure gradient

$$-\frac{1}{\rho}\frac{dP}{dx} = v_e\frac{dv_e}{dx} = m\,d^2x^{2m-1} \tag{15-27}$$

will be used in Section 25.6 for the description of the viscous boundary layer that forms beneath the inviscid flow accelerated by the wedge.

15.6 PROBLEMS

15-1 Consider an incompressible potential flow defined by placing a line vortex a distance d away from a wall. Determine the pressure distribution on the wall. How will the vortex move if let loose? What is the force on the vortex?

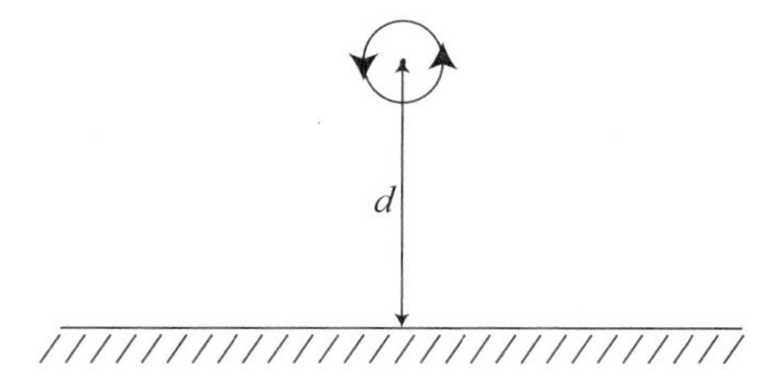

15-2 Derive the stream functions for a source flow and a vortex flow. Show that superposition of a source and a sink separated by a distance ε gives the stream function for a doublet when $\varepsilon \to 0$.

15-3 A vacuum could be modeled as a line sink adjacent to a wall. Roughly, where is the greatest vacuum (lowest pressure) along the floor? Justify your answer based on arguments that can be made from the illustration. Roughly, where is the poorest vacuum (highest pressure) along the floor?

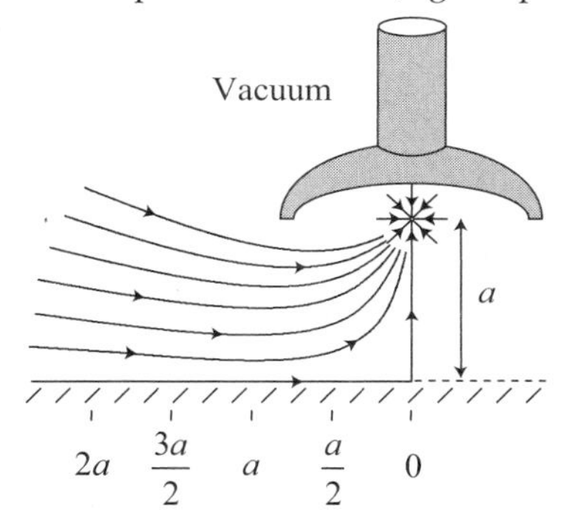

Find an expression for

$$\frac{P_\infty - P(x)}{\rho},$$

where $P(x)$ is the pressure along the floor, as a function of position x. Verify the location of lowest pressure along the floor. Find the acceleration of the fluid as it passes the point of lowest pressure along the floor.

15-4 Show that the stream function for flow over a rotating cylinder having a radius "a" is given by

$$\psi = U_x \sin\theta\left(r - \frac{a^2}{r}\right) + \frac{-m}{2\pi}\ln(r).$$

Determine the angular location of the stagnation points. Show that the lift force on the cylinder is $Lift = (-m)\rho U_x$.

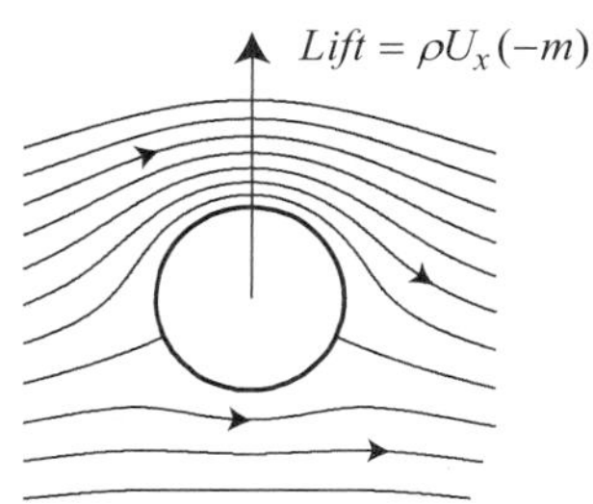

Chapter 16

Complex Variable Methods

16.1 Brief Review of Complex Numbers

16.2 Complex Representation of Potential Flows

16.3 The Joukowski Transform

16.4 Joukowski Symmetric Airfoils

16.5 Joukowski Cambered Airfoils

16.6 Heat Transfer between Nonconcentric Cylinders

16.7 Transport with Temporally Periodic Conditions

16.8 Problems

Two examples of complex variable methods are illustrated in this chapter. The first addresses the use of conformal mapping, which is a mathematical technique used to convert (or map) one mathematical problem and solution into another. One noteworthy application of this technique is in extending the application of potential flow theory to solving aerodynamic airfoil design. The second complex variable method discussed in this chapter seeks to establish quasi-steady-state solutions to transport problems with periodic temporal conditions. The method, called complex combination, is used to transform a governing partial differential equation into an ordinary differential equation governing a complex dependent variable. The method capitalizes on the ability of the complex dependent variable to describe both the phase and amplitude of a periodic response. Since both methods rely on the use of the complex plane, this chapter begins with a brief review of complex numbers.

16.1 BRIEF REVIEW OF COMPLEX NUMBERS

The imaginary unit number "i" is a constant with the special property $i^2 = -1$. An imaginary number is formed by multiplying the imaginary unit number by a real number, such as $i\,y$. A complex number is formed by adding a real number to an imaginary number, such as $z = x + i\,y$. Complex numbers can be added or subtracted by adding or subtracting the real and imaginary parts. For example,

$$z_1 - z_2 = (x_1 + iy_1) - (x_2 + iy_2) = (x_1 - x_2) + i(y_1 - y_2). \tag{16-1}$$

Complex numbers can be multiplied according to the usual rules of algebra. For example,

$$\begin{aligned} z_1 z_2 &= (x_1 + iy_1)(x_2 + iy_2) = x_1x_2 + iy_1x_2 + iy_2x_1 + i^2y_1y_2 \\ &= (x_1x_2 - y_1y_2) + i(y_1x_2 + y_2x_1), \end{aligned} \tag{16-2}$$

in which the rule $i^2 = -1$ has been invoked.

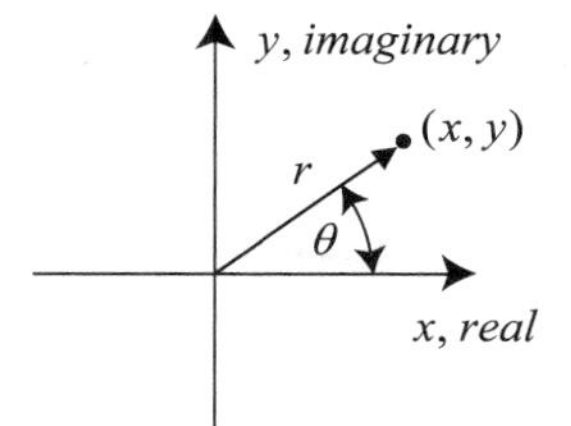

Figure 16-1 The complex plane.

Complex numbers can be presented graphically on a plane, with the real part representing the abscissa and the imaginary part the ordinate. Therefore, the Cartesian point (x, y) identifies the complex number $x + i\,y$, as shown in Figure 16-1. Points on the complex plane can also be identified with polar coordinates. Using *Euler's formula,**

$$e^{i\theta} = \cos\theta + i\sin\theta, \tag{16-3}$$

a complex number can be expressed in polar form as

$$z = x + iy = r(\cos\theta + i\sin\theta) = re^{i\theta} \tag{16-4}$$

where

$$x = r\cos\theta, \quad y = r\sin\theta \tag{16-5}$$

and

$$r = \sqrt{x^2 + y^2}, \quad \theta = \tan^{-1}(y/x). \tag{16-6}$$

Multiplication of complex numbers is easier to perform with polar variables:

$$z_1 z_2 = r_1 e^{i\theta_1} r_2 e^{i\theta_2} = r_1 r_2 e^{i(\theta_1+\theta_2)}. \tag{16-7}$$

However, addition and subtraction is easier to perform in Cartesian coordinates.

The *modulus* (*or absolute value*) of a complex number is $|z| = |re^{i\theta}| = r$, while the *argument* (*or phase*) of a complex number is defined by $\arg z = \arg(re^{i\theta}) = \theta$. The *complex conjugate* z^* of a complex number $z = x + iy = re^{i\theta}$ has the same real part and opposite sign for the imaginary part or, equivalently, the same magnitude but opposite sign for the phase: $z^* = x - iy = re^{-i\theta}$. Notice that $zz^* = |z|^2$, which is the modulus squared.

16.2 COMPLEX REPRESENTATION OF POTENTIAL FLOWS

The Laplace equation can be used to describe steady potential flows as discussed in Chapters 14 and 15 and steady diffusion transport as discussed in Chapter 8. Conformal mapping is a method used to extend the application of known solutions of the Laplace equation. Conformal mapping allows simple solutions to be shaped into descriptions of more complicated and potentially important problems. However, to apply conformal mapping requires that solutions given in terms of potential functions be expressed in a complex variable form.

To describe a plane flow, the *complex potential* $F(z)$ is defined with respect to the velocity components such that

*Euler's formula is named after the Swiss mathematician and physicist Leonhard Euler (1707–1783).

$$\frac{dF}{dz} = v_x - i\, v_y. \tag{16-8}$$

Let the real and imaginary parts of the complex potential be Φ and ψ, such that

$$F = \Phi + i\, \psi. \tag{16-9}$$

Then by definition, $dF = (v_x - iv_y)dz$ is

$$d(\Phi + i\psi) = (v_x - iv_y)d(x + iy) = (v_x dx + v_y dy) + i(v_x dy - v_y dx). \tag{16-10}$$

From this result, the relation between the velocity field (v_x, v_y) and the potentials (Φ, ψ) becomes evident:

$$v_x = \frac{\partial \Phi}{\partial x} = +\frac{\partial \psi}{\partial y} \quad \text{and} \quad v_y = \frac{\partial \Phi}{\partial y} = -\frac{\partial \psi}{\partial x}. \tag{16-11}$$

Equation (16-11) expresses the conventional definitions for the *velocity potential* Φ and the *stream function* ψ. Additionally, the relations between the derivatives of the real and imaginary parts of F expressed in Eq. (16-11) are known as the *Cauchy-Riemann equations*.† The driving force behind many applications of complex analysis is the fact that the real and imaginary parts of any complex potential $F(z)$ are automatic solutions to the Laplace equation of two variables. Using the Cauchy-Riemann equations expressed in Eq. (16-11), this fact can be demonstrated:

$$\begin{aligned} \nabla^2 \Phi &= \frac{\partial}{\partial x}\left(\frac{\partial \Phi}{\partial x}\right) + \frac{\partial}{\partial y}\left(\frac{\partial \Phi}{\partial y}\right) = \frac{\partial}{\partial x}\left(\frac{\partial \psi}{\partial y}\right) + \frac{\partial}{\partial y}\left(-\frac{\partial \psi}{\partial x}\right) = 0 \\ \nabla^2 \psi &= \frac{\partial}{\partial x}\left(\frac{\partial \psi}{\partial x}\right) + \frac{\partial}{\partial y}\left(\frac{\partial \psi}{\partial y}\right) = \frac{\partial}{\partial x}\left(-\frac{\partial \Phi}{\partial y}\right) + \frac{\partial}{\partial y}\left(\frac{\partial \Phi}{\partial x}\right) = 0 \end{aligned} \tag{16-12}$$

A most useful consequence of this arises from the assurance that when conformal mapping is used to create a new complex potential function, the new function will also be a solution to the Laplace equation.

The basic flows used in potential flow theory, such as uniform flow, source, sink, doublet, and vortex, can all be represented with the complex potential $F(z) = \Phi + i\psi$, as shown in Table 16-1.

16.3 THE JOUKOWSKI TRANSFORM

Conformal mapping is performed through the transformation of a complex function from one coordinate system to another. A transformation function is applied to the original function to perform this mapping. For airfoil design, the Joukowski transform [1] is an important function:

$$w = z + \frac{b^2}{z}, \tag{16-13}$$

†The Cauchy-Riemann differential equations in complex analysis are named after the French mathematician Augustin-Louis Cauchy (1789–1857) and the German mathematician Georg Friedrich Bernhard Riemann (1826–1866).

Table 16-1 Catalog of complex potential functions for planar flows

	uniform	$F(z) = (U_x - i\,U_y)\,z$
	vortex	$F(z) = i\dfrac{-m}{2\pi}\ln(z) = i\dfrac{-m}{2\pi}\ln(re^{i\theta}) = \dfrac{m}{2\pi}\theta + i\dfrac{-m}{2\pi}\ln(r)$
source/sink	or	$F(z) = \dfrac{m}{2\pi}\ln(z) = \dfrac{m}{2\pi}\ln(re^{i\theta}) = \dfrac{m}{2\pi}\ln(r) + i\dfrac{m}{2\pi}\theta$
doublet		$F(z) = \dfrac{m}{2\pi}\dfrac{1}{z} = \dfrac{m}{2\pi}\dfrac{1}{re^{i\theta}} = \dfrac{m}{2\pi}\dfrac{\cos\theta}{r} - i\dfrac{m}{2\pi}\dfrac{\sin\theta}{r}$

where b is a constant. Here the conformal mapping will transform a complex plane in z ($z = x + iy$) onto a complex plane in the new variable w ($w = \xi + i\varsigma$). This mapping is illustrated in Figure 16-2. Notice that the local orthogonality of lines drawn in the *z-plane* is preserved in the *w-plane*. This ability to preserve angles at noncritical points is a property associated with conformal mapping.

Any function described in the *z-plane* can be transformed into a related function in the *w-plane*. As an example, consider a circle drawn in the *z-plane* defined by $z = be^{i\theta}$. The Joukowski transform maps a circle of radius b into a straight line in the *w-plane* that lies entirely on the real axis between $-2b$ and $2b$, as demonstrated by

$$w = z + \frac{b^2}{z} = be^{i\theta} + \frac{b^2}{be^{i\theta}} = be^{i\theta} + be^{-i\theta} = 2b\cos(\theta) + i0 \tag{16-14}$$

and as illustrated in Figure 16-3. If a flow had been drawn over the circle, the transform (16-13) could map that flow into a flow over a flat plate in the *w-plane*.

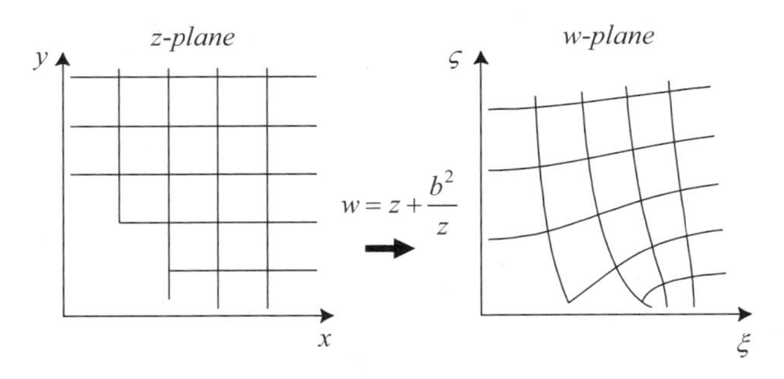

Figure 16-2 Illustration of conformal mapping.

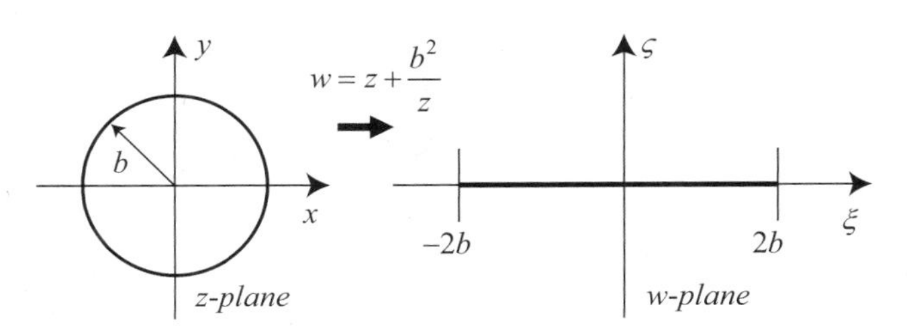

Figure 16-3 Joukowski transform mapping a circle into a flat plate.

If the circle in the *z-plane* originally had a radius slightly larger than the transform constant b, $z = ae^{i\theta}$, with $a > b$, the circle would have formed an ellipse instead of the flat plate:

$$w = z + \frac{b^2}{z} = ae^{i\theta} + \frac{b^2}{ae^{i\theta}} = \left(a + \frac{b^2}{a}\right)\cos(\theta) + i\left(a - \frac{b^2}{a}\right)\sin(\theta) = \xi + i\varsigma. \tag{16-15}$$

The transformed equation can be written in the usual form of an ellipse in the *w-plane*:

$$\frac{\xi^2}{\left(a + \frac{b^2}{a}\right)^2} + \frac{\varsigma^2}{\left(a - \frac{b^2}{a}\right)^2} = 1. \tag{16-16}$$

A flow over the circle could now be transformed into a flow over an ellipse.

16.4 JOUKOWSKI SYMMETRIC AIRFOILS

An important application of the Joukowski transform is to an offset circle. If we consider a circle slightly offset from the origin along the negative real axis, one obtains a symmetric Joukowski airfoil, as illustrated in Figure 16-4.

The equation of the offset circle is $z = ae^{i\theta} - \varepsilon$, where the constant $\varepsilon = a - b$ (see Figure 16-4) is typically a small number. If a flow had been drawn over the circle, the Joukowski transform could map that flow over a symmetric airfoil in the *w-plane*. For example, the flow past a rotating offset cylinder of radius $a = b + \varepsilon$ is described by the complex potential (see Problem 16-1):

$$F(z) = U\left(z + \varepsilon + \frac{a^2}{z + \varepsilon}\right) - i\frac{\Gamma}{2\pi}\ln(z + \varepsilon). \tag{16-17}$$

When this flow field is mapped over the symmetric airfoil using $b = 1$, $\varepsilon = 0.2$, $U = 1$, and $\Gamma = -2\pi$, the result is illustrated in Figure 16-5.

Although the flow illustrated is mathematically possible, the stagnation points on the cylinder map to locations on the airfoil that are not physically realistic. A *stagnation point* occurs where a streamline ends on a surface in the flow. To make a realistic flow requires the *Kutta condition*,[‡] which stipulates that the rear stagnation point attaches to the trailing edge of the airfoil. This causes the flow to depart smoothly from the airfoil. Having a rear stagnation point at any other location causes the velocity near the trailing edge to tent to infinity as streamlines wrap around the rear of the airfoil.

The Kutta condition can be satisfied by adjusting the vorticity strength Γ, such that the rear stagnation point on the airfoils resides on the trailing edge of the airfoil.

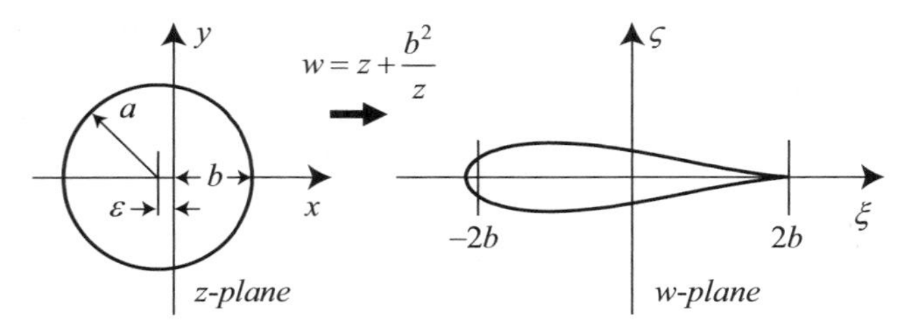

Figure 16-4 Joukowski transform mapping of a circle into a symmetric airfoil.

[‡]The Kutta condition is named after the German mathematician Martin Wilhelm Kutta (1867–1944).

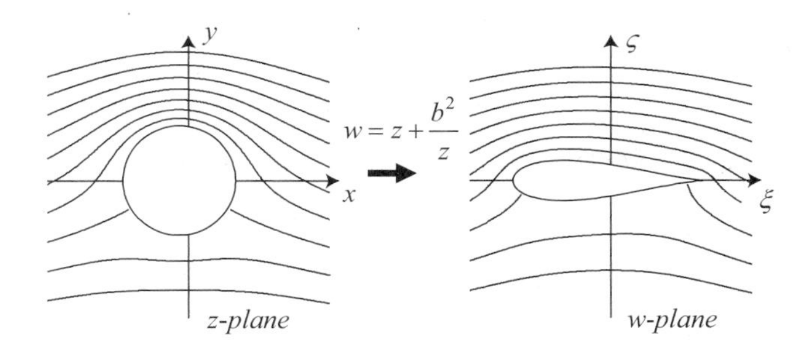

Figure 16-5 Joukowski transform mapping a circulating flow over a cylinder into a circulating flow over a symmetric airfoil.

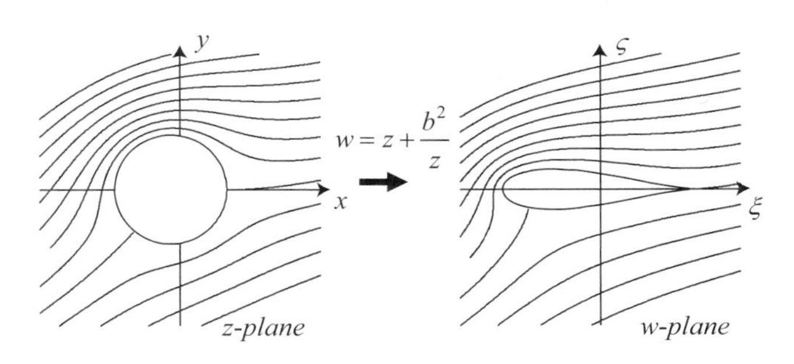

Figure 16-6 Joukowski transform of flow field shown in Figure 16-5 rotated to satisfy the Kutta condition.

Equivalently, the value of Γ is adjusted such that the rear stagnation point on the cylinder (in the *z-plane*) resides at the cylinder's intercept with the real axis (since this point maps to the trailing edge of the airfoil). Alternatively, the Kutta condition can be satisfied by adjusting the angle of attack of the uniform flow when $\Gamma \neq 0$. The flow in the *z-plane* is rotated by an angle α about the center of the cylinder by replacing $z + \varepsilon$ with $(z + \varepsilon)e^{-i\alpha}$, yielding

$$F(z) = U\left((z+\varepsilon)e^{-i\alpha} + \frac{a^2}{(z+\varepsilon)e^{-i\alpha}}\right) - i\frac{\Gamma}{2\pi}\ln\left[(z+\varepsilon)e^{-i\alpha}\right]. \tag{16-18}$$

The velocity at the stagnation point $z = b + i0$ is evaluated from the derivative of the complex potential function:

$$\left.\frac{dF}{dz}\right|_{z=b} = U\left(e^{-i\alpha} - \frac{1}{e^{-i\alpha}}\right) - i\frac{\Gamma}{2\pi a} = 0 - i\left(2U\sin(\alpha) + \frac{\Gamma}{2\pi a}\right). \tag{16-19}$$

This velocity will evaluate to zero at the stagnation point only if

$$\alpha = \sin^{-1}\left(\frac{-\Gamma}{4\pi a U}\right). \tag{16-20}$$

This yields the required angle of attack for the symmetric Joukowski airfoil to satisfy the Kutta condition. Appling this rotation to the flow shown previously in Figure 16-5 yields the physically correct flow over the symmetric airfoil, as illustrated in Figure 16-6.

The lift force generated by the lifting flow over the cylinder is proportional to the circulation about the cylinder imposed by the added vortex flow. The lifting force on the resulting Joukowski airfoil is the same because both flows have the same circulation $-\Gamma$. The lift on the airfoil (and cylinder) is given by

$$Lift = \rho U(-\Gamma) = 4\pi a U \sin(\alpha), \tag{16-21}$$

and is perpendicular to the direction of the rotated flow. Notice that the lift does not depend on the constant $\varepsilon = a - b$, which gives the shape of the symmetric airfoil. Indeed, only for asymmetric airfoils does the actual shape influence the lifting force through implementation of the Kutta condition.

16.5 JOUKOWSKI CAMBERED AIRFOILS

If the cylinder is displaced slightly along the complex axis as well as the real axis, one obtains a cambered (asymmetric) airfoil, as illustrated in Figure 16-7. If a lifting flow about the original circle is imposed, the Joukowski transformation will generate a lifting flow about the Joukowski cambered airfoil. However, to create a realistic flow requires imposing the Kutta condition, such that the rear stagnation point attaches to the trailing edge of the airfoil. The lifting flow around the offset circle is now described by the complex potential:

$$F(z) = U\left((z+\varepsilon-i\delta)e^{-i\alpha} + \frac{a^2}{(z+\varepsilon-i\delta)e^{-i\alpha}}\right) - i\frac{\Gamma}{2\pi}\ln\left[(z+\varepsilon-i\delta)e^{-i\alpha}\right]. \tag{16-22}$$

The derivative of the complex potential function yields the velocity field:

$$\frac{dF}{dz} = U\left(e^{-i\alpha} - \frac{a^2}{(z+\varepsilon-i\delta)^2 e^{-i\alpha}}\right) - i\frac{\Gamma}{2\pi}\frac{1}{z+\varepsilon-i\delta} = v_x - iv_y. \tag{16-23}$$

Letting $b+\varepsilon-i\delta = ae^{-i\beta}$, where $\beta = \tan^{-1}(\delta/(b+\varepsilon))$, the velocity at the stagnation point $z = b + i0$ can be evaluated from:

$$\left.\frac{dF}{dz}\right|_{z=b} = U\left(e^{-i\alpha} - e^{i(\alpha+2\beta)}\right) - i\frac{\Gamma}{2\pi a}e^{i\beta}, \tag{16-24}$$

or

$$\left.\frac{dF}{dz}\right|_{z=b} = U\left(2\sin(\alpha+\beta) + \frac{\Gamma}{2\pi aU}\right)\sin(\beta) - iU\left(2\sin(\alpha+\beta) + \frac{\Gamma}{2\pi aU}\right)\cos(\beta). \tag{16-25}$$

Requiring the stagnation point velocity to be zero yields the relation

$$\sin(\alpha+\beta) = \frac{-\Gamma}{4\pi aU}. \tag{16-26}$$

Therefore, the lifting force on the resulting cambered Joukowski airfoil is

$$Lift = \rho U(-\Gamma) = 4\pi a\rho U^2 \sin(\alpha+\beta). \tag{16-27}$$

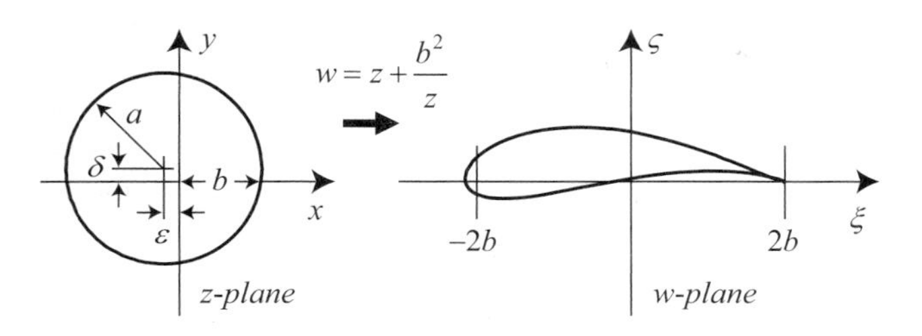

Figure 16-7 Joukowski transform mapping of a circle into a cambered airfoil.

Code 16-1 Potential flow over a Joukowski airfoil. (C++)

```
#include <iostream>
#include <fstream>
#include <complex>

using namespace std;

#define  Cmplx complex<double>
#define  _i Cmplx(0.,1.)

int main()
{
   Cmplx z,zed,F;
   double b=1.,epsln=0.2,delta=0.2;
   double a=sqrt((b+epsln)*(b+epsln)+delta*delta);
   double beta=atan(delta/(b+epsln));

   double alpha=10.*M_PI/180.;
   double K=2.*a*sin(alpha+beta);

   Cmplx C=a*exp(-_i*alpha)+a*a/a/exp(-_i*alpha)+_i*K*log(a*exp(-_i*alpha));

   ofstream outt;
   outt.open("JoukAir.dat");
   for (double x=-3.5;x<=3.501;x+=.02) {
      for (double y=-3.5;y<=3.501;y+=.02) {
         z=x+_i*(x*x+y*y >= a*a ? y : sqrt(a*a-x*x)*y/fabs(y));
         F=z*exp(-_i*alpha)+a*a/z/exp(-_i*alpha)+_i*K*log(z*exp(-_i*alpha));
         F-=C;
         z+=Cmplx(-epsln,delta);
         zed=(z+b*b/z);
         zed*=exp(-_i*alpha);
         outt << real(zed) <<' '<< imag(zed) <<' '<< imag(F) << endl;
      }
   }
   outt.close();

   cout << "\nCL=" << 2.*M_PI*a*sin(alpha+beta)/b << endl;
   return 0;
}
```

For small ε, the *cord length* c of the Joukowski airfoil can be approximated as $4b$. Therefore, the *lift coefficient* can be expressed as

$$C_L = \frac{Lift}{\frac{1}{2}\rho U^2 c} \approx \frac{\rho U(-\Gamma)}{\frac{1}{2}\rho U^2 4b} = \frac{-\Gamma}{2U^2 b} = \frac{2\pi a}{b}\sin(\alpha+\beta). \tag{16-28}$$

With the assumption that $b \approx a$,

$$C_L \approx 2\pi \sin(\alpha+\beta). \tag{16-29}$$

Notice that two geometric factors influence the lift of the asymmetric Joukowski airfoil: the angle of attack α and the shape factor β that gives camber to the airfoil. Code 16-1 calculates the flow field around a Joukowski airfoil defined by the transform variables $b = 1.0$ and $\delta = \varepsilon = 0.2$, and $U = 1.0$. Figure 16-8 illustrates the flow fields around the airfoil for three angles of attack, $\alpha = 0^\circ$, 10°, and 20°, and reports the coefficient of lift calculated from Eq. (16-29).

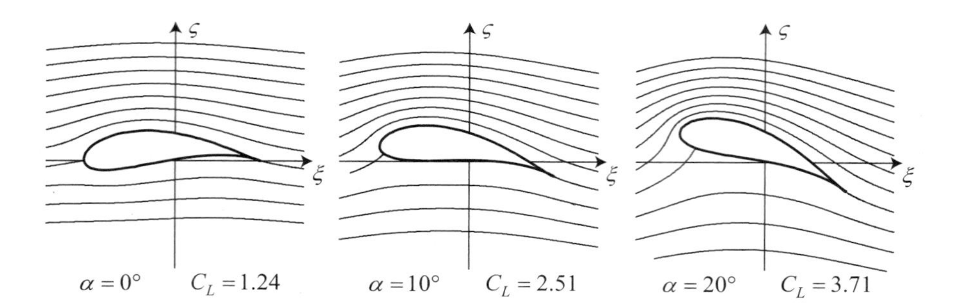

Figure 16-8 Streamlines and coefficient of lift for a cambered Joukowski airfoil at three angles of attack.

16.6 HEAT TRANSFER BETWEEN NONCONCENTRIC CYLINDERS

Conformal mapping of solutions to the Laplace equation is as useful to the description of diffusion transport as it is to ideal potential flow (advection transport). For example, the complex potential may be reinterpreted to describe a planar temperature field when written as

$$F(z) = T + i\varphi. \tag{16-30}$$

Now T describes the isotherms (analogous to constant potential lines), and φ describes the heat transfer lines (analogous to the stream function). Since the real and imaginary parts of any complex potential $F(z)$ are automatic solutions to the Laplace equation, we may use complex potentials to describe heat transfer solutions to the steady-diffusion equation $\nabla^2 T = 0$.

In analogy to Eq. (16-8), the derivative of the complex potential can be shown (see Problem 16-6) to be

$$\frac{dF}{dz} = \frac{\partial T}{\partial x} - i\frac{\partial T}{\partial y}. \tag{16-31}$$

Suppose one is interested in solving the steady heat diffusion equation $\nabla^2 T = 0$ between nonconcentric cylindrical surfaces, each at a constant but different temperature. Interest in this problem might arise from the desire to quantify the net heat transfer between these nonconcentric surfaces. The complex geometry of the domain makes imposing boundary conditions difficult. This can be remedied by solving $\nabla^2 T = 0$ on a simpler domain that can be mapped into the more complicated domain. Consider mapping between concentric cylinders and nonconcentric cylinders performed with the transform:

$$w = \frac{z - \alpha}{\alpha^* z - 1}. \tag{16-32}$$

This transform function maps a unit disk onto itself, but the origin $z = 0$, in the *z-plane*, is moved to the point $w = \alpha$ in the *w-plane*, as illustrated in Figure 16-9. Consequently, circles in the *w-plane* are forced to be nonconcentric.

Suppose one is interested in a nonconcentric region in the *w-plane* where $w < 1$ and $|w - c| > c$, as illustrated in Figure 16-10. The inner cylinder is centered on the real axis at $c + i0$, and has a radius of c. The outer cylinder is centered at $0 + i0$ and has a radius of 1. The inner cylinder is held at a temperature $T = T_c$, while the outer is held at $T = 0$.

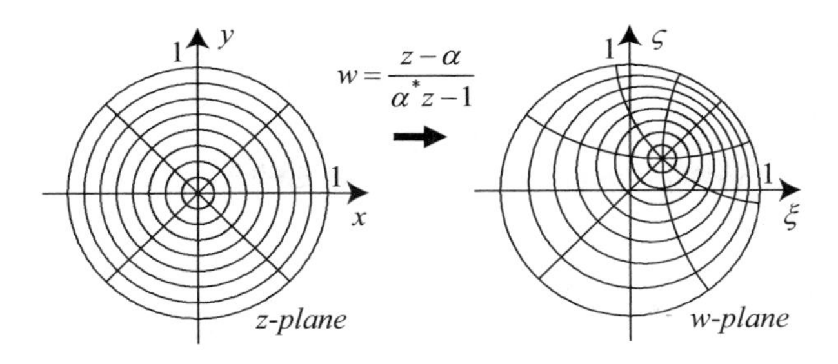

Figure 16-9 Transform illustrated with $\alpha=(1+i)/4$.

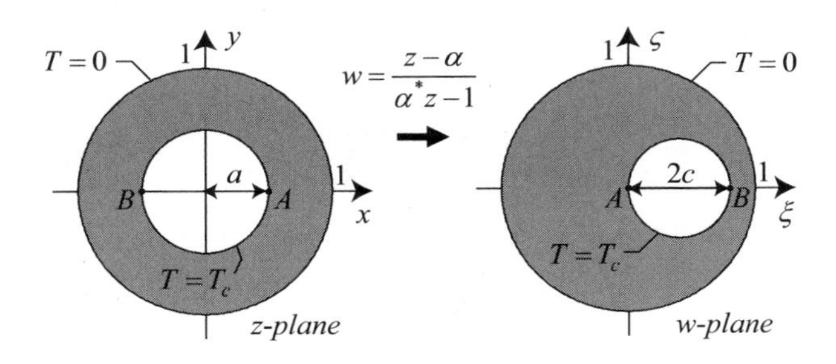

Figure 16-10 Transform illustrated with $a=\alpha=1/2$ and $c=2/5$.

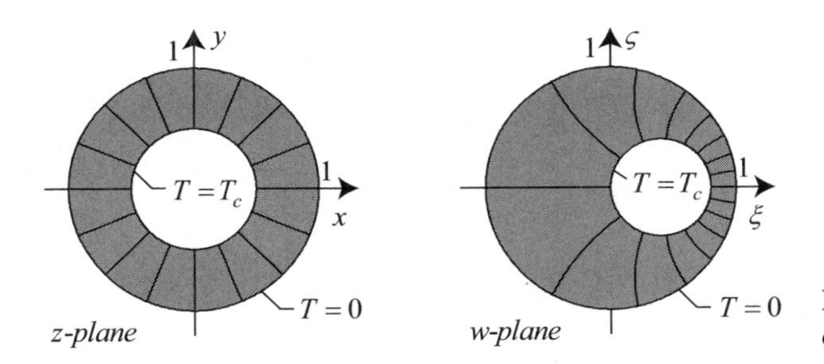

Figure 16-11 Heat lines between cylinders of constant temperature.

The region $a < |z| < 1$ in the *z-plane* can be mapped into the nonconcentric region in the *w-plane* using the transformation function given by Eq. (16-32). Note the mapping of points A and B in the *z-plane* to their respective positions in the *w-plane*, as illustrated in Figure 16-10. Using the transformation function, it is straightforward to show that $\alpha = a + i0$ is required for point A at $z = a + i0$ to map to $w = 0 + i0$. For point B at $z = -a + i0$ to map to $w = 2c + i0$ requires $a = (1 - \sqrt{1 - 4c^2})/(2c)$.

The temperature field between the concentric cylinders having $T(r = a) = 1$ and $T(r = 1) = 0$ is easily determined. That solution, expressed in the complex-number plane, is given by

$$F(z) = \frac{T_c}{\ln(a)}\ln(z) = \frac{T_c}{\ln(a)}\ln(r) + i\frac{\theta}{\ln(a)} = T + i\varphi. \tag{16-33}$$

Figure 16-11 shows the conformal mapping of heat lines, φ, between the concentric cylinders in the *z-plane*, to the region between nonconcentric cylinders in the *w-plane*. Notice that there is a concentration of heat transfer in the region where the inner and outer cylinders are closest.

For the concentric cylinders in the *z-plane*, the heat flux delivered to the outer surface at $r = 1$ is given by

$$q(z)_{r=1} = -k\frac{\partial T(z)}{\partial r}\bigg|_{r=1} = -k\,\mathrm{Re}\left\{\frac{\partial F(z)}{\partial r}\bigg|_{r=1}\right\} = -\frac{kT_c}{\ln(a)}. \tag{16-34}$$

The total heat loss to the outer cylinder (per unit length) is

$$Q_z/L = \int_0^{2\pi} q(z)_{r=1} r d\theta = -\frac{2\pi kT_c}{\ln(a)}. \tag{16-35}$$

For the nonconcentric cylinders, in the *w-plane*, the heat flux delivered to the outer surface at $r = 1$ is given by

$$q(w)_{r=1} = -k\frac{\partial T(w)}{\partial r}\bigg|_{r=1} = -k\,\mathrm{Re}\left\{\frac{\partial F(w)}{\partial r}\bigg|_{r=1}\right\} = -k\,\mathrm{Re}\left\{\frac{dz}{dw}\frac{\partial F(z)}{\partial r}\bigg|_{r=1}\right\}. \tag{16-36}$$

With

$$\frac{dz}{dw} = \frac{(\alpha^* z - 1)^2}{|\alpha|^2 - 1}, \tag{16-37}$$

found by differentiating the transformation function, Eq. (16-32), the flux delivered to the outer surface at $r = 1$ can be evaluated:

$$q(w)_{r=1} = -\frac{kT_c}{\ln(a)}\left|\frac{(\alpha^* z - 1)^2}{(|\alpha|^2 - 1)z}\right|_{r=1} = -\frac{kT_c}{\ln(a)}\frac{e^{-i\theta}(e^{i\theta} - \alpha)(1 - \alpha^* e^{i\theta})}{1 - |\alpha|^2}. \tag{16-38}$$

The total heat loss to the outer cylinder (per unit length) is then found to be

$$Q_w/L = \int_0^{2\pi} q(w)_{r=1} r d\theta = -\frac{2\pi kT_c}{\ln(a)}\frac{1 + |\alpha|^2}{1 - |\alpha|^2}. \tag{16-39}$$

Comparing the heat loss between the concentric cylinders and the heat loss between the nonconcentric cylinders gives

$$\frac{Q_w}{Q_z} = \frac{1 + |\alpha|^2}{1 - |\alpha|^2}. \tag{16-40}$$

16.7 TRANSPORT WITH TEMPORALLY PERIODIC CONDITIONS

Among other utilities, complex numbers are helpful in solving transport problems having a temporally periodic nature—for example, problems with a boundary condition or source term that is temporally periodic. Partial differential equations govern such problems. The desired solution may describe a *quasi-steady-state*, in that there is no change in its periodic nature with time. For such problems, only the amplitude and phase of the solution will vary from one point to another in the domain. It is easy to represent the amplitude and phase attributes of a solution when the dependent variable is complex. With this transformation of the dependent variable to a complex variable, the quasi-steady-state solution may be cast into an ordinary differential equation.

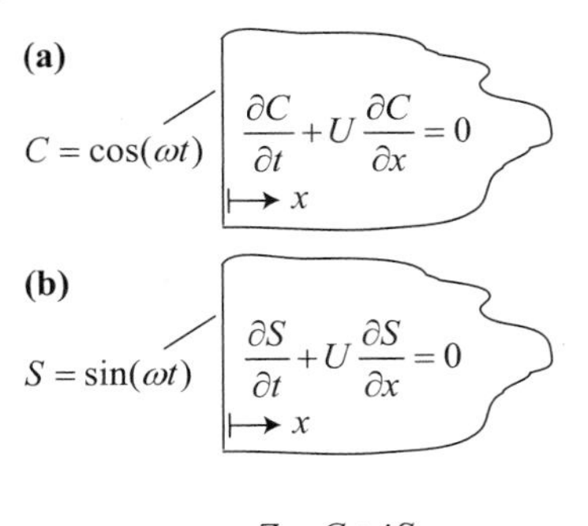

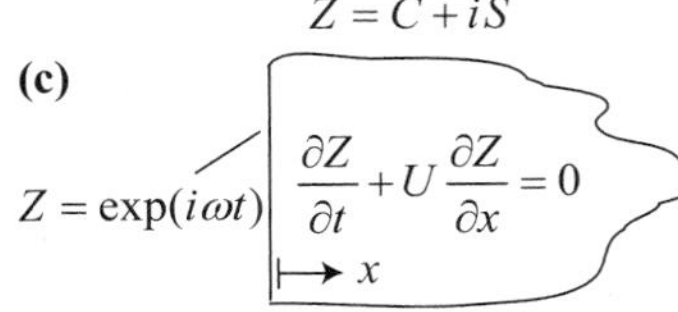

Figure 16-12 Construction of a complex problem for $Z(t,x)$ from original problem for $C(t,x)$.

To illustrate the solution approach, consider the transient advection of a species in a one-dimensional flow, as illustrated mathematically in Figure 16-12(a). The concentration variable is made nondimensional through the definition

$$C = \frac{c - c_o}{A}, \tag{16-41}$$

where c_o is the average concentration at the boundary of the semi-infinite body, and A is the amplitude of periodic concentration excursions at this boundary. In other words, the dimensional concentration at the surface of the semi-infinite body is given by

$$c(t, x = 0) = c_o + A\cos(\omega t). \tag{16-42}$$

A solution for the downstream species concentration $C(t, x)$ can be sought by adding an imaginary component to the original problem, such that $Z(t, x) = C(t, x) + i\, S(t, x)$ as shown in Figure 16-12. This method is known as complex combination. Notice that in this example the imaginary problem described by $S(t, x)$ compliments the boundary condition for $C(t, x)$ such that the boundary condition for $Z(t, x)$ becomes

$$Z(x = 0) = C(x = 0) + i\, S(x = 0) = \cos(\omega t) + i\, \sin(\omega t) = \exp(i\, \omega t) \tag{16-43}$$

with the use of Euler's formula. The desired solution for C is related to the real part of the new problem for Z. In other words, $C = \text{Re}\{Z\}$. The new problem for Z can be solved by assuming a solution with the form

$$Z(t, x) = \exp(i\, \omega t)\, \tilde{C}(x), \tag{16-44}$$

where the temporal part of the solution is described by $\exp(i\, \omega t)$. The complex concentration field $\tilde{C}(x)$ describes the phase and amplitude of the species concentration downstream, and is not a function of time. Substituting the assumed form of the solution into the governing equation and boundary condition for Z yields:

$$\frac{\partial Z}{\partial t} + U\frac{\partial Z}{\partial x} = 0 \qquad \rightarrow \qquad \exp(i\omega t)\left\{ i\omega\, \tilde{C}(x) + U\frac{d\tilde{C}(x)}{dx} \right\} = 0 \tag{16-45}$$

$$Z(x=0) = \exp(i\omega t) \quad \rightarrow \quad \exp(i\omega t)\,\tilde{C}(x=0) = \exp(i\omega t) \tag{16-46}$$

Therefore, complex concentration field $\tilde{C}(x)$ is described by the solution to the ordinary differential equation:

$$\frac{d\tilde{C}}{dx} + \frac{i\omega}{U}\tilde{C} = 0 \quad \text{with} \quad \tilde{C}(0) = 1. \tag{16-47}$$

From this point integration is straightforward and the solution for the complex concentration is given by

$$\tilde{C}(x) = \exp(-i\omega x/U). \tag{16-48}$$

Therefore, for the original problem, the solution is constructed from

$$\begin{aligned} C(t,x) &= \mathrm{Re}\{\exp(i\omega t)\exp(-i\omega x/U)\} = \mathrm{Re}\{\exp(i\omega t - i\omega x/U)\} \\ &= \cos(\omega t - \omega x/U), \end{aligned} \tag{16-49}$$

where Euler's formula was used to put the result into the final form. In Problem 16-9, the current problem is extended to include the effect of diffusion transport.

16.8 PROBLEMS

16-1 Show that the complex potential function for flow over a rotating cylinder of radius a is given by

$$F(z) = U\left(z + \frac{a^2}{z}\right) - i\frac{\Gamma}{2\pi}\ln(z)$$

16-2 Identify the transformations needed to plot ideal plane flow over a unit high wall as illustrated. Contour plot the streamlines for this flow.

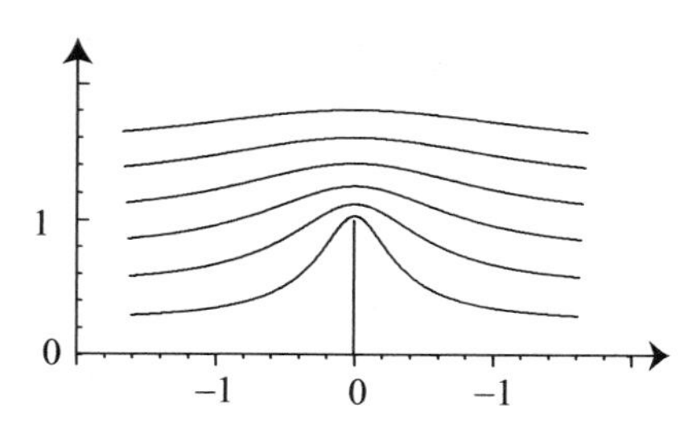

16-3 A Joukowski airfoil is formed by displacing a circle of radius 1 by $\Delta x = -0.16$ (real axis) and $\Delta y = 0.10$ (imaginary axis). Plot the shape of the airfoil. Find the vortex strength Γ if $\alpha = 0°$ and $U = 10$ m/s. Find C_L for $\alpha = 0°$ and $\alpha = 10°$, using the exact cord length.

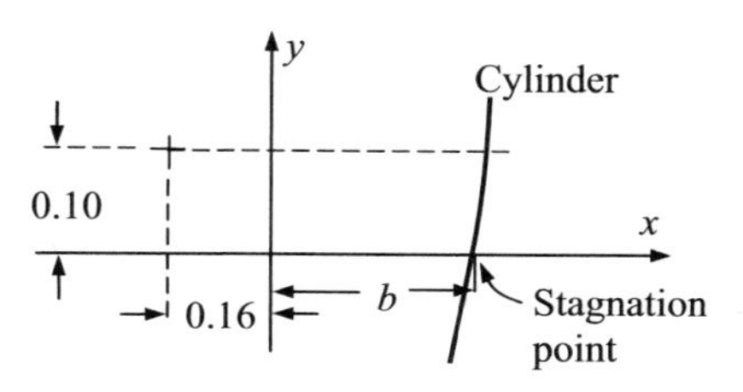

16-4 Using the Kutta condition, plot the streamlines for a flow over a flat plate with a cord length of 4, rotated at an angle of $\alpha = 10^\circ$.

16-5 The Joukowski transform $w = z + 1/z$ can be decomposed into three steps:

Step 1, z to the *u-plane*: $u = \dfrac{z-1}{z+1}$

Step 2, u to the *v-plane*: $v = u^2$

Step 3, v to the *w-plane*: $w = \dfrac{2+2v}{1-v}$

Show that these three steps yield the Joukowski transform $w = z + 1/z$. Consider a modification of Step 2, where $v = u^{1.925}$. For a Joukowski airfoil defined by $b = 1.0$ $\varepsilon = 0.1$ and $\delta = 0.2$, plot the shape of the airfoil given by the modified transform and contrast it with the shape given by the traditional transform $w = z + 1/z$. The advantage of the modified transform is that the sides of the airfoil at the trailing edge form an angle of 0.15π radians, or 27°, which is more realistic than the angle of 0° yielded by the traditional Joukowski transform.

16-6 Show that for the complex temperature potential $F(z) = T + i\varphi$,

$$\frac{dF}{dz} = \frac{\partial T}{\partial x} - i\,\frac{\partial T}{\partial y}.$$

16-7 Demonstrate the sin (z) transformation mapping:

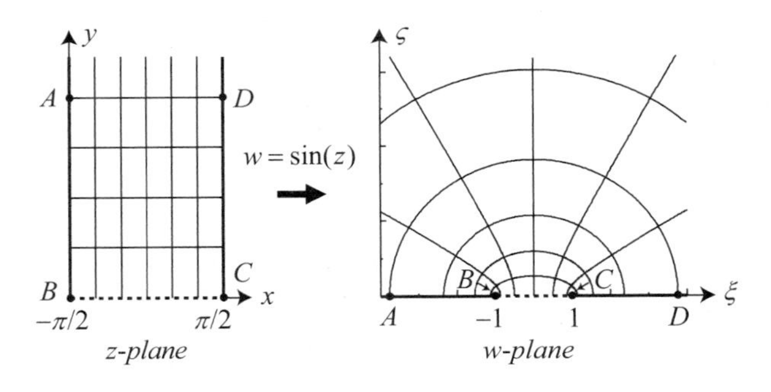

Consider a heat transfer problem in the *w-plane*, where the surface defined by $\varsigma = 0$ and $\xi < -1$ has a temperature T_1, the surface defined by $\varsigma = 0$ and $\xi > +1$ has a temperature T_2, and the surface defined by $\varsigma = 0$ and $-1 < \xi < +1$ is adiabatic. Find the heat transfer rate between T_1 and T_2.

16-8 A semi-infinite fluid is bounded by a large plate. The plate is given a periodic motion such that the fluid in contact with the plate moves with the speed $v_x(y = 0, t) = v_o \cos(\omega t)$. Determine the quasi-steady velocity profile $v_x(y, t)$ in the fluid overlying the plate resulting from diffusion transport. Plot the solution at time intervals of $\omega t = 0$, $\pi/8$, $\pi/4$, $3\pi/8$, $\pi/2$, $5\pi/8$, $3\pi/4$, and π.

y
Fluid
x
$v_x(y = 0, t) = v_o \cos(\omega t)$

16-9 Consider the transient advection and diffusion of a dilute species in a one-dimensional flow, as illustrated. At $x = 0$, the concentration of the species changes periodically in time. Solve for the quasi-steady downstream concentration resulting from advection and diffusion transport. Plot the amplitude of downstream concentration oscillations.

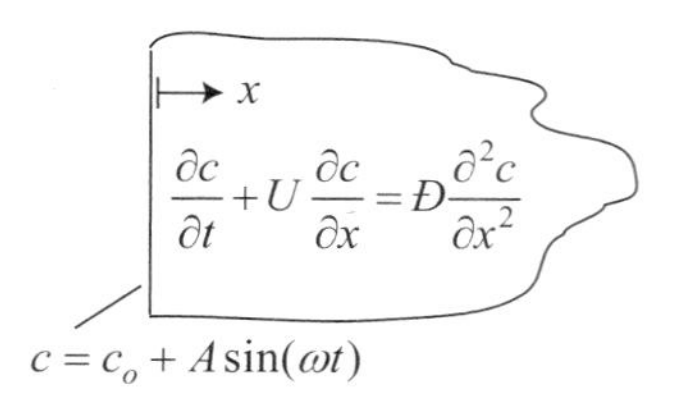

16-10 A long horizontal pipe is filled with a fluid. The fluid is subject to an oscillatory pressure gradient $dP/dz = A\cos(\omega t)$. Show that the quasi-steady-state solution for the fluid velocity in the pipe is given by

$$\frac{v_z(t,r)}{r_o^2 A/\mu} = \mathrm{Re}\left\{\exp(i\omega t)\sum_{n=1}^{\infty}\frac{-2J_0(\lambda_n r/r_o)/J_1(\lambda_n)}{\lambda_n\left(\lambda_n^2 + i\omega r_o^2/\nu\right)}\right\}$$

where λ_n are the roots of $J_0(\lambda_n) = 0$.

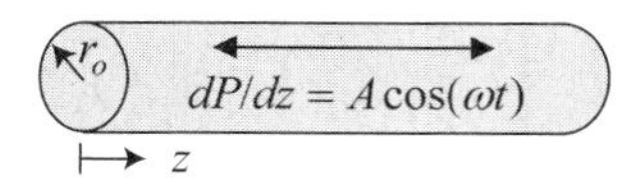

REFERENCE

[1] N. E. Joukovskii, "De la Chute dans l'Air de Corps Légers de Forme Allongée, Animés d'un Mouvement Rotatoire," *Bulletin de l'Institut Aérodynamique de Koutchino*, **1**, 51 (1906).

Chapter 17

MacCormack Integration

17.1 Flux-Conservative Equations

17.2 MacCormack Integration

17.3 Transient Convection

17.4 Steady-State Solution of Coupled Equations

17.5 Problems

The numerical task of solving the transient advection equation is addressed in this chapter. Since problems for which analytic solutions to the advection equation exist are rather limited, using a numerical technique is quite beneficial. However, numerical integration requires some attention to details like numerical stability and accuracy of solutions, which alone can be the subject of extensive study. The present treatment is very cursory in this respect, and is designed mainly to demonstrate the accessibility of numerical solutions. Books devoted to the subject of numerical methods include those given in references [1], [2], and [3].

17.1 FLUX-CONSERVATIVE EQUATIONS

Conservation equations can be written in *flux-conservative* and *nonconservative* forms. Because the two forms are analytically equivalent, no attention was previously given to the distinction. However, for numerical integration, the non-conservative forms of the governing equations can yield erroneous solutions for flows exhibiting discontinuities, such as the hydraulic jump treated in Chapter 18 and the shock wave analyzed in Chapter 20.

Partial differential equations written in flux-conservative form exhibit derivative terms with constant coefficients, or, if variable, derivatives of these coefficients must not appear elsewhere in the equation. The prototypical form of transport equations written in flux-conservative or *conservation form* is

$$\partial_o \phi + \partial_j F_j(\phi) = (sources) \tag{17-1}$$

where ϕ is the conserved property of interest and $F_j(\phi)$ describes the fluxes of ϕ. For example, if the governing equation describes conservation of momentum in the x-direction, then $\phi = \rho v_x$ is the momentum content and $F_j(\phi) = \rho v_x v_j + M_{jx}$ are the advection and diffusion fluxes of momentum. For an inviscid ($M_{jx} = 0$) unidirectional flow, the conservative form of the momentum equation is

$$\text{conservation form:} \; \frac{\partial}{\partial t}(\rho v_x) + \frac{\partial}{\partial x}(\rho v_x^2) = -\frac{\partial P}{\partial x}. \tag{17-2}$$

A nonconservative form of this equation is obtained by expanding derivatives on the left-hand side and eliminating the appearance of the continuity statement. The result is given by

$$\text{nonconservation form: } \rho\frac{\partial v_x}{\partial t} + \rho v_x \frac{\partial v_x}{\partial x} = -\frac{\partial P}{\partial x}. \tag{17-3}$$

The nonconservation form has the "usual" appearance of transport equations derived in earlier chapters. Notice that the nonconservation form contains a nonconstant coefficient v_x that also appears in derivative form in the same equation. Again, the conservation and nonconservation forms are analytically equivalent, but perform differently under numerical integration when discontinuities are exhibited in the flow solution.

17.2 MACCORMACK INTEGRATION

There exist a large number of numerical methods with which transport equations can be integrated. All offer tradeoffs that should be considered if the "best" method is to be selected for a specific problem. However, such an activity is beyond the scope of this text. For numerical integration of the transient advection equation, the MacCormack method [4] has been selected for its relatively simple predictor–corrector scheme. In this scheme, the predictor step estimates the change in the dependent variable using a first-order forward-differencing representation of spatial gradients in the transport equation. After the predictor step is completed for the entire domain, corrector values are generated. However, the corrector step makes use of predictor values in the discretized equation, and uses a first-order backward-differencing representation of spatial gradients. When the predictor and the corrector values are averaged, a second-order integration scheme is realized.*

MacCormack integration is illustrated with the inviscid Burgers' equation† [5], which can be written in two forms:

$$\text{conservative form: } \frac{\partial v}{\partial t} + \frac{\partial}{\partial x}\left(\frac{v^2}{2}\right) = 0, \tag{17-4}$$

$$\text{nonconservative form: } \frac{\partial v}{\partial t} + v\frac{\partial v}{\partial x} = 0. \tag{17-5}$$

In the nonconservative form, Burgers' equation is seen to have the prototypical form of transient advection of momentum. When MacCormack integration is applied to the conservative equation (17-4), the finite differencing representation for the predictor step is written as

$$\frac{\{v_n^{t+\Delta t}\}_p - v_n^t}{\Delta t} = -\left[\frac{(v_{n+1}^t)^2/2 - (v_n^t)^2/2}{\Delta x}\right]. \tag{17-6}$$

Notice that forward differencing is used to evaluate the spatial gradient. Since the dependent variable is known at time t, the only unknown in this equation is $v_n^{t+\Delta t}$. This is a feature of *explicit numerical schemes.* The unknown velocity is denoted by $\{v_n^{t+\Delta t}\}_p$ as a reminder that this is an estimate of the solution at $t + \Delta t$ based on the predictor step. The above equation is solved for $\{v_n^{t+\Delta t}\}_p$:

---

*The MacCormack scheme may also be written with backward differencing in the predictor step and forward differencing in the corrector step.

†The viscid Burgers' equation was studied by Johannes Martinus Burgers in 1948 as a simplification of the Navier-Stokes equations.

$$\text{Predictor step: } \{v_n^{t+\Delta t}\}_p = v_n^t - \frac{\Delta t}{2\Delta x}\left[(v_{n+1}^t)^2 - (v_n^t)^2\right]. \tag{17-7}$$

Equation (17-7) for the predictor step can be evaluated for every node in the domain, except the last. The last node must be handled differently, since forward differencing used in the predictor step would extend outside the discretized domain. Therefore, one resorts to backward differencing for the final node.

For the corrector step, the finite differencing representation of equation (17-4) is written as

$$\frac{\{v_n^{t+\Delta t}\}_c - v_n^t}{\Delta t} = -\left[\frac{\{v_n^{t+\Delta t}\}_p^2/2 - \{v_{n-1}^{t+\Delta t}\}_p^2/2}{\Delta x}\right]. \tag{17-8}$$

Solving the above equation for the only unknown, $\{v_n^{t+\Delta t}\}_c$, yields

$$\text{Corrector step: } \{v_n^{t+\Delta t}\}_c = v_n^t - \frac{\Delta t}{2\Delta x}\left[\{v_n^{t+\Delta t}\}_p^2 - \{v_{n-1}^{t+\Delta t}\}_p^2\right]. \tag{17-9}$$

Notice that backward differencing was used to evaluate the spatial gradient in the corrector step. Additionally, the spatial part of the differential equation has been evaluated with the predictor step results for $\{v^{t+\Delta t}\}_p$. Since $\{v^{t+\Delta t}\}_p$ is everywhere known (after the predictor step), the corrector step can be evaluated for the only unknown $\{v_n^{t+\Delta t}\}_c$. Equation (17-9) is evaluated for every node in the domain except the first, where forward differencing is applied to avoid extending outside the discretized domain.

Both the predictor step and corrector step estimate the solution at $t + \Delta t$. The accepted value of $v_n^{t+\Delta t}$ at the end of the time step is calculated from the average result of the predictor and corrector steps:

$$\text{MacCormack scheme: } v_n^{t+\Delta t} = \frac{\{v_n^{t+\Delta t}\}_p + \{v_n^{t+\Delta t}\}_c}{2}. \tag{17-10}$$

Once the MacCormack scheme has been used to evaluate the dependent variable at $t + \Delta t$, the time index can be incremented forward and the predictor–corrector steps repeated. In this way, the numerical solution is marched forward in time.

17.2.1 Stability of Numerical Integration

Like other explicit methods, numerical integration becomes unstable if the size of the time step Δt is too large. For stable integration, the time step is bounded by the *Courant condition* [6]:

$$0 < \frac{\{|v|\}_{\max}\Delta t}{\Delta x} \leq 1. \tag{17-11}$$

This requirement can be understood by inspecting the nonconservative form of the inviscid Burgers' equation (17-5). If Δt is larger than dictated by Eq. (17-11), the flow speed v will carry information (by advection) further than the distance Δx in one time step. However, with explicit finite differencing, the governing equation is only correctly enforced over a distance of $\pm\Delta x$. Therefore, it should come as no surprise that integration becomes unstable if Δt exceeds the bounds stipulated by Eq. (17-11). For the open channel flows treated in Chapters 18 and 19, information propagates with the combined speed of the flow v and the wave speed c. In this situation, the Courant condition becomes

$$0 < \frac{\{|v| + c\}_{\max} \Delta t}{\Delta x} \leq 1. \tag{17-12}$$

For the compressible flows treated in Chapters 20 through 22, information propagates with the combined speed of the flow v and the speed of sound a. In this situation, the Courant condition becomes

$$0 < \frac{\{|v| + a\}_{\max} \Delta t}{\Delta x} \leq 1. \tag{17-13}$$

It is undesirable to make Δt too small when satisfying the Courant condition. All finite differencing representations of equations introduce an artificial diffusion term that scales as $(\Delta x)^2/\Delta t$. Therefore, in addition to slowing down the process of integration, making Δt too small can introduce an unacceptable amount of artificial diffusion into the solution.

17.2.2 Addition of Viscosity for Numerical Stability

One difficulty in numerically solving the advection equation is that oscillations (or disturbances) in the solution have an opportunity to grow and cause integration to become unstable. This issue is particularly acute near a jump or discontinuity in the solution that provides a strong source for disturbances. Artificial viscosity that is implicit to the discretized form of the governing equation helps dampen unwanted oscillations, but is not always sufficient to keep the solution stable. Oscillations in a solution can be further reduced by explicitly adding a diffusion term to the (inviscid) transport equation. With a diffusion term added, Burgers' equation becomes

$$\text{conservative form:} \quad \frac{\partial v}{\partial t} + \frac{\partial}{\partial x}\left(\frac{v^2}{2}\right) = \nu_a \frac{\partial^2 v}{\partial x^2}. \tag{17-14}$$

The finite differencing form of Burgers' equation can be modified to include the diffusion term:

$$v_n^{t+\Delta t} - v_n^t = (\cdots) + \underbrace{\Delta t \, \nu_a \frac{v_{n+1} - 2v_n + v_{n-1}}{\Delta x^2}}_{\textit{diffusion}}. \tag{17-15}$$

The value of the added viscosity ν_a is somewhat arbitrary; the intent is to introduce sufficient diffusion into the equation to dampen unwanted numerical oscillations, but without changing the dominance of advection transport over the flow physics. Too high a value of ν_a will cause excessive diffusion in the solution (and possible stability issues, as discussed in Section 17.3), while too low a value will exhibit numerical oscillations near discontinuities or allow integration to become unstable and "blow up." An appropriate magnitude for ν_a can be deduced from Burgers' equation. Using the scales $t \sim \Delta t$, $x \sim \Delta x$, and $v \sim \{v\}_{\max}$, scaling of Burgers' equation with the artificial diffusion term reveals

$$\frac{\{v\}_{\max}}{\Delta t} \text{ "+" } \frac{\{v\}^2_{\max}}{\Delta x} \sim \nu_a \frac{\{v\}_{\max}}{\Delta x^2}. \tag{17-16}$$

To ensure that the added diffusion has a negligible effect on the solution requires

$$\frac{diffusion}{advection} \sim \frac{\nu_a \Delta t/\Delta x^2}{\{v\}_{\max}\Delta t/\Delta x} \ll 1. \tag{17-17}$$

However, the Courant condition necessitates that $\{v\}_{\max}\Delta t/\Delta x \sim 1$. Therefore, Eq. (17-17) reduces to the requirement that $\nu_a \Delta t/\Delta x^2 \ll 1$. The recommended magnitude of artificial viscosity is specified by

$$\nu_a \approx \frac{A}{\Delta t/\Delta x^2} < \frac{0.5}{\Delta t/\Delta x^2}, \tag{17-18}$$

where constant $A = 0.01 - 0.1$ is sized to provide enough diffusion to damp unwanted oscillations, but without making diffusion transport significant to the physics of the solution. The value of artificial viscosity recommended in Eq. (17-18) ensures that the magnitude of the explicit diffusion added for numerical stability is about $\sim(100 \times A)\%$ of the magnitude of the advection term in the transport equation. Using $A > 0.5$ results in unstable integration, as will be discussed in Section 17.3.

The Burgers' equation, with added artificial diffusion, is cast into the finite differencing forms of the predictor and corrector steps used in the MacCormack scheme:

$$\begin{aligned}\text{Predictor: } \{v_n^{t+\Delta t}\}_p &= v_n^t - \frac{\Delta t}{2\Delta x}\left[(v_{n+1}^t)^2 - (v_n^t)^2\right] \\ &+ \frac{\nu_a \Delta t}{\Delta x^2}\left[v_{n+1}^t - 2v_n^t + v_{n-1}^t\right] \leftarrow \textit{for stability}\end{aligned} \tag{17-19}$$

$$\begin{aligned}\text{Corrector: } \{v_n^{t+\Delta t}\}_c &= v_n^t - \frac{\Delta t}{2\Delta x}\left[\{v_n^{t+\Delta t}\}_p^2 - \{v_{n-1}^{t+\Delta t}\}_p^2\right] \\ &+ \frac{\nu_a \Delta t}{\Delta x^2}\left[\{v_{n+1}^{t+\Delta t}\}_p - 2\{v_n^{t+\Delta t}\}_p + \{v_{n-1}^{t+\Delta t}\}_p\right] \leftarrow \textit{for stability}.\end{aligned} \tag{17-20}$$

At boundary nodes, it is impossible to evaluate the diffusion terms with the expressions given without extending beyond the physical domain. However, the diffusion terms may be omitted from the predictor and corrector equations at boundary nodes without a noticeable effect on stability.

17.2.3 Numerical Solution to Burgers' Equation

To illustrate MacCormack integration, consider a problem governed by the Burgers' equation, where artificial diffusion is added for numerical stability:

$$\frac{\partial v}{\partial t} + \frac{\partial}{\partial x}\left(\frac{v^2}{2}\right) = \nu_a \frac{\partial^2 v}{\partial x^2}. \tag{17-21}$$

A domain $0 \le x \le 1$ (m) is subject to the initial condition:

$$v(t=0,x) = \begin{cases} 1 \text{ m/s} & 0 \le x \le 0.25 \text{ (m)} \\ 0 & 0.25 < x \le 1 \text{ (m)}. \end{cases} \tag{17-22}$$

Burgers' equation is integrated with Code 17-1 over the time interval: $0 \le t \le 1$ (s). The code uses a mesh of $N = 501$ points and a time step of $\Delta t = 0.0016$ (s). The Courant

Code 17-1 Solution to the inviscid Burgers' equation

```
#include <stdio.h>
#include <math.h>

int MacC(int N,double *u,double dt,double dx)
{
   int n;
   double nu=0.001;
   double Pu[N],Cu[N];

   for (n=0;n<N-1;++n) { // perform predictor step
      Pu[n]=u[n]-(dt/dx/2.)*( u[n+1]*u[n+1] - u[n]*u[n] );
      if (n>0) Pu[n]+=(dt/dx/dx)*nu*(u[n+1]-2.*u[n]+u[n-1]);
   }
   Pu[n]=u[n]-(dt/dx/2.)*( u[n]*u[n] - u[n-1]*u[n-1] );

   for (n=1;n<N;++n) { // perform corrector step
     Cu[n]=u[n]-(dt/dx/2.)*( Pu[n]*Pu[n] - Pu[n-1]*Pu[n-1] );
      if (n<N-1) Cu[n]+=(dt/dx/dx)*nu*(Pu[n+1]-2.*Pu[n]+Pu[n-1]);
   }

   // average predictor and corrector results (interior nodes)
   for (n=1;n<N-1;++n) {
      u[n]=0.5*(Pu[n]+Cu[n]);
      if (isnan(u[n])) {
         printf("\nsoln BLEW!\n");
         return 0;
      }
   }
   return 1;
}

int main(void)
{
   int cnt=0,N=501;
   int n,n0=(N-1)/4;
   double speed,u[N];
   double umax=1.,dx=1./(N-1);
   double dt=.8*dx/umax,tend=1.;
   FILE *fp;

   for (n=0;n<=n0;++n) u[n]=umax;  // initial distribution
   for ( ;n<N;++n)     u[n]=0.0;

   do { // integrate forward in time
      if ( !MacC(N,u,dt,dx) ) break;
   } while (++cnt*dt<tend);
   printf("\ncnt=%d  t=%e",cnt,cnt*dt);

   for (n=0;n<N;++n) if (u[n]<.8) break;
   speed=dx*(n-1+(0.5-u[n-1])/(u[n]-u[n-1])-n0)/cnt/dt;
   printf("\nspeed=%e",speed);

   fp=fopen("out.dat","w");
   for (n=0;n<N;++n)
      fprintf(fp,"%e %e\n",n*dx,u[n]);

   fclose(fp);
   return 1;
}
```

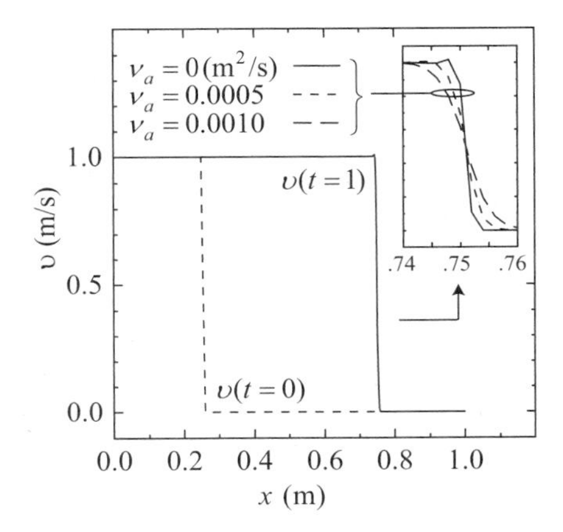

Figure 17-1 Wave propagation shown at $t = 1$ for Courant number of 0.8 and different added viscosity.

number for integration is $\{|\upsilon|\}_{\max}\Delta t/\Delta x = 0.8$. At $t = 1$ (s) the solution for $\upsilon(t = 1, x)$ is shown in Figure 17-1 for three choices of added viscosity ($0 \le \nu_a \le 0.001$). The wave front (where υ transitions between 1 and 0) is initially at $x = 0.25$ (m) and propagates to $x = 0.75$ (m) in one second, for a wave speed of 0.50 (m/s). Small numerical oscillations occur at the wave front. With increasing added viscosity, the wave front broadens by diffusion and numerical oscillations are dampened, as shown in the inset of Figure 17-1. For $\nu_a = 0.002$, integration becomes unstable for a reason that is explained in the next section.

The numerical oscillations in the solution to Burgers' equation are relatively small. However, larger-magnitude numerical oscillations will occur for equations governing the hydraulic jump treated in Chapter 18 and the shock wave discussed in Chapter 20. Numerical integration of these flows will benefit more from added viscosity than what is observed here for Burgers' equation.

17.3 TRANSIENT CONVECTION

When advection and diffusion are both physically important to transport, the problem is classified as convection. The viscid Burgers' equation is an example of a transient convection equation:

$$\text{conservative form: } \frac{\partial \upsilon}{\partial t} + \frac{\partial}{\partial x}\left(\frac{\upsilon^2}{2}\right) = \nu \frac{\partial^2 \upsilon}{\partial x^2} \tag{17-23}$$

$$\text{nonconservative form: } \frac{\partial \upsilon}{\partial t} + \upsilon \frac{\partial \upsilon}{\partial x} = \nu \frac{\partial^2 \upsilon}{\partial x^2}. \tag{17-24}$$

The viscid Burgers' equation can be solved by MacCormack integration, as was discussed in Section 17.2, but with the understanding that ν is now a physically important viscosity. The MacCormack integration predictor and corrector steps are the same:

$$\begin{aligned}\text{Predictor: } \{\upsilon_n^{t+\Delta t}\}_p &= \upsilon_n^t - \frac{\Delta t}{2\Delta x}\left[(\upsilon_{n+1}^t)^2 - (\upsilon_n^t)^2\right] \\ &+ \frac{\nu\,\Delta t}{\Delta x^2}\left[\upsilon_{n+1}^t - 2\upsilon_n^t + \upsilon_{n-1}^t\right]\end{aligned} \tag{17-25}$$

$$\begin{aligned}\text{Corrector: } \{\upsilon_n^{t+\Delta t}\}_c &= \upsilon_n^t - \frac{\Delta t}{2\Delta x}\left[\{\upsilon_n^{t+\Delta t}\}_p^2 - \{\upsilon_{n-1}^{t+\Delta t}\}_p^2\right] \\ &+ \frac{\nu\,\Delta t}{\Delta x^2}\left[\{\upsilon_{n+1}^{t+\Delta t}\}_p - 2\{\upsilon_n^{t+\Delta t}\}_p + \{\upsilon_{n-1}^{t+\Delta t}\}_p\right].\end{aligned} \tag{17-26}$$

However, since diffusion is a physically significant transport term in the equation, a new stability requirement exists on the numerical time step:

$$0 < \frac{\{|v|\}_{\max}\Delta t}{\Delta x} \le \frac{\sqrt{2\nu\,\Delta t}}{\Delta x} \le 1. \tag{17-27}$$

This stability requirement includes consideration of a diffusion length scale $\sqrt{2\nu\,\Delta t}$, and stipulates that this length scale be smaller than Δx. This necessity follows from the same logic used to justify Eq. (17-11); if in a single time step Δt information is carried a distance greater than Δx, the governing equation in explicit finite differencing form cannot be adequately enforced. In general, *von Neumann stability analysis* [1] can be used in numerical analysis to determine the stability requirements for finite difference schemes.

Since diffusion is physically important to the solution, it is unsatisfactory to omit the diffusion term in the governing equation at the boundaries, as was done when diffusion was added in Section 17.2 solely for numerical stability. To describe diffusion accurately at the boundary nodes, without extending differencing operations beyond the bounds of the domain, requires alternate forms of the diffusion terms. At the first node of the domain, the finite differencing equation can use a diffusion term expressed by

$$n = \text{first nodes: } v_n^{t+\Delta t} - v_n^t = (\cdots) + \underbrace{\Delta t\,\nu\frac{2v_n - 5v_{n+1} + 4v_{n+2} - v_{n+3}}{\Delta x^2}}_{diffusion}. \tag{17-28}$$

At the last node of the domain, the diffusion term can be expressed by

$$n = \text{last node: } v_n^{t+\Delta t} - v_n^t = (\cdots) + \underbrace{\Delta t\,\nu\frac{-v_{n-3} + 4v_{n-2} - 5v_{n-1} + 2v_n}{\Delta x^2}}_{diffusion}. \tag{17-29}$$

Both forms of the diffusion term retain the same second-order accuracy provided by the center differencing employed for internal nodes:

$$n = \text{interior nodes: } v_n^{t+\Delta t} - v_n^t = (\cdots) + \underbrace{\Delta t\,\nu\frac{v_{n+1} - 2v_n + v_{n-1}}{\Delta x^2}}_{diffusion}. \tag{17-30}$$

The numerical solution of Burgers' viscid equation is illustrated in the next section.

17.3.1 Groundwater Contamination

Consider the problem illustrated in Figure 17-2 of transient bacteria transport in groundwater, which was solved by separation of variables in Section 7.4. Here the problem is resolved by MacCormack time integration. The transient bacteria concentration in the groundwater is governed by the transport equation:

$$\frac{\partial c_B}{\partial t} - U\frac{\partial c_B}{\partial x} = D_B\frac{\partial^2 c_B}{\partial x^2}. \tag{17-31}$$

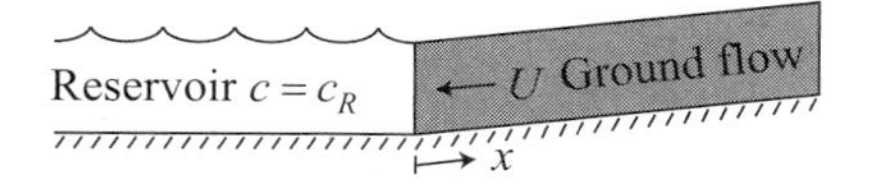

Figure 17-2 Groundwater flow to a reservoir.

The groundwater flows with a uniform velocity $v_x = -U$ and empties into a reservoir with a concentration c_R of bacteria. The groundwater is initially uncontaminated. Therefore, the conditions to be imposed on the solution are:

$$t = 0: \quad c_B\,(t = 0, x) = 0 \tag{17-32}$$

$$x = 0: \quad c_B\,(x = 0) = c_R \tag{17-33}$$

$$x \rightarrow \infty: \quad c_B\,(x \rightarrow \infty) = 0. \tag{17-34}$$

Letting $\eta = Ux/\,Đ_B$, $\tau = t\,U^2/\,Đ_B$, and $\theta = c_B/c_R$, the problem statement can be non-dimensionalized. The governing equation becomes

$$\frac{\partial \theta}{\partial \tau} = \frac{\partial \theta}{\partial \eta} + \frac{\partial^2 \theta}{\partial \eta^2}, \tag{17-35}$$

with the following conditions to be imposed on the solution:

$$\tau = 0: \quad \theta(\tau = 0, \eta) = 0 \tag{17-36}$$

$$\eta = 0: \quad \theta(\eta = 0) = 1 \tag{17-37}$$

$$\eta \rightarrow \infty: \quad \theta(\eta \rightarrow \infty) = 0. \tag{17-38}$$

Equation (17-35) can be integrated with the MacCormack scheme to determine the groundwater bacteria concentration as a function of time. Casting Eq. (17-35) into the finite differencing forms for the predictor and corrector steps yields:

$$\text{Predictor:} \quad \{\theta_n^{t+\Delta t}\}_p = \theta_n^t + \frac{\Delta\tau}{\Delta\eta}\left[\theta_{n+1}^t - \theta_n^t\right] + \frac{\Delta\tau}{\Delta\eta^2}\left[\theta_{n+1}^t - 2\theta_n^t + \theta_{n-1}^t\right] \tag{17-39}$$

$$\text{Corrector:} \quad \{\theta_n^{t+\Delta t}\}_c = \theta_n^t + \frac{\Delta\tau}{\Delta\eta}\left[\{\theta_n^{t+\Delta t}\}_p - \{\theta_{n-1}^{t+\Delta t}\}_p\right] + \frac{\Delta\tau}{\Delta\eta^2}\left[\{\theta_{n+1}^{t+\Delta t}\}_p - 2\{\theta_n^{t+\Delta t}\}_p + \{\theta_{n-1}^{t+\Delta t}\}_p\right]. \tag{17-40}$$

The presence of physical diffusion negates any need for introducing artificial diffusion into the equation. For numerical stability let $\Delta\tau = \Delta\eta^2/2$, which will satisfy the condition

$$0 \leq \frac{\Delta\tau}{\Delta\eta} \leq \sqrt{\frac{2\,\Delta\tau}{\Delta\eta^2}} \leq 1 \tag{17-41}$$

as long as $\Delta\eta \leq 2$. Code 17.2 integrates the predictor–corrector equations (17-39) and (17-40) in time. The diffusion terms in the predictor and corrector steps are modified in Code 17-2 for the boundary nodes as indicated by Eqs. (17-28) and (17-29). The accuracy of the numerical integration can be confirmed by a comparison with the exact solution derived in Section 7.4.

Code 17-2 Transient mass transport of bacteria in groundwater

```
#include <stdio.h>
#include <math.h>

int MacC(int N,double *c,double dt,double dx)
{
   int n;
   double Pc[N],Cc[N];

   for (n=0;n<N-1;++n) { // predictor step
      Pc[n]=c[n]+(dt/dx)*( c[n+1] - c[n] );
      if (n>0) Pc[n]+=(dt/dx/dx)*(c[n+1]-2.*c[n]+c[n-1]);
      else Pc[n]+=(dt/dx/dx)*(-c[n+3]+4.*c[n+2]-5.*c[n+1]+2.*c[n]);
   }
   Pc[n]=c[n]+(dt/dx)*( c[n] - c[n-1] );
   Pc[n]+=(dt/dx/dx)*(-c[n-3]+4.*c[n-2]-5.*c[n-1]+2.*c[n]);

   for (n=1;n<N;++n) { // corrector step
      Cc[n]=c[n]+(dt/dx)*( Pc[n] - Pc[n-1] );
      if (n<N-1) Cc[n]+=(dt/dx/dx)*(Pc[n+1]-2.*Pc[n]+Pc[n-1]);
      else Cc[n]+=(dt/dx/dx)*(-Pc[n-3]+4.*Pc[n-2]-5.*Pc[n-1]+2.*Pc[n]);
   }

   for (n=1;n<N-1;++n) { // average predictor and corrector
      c[n]=0.5*(Pc[n]+Cc[n]);
      if (isnan(c[n])) {
         printf("\nsoln BLEW!\n");
         return 0;
      }
   }
   return 1;
}
```

```
int main(void)
{
   int n,cnt=0,N=251;
   double c[N];
   FILE *fp;
   double L=8., dx=L/(N-1);
   double dt=dx*dx/2.0;
   double tend=10.;
   printf("\ndt=%e",dt);

   c[0]=1.0; // initial distribution
   for (n=1;n<N;++n) c[n]=0.;

   fp=fopen("out.dat","w");
   double log_out=-2.;
   do { // integrate forward in time
      if (log10(cnt*dt)>=log_out) {
         for (n=0;n<N;++n)          // output
            fprintf(fp,"%e %e\n",n*dx,c[n]);
         fprintf(fp,"\n");
         printf("\ncnt=%d t=%e",cnt,cnt*dt);
         log_out+=1.;
      }
      if ( !MacC(N,c,dt,dx) ) break;
   } while ((cnt++)*dt < tend);
   fclose(fp);
   return 1;
}
```

17.4 STEADY-STATE SOLUTION OF COUPLED EQUATIONS

For some problems, a steady-state solution to advection transport equations is sought. One strategy for obtaining such a solution is to integrate the transient equations forward in time until the steady-state solution is reached. This approach is illustrated with MacCormack integration for a flow described by the dependent variables $u(x,t)$ and $\eta(x,t)$, which are governed by the coupled equations

$$\frac{\partial \eta}{\partial t} + \frac{\partial}{\partial x}(\eta u) = 0 \tag{17-42}$$

and

$$\frac{\partial(\eta u)}{\partial t} + \frac{\partial}{\partial x}((\eta u)u) = \eta. \tag{17-43}$$

Notice that both equations have the form of advection transport (where u is related to the flow velocity) and both equations are written in a conservative form. The second governing equation is a "made-up" equation describing the transport of a flow quantity (ηu), where η is an undisclosed property of the flow.

Suppose a steady-state solution to Eqs. (17-42) and (17-43) is desired for the domain $0 \le x \le 1$, and is subject to the boundary conditions

$$\eta(x=0) = 1 \quad \text{and} \quad u(x=0) = 1. \tag{17-44}$$

Since the governing equations are first order, only one boundary condition for each dependent variable can be specified. Therefore, values of the dependent variables at the second boundary must be part of the unknown solution to the transport equations.

To integrate the governing equations forward in time also requires an initial condition. Since any initial condition will integrate to the same steady-state solution, the only stipulation on the initial condition is that it should satisfy the known boundary conditions. For example, in the present problem it suffices to use the initial conditions

$$\eta(t=0,x) = 1 \quad \text{and} \quad u(t=0,x) = 1. \tag{17-45}$$

The governing equations (17-42) and (17-43) can be solved by MacCormack integration for the variables η and (ηu). Artificial viscosity is added to these equations for improved stability. The predictor step calculations for the governing equations are given by

$$\begin{aligned}\{\eta_n^{t+\Delta t}\}_p &= \eta_n^t - \frac{\Delta t}{\Delta x}\left[\eta_{n+1}^t u_{n+1}^t - \eta_n^t u_n^t\right] \\ &+ \frac{\nu_a \Delta t}{\Delta x^2}\left[\eta_{n+1}^t - 2\eta_n^t + \eta_{n-1}^t\right] \leftarrow \textit{for stability}\end{aligned} \tag{17-46}$$

and

$$\begin{aligned}\{(\eta u)_n^{t+\Delta t}\}_p &= (\eta u)_n^t - \frac{\Delta t}{\Delta x}\left[(\eta u)_{n+1}^t u_{n+1}^t - (\eta u)_n^t u_n^t\right] + \Delta t\, \eta_n^\tau \\ &+ \frac{\mu_a \Delta t}{\Delta x^2}\left[(\eta\, u)_{n+1}^t - 2(\eta\, u)_n^t + (\eta\, u)_{n-1}^t\right] \leftarrow \textit{for stability}.\end{aligned} \tag{17-47}$$

Predictor step values for velocity $\{u_n^{t+\Delta t}\}_p$ are obtained from $\{(\eta\, u)_n^{t+\Delta t}\}_p$ using

$$\{u_n^{t+\Delta t}\}_p = \frac{\{(\eta u)_n^{t+\Delta t}\}_p}{\{\eta_n^{t+\Delta t}\}_p}. \tag{17-48}$$

The corrector step calculations for the governing equations (17-42) and (17-43) are given by

$$\begin{aligned}\{\eta_n^{t+\Delta t}\}_c &= \eta_n^t - \frac{\Delta t}{\Delta x}\left[\{\eta_n^{t+\Delta t}\}_p\{u_n^{t+\Delta t}\}_p - \{\eta_{n-1}^{t+\Delta t}\}_p\{u_{n-1}^{t+\Delta t}\}_p\right] \\ &+ \frac{\nu_a \Delta t}{\Delta x^2}\left[\{\eta_{n+1}^{t+\Delta t}\}_p - 2\{\eta_n^{t+\Delta t}\}_p + \{\eta_{n-1}^{t+\Delta t}\}_p\right] \leftarrow \textit{for stability}\end{aligned} \tag{17-49}$$

and

$$\begin{aligned}\{(\eta u)_n^{t+\Delta t}\}_c &= (\eta u)_n^t - \frac{\Delta t}{\Delta x}\left[\{(\eta u)_n^{t+\Delta t}\}_p\{u_n^{t+\Delta t}\}_p - \{(\eta u)_{n-1}^{t+\Delta t}\}_p\{u_{n-1}^{t+\Delta t}\}_p\right] + \Delta t\{\eta_n^{t+\Delta t}\}_p \\ &+ \frac{\mu_a \Delta t}{\Delta x^2}\left[\{(\eta u)_{n+1}^{t+\Delta t}\}_p - 2\{(\eta u)_n^{t+\Delta t}\}_p + \{(\eta u)_{n-1}^{t+\Delta t}\}_p\right] \leftarrow \textit{for stability}.\end{aligned} \tag{17-50}$$

Corrector step values for velocity $\{u_n^{t+\Delta t}\}_c$ are obtained from $\{(\eta u)_n^{t+\Delta t}\}_c$ using

$$\{u_n^{t+\Delta t}\}_c = \frac{\{(\eta u)_n^{t+\Delta t}\}_c}{\{\eta_n^{t+\Delta t}\}_c}. \tag{17-51}$$

The predictor step and corrector step values are averaged to obtain the end of time step values:

$$\eta_n^{t+\Delta t} = \frac{\{\eta_n^{t+\Delta t}\}_p + \{\eta_n^{t+\Delta t}\}_c}{2}, \tag{17-52}$$

$$(\eta u)_n^{t+\Delta t} = \frac{\{(\eta u)_n^{t+\Delta t}\}_p + \{(\eta u)_n^{t+\Delta t}\}_c}{2}. \tag{17-53}$$

and

$$u_n^{t+\Delta t} = \frac{(\eta u)_n^{t+\Delta t}}{\eta_n^{t+\Delta t}}. \tag{17-54}$$

One simple numerical way to handle unspecified boundary values is to let them "float." The floating condition linearly extrapolates boundary values from internal points of the flow solution. In this manner, the requirements of the transport equations are allowed to propagate to the boundary. For example, the floating boundary conditions for η and (ηu) are

$$\text{at } x = 1: \quad \text{for } n = \text{last node}: \begin{cases} \eta_n^{t+\Delta t} = 2\eta_{n-1}^{t+\Delta t} - \eta_{n-2}^{t+\Delta t} \\ (\eta u)_n^{t+\Delta t} = 2(\eta u)_{n-1}^{t+\Delta t} - (\eta u)_{n-2}^{t+\Delta t} \end{cases}. \tag{17-55}$$

Alternatively, boundary values can be calculated explicitly from the governing equations. However, this will require the direction of finite differencing to be reversed at boundary

Code 17-3 Integration to steady-state solution of coupled equations

```
#include <stdio.h>
#include <math.h>

inline double Eq1_diff(char t,int n,double *y,double *v) {
   return t=='p' ? y[n+1]*v[n+1] - y[n]*v[n] :
                   y[n]*v[n] - y[n-1]*v[n-1] ;          }

inline double Eq2_diff(char t,int n,double *y,double *y_v,double *v) {
   return t=='p' ? y_v[n+1]*v[n+1] - y_v[n]*v[n] :
                   y_v[n]*v[n] - y_v[n-1]*v[n-1] ;                 }

double MacC(int N,double *y,double *y_v,double *v,double *dt_,double dx)
{
   int n,step;
   double dt=*dt_,Py[N],Cy[N],Py_v[N],Cy_v[N],Pv[N],Cv[N];
   double C,Cmax=0,Ctarg=0.95,dtdx=dt/dx,dtddx=dt/dx/dx,ma,na;
   na=ma=0.04/dtddx;

   for (n=0;n<N;++n) {              // perform predictor step
      step=(n<N-1 ? 'p' : 'c');
      Py[n]=y[n]-dtdx*Eq1_diff(step,n,y,v);
      if (n>0 && n<N-1) Py[n]+=dtddx*na*(y[n+1]-2.*y[n]+y[n-1]);
      Py_v[n]=y_v[n]-dtdx*Eq2_diff(step,n,y,y_v,v) + dt*y[n];
      if (n>0 && n<N-1) Py_v[n]+=dtddx*ma*(y_v[n+1]-2.*y_v[n]+y_v[n-1]);
      Pv[n]=Py_v[n]/Py[n];
      C=dt*(fabs(Pv[n]))/dx;        // dt size checking
      if (C>Cmax) Cmax=C;
   }
   if (Cmax>1. || Cmax<2.*Ctarg-1.) { // integrate with different time step
      *dt_=0.;
      return dt*Ctarg/Cmax;
   }
   for (n=0;n<N;++n) {              // perform corrector step
      step=(n>0 ? 'c' : 'p');
      Cy[n]=y[n]-dt/dx*Eq1_diff(step,n,Py,Pv);
      if (n>0 && n<N-1) Cy[n]+=dtddx*na*(Py[n+1]-2.*Py[n]+Py[n-1]);
      Cy_v[n]=y_v[n]-dtdx*Eq2_diff(step,n,Py,Py_v,Pv) + dt*Py[n];
      if (n>0 && n<N-1) Cy_v[n]+=dtddx*ma*(Py_v[n+1]-2.*Py_v[n]+Py_v[n-1]);
      Cv[n]=Cy_v[n]/Cy[n];
      if (isnan(Cv[n])) {
         printf("\nsoln BLEW! dt=%e (Cmax=%e)\n",dt,Cmax);
         return 0.;
      }
   }
   for (n=1;n<N;++n) { // average predictor and corrector results
      y[n]=0.5*(Py[n]+Cy[n]);
      y_v[n]=0.5*(Py_v[n]+Cy_v[n]);
      v[n]=y_v[n]/y[n];
   }
   v[N-1]=(y_v[N-1]=2.*y_v[N-2]-y_v[N-3])/y[N-1]; // float exit u
   y[N-1]=2.*y[N-2]-y[N-3];                       // float exit y

   return dt*Ctarg/Cmax;
}

int main(void)
{
   int n,cnt=0,N=1201,ConvergeCnt=0;;
   double y[N],y_u[N],u[N],u_last=0.,soln;
   double t=0,L=1.,dx=L/(N-1),dt=0.95*dx,dt_next=dt;
   FILE *fp;
   for (n=0;n<N;++n) // initial distributions
      y[n]=y_u[n]=u[n]=1.;

   do {
      if (!((cnt++)%500)) { // check every 500 dt
         if (fabs((u[N-1]-u_last))<5.e-6) ++ConvergeCnt;
         else ConvergeCnt=0;
         printf("\nt=%e Exit_u=%e",t,u[N-1]);
         u_last=u[N-1];
      }
      dt=dt_next;
      if ( !(dt_next=MacC(N,y,y_u,u,&dt,dx)) ) break;
   } while ((t+=dt)<100. && ConvergeCnt<5);

   fp=fopen("out.dat","w");
   for (n=0;n<N;++n) {
      soln=sqrt(2.*n*dx+u[0]*u[0]);
      fprintf(fp,"%e %e %e %e %e\n",n*dx,y[n],u[n],soln,1./soln);
   }
   fclose(fp);

   return 1;
}
```

nodes, for either the predictor or corrector step, to prevent the operation from extending beyond the numerical domain. In the current problem, finite differencing during the predictor step must be reversed at $x = 1$. Using the governing equations to calculate unknown boundary values is desirable when the boundary is physical (like a wall) as opposed to being imaginary. Imaginary boundaries are used in steady-state problems at locations where the flow enters or exits the computation domain.

Code 17-3 integrates the predictor–corrector equations forward in time. Each time the MacCormack subroutine is called, the solutions to the temporal equations are advanced by Δt. Notice that the integration time step Δt is dynamically adjusted in the MacCormack subroutine to satisfy the condition

$$\frac{\Delta t\{u\}_{\max}}{\Delta x} \approx C_{targ} = 0.95. \tag{17-56}$$

This is a useful strategy for satisfying the Courant condition, while maintaining as large an integration time step as possible. Following Eq. (17-18), values of the artificial diffusivities added to the governing equations are also dynamically adjusted to the value

$$\nu_a = \mu_a = \frac{0.04}{\Delta t/\Delta x^2}. \tag{17-57}$$

This ensures that the magnitude of the explicit diffusion added for numerical stability is minimally sufficient to dampen unwanted oscillations.

Integration to steady-state requires an assessment of when the solution stops changing in time. The most robust assessment is to look at every number that is calculated, to determine if there is any change over the current time step. A simpler approach is to check a single result for change. For example, the exit velocity is being polled in Code 17-3 for convergence to steady-state. However, some care is require to ensure that one instant in which the exit velocity appears to have stopped changing is not a prelude to persisting upstream transients that are still being swept toward the exit.

The exit conditions are allowed to float in Code 17-3. However, by commenting out the lines of code that perform the floating (after the predictor and corrector averaging step), the boundary values at $x = 1$ can be calculated explicitly from the governing equations. Performing this change will demonstrate that both approaches yield identical results. The accuracy of numerical integration can be confirmed by comparison of results with the exact steady-state solution:

$$u = \sqrt{2x+1} \quad \text{and} \quad \eta = 1/\sqrt{2x+1}. \tag{17-58}$$

17.5 PROBLEMS

17-1 Derive the conservation form of the governing equations for an inviscid adiabatic flow:

$$\begin{aligned}
&\text{Continuity:} && \partial_o\rho + \partial_j(\rho v_j) = 0 \\
&\text{Momentum:} && \partial_o(\rho v_i) + \partial_j(\rho v_j v_i) + \partial_i P = 0 \\
&\text{Energy:} && \partial_o(\rho e) + \partial_j(\rho e\, v_j + P v_j) = 0
\end{aligned}$$

where $e = u + v_k v_k/2$.

17-2 Investigate the numerical integration of Burgers' equation, as described in Section 17.2.3. (A) Using $N = 501$ mesh points and $\nu_a = 0.0$, try integrating the conservation form of Burgers' equation for three different Courant numbers $\{|v|\}_{\max}\Delta t/\Delta x = 0.4$, 0.8, and 1.0. What happens? (B) Try integrating the conservation form of Burgers' equation for the same set of three

Courant numbers when $\nu_a = 0.0005$. What happens? (C) Try integrating the *non*conservation form of Burgers' equation for the same set of three Courant numbers and $\nu_a = 0.0005$. What happens? Comment on the stability requirements for these three cases.

17-3 Solve Burgers' equation with added viscosity for stability for the domain $0 \leq x \leq 1$ (m), using the initial and boundary conditions given:

$$v(t = 0, x) = 0.5 \sin(\pi x) + \sin(2\pi x) \text{ (m/s)}$$

$$v(t, x = 0) = 0 \quad \text{and} \quad v(t, x = 1) = 0 \text{ (m/s)}$$

Plot the solution at times $t = 0.0$, 0.2, 0.6, 1.0, 1.4, and 2.0 (s).

17-4 Solve the viscid Burgers' equation:

$$\frac{\partial v}{\partial t} + \frac{\partial}{\partial x}\left(\frac{v^2}{2}\right) = \nu \frac{\partial^2 v}{\partial x^2}$$

for $0 \leq x \leq 1$ and the initial and boundary conditions given by

$$v(t = 0, x) = \begin{cases} +1 & x = 0 \\ 0 & 0 < x < 1 \text{ (m/s)} \\ -1 & x = 1 \end{cases}$$

$$v(t, x = 0) = +1 \quad \text{and} \quad v(t, x = 1) = -1 \text{ (m/s)}.$$

Show that numerical stability requires a time step of $\Delta t \leq \Delta x^2/(2\nu)$. Numerically integrate the viscid Burgers' equation for two values of diffusivity: $\nu = 0.02$ and $\nu = 0.2(\text{m}^2/\text{s})$. For the period $0 \leq t \leq 1.5$(s), plot $v(x)$ at time intervals of 0.1 (s) for both viscosity cases. Is there a steady-state solution to this problem?

17-5 The illustrated problem for transient advection and diffusion of species in a one-dimensional flow was solved with complex variable definitions in Problem 16-9. The quasi-steady solution was found to be

$$\frac{c(x,t) - c_o}{A} = \text{Im}\left\{\exp\left[\left(i\,B\frac{t}{Đ/U^2} + \frac{1 - \sqrt{1 + i\,4B}}{2}\frac{x}{Đ/U}\right)\right]\right\} \text{ where } B = \omega Đ/U^2.$$

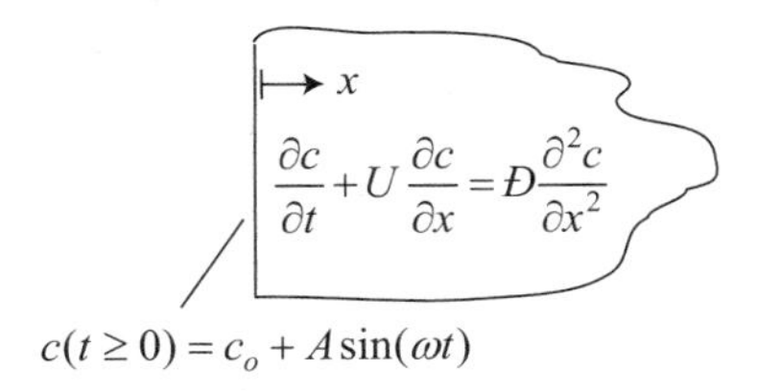

For the current problem, let $c(x, t < 0) = c_o$ and $c(x = 0, t \geq 0) = c_o + A\sin(\omega t)$. Recast the problem statement into the variables $C = (c - c_o)/A$, $\eta = Ux/Đ$, and $\tau = U^2 t/Đ$, and numerically solve for the case where $B = \omega Đ/U^2 = 1$. Use the computational domain $0 \leq \eta \leq 20$, and float the concentration at $\eta = 20$. Integrate forward in time while $\tau \leq 12\pi$. For both the numerical solution and analytic quasi-steady solution, plot $C(\eta = 10, \tau)$ and $C(\eta, \tau = 12\pi)$.

REFERENCES

[1] D. A. Anderson, J.C. Tannehill, and R. H. Pletcher, *Computational Fluid Mechanics and Heat Transfer.* New York, NY: Hemisphere Publishing Co., 1984.
[2] S. V. Patankar, *Numerical Heat Transfer and Fluid Flow.* New York, NY: Hemisphere Publishing Co., 1980.
[3] R. Peyret, *Handbook of Computational Fluid Mechanics*. San Diego, CA: Academic Press, 1996.
[4] R. W. MacCormack, "The Effect of Viscosity in Hypervelocity Impact Cratering." *American Institute of Aeronautics and Astronautics Paper,* 69–354 (1969).
[5] J. Burgers "A Mathematical Model Illustrating the Theory of Turbulence." Advances in Applied Mechanics, **1**, 171 (1948).
[6] R. Courant, K. O. Friedrichs and H. Lewy, "Üeber die Partiellen Differenzgleichungen der Mathematische Physik." *Mathematische Annalen*, **100**, 32 (1928).

Chapter 18

Open Channel Flow

18.1 Analysis of Open Channel Flows
18.2 Simple Surface Waves
18.3 Depression and Elevation Waves
18.4 The Hydraulic Jump
18.5 Energy Conservation
18.6 Dam-Break Example
18.7 Tracer Transport in the Dam-Break Problem
18.8 Problems

Hydrology is the study of the distribution, movement, and properties of the waters of the earth and their relation to the environment. An important problem arising in hydrology concerns the transport of water in open channels, which is governed by the forces of gravity and inertia. One interesting feature of open channel flows is that unlike enclosed flows, the cross-sectional area of the flow is a variable governed by the transport equations. This additional degree of freedom gives rise to multiple solutions to the nonlinear transport equations describing the state and behavior of the flow.* Revealing the consequences of this, and developing strategies for solving the equations governing advection transport in open channels, are the main topics treated in this chapter and Chapter 19. Textbooks devoted to open channel hydraulics, such as references [1] and [2], will provide a wider range of applied topics.

18.1 ANALYSIS OF OPEN CHANNEL FLOWS

Open channel flows of incompressible fluids are driven by gravity. To a first approximation, open channel flows may often be treated as inviscid and irrotational. Consequently, with diffusion terms omitted and the advection term written in irrotational form, the momentum equation reduces to Bernoulli's equation,

$$\partial_o \Phi + v^2/2 + P/\rho + gh = C(t), \tag{18-1}$$

as was demonstrated in Section 14.2. In Bernoulli's equation the velocity potential function Φ is defined such that

$$v_i = \partial_i \Phi. \tag{18-2}$$

*This consequence of additional degrees of freedom will arise again in the compressibility of fluids, as treated in Chapters 21 through 23.

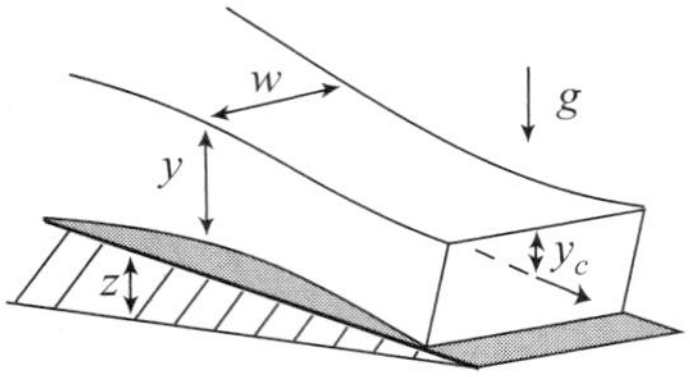

Figure 18-1 Illustration for open channel flow.

For analysis, consider a streamline through the center of the flow, as shown in Figure 18-1. Treating the flow as quasi-one-dimensional, with the x-direction as the direction of motion, Bernoulli's equation can be differentiated with respect to change along the streamline:

$$\frac{\partial v}{\partial t}+\frac{\partial}{\partial x}\left(\frac{v^2}{2}+\frac{P}{\rho}+gh\right)=0 \quad\rightarrow\quad \frac{\partial v}{\partial t}+\frac{\partial}{\partial x}\left(\frac{v^2}{2}+\frac{P_o+\rho g y_c}{\rho}+g(z+y-y_c)\right)=0. \quad (18\text{-}3)$$

The x-direction velocity component is denoted by $v_x \rightarrow v$ for notational simplicity and P_o is the ambient pressure overlying the flow. The streamline passing through the centroid of the channel characterizes the average pressure and elevation of the flow cross-section. The hydrostatic pressure at the centroid is $P = P_o + \rho g y_c$ and the elevation of the flow at the centroid is $h = z + y - y_c$, where y_c is the vertical distance to the centroid from the free surface of the flow. However, notice that the quantity $P/\rho + gh$ is independent of the value of y_c (or choice of streamline). Completing the task of differentiation, the momentum equation becomes

$$\frac{\partial v}{\partial t}+v\frac{dv}{dx}+g\left(\frac{dy}{dx}+\frac{dz}{dx}\right)=0. \quad (18\text{-}4)$$

The momentum equation now contains two dependent variables describing the flow: the velocity v and the flow thickness y. Continuity provides the additional equation required for closure. It is left as an exercise (see Problem 18-1) to show that the transient continuity equation for a fluid in an open channel can be written as

$$\frac{\partial y}{\partial t}=-\frac{1}{w}\frac{\partial}{\partial x}(vyw), \quad (18\text{-}5)$$

where w is the width of the flow. The momentum equation (18-4) and continuity equation (18-5) govern advection of a constant density fluid in a frictionless open channel of rectangular cross-section.

Open channel flows on an inclined bed exhibit interesting characteristics that can be illustrated with a steady-state flow ($\partial v/\partial t = 0$) through a constant-width channel ($\partial w/\partial x = 0$). With these restrictions, the momentum (18-4) and continuity (18-5) equations combine to yield

$$\frac{dy}{dx}=-\frac{dz}{dx}\left(1-\frac{v^2}{gy}\right)^{-1} \quad \text{(steady flow through a channel of constant width).} \quad (18\text{-}6)$$

This result indicates that the flow thickness may increase ($dy/dx > 0$) or decrease ($dy/dx < 0$) with a change in the bed height dz/dx as the fluid advances depending on the value of the local *Froude number*† [3]:

†The Froude number is named after the English engineer, hydrodynamicist, and naval architect William Froude (1810–1879).

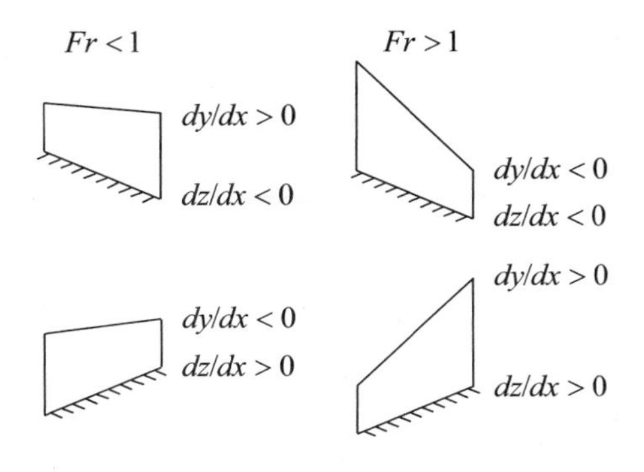

Figure 18-2 Illustration of relation between *Fr*, dz/dx, and dy/dx for a frictionless channel.

$$Fr = \frac{v}{\sqrt{gy}}. \tag{18-7}$$

If $Fr < 1$, the flow is *subcritical*, and the change in flow thickness dy/dx given by Eq. (18-6) has a sign opposite to the slope of the channel bottom dz/dx. In contrast, if $Fr > 1$, the flow is *supercritical*, and the change in flow thickness dy/dx has the same sign as the slope of the channel bottom dz/dx. Figure 18-2 illustrates how a constant-width flow thins or thickens in response to a change in bed height z when $Fr < 1$ and when $Fr > 1$. The demarcation for these two distinct behaviors is the *critical flow* condition when $Fr = 1$.

As will be demonstrated in the next section, simple surface waves propagate with a speed of $c = \sqrt{gy}$. Therefore, the Froude number is interpreted as the ratio of the flow speed to the wave speed. Information about the flow propagates through the fluid at the wave speed. Therefore, when a flow is subcritical, $Fr < 1$, information concerning downstream conditions can propagate upstream to the oncoming flow. However, when the flow is supercritical, $Fr > 1$, information concerning downstream conditions cannot propagate upstream. Consequently, supercritical flows are sometimes forced to make dynamic adjustments to downstream conditions that are unknown in advance of arrival. This is facilitated by the nonlinear nature of the momentum equation that permits two finitely different flow thickness and speed combinations to exist that identically satisfy momentum and continuity (when $Fr \neq 1$). This allows supercritical flows to transition into subcritical flows through a *hydraulic jump*, as will be shown in Section 18.4. This degree of freedom allows supercritical flows to adjust instantaneously to downstream conditions.

18.2 SIMPLE SURFACE WAVES

Conservation laws can be used to determine the speed of surface waves "c." For this analysis, the flow is taken to be traveling to the right on a horizontal bed with a speed v and flow thickness y. Surface waves can propagate in the same or opposing direction to the flow, as shown in Figure 18-3. A wave propagating in the same direction travels with a speed of $v + c$, while a wave propagating in the opposing direction travels with a speed of $v - c$.

There is an altered state in the wake of a surface wave described by $v + dv$ and $y + dy$. It is assumed that this altered state is only differentially different from the original state. This distinguishes the effect of a *simple surface wave* from waves that have finite magnitude. Figure 18-3 illustrates the control surface surrounding the wave that will be used in the following analysis. The frame of reference for the control surface is fixed to the surface wave, such that the flow approaching the wave location has a speed of $\pm c$. Surfaces "a" and "b" shown in Figure 18-3 are separated by a vanishingly small distance.

Since integration across the wave does not involve a change in the channel bed height ($dz = 0$) or a change in channel width ($dw = 0$), the differential forms of the steady continuity and momentum equations become

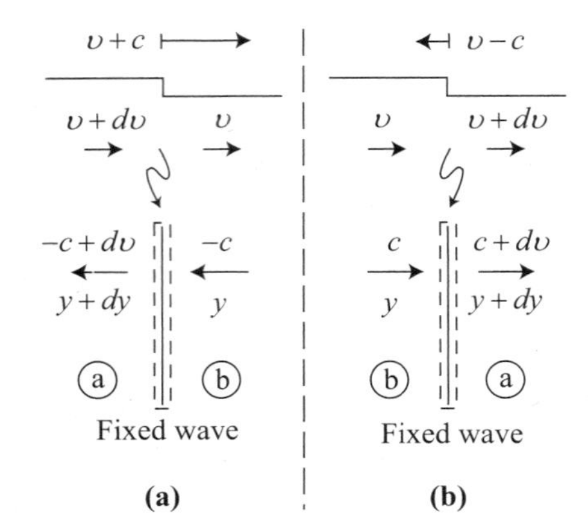

Figure 18-3 Propagation of simple surface waves in the same (a) and opposing (b) direction as the flow.

$$\text{Continuity:} \quad d(vyw) = 0 \quad \rightarrow \quad vdy + ydv = 0 \tag{18-8}$$

$$\text{Momentum:} \quad d\left(\frac{v^2}{2} + \frac{P_o}{\rho} + g(z+y)\right) = 0 \quad \rightarrow \quad vdv + gdy = 0. \tag{18-9}$$

It is important to remember that the differential change described by equations (18-8) and (18-9) is made from the frame of reference of the moving wave. In this reference frame, the flow velocity can only be the positive or negative value of the wave speed $v = \pm c$, as shown in Figure 18-3. Consequently, the differential forms of the continuity (18-8) and momentum (18-9) equations for a simple surface wave can be written in the form

$$\text{Continuity:} \quad (\pm c)dy + ydv = 0 \tag{18-10}$$

$$\text{Momentum:} \quad (\pm c)dv + gdy = 0. \tag{18-11}$$

Combining the continuity and momentum equations to eliminate dv gives the result:

$$\frac{gy\,dy}{(\pm c)} = (\pm c)dy \quad \text{or} \quad c^2 = gy. \tag{18-12}$$

This result dictates that simple surface waves propagate with a speed dependent on the flow thickness:

$$c = \sqrt{gy}. \tag{18-13}$$

18.3 DEPRESSION AND ELEVATION WAVES

Waves interact in distinct ways depending on whether they are propagating into regions of higher or lower water levels. As discussed in the previous section, there is an altered state in the wake of a wave. This state differentiates two types of waves: depression and elevation waves. The water level thins as a flow passes through a *depression wave* and thickens as it passes through an *elevation wave*. These two types of waves are shown schematically in Figure 18-4. Depicted in this figure is an initially stationary fluid that experiences a series of discrete changes in velocity imposed by the left-hand boundary. The boundary motion sets up a series of discrete waves that propagate into the fluid. Panel (a) of Figure 18-4 illustrates a series of depression waves, while panel (b) shows a

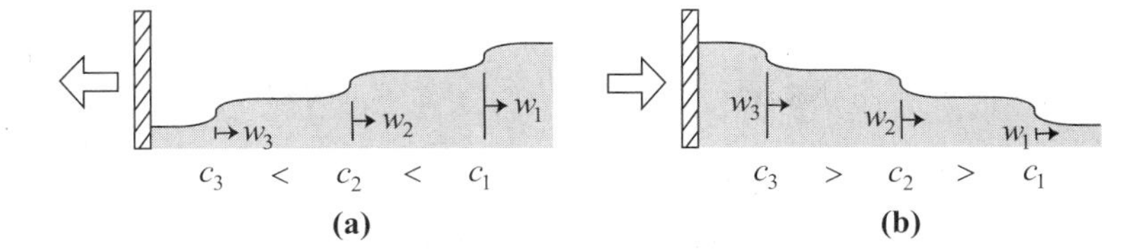

Figure 18-4 A series of depression waves (a) and elevation waves (b) created by the motion of a bounding wall.

series of elevation waves. All waves propagate away from the disturbance with the local value of the surface wave speed $c = \sqrt{gy}$. Since the water level drops behind each depression wave, the next wave in the series propagates with a lower speed, such that $c_3 < c_2 < c_1$. Therefore, the series of depression waves tends to spread further apart with time. However, there is a water level rise behind each elevation wave. Therefore, the next wave in the series propagates with a higher speed, such that $c_3 > c_2 > c_1$. Consequently, the spacing between the series of elevation waves tends to diminish with time. As the elevation waves coalesce, a single *hydraulic jump* forms.

Recalling the simple surface wave continuity equation (18-10), a relation between the change in velocity and water level in the wake of the wave can be established using the wave speed $c = \sqrt{gy}$. With $dy = 2y dc/c$, the simple surface wave continuity equation (18-10) becomes

$$dv + (\pm c)\frac{dy}{y} = 0 \quad \rightarrow \quad dv \pm 2\, dc = 0. \tag{18-14}$$

A simple surface wave can propagate in two directions relative to the flow, as indicated with $\pm c$. The wave traveling in the same direction as v utilizes the $(-)$ sign (see Figure 18-3) and the $(+)$ sign denotes a wave traveling in the opposite direction. The simple surface wave equation (18-14) can be integrated for the result

$$dv \pm 2dc = 0 \quad \rightarrow \quad \begin{cases} v + 2c = C_1 & v \text{ and } c \text{ in opposing directions} \qquad (18\text{-}15a) \\ v - 2c = C_2 & v \text{ and } c \text{ in the same direction.} \qquad (18\text{-}15b) \end{cases}$$

The depression wave illustrated in Figure 18-4(a) shows that surface waves are propagating in the reverse direction of the flow. Because the effect of spreading a simple surface wave is everywhere differential, the integrated effect describing a finite depression wave is governed by Eq. (18-15a). In contrast, elevation waves illustrated in Figure 18-4(b) coalesce into a finite (nondifferential) hydraulic jump that cannot be described by the result of Eq. (18-15b). The conservation laws across a hydraulic jump should be addressed in integral form, rather than differential form.

18.4 THE HYDRAULIC JUMP

An interesting feature of open channel flow is that two finitely different flow thickness and flow speed combinations exist that identically satisfy momentum and continuity (when $Fr \neq 1$). This can be demonstrated by evaluating the momentum and continuity equations between two surface heights, y_a and y_b, in a steady flow, as illustrated in Figure 18-5. The nonconservative differential form of the continuity equation and momentum equation (Bernoulli's equation) cannot describe a discontinuous change in the flow. Therefore, it is appropriate to start from an integral form of these two conservation equations:

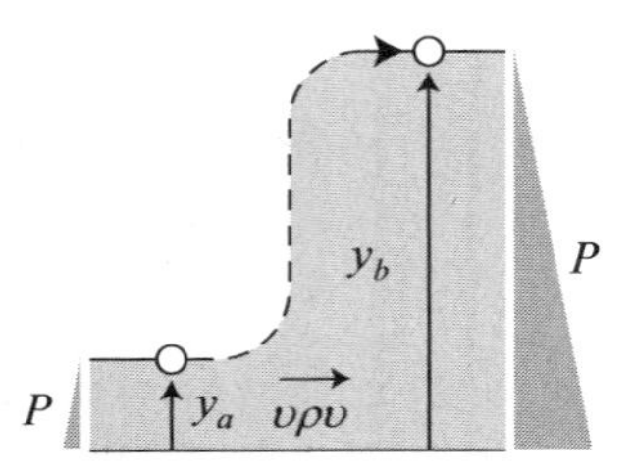

Figure 18-5 A stationary hydraulic jump.

$$\text{Continuity:}\ 0 = \int_A -\upsilon_j(\rho)n_j dA \qquad \rightarrow [\upsilon\rho A]_a = [\upsilon\rho A]_b \tag{18-16}$$

$$\text{Momentum:}\ 0 = \int_A \left[-\upsilon_j(\rho\upsilon_i) - P\delta_{ji}\right]n_j dA \quad \rightarrow [(\upsilon\rho\upsilon + P)A]_a = [(\upsilon\rho\upsilon + P)A]_b. \tag{18-17}$$

The average hydrostatic pressure is used to evaluate pressure terms in the momentum statement, such that $PA = (\rho g y/2)(y \cdot 1)$, where density is constant and the width of the channel is taken as unity. It is left as an exercise (see Problem 18-3) to show that the momentum and continuity equations can be combined to eliminate υ_b for the result

$$(y_a - y_b)\left[y_a + y_b - 2(y_a^2/y_b)Fr_a^2\right] = 0, \tag{18-18}$$

where $Fr_a = \upsilon_a/\sqrt{g y_a}$. Equation (18-18) may be satisfied in two physically meaningful ways. When $y_a \neq y_b$, Eq. (18-18) stipulates that

$$\frac{y_b^2}{y_a^2} + \frac{y_b}{y_a} - 2Fr_a^2 = 0 \quad \rightarrow \quad \frac{y_b}{y_a} = \frac{1}{2}\left(\sqrt{1 + 8Fr_a^2} - 1\right). \tag{18-19}$$

This solution to Eq. (18-18) corresponds to a hydraulic jump when $y_b > y_a$, which is possible when $Fr_a > 1$. Momentum conservation and continuity dictate that if $Fr_a > 1$, then $Fr_b < 1$ after the hydraulic jump. For $Fr_a < 1$, Eq. (18-18) suggests $y_b/y_a < 1$ is a possibility. However, this reversal of the hydraulic jump is not physically possible in a steady-state flow, since such a state would transiently spread into an infinite number of depression waves (see Section 18.3). Additionally, it is shown in Section 18.5 that $y_b/y_a < 1$ violates the second law of thermodynamics. The quadratic equation (18-18) for y_b/y_a also has a negative root that has no physical meaning. Therefore, the final significant solution to Eq. (18-18) is when $y_a = y_b$, corresponding to the trivial case in which there is no hydraulic jump.

Another interesting scenario to consider is the steady motion of a hydraulic jump moving into a stationary fluid. By changing the coordinate system to move with the hydraulic jump, as illustrated in Figure 18-6, the speed of the hydraulic jump can

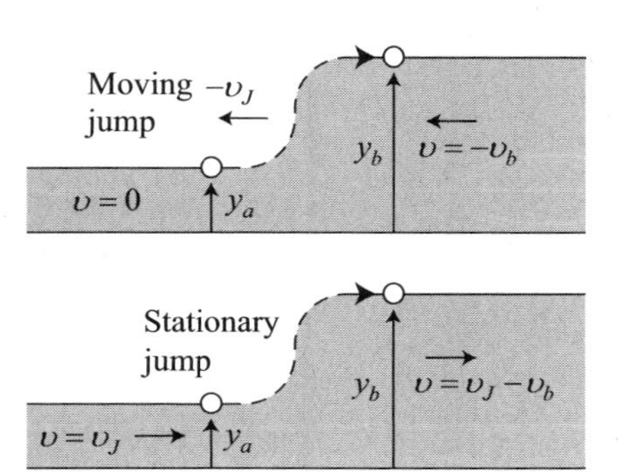

Figure 18-6 A moving hydraulic jump.

be analyzed with the results for the stationary jump problem. An extension of this analysis considers the limiting case of $y_b - y_a \rightarrow 0$, in which case the infinitesimal hydraulic jump becomes a simple surface wave. It is left as an exercise (see Problem 18-4) to show that Eq. (18-18), in the limit that $y_b - y_a \rightarrow 0$, yields the simple surface wave speed: $v_J(y_b - y_a \rightarrow 0) = \sqrt{gy}$.

18.5 ENERGY CONSERVATION

For a compressible viscid flow, kinetic energy can be converted to thermal energy by both compression (Section 5.3) and viscous heating (Section 5.6). For inviscid incompressible flows, there is typically no way for energy in the fluid to be converted between mechanical and thermal forms. However, the hydraulic jump proves to be an exception to this statement. To demonstrate, the energy transport equation is derived for open channel flow.

Figure 18-7 details the terms needed for an energy conservation statement, where $e = (u + v^2/2)$ consists of the internal energy (u) and kinetic energy $(v^2/2)$ on a per-unit-mass basis. The conservation statement is formulated for a control volume that is bounded by the cross-sectional area of the flow and is differential with respect to the direction of the flow. Conservation of energy requires that the rate of change of energy in the control volume be balanced by advection transport of energy into the control volume plus the net effect of the sources of energy acting on the control volume. Forces shown in Figure 18-7 that transmit mechanical energy to (and from) the flow are both related to gravity: first, acting on the fluid mass within the control volume and, second, acting through hydrostatic pressure on the surfaces of the control volume. Expressed in integral form, the energy equation is

$$\begin{array}{ccccc} (increase) & = & (in - out) & + & (sources) \\ \downarrow & & \downarrow & & \downarrow \end{array}$$

$$\int_{\Delta x} \partial_o(\rho e A)dx = -\int_A (\rho e v_j)n_j dA + \left(-\int_A (P\delta_{ji}v_i)n_j dA + \int_{\Delta x} (\rho g_j v_j A)dx \right), \tag{18-20}$$

where the area integrals are comprised of the three surfaces shown in Figure 18-7. Notice that although pressure transmits a force through surface (3), work is not performed without the motion of a flow through the surface. Therefore, surface (3) does not contribute to the area integrals. Integrating the energy equation over the control volume yields

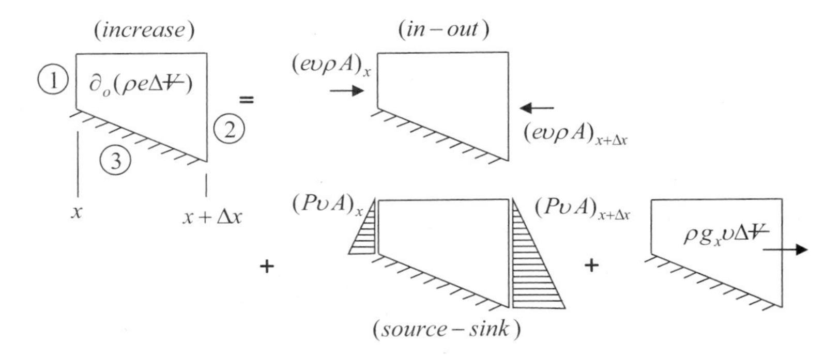

Figure 18-7 Elements of energy conservation for open channel flow without friction.

$$\frac{\partial}{\partial t}(\rho e A)\Delta x = \overbrace{(\rho e v A)_x}^{\text{surface 1}} + \overbrace{-(\rho e v A)_{x+\Delta x}}^{\text{surface 2}}$$

$$+ \underbrace{(PvA)_x}_{\text{surface 1}} + \underbrace{-(PvA)_{x+\Delta x}}_{\text{surface 2}} + \underbrace{\rho g_x v A \Delta x}_{\text{volume}}, \tag{18-21}$$

where g_x is the component of $\vec{g}$ aligned in the direction of the flow. Dividing by Δx, and considering the differential limit in which $\Delta x \to dx$, one obtains

$$\frac{\partial}{\partial t}(\rho e A) = -\frac{\partial}{\partial x}(\rho e v A) - \frac{\partial}{\partial x}(PvA) + \rho g_x v A. \tag{18-22}$$

Using continuity, this result simplifies to

$$\rho A\frac{\partial e}{\partial t} + \underbrace{e\frac{\partial}{\partial t}(\rho A) = -e\frac{\partial}{\partial x}(\rho v A)}_{\text{continuity}=0} - \rho v A\frac{\partial e}{\partial x} - \frac{\partial}{\partial x}(PvA) + \rho g_x v A, \tag{18-23}$$

yielding the result

$$\frac{De}{Dt} = -\frac{1}{\rho A}\frac{\partial}{\partial x}(PvA) + g_x v \tag{18-24}$$

or

$$\rho\frac{D(e + P/\rho)}{Dt} - \rho\frac{D(P/\rho)}{Dt} = -\frac{1}{A}\frac{\partial}{\partial x}(PvA) + \rho g_x v. \tag{18-25}$$

Continuity can be used again to shown that

$$\rho\frac{D(P/\rho)}{Dt} = \frac{\partial P}{\partial t} + \frac{1}{A}\frac{\partial}{\partial x}(PvA). \tag{18-26}$$

Therefore, the energy equation becomes

$$\rho\frac{D(e + P/\rho)}{Dt} = \frac{\partial P}{\partial t} + \rho g_x v. \tag{18-27}$$

Using the arguments of Section 14.2, the effect of gravity can be written as a potential function, $g_x = -g(\partial h/\partial x)$, where the variable h is a measure of distance along the direction of $-\vec{g}$. The scalar g is the magnitude of $\vec{g}$. In terms of the gravity potential function, the energy equation becomes

$$\rho\frac{D(e + P/\rho)}{Dt} + \rho g\left(v\frac{\partial h}{\partial x}\right) = \frac{\partial P}{\partial t} \tag{18-28}$$

or

$$\frac{D}{Dt}(e + P/\rho + gh) = \frac{\partial}{\partial t}(P/\rho + gh). \tag{18-29}$$

Although both the hydrostatic pressure P/ρ and gravity potential gh vary with vertical position in the flow, their sum is equal to a constant, $P/\rho + gh = g(y+z) + P_o/\rho$, where P_o is the ambient pressure overlying the flow. Making use of this observation, the energy equation becomes

$$\frac{D}{Dt}\left[u + v^2/2 + g(y+z)\right] = g\frac{\partial}{\partial t}(y+z). \tag{18-30}$$

The energy equation can also be written in a form that reveals the momentum equation:

$$\frac{Du}{Dt} + v\overbrace{\left[\frac{Dv}{Dt} + g\left(\frac{\partial y}{\partial x} + \frac{\partial z}{\partial x}\right)\right]}^{momentum=0} = 0. \tag{18-31}$$

In this final form, the energy equation illustrates that generally $Du/Dt = 0$ must be satisfied when the differential form of the momentum equation is satisfied. However, across a hydraulic jump, v and y are discontinuous and the result that $Du/Dt = 0$ is no longer valid because the nonconservative differential form of the momentum equation is not satisfied. Equation (18-30) reveals that, for a steady-state flow, the energy content $[u + v^2/2 + g(y+z)]$ is constant for a material fluid element traveling with the flow. Therefore, equating this energy content between two states across a hydraulic jump yields

$$\left[u + v^2/2 + g(y+z)\right]_a = \left[u + v^2/2 + g(y+z)\right]_b. \tag{18-32}$$

Or, with $z_a = z_b$,

$$u_b - u_a = (v_a^2 - v_b^2)/2 + g(y_a - y_b). \tag{18-33}$$

Using continuity across the hydraulic jump, $v_a y_a = v_b y_b$, the change in internal energy is found to be

$$\frac{u_b - u_a}{g y_a} = \frac{Fr_a^2}{2}\left[1 - \left(\frac{y_a}{y_b}\right)^2\right] + \left(1 - \frac{y_b}{y_a}\right). \tag{18-34}$$

Using Eq. (18-19) to eliminate Fr_a^2 yields

$$\frac{u_b - u_a}{g y_a} = \frac{(y_b/y_a - 1)^3}{4 y_b/y_a}. \tag{18-35}$$

This result reveals that mechanical energy is dissipated into thermal energy by the hydraulic jump, since $y_b/y_a > 1$. This result alludes to the "viscous" nature of a hydraulic jump that is accompanied by dissipation of mechanical energy into heat ($u_b > u_a$). Since $ds = du/T$ for an incompressible fluid (Table 1-1), it is clear that $ds > 0$ through the hydraulic jump and that the flow is irreversible. Arguing for the possibility that $y_b/y_a < 1$ requires $u_b < u_a$ and $ds < 0$, which violates the second law of thermodynamics.

18.6 DAM-BREAK EXAMPLE

The dam-break problem provides a good illustration of the wave nature of open channel flows in a transient one-dimensional problem that can be solved both analytically and numerically. Consider an open channel that is partitioned into a high-level water region

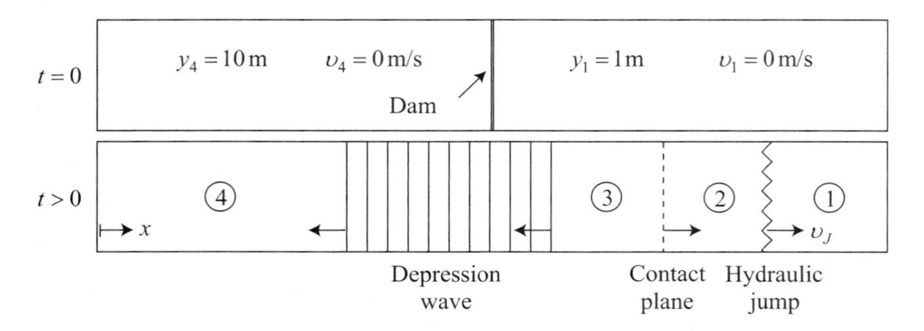

Figure 18-8 Dam-break illustration.

and a low-level water region by a dam, as illustrated in Figure 18-8. The high and low level regions in the channel are finite in length, and the water is initially stationary. At $t = 0$ the dam breaks and the high-water region moves into the low-water region with two resulting waves. One is a hydraulic jump moving to the right into the low-water region. The second is a depression wave moving to the left into the high-water region. Both waves originate from a plane of contact that divides the water originating from the high-level region from the water originating from the low-level region. This plane of contact exists for all time (since the water from the two regions cannot comingle without diffusion). However, the water advancing from the high-level region pushes the contact plane to the right. The depression wave spreads as it advances to the left. In contrast, the hydraulic jump remains discrete as it moves to the right. The two waves and contact plane define four regions of constant flow conditions in the open channel, as labeled in Figure 18-8. Region 1 is the stationary low-level region in advance of the hydraulic jump. Region 2 is the water advancing behind the hydraulic jump. Region 4 is the stationary high-level region in advance of the depression wave. Region 3 is the flow behind the depression wave. Regions 2 and 3 are separated by the moving contact plane.

18.6.1 Analytic Description

To solve the dam-break problem analytically, conditions on the far right end of the channel must be related to conditions at the far left end through the two waves. This analysis starts with the hydraulic jump. Figure 18-9 summarizes the flow variables on either side of the hydraulic jump with respect to coordinates that are moving with the jump. The conservation laws across the jump (continuity and momentum) given by Eqs. (18-16) and (18-17) are applied to yield

Hydraulic jump equations:

$$\text{Continuity:}\ \left[(v_2 - v_J)y_2\right]_b = \left[(-v_J)y_1\right]_a \tag{18-36}$$

$$\text{Momentum:}\ [y_2(v_2 - v_J)^2 + gy_2^2/2]_b = [y_1(-v_J)^2 + gy_1^2/2]_a. \tag{18-37}$$

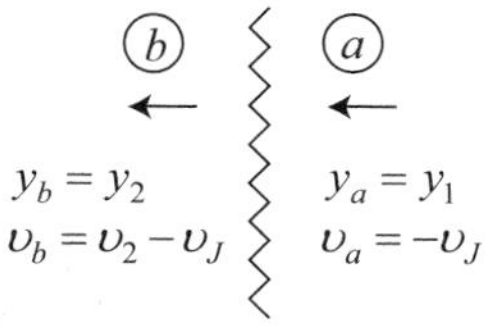

(b) (a)

$y_b = y_2$ $y_a = y_1$

$v_b = v_2 - v_J$ $v_a = -v_J$

Figure 18-9 Variables across the hydraulic jump moving into the low-level water.

The hydraulic jump equations make use of $PA = (\rho g y/2)(y \cdot 1)$ for the average hydrostatic pressure terms in the momentum equation, and the constant density is factored out of both the continuity and momentum equations.

Analysis of the flow in the vicinity of the contact plane and across the depression wave is needed to relate the hydraulic jump equations to the conditions at the far left end of the open channel. Therefore, the contact plane and depression wave contribute these equations:

$$\text{Contact plane:} \qquad y_3 = y_2 \tag{18-38}$$

$$v_3 = v_2 \tag{18-39}$$

Depression wave equation:

$$v + 2c = C_1: \qquad v_4 + 2\sqrt{g y_4} = v_3 + 2\sqrt{g y_3}. \tag{18-40}$$

Equation (18-40) was derived in Section (18.3) and is a consequence of satisfying both continuity and momentum throughout a depression wave.

The unknowns in this problem are v_2, y_2, v_3, y_3, and v_J. Since there are five unknowns and five equations (18-36 through 18-40), the remaining steps in solving this problem are algebraic.

The momentum equation (18-37) can be written in the form

$$g\frac{y_1^2 - y_2^2}{2} = y_2(v_2 - v_J)^2 - y_1 v_J^2 = -v_2 v_J y_1 = -v_J^2(1 - y_1/y_2)y_1, \tag{18-41}$$

in which v_2 is eliminated with the continuity equation (18-36). Equation (18-41) can be solved for the hydraulic jump velocity:

$$v_J = \sqrt{g\frac{y_2}{y_1}\frac{y_2 + y_1}{2}}, \tag{18-42}$$

or, using Eq. (18-41) again, the flow speed v_2:

$$v_2 = \left(\frac{y_2}{y_1} - 1\right)\sqrt{g\frac{y_1}{y_2}\frac{y_2 + y_1}{2}}. \tag{18-43}$$

With $v_4 = 0$, and the contact plane relations $y_3 = y_2$ and $v_3 = v_2$, the depression wave equation (18-40) can be written as

$$2\sqrt{g y_4} = v_2 + 2\sqrt{g y_2}. \tag{18-44}$$

Substituting Eq. (18-43) for v_2 into the last equation yields

$$\sqrt{\frac{y_4}{y_1}} = \frac{1}{2}\left(\frac{y_2}{y_1} - 1\right)\sqrt{\frac{1}{2}\left(1 + \frac{y_1}{y_2}\right)} + \sqrt{\frac{y_2}{y_1}}. \tag{18-45}$$

For a given initial water level ratio y_4/y_1, Eq. (18-45) can be solved for the water level ratio across the hydraulic jump y_2/y_1.

To illustrate some results of this analysis, consider a case in which $y_4/y_1 = 10$. The solution to Eq. (18-45) yields $y_2/y_1 = 3.962$. With $g = 9.807\ \text{m/s}^2$, it is found from Eq. (18-43) that $v_3 = v_2 = 7.340$ m/s, and from Eq. (18-42) that $v_J = 9.818$ m/s. Although the conditions internal to the depression wave change as a function of time and position, the leading edge of the depression wave has a constant velocity of $v_4 - c_4 = 0 - \sqrt{gy_4} = -9.903$ m/s (traveling to the left). The trailing edge of the depression wave also has a constant velocity equal to $v_3 - c_3$. With $c_3 = \sqrt{gy_3} = \sqrt{gy_2} = 6.233$ m/s, the depression wave trailing edge velocity is found to be $v_3 - c_3 = +1.106$ m/s (traveling to the right).

18.6.2 Numerical Description

The equations governing momentum (18-4) and continuity (18-5) in the open channel can be numerically integrated when formulated in conservation form, as discussed in Chapter 17. As determined in Problem 18-7, the conservation forms of the governing equations applied to the dam-break problem are:

$$\text{Continuity:} \quad \frac{\partial}{\partial t}(y) + \frac{\partial}{\partial x}(yv) = 0 \tag{18-46}$$

$$\text{Momentum:} \quad \frac{\partial}{\partial t}(yv) + \frac{\partial}{\partial x}\left((yv)v + g\frac{y^2}{2}\right) = 0. \tag{18-47}$$

Again, the x-direction velocity component is denoted by $v_x \to v$ for notational simplicity. The momentum equation (18-47) will be integrated for the dependent variable (yv), rather than velocity alone. The initial conditions in the channel are described by

$$t = 0 : x \le L/2 : \begin{cases} y = y_4 \\ yv = 0 \end{cases} \quad \text{and} \quad x > L/2 : \begin{cases} y = y_1 \\ yv = 0 \end{cases} \tag{18-48}$$

and the boundary conditions are

$$yv(t, x = 0) = 0 \quad \text{and} \quad yv(t, x = L) = 0. \tag{18-49}$$

The equations governing the open channel flow can be integrated for y and yv with the MacCormack scheme described in Chapter 17. To dampen numerical oscillations, diffusion terms are added to the finite differencing form of the governing equations, as discussed in Section 17.2.2. The continuity equation (18-46) and momentum equation (18-47) are written for the predictor and corrector steps discussed in Section 17.2. The predictor step calculations are given by

$$\begin{aligned} \{y_n^{t+\Delta t}\}_p = y_n^t - \frac{\Delta t}{\Delta x}\left[y_{n+1}^t v_{n+1}^t - y_n^t v_n^t\right] \\ + \frac{\nu_a \Delta t}{\Delta x^2}\left[y_{n+1}^t - 2y_n^t + y_{n-1}^t\right] \leftarrow \textit{for stability} \end{aligned} \tag{18-50}$$

$$\begin{aligned} \{(yv)_n^{t+\Delta t}\}_p = (yv)_n^t - \frac{\Delta t}{\Delta x}\left[\left((yv)_{n+1}^t v_{n+1}^t + g(y_{n+1}^t)^2/2\right) - \left((yv)_n^t v_n^t + g(y_n^t)^2/2\right)\right] \\ + \frac{\mu_a \Delta t}{\Delta x^2}\left[(yv)_{n+1}^t - 2(yv)_n^t + (yv)_{n-1}^t\right] \leftarrow \textit{for stability.} \end{aligned} \tag{18-51}$$

The predictor step values for velocity $\{v_n^{t+\Delta t}\}_p$ are obtained from $\{(yv)_n^{t+\Delta t}\}_p$ using

$$\{v_n^{t+\Delta t}\}_p = \frac{\{(yv)_n^{t+\Delta t}\}_p}{\{y_n^{t+\Delta t}\}_p}. \tag{18-52}$$

The corrector step calculations are given by

$$\{y_n^{t+\Delta t}\}_c = y_n^t - \frac{\Delta t}{\Delta x}\left[\{y_n^{t+\Delta t}\}_p\{v_n^{t+\Delta t}\}_p - \{y_{n-1}^{t+\Delta t}\}_p\{v_{n-1}^{t+\Delta t}\}_p\right]$$

$$+\frac{\nu_a \Delta t}{\Delta x^2}\left[\{y_{n+1}^{t+\Delta t}\}_p - 2\{y_n^{t+\Delta t}\}_p + \{y_{n-1}^{t+\Delta t}\}_p\right] \leftarrow \textit{for stability} \tag{18-53}$$

$$\{(yv)_n^{t+\Delta t}\}_c = (yv)_n^t - \frac{\Delta t}{\Delta x}\left[\left(\{(yv)_n^{t+\Delta t}\}_p\{v_n^{t+\Delta t}\}_p + g\{y_n^{t+\Delta t}\}_p^2/2\right)\right.$$

$$\left. -\left(\{(yv)_{n-1}^{t+\Delta t}\}_p\{v_{n-1}^{t+\Delta t}\}_p + g\{y_{n-1}^{t+\Delta t}\}_p^2/2\right)\right]$$

$$+\frac{\mu_a \Delta t}{\Delta x^2}\left[\{(yv)_{n+1}^{t+\Delta t}\}_p - 2\{(yv)_n^{t+\Delta t}\}_p + \{(yv)_{n-1}^{t+\Delta t}\}_p\right] \leftarrow \textit{for stability}. \tag{18-54}$$

The corrector step values for velocity $\{v_n^{t+\Delta t}\}_c$ are obtained from $\{(yv)_n^{t+\Delta t}\}_c$ using

$$\{v_n^{t+\Delta t}\}_c = \frac{\{(yv)_n^{t+\Delta t}\}_c}{\{y_n^{t+\Delta t}\}_c}. \tag{18-55}$$

As was noted in Chapter 17, the direction of finite differencing may need to be reversed for either the predictor or corrector step at boundary nodes to prevent the finite differencing operation from extending beyond the physical domain. For the same reason, the diffusion terms, added to the governing equations for numerical stability, are excluded at the boundary nodes.

The predictor step and corrector step values are averaged at the end of each time step to obtain

$$y_n^{t+\Delta t} = \frac{\{y_n^{t+\Delta t}\}_p + \{y_n^{t+\Delta t}\}_c}{2} \tag{18-56}$$

$$(yv)_n^{t+\Delta t} = \frac{\{(yv)_n^{t+\Delta t}\}_p + \{(yv)_n^{t+\Delta t}\}_c}{2}. \tag{18-57}$$

Code 18-1 implements the numerical integration of the continuity and momentum equations, as described above. Since the MacCormack scheme is explicit, stability of integration requires that each time step be selected such that the Courant condition is satisfied:

$$\frac{\Delta t\{|v| + c\}_{\max}}{\Delta x} = \frac{\Delta t\{|v| + \sqrt{gy}\}_{\max}}{\Delta x} \le 1, \tag{18-58}$$

as discussed in Section 17.2.1. As a practical approach, each time step can be selected to achieve a specified target Courant number:

$$\frac{\Delta t\{|v| + \sqrt{gy}\}_{\max}}{\Delta x} \approx C_{targ} = 0.9, \tag{18-59}$$

as implemented in Code 18-1.

The values of artificial viscosities (ν_a and μ_a) used in numerical integration are somewhat arbitrary. However, too little artificial viscosity can result in large oscillations near the hydraulic jump, and undesired consequences. On the other hand, too much artificial viscosity can result in the appearance of excessive diffusion in the solution, or cause instability if the conditions

$$\sqrt{\frac{2\nu_a \Delta t}{\Delta x^2}} \leq 1 \quad \text{and} \quad \sqrt{\frac{2\mu_a \Delta t}{\Delta x^2}} \leq 1 \tag{18-60}$$

are not satisfied, as discussed in Section 17.3. Code 18-1 implements integration with artificial diffusivities dynamically selected such that

$$\nu_a = \mu_a = \frac{0.1}{\Delta t/\Delta x^2}. \tag{18-61}$$

As discussed in Section 17.2.2, the artificial diffusivities assigned by Eq. (18-61) yield a magnitude of explicit diffusion added for numerical stability that is at most ~10% of the magnitude of the advection term in the transport equations. This may be reduced by lowering the value of the constant in the numerator.

With $y_4 = 10$ m, $y_1 = 1$ m, and $L = 400$ m, the numerical solution can be integrated forward in time to compare with the analytical results investigated in Section 18.6.1. Figure 18-10 shows the numerical results for flow thickness and velocity in the open channel at $t = 12$ s. Superimposed on the figure are the analytical results calculated for $y_2(= y_3)$ and $v_2(= v_3)$. The numerical results compare very well with the analytic results. Additionally, the location of the depression wave, contact plane, and hydraulic jump are calculated from flow speeds and wave speeds determined in Section 18.6.1. These locations are also identified at 12 s in Figure 18-10, and are in good agreement with the numerical results. The depression wave is slightly broadened in the numerical results, which is caused by the diffusion terms added to the governing equations for stability. For the same reason, the discontinuities of flow conditions across the hydraulic jump are also slightly broadened.

After longer times, the hydraulic jump reflects off the right-hand boundary of the channel and advances back into the depression wave. The resulting interactions of waves become more complex to describe analytically, but may be easily explored with Code 18-1.

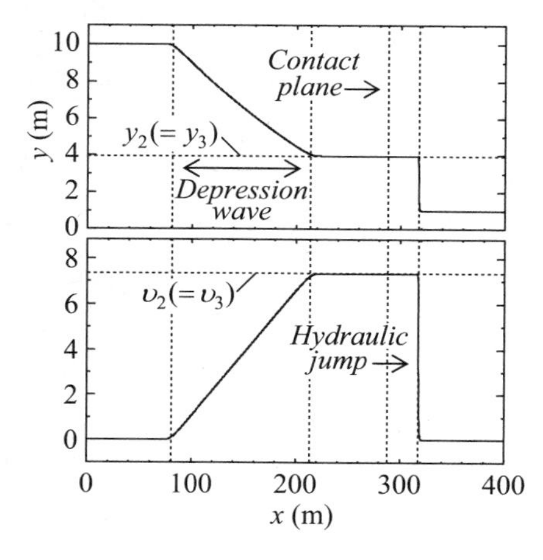

Figure 18-10 Flow conditions in the open channel at $t = 12$ s.

Code 18-1 Transient open channel flow

```
#include <stdio.h>
#include <string.h>
#include <math.h>

double cont_diff(char t,int n,double *y,double *v) {
   return t=='p' ? y[n+1]*v[n+1] - y[n]*v[n] :
                   y[n]*v[n] - y[n-1]*v[n-1] ;     }

double mom_diff(char t,int n,double *y,double *y_v,double *v,double g) {
   return t=='p' ?
   (y_v[n+1]*v[n+1]+g*y[n+1]*y[n+1]/2.) - (y_v[n]*v[n]+g*y[n]*y[n]/2.) :
   (y_v[n]*v[n]+g*y[n]*y[n]/2.) - (y_v[n-1]*v[n-1]+g*y[n-1]*y[n-1]/2.) ; }

double MacC(int N,double *y,double *y_v,double *v,
            double *dt_,double dx,double g)
{
   int n,step;
   double C,Cmax=0,dt=*dt_,Py[N],Cy[N],Py_v[N],Cy_v[N],Pv[N],Cv[N];
   double na,ma,Ctarg=0.9,dtdx=dt/dx,dtddx=dt/dx/dx;
   na=ma=0.1/dtddx;

   for (n=0;n<N;++n) {              // perform predictor step
      step=(n<N-1 ? 'p' : 'c');
      Py[n]=y[n]-dtdx*cont_diff(step,n,y,v);
      if (n>0 && n<N-1) Py[n]+=dtddx*na*(y[n+1]-2.*y[n]+y[n-1]);
      Py_v[n]=y_v[n]-dtdx*mom_diff(step,n,y,y_v,v,g);
      if (n>0 && n<N-1) Py_v[n]+=dtddx*ma*(y_v[n+1]-2.*y_v[n]+y_v[n-1]);
      Pv[n]=Py_v[n]/Py[n];
      C=dt*(fabs(Pv[n])+sqrt(g*Py[n]))/dx; // dt size checking
      if (C>Cmax) Cmax=C;
   }
   if (Cmax>1.) { // integrate with a smaller time step
      *dt_=0.;
      return dt*Ctarg/Cmax;
   }

   for (n=0;n<N;++n) { // perform corrector step
      step=(n>0 ? 'c' : 'p');
      Cy[n]=y[n]-dtdx*cont_diff(step,n,Py,Pv);
      if (n>0 && n<N-1) Cy[n]+=dtddx*na*(Py[n+1]-2.*Py[n]+Py[n-1]);
      Cy_v[n]=y_v[n]-dtdx*mom_diff(step,n,Py,Py_v,Pv,g);
      if (n>0 && n<N-1) Cy_v[n]+=dtddx*ma*(Py_v[n+1]-2.*Py_v[n]+Py_v[n-1]);
      Cv[n]=Cy_v[n]/Cy[n];
      if (isnan(Cv[n])) {
         printf("\nsoln BLEW! dt=%e ma=%e (Cmax=%e)\n",dt,ma,Cmax);
         return 0;
      }
   }
   for (n=0;n<N;++n) { // average predictor and corrector results
      y[n]=0.5*(Py[n]+Cy[n]);
      if (y[n]<0) y[n]=.01;
      y_v[n]=0.5*(Py_v[n]+Cy_v[n]);
      v[n]=y_v[n]/y[n];
   }
   y_v[0]=y_v[N-1]=0.; // enforce BC

   return dt*Ctarg/Cmax;
}

int main(void)
{

   int n,cnt=0,N=4001;
   double y[N],y_vel[N],vel[N],y4=10.0,y1=1.0;
   double g=9.807,L=400.0,dx=L/(N-1),t=0.,dt=5.0e-3,dt_next=7.0e-3,tend=12.;
   char file[99];
   FILE *fp;
   for (n=0;n<N;++n) { // initial distributions through tube
      y[n]=(n<=(N-1)/2 ? y4 : y1);
      y_vel[n]=vel[n]=0.0;
   }
   y[(N-1)/2]=(y4+y1)/2.;
   do {
      if (t+(dt=dt_next)>tend) dt=tend-t;
      if ( !(dt_next=MacC(N,y,y_vel,vel,&dt,dx,g)) ) break;
      t+=dt;
      if ( !(++cnt%100) ) printf("\ncnt=%d t=%f dt=%e\n",cnt,t,dt_next);

   } while (t<tend);

   sprintf(file,"out%5dms.dat",(int)(t*1.e3+.5));
   while (strstr(file," ")) *(strstr(file," "))='0';
   printf(" writing %s ... ",file);
   fp=fopen(file,"w");
   for (n=0;n<N;++n) fprintf(fp,"%e %e %e\n",n*dx,y[n],vel[n]);
   fclose(fp);

   return 1;
}
```

18.7 TRACER TRANSPORT IN THE DAM-BREAK PROBLEM

For the dam-break problem discussed in Section 18.6, the moving contact plane, illustrated in Figure 18-8, separates two regions of the fluid that were originally on either side of the dam. However, after the dam break there is no discernable difference in the flow immediately behind and immediately ahead of the contact plane, since $y_3 = y_2$ and $v_3 = v_2$ across the contact plane. This situation exists because the intensive properties of the fluids initially on either side of the dam are indistinguishable. However, suppose that the fluid on one side of the dam was doped with a small concentration of some "tracer" element to track the flow. Then the contact plane becomes an identifiable position in the flow, as distinguished by a step change in the tracer concentration. If the dam-break problem is solved numerically, a transport equation for the tracer concentration must be solved simultaneously with the continuity equation (18-46) and momentum equation (18-47) (see Problem 18-8). In the absence of sources and sinks, the tracer element concentration for each material element of the flow is swept along unaltered. Accordingly, the substantial derivative of the tracer element concentration must be zero. Therefore, the transport equation for the tracer concentration c is simply

$$\frac{Dc}{Dt} = \frac{\partial c}{\partial t} + v\frac{\partial c}{\partial x} = 0, \tag{18-62}$$

or in conservation form:

$$\frac{\partial}{\partial t}(cy) + \frac{\partial}{\partial x}((cy)v) = 0. \tag{18-63}$$

Observe that the conservation form of the transport equation (18-63) for c combined with continuity (18-46) gives back the requirement that $Dc/Dt = 0$. Since neither the momentum or continuity equations depends on c, the dynamics of the open channel flow are unaltered by the tracer element transport. However, Problem 18-11 addresses the dam-break problem when the dam initially partitions fluids of different density. The requirement that $D\rho/Dt = 0$ must be imposed on the numerical solution to this problem, and the momentum and continuity equations must be derived to demonstrate explicitly their dependency on density (see Problem 18-13).

18.8 PROBLEMS

18-1 Show that for a $\rho = const.$ flow in an open channel, the transient continuity equation can be written as

$$\frac{\partial y}{\partial t} = -\frac{1}{w}\frac{\partial}{\partial x}(vy\,w)$$

where y and w are the thickness and width of the flow.

18-2 Consider a frictionless flat-bottomed open channel of rectangular cross-section. The width of the channel is variable. Determine whether the flow thins or thickens as a function of the local Froude number and the change in channel width.

18-3 Show that the combination of momentum conservation and continuity across the hydraulic jump results in the expression $(y_a - y_b)\left[y_a + y_b - 2(y_a^2/y_b)Fr_a^2\right] = 0$, as discussed in Section 18.4.

18-4 Consider the limiting case of an infinitesimal hydraulic jump where $y_b - y_a \to 0$, as discussed in Section 18.4. Show that in this limit the jump speed is dependent on the flow thickness y and is given by $v_J(y_b - y_a \to 0) = \sqrt{gy}$.

18-5 Consider the steady flow after a sluice gate where the length and slope of the spillway is $L/y_1 = 25$ and $dz/dx = -0.02$. The velocity of flow under the sluice gate is supercritical, with $v_x(x = 0)/\sqrt{gy_1} = 1.1$. Assume that the spillway is frictionless. Formulate a set of analytic equations that describe all physically possible backwater heights y_2/y_1. What is the possible range of backwater heights?

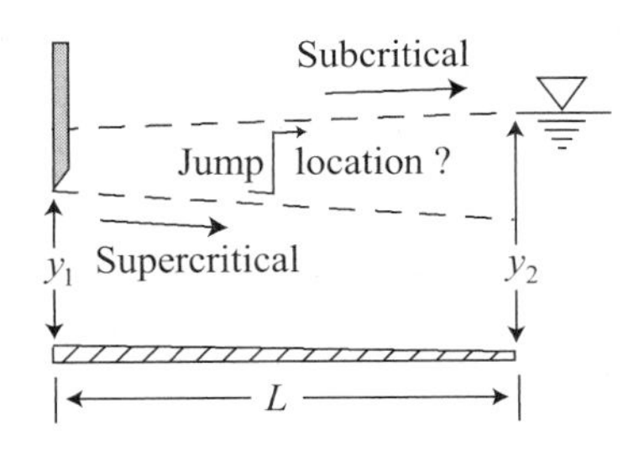

18-6 Consider the dam-break problem of Section 18.6, for the same initial condition prior to the dam break ($y_4 = 10$ m and $y_1 = 1$ m). Determine (analytically) the flow state in region 1 after the hydraulic jump has reflected from the wall at $L = 400$ m.

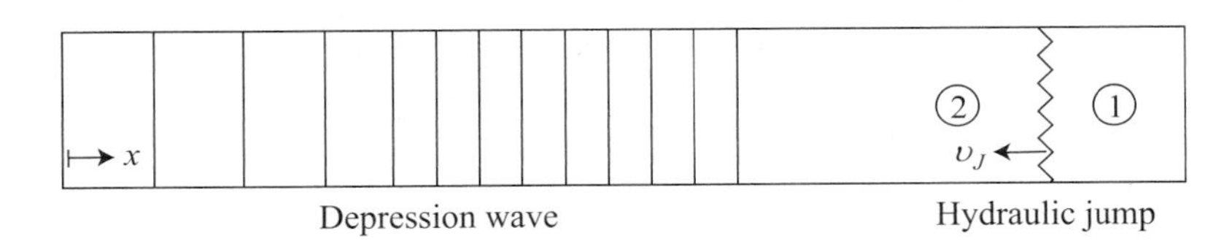

When does the contact plane first stop moving?

18-7 Derive the conservation form of the frictionless open channel momentum equation used for the dam-break problem described in Section 18.6:

$$\frac{\partial}{\partial t}(yv) + \frac{\partial}{\partial x}\left((yv)v + g\frac{y^2}{2}\right) = 0.$$

The conservation form of the momentum equation can be derived from the nonconservation form using continuity. How does the conservation form of the momentum equation change when the width of the channel is not constant?

18-8 Numerically solve the dam-break problem described in Section 18.6, including transport of a tracer concentration to reveal the location of the contact plane, as discussed in Section 18.7. Take the initial concentration of the tracer element to be $c_4 = 0.01$ in the high-water region and $c_1 = 0.0$ in the low-water region. Compare the numerical result for the contact plane location at 12 s with the analytic result.

18-9 Consider the dam-break problem of Section 18.6, with the channel open at one end to a large reservoir.

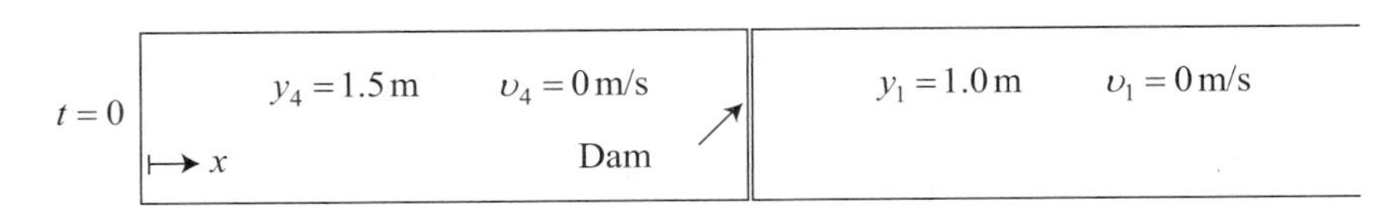

Let $y_4 = 1.5$ m and $y_1 = 1.0$ m. When the hydraulic jump reaches the open end of the channel, emptying into the reservoir, a depression wave is reflected back. Assuming that the channel is frictionless, determine the velocity of the leading and trailing edges of the depression wave that moves back through the channel, making note of significant assumptions. Is there a value y_4 for which the entirety of the depression wave will not be reflected back into the channel?

18-10 Consider the dam-break problem illustrated. The water level at the far left wall is $y_4 = 3$ m and at $L = 30$ m to the far right, the water level is $y_1 = 1$ m at the wall. The floor of the basin is sloped such that $z_4 = y_1 + z_1$. At $t = 0$ the dam breaks. Derive the conservation form of the frictionless open channel momentum equation from the nonconservation form using continuity. Numerically integrate the open channel flow equations. Determine the time it takes for the hydraulic jump to travel halfway to the right-hand wall. Plot the $y + z$ and v distributions at this time.

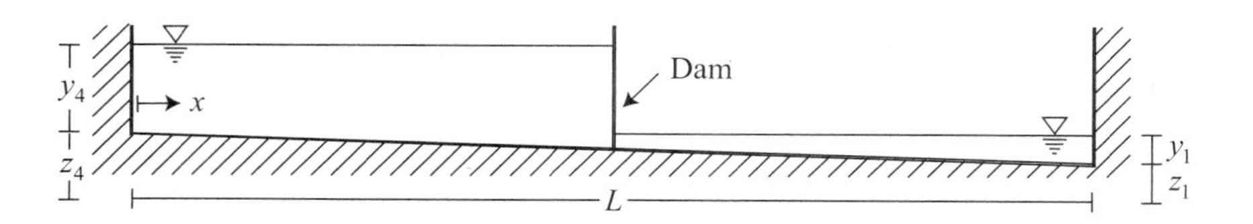

18-11 Consider the dam-break problem illustrated. In addition to the imbalance in the fluid levels on either side of the dam, the fluid on the high-level side has a density that is smaller than the fluid on the low-level side. Analytically determine the flow conditions a short time after the dam breaks, as illustrated for $t > 0$. Specifically, determine v_2, y_2, v_3, y_3, v_J, and the velocities of the leading and trailing edges of the depression wave. Assume that the channel is frictionless.

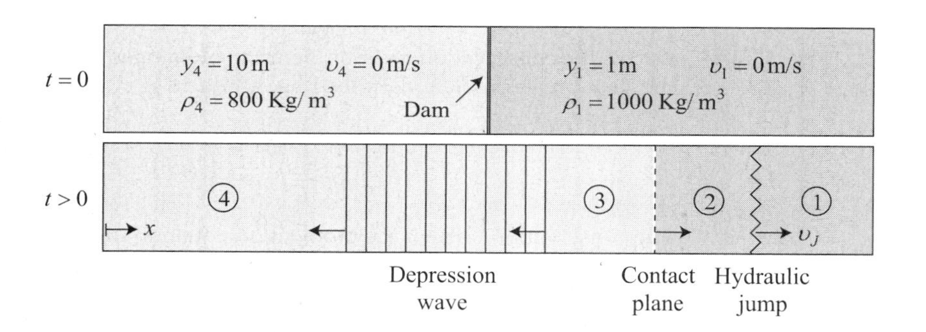

18-12 Consider open channel flow when the fluid density changes with downstream distance, although the flow is still incompressible. Assume that the flow area has a rectangular cross section. Using a control volume approach, derive the frictionless open channel momentum equation for the variable density flow:

$$\frac{\partial v}{\partial t} + v\frac{\partial v}{\partial x} + g\left(\frac{\partial y}{\partial x} + \frac{\partial z}{\partial x} + \frac{y}{2\rho}\frac{\partial \rho}{\partial x}\right) = 0.$$

When $\partial\rho/\partial x = 0$, this result becomes the same as Eq. (18-4).

18-13 Consider open channel flow of an incompressible fluid of nonconstant density. Derive the conservation form of the three governing equations for y, ρy, and $\rho y v$ in a frictionless channel of constant width. Show that these equations can be expressed in the form

$$\frac{\partial y}{\partial t} + \frac{\partial}{\partial x}(yv) = 0$$

$$\frac{\partial}{\partial t}(\rho y) + \frac{\partial}{\partial x}((\rho y)v) = 0$$

$$\frac{\partial}{\partial t}(\rho v y) + \frac{\partial}{\partial x}\left((\rho y v)v + g\frac{(\rho y)y}{2}\right) = 0.$$

18-14 Numerically solve Problem 18-11 using the conservation form of the governing equations derived in Problem 18-13. Choose your dependent variables to be y, ρy, and $\rho y v$. Add diffusion terms to the governing equations to smooth numerical oscillations:

$$\frac{\partial y}{\partial t} = \cdots + \lambda_a \frac{\partial^2 y}{\partial x^2}$$

$$\frac{\partial(\rho y)}{\partial t} = \cdots + \nu_a \frac{\partial^2(\rho y)}{\partial x^2}$$

$$\frac{\partial(\rho y v)}{\partial t} = \cdots + \mu_a \frac{\partial^2(\rho y v)}{\partial x^2}.$$

Plot $\rho(x)$, $y(x)$, and $v(x)$ at $t = 15$ s. Using the analytic solution developed in Problem 18-11, indicate on these plots the locations of the leading and trailing edges of the depression wave, and the locations of the contact plane and the hydraulic jump.

REFERENCES

[1] V. T. Chow, *Open Channel Hydraulics*. New York, NY: McGraw-Hill, 1959.

[2] A. O. Akan, *Open Channel Hydraulics*. Burlington, VT: Butterworth-Heinemann/Elsevier, 2006.

[3] W. Froude, "On Useful Displacement as Limited by Weight of Structure and of Propulsive Power." *Transactions of the Institution of Naval Architects*, **15**, 148 (1874).

Chapter 19

Open Channel Flow with Friction

19.1 The Saint-Venant Equations
19.2 The Friction Slope
19.3 Flow through a Sluice Gate
19.4 Problems

Neglecting friction in the description of an open channel flow becomes less realistic as the length of the channel increases. Therefore, it may become desirable to include the first-order effect of shear stresses acting on the containing boundaries of the flow. However, the actual wall shear stresses cannot be revealed by the flow solution without increasing the dimensionality of the flow description (considering two or more spatial dimensions). Consequently, empirical relations for the boundary friction are commonly employed to capture the first-order effect of drag, without increasing the dimensionality of the flow description.

19.1 THE SAINT-VENANT EQUATIONS

The momentum transport equation needs to be formulated to include boundary friction as a sink of momentum. To this end, consider the control volume illustrated in Figure 19-1 that is differential with respect to the direction of fluid motion but considers the entire cross-sectional area of the flow. The elevation of the flow bed is denoted by z, and is a function of the downstream distance x. The horizontal surfaces, containing the flow of depth y, and the bottom surface, of width w, are the flow boundaries that are subject to friction. The top (free) surface is the only boundary not subject to frictional effects.

Figure 19-2 identifies the elements appearing in the momentum conservation statement. Forces that transmit momentum to (and from) the flow include gravity and boundary friction. Gravity acts on the flow both through hydrostatic pressure and as a

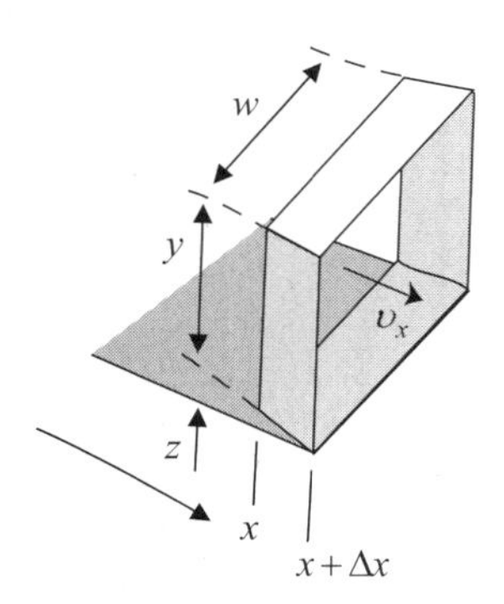

Figure 19-1 Differential volume of open channel flow with friction.

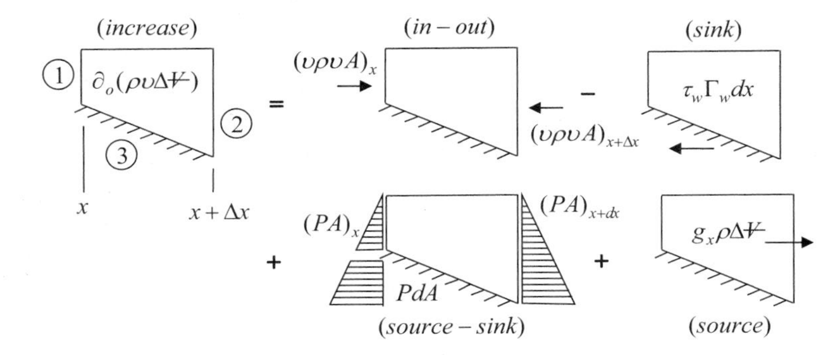

Figure 19-2 Elements of momentum conservation for open channel flow with friction.

body force. The wall shear stress τ_w acts on the wetted circumference Γ_w of the bounding surfaces. For a flow of rectangular cross-sectional area, as shown in Figure 19-1, the wetted circumference is $\Gamma_w = 2y + w$, where y is the flow depth and w is the width of the channel.

Expressed in integral form, the x-direction momentum equation is

$$\underbrace{\int_{\Delta x} \partial_o(\rho v_x A)dx}_{(increase)} = \underbrace{-\int_A (\rho v_x v_j + M_{jx})n_j dA}_{(in\,-\,out)} + \underbrace{\left(-\int_A (P\delta_{jx})n_j dA + \int_{\Delta x} (\rho g_x A)dx,\right)}_{(sources)} \tag{19-1}$$

where the area integrals are comprised of the three surfaces shown in Figure 19-2. One surface intersecting the flow at x (surface 1), another intersecting the flow at $x + \Delta x$ (surface 2), and the final surface is the area of the wetted circumference over the differential distance Δx (surface 3). Integrating the momentum equation over the differential volume yields

$$\frac{\partial}{\partial t}(\rho v A)\Delta x = \overbrace{(\rho v^2 A)_x}^{\text{surface 1}} + \overbrace{-(\rho v^2 A)_{x+\Delta x}}^{\text{surface 2}} + \overbrace{-\tau_w \Gamma_w \Delta x}^{\text{surface 3}}$$

$$+ \underbrace{(PA)_x}_{\text{surface 1}} + \underbrace{-(PA)_{x+\Delta x}}_{\text{surface 2}} + \underbrace{PdA}_{\text{surface 3}} + \underbrace{\rho g A \Delta x}_{\text{volume}}, \tag{19-2}$$

where the x-direction velocity component is denoted with $v_x \to v$ for notational simplicity. The magnitude of the momentum flux to the walls ($M_{jx}n_j$ evaluated for surface 3) equals the wall shear stress τ_w acting on the area of the wetted perimeter $\Gamma_w \Delta x$. Additionally, although the flow does not penetrate surface 3, momentum is still introduced through this surface by pressure. Dividing through by Δx, and considering the differential limit in which $\Delta x \to dx$, one obtains

$$\frac{\partial}{\partial t}(\rho v A) + \frac{\partial}{\partial x}(\rho v^2 A) = -A\frac{\partial P}{\partial x} - \tau_w \Gamma_w + g_x \rho A. \tag{19-3}$$

With the body force of gravity expressed as a potential function: $g_x = -g(\partial h/\partial x)$ (as discussed in Section 14.2), the momentum equation can be rewritten for constant ρ in the form

$$\frac{\partial}{\partial t}(vA) + \frac{\partial}{\partial x}(v^2 A) = -A\frac{\partial}{\partial x}\left(\frac{P}{\rho} + gh\right) - \frac{\tau_w \Gamma_w}{\rho}. \tag{19-4}$$

Although both the hydrostatic pressure P/ρ and the gravity potential gh vary with vertical position in the flow, their sum is a constant equal to $P/\rho + gh = g(y+z) + P_o/\rho$. P_o is the ambient pressure overlying the open channel. Making use of this observation, the momentum equation becomes

$$\frac{\partial}{\partial t}(vA) + \frac{\partial}{\partial x}(v^2 A) = -gA\left(\frac{\partial y}{\partial x} + \frac{\partial z}{\partial x} + S_f\right) \quad \text{with} \quad S_f = \frac{\tau_w \Gamma_w}{\rho g A}. \tag{19-5}$$

The *friction slope* S_f is introduced to express the effect of boundary friction on the momentum transport equation, as is customary in the literature. The continuity equation could be applied to Eq. (19-5) to put the momentum equation into a form similar to Eq. (18-1), which was derived from Bernoulli's equation. However, to facilitate numerical integration, it is desirable to leave the momentum equation in a conservation form (as discussed in Section 17.1). For a flow of rectangular cross section, $A = wy$ and

$$A\frac{\partial y}{\partial x} = wy\frac{\partial y}{\partial x} = \frac{\partial}{\partial x}\frac{y^2 w}{2} - \frac{y^2}{2}\frac{\partial w}{\partial x} = \frac{\partial}{\partial x}\frac{A^2}{2w} - \frac{A^2}{2w^2}\frac{\partial w}{\partial x}. \tag{19-6}$$

Therefore, the momentum equation (19-5) can be written as

$$\text{Momentum:} \quad \frac{\partial}{\partial t}(vA) + \frac{\partial}{\partial x}\left(v^2 A + \frac{gA^2}{2w}\right) = \frac{gA^2}{2w^2}\frac{\partial w}{\partial x} - gA\left(\frac{\partial z}{\partial x} + S_f\right). \tag{19-7}$$

Combined with the continuity equation, also written in conservation form:

$$\text{Continuity:} \quad \frac{\partial}{\partial t}(A) + \frac{\partial}{\partial x}(vA) = 0, \tag{19-8}$$

these equations (19-7) and (19-8) are known as the *Saint-Venant equations* (for a flow of rectangular cross section) [1]. Notice that the form of the continuity equation (19-8) is unaffected by the consideration of wall friction.

In addition to describing a spectrum of transient open channel flows, solutions to the Saint-Venant equations for steady-state problems are of interest. Therefore, attention to the number of constraints that can be imposed on a steady-state solution is warranted. Open channel flows are governed by two first-order differential equations ((19-7) and (19-8)) that can be expressed in terms of two unknown variables, the flow depth y and the flow speed v. Integration of these two equations requires two boundary conditions. However, given a flow with inlet and exit conditions, there are a total of four boundary values between y and v. Generally, only two of these boundary values can be specified independently, and the conservation laws governing the flow dictate the remaining two. However, when supercritical ($Fr > 1$), a flow (or a section of the flow) loses the ability to communicate upstream the influence of downstream conditions (as discussed in Section 18.1). Therefore, an additional downstream boundary condition is required to trigger selection of a subcritical exit state attained through a hydraulic jump. There is a finite range of subcritical exit conditions that, by influencing the location of the hydraulic jump, can satisfy both the upstream boundary conditions and the governing equations.

19.2 THE FRICTION SLOPE

The momentum lost to the boundaries of a channel depends on the shear stress at these surfaces. The shear stress can be characterized in terms of the *coefficient of friction* or Fanning friction factor*:

*Named after the American engineer John Thomas Fanning (1837–1911).

$$c_f = \frac{2\tau_w}{\rho v^2}. \qquad (19\text{-}9)$$

Therefore, in terms of the coefficient of friction, the friction slope becomes

$$S_f = \frac{c_f v^2 \Gamma_w}{2gA} = \frac{c_f v^2}{2gR_w}, \qquad \text{where} \quad R_w = \frac{A}{\Gamma_w} \qquad (19\text{-}10)$$

is the hydraulic radius of a channel defined by the ratio of its cross-sectional area A to its wetted perimeter Γ_w. However, a slightly different empirical expression, known as *Manning's formula*† [2], is more commonly used to evaluate the friction slope:

$$S_f = \frac{n_m^2 v^2}{R_w^{4/3}}. \qquad (19\text{-}11)$$

In Manning's formula, n_m is a dimensional roughness coefficient with units of $(m^{-1/3}s)$. The roughness coefficient may be thought of as an index for the features of channel roughness that contribute to the dissipation of flow momentum. Correlations for the roughness coefficient are generally dependent on the roughness of the channel bed material. For example, Limerinos [3] related n_m to the hydraulic radius of the channel R_w and the size of particles in the surface of the channel bed:

$$n_m = \frac{0.8204\, R_w^{1/6}}{1.16 + 2\log(R_w/d_{84})}. \qquad (19\text{-}12)$$

In Limerinos' correlation, the channel bed particle diameter d_{84} is the size that equals or exceeds the diameter of 84 percent of the particles. The data used in this correlation were limited to $1.5 \leq d_{84} \leq 250$ mm and $30 \leq R_w \leq 180$ cm.

19.3 FLOW THROUGH A SLUICE GATE

Consider the open channel flow after a sluice gate, as illustrated in Figure 19-3. The water reservoir behind the gate has a depth $y_r = 1.22$ m, and is sufficiently expansive that y_r does

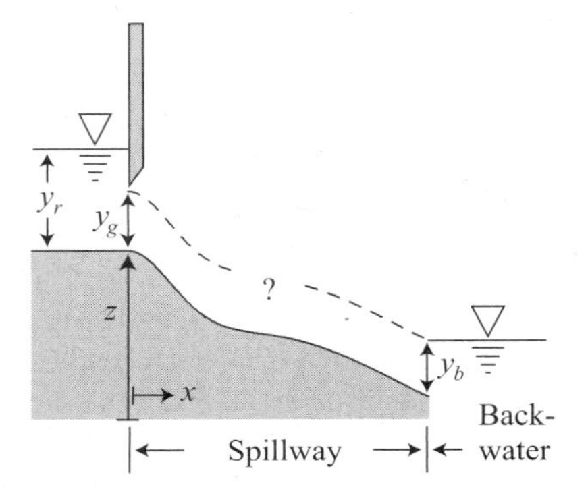

Figure 19-3 Transition to backwater height after sluice gate.

†Named after the Irish engineer Robert Manning (1816-1897).

not fall when the sluice gate is opened. The gate is opened to achieve an initial flow depth $y_g = 0.70$ m, and the flow travels down a spillway of length $L = 100$ m into a backwater of depth y_b. The spillway is characterized with a roughness coefficient of $n_m = 0.03\ \mathrm{m}^{-1/3}\mathrm{s}$. The backwater travels away from the spillway with a speed v_b. As will be established in the analysis to follow, only certain values of the backwater depth y_b and speed v_b admit a steady-state solution for the flow on the spillway.

The flow velocity under the sluice gate v_g is dictated by the reservoir water level y_r through the use of Bernoulli's equation:

$$\frac{v_g^2 - \overbrace{v_r^2}^{=0}}{2} = g(y_r - y_g). \tag{19-13}$$

Assuming that the free surface of the reservoir far behind the sluice gate experiences negligible flow ($v_r \approx 0$), the flow speed under the sluice gate can be determined: $v_g = \sqrt{2g(y_r - y_g)} = 3.194\ \mathrm{m/s}$. From this result it is found that the inlet flow onto the spillway is supercritical: $Fr_g = v_g/\sqrt{gy_g} = 1.219$. When the spillway flow entering the backwater is also supercritical, no additional exit boundary conditions can be specified. However, if the flow entering the backwater is subcritical, one more boundary condition at the exit is needed, such as the backwater depth y_b.

An investigation is made into how the flow transitions from the sluice gate conditions to the backwater conditions for the case where the width of the spillway is constant and much greater than the depth of the flow ($w \gg y$). This condition dictates that $R_w = A/\Gamma_w = y$. The elevation of the spillway surface is described by the function:

$$z(x) = \exp\left\{-\left(\frac{x}{20}\right)^2\right\} + \exp\left\{-\left(\frac{x}{200}\right)^2\right\} + \frac{1}{2}\sin^2\left(\frac{x}{40}\right) \quad \text{for} \quad 0 \le x \le 100\ \mathrm{m}. \tag{19-14}$$

The spillway flow is governed by the Saint-Venant equations developed in Section 19-1. The conservation form of the governing equations (19-7) and (19-8) are rewritten for the case of constant channel width and shallow flow depth ($y \ll w = const.$):

$$\text{Continuity}: \quad \frac{\partial y}{\partial t} = -\frac{\partial}{\partial x}(yv). \tag{19-15}$$

$$\text{Momentum}: \quad \frac{\partial}{\partial t}(yv) = -\frac{\partial}{\partial x}\left((yv)v + \frac{gy^2}{2}\right) - g\left(y\frac{\partial z}{\partial x} + \frac{(n_m v)^2}{y^{1/3}}\right). \tag{19-16}$$

Manning's formula is used to assign frictional losses on the spillway in the momentum equation (19-16).

19.3.1 Numerical Solution to the Spillway Flow

Although a steady-state description of the flow is sought, integration of the transient equations in time provides a straightforward approach to the solution, as discussed in Section 17.4. Some kind of iterative approach is generally needed to solve nonlinear equations numerically, and here the iterative process is performed in time.

To solve for the flow on the spillway, the Saint-Venant equations are solved by MacCormack integration. As discussed in Chapter 17, MacCormack integration requires the averaging of two estimates of the dependent variables at the end of each time step. These estimates are derived from the predictor and corrector steps. Following the procedure described in Chapter 17, the predictor step calculations of the continuity equation (19-15) and the momentum equation (19-16) are given by

$$\{y_n^{t+\Delta t}\}_p = y_n^t - \frac{\Delta t}{\Delta x}\left[y_{n+1}^t v_{n+1}^t - y_n^t v_n^t\right] + \frac{\nu_a \Delta t}{\Delta x^2}\left[y_{n+1}^t - 2y_n^t + y_{n-1}^t\right] \leftarrow \textit{for stability} \tag{19-17}$$

$$\left\{(yv)_n^{t+\Delta t}\right\}_p = (yv)_n^t - \frac{\Delta t}{\Delta x}\left\{\left[\left((yv)_{n+1}^t v_{n+1}^t + g(y_{n+1}^t)^2/2\right) - \left((yv)_n^t v_n^t + g(y_n^t)^2/2\right)\right] + g y_n^t [z_{n+1} - z_n]\right\} - \Delta t g (n_m v_n^t)^2/(y_n^t)^{1/3} + \frac{\mu_a \Delta t}{\Delta x^2}\left[(yv)_{n+1}^t - 2(yv)_n^t + (yv)_{n-1}^t\right] \leftarrow \textit{for stability.} \tag{19-18}$$

Predictor step values for velocity $\{v_n^{t+\Delta t}\}_p$ are obtained from $\{(yv)_n^{t+\Delta t}\}_p$ using

$$\{v_n^{t+\Delta t}\}_p = \frac{\left\{(yv)_n^{t+\Delta t}\right\}_p}{\{y_n^{t+\Delta t}\}_p}. \tag{19-19}$$

The corrector step calculations for the continuity and momentum equations are given by

$$\{y_n^{t+\Delta t}\}_c = y_n^t - \frac{\Delta t}{\Delta x}\left[\{y_n^{t+\Delta t}\}_p\{v_n^{t+\Delta t}\}_p - \{y_{n-1}^{t+\Delta t}\}_p\{v_{n-1}^{t+\Delta t}\}_p\right] + \frac{\nu_a \Delta t}{\Delta x^2}\left[\{y_{n+1}^{t+\Delta t}\}_p - 2\{y_n^{t+\Delta t}\}_p + \{y_{n-1}^{t+\Delta t}\}_p\right] \leftarrow \textit{for stability} \tag{19-20}$$

$$\begin{aligned}\left\{(yv)_n^{t+\Delta t}\right\}_c = (yv)_n^t &- \frac{\Delta t}{\Delta x}\left\{\left[\left(\left\{(yv)_n^{t+\Delta t}\right\}_p\{v_n^{t+\Delta t}\}_p + g\{y_n^{t+\Delta t}\}_p^2/2\right)\right.\right.\\ &\left.\left.- \left(\left\{(yv)_{n-1}^{t+\Delta t}\right\}_p\{v_{n-1}^{t+\Delta t}\}_p + g\{y_{n-1}^{t+\Delta t}\}_p^2/2\right)\right] + g\{y_n^{t+\Delta t}\}_p[z_n - z_{n-1}]\right\}\\ &- \Delta t g\left(n_m\{v_n^{t+\Delta t}\}_p\right)^2 / \left(\{y_n^{t+\Delta t}\}_p\right)^{1/3}\\ &+ \frac{\mu_a \Delta t}{\Delta x^2}\left[\left\{(yv)_{n-1}^{t+\Delta t}\right\}_p - 2\left\{(yv)_n^{t+\Delta t}\right\}_p + \left\{(yv)_{n-1}^{t+\Delta t}\right\}_p\right] \leftarrow \textit{for stability.}\end{aligned} \tag{19-21}$$

Corrector step values for velocity $\{v_n^{t+\Delta t}\}_c$ are obtained from $\left\{(yv)_n^{t+\Delta t}\right\}_c$ using

$$\{v_n^{t+\Delta t}\}_c = \frac{\left\{(yv)_n^{t+\Delta t}\right\}_c}{\{y_n^{t+\Delta t}\}_c}. \tag{19-22}$$

As was noted in Chapter 17, the direction of finite differencing may need to be reversed for either the predictor or corrector step at boundary nodes to prevent the differencing operation from extending beyond the physical domain. For the same reason, the artificial diffusion terms, added to the predictor and corrector step equations for numerical stability, are not evaluated at the boundary nodes.

To march MacCormack integration forward, the predictor and corrector step values are averaged to obtain the end of time step values:

$$y_n^{t+\Delta t} = \frac{\{y_n^{t+\Delta t}\}_p + \{y_n^{t+\Delta t}\}_c}{2} \tag{19-23}$$

$$(yv)_n^{t+\Delta t} = \frac{\left\{(yv)_n^{t+\Delta t}\right\}_p + \left\{(yv)_n^{t+\Delta t}\right\}_c}{2} \tag{19-24}$$

and

$$v_n^{t+\Delta t} = \frac{(yv)_n^{t+\Delta t}}{y_n^{t+\Delta t}}. \tag{19-25}$$

When the spillway exit flow is subcritical, the backwater depth y_b may be specified for a unique steady-state solution. However, the second exit flow variable $(yv)_b$ may not be freely specified at the exit, and is dictated by the solution to the transport equations. One way to handle unspecified boundary values numerically is to let them "float," as discussed in Section 17.4. The floating boundary condition for (yv) is

$$\text{Floating exit:} \qquad \text{at } n = \text{last node: } (yv)_n^{t+\Delta t} = 2(yv)_{n-1}^{t+\Delta t} - (yv)_{n-2}^{t+\Delta t}. \tag{19-26}$$

If the spillway exit flow is supercritical, the backwater conditions are fully prescribed by the upstream state of the flow. In this case, the value of the backwater depth must also be floated:

$$\text{Floating exit:} \qquad \text{at } n = \text{last node: } y_n^{t+\Delta t} = 2y_{n-1}^{t+\Delta t} - y_{n-2}^{t+\Delta t}. \tag{19-27}$$

In this way, the governing equations can propagate all the requirements of the conservation laws to the exit condition.

Code 19-1 is used to determine the flow solution on the spillway for known inlet conditions y_g and $(yv)_g$. The special case of supercritical exit conditions employs the floating exit boundary condition on both y_b and $(yv)_b$, as described above. Otherwise, the exit flow depth y_b is specified for a unique solution. The nodes between the boundaries are integrated in time to the steady-state solution. A linear interpolation between inlet and exit conditions is used as an initial condition, where steady-state continuity is used to initialize the exit velocity $v_b = (yv)_g / y_b$. The integration time step is dynamically adjusted in Code 19-1 to satisfy

$$\frac{\Delta t\{|v| + c\}_{\max}}{\Delta x} \approx C_{targ} = 0.9, \tag{19-28}$$

where the local wave speed is known from $c = \sqrt{gy}$. This satisfies the Courant condition required for numerical stability, as discussed in Section 17.2. The artificial diffusivities are specified dynamically by the condition

$$\nu_a = \mu_a = \frac{0.1}{\Delta t / \Delta x^2}. \tag{19-29}$$

The coefficient in the numerator may be changed for varied amounts of diffusion in the solution, as long as the stability requirement given by Eq. (18-60) is observed.

Integration to steady-state requires an assessment of when the solution stops changing in time. The most robust assessment is to look at every number that is calculated, to determine if there is any change over a time step. A simpler approach is to check a single result for change. In Code 19-1 the exit velocity v_b alone is polled for convergence to a steady-state value.

19.3.2 Solutions to the Spillway Flow

Figure 19-4 shows the numerical solution corresponding to a supercritical state entering the backwater. By floating both exit conditions, the supercritical exit speed is determined to be $v_b = 3.344$ m/s and flow depth to be $y_b = 0.6684$ m. This corresponds to a Froude number of $Fr_b = v_b/\sqrt{gy_b} = 1.306$. It is seen in Figure 19-4 that a hydraulic jump occurs at a distance of $x = 40$ m down the spillway. Approaching the hydraulic jump, the flow is everywhere supercritical. For frictionless flow, the observations made in Section 18.1 suggest that a supercritical flow should thin on the downward slope of the spillway. However, here, in the presence of friction, the supercritical flow is seen to thicken on the less steep sections of the spillway, such as immediately after the sluice gate and immediately before the hydraulic jump. It is left as an exercise (see Problem 19-1) to investigate the relation between the spillway slope and friction on changes in flow depth.

Other possible steady-state solutions exist when the flow is forced to approach the backwater in a subcritical state. For example, suppose the flow undergoes a second hydraulic jump at the end of the spillway, at $x = 100$ m. Using Eq. (18-19), the new backwater flow is now required to have a depth of

$$\frac{y_b}{0.6684} = \frac{1}{2}\left(\sqrt{1 + 8(1.306)^2} - 1\right) \tag{19-30}$$

or

$$y_b = 0.945\,\text{m}.$$

Solutions also exist for greater backwater depths, $y_b > 0.945$ m, which result in the second hydraulic jump being pushed upstream toward the sluice gate. Figure 19-5 illustrates the range of steady-flow solutions that exist for the specified flow entering under the sluice gate. All steady-state flows with a subcritical approach to the backwater are attained through one or two upstream hydraulic jumps. The range of possible subcritical flows on

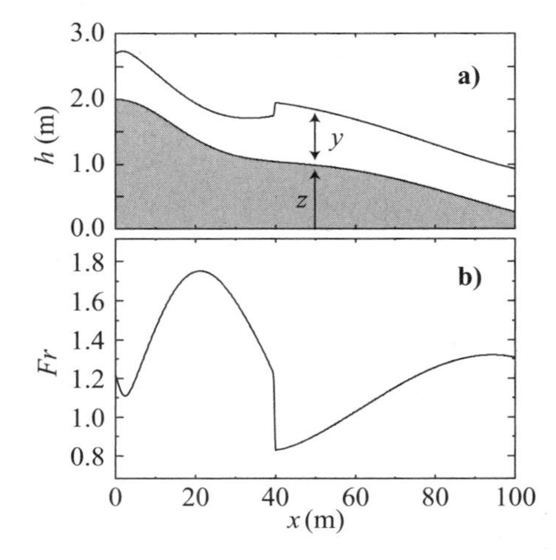

Figure 19-4 Flow over spillway: (a) depth + spillway height and (b) Froude number.

Code 19-1 Steady-flow on spillway

```
#include <stdio.h>
#include <math.h>

double cont_diff(char t,int n,double *y,double *v) {
   return t=='p' ? y[n+1]*v[n+1] - y[n]*v[n] :
                   y[n]*v[n] - y[n-1]*v[n-1] ;       }

double mom_diff(char t,int n,
   double g,double *y,double *y_v,double *v,double *z) {
   double val= t=='p' ?
   (y_v[n+1]*v[n+1]+g*y[n+1]*y[n+1]/2.) - (y_v[n]*v[n]+g*y[n]*y[n]/2.) :
   (y_v[n]*v[n]+g*y[n]*y[n]/2.) - (y_v[n-1]*v[n-1]+g*y[n-1]*y[n-1]/2.) ;
   return val + g*y[n]*(t=='p' ? z[n+1]-z[n] : z[n]-z[n-1]);          }

double MacC(int N,double *z,double *y,double *y_v,double *v,
                 double *dt_,double dx,double g)

{
   int n,step;
   double dt=*dt_,Py[N],Cy[N],Py_v[N],Cy_v[N],Pv[N],Cv[N];
   double nm=0.03,C,Cmax=0,Ctarg=0.9,dtdx=dt/dx,dtddx=dt/dx/dx,na,ma;
   na=ma=0.1/dtddx;

   for (n=0;n<N;++n) {                // perform predictor step
      step=(n<N-1 ? 'p' : 'c');
      Py[n]=y[n]-dtdx*cont_diff(step,n,y,v);
      if (n>0 && n<N-1) Py[n]+=dtddx*na*(y[n+1]-2.*y[n]+y[n-1]);
      Py_v[n]=y_v[n]-dtdx*mom_diff(step,n,g,y,y_v,v,z);
      Py_v[n]-=dt*g*pow(nm*v[n]/pow(y[n],0.16666666667),2.);
      if (n>0 && n<N-1) Py_v[n]+=dtddx*ma*(y_v[n+1]-2.*y_v[n]+y_v[n-1]);
      Pv[n]=Py_v[n]/Py[n];
      C=dt*(fabs(Pv[n])+sqrt(g*Py[n]))/dx; // dt size checking
      if (C>Cmax) Cmax=C;
   }
   if (Cmax>1.) { // integrate with a smaller time step
      *dt_=0.;
      return dt*Ctarg/Cmax;
   }
   for (n=0;n<N;++n) {                // perform corrector step
      step=(n>0 ? 'c' : 'p');
      Cy[n]=y[n]-dt/dx*cont_diff(step,n,Py,Pv);
        if (n>0 && n<N-1) Cy[n]+=dtddx*na*(Py[n+1]-2.*Py[n]+Py[n-1]);
      Cy_v[n]=y_v[n]-dtdx*mom_diff(step,n,g,Py,Py_v,Pv,z);
      Cy_v[n]-=dt*g*pow(nm*Pv[n]/pow(Py[n],0.16666666667),2.);
     if (n>0 && n<N-1) Cy_v[n]+=dtddx*ma*(Py_v[n+1]-2.*Py_v[n]+Py_v[n-1]);
      Cv[n]=Cy_v[n]/Cy[n];
      if (isnan(Cv[n])) {
         printf("\nsoln BLEW! dt=%e ma=%e (Cmax=%e)\n",dt,ma,Cmax);
         return 0;
      }
   }

   for (n=1;n<N-1;++n) { // average predictor and corrector results
      y[n]=0.5*(Py[n]+Cy[n]);
      if (y[n]< .1) y[n]=.1;
      y_v[n]=0.5*(Py_v[n]+Cy_v[n]);
      v[n]=y_v[n]/y[n];
   }
//   y[N-1]=2.*y[N-2]-y[N-3]; // for supercrit. exit
   v[N-1]=(y_v[N-1]=2.*y_v[N-2]-y_v[N-3])/y[N-1]; // float exit velocity

   return dt*Ctarg/Cmax;
}

int main(void)
{
   int n,cnt=0,N=1201,ConvergeCnt=0;;
   double y[N],y_vel[N],vel[N],z[N],LastExitVel;
   double x,t=0,g=9.807,L=100.0,dx=L/(N-1),dt,dt_next=0.01;
   FILE *fp;

   y_vel[0]=(vel[0]=3.1936)*(y[0]=0.7); // inlet conditions
   y[N-1]=1.0;                          // exit boundary condition
   LastExitVel=vel[N-1]=(y_vel[N-1]=y_vel[0])/y[N-1];
   for (n=0;n<N;++n) { // initial distributions through channel
      y[n]=        y[0]+n*(    y[N-1]-    y[0])/(N-1);
      y_vel[n]=y_vel[0]+n*(y_vel[N-1]-y_vel[0])/(N-1);
      vel[n]=y_vel[n]/y[n];
      x=n*dx;
      z[n]=exp(-x*x/400)+exp(-x*x/4000)+.5*sin(x/40.)*sin(x/40.);
   }

   do {
      if (!((cnt++)%100)) { // check every 100 dt
         if (fabs((vel[N-1]-LastExitVel))<5.e-6) ++ConvergeCnt;
         else ConvergeCnt=0;
         LastExitVel=vel[N-1];
         printf("\nt=%e ExitVel=%e ",t,LastExitVel);
      }
      dt=dt_next;
      if ( !(dt_next=MacC(N,z,y,y_vel,vel,&dt,dx,g)) ) break;
   } while ((t+=dt)<1000. && ConvergeCnt<10);

   fp=fopen("out.dat","w");
   for (n=0;n<N;++n)
      fprintf(fp,"%e %e %e %e %e\n",
         n*dx,z[n],y[n],vel[n],vel[n]/sqrt(g*y[n]));
   fclose(fp);

   return 1;
}
```

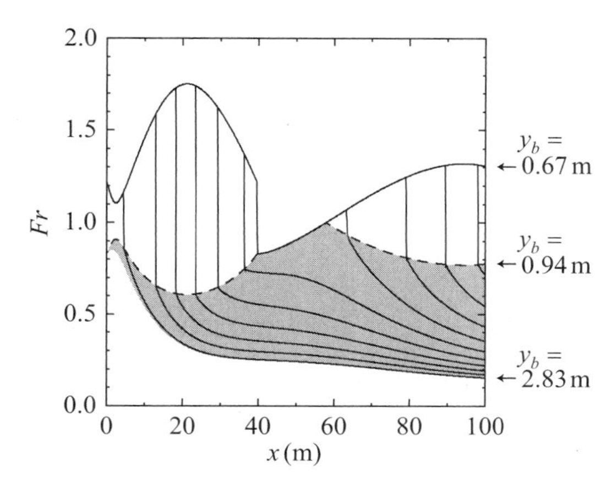

Figure 19-5 Admissible solutions to spillway flow.

the spillway is indicated by the shaded region in Figure 19-5. Solid lines are used to show some of the steady-state flow solutions that lead to various backwater depths. Steady-state solutions are shown in Figure 19-5 for various backwater heights y_b, starting with $y_b = 1.0$ m and ending at $y_b = 2.8$ m, with intervals of 0.2 m between each case. The greatest possible backwater depth is achieved when the hydraulic jump occurs immediately at the sluice gate, such that the flow is subcritical at the start of the spillway. In this case, the initial flow state (after the hydraulic jump) is specified by Eq. (18-19) to yield

$$\frac{y_g}{0.7} = \frac{1}{2}\left(\sqrt{1 + 8(1.219)^2} - 1\right) \quad \text{or} \quad y_g = 0.9064\,\text{m}. \tag{19-31}$$

And, from continuity,

$$v_g = \frac{(3.194\ \text{m/s})(0.7\ \text{m})}{0.9064\ \text{m}} = 2.466\ \text{m/s}. \tag{19-32}$$

When the flow is integrated with these subcritical inlet conditions, the backwater flow depth is determined to be $y_b = 2.83$ m.

19.4 PROBLEMS

19-1 Without friction, a flow will thicken or thin on an incline, depending on whether the flow is subcritical or supercritical. However, the situation becomes more complicated with friction. For a "wide" open channel (where the wetted circumference divided by the cross-sectional area is $\Gamma_w/A = 1/y$) show that the steady-state momentum and continuity equations can be combined for the result

$$\frac{\partial y}{\partial x} = -\frac{dz/dx + c_f Fr^2/2}{1 - Fr^2}.$$

Determine the slope required for a constant flow depth to occur on an inclined surface in the presence of friction.

19-2 Consider the flow after a sluice gate where the length and slope of the spillway is $x_2/y_1 = 25.0$ and $dz/dx = -0.02$, and the depth of the backwater is $y_2/y_1 = 1.7$. The velocity of flow under the sluice gate is $v(x = 0)/\sqrt{gy_1} = 1.1$. The spillway has a friction coefficient $c_f = 0.01$. Non-dimensionalize the conservative form of the governing equations, using

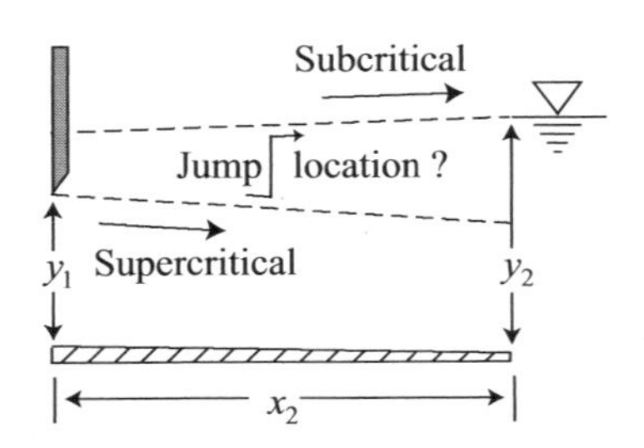

$$u = v/\sqrt{gy_1}, \quad \eta = y/y_1, \quad \xi = x/y_1, \quad \zeta = z/y_1, \quad \text{and} \quad \tau = t/\sqrt{y_1/g}.$$

Add diffusion terms to the continuity and momentum equations to smooth numerical oscillations:

$$\frac{\partial \eta}{\partial \tau} = \cdots + \nu_a \frac{\partial^2 \eta}{\partial \xi^2} \quad \text{and} \quad \frac{\partial(\eta\, u)}{\partial \tau} = \cdots + \mu_a \frac{\partial^2 (\eta\, u)}{\partial \xi^2}.$$

Numerically integrate the governing equations to determine the flow depth and local Froude number as a function of the distance down the spillway.

19-3 Consider the flow described in Problem 19-2. For what range of backwater depths are there physical solutions to this problem?

19-4 Consider flow under the sluice gate shown. The flow leaving the gate spills down a spillway into a backwater of depth y_4. The wide basin has three sloped regions. In the regions between x_1 and x_2, and between x_3 and x_4, the basin has a shallow slope $dz/dx = -0.002$. In the region between x_2 and x_3, the basin has a steeper slope $dz/dx = -0.009$. The basin has a friction coefficient $c_f = 0.01$. Nondimensionalize the conservative form of the governing equations, using

$$u = v/\sqrt{gy_1}, \quad \eta = y/y_1, \quad \xi = x/y_1, \quad \zeta = z/y_1, \quad \text{and} \quad \tau = t/\sqrt{y_1/g}.$$

Add diffusion terms to the continuity and momentum equations to smooth numerical oscillations:

$$\frac{\partial \eta}{\partial \tau} = \cdots + \nu_a \frac{\partial^2 \eta}{\partial \xi^2} \quad \text{and} \quad \frac{\partial(\eta\, u)}{\partial \tau} = \cdots + \mu_a \frac{\partial^2 (\eta\, u)}{\partial \xi^2}.$$

Numerically integrate the governing equations to determine the flow depth and local Froude number as a function of the distance down the spillway.

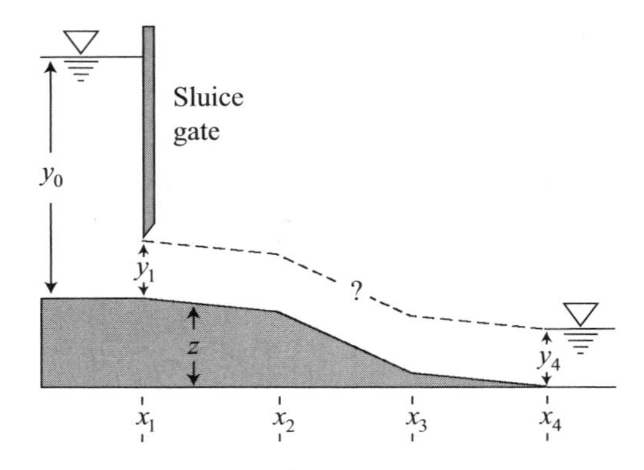

Distances along the spillway are $\xi(x_1) = 0$, $\xi(x_2) = 150$, $\xi(x_3) = 300$, and $\xi(x_4) = 450$. The flow depth of the backwater equals $\eta(x_4) = 1.5$. The Froude number for the flow under the sluice gate is $Fr(0) = 0.657$. Determine the steady-state flow depth from the gate condition to the backwater condition. Superimpose plots of $\zeta + \eta$ and ζ. Plot the local value of the Froude number as a function of distance. Discuss the behavior of the flow in every region of the spillway. Does the flow thin or thicken as expected leaving the gate? Explain your expectations in the context of Problem 19-1. Try to solve this problem for $Fr(0) = 0.8$. Explain what happens to the solution.

19-5 Consider the open channel flow discussed in Problem 19-4. Now let the backwater flow depth be $\eta(x_4) = 1.0$. Determine the velocity of flow under the sluice gate $u(x_1) = Fr(0)$. *Hint*: Let the entrance velocity be transiently dictated by steady-state continuity, $u(x_1) = u(x_4)\eta(x_4)$, while floating the exit velocity. How does this entrance velocity compare with the case when $\eta(x_4) = 1.5$? Explain this comparison.

19-6 Consider a 30-m-long open channel of rectangular cross section and flat bottom. The width of the channel is variable: $w(x)/w_t = 1 + (x/15 - 1)^2/2$. The throat of the channel is $w_t = 10$ m. At the inlet ($x = 0$), the flow depth is $y^i = 2$ m and flow speed is $v^i = 15$ m/s. The walls and the bottom of the channel are characterized by a Manning's roughness coefficient of $n_m = 0.01$ m$^{-1/3}$s. Determine the supercritical flow profile through the channel. Plot the flow depth and Froude number over the length of the channel. Find the location where the flow depth is a minimum and flow speed is a maximum.

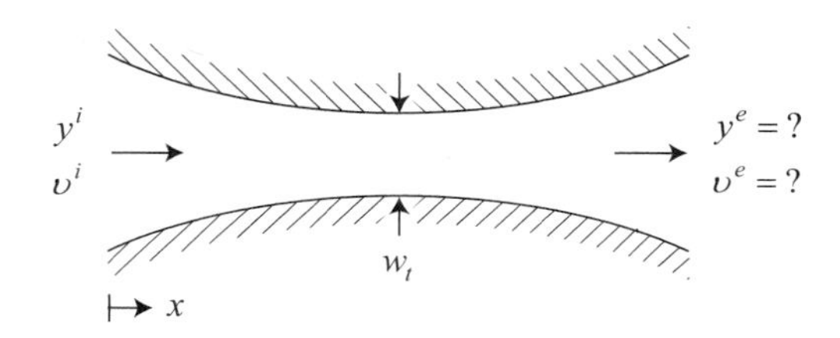

REFERENCES

[1] A. J. C. de Saint-Venant, "Théorie du Mouvement Non-Permanent des Eaux avec Application aux Crues des Rivières et à l'Introduction des Marées dans leur Lit." *C. R. Acad. Sc. Paris*, **73**, 147 (1871).

[2] R. Manning, "On the Flow of Water in Open Channels and Pipes." *Transactions of the Institution of Civil Engineers of Ireland*, **20**, 161 (1891).

[3] J. T. Limerinos, "Determination of the Manning Coefficient from Measured Bed Roughness in Natural Channels." U.S. Geological Survey Water Supply paper, 1898-B, **47** (1970).

Chapter 20

Compressible Flow

20.1 General Equations of Momentum and Energy Transport

20.2 Reversible Flows

20.3 Sound Waves

20.4 Propagation of Expansion and Compression Waves

20.5 Shock Wave (Normal to Flow)

20.6 Shock Tube Analytic Description

20.7 Shock Tube Numerical Description

20.8 Shock Tube Problem with Dissimilar Gases

20.9 Problems

In most chapters of this text, energy conservation is subdivided into mechanical and thermal components. Satisfying the momentum equation ensures conservation of mechanical energy in a flow, and, in many instances, the mechanical energy content is uncoupled from the thermal energy. However, some exceptions, such as natural convection (Chapter 26) and problems for which fluid diffusivities are temperature dependent (Chapter 29), will be considered in later chapters. Additionally, flows of particularly high speed tend to couple mechanical energy and thermal energy in ways that cannot be avoided. For example, close to flow boundaries, mechanical energy is lost to heat through viscous effects, as is accounted for by the momentum equation. However, the reappearance of this energy in the heat equation is usually neglected at lower flow speeds, as discussed in Chapter 5, but may become important at high speeds, as is considered in Section 25.7.

This chapter focuses on the coupling between mechanical and thermal energy that arises in advection transport due to the compressibility of gases. At sufficiently low speeds even compressible gases tend to flow as though incompressible. However, when flow speeds become comparable to the speed of sound, the compressibility of gases can no longer be ignored. In such cases, flows are incapable of fully adjusting for downstream conditions to avoid highly dynamic accelerations that invoke gas compressibility. Therefore, consequences of gas compressibility on the transport equations describing high-speed flows are investigated in this chapter. This is the subject of *gas dynamics* [1].

20.1 GENERAL EQUATIONS OF MOMENTUM AND ENERGY TRANSPORT

The general conservation equations were developed in Chapter 5, and are summarized here for reference:

$$\text{Continuity:} \qquad \frac{D\rho}{Dt} = -\rho\partial_j v_j \tag{20-1}$$

$$\text{Momentum:} \quad \rho \frac{Dv_i}{Dt} = -\partial_j M_{ji} - \partial_i P + \rho g_i \tag{20-2}$$

$$\text{Energy:} \quad \rho \frac{Dh_o}{Dt} = \partial_o P - \partial_j q_j - \partial_j (M_{ji} v_i) + \rho g_j v_j. \tag{20-3}$$

It is left as an exercise (see Problem 20-1) to derive the current form of the energy equation, which is expressed in terms of the stagnation enthalpy:

$$h_o = h + v^2/2. \tag{20-4}$$

The diffusion laws needed in the general transport equations are

$$M_{ji} = -\mu(\partial_j v_i + \partial_i v_j) + (2/3)\mu \delta_{ji} \partial_k v_k \quad \text{– Newton's momentum diffusion law} \tag{20-5}$$

$$q_i = -k \partial_i T \quad \text{– Fourier's heat diffusion law.} \tag{20-6}$$

20.1.1 Flow Equations Far from Boundaries

The flow equations that describe advection transport far from boundaries were developed in Chapter 14. These equations simplify by considering the dominance of advection transport over diffusion processes in the flow. Additionally, the influence of body forces are generally small compared with the inertial forces imposed at high speeds. Consequently, the following simplifications are typical of problems in gas dynamics:

1. Inviscid – drop terms related to momentum diffusion M_{ji}.
2. Adiabatic – drop terms related to heat diffusion q_i.
3. Small body forces – drop terms related to g_i.

With these assumptions, it is easily seen that the governing equations (20-1) through (20-3) simplify to

$$\text{Continuity:} \quad \frac{D\rho}{Dt} = -\rho \partial_j v_j \tag{20-7}$$

$$\text{Momentum:} \quad \rho \frac{Dv_i}{Dt} = -\partial_i P \tag{20-8}$$

$$\text{Energy:} \quad \rho \frac{Dh_o}{Dt} = \partial_o P. \tag{20-9}$$

These equations, which govern inviscid flow, are the *Euler equations** [2]. Notice that there are three equations in terms of four unknown properties ρ, v, P, and h. Therefore, a fourth equation of state is required. Taking the ideal gas law as suitable for problems in gas dynamics, the fourth governing equation is

$$\text{Ideal gas:} \quad P = \rho R T. \tag{20-10}$$

---

*The Euler equations are named after the Swiss mathematician and physicist Leonhard Euler (1707–1783).

Table 20-1 Thermodynamic relations for an ideal gas

Energy	Entropy
$dh = C_p dT$	$ds = C_v dT/T - R d\rho/\rho$
$du = C_v dT$	$ds = C_p dT/T - R dP/P$
	$ds = C_v dP/P - C_p d\rho/\rho$

Since for an ideal gas enthalpy $h(T)$ is only a function of temperature, these four equations can be expressed in terms of four unknown properties ρ, v, P, and T.

For an ideal gas, the thermodynamic relations given in Table 20-1 (which is excerpted from Table 1-1) are applicable for determining changes in the energy quantities (u or h) and entropy (s).

20.2 REVERSIBLE FLOWS

As discussed in Section 1.1, a flow is reversible if no entropy generation occurs. Such a flow is also isentropic because of the constant entropy value that each control mass in the flow carries. This definition does not require that entropy be a spatial constant, except along steady-state streamlines. Reversible flows necessarily disallow diffusion transport, which results in entropy generation as discussed in Section 5.7. This is not an added limitation to the topics of this chapter, since attention has already been restricted to flows without diffusion. However, irreversible discontinuous changes can sometimes occur when a compressible flow transitions between two solutions of the governing equations. In gas dynamics, this occurs through a shock wave, as will be discussed in Section 20.5. Since shocks produce entropy, their occurrence is prohibited in the discussion of reversible flows.

An interesting simplification to the energy equation for isentropic (reversible) flows can be observed when the energy equation (20-9) is rewritten (using $h_o = h + v^2/2$) in the form

$$\text{Energy:} \quad \rho\frac{D}{Dt}\left(\frac{v^2}{2}\right) = \partial_o P - \rho\frac{Dh}{Dt} \quad \text{or} \quad v_i\rho\frac{Dv_i}{Dt} = \partial_o P - \rho\frac{Dh}{Dt}. \tag{20-11}$$

When a flow is isentropic ($ds = 0$), the fundamental thermodynamic equation $dh = Tds + dP/\rho$ (Eq.(1-8)) requires that $dh = dP/\rho$. Therefore, for an isentropic flow,

$$\frac{Ds}{Dt} = 0 \rightarrow \rho\frac{Dh}{Dt} = \frac{DP}{Dt} \quad \text{(isentropic flow)}. \tag{20-12}$$

With this restriction, the energy equation (20-11) becomes

$$\text{Energy:} \quad v_i\rho\frac{Dv_i}{Dt} = \partial_o P - \frac{DP}{Dt} = -v_i\partial_i P$$

$$\text{or} \quad v_i\left(\rho\frac{Dv_i}{Dt} + \partial_i P\right) = 0 \quad \text{(isentropic flow)}. \tag{20-13}$$

In this final form, it is clear that the energy equation is satisfied by the momentum equation (20-8) when the flow is reversible (isentropic).

20.3 SOUND WAVES

Flows that are comparable in speed to sound waves define the meaning of "high speed" in gas dynamics. Sound waves dictate the speed at which information propagates within a flow. When insufficient information can propagate upstream, flows are incapable of sufficiently adjusting for downstream conditions to avoid dynamic accelerations that invoke gas compressibility.

The conservation laws can be used to determine the speed of sound "a" in a gas. For this analysis, the flow is taken to be traveling to the right with properties v, ρ, and P. Sound waves can propagate in the same or opposing direction as the flow, as shown in Figure 20-1. Therefore, a sound wave propagating in the same direction as the flow travels with a speed of $v + a$, while a wave propagating in the opposing direction travels with a speed of $v - a$. There is an altered state in the wake of a sound wave described by $v + dv$, $\rho + d\rho$, and $P + dP$. It is assumed that this altered state is only differentially different from the original state and that the change is reversible ($ds = 0$). This distinguishes the effect of a sound wave from other types of waves that have finite magnitude and may not be reversible (such as shock waves). Figure 20-1 illustrates the control surface surrounding the sound wave that will be used in analysis. The control surface is fixed to the sound wave, such that the flow approaching the wave location has a speed of $\pm a$. Surfaces A and B are separated by a vanishingly small distance.

For a reversible differential change in state, the flow is isentropic ($ds = 0$), and energy conservation does not need to be considered separately from momentum conservation. The differential forms of the remaining equations governing a steady flow along a streamline across the wave require

$$\text{Continuity:} \quad d(v\rho) = 0 \rightarrow \rho dv + v d\rho = 0 \tag{20-14}$$

$$\text{Momentum:} \quad \rho v dv = -dP. \tag{20-15}$$

Equations (20-14) and (20-15) are conservation requirements descriptive of the differential change in state across a sound wave made from a frame of reference moving with the wave. In this reference frame, the flow velocity can only be the positive or negative value of the speed of sound $v = \pm a$, as shown in Figure 20-1. Consequently, the differential forms of the continuity and momentum equations for the sound wave become

$$\text{Continuity:} \quad \rho dv + (\pm a) d\rho = 0 \tag{20-16}$$

$$\text{Momentum:} \quad (\pm a)\rho dv = -dP. \tag{20-17}$$

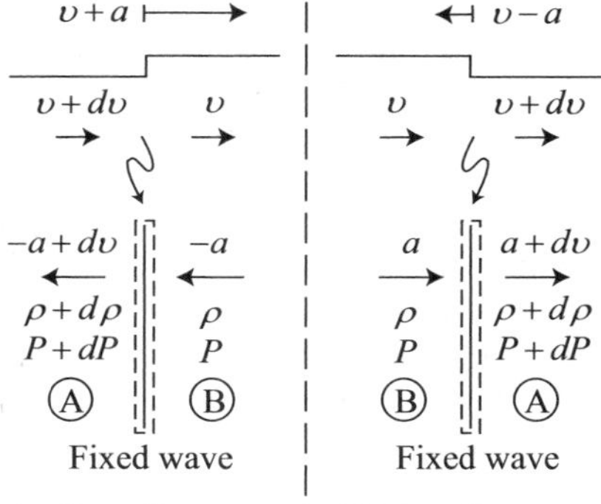

Figure 20-1 Propagation of sound waves in the same (a) and opposing (b) direction as the flow.

Combining the continuity and momentum equations to eliminate ρdv gives the result

$$\frac{dP}{(\pm a)} = (\pm a)d\rho \quad \text{or} \quad a^2 = \frac{dP}{d\rho}. \tag{20-18}$$

This result can be used to evaluate the sound wave speed. The assumed reversible change in state across the sound wave imposes the constraint that $ds = 0$. Therefore, the speed of sound is formally written as

$$a = \sqrt{\left.\frac{\partial P}{\partial \rho}\right|_{ds=0}}. \tag{20-19}$$

For an ideal gas,

$$ds = C_v dP/P - C_p d\rho/\rho, \tag{20-20}$$

such that

$$\partial P/\partial \rho|_{ds=0} = \gamma P/\rho = \gamma RT. \tag{20-21}$$

Therefore, the speed of sound for an ideal gas evaluates to

$$a = \sqrt{\gamma RT} \quad \text{(ideal gas)}. \tag{20-22}$$

Since the speed of sound is an important measure of how fast a flow is moving, the *Mach number*† defined as the ratio of the flow speed to the speed of sound

$$\mathrm{M} = \frac{v}{a}, \tag{20-23}$$

is of considerable significance to the behavior of a flow. When a flow is *subsonic*, the flow speed is less than the local speed of sound in the fluid, $\mathrm{M} < 1$. When a flow is *supersonic*, the flow speed is greater than the local speed of sound, $\mathrm{M} > 1$. Finally, when a flow is *sonic*, the flow speed is equal to the local speed of sound, $\mathrm{M} = 1$. As a rule of thumb, compressibility effects begin to become important in gas flows at speeds where $\mathrm{M} = 0.3$.

20.4 PROPAGATION OF EXPANSION AND COMPRESSION WAVES

As discussed in the last section, there is an altered state in the wake of a wave, the nature of which differentiates two types of waves, expansion and compression waves. A material fluid element expands as it passes through an *expansion wave* (also called a *rarefaction wave*) and is compressed as it passes through a *compression wave*. These two types of waves are shown schematically in Figure 20-2, where the incremental motion of a bounding wall is used to create a series of waves. The left-hand illustration shows a series of expansion waves, while the right-hand side shows a series of compression waves. All waves propagate with the local speed of sound, which was demonstrated in Section 20.3 to be related to the temperature of an ideal gas by $a = \sqrt{\gamma RT}$. Since there is a temperature drop behind each expansion wave, the next wave in the series propagates with a lower

†The Mach number is named after Austrian physicist and philosopher Ernst Mach (1838–1916).

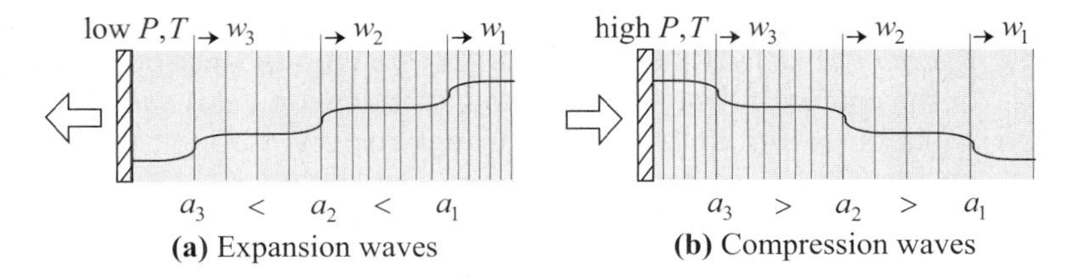

Figure 20-2 A series of (a) expansion waves and (b) compression waves created by the motion of a bounding wall.

speed, such that $a_3 < a_2 < a_1$. Consequently, a series of expansion waves tends to spread further apart with time. However, there is a temperature rise behind each compression wave. Therefore, the next wave in the series propagates with a higher speed, such that $a_3 > a_2 > a_1$, and the spacing between a series of compression waves tends to diminish with time. As compression waves coalesce, a single finite-magnitude *shock* wave is formed. Flow through a finite shock wave is no longer differentially different from the original state, and is irreversible ($ds > 0$). A similar discontinuous transition was observed in the hydraulic jump of open channel incompressible flows, discussed in Chapter 18. Shock waves and hydraulic jumps are both examples of transitions permitted by multiple solutions to nonlinear governing equations.

Recalling the sound wave continuity equation (20-16), a relation between the change in velocity and density in the wake of the wave can be established:

$$\rho\, dv + (\pm a)d\rho = 0 \rightarrow \frac{dv}{a} \pm \frac{d\rho}{\rho} = 0. \tag{20-24}$$

Sound can propagate in two directions relative to the flow, as indicated with $\pm a$. A sound wave traveling in the same direction as v utilizes the $(-)$ sign (see Figure 20-1) and the $(+)$ sign denotes a sound wave traveling in the opposite direction. For an ideal gas, changes in the speed of sound are regulated by changes in temperature of the flow by

$$\frac{d(a^2)}{a^2} = \frac{d(\gamma RT)}{\gamma RT} \rightarrow 2\frac{da}{a} = \frac{dT}{T}. \tag{20-25}$$

Furthermore, since sound waves propagate isentropically, from the entropy relation $ds = C_v dT/T - Rd\rho/\rho$ for an ideal gas it is found that

$$ds = 0 \rightarrow \frac{dT}{T} = (\gamma - 1)\frac{d\rho}{\rho}. \tag{20-26}$$

Combining Eqs. (20-25) and (20-26) yields the result

$$\frac{d\rho}{\rho} = \frac{2}{\gamma - 1}\frac{da}{a}, \tag{20-27}$$

which applied to the sound wave continuity equation (20-24) yields

$$dv \pm \frac{2}{\gamma - 1} da = 0 \rightarrow \begin{cases} v + \dfrac{2}{\gamma - 1}a = C_1 & v \text{ and } a \text{ in opposing directions} \quad \text{(20-28a)} \\ v - \dfrac{2}{\gamma - 1}a = C_2 & v \text{ and } a \text{ in the same direction.} \quad \text{(20-28b)} \end{cases}$$

The constants C_1 and C_2 associated with the integrated form of Eq. (20-28) are known as *Riemann variables* [3]. Equation (20-28) is a consequence of enforcing the differential forms of the continuity and momentum equations for a wave in an ideal gas. (The momentum equation is introduced in the current analysis through the speed of sound relation for an ideal gas.) Energy is conserved implicitly by the isentropic constraint. Therefore, Eq. (20-28) is useful for relating v and a over a finite distance only if the flow is isentropic. The expansion wave illustrated in Figure 20-2(a) shows that sound waves are propagating in the reverse direction of the flow. Therefore, Eq. (20-28a) dictates that v and a are related over the extent of the expansion wave by

$$v + 2a/(\gamma - 1) = C_1 \quad \text{(expansion wave)}. \tag{20-29}$$

In contrast, compression waves coalescing into a finite irreversible shock wave cannot be described by the result of Eq. (20-28b). Instead, the shock wave must be analyzed with conservation of energy providing an additional constraint to the momentum and continuity equations for an irreversible flow. Additionally, the conservation equations should be evaluated in integral form, since the nonconservative differential forms are invalid for describing discontinuous change.

20.5 SHOCK WAVE (NORMAL TO FLOW)

Figure 20-3 illustrates the control surface surrounding a normal shock that will be used for the following analysis. A normal shock has a surface perpendicular to the flow (in contrast to an oblique shock). Surfaces A and B surrounding the shock are separated by a vanishingly small distance; the volume encompassed by the control surface can be taken as zero. Let the frame of reference for the control surfaces be fixed to the shock wave. The continuity equation, momentum equation, and energy equation can be written in integral form to evaluate changes in the flow across the shock wave:

$$\text{Continuity:}\ \ 0 = \int_A -v_j(\rho)n_j dA \quad \rightarrow \quad [v\rho A]_A = [v\rho A]_B \tag{20-30}$$

$$\text{Momentum:}\ \ 0 = \int_A \left[-v_j(\rho v_i) - P\delta_{ji}\right] n_j dA \ \rightarrow\ [(v\rho v + P)A]_A = [(v\rho v + P)A]_B \tag{20-31}$$

$$\text{Energy:}\ \ 0 = \int_A \left[-v_j(\rho e) - P\delta_{ji} v_i\right] n_j dA \ \rightarrow\ [(e + P/\rho)v\rho A]_A = [(e + P/\rho)v\rho A]_B \tag{20-32}$$

In the energy equation, $e = u + v^2/2$ accounts for the internal energy and the kinetic energy of the flow. Unlike the situation considered for a sound wave in Section 20.4, the energy equation is indispensable in the description of an irreversible flow. The shock wave equations use the subscript "A" to reference the state of the flow crossing surface

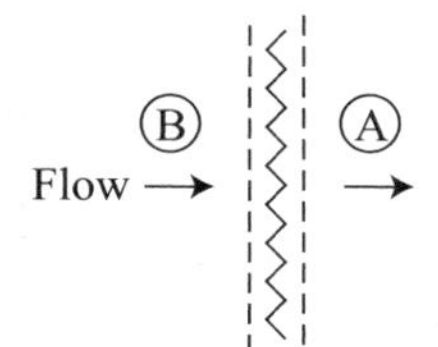

Figure 20-3 Control surfaces for normal shock analysis.

A (after the wave), and subscript "B" to reference the state of the flow crossing surface B (before the wave), as shown in Figure 20-3. Velocities are measured with respect to coordinates fixed to the shock wave. The normal shock equations are simplified by factoring out the common area $A_A = A_B$ from the continuity and momentum equations, and by factoring out the mass flux $[\upsilon\rho A]_A = [\upsilon\rho A]_B$ from the energy equation:

$$\text{Continuity:} \quad [\upsilon\rho]_A = [\upsilon\rho]_B \tag{20-33}$$

$$\text{Momentum:} \quad [\rho\upsilon^2 + P]_A = [\rho\upsilon^2 + P]_B \tag{20-34}$$

$$\text{Energy:} \quad \left[h + \upsilon^2/2\right]_A = \left[h + \upsilon^2/2\right]_B. \tag{20-35}$$

Notice that a change in variables $(e + P/\rho) \rightarrow (h + \upsilon^2/2)$ has been utilized in the energy equation (20-35). To solve these three equations, they will be expressed in terms of three unknowns T, P, and M. With velocities expressed as the product of the Mach number and the speed of sound, $\upsilon^2 = \mathrm{M}^2 a^2 = \mathrm{M}^2\gamma RT$, the energy equation (20-35) for an ideal gas can be written as

$$C_p T_A + \mathrm{M}_A^2 \gamma R T_A/2 = C_p T_B + \mathrm{M}_B^2 \gamma R T_B/2. \tag{20-36}$$

Using $R = C_p - C_\upsilon$ and $\gamma = C_p/C_\upsilon$, the energy equation can be rewritten as

$$\text{Energy:} \quad T_A\left(1 + \mathrm{M}_A^2(\gamma - 1)/2\right) = T_B\left(1 + \mathrm{M}_B^2(\gamma - 1)/2\right). \tag{20-37}$$

Similarly, with $\upsilon^2 = \mathrm{M}^2\gamma RT$ and $P = \rho RT$ the momentum equation (20-34) and continuity equation (20-33) for an ideal gas become

$$\text{Momentum:} \quad P_A(\mathrm{M}_A^2\gamma + 1) = P_B(\mathrm{M}_B^2\gamma + 1) \tag{20-38}$$

$$\text{Continuity:} \quad (\mathrm{M}_A/\mathrm{M}_B)^2 (P_A/P_B)^2 = T_A/T_B. \tag{20-39}$$

Using the energy equation (20-37) and the momentum equation (20-38) to eliminate T_A/T_B and P_B/P_A from the continuity equation (20-39) yields

$$\left(\frac{\mathrm{M}_A}{\mathrm{M}_B}\right)^2 \left(\frac{\mathrm{M}_B^2\gamma + 1}{\mathrm{M}_A^2\gamma + 1}\right)^2 = \frac{1 + \mathrm{M}_B^2(\gamma - 1)/2}{1 + \mathrm{M}_A^2(\gamma - 1)/2}. \tag{20-40}$$

This result can be solved for the Mach number after the normal shock:

$$\mathrm{M}_A^2 = \frac{\mathrm{M}_B^2 + \dfrac{2}{\gamma - 1}}{\dfrac{2\gamma}{\gamma - 1}\mathrm{M}_B^2 - 1}. \tag{20-41}$$

Once Eq. (20-41) is used to evaluate the flow Mach number following a shock, the energy equation (20-37) can be revisited to determine the change in temperature across the shock, and the momentum equation (20-38) can be revisited to determine the change in pressure.

20.6 SHOCK TUBE ANALYTIC DESCRIPTION

The shock tube problem provides a good illustration of a transient one-dimensional gas dynamic problem that can be solved analytically. A shock tube is comprised of an enclosure that is partitioned into a high-pressure region and a low-pressure region, as illustrated in Figure 20-4. At $t = 0$, the partition is ruptured, and the high-pressure gas moves into the low-pressure region with two resulting waves. One is a shock wave moving to the right into the low-pressure gas. The second is an expansion wave moving to the left into the high-pressure gas. Both waves originate from a plane of contact between the two pressure regions; the plane of contact exists for all time (since the two gases cannot comingle without diffusion), although it is pushed to the right by the high-pressure gas. The expansion wave is spreading as it advances to the left. In contrast, the shock wave remains as a discrete plane moving to the right. The two waves and contact plane define four regions in the shock tube, as labeled in Figure 20-4. Region 1 is the initial low-pressure gas state in advance of the shock. Region 2 is the gas state behind the shock wave. Region 4 is the initial high-pressure gas state in advance of the expansion wave. Region 3 is the gas state behind the expansion wave. Regions 2 and 3 are separated by the moving contact plane. The pressure and speed of the gases on both sides of the contact plane must be balanced. However, there is a temperature jump across the contact plane that results from the compression of gas on one side caused by the expansion of gas on the other side. The process by which heat could be transferred between regions 2 and 3 is diffusion, which has been disallowed in the present analysis.

Figure 20-5 summarizes the flow variables in the vicinity of the shock with respect to coordinates moving with the wave. The conservation laws across the shock wave (continuity, momentum, and energy) given by Eqs. (20-33) through (20-35) are evaluated to yield

Shock wave equations:

$$\text{Continuity:} \quad [(\upsilon_2 - \upsilon_s)\rho_2]_A = [(-\upsilon_s)\rho_1]_B \tag{20-42}$$

$$\text{Momentum:} \quad [\rho_2(\upsilon_2 - \upsilon_s)^2 + P_2]_A = [\rho_1(-\upsilon_s)^2 + P_1]_B \tag{20-43}$$

$$\text{Energy:} \quad \left[C_pT_2 + (\upsilon_2 - \upsilon_s)^2/2\right]_A = \left[C_pT_1 + (-\upsilon_s)^2/2\right]_B \tag{20-44}$$

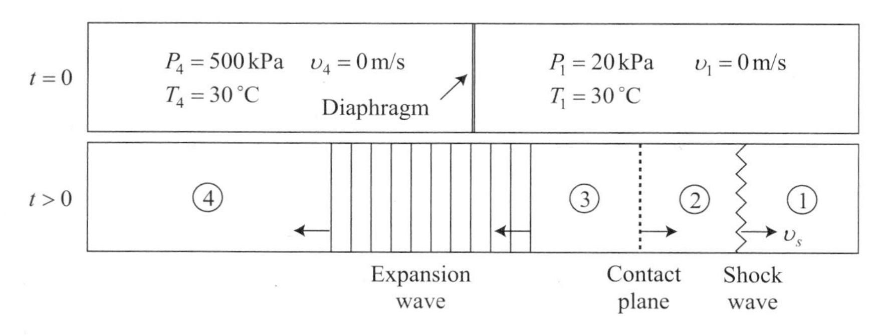

Figure 20-4 Shock tube illustration.

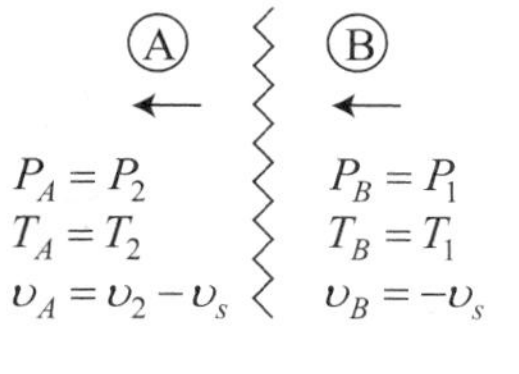

A | B

$P_A = P_2$ $P_B = P_1$

$T_A = T_2$ $T_B = T_1$

$\upsilon_A = \upsilon_2 - \upsilon_s$ $\upsilon_B = -\upsilon_s$

Figure 20-5 Variables across shock moving into low-pressure gas.

Analysis of the flow in the vicinity of the contact plane and across the expansion wave is needed to relate the shock equations to the conditions at the far left end of the shock tube:

Contact plane:

$$P_3 = P_2 \tag{20-45}$$

$$v_3 = v_2 \tag{20-46}$$

Expansion wave equations:

$$ds = 0: \quad T_3/T_4 = (P_3/P_4)^{(\gamma-1)/\gamma} \tag{20-47}$$

$$v + 2a/(\gamma - 1) = C_1: \quad v_4 + 2a_4/(\gamma - 1) = v_3 + 2a_3/(\gamma - 1) \tag{20-48}$$

The first expansion wave equation (20-47) is simply a consequence of isentropic flow, for which $ds = C_p dT/T - RdP/P = 0$ (see Table 20-1) may be integrated between states 3 and 4, assuming a calorically perfect gas. The second equation (20-48) was derived in Section (20.5) from the continuity and momentum equations for an isentropic expansion wave propagating through an ideal gas.

The unknowns in this problem are v_2, P_2, T_2, v_3, P_3, T_3, and v_s. Not included in the list of unknowns are densities $\rho = P/(RT)$ and speeds of sound $a = \sqrt{\gamma RT}$, since both can be expressed with ideal gas relations in terms of P and T. Since there are seven unknowns and seven equations, (20-42) through (20-48), the remaining steps in solving this problem are "simply" algebraic.

The energy equation (20-44) can be rewritten as

$$C_p(T_2 - T_1) = \frac{a_2^2 - a_1^2}{\gamma - 1} = v_s v_2 - \frac{v_2^2}{2}. \tag{20-49}$$

Then the momentum equation (20-43) is written in the form

$$P_1 - P_2 = \rho_2 (v_s - v_2)^2 - \rho_1 v_s^2 \tag{20-50}$$

and combined with continuity (20-42) to yield

$$P_1 - P_2 = \rho_1 v_s (v_s - v_2) - \rho_1 v_s^2 = -\rho_1 v_s v_2. \tag{20-51}$$

With pressures expressed as $P = \rho RT = \rho a^2/\gamma$, Eq. (20-51) becomes

$$\rho_1 a_1^2 - \rho_2 a_2^2 = -\gamma \rho_1 v_s v_2. \tag{20-52}$$

Or, using continuity (20-42) to eliminate $\rho_2 = \rho_1 v_s/(v_s - v_2)$, yields

$$a_1^2 - \frac{v_s a_2^2}{v_s - v_2} = -\gamma v_s v_2. \tag{20-53}$$

Combining this last result with the energy equation (20-49) to eliminate a_2 yields

$$v_2 = \frac{2a_1}{\gamma + 1}\left(\frac{v_s}{a_1} - \frac{a_1}{v_s}\right). \tag{20-54}$$

Upon substitution of this expression for v_2 into Eq. (20-51) yields

$$P_2 - P_1 = \frac{2\rho_1 v_s}{\gamma+1}\left(v_s - \frac{a_1^2}{v_s}\right) \quad \text{or} \quad \frac{P_2}{P_1} - 1 = \frac{2}{\gamma+1}\frac{v_s}{RT_1}\left(v_s - \frac{a_1^2}{v_s}\right) = \frac{2\gamma}{\gamma+1}\left[\left(\frac{v_s}{a_1}\right)^2 - 1\right]. \tag{20-55}$$

Solving this last expression for v_s/a_1 yields

$$\frac{v_s}{a_1} = \sqrt{\frac{\gamma+1}{2\gamma}\left(\frac{P_2}{P_1} - 1\right) + 1}. \tag{20-56}$$

Substituting Eq. (20-56) back into Eq. (20-54) yields

$$\frac{v_2}{a_1} = \frac{\dfrac{P_2}{P_1} - 1}{\gamma\sqrt{\dfrac{\gamma+1}{2\gamma}\left(\dfrac{P_2}{P_1} - 1\right) + 1}}. \tag{20-57}$$

With $v_4 = 0$, $v_3 = v_2$, and $a_4 = a_1$, the second expansion wave equation (20-48) may be written in the form

$$\frac{v_2}{a_1} = \frac{2}{\gamma-1}\left(1 - \frac{a_3}{a_4}\right) \quad \text{or} \quad \frac{v_2}{a_1} = \frac{2}{\gamma-1}\left[1 - \left(\frac{P_3}{P_4}\right)^{\frac{\gamma-1}{2\gamma}}\right]. \tag{20-58}$$

The latter form makes use of the first expansion wave equation (20-47). Combining this last result with Eq. (20-57) yields

$$\frac{2}{\gamma-1}\left[1 - \left(\frac{P_3}{P_4}\right)^{\frac{\gamma-1}{2\gamma}}\right] = \frac{\dfrac{P_2}{P_1} - 1}{\gamma\sqrt{\dfrac{\gamma+1}{2\gamma}\left(\dfrac{P_2}{P_1} - 1\right) + 1}}. \tag{20-59}$$

With $P_3 = P_2$, this can be rearranged for the final result:

$$\frac{P_1}{P_4} = \frac{P_1}{P_2}\left[1 - \frac{\dfrac{\gamma-1}{2}\left(\dfrac{P_2}{P_1} - 1\right)}{\gamma\sqrt{\dfrac{\gamma+1}{2\gamma}\left(\dfrac{P_2}{P_1} - 1\right) + 1}}\right]^{\frac{2\gamma}{\gamma-1}}. \tag{20-60}$$

For a given initial pressure ratio P_1/P_4, this equation can be solved for the pressure ratio across the shock P_2/P_1.

To illustrate some results of this analysis, consider a case in which $P_1/P_4 = 20/500$, with $R = 287.0$ J/kg/K and $\gamma = 1.4$ (air). The solution to Eq. (20-60) yields $P_2/P_1 = 4.047$. From Eq. (20-57) it is found that $v_2/a_1 = 1.145$, and from Eq. (20-56) it is found that $v_s/a_1 = 1.90$. Therefore, $v_s = (v_s/a_1)\sqrt{\gamma RT_1} = 663.3$ m/s and $v_2 = (v_2/a_1)\sqrt{\gamma RT_1} = 399.7$ m/s. From Eq. (20-49) it is determined that $a_2 = \sqrt{(\gamma-1)(v_s v_2 - v_2^2/2) + a_1^2} = 442.6$ m/s.

Since $a_2 = \sqrt{\gamma R T_2}$, it can be concluded that $T_2 = 487.5$ K $= 214.4$ °C. Furthermore, from Eq. (20-47) with $P_3 = P_2 = 4.047 P_1$, it is found that $T_3 = 180.2$ K $= -92.97$ °C.

Although the conditions internal to the expansion wave change as a function of time, the leading edge of the expansion wave has a constant velocity of $v_4 - a_4 = 0 - \sqrt{\gamma R T_4} = -349.0$ m/s. The trailing edge of the expansion wave also has a constant velocity equal to $v_3 - a_3$. With $a_3 = \sqrt{\gamma R T_3} = 269.1$ m/s, the expansion wave trailing edge velocity is found to be $v_3 - a_3 = +130.6$ m/s.

20.7 SHOCK TUBE NUMERICAL DESCRIPTION

Although the shock tube flow has been solved analytically in the previous section, it is useful to seek a numerical solution that can broaden the shock tube investigation for more complex situations. The governing equations (20-7) through (20-9) can be rewritten in one-dimensional form for application to the inviscid shock tube problem as follows:

$$\text{Continuity:} \quad \frac{\partial}{\partial t}(\rho) + \frac{\partial}{\partial x}[\rho v] = 0 \tag{20-61}$$

$$\text{Momentum:} \quad \frac{\partial}{\partial t}(\rho v) + \frac{\partial}{\partial x}[(\rho v)v + P] = 0 \tag{20-62}$$

$$\text{Energy:} \quad \frac{\partial}{\partial t}(\rho e) + \frac{\partial}{\partial x}[(\rho e + P)v] = 0 \tag{20-63}$$

The x-direction velocity component is denoted by $v_x \to v$ for notational simplicity. Additionally, the energy equation has been expressed in terms of the specific energy $e = (u + v^2/2)$. If the fluid is assumed to be an ideal gas, where $P = \rho R T$, and calorically perfect, such that C_p and C_v are constant, then $u = C_v T$, and the specific energy can be evaluated from

$$e = \frac{RT}{\gamma - 1} + \frac{v^2}{2}, \tag{20-64}$$

where $C_v = R/(\gamma - 1)$.

Equations (20-61) through (20-63) can be solved for the flow variables ρ, ρv, and ρe, from which

$$v = (\rho v)/\rho \quad \text{and} \quad P = (\gamma - 1)[\rho e - (\rho v)^2/(2\rho)] \tag{20-65}$$

are calculated.

The initial conditions in the shock tube are

$$t = 0: \quad x \leq L/2: \quad \begin{cases} \rho = P_4/(RT_4) \\ \rho v = 0 \\ \rho e = P_4/(\gamma - 1) \end{cases} \quad \text{and} \quad x > L/2: \quad \begin{cases} \rho = P_1/(RT_1) \\ \rho v = 0 \\ \rho e = P_1/(\gamma - 1) \end{cases} \tag{20-66}$$

and the boundary conditions are

$$v(t, x = 0) = 0 \quad \text{and} \quad v(t, x = L) = 0. \tag{20-67}$$

The equations (20-61) through (20-63) governing the gas dynamics in the shock tube can be integrated with the MacCormack scheme described in Chapter 17. To dampen numerical oscillations, diffusion terms are added to the finite differencing equations. The predictor step calculations for the MacCormack scheme are given by

$$\{(\rho)_n^{t+\Delta t}\}_p = (\rho)_n^t - \frac{\Delta t}{\Delta x}\left[(\rho v)_{n+1}^t - (\rho v)_n^t\right] + \frac{\nu_a \Delta t}{\Delta x^2}\left[\rho_{n+1}^t - 2\rho_n^t + \rho_{n-1}^t\right] \leftarrow \textit{for stability} \tag{20-68}$$

$$\{(\rho v)_n^{t+\Delta t}\}_p = (\rho v)_n^t - \frac{\Delta t}{\Delta x}\left[\left((\rho v)_{n+1}^t v_{n+1}^t + P_{n+1}^t\right) - \left((\rho v)_n^t v_n^t + P_n^t\right)\right] + \frac{\mu_a \Delta t}{\Delta x^2}\left[(\rho v)_{n+1}^t - 2(\rho v)_n^t + (\rho v)_{n-1}^t\right] \leftarrow \textit{for stability} \tag{20-69}$$

$$\{(\rho e)_n^{t+\Delta t}\}_p = (\rho e)_n^t - \frac{\Delta t}{\Delta x}\left[\left((\rho e)_{n+1}^t + P_{n+1}^t\right)v_{n+1}^t - \left((\rho e)_n^t + P_n^t\right)v_n^t\right] + \frac{\kappa_a \Delta t}{\Delta x^2}\left[(\rho e)_{n+1}^t - 2(\rho e)_n^t + (\rho e)_{n-1}^t\right] \leftarrow \textit{for stability} \tag{20-70}$$

Predictor step values for velocity $\{v_n^{t+\Delta t}\}_p$ and pressure $\{P_n^{t+\Delta t}\}_p$ are obtained from $\{\rho_n^{t+\Delta t}\}_p$, $\{(\rho v)_n^{t+\Delta t}\}_p$, and $\{(\rho e)_n^{t+\Delta t}\}_p$ using

$$\{v_n^{t+\Delta t}\}_p = \frac{\{(\rho v)_n^{t+\Delta t}\}_p}{\{\rho_n^{t+\Delta t}\}_p} \quad \text{and} \quad \{P_n^{t+\Delta t}\}_p = (\gamma - 1)\left[\{(\rho e)_n^{t+\Delta t}\}_p - \frac{\left(\{(\rho v)_n^{t+\Delta t}\}_p\right)^2}{2\{\rho_n^{t+\Delta t}\}_p}\right]. \tag{20-71}$$

The corrector step calculations for the MacCormack scheme are given by

$$\{(\rho)_n^{t+\Delta t}\}_c = (\rho)_n^t - \frac{\Delta t}{\Delta x}\left[\{(\rho v)_n^{t+\Delta t}\}_p - \{(\rho v)_{n-1}^{t+\Delta t}\}_p\right] + \frac{\nu_a \Delta t}{\Delta x^2}\left[\{\rho_{n+1}^{t+\Delta t}\}_p - 2\{\rho_n^{t+\Delta t}\}_p + \{\rho_{n-1}^{t+\Delta t}\}_p\right] \leftarrow \textit{for stability} \tag{20-72}$$

$$\begin{aligned}\{(\rho v)_n^{t+\Delta t}\}_c = (\rho v)_n^t &- \frac{\Delta t}{\Delta x}\Big[\left(\{(\rho v)_n^{t+\Delta t}\}_p\{v_n^{t+\Delta t}\}_p + \{P_n^{t+\Delta t}\}_p\right) \\ &- \left(\{(\rho v)_{n-1}^{t+\Delta t}\}_p\{v_{n-1}^{t+\Delta t}\}_p + \{P_{n-1}^{t+\Delta t}\}_p\right)\Big] \\ &+ \frac{\mu_a \Delta t}{\Delta x^2}\left[\{(\rho v)_{n+1}^{t+\Delta t}\}_p - 2\{(\rho v)_n^{t+\Delta t}\}_p + \{(\rho v)_{n-1}^{t+\Delta t}\}_p\right] \leftarrow \textit{for stability}\end{aligned} \tag{20-73}$$

$$\{(\rho e)_n^{t+\Delta t}\}_c = (\rho e)_n^t - \frac{\Delta t}{\Delta x}\Big[\Big(\{(\rho e)_n^{t+\Delta t}\}_p + \{P_n^{t+\Delta t}\}_p\Big)\{v_n^{t+\Delta t}\}_p$$
$$-\Big(\{(\rho e)_{n-1}^{t+\Delta t}\}_p + \{P_{n-1}^{t+\Delta t}\}_p\Big)\{v_{n-1}^{t+\Delta t}\}_p\Big]$$
$$+\frac{\kappa_a \Delta t}{\Delta x^2}\Big[\{(\rho e)_{n+1}^{t+\Delta t}\}_p - 2\{(\rho e)_n^{t+\Delta t}\}_p + \{(\rho e)_{n-1}^{t+\Delta t}\}_p\Big] \leftarrow \textit{for stability} \quad (20\text{-}74)$$

Corrector step values for velocity $\{v_n^{t+\Delta t}\}_c$ and pressure $\{P_n^{t+\Delta t}\}_c$ are obtained from $\{\rho_n^{t+\Delta t}\}_c$, $\{(\rho v)_n^{t+\Delta t}\}_c$, and $\{(\rho e)_n^{t+\Delta t}\}_c$ using

$$\{v_n^{t+\Delta t}\}_c = \frac{\{(\rho v)_n^{t+\Delta t}\}_c}{\{\rho_n^{t+\Delta t}\}_c} \quad \text{and} \quad \{P_n^{t+\Delta t}\}_c = (\gamma - 1)\left[\{(\rho e)_n^{t+\Delta t}\}_c - \frac{\Big(\{(\rho v)_n^{t+\Delta t}\}_c\Big)^2}{2\{\rho_n^{t+\Delta t}\}_c}\right]. \quad (20\text{-}75)$$

As was noted in Chapter 17, the direction of finite differencing may need to be reversed for either the predictor or corrector step at boundary nodes to prevent the operation from extending beyond the physical domain. For the same reason, the diffusion terms, added for numerical stability, are not included at the boundary nodes.

In implementing the MacCormack scheme, the predictor step and corrector step values are averaged to obtain the end of the time step values:

$$\rho_n^{t+\Delta t} = \frac{\{\rho_n^{t+\Delta t}\}_p + \{\rho_n^{t+\Delta t}\}_c}{2} \quad (20\text{-}76)$$

$$(\rho v)_n^{t+\Delta t} = \frac{\{(\rho v)_n^{t+\Delta t}\}_p + \{(\rho v)_n^{t+\Delta t}\}_c}{2} \quad (20\text{-}77)$$

$$(\rho e)_n^{t+\Delta t} = \frac{\{(\rho e)_n^{t+\Delta t}\}_p + \{(\rho e)_n^{t+\Delta t}\}_c}{2}. \quad (20\text{-}78)$$

Code 20-1 implements the numerical integration of Eq. (20-61) through (20-63), as described above. Since the MacCormack scheme is explicit, stability of integration requires that the time step be bounded by the Courant condition, as discussed in Section 17.2.1. As a practical approach, each time step can be selected to achieve a specified target Courant number:

$$\frac{\Delta t(|v| + a)}{\Delta x} \approx C_{targ} = 0.9. \quad (20\text{-}79)$$

The artificial diffusivities added to the transport equations for numerical stability are dynamically selected such that

$$\nu_a = \mu_a = \kappa_a = \frac{0.04}{\Delta t/\Delta x^2}. \quad (20\text{-}80)$$

As discussed in Section 17.2.2, the artificial diffusivities assigned by Eq. (20-80) yield a magnitude of diffusion that is at most $\sim$4% of the magnitude of the advection term in the transport equations.

With $L = 15$ m, the governing equations, as described with the MacCormack scheme, are integrated to $t = 8$ ms for comparison with the analytical results determined in

Code 20-1 Transient shock tube

```
#include <stdio.h>
#include <string.h>
#include <math.h>

inline double enrgy_diff(char t,int n,double *d_e,double *v,double *p) {
   return t=='p' ?  (d_e[n+1] + p[n+1])*v[n+1] - (d_e[n] + p[n])*v[n] :
       /* t=='c' */ (d_e[n] + p[n])*v[n] - (d_e[n-1] + p[n-1])*v[n-1] ; }

inline double mom_diff(char t,int n,double *d_v,double *v,double *p) {
   return t=='p' ?  (d_v[n+1]*v[n+1] + p[n+1]) - (d_v[n]*v[n] + p[n]) :
       /* t=='c' */ (d_v[n]*v[n] + p[n]) - (d_v[n-1]*v[n-1] + p[n-1]) ; }

inline double cont_diff(char t,int n,double *d,double *v) {
   return t=='p' ?  d[n+1]*v[n+1] - d[n]*v[n] :
       /* t=='c' */ d[n]*v[n] - d[n-1]*v[n-1] ;          }

double MacC(int N,double *d,double *d_v,double *d_e,double *p,
        double *v,double *dt_,double dx,double gam)
{
   int n,step;
   double dt=*dt_,Pd[N],Cd[N],Pd_v[N],Cd_v[N],Pd_e[N],Cd_e[N];
   double Pv[N],Cv[N],Pp[N],Cp[N];
   double C,Cmax=0,Ctarg=0.9,dtdx=dt/dx,dtddx=dt/dx/dx,na,ma,ka;
   na=ma=ka=0.04/dtddx;

   for (n=0;n<N;++n) { // perform predictor step
      step=(n<N-1 ? 'p':'c');
      Pd[n]=d[n]-dtdx*cont_diff(step,n,d,v);
      if (n>0 && n<N-1) Pd[n]+=dtddx*na*(d[n+1]-2.*d[n]+d[n-1]);
      Pd_v[n]=d_v[n]-dtdx*mom_diff(step,n,d_v,v,p);
      if (n>0 && n<N-1) Pd_v[n]+=dtddx*ma*(d_v[n+1]-2.*d_v[n]+d_v[n-1]);
      Pv[n]=Pd_v[n]/Pd[n];
      Pd_e[n]=d_e[n]-dtdx*enrgy_diff(step,n,d_e,v,p);
      if (n>0 && n<N-1) Pd_e[n]+=dtddx*ka*(d_e[n+1]-2.*d_e[n]+d_e[n-1]);
      Pp[n]=(gam-1.)*(Pd_e[n]-Pd_v[n]*Pd_v[n]/2./Pd[n]);
      C=dt*(fabs(Pv[n])+sqrt(gam*Pp[n]/Pd[n]))/dx; // dt size checking
      if (C>Cmax) Cmax=C;
   }
   if (Cmax>1.) { // integrate with a smaller time step
      *dt_=0.;
      return dt*Ctarg/Cmax;
   }
   for (n=0;n<N;++n) { // perform corrector step
      step=(n>0 ? 'c':'p');
      Cd[n]=d[n]-dtdx*cont_diff(step,n,Pd,Pv);
      if (n>0 && n<N-1) Cd[n]+=dtddx*na*(Pd[n+1]-2.*Pd[n]+Pd[n-1]);
      Cd_v[n]=d_v[n]-dtdx*mom_diff(step,n,Pd_v,Pv,Pp);
      if (n>0 && n<N-1) Cd_v[n]+=dtddx*ma*(Pd_v[n+1]-2.*Pd_v[n]+Pd_v[n-1]);
      Cv[n]=Cd_v[n]/Cd[n];
      Cd_e[n]=d_e[n]-dtdx*enrgy_diff(step,n,Pd_e,Pv,Pp);
      if (n>0 && n<N-1) Cd_e[n]+=dtddx*ka*(Pd_e[n+1]-2.*Pd_e[n]+Pd_e[n-1]);
      Cp[n]=(gam-1.)*(Cd_e[n]-Cd_v[n]*Cd_v[n]/2./Cd[n]);
      if (isnan(Cp[n])) { printf("\nsoln BLEW!\n"); return 0; }
   }
   for (n=0;n<N;++n) { // average predictor and corrector results
      d[n]=0.5*(Pd[n]+Cd[n]);
      d_v[n]=0.5*(Pd_v[n]+Cd_v[n]);
      v[n]=0.5*(Pv[n]+Cv[n]);
      d_e[n]=0.5*(Pd_e[n]+Cd_e[n]);
      p[n]=0.5*(Pp[n]+Cp[n]);
   }
   v[0]=d_v[0]=v[N-1]=d_v[N-1]=0.; // enforce BC
   return dt*Ctarg/Cmax;
}

int main(void)
{
   int n,cnt=0,N=1501;
   double den[N],den_vel[N],den_enrgy[N],pres[N],vel[N];
   double T1_4=303.15,P4=500000.0,P1=20000.0,R=287.0,gam=1.4;
   double L=15.0,dx=L/(N-1),t=0.,dt,dt_next=5.0e-6,tend=0.008;
   FILE *fp;

   for (n=0;n<(N-1)/2;++n) { // initial distributions through tube
      den[n]=((pres[n]=P4)/T1_4)/R;
      den_enrgy[n]=P4/(gam-1);
      den_vel[n]=vel[n]=0.0;
   }
   den[n]=((pres[n]=(P4+P1)/.2)/T1_4)/R;
   den_enrgy[n]=pres[n]/(gam-1);
   den_vel[n]=vel[n]=0.0;
   for ( ;n<N;++n) {
      den[n]=((pres[n]=P1)/T1_4)/R;
      den_vel[n]=vel[n]=0.0;
      den_enrgy[n]=P1/(gam-1);
   }

   do {
      if (t+(dt=dt_next)>tend) dt=tend-t;
      dt_next=MacC(N,den,den_vel,den_enrgy,pres,vel,&dt,dx,gam);
      if ( !dt_next ) break;
      t+=dt;
      if ( !(++cnt%100) ) printf("\ncnt=%d t=%f dt=%e\n",cnt,t,dt_next);
   } while (t<tend);

   fp=fopen("out.dat","w");
   for (n=0;n<N;++n)
      fprintf(fp,"%e %e %e %e %e\n",
              n*dx,den[n],vel[n],pres[n],pres[n]/den[n]/R);

   fclose(fp);
   return 1;
}
```

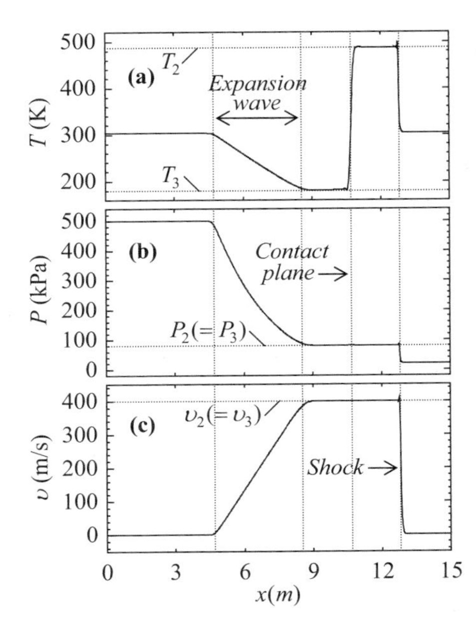

Figure 20-6 Gas conditions in shock tube at $t = 8\,\text{ms}$.

Section 20.7. Figure 20-6 shows the numerical results for temperature, pressure, and velocity of the gas in the shock tube. Superimposed on the figure are the analytical results for T_2, T_3, $P_2(= P_3)$, and $v_2(= v_3)$. The numerical results compare very well with these analytic results.

Additionally, the location of the expansion wave, contact plane, and shock wave can be calculated from flow speeds and wave speeds determined from the analytic solution. These locations in the flow are also identified at 8 ms in Figure 20-6. The expansion wave is slightly broadened in the numerical results, which is caused by the diffusion terms added to the governing equations for stability. For the same reason, the discontinuity in temperature at the contact plane, and jump in properties across the shockwave, are also slightly broadened.

After longer times, the shock wave reflects off the right-hand side of the tube and advances back into the expansion wave. The resulting interactions of waves is too complex to describe analytically, but may be explored numerically with Code 20-1.

20.8 SHOCK TUBE PROBLEM WITH DISSIMILAR GASES

Consider a shock tube with a high-pressure gas of helium and low-pressure gas of argon. The gas constants for helium R_{He} and argon R_{Ar} are different, but their ratio of specific heats are the same, $\gamma_{He} = \gamma_{Ar}$. The gas constants are needed to evaluate pressure $P = \rho RT$ and energy $e = RT/(\gamma - 1) + v^2/2$ in the transport equations. However, without diffusion or temperature dependence, the gas constants are properties of the fluid that remain constant for each material element in the flow. Therefore, enforcing the transport equation $DR/Dt = 0$, or $\partial R/\partial t + v(\partial R/\partial x) = 0$, allows the ideal gas constant $R(x,t)$ to be determined as part of the shock tube solution. Utilizing the continuity equation, the transport equation for $R(x,t)$ can be written in conservation form:

$$\text{Gas constant:} \quad \frac{\partial}{\partial t}(\rho R) + \frac{\partial}{\partial x}(\rho R v) = 0. \tag{20-81}$$

Equations (20-61) through (20-63) and Eq. (20-81) can now be solved for the flow variables ρ, ρv, ρe, and ρR, as addressed in Problem 20-11.

20.9 PROBLEMS

20-1 What physical effect that is important at high speeds may be unimportant at low speeds? What is the correct dimensionless number to consult as to whether this high-speed effect may be important for a given flow? Justify the physical significance of this dimensionless number.

20-2 Derive the energy equation to have the form of Eq. (20-3), expressed in terms of the stagnation enthalpy $h_o = h + v^2/2$. See Section 5.8.3 for the derivation of the energy equation in terms of $e = (u + v^2/2)$.

20-3 Under a certain operating condition, the piston speed in an auto engine is 10 m/s. Approximate engine "knock" as the occurrence of a normal shock wave traveling downward, into the unburned mixture at 700 kPa and 500 K. Determine the pressure acting on the piston face after the shock reflects from it. Assume that the gas has the properties of air and acts as a perfect gas, with $\gamma = 1.4$.

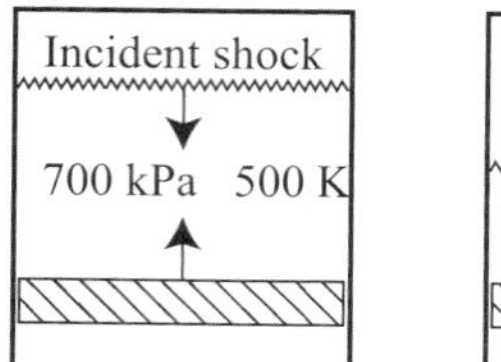

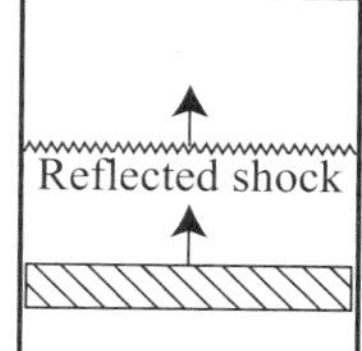

20-4 Water vapor passes through a normal shock. After the shock, the water is saturated ($x_2 = 1$) at a pressure $P_2 = 320$ kPa and is traveling at a velocity of $v_2 = 345$ m/s. Is the water approaching the shock in a two-phase or superheated state? Find the velocity and thermodynamic state (two independent thermodynamic properties) of the water before the shock.

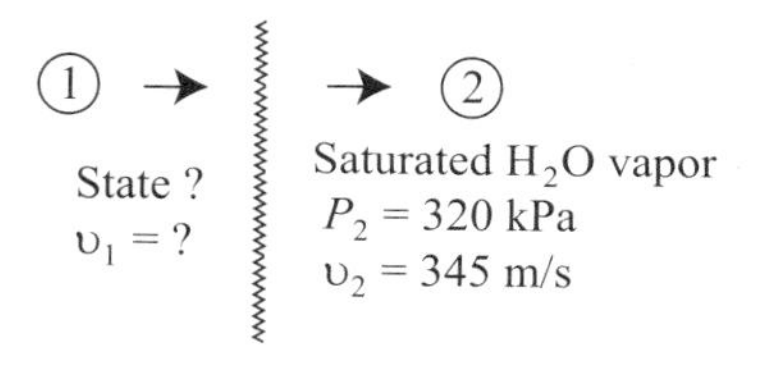

20-5 Consider the shock tube problem discussed in Section 20.7, for the same initial condition ($P_1/P_4 = 20/500$, $T_1 = T_4 = 30\,°C$). Analytically determine the air temperature at the right-hand wall immediately after the shock wave is reflected.

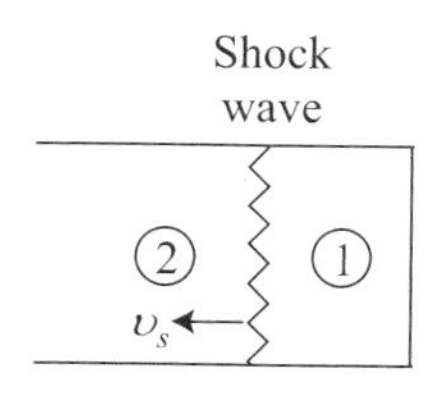

20-6 Numerically solve Problem 20-5. Immediately after the shock wave is reflected, is the air temperature at the right-hand wall at its highest value? Why?

20-7 Consider the shock tube problem discussed in Section 20.7, for the same initial condition ($P_1/P_4 = 20/500$, $T_1 = T_4 = 30\,°C$). After the shock has reflected from the right-hand wall (see Problem 20-5), the shock wave approaches the contact plane. When the shock wave meets the contact plane, it is partially reflected and partially transmitted, as shown. Analytically determine the state of the flow immediately after the shock passes through the contact plane.

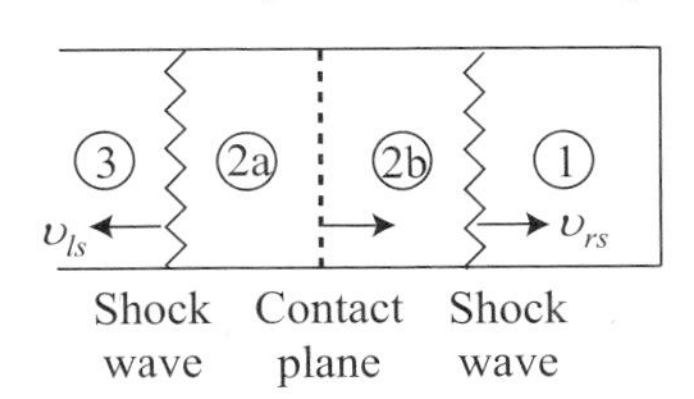

20-8 A normal shock moves down an open-ended tube with a velocity $v_s = 400$ m/s. In advance of the normal shock, the static air in the tube is at the ambient conditions: $P_a = 101$ kPa and $T_a = 25\,°\text{C}$. When the shock reaches the open end, an expansion wave is reflected back. Determine the velocity of the leading and trailing edges of the expansion wave moving back through the tube. What happens to the expansion wave when $v_s = 600$ m/s and when $v_s = 800$ m/s? Use the gas properties $R = 287$ J/kg/K and $\gamma = 1.40$ for air.

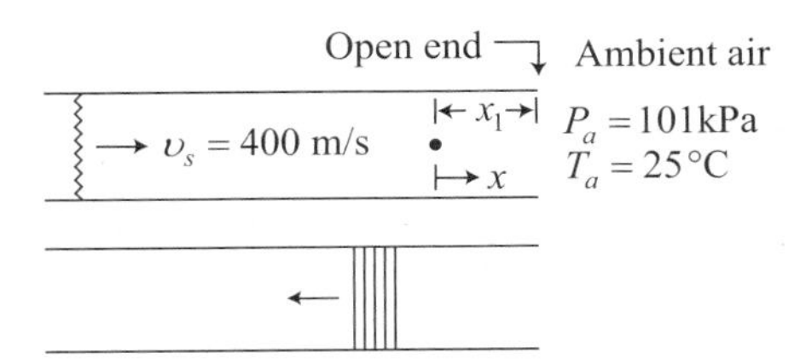

20-9 Consider the flow described in Problem 20-8. Derive a set of governing differential equations that describe the transient flow in terms of $\rho(x,t)$, $v(x,t)$, and $T(x,t)$. Present these equations in nonconservation form. (The shock wave described in Problem 20-8 is no longer in the domain of the solution.) Specify the initial and boundary conditions required to solve the governing equations for $t > 0$, when at $t = 0$ the shock passes through the open end. Numerically integrate the governing equations over a length of $L = 1$ m in the tube. Add diffusion terms to the momentum and energy equations to smooth numerical oscillations:

$$\frac{\partial v}{\partial t} = \ldots + \frac{\mu_a}{\rho}\frac{\partial^2 v}{\partial x^2} \quad \text{and} \quad \frac{\partial T}{\partial t} = \ldots + \frac{\kappa_a}{\rho}\frac{\partial^2 T}{\partial x^2}.$$

Let $\mu_a = \kappa_a = 0.02$. Plot $T(x,t)$ for three different times: $t_1 = 1$ ms, $t_2 = 2$ ms, and $t_3 = 3$ ms. On these plots, indicate where the leading and trailing edges of the expansion wave should be, using your analytic solution. What happens to this comparison when $\mu_a = \kappa_a = 0.2$ and when $\mu_a = \kappa_a = 0.0$?

20-10 Consider the shock tube problem described in Section 20.6 for the same initial conditions ($P_1/P_4 = 20/500$, $T_1 = T_4 = 30\,°\text{C}$) but with helium as the high-pressure gas and argon as the low-pressure gas. The gas constants for these gases are $R_{He} = 2077$ J/kg/K, $R_{Ar} = 208.1$ J/kg/K, and $\gamma_{He} = \gamma_{Ar} = 5/3$. Analytically determine P_2, T_2, v_2, P_3, T_3, v_3, and v_s before the shock wave reaches the right-hand wall. Also, determine the leading edge and trailing edge velocities of the expansion wave.

20-11 Numerically solve the shock tube flow described in Problem 20-10. Add a diffusion term to the gas constant transport equation to smooth numerical oscillations:

$$\frac{\partial(\rho R)}{\partial t} = \ldots + \lambda_a \frac{\partial^2(\rho R)}{\partial x^2}.$$

Plot v, P, and T after $t = 8$ ms. Determine the highest gas temperature and density achieved in the tube within the first 50 ms. At what location does the highest temperature occur? What numerical effect will have the largest effect on the accuracy of this result?

REFERENCES

[1] J. D. Anderson, *Modern Compressible Flow with Historical Perspective*, Third Edition. New York, NY: McGraw-Hill, 2002.

[2] L. Euler, "Principes generaux de l'etat d'equilibre des fluids." *Mémoires de l'Academie des Sciences de Berlin*, **11**, 217 (1757).

[3] B. Riemann, "Über die Fortpflanzung ebener Luftwellen von endlicher Schwingungsweite." *Abhandlungen der Gesellschaft der Wissenschaften zu Gottingen, Mathematisch-physikalische Klasse*, **8**, 43 (1860).

Chapter 21

Quasi-One-Dimensional Compressible Flows

21.1 Quasi-One-Dimensional Flow Equations

21.2 Quasi-One-Dimensional Steady Flow Equations without Friction

21.3 Numerical Solution to Quasi-One-Dimensional Steady Flow

21.4 Problems

The multidirectional equations of gas dynamics require difficult numerical solutions. However, relatively more simple quasi-one-dimensional flow solutions can adequately describe many interesting problems in gas dynamics. Such flows are called quasi-one-dimensional because although the flow is considered unidirectional, the cross-sectional area of the flow is not constant. For these problems, the question is not "where does the flow go?" but rather "how does flow state (v, ρ, P, T) change?" Changes in cross-sectional area combined with inertial consequences of the flow dynamics give rise to compressibility effects. The quasi-one-dimensional flow equations find utility in the description of propulsion and turbomachinery subsystems [1, 2].

21.1 QUASI-ONE-DIMENSIONAL FLOW EQUATIONS

To derive the quasi-one-dimensional differential equations of transport, a control volume is considered as shown in Figure 21-1. The control volume is differential only with respect to the flow direction (x-direction), and has a finite cross-sectional area A with circumference Γ. Advection transport occurs across the front and back surfaces labeled 1 and 2, respectively, that are separated by a distance of Δx. Transport variables are averaged with respect to the cross-sectional area. A unique attribute of the quasi-one-dimensional description is the peripheral surface labeled 3. Although one is not interested in considering traditional transport through this surface (thereby rendering a multidimensional problem), the

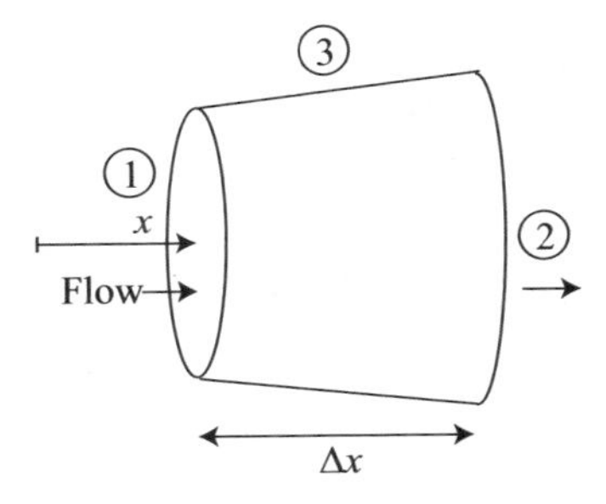

Figure 21-1 Differential control volume for quasi-one-dimensional flow in x-direction.

peripheral surface can cause important "squeezing" or "expansion" effects on transport in the primary direction.

21.1.1 Continuity Equation

Expressed in integral form, the continuity equation is

$$\int_{\forall} \frac{\partial}{\partial t}(\rho)d\forall = \int_{A} \overbrace{-v_j(\rho)}^{\text{advection}} n_j dA\,, \tag{21-1}$$

where the area integral is comprised of the three surfaces shown in Figure 21-1 surrounding the stationary control volume $\forall$. The rate of change of mass in the volume $\forall$ is balance by the net advection of mass through surfaces 1 and 2. Integrating Eq. (21-1) over the control volume yields

$$\frac{\partial}{\partial t}(\rho A)\Delta x = \overbrace{(\rho v A)_x}^{\text{surface 1}} + \overbrace{-(\rho v A)_{x+\Delta x}}^{\text{surface 2}} + \overbrace{0}^{\text{surface 3}}\,, \tag{21-2}$$

where v is used to denote v_x, since it is understood that the flow is in the x-direction. Dividing by Δx, and considering the differential limit in which $\Delta x \rightarrow dx$, one obtains the quasi-one-dimensional continuity equation:

$$\frac{\partial(\rho A)}{\partial t} + \frac{\partial}{\partial x}(\rho v A) = 0. \tag{21-3}$$

Notice that Eq. (21-3) is quite different from simply expressing the general continuity equation in one dimension, which would not account for the change in cross-sectional area A with downstream distance.

21.1.2 Momentum Equation

Expressed in integral form, the momentum equation is

$$\int_{\forall} \frac{\partial}{\partial t}(\rho v_x)d\forall = \int_{A} \left[\overbrace{-v_j(\rho v_x)}^{\text{advection}} \overbrace{- M_{jx} - P\delta_{jx}}^{\text{(sources + sinks)}}\right] n_j dA, \tag{21-4}$$

where the area integral is again comprised of the three surfaces shown in Figure 21-1. Conservation of momentum requires that the net advection of momentum into the control volume plus the net effect of the sources and sinks of momentum acting on the control volume balance the rate of change of momentum in the control volume. In addition to the usual contribution of pressure, the momentum diffusion flux to the wall M_{jx} is included for transmitting drag on the flow through the peripheral surface. Integrating Eq. (21-4) over the control volume yields

$$\begin{aligned}\frac{\partial}{\partial t}(\rho v A)\Delta x = {} & \overbrace{(\rho v^2 A)_x}^{\text{surface 1}} + \overbrace{-(\rho v^2 A)_{x+\Delta x}}^{\text{surface 2}} + \overbrace{-\tau_w \Gamma \Delta x}^{\text{surface 3}} \\ & + \overbrace{(PA)_x}^{\text{surface 1}} + \overbrace{-(PA)_{x+\Delta x}}^{\text{surface 2}} + \overbrace{P_x \Delta A}^{\text{surface 3}}\,,\end{aligned} \tag{21-5}$$

where the x-direction velocity component is denoted by $v_x \rightarrow v$ for notational simplicity. The magnitude of the momentum flux to the walls ($M_{jx}n_j$ evaluated for surface 3) equals the wall shear stress τ_w acting on the peripheral area $\Gamma\Delta x$. The change in area is $\Delta A = A_2 - A_1$. Dividing by Δx, and considering the differential limit in which $\Delta x \to dx$ and $\Delta A \to dA$, one obtains

$$\frac{\partial}{\partial t}(\rho v A) = -\frac{\partial}{\partial x}(\rho v^2 A) - \tau_w \Gamma - \frac{\partial}{\partial x}(PA) + P\frac{dA}{dx}. \tag{21-6}$$

Expanding the derivatives

$$\rho A\frac{\partial v}{\partial t} + \underbrace{v\frac{\partial}{\partial t}(\rho A) = -v\frac{\partial}{\partial x}(\rho v A)}_{= 0 \text{ by continuity}} - \rho v A\frac{\partial v}{\partial x} - \tau_w \Gamma - A\frac{\partial P}{\partial x} \underbrace{- P\frac{\partial A}{\partial x} + P\frac{dA}{dx}}_{= 0}, \tag{21-7}$$

reveals continuity, and allows the final form of the quasi-one-dimensional momentum equation to simplify to

$$\frac{\partial v}{\partial t} + v\frac{\partial v}{\partial x} = -\frac{\tau_w}{\rho}\frac{\Gamma}{A} - \frac{1}{\rho}\frac{\partial P}{\partial x}. \tag{21-8}$$

21.1.3 Energy Equation

Expressed in integral form, the energy equation is

$$\int_{\forall} \frac{\partial}{\partial t}(\rho e) d\forall = \int_A \Big[\overbrace{-v_j(\rho e)}^{\text{advection}} \quad \overbrace{-M_{jx}v_x - P\partial_{jx}U_x}^{\text{(sources+sinks)}}\Big] n_j dA, \tag{21-9}$$

where the area integral is again comprised of the three surfaces shown in Figure 21-1. The energy variable $e = u + v^2/2$ accounts for both thermal and kinetic forms. Conservation of energy requires that the net advection of energy into the control volume plus the net effect of the sources and sinks of energy acting on the control volume balance the rate of change of energy in the control volume. In addition to the usual contribution of the rate of pressure work being done, shear stresses transmitted to the walls perform work on the flow at a rate of $M_{jx}v_x n_j$, as evaluated over surface 3. Letting $v_x \to v$ and $M_{jx}n_j \to \tau_w$, Eq. (21-9) is integrated over the control volume to yield

$$\frac{\partial}{\partial t}(\rho\, e\, A)\Delta x = \overbrace{(\rho e v A)_x}^{\text{surface 1}} + \overbrace{-(\rho e v A)_{x+\Delta x}}^{\text{surface 2}} + \overbrace{-\tau_w v\, \Gamma \Delta x}^{\text{surface 3}}$$

$$+ \overbrace{(P v A)_x}^{\text{surface 1}} + \overbrace{-(P v A)_{x+\Delta x}}^{\text{surface 2}}. \tag{21-10}$$

Notice that although pressure transmits a force to surface 3, work cannot be performed by pressure without motion of the surface. Dividing by Δx, and considering the differential limit in which $\Delta x \to dx$, one obtains

$$\frac{\partial}{\partial t}(\rho e A) = -\frac{\partial}{\partial x}(\rho e v A) - \tau_w v \Gamma - \frac{\partial}{\partial x}(P v A). \tag{21-11}$$

This result simplifies through the use of continuity to

$$\rho A\frac{\partial e}{\partial t}+\underbrace{e\frac{\partial}{\partial t}(\rho A)=-e\frac{\partial}{\partial x}(\rho vA)}_{=0\text{ by continuity}}-\rho vA\frac{\partial e}{\partial x}-\tau_w v\Gamma-\frac{\partial}{\partial x}(PvA). \tag{21-12}$$

Changing the dependent variable to the stagnation enthalpy $h_o=e+P/\rho=h+v^2/2$, the energy equation becomes

$$\rho A\frac{De}{Dt}=\rho A\frac{Dh_o}{Dt}-\rho A\frac{D(P/\rho)}{Dt}=-\tau_w v\Gamma-\frac{\partial}{\partial x}(PvA). \tag{21-13}$$

However, with continuity it can be shown that

$$\rho A\frac{D(P/\rho)}{Dt}-\frac{\partial}{\partial x}(PvA)=A\frac{\partial P}{\partial t}+P\frac{\partial A}{\partial t}. \tag{21-14}$$

Therefore, the final form of the quasi-one-dimensional energy equation can be written as

$$\frac{\partial h_o}{\partial t}+v\frac{\partial h_o}{\partial x}=-\frac{\tau_w}{\rho}v\frac{\Gamma}{A}+\frac{1}{\rho}\frac{\partial P}{\partial t}+\frac{P}{\rho A}\frac{\partial A}{\partial t}. \tag{21-15}$$

The energy equation (21-15) becomes particularly simple when time derivatives are dropped in steady-state problems. However, for the transient problems, as discussed in Section 21.3, the energy equation (21-11) is better suited for numerical integration.

21.2 QUASI-ONE-DIMENSIONAL STEADY FLOW EQUATIONS WITHOUT FRICTION

The quasi-one-dimensional equations for steady-state flows without wall friction ($\tau_w=0$) are easily written from the more general equations (21-3), (21-8), and (21-15) of the previous section:

$$\text{Continuity:}\quad d(\rho vA)=0 \tag{21-16}$$

$$\text{Momentum:}\quad \rho v dv+dP=0 \tag{21-17}$$

$$\text{Energy:}\quad dh+v dv=0 \quad (\text{or}\quad dh_o=0) \tag{21-18}$$

Explicit reference to a spatial variable (i.e., x) has been dropped in these equations. Notice that there are three equations in terms of four unknown properties, ρ, v, P, and h. Therefore, an equation of state is required, which is taken to be the ideal gas law:

$$\text{Ideal gas:}\quad P=\rho RT. \tag{21-19}$$

A quasi-one-dimensional flow is governed by three first-order differential equations that can be expressed in terms of three unknown variables, using the equation of state. Integration of these three equations requires three boundary conditions. However, given a steady-state flow with inlet and exit conditions, there are a total of six boundary values from which only three can generally be independently specified. Conservation laws govern the remaining three. However, if the entrance condition is supersonic, an additional downstream boundary condition is required to select between possible supersonic or subsonic flow conditions at the exit (as discussed in Section 21.2.2). If the flow inlet condition is subsonic, and the flow becomes supersonic downstream, then there is one less inlet condition that can be independently specified, as will be discussed in the example of Section 21.3.4. In this case, the total number of independent boundary conditions remains three.

21.2.1 Isentropic Flows

Some interesting characteristics of gas dynamics can be learned from steady isentropic flows. When a flow is isentropic ($ds = 0$), the fundamental thermodynamic equation, $dh = Tds + dP/\rho$, requires $dh = dP/\rho$. With this restriction, the energy equation (21-18) becomes identical to the momentum equation (21-17) for steady quasi-one-dimensional flows and, therefore, is no longer an independent equation (as expected from the discussion in Section 20.3). The momentum equation may be written as

$$-vdv = \frac{dP}{\rho} = \frac{dP}{d\rho}\frac{d\rho}{\rho} \tag{21-20}$$

and constrained for an ideal gas by $ds = C_v dP/P - C_p d\rho/\rho = 0$ (see Table 20-1) to an isentropic path, such that

$$\left(\frac{\partial P}{\partial \rho}\right)_s = \frac{C_p}{C_v}\frac{P}{\rho} = \gamma\frac{P}{\rho} = \gamma RT = a^2. \tag{21-21}$$

The term $a = \sqrt{\gamma RT}$ is the speed of sound of an ideal gas, as was demonstrated in Section 20.4. With Eq. (21-21), the momentum equation (21-20) may be written for an isentropic flow as

$$\frac{d\rho}{\rho} = -\left(\frac{v}{a}\right)^2\frac{dv}{v} = -M^2\frac{dv}{v}, \tag{21-22}$$

where the local *Mach number* is defined as the ratio of the flow speed to the speed of sound in the fluid:

$$M = \frac{v}{a}. \tag{21-23}$$

The continuity equation (21-16) written in the form

$$\frac{dA}{A} + \frac{dv}{v} + \frac{d\rho}{\rho} = 0 \tag{21-24}$$

can be used to eliminate $d\rho/\rho$ from the momentum equation (21-22) to yield

$$\frac{dA}{A} = \left(M^2 - 1\right)\frac{dv}{v}. \tag{21-25}$$

Equation (21-25) describes the relation between changes in the speed of an isentropic flow and changes in the cross-sectional area. If the cross-sectional area is increasing, $dA/A > 0$, subsonic flows $M < 1$ will decrease in speed $dv/v < 0$, while supersonic flows $M > 1$ will increase in speed, $dv/v > 0$. If the cross-sectional area is decreasing, $dA/A < 0$, subsonic flows $M < 1$ will increase in speed, $dv/v > 0$, while supersonic flows $M > 1$ will decrease in speed, $dv/v < 0$. These contrasting behaviors are summarized in Figure 21-2. It is also interesting to observe that every flow, whether it is subsonic or supersonic, will approach the sonic condition when "squeezed." Conversely, a flow will move further from the sonic condition when allowed to expand. Furthermore, the only isentropic path

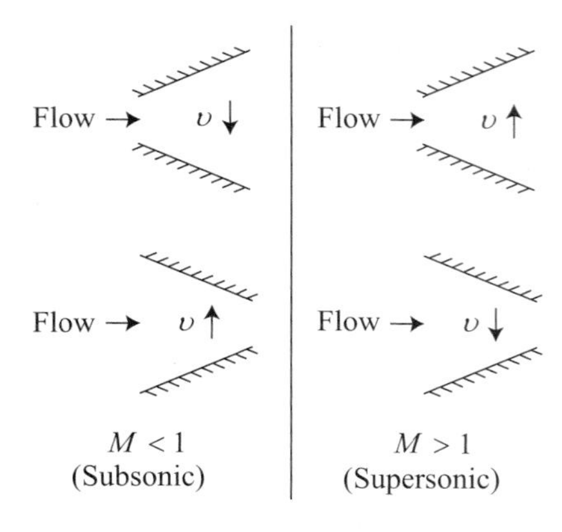

Figure 21-2 Contrasting subsonic and supersonic flows in converging and diverging channels.

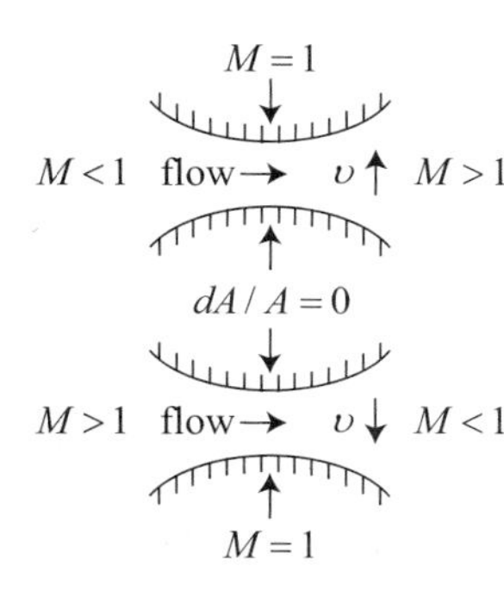

Figure 21-3 Isentropic transitions between subsonic and supersonic flows.

for a flow to transition between subsonic and supersonic states requires squeezing the flow to the sonic condition before allowing it to expand, as illustrated in Figure 21-3. Equation (21-25) dictates that the sonic condition $M = 1$ can only occur when $dA/A = 0$.

To proceed further with analytic analysis, it is useful to define two reference states:

Stagnation: $v \to 0$, denoted with subscript "o"
Sonic: $M \to 1$, denoted with superscript "*"

Consider a flow in a state described by a local value of T and v. With $dh = C_p dT$ for an ideal gas, the energy equation (21-18) may be expressed as

$$\text{Energy:} \quad C_p dT + v dv = 0. \tag{21-26}$$

To facilitate integration of the energy equation, it is convenient to assume a calorically "perfect" gas, for which C_p and C_v are constants. Then the energy equation may be trivially integrated between the current state (at T and v) and the stagnation state (where $v_o = 0$), for the result

$$\frac{v^2}{2} = C_p(T_o - T). \tag{21-27}$$

Since for steady-state $dh_o = 0$, the stagnation temperature T_o is a constant property of the flow. Knowing T_o permits Eq. (21-27) to be used to calculate temperature from the speed of the flow, or to calculate speed from the temperature of the flow. Equation (21-27) may be expressed in terms of the Mach number using the ideal gas relations $a^2 = \gamma RT$ and $R = C_p - C_v$, and the definitions $\gamma = C_p/C_v$ and $M = v/a$ for the result

$$\frac{T_o}{T} = 1 + \frac{M^2}{2}(\gamma - 1). \tag{21-28}$$

Next, consider a flow in a state described by T and v that is squeezed to the sonic state where $M^* = 1$. By continuity,

$$\rho v A = \rho^* v^* A^*. \tag{21-29}$$

This result can be used to relate the current flow area A to the value A^* at which sonic conditions are achieved:

$$\frac{A}{A^*} = \frac{\rho^*}{\rho}\frac{v^*}{v}. \tag{21.30}$$

Using Eq. (21-27), the current flow state can be related to the sonic state through

$$\left(\frac{v^*}{v}\right)^2 = \frac{1 - T^*/T_o}{1 - T/T_o} = \frac{1 - \left[1 + (\gamma - 1)\,\mathrm{M}^{*2}/2\right]^{-1}}{1 - \left[1 + (\gamma - 1)\,\mathrm{M}^2/2\right]^{-1}} = \frac{1 + (\gamma - 1)\,\mathrm{M}^2/2}{(\gamma + 1)\,\mathrm{M}^2/2}, \tag{21-31}$$

where a substitution is made using Eq. (21-28) to express the result in terms of the Mach number. Furthermore, if the flow is constrained to an isentropic path, then integration of $ds = C_v dT/T - R d\rho/\rho = 0$ (see Table 20-1) for a calorically perfect ideal gas yields the following result:

$$\frac{\rho^*}{\rho} = \left(\frac{T^*}{T}\right)^{\frac{1}{\gamma-1}} = \left(\frac{T_o}{T}\frac{T^*}{T_o}\right)^{\frac{1}{\gamma-1}} = \left(\frac{1 + \mathrm{M}^2(\gamma - 1)/2}{1 + (\gamma - 1)/2}\right)^{\frac{1}{\gamma-1}} \quad (\text{when} \quad ds = 0). \tag{21-32}$$

The last substitution in Eq. (21-32) is made with Eq. (21-28) to express the result in terms of the Mach number. Applying the results of Eqs. (21-31) and (21-32) to Eq. (21-30), the Mach number area relation for an isentropic flow becomes

$$\left(\frac{A}{A^*}\right)^2 = \left(\frac{1 + \mathrm{M}^2(\gamma - 1)/2}{1 + (\gamma - 1)/2}\right)^{\frac{2}{\gamma-1}} \frac{1 + (\gamma - 1)\,\mathrm{M}^2/2}{(\gamma + 1)\,\mathrm{M}^2/2}. \tag{21-33}$$

Or, after simplifying,

$$\left(\frac{A}{A^*}\right)^2 = \frac{1}{\mathrm{M}^2}\left[\frac{2}{\gamma + 1}\left(1 + \frac{\gamma - 1}{2}\mathrm{M}^2\right)\right]^{\frac{\gamma+1}{\gamma-1}} \quad (\text{when} \quad ds = 0). \tag{21-34}$$

Knowing the cross-sectional area required to squeeze the flow to sonic conditions A^*, the Mach number anywhere in the flow can be calculated from the local value of A using the Mach number area relation (21.34). Once the Mach number is known, Eq. (21-28) can be used to find the local value of T from the stagnation temperature. Once the local temperature is known, the isentropic ideal gas relations (see Table 20-1) can be integrated for a calorically perfect gas to determine the local values of P and ρ from their respective stagnation values:

$$\frac{T}{T_o} = \left(\frac{P}{P_o}\right)^{\frac{\gamma-1}{\gamma}} = \left(\frac{\rho}{\rho_o}\right)^{\gamma - 1}. \tag{21-35}$$

21.2.2 Flow through a Converging-Diverging Nozzle

Consider the flow through the converging-diverging nozzle illustrated in Figure 21-4. The flow is fed from a stagnation state of P_o and T_o. The mass flow rate is increase by lowering the nozzle exit pressure (back pressure), $P^e < P_o$. As the flow rate increases, the Mach number at the throat increases until it reaches a value of $M = 1$, as illustrated in Figure 21-4. Once the flow becomes sonic at the throat, it is impossible to communicate the exit conditions upstream of the throat. Further lowering of the exit pressure will not increase the mass flow rate, and in this condition, the nozzle is *choked*. All exit pressures higher than the choking value result in a subsonic expansion of the gas through the diverging section of the nozzle (shaded grey in Figure 21-4). All exit pressures lower than the choking value result in the flow going supersonic in the diverging section of the nozzle. However, only one specific exit pressure will yield an isentropic solution with supersonic expansion throughout the diverging section of the nozzle. Other exit pressures below the choking value will require a shock to transition the flow to subsonic conditions somewhere in the diverging section of the nozzle.

Although the process is cumbersome, it is possible to semi-analytically describe a shock containing steady-state flow through the nozzle illustrated in Figure 21-4. Two isentropic paths can be drawn through the expansion region of the nozzle, as shown in Figure 21-5. One path is traced forward from the throat along a supersonic path using the Mach number area relation (21-34), where A^* is the area of the throat. The second path, which is subsonic, is traced backward from the exit condition, again using Eq. (21-34). The reverse path uses a different value of A^*, since the shock will alter this property of the flow. The second post-shock value of A^* is found by applying Eq. (21-34) to the known exit conditions. (The second value of A^* is a hypothetical area that would be required to squeeze the post-shock flow back to the sonic condition.) The isentropic paths traced forward and backward are bridged by a normal shock. However, the change in Mach number across the shock, $(M_B - M_A)$, must satisfy the shock relation (see Section 20.6):

$$M_A^2 = \frac{M_B^2 + \dfrac{2}{\gamma - 1}}{\dfrac{2\gamma}{\gamma - 1} M_B^2 - 1}. \tag{21-36}$$

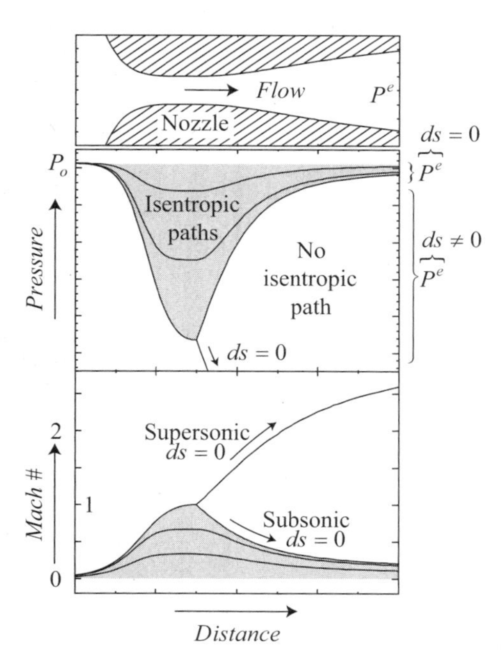

Figure 21-4 Flow through a converging-diverging nozzle.

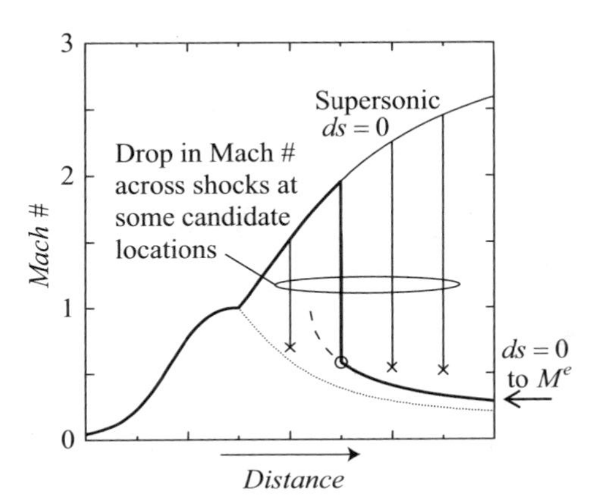

Figure 21-5 Searching for shock location.

Figure 21-5 illustrates four candidate shock transitions from the isentropic supersonic path. However, only at one location (or value of A) will the shock relation (21-36) exactly span the two isentropic paths. Once the correct shock location is found, the entire flow solution through the diverging section of the nozzle is described.

21.3 NUMERICAL SOLUTION TO QUASI-ONE-DIMENSIONAL STEADY FLOW

An analytic solution imposes several limitations that a numerical approach can overcome. Analysis thus far has been restricted to quasi-one-dimensional flows without friction or sources of heat of a calorically perfect ideal gas. In this section, a numerical approach based on MacCormack integration is developed for quasi-one-dimensional flows. This approach can be extended to flows with heating and friction, and to gases that obey a more complicated equation of state (than the ideal gas law) and that have variable specific heats (not calorically perfect). In this section, the governing equations are integrated forward in time to a steady-state solution of the problem described analytically in the previous section. In Chapter 22, MacCormack integration is used to obtain a fully two-dimensional description of supersonic expansion through a diverging nozzle.

21.3.1 Unsteady Flow Equations without Friction

The quasi-one-dimensional unsteady equations for a flow without wall friction ($\tau_w = 0$) are written for a calorically perfect ideal gas in conservation form (see Problem 21.1):

$$\text{Continuity:} \quad \frac{\partial}{\partial t}(\rho) + \frac{1}{A}\frac{\partial}{\partial x}(\rho v A) = 0 \tag{21-37}$$

$$\text{Momentum:} \quad \frac{\partial}{\partial t}(\rho v) + \frac{1}{A}\frac{\partial}{\partial x}[(\rho v)vA] + \frac{\partial P}{\partial x} = 0 \tag{21-38}$$

$$\text{Energy:} \quad \frac{\partial}{\partial t}(\rho e) + \frac{1}{A}\frac{\partial}{\partial x}[(\rho e + P)vA] = 0 \tag{21-39}$$

$$\text{where} \quad v = \rho v/\rho \quad \text{and} \quad P = (\gamma - 1)[\rho e - (\rho v)^2/2\rho]. \tag{21-40}$$

It is assumed that geometric constraints on the flow are static ($\partial A/\partial t = 0$). Equations (21-37) through (21-39) are written in a form where continuity is solved for ρ, momentum is solved for ρv, and energy is solved for ρe. For solutions to steady-state problems, Eqs. (21-37) through (21-39) may be integrated forward in time, from any suitable initial condition, until steady-state is achieved.

The governing equations can be cast into a finite differencing form using the MacCormack scheme. As discussed in Chapter 17, MacCormack integration requires the averaging of two estimates for the value of the dependent variables at the end of each time step. These estimates are derived from a predictor and a corrector step. The predictor step calculations are given by

$$\{\rho_n^{t+\Delta t}\}_p = \rho_n^t - \frac{\Delta t}{\Delta x}\left[\frac{\rho_{n+1}^t v_{n+1}^t A_{n+1} - \rho_n^t v_n^t A_n}{A_n}\right] + \frac{\nu_a \Delta t}{\Delta x^2}\left[\rho_{n+1}^t - 2\rho_n^t + \rho_{n-1}^t\right] \leftarrow \textit{for stability} \tag{21-41}$$

$$\{(\rho v)_n^{t+\Delta t}\}_p = (\rho v)_n^t - \frac{\Delta t}{\Delta x}\left[\frac{(\rho v)_{n+1}^t v_{n+1}^t A_{n+1} - (\rho v)_n^t v_n^t A_n}{A_n} + P_{n+1}^t - P_n^t\right] + \frac{\mu_a \Delta t}{\Delta x^2}\left[(\rho v)_{n+1}^t - 2(\rho v)_n^t + (\rho v)_{n-1}^t\right] \leftarrow \textit{for stability} \tag{21-42}$$

$$\{(\rho e)_n^{t+\Delta t}\}_p = (\rho e)_n^t - \frac{\Delta t}{\Delta x}\left[\frac{((\rho e)_{n+1}^t + P_{n+1}^t) v_{n+1}^t A_{n+1} - ((\rho e)_n^t + P_n^t) v_n^t A_n}{A_n}\right] + \frac{k_a \Delta t}{\Delta x^2}\left[(\rho e)_{n+1}^t - 2(\rho e)_n^t + (\rho e)_{n-1}^t\right] \leftarrow \textit{for stability} \tag{21-43}$$

Predictor step values for velocity $\{v_n^{t+\Delta t}\}_p$ and pressure $\{P_n^{t+\Delta t}\}_p$ are obtained from $\{\rho_n^{t+\Delta t}\}_p$, $\{(\rho v)_n^{t+\Delta t}\}_p$, and $\{(\rho e)_n^{t+\Delta t}\}_p$ using

$$\{v_n^{t+\Delta t}\}_p = \frac{\{(\rho v)_n^{t+\Delta t}\}_p}{\{\rho_n^{t+\Delta t}\}_p} \quad \text{and} \quad \{P_n^{t+\Delta t}\}_p = (\gamma - 1)\left[\{(\rho e)_n^{t+\Delta t}\}_p - \frac{\{(\rho v)_n^{t+\Delta t}\}_p^{\,2}}{2\rho_n^{t+\Delta t}}\right]. \tag{21-44}$$

The corrector step calculations are given by

$$\{\rho_n^{t+\Delta t}\}_c = \rho_n^t - \frac{\Delta t}{\Delta x}\left[\frac{\{\rho_n^{t+\Delta t}\}_p\{v_n^{t+\Delta t}\}_p A_n - \{\rho_{n-1}^{t+\Delta t}\}_p\{v_{n-1}^{t+\Delta t}\}_p A_{n-1}}{A_n}\right] + \frac{\nu_a \Delta t}{\Delta x^2}\left[\{\rho_{n+1}^{t+\Delta t}\}_p - 2\{\rho_n^{t+\Delta t}\}_p + \{\rho_{n-1}^{t+\Delta t}\}_p\right] \leftarrow \textit{for stability} \tag{21-45}$$

$$\{(\rho v)_n^{t+\Delta t}\}_c = (\rho v)_n^t - \frac{\Delta t}{\Delta x}\left[\frac{\{(\rho v)_n^{t+\Delta t}\}_p\{v_n^{t+\Delta t}\}_p A_n - \{(\rho v)_{n-1}^{t+\Delta t}\}_p\{v_{n-1}^{t+\Delta t}\}_p A_{n-1}}{A_n} + \{P_n^{t+\Delta t}\}_p - \{P_{n-1}^{t+\Delta t}\}_p\right] + \frac{\mu_a \Delta t}{\Delta x^2}\left[\{v_{n+1}^{t+\Delta t}\}_p - 2\{v_n^{t+\Delta t}\}_p + \{v_{n-1}^{t+\Delta t}\}_p\right] \leftarrow \textit{for stability} \tag{21-46}$$

$$\{(\rho e)_n^{t+\Delta t}\}_c = (\rho e)_n^t - \frac{\Delta t}{\Delta x}\left[\frac{\left(\{(\rho e)_n^{t+\Delta t}\}_p + \{P_n^{t+\Delta t}\}_p\right)\{v_n^{t+\Delta t}\}_p A_n - \left(\{(\rho e)_{n-1}^{t+\Delta t}\}_p + \{P_{n-1}^{t+\Delta t}\}_p\right)\{v_{n-1}^{t+\Delta t}\}_p A_{n-1}}{A_n}\right] + \frac{k_a \Delta t}{\Delta x^2}\left[(\rho e)_{n+1}^t - 2(\rho e)_n^t + (\rho e)_{n-1}^t\right] \leftarrow \textit{for stability} \tag{21-47}$$

Corrector step values for velocity $\{v_n^{t+\Delta t}\}_c$ and pressure $\{P_n^{t+\Delta t}\}_c$ are obtained from $\{\rho_n^{t+\Delta t}\}_c$, $\{(\rho v)_n^{t+\Delta t}\}_c$, and $\{(\rho e)_n^{t+\Delta t}\}_c$ using

$$\{v_n^{t+\Delta t}\}_c = \frac{\{(\rho v)_n^{t+\Delta t}\}_c}{\{\rho_n^{t+\Delta t}\}_c} \quad \text{and} \quad \{P_n^{t+\Delta t}\}_c = (\gamma - 1)\left[\{(\rho e)_n^{t+\Delta t}\}_c - \frac{\{(\rho v)_n^{t+\Delta t}\}_c^2}{2\rho_n^{t+\Delta t}}\right]. \tag{21-48}$$

As was noted in Chapter 17, the direction of finite differencing may need to be reversed for either the predictor or corrector step at boundary nodes to prevent the operation from extending beyond the physical domain. For the same reason, the diffusion terms, added for numerical stability, are not included at boundary nodes.

The predictor and corrector step values are averaged to obtain the end of the time step values:

$$\rho_n^{t+\Delta t} = \frac{\{\rho_n^{t+\Delta t}\}_p + \{\rho_n^{t+\Delta t}\}_c}{2} \tag{21-49}$$

$$(\rho v)_n^{t+\Delta t} = \frac{\{(\rho v)_n^{t+\Delta t}\}_p + \{(\rho v)_n^{t+\Delta t}\}_c}{2} \tag{21-50}$$

$$(\rho e)_n^{t+\Delta t} = \frac{\{(\rho e)_n^{t+\Delta t}\}_p + \{(\rho e)_n^{t+\Delta t}\}_c}{2}. \tag{21-51}$$

Since the MacCormack scheme is explicit, stability of integration requires the time step be bounded by the Courant condition, as discussed in Section 17.2.1. As a practical approach, each time step can be selected to achieve a specified target Courant number:

$$\frac{\Delta t(|v| + a)}{\Delta x} \approx C_{targ} = 0.9. \tag{21-52}$$

The artificial diffusivities added to the transport equations for numerical stability are dynamically selected, such that

$$\nu_a = \mu_a = \kappa_a = \frac{0.1}{\Delta x^2/\Delta t}. \tag{21-53}$$

The coefficient in the numerator may be changed for varied amounts of diffusion in the solution, as long as the stability requirements discussed in Section 17.3 for diffusion are observed.

21.3.2 Boundary Conditions

With both inlet and exit conditions, there are a total of six boundary values for the variables ρ, ρv, and ρe, or related variables. Only three are boundary conditions that must be specified for a unique solution. The conservation laws governing the flow dictate the remaining three boundary values. One simple way to satisfy this requirement numerically is to let the unspecified boundary values "float," as discussed in Section 17.4. The floating condition extrapolates boundary values from internal points of the flow. In this manner, the governing equations can propagate the requirements of the conservation laws to the boundaries. For example, suppose the required boundary conditions for a specific problem are given by the inlet conditions ρ^i and P^i, and the exit condition P^e.

The first and last nodes in the numerical solution can employ floating boundary values for the unknown inlet value of $(\rho v)^i$, and exit values of ρ^e and $(\rho v)^e$ using

$$\text{at} \quad n = \text{first node}: \qquad (\rho v)_n^{t+\Delta t} = 2(\rho v)_{n+1}^{t+\Delta t} - (\rho v)_{n+2}^{t+\Delta t} \tag{21-54}$$

$$\text{at} \quad n = \text{last node}: \qquad \rho_n^{t+\Delta t} = 2\rho_{n-1}^{t+\Delta t} - \rho_{n-2}^{t+\Delta t} \tag{21-55}$$

$$\text{at} \quad n = \text{last node}: \qquad (\rho v)_n^{t+\Delta t} = 2(\rho v)_{n-1}^{t+\Delta t} - (\rho v)_{n-2}^{t+\Delta t}. \tag{21-56}$$

The inlet and exit energy fluxes are evaluated from

$$\rho e = P/(\gamma - 1) + (\rho v)^2/2\rho. \tag{21-57}$$

In the current example, the inlet momentum flux is determined by floating, and the inlet density and pressure are specified, while at the outlet the density and momentum fluxes are determined by floating, and the outlet pressure is specified.

21.3.3 Initial Conditions and Convergence

Any physically sound initial state can be used for the purpose of integrating forward in time to the steady-state solution. The simplest initial condition is to specify a stagnant gas at the thermodynamic conditions of the inlet. Only the exit condition is specified otherwise, using the required exit pressure:

$$t = 0: \quad 0 \le x < L: \quad \begin{cases} \rho = \rho^i \\ \rho v = v = 0 \\ \rho e = P^i/(\gamma - 1) \end{cases} \quad \text{and} \quad x = L: \quad \begin{cases} \rho = \rho^i \\ \rho v = v = 0 \\ \rho e = P^e/(\gamma - 1) \end{cases} \tag{21-58}$$

These initial conditions can be integrated forward in time to achieve the steady-state solution. As an illustration of the transient development of a supersonic flow, the progression of the pressure distribution in a nozzle with time is shown in Figure 21-6.

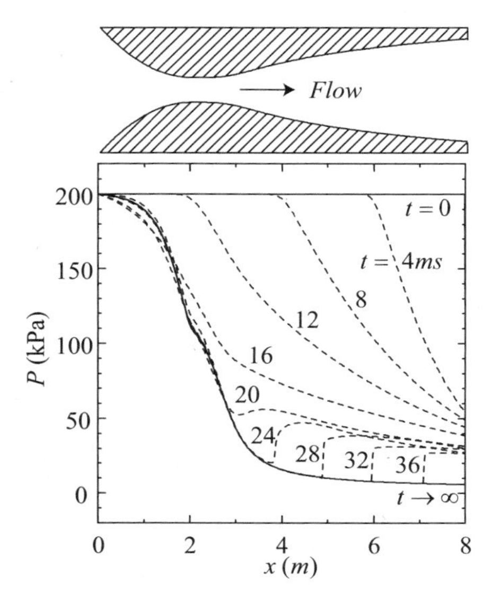

Figure 21-6 Time evolution of pressure in a nozzle from constant initial condition.

However, to achieve the supersonic exit condition, the exit pressure must be allowed to float. Therefore, the exit pressure specified for the initial time step is only required to be sufficiently low to cause the development of a supersonic flow.

Numerical integration requires an assessment of when the solution has achieved steady-state. The most robust assessment is to look at every number that is calculated to determine if there is any change over a time step. A simpler approach may poll a single result for change. One method, following the second approach, is to calculate the net rate of entropy leaving the nozzle:

$$\frac{(\dot{m}s)^e - (\dot{m}s)^i}{C_{\bar{v}}} = \rho^e v^e A^e[\ln(P^e) - \gamma\ln(\rho^e)] - \rho^i v^i A^i\left[\ln(P^i) - \gamma\ln(\rho^i)\right]. \tag{21-59}$$

When this number stops changing, integration has presumably reached steady-state. Equation (21-59) equates to the steady-state rate of entropy generation in the nozzle. In the absence of a shock, entropy generation is theoretically zero. However, numerical integration always introduces artificial diffusion into the solution (plus the diffusion terms added explicitly for numerical stability). Therefore, even in a shock-free flow some entropy generation always exists in the numerical solution.

21.3.4 Converging-Diverging Nozzle Example

Several problem scenarios can be addressed for a flow through a converging-diverging nozzle. If the flow is everywhere subsonic, then the flow solution is constrained by three independent variables specified anywhere in the flow. For example, boundary conditions could be fully specified at the inlet, ρ^i, v^i, and P^i, or a combination of inlet and exit conditions, ρ^i, P^i, and P^e, could be specified. However, when the flow is choked, three inlet conditions can no longer be independently assigned because of the added constraint that $\mathrm{M} = 1$ at the throat. For example, if ρ^i and P^i are specified, only one value of v^i will yield the choking condition. Lower values will not allow the flow to attain sonic conditions at the throat, and higher values would imply the flow goes sonic before reaching the throat, which violates the physical laws of a steady-flow solution (see Section 21.2.1). Therefore, the correct value of v^i is part of the unknown solution.

When the flow through the nozzle is choked, one additional boundary condition is required to identify the nature of the flow after the sonic throat conditions, such as P^e. If the choked flow is isentropic, only two possibilities for P^e exist, corresponding to the subsonic and supersonic expansions through the nozzle. All exit pressures falling between the high subsonic expansion P^e and the low supersonic expansion P^e yield a nonisentropic shock somewhere downstream of the throat. It is important to note that the value of P^e does not influence the flow entering the nozzle (i.e., the value of v^i) when the flow is choked, since no downstream information can propagate upstream through sonic or supersonic conditions. However, the value of P^e will dictate the location of a shock appearing in the diverging section of the nozzle.

Consider a choked flow specified by $\rho^{i,} P^i$, and P^e. The unspecified boundary values v^i, ρ^e, and v^e are part of the solution. To impose the choked condition on the solution, Eq. (21-34) describing the steady-state isentropic relation between the flow Mach number and area could be used to determine the inlet Mach number M^i, from which the inlet velocity $v^i = \mathrm{M}^i\sqrt{\gamma(P^i/\rho^i)}$ can be found. Additionally, the steady-state values of ρ^e and v^e could be calculated from continuity and conservation of energy applied between the inlet and exit conditions. However, in the spirit of letting the governing differential equations dictate the solution, unknown boundary values of $(\rho v)^i$, ρ^e, and $(\rho v)^e$ can also be allowed to float as a part of the numerical solution, as described in Section 21.3.2.

Code 21-1 was written to integrate the equations of continuity (21-37), momentum (21-38), and energy (21-39) for flow through a converging-diverging nozzle. The geometry of the nozzle is given by

$$\frac{A(x)}{A_t} = \begin{cases} 1 + x^2 & (-2 \le x \le 0) \\ 1 + \log(1 + x^4) & (0 < x \le 6) \end{cases}, \tag{21-60}$$

where A_t is the area of the throat. For this example, the nozzle is fed by a gas of $\rho^i = 1.2\ \mathrm{kg}/m^3$ and $P^i = 200\mathrm{kPa}$, and the nozzle exit pressure is $P^e = 100\mathrm{kPa}$. The exit pressure is sufficiently low to choke the flow at the nozzle throat. Viscosity is added to the equations for numerical stability, as described in Section 21.3.1. The boundary conditions are handled as discussed in Section 21.3.2, and the domain is initialized as discussed in Section 21.3.3.

Figure 21-7 plots the Mach number through the nozzle and contrasts the numerical solution with the exact analytic solution (the dashed line) discussed in Section 21.2.2. The agreement is good, although some smoothing of the numerical solution occurs around the rapid transition seen near the throat. The shock also exhibits small oscillations in the numerical solution. On a cautionary note: Underdamped numerical oscillations near the shock can lead to a misplacement of the shock in the numerical solution. The shock is a bridge between supersonic and subsonic isentropic branches of the solution. When the separating distance between the two isentropic branches becomes misrepresented by numerical oscillations, the shock can be misplaced. Therefore, it is recommended to use sufficiently high artificial viscosity to dampen most of the numerical oscillations near the shock.

Code 21-1 can also be used to find the isentropic solution corresponding to a supersonic exit flow. This is accomplished by forcing P^e below the subsonic exit pressure for a choked isentropic flow, but otherwise letting P^e float. If P^e is allowed to float (but not to any of the subsonic exit pressures), it will float to the supersonic exit pressure for an isentropic solution. It will not float to any values associated with a shock-containing solution. The only way to pin a shock into the solution is by fixing the exit pressure at an appropriate value.

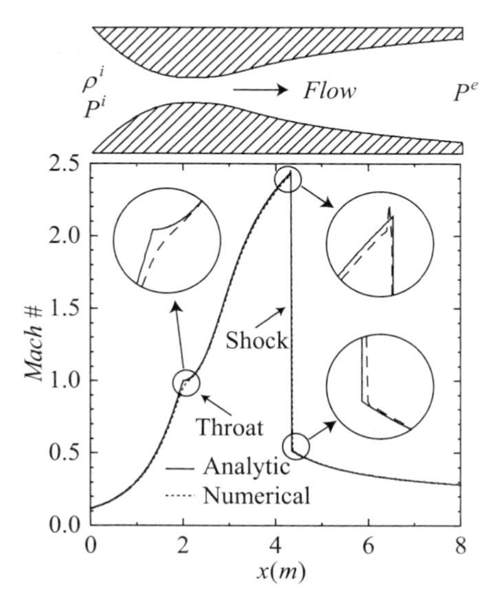

Figure 21-7 Comparison of exact (solid line) and numerical solutions through the nozzle.

Code 21-1 Flow through converging–diverging nozzle

```
#include <stdio.h>
#include <math.h>

double cont_diff(char t,double *d,double *v,double *A) {
    return (t=='p' ?  d[+1]*v[+1]*A[+1] - *d**v**A :
         /* t=='c' */ *d**v**A - d[-1]*v[-1]*A[-1] )/(*A);  }

double mom_diff(char t,double *d_v,double *v,double *p,double *A) {
   double val= (t=='p' ?  d_v[+1]*v[+1]*A[+1] - *d_v**v**A :
             /* t=='c' */ *d_v**v**A - d_v[-1]*v[-1]*A[-1] )/(*A);
   return val+=(t=='p' ? p[+1] - *p : *p - p[-1]);         }

double enrgy_diff(char t,double *d_e,double *v,double *p,double *A) {
   return (t=='p' ?  (d_e[+1]+p[+1])*v[+1]*A[+1]-(*d_e+*p)**v**A :
 /* t=='c' */ (*d_e+*p)**v**A - (d_e[-1]+p[-1])*v[-1]*A[-1] )/(*A); }

double MacC(int N,double *dt_,double dx,double gam,
              double *d,double *d_v,double *d_e,double *v,double *p,double *A)
{
   int n,step;
   double Pd[N],Cd[N],Pd_v[N],Cd_v[N],Pd_e[N],Cd_e[N],Pv[N],Cv[N],Pp[N],Cp[N];
   double C,Cmax=0,Ctarg=0.9,dt=*dt_,dtdx=dt/dx,dtddx=dt/dx/dx,na,ma,ka;
   na=ma=ka=0.1/dtddx;
   for (n=0;n<N;++n) {                       // perform predictor step
      step=(n<N-1 ? 'p' : 'c');
      Pd[n]=d[n]-dtdx*cont_diff(step,d+n,v+n,A+n);
      if (n>0 && n<N-1) Pd[n]+=dtddx*na*(d[n+1]-2.*d[n]+d[n-1]);
      Pd_v[n]=d[n]*v[n]-dtdx*mom_diff(step,d_v+n,v+n,p+n,A+n);
      if (n>0 && n<N-1) Pd_v[n]+=dtddx*ma*(d_v[n+1]-2.*d_v[n]+d_v[n-1]);
      Pv[n]=Pd_v[n]/Pd[n];
      Pd_e[n]=d_e[n]-dtdx*enrgy_diff(step,d_e+n,v+n,p+n,A+n);
      if (n>0 && n<N-1) Pd_e[n]+=dtddx*ka*(d_e[n+1]-2.*d_e[n]+d_e[n-1]);
      Pp[n]=(gam-1.)*(Pd_e[n]-Pd_v[n]*Pd_v[n]/2./Pd[n]);
      C=dt*(fabs(Pv[n])+sqrt(gam*Pp[n]/Pd[n]))/dx; // dt size checking
      if (C>Cmax) Cmax=C;
   }

   if (Cmax>1.) { // integrate with a smaller time step
      *dt_=0.;
      return dt*Ctarg/Cmax;
   }
   for (n=0;n<N;++n) {                       // perform corrector step
      step=(n>0 ? 'c' : 'p');
      Cd[n]=d[n]-dtdx*cont_diff(step,Pd+n,Pv+n,A+n);
      if (n>0 && n<N-1) Cd[n]+=dtddx*na*(Pd[n+1]-2.*Pd[n]+Pd[n-1]);
      Cd_v[n]=d_v[n]-dtdx*mom_diff(step,Pd_v+n,Pv+n,Pp+n,A+n);
      if (n>0 && n<N-1) Cd_v[n]+=dtddx*ma*(Pd_v[n+1]-2.*Pd_v[n]+Pd_v[n-1]);
      Cv[n]=Cd_v[n]/Cd[n];
      Cd_e[n]=d_e[n]-dtdx*enrgy_diff(step,Pd_e+n,Pv+n,Pp+n,A+n);
      if (n>0 && n<N-1) Cd_e[n]+=dtddx*ka*(Pd_e[n+1]-2.*Pd_e[n]+Pd_e[n-1]);
      Cp[n]=(gam-1.)*(Cd_e[n]-Cd_v[n]*Cd_v[n]/2./Cd[n]);
      if (isnan(Cp[n])) { printf("\nsoln BLEW!\n"); return 0.; }
   }
   for (n=1;n<N-1;++n) {        // average predictor and corrector results
      d[n]=0.5*(Pd[n]+Cd[n]);
      d_v[n]=0.5*(Pd_v[n]+Cd_v[n]);
      d_e[n]=0.5*(Pd_e[n]+Cd_e[n]);
      v[n]=d_v[n]/d[n];
      p[n]=(gam-1.)*(d_e[n]-d_v[n]*d_v[n]/2./d[n]);
      if (isnan(p[n])) { printf("\nsoln BLEW!\n"); return 0; }
   }
   v[0]=(d_v[0]=2.*d_v[1]-d_v[2])/d[0];  // float entrance
   d_e[0]=p[0]/(gam-1.)+d_v[0]*d_v[0]/2./d[0];
   d[N-1]=2.*d[N-2]-d[N-3];              // float exit
   v[N-1]=(d_v[N-1]=2.*d_v[N-2]-d_v[N-3])/d[N-1];
   d_e[N-1]=p[N-1]/(gam-1.)+d_v[N-1]*d_v[N-1]/2./d[N-1];
   return dt*Ctarg/Cmax;
}

int main(void)
{
   FILE *fp;
   int n,cnt=0,N=601,ConvergeCnt=0;
   double den[N],den_vel[N],den_enrgy[N],vel[N],pres[N],A[N];
   double t=0,x,M,Sgen,SgenLast=0.,delSgen;
   double gam=1.4,At=0.1,L=8.0,dx=L/(N-1),dt,dt_next=1.0e-5;
   double den_in=1.2,pres_in=200.0e3,pres_exit=100.0e3;
   for (x=-2.,n=0;n<N;++n,x+=dx)
      A[n]=At*( x<=0. ? 1.+x*x : 1.+log10(1.+x*x*x*x));
   for (n=0;n<N;++n) {      // initial distributions
      den[n]=den_in;
      den_vel[n]=vel[n]=0.;
      den_enrgy[n]=(pres[n]=pres_in)/(gam-1.);
   }
   den_enrgy[N-1]=(pres[N-1]=pres_exit)/(gam-1.);
   do {                        // step soln by dt
      if (!((cnt++)%100)) { // check every 100 dt
         Sgen=den_vel[N-1]*A[N-1]*(log(pres[N-1])-gam*log(den[N-1]))
               -den_vel[0] * A[0]  *(log(pres[0]) - gam*log(den[0])  );
         if ((delSgen=fabs((Sgen-SgenLast)))< 5.e-5) ++ConvergeCnt;
         else ConvergeCnt=0;
         printf("\nt=%e Sgen=%e delSgen=%e",t,Sgen,delSgen);
         SgenLast=Sgen;
      }
      dt=dt_next;
      dt_next=MacC(N,&dt,dx,gam,den,den_vel,den_enrgy,vel,pres,A);
      if ( !dt_next ) break;
//    pres[N-1]=2.*pres[N-2]-pres[N-3]; // float exit P for M>1 at exit
   } while ((t+=dt) < 1. && ConvergeCnt<10);
   fp=fopen("out.dat","w");
   for (n=0;n<N;++n) {
      M=vel[n]/sqrt(gam*pres[n]/den[n]);
      fprintf(fp,"%e %e %e %e %e\n",n*dx,M,den[n],vel[n],pres[n]/1000.);
   }
   fclose(fp);
   return 1;
}
```

21.4 PROBLEMS

21-1 Assuming a calorically perfect ideal gas, derive the governing quasi-one-dimensional unsteady equations in conservation form for a flow without wall friction ($\tau_w = 0$):

Continuity: $$\frac{\partial}{\partial t}(\rho) + \frac{1}{A}\frac{\partial}{\partial x}(\rho v A) = 0$$

Momentum: $$\frac{\partial}{\partial t}(\rho v) + \frac{1}{A}\frac{\partial}{\partial x}(\rho v^2 A) + \frac{\partial P}{\partial x} = 0$$

Energy: $$\frac{\partial}{\partial t}(\rho e) + \frac{1}{A}\frac{\partial}{\partial x}[(\rho e + P)vA] = 0 \quad \text{where} \quad e = \frac{P}{\rho(\gamma - 1)} + \frac{v^2}{2}$$

21-2 Consider a quasi-one-dimensional compressible flow through a converging-diverging nozzle. Let inlet conditions be denoted by the superscript "i" and exit conditions be denoted by the superscript "e". The inlet flow is subsonic, $M^i < 1$. Which of the following sets of boundary conditions are appropriate for the flow conditions described?

$\rho^i, P^i, T^i,$ $v^i, M^i < 1$ flow → $\rho^e, P^e, T^e,$ v^e, M^e

$dA/A = 0$

1. ρ^i, P^i, and P^e.
2. ρ^i, P^i, and T^i for a flow that is subsonic everywhere.
3. ρ^e, P^e, and M^e for a flow that is subsonic everywhere.
4. ρ^i, P^i, and v^i for a flow that is choked (sonic at the throat).
5. ρ^i and P^i for an isentropic flow with supersonic exit.
6. ρ^i and P^i for an isentropic flow with subsonic exit.
7. ρ^i, P^i, and P^e for an isentropic flow with supersonic exit.

21-3 A converging-diverging nozzle is designed to operate isentropically with an exit Mach number of 1.5. The nozzle is supplied from an air reservoir ($\gamma = 1.4$ and $R = 0.2870$ kJ/kgK) in which the pressure is 500 kPa; the temperature is 500 K. The nozzle throat is 5 cm^2. Assume air to behave as an ideal and calorically perfect gas. Determine: (A) the ratio of exit area to throat area; (B) the back pressure below which the nozzle is choked (where a decrease in back pressure does not increase the flow rate); (C) the mass flow rate for a back pressure of 450 kPa; and (D) the mass flow rate for a back pressure of 0 kPa.

21-4 Air moves through a converging-diverging nozzle with the conditions illustrated. The circular area of the nozzle is $A/A_t = 1 + 2.2(x - 1.5)^2$. At the inlet $P_i = 1$ atm, $T_i = 0\,°\text{C}$, and $M_i = 2$. For air: $\gamma = 1.4$ and $R = 0.2870$ kJ/kgK.

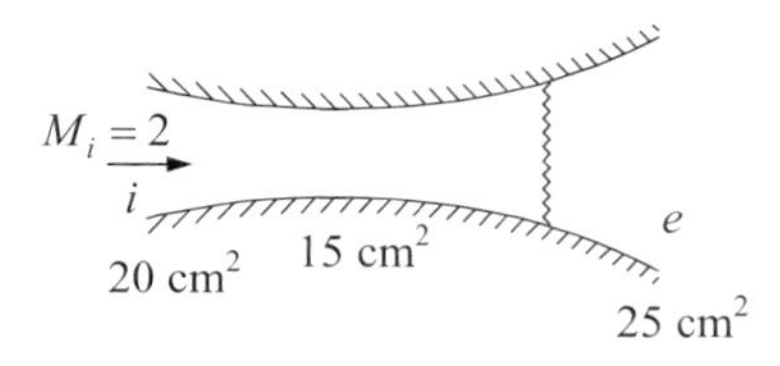

Analyze the flow analytically. For what back pressure P_e is there no shock? For what range of back pressures is there a shock in the nozzle? Is there a back pressure for which the shock can appear in the converging section of the nozzle? Explain. What happens when $P_e = 700$ kPa?

21-5 A 3-m-long converging-diverging nozzle has a cross-sectional area given by:

$$\frac{A}{A_t} = 1 + 2.2\,(x - 1.5)^2,$$

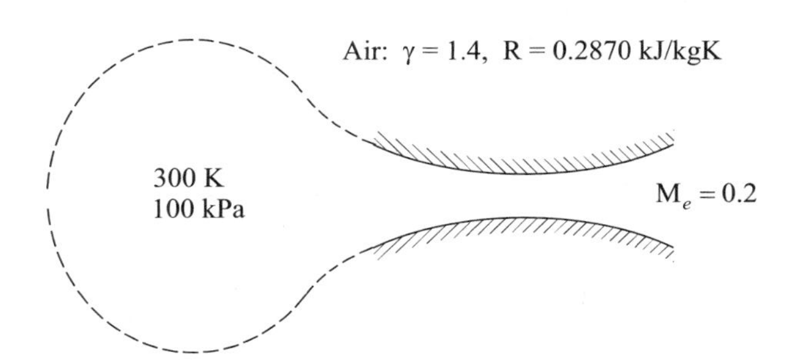

with a throat area of $A_t = 0.1\ m^2$. Sonic conditions exist at the throat of the nozzle, and the inlet stagnation temperature and stagnation pressure of the air flowing through the nozzle are $T_{o,i} = 300\ K$ and $P_{o,i} = 100\ kPa$, respectively. Analyze the isentropic flow analytically. Plot T, P, and M over the length of the nozzle (A) assuming the flow goes supersonic after the throat, and then (B) assuming the flow remains subsonic after the throat.

21-6 Plot the "analytic solution" for T, P, and M over the length of the nozzle described in Problem 21-5 for the nonisentropic flow with $P_e = 80kPa$.

21-7 Numerically solve Problem 21-6. Compare the numerical solution for T, P, and M with the distributions obtained from the analytic solution over the length of the nozzle.

21-8 Consider the nozzle discussed in Section 21.3, which is fed by a gas of $\rho^i = 1.2\ \text{kg}/m^3$ and $P^i = 200\text{kPa}$. Let the flow go supersonic to the exit of the diverging section of the nozzle. Solve this problem numerically. Let P^e float to the correct isentropic value. Evaluate the shock-free entropy generation in the nozzle. How is entropy generation effected by the value of artificial viscosity? Contrast entropy generation in the shock-free flow with the case where $P^e = 100\text{kPa}$. Comment on this comparison.

21-9 Consider the nonisentropic effect of heat transfer to the air for the nozzle conditions described in Problem 21-5. Suppose the nozzle cross-section is circular and a heat flux $q_w = 500\ \text{kW}/m^2$ passes through the wall of the nozzle. Modify the numerical treatment to account for this, and reinvestigate the supersonic exit condition for a shock-free flow. Compare the new results for M with the distribution obtained for the adiabatic problem. What effect does the added heat have on the rate of mass flow through the nozzle?

21-10 Air moves through a converging-diverging nozzle with the conditions illustrated. The flow experiences wall friction, as characterized by $\tau_w = c_f \rho v^2/2$. The circular area of the nozzle is $A/A_t = 1 + 2.2{*}(x - 1.5)^2$. At the inlet $P_i = 1$ atm and $T_i = 0\,°\text{C}$, and at the exit the pressure is $P_e = 4.5$ atm. The fluid is air: $\gamma = 1.4$ and $R = 0.2870$ kJ/kgK.

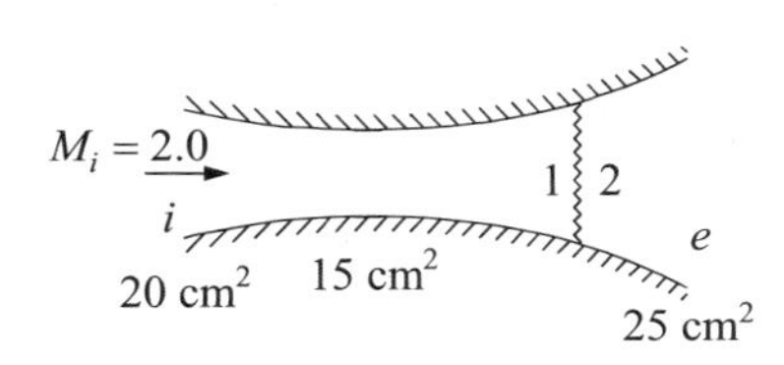

Plot the Mach number and pressure as a function of distance through the nozzle, considering the skin friction coefficient to be $c_f = 0.0$, $c_f = 0.0025$, and $c_f = 0.005$. What is the exit temperature for each case?

REFERENCES

[1] S. S. Penner, *Chemistry Problems in Jet Propulsion*. New York, NY: Pergamon, 1957.
[2] Erian A. Baskharone, *Principles of turbomachinery in air-breathing engines. Cambridge Aerospace Series* (No. 18), Cambridge University Press, 2006.

Chapter 22

Two-Dimensional Compressible Flows

22.1 Flow through a Diverging Nozzle
22.2 Problems

In this chapter, the governing equations for a two-dimensional inviscid reversible flow are solved numerically for a diverging nozzle. This exercise will illustrate the effort involved in abandoning the quasi-one-dimensional equations of Chapter 21 in favor of an exact two-dimensional description. Additionally, the exact two-dimensional solution can be contrasted with the approximate quasi-one-dimensional results for an indication of the degree of information lost.

22.1 FLOW THROUGH A DIVERGING NOZZLE

Consider a sonic flow entering the diverging section of a nozzle, as shown in Figure 22-1. Assume that the gas expands supersonically without a shock. For this adiabatic reversible flow, entropy transport requires that

$$Ds/Dt = 0. \tag{22-1}$$

Description of the entropy content in the flow is simplified by restricting analysis to flows of a calorically perfect ideal gas. In this situation, $ds = C_v dP/P - C_p d\rho/\rho = 0$ (see Table 20-1) may be integrated for the result

$$s - s^* = C_v \ln(P/P^*) - C_p \ln(\rho/\rho^*), \tag{22-2}$$

in which the superscript "*" denotes the sonic inlet conditions, which are used as a reference state. Therefore, a convenient dimensionless entropy variable is defined by

$$\sigma = \frac{s - s^*}{C_v} = \ln(P/P^*) - \gamma \ln(\rho/\rho^*). \tag{22-3}$$

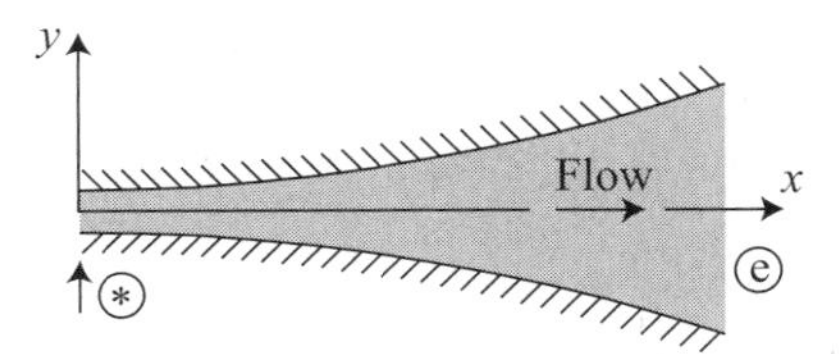

Figure 22-1 Flow through a two-dimensional diverging nozzle.

Isentropic flows satisfying the momentum equation also satisfy the energy equation, as discussed in Section 20.3. Therefore, the governing equations sufficient for describing a reversible flow are

$$\text{Continuity:} \quad \frac{D(\ln \rho)}{Dt} = -\partial_j v_j \tag{22-4}$$

$$\text{Momentum:} \quad \rho \frac{Dv_i}{Dt} = -\partial_i P \tag{22-5}$$

$$\text{Entropy:} \quad \frac{D\sigma}{Dt} = 0. \tag{22-6}$$

Notice that the density variable has been expressed in terms of $\ln \rho$, where $d(\ln \rho) = d\rho/\rho$, and the entropy variable σ adopts the definition given by Eq. (22-3). Although the three transport equations are expressed in terms of four unknowns ($\ln \rho$, v_i, P, and σ), the problem is closed by the entropy variable definition (22-3) that enforces the ideal gas law as an equation of state. The governing equations may be expanded for a two-dimensional problem in Cartesian coordinates:

$$\text{Continuity:} \quad \frac{\partial \ln\rho}{\partial t} = -\left[v_x \frac{\partial \ln\rho}{\partial x} + v_y \frac{\partial \ln\rho}{\partial y} + \frac{\partial v_x}{\partial x} + \frac{\partial v_y}{\partial y}\right] \tag{22-7}$$

$$x\text{-dir.Mom:} \quad \frac{\partial v_x}{\partial t} = -\left[v_x \frac{\partial v_x}{\partial x} + v_y \frac{\partial v_x}{\partial y} + \frac{1}{\rho}\frac{\partial P}{\partial x}\right] \tag{22-8}$$

$$y\text{-dir.Mom:} \quad \frac{\partial v_y}{\partial t} = -\left[v_x \frac{\partial v_y}{\partial x} + v_y \frac{\partial v_y}{\partial y} + \frac{1}{\rho}\frac{\partial P}{\partial y}\right] \tag{22-9}$$

$$\text{Entropy:} \quad \frac{\partial \sigma}{\partial t} = -\left[v_x \frac{\partial \sigma}{\partial x} + v_y \frac{\partial \sigma}{\partial y}\right]. \tag{22-10}$$

Notice that these equations are not expressed in conservation form. However, this is acceptable for numerical integration of isentropic flows since the dependent variables are everywhere differentiable.

The geometric domain must be discretized for a numerical solution. To fill a two-dimensional domain, it is desired to distort the mesh to conform to physical boundaries. This requires transformation of the governing equations to map onto a regular computational domain. To illustrate, the gas flow above the centerline of the nozzle is discretized as shown schematically in Figure 22-2. The vertical spacing in the computational domain is stretched by the local half-width of the nozzle. Therefore, new dimensionless spatial variables can be defined by

$$Y = y/W(x) \quad \text{and} \quad X = x/W^*, \tag{22-11}$$

where $W(x)$ is the half-width of the nozzle and $W^* = W(x = 0)$ is the inlet dimension of the nozzle.

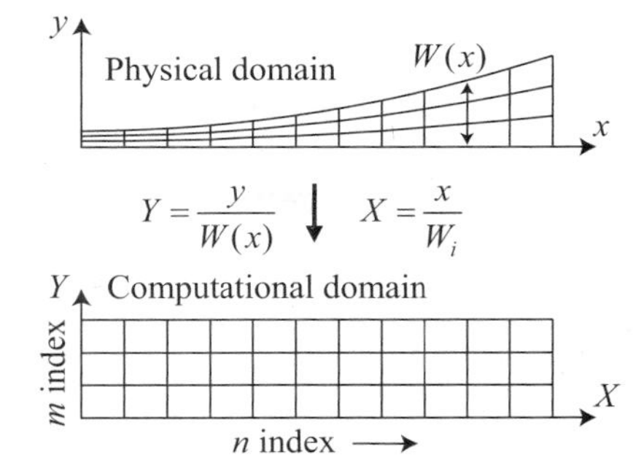

Figure 22-2 Mapping nozzle onto a regular computational domain.

To solve the governing equations on the computational domain, they must be transformed from functions of x and y into functions of X and Y. Transforming the derivatives requires

$$\frac{\partial(\,)}{\partial x} = \frac{\partial X}{\partial x}\frac{\partial(\,)}{\partial X} + \frac{\partial Y}{\partial x}\frac{\partial(\,)}{\partial Y} = \frac{1}{W^*}\frac{\partial(\,)}{\partial X} - \frac{Y}{W}\frac{\partial W}{\partial x}\frac{\partial(\,)}{\partial Y} = \frac{1}{W^*}\frac{\partial(\,)}{\partial X} + \frac{\chi}{W^*}\frac{\partial(\,)}{\partial Y}, \tag{22-12}$$

where $\chi = -(Y/w)(\partial w/\partial X)$ and $w = W(x)/W^*$.

Similarly,

$$\frac{\partial(\,)}{\partial y} = \frac{\partial X}{\partial y}\frac{\partial(\,)}{\partial X} + \frac{\partial Y}{\partial y}\frac{\partial(\,)}{\partial Y} = \frac{1}{W}\frac{\partial(\,)}{\partial Y}. \tag{22-13}$$

It is convenient to define the following dimensionless variables:

$$u = v_x/a^*, \quad v = v_y/a^*, \quad r = \rho/\rho^*, \quad \pi = P/(\rho^*(a^*)^2),$$

$$\tau = t\,a^*/W^*, \quad \text{and} \quad \beta = \chi\, u + v/w, \tag{22-14}$$

where a^* and ρ^* are the speed of sound and the gas density, respectively, at the sonic inlet of the nozzle. The governing equations can be transformed into the new variables with the result

$$\text{Continuity:} \quad \frac{\partial \ln r}{\partial \tau} = -\left[u\frac{\partial \ln r}{\partial X} + \frac{\partial u}{\partial X} + \beta\frac{\partial \ln r}{\partial Y} + \chi\frac{\partial u}{\partial Y} + \frac{1}{w}\frac{\partial v}{\partial Y}\right] \tag{22-15}$$

$$x\text{-dir.Mom.:} \quad \frac{\partial u}{\partial \tau} = -\left[u\frac{\partial u}{\partial X} + \beta\frac{\partial u}{\partial Y} + \exp(-\ln r)\left(\frac{\partial \pi}{\partial X} + \chi\frac{\partial \pi}{\partial Y}\right)\right] \tag{22-16}$$

$$y\text{-dir.Mom.:} \quad \frac{\partial v}{\partial \tau} = -\left[u\frac{\partial v}{\partial X} + \beta\frac{\partial v}{\partial Y} + \frac{\exp(-\ln r)}{w}\frac{\partial \pi}{\partial Y}\right] \tag{22-17}$$

$$\text{Entropy:} \quad \frac{\partial \Sigma}{\partial \tau} = -\left[u\frac{\partial \Sigma}{\partial X} + \beta\frac{\partial \Sigma}{\partial Y}\right]. \tag{22-18}$$

where

$$\pi = \exp(\Sigma + \gamma \ln r) \quad (\Sigma = \ln \pi - \gamma \ln r). \tag{22-19}$$

For convenience, the dimensionless entropy variable has been redefined by

$$\Sigma = \sigma - \ln(\gamma). \tag{22-20}$$

The transport equations (22-15) through (22-18) may be simultaneously solved using MacCormack integration, as discussed in Chapter 17. Unlike the quasi-one-dimensional formulation, the containing walls of the flow provide boundaries off of which waves can be reflected. However, at the boundaries of the domain, the standard rules for forward and backward differencing used in the predictor and corrector steps may not be possible. For example, when m is a node lying on the upper wall of the nozzle, it is impossible to evaluate flow properties at $m+1$ in the standard predictor step. Therefore, exceptions are handled by reversing the predictor and corrector differencing rules, when necessary.

To make presentation of the predictor and corrector finite difference equations more compact, the following finite differencing operators are defined:

$$\Delta^{\tau}_{X,p}\{\ (\ )\ \} = \begin{cases} [\ (\ )^{\tau}_{m,n+1} - (\ )^{\tau}_{m,n}\ ]/\Delta X & \text{(for } n < \text{last node)} \\ [\ (\ )^{\tau}_{m,n} - (\ )^{\tau}_{m,n-1}\ ]/\Delta X & \text{(for } n = \text{last node)} \end{cases} \tag{22-21}$$

$$\Delta^{\tau}_{Y,p}\{\ (\ )\ \} = \begin{cases} [\ (\ )^{\tau}_{m+1,n} - (\ )^{\tau}_{m,n}\ ]/\Delta Y & \text{(for } m < \text{last node)} \\ [\ (\ )^{\tau}_{m,n} - (\ )^{\tau}_{m-1,n}\ ]/\Delta Y & \text{(for } m = \text{last node)} \end{cases} \tag{22-22}$$

$$\Delta^{\tau+\Delta\tau}_{X,c}\{\ (\ )\ \} = \begin{cases} [\ \{(\ )^{\tau+\Delta\tau}_{m,n}\}_p - \{(\ )^{\tau+\Delta\tau}_{m,n-1}\}_p\ ]/\Delta X & \text{(for } n > \text{first node)} \\ [\ \{(\ )^{\tau+\Delta\tau}_{m,n+1}\}_p - \{(\ )^{\tau+\Delta\tau}_{m,n}\}_p\ ]/\Delta X & \text{(for } n = \text{first node)} \end{cases} \tag{22-23}$$

$$\Delta^{\tau+\Delta\tau}_{Y,c}\{\ (\ )\ \} = \begin{cases} [\ \{(\ )^{\tau+\Delta\tau}_{m,n}\}_p - \{(\ )^{\tau+\Delta\tau}_{m-1,n}\}_p\ ]/\Delta Y & \text{(for } m > \text{first node)} \\ [\ \{(\ )^{\tau+\Delta\tau}_{m+1,n}\}_p - \{(\ )^{\tau+\Delta\tau}_{m,n}\}_p\ ]/\Delta Y & \text{(for } m = \text{first node).} \end{cases} \tag{22-24}$$

To illustrate, in the predictor differencing $\Delta^{\tau}_{X,p}\{u\}$ is evaluated as $[u^{\tau}_{m,n+1} - u^{\tau}_{m,n}]/\Delta X$ when $n <$ *last node* and is evaluated as $[u^{\tau}_{m,n} - u^{\tau}_{m,n-1}]/\Delta X$ when $n =$ *last node*.

The dependent variables are integrated forward in time using the average of the predictor and corrector calculations:

$$\ln r^{\tau+\Delta\tau}_{m,n} = \left[\ \{\ln r^{\tau+\Delta\tau}_{m,n}\}_p + \{\ln r^{\tau+\Delta\tau}_{m,n}\}_c\ \right]/2 \tag{22-25}$$

$$u^{\tau+\Delta\tau}_{m,n} = \left[\ \{u^{\tau+\Delta\tau}_{m,n}\}_p + \{u^{\tau+\Delta\tau}_{m,n}\}_c\right]/2 \tag{22-26}$$

$$v^{\tau+\Delta\tau}_{m,n} = \left[\ \{v^{\tau+\Delta\tau}_{m,n}\}_p + \{v^{\tau+\Delta\tau}_{m,n}\}_c\ \right]/2 \tag{22-27}$$

$$\Sigma^{\tau+\Delta\tau}_{m,n} = \left[\ \{\Sigma^{\tau+\Delta\tau}_{m,n}\}_p + \{\Sigma^{\tau+\Delta\tau}_{m,n}\}_c\ \right]/2. \tag{22-28}$$

The two estimates of the dependent variables at the end of each time step are obtained from the predictor $\{\cdots\}_p$ and corrector $\{\cdots\}_c$ forms of the governing equations:

Continuity:

$$\{\ln r_{m,n}^{\tau+\Delta\tau}\}_p = \ln r_{m,n}^{\tau} - \Delta\tau\Big[u_{m,n}^{\tau}\Delta_{X,p}^{\tau}\{\ln r\} + \Delta_{X,p}^{\tau}\{u\} + \beta_{m,n}^{\tau}\Delta_{Y,p}^{\tau}\{\ln r\} + C_{m,n}^{\tau}\Delta_{Y,p}^{\tau}\{u\} + w_n^{-1}\Delta_{Y,p}^{\tau}\{v\}\Big] \tag{22-29}$$

$$\{\ln r_{m,n}^{\tau+\Delta\tau}\}_c = \ln r_{m,n}^{\tau} - \Delta\tau\Big[\{u_{m,n}^{\tau+\Delta\tau}\}_p\Delta_{X,c}^{\tau+\Delta\tau}\{\ln r\} + \Delta_{X,c}^{\tau+\Delta\tau}\{u\} + \{\beta_{m,n}^{\tau+\Delta\tau}\}_p\Delta_{Y,c}^{\tau+\Delta\tau}\{\ln r\} + \{C_{m,n}^{t+\Delta t}\}_p\Delta_{Y,c}^{\tau+\Delta\tau}\{u\} + w_n^{-1}\Delta_{Y,c}^{\tau+\Delta\tau}\{v\}\Big] \tag{22-30}$$

x-dir. Momentum:

$$\{u_{m,n}^{\tau+\Delta\tau}\}_p = u_{m,n}^{\tau} - \Delta\tau\Big[u_{m,n}^{\tau}\Delta_{X,p}^{\tau}\{u\} + \beta_{m,n}^{\tau}\Delta_{Y,p}^{\tau}\{u\} + \exp(-\ln r_{m,n}^{\tau})(\Delta_{X,p}^{\tau}\{\pi\} + C_{m,n}^{\tau}\Delta_{Y,p}^{\tau}\{\pi\})\Big] \tag{22-31}$$

$$\{u_{m,n}^{\tau+\Delta\tau}\}_c = u_{m,n}^{\tau} - \Delta\tau\Big[\{u_{m,n}^{\tau+\Delta\tau}\}_p\Delta_{X,c}^{\tau+\Delta\tau}\{u\} + \{\beta_{m,n}^{\tau+\Delta\tau}\}_p\Delta_{Y,c}^{\tau+\Delta\tau}\{u\} + \exp(-\{\ln r_{m,n}^{\tau+\Delta\tau}\}_p)(\Delta_{X,c}^{\tau+\Delta\tau}\{\pi\} + \{C_{m,n}^{t+\Delta t}\}_p\Delta_{Y,c}^{\tau+\Delta\tau}\{\pi\})\Big] \tag{22-32}$$

y-dir. Momentum:

$$\{v_{m,n}^{\tau+\Delta\tau}\}_p = v_{m,n}^{\tau} - \Delta\tau\Big[u_{m,n}^{\tau}\Delta_{X,p}^{\tau}\{v\} + \beta_{m,n}^{\tau}\Delta_{Y,p}^{\tau}\{v\} + w_n^{-1}\exp(-\ln r_{m,n}^{\tau})\Delta_{Y,p}^{\tau}\{\pi\}\Big] \tag{22-33}$$

$$\{v_{m,n}^{\tau+\Delta\tau}\}_c = v_{m,n}^{\tau} - \Delta\tau\Big[\{u_{m,n}^{\tau+\Delta\tau}\}_p\Delta_{X,c}^{\tau+\Delta\tau}\{v\} + \{\beta_{m,n}^{\tau+\Delta\tau}\}_p\Delta_{Y,c}^{\tau+\Delta\tau}\{v\} + w_n^{-1}\exp(-\{\ln r_{m,n}^{\tau+\Delta\tau}\}_p)\Delta_{Y,c}^{\tau+\Delta\tau}\{\pi\}\Big] \tag{22-34}$$

Entropy:

$$\{\Sigma_{m,n}^{\tau+\Delta\tau}\}_p = \Sigma_{m,n}^{\tau} - \Delta\tau\Big[u_{m,n}^{\tau}\Delta_{X,p}^{\tau}\{\Sigma\} + \beta_{m,n}^{\tau}\Delta_{Y,p}^{\tau}\{\Sigma\}\Big] \tag{22-35}$$

$$\{\Sigma_{m,n}^{\tau+\Delta\tau}\}_c = \Sigma_{m,n}^{\tau} - \Delta\tau\Big[\{u_{m,n}^{\tau+\Delta\tau}\}_p\Delta_{X,c}^{\tau+\Delta\tau}\{\Sigma\} + \{\beta_{m,n}^{\tau+\Delta\tau}\}_p\Delta_{Y,c}^{\tau+\Delta\tau}\{\Sigma\}\Big] \tag{22-36}$$

Since the MacCormack scheme is explicit, stability of integration requires the time step to be bounded by

$$\Delta\tau < \frac{\Delta X}{u + \sqrt{\gamma\pi/r}} \quad \text{and} \quad \Delta\tau < \frac{\Delta Y}{v + \sqrt{\gamma\pi/r}}, \tag{22-37}$$

where $\sqrt{\gamma\pi/r}$ is the dimensionless local speed of sound (a/a^*). If $d\tau$ is larger than permitted by Eq. (22-37), the flow carries information further than the distance between adjacent nodes in one time step, and integration becomes unstable.

22.1.1 Boundary Conditions and Initial Condition

For the sonic inlet $M^* = 1$ condition, all the dimensionless inlet values are specified:

$$\text{at } X = 0: \qquad \pi = 1/\gamma, \quad \ln r = 0, \quad u = 1, \quad v = 0, \quad \Sigma = -\ln(\gamma). \tag{22-38}$$

All the exit conditions will be governed by the transport equations explicitly, rather than using floating boundary conditions. The supersonic expansion of the gas downstream is dictated by the geometry of the diverging nozzle. For illustration, suppose the geometry of the nozzle is given by

$$w(X) = 1 + 0.025X^2 \quad \text{for} \quad 0 \leq X \leq 10. \tag{22-39}$$

The centerline and the top wall of the diverging nozzle impose the boundary conditions on v:

$$\text{at } Y = 0: \qquad v = 0 \qquad \text{(for centerline symmetry)} \tag{22-40}$$

$$\text{at } Y = 1: \qquad v = u\,(dw/dX) \qquad \text{(for an impenetrable outer wall)}. \tag{22-41}$$

The domain can be initialized with a linear interpolation between the known inlet conditions and a reasonable guess at the exit conditions. A guess at the exit conditions can be made from the quasi-one-dimensional model. With Eq. (21-24), the exit Mach number M^e is estimated from the supersonic root of

$$w^e = \frac{1}{M^e}\left[\frac{2}{\gamma+1}\left(1 + \frac{\gamma-1}{2}(M^e)^2\right)\right]^{\frac{\gamma+1}{2(\gamma-1)}}, \tag{22-42}$$

where $w^e = W(x = L)/W^*$ is the dimensionless half-width of the nozzle exit. Then, with M^e and Eq. (21-22), the exit velocity can be estimated from

$$u^e = M^e\sqrt{\frac{\gamma+1}{2 + (M^e)^2(\gamma-1)}}. \tag{22-43}$$

Finally, with this result, continuity, and the ideal gas law, the exit pressure can be estimated from

$$\pi^e = \frac{1}{\gamma w^e M^e}\frac{u^e}{M^e} = \frac{1}{\gamma w^e M^e}\sqrt{\frac{2 + (\gamma-1)}{2 + (M^e)^2(\gamma-1)}}. \tag{22-44}$$

The two-dimensional domain is initialized with

$$\pi(\tau = 0, X, Y) = \pi^* + \frac{X}{L/W^*}(\pi^e - \pi^*) \tag{22-45}$$

$$u(\tau = 0, X, Y) = u^* + \frac{X}{L/W^*}(u^e - u^*) \tag{22-46}$$

$$v(\tau = 0, X, Y) = \left(Y\frac{dw}{dX}\right)u(\tau = 0, X, Y) \tag{22-47}$$

$$\Sigma(\tau = 0, X, Y) = -\ln(\gamma) \tag{22-48}$$

$$\ln r(\tau = 0, X, Y) = \frac{\ln \pi(\tau = 0, X, Y) + \ln(\gamma)}{\gamma} \tag{22-49}$$

where $\pi^* = 1/\gamma$ and $u^* = 1$. Notice that entropy is initialized everywhere to equal the inlet value $\Sigma = -\ln(\gamma)$, which is the steady-state requirement. Also, notice that $v(X, Y)$ is initialized in a way that satisfies both boundary conditions at $Y = 0$ and $Y = 1$.

22.1.2 Illustrative Result

Code 22-1 was written to integrate the continuity, (22-15), and momentum, (22-16) and (22-17), to steady-state using the initial conditions described in Section 22.1.1. The flow is comprised of air for which $\gamma = 1.4$. Since entropy is everywhere initialized to a constant, $\Sigma(\tau = 0, X, Y) = -\ln(\gamma)$, time integration of the entropy transport equation yields the trivial result: $\Sigma(\tau > 0, X, Y) = -\ln(\gamma)$. Therefore, Code 22-1 does not integrate the entropy transport equation. Instead, pressure required for the momentum equation can be evaluated from Eq. (22-19) with $\Sigma = -\ln(\gamma)$ such that

$$\pi = \exp[\gamma \ln r - \ln(\gamma)]. \tag{22-50}$$

Figure 22-3 plots the Mach number through the nozzle and contrasts the exact two-dimensional numerical solution with the analytic result from the quasi-one-dimensional solution (dashed lines). The Mach lines, starting after M = 1 at the throat, are contoured at 0.2 intervals downstream. The two-dimensional Mach lines have increasing curvature with distance downstream. Although the quasi-one-dimensional formulation loses spatial information across the width of the flow, there is good agreement between the two solutions when the two-dimensional results are averaged across the width of the nozzle.

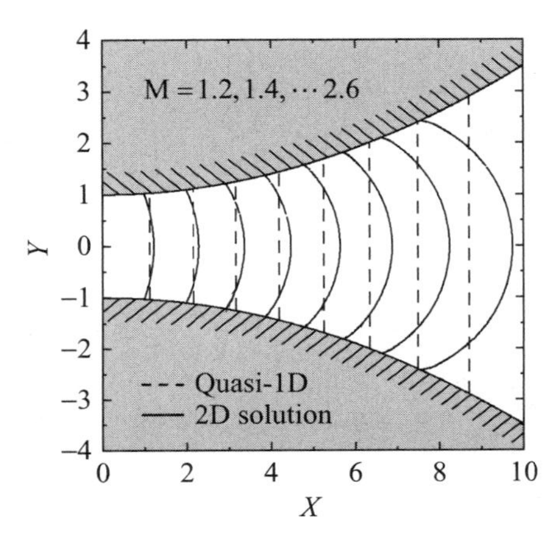

Figure 22-3 Comparison of two-dimensional and quasi-one-dimensional solutions.

Code 22-1 Two-dimensional flow through diverging nozzle

```
#include <stdio.h>
#include <math.h>
#define NN 301
#define MM 21
#define _delXp(_v)  (n==NN-1 ? _v[m][n]-_v[m][n-1]  : _v[m][n+1]-_v[m][n])/dX
#define _delYp(_v)  (m==MM-1 ? _v[m][n]-_v[m-1][n]  : _v[m+1][n]-_v[m][n])/dY
#define _delXc(_v)  (n==  0  ? _v[m][n+1]-_v[m][n]  : _v[m][n]-_v[m][n-1])/dX
#define _delYc(_v)  (m==  0  ? _v[m+1][n]-_v[m][n]  : _v[m][n]-_v[m-1][n])/dY
double   lnr[MM][NN],u[MM][NN],v[MM][NN],pi[MM][NN],w[NN];
double MacC2D(double *dt_,double dX,double dY,double gam)
{
        int n,m;
        extern double lnr[MM][NN],u[MM][NN],v[MM][NN],pi[MM][NN],w[NN];
        double den,Pden,B,C,Plnr[MM][NN],Clnr[MM][NN],Pu[MM][NN],Cu[MM][NN];
        double Ppi[MM][NN],Pv[MM][NN],Cv[MM][NN],Cx,Cy,Cmax=0.,dt=*dt_;
        for (n=0;n<NN;++n)                        /* perform predictor step */
          for (m=0;m<MM;++m) {
             den=exp(lnr[m][n]);
             Cx=dt*(u[m][n]+sqrt(gam*pi[m][n]/den))/dX;  /* dt size checking */
             Cmax=(Cx>Cmax ? Cx : Cmax);
             Cy=dt*(v[m][n]+sqrt(gam*pi[m][n]/den))/dY;  /* dt size checking */
             Cmax=(Cy>Cmax ? Cy : Cmax);
             C= -m*dY*(n==NN-1 ? w[n]-w[n-1] : w[n+1]-w[n])/dX/w[n];
             B=C*u[m][n]+v[m][n]/w[n];
             Plnr[m][n]=lnr[m][n]-dt*( u[m][n]*_delXp(lnr)+ _delXp(u)
                + B*_delYp(lnr) + C*_delYp(u) + _delYp(v)/w[n] );
             Pu[m][n]=u[m][n]-dt*( u[m][n]*_delXp(u) + B*_delYp(u)
                + (_delXp(pi) + C*_delYp(pi))/den );
             Pv[m][n]=v[m][n]-dt*( u[m][n]*_delXp(v) + B*_delYp(v)
                + _delYp(pi)/w[n]/den );
             Ppi[m][n]=exp(gam*Plnr[m][n]-log(gam));
       }
   if (Cmax>1. || Cmax<0.8) {     /* integrate with a different time step */
        *dt_=dt*0.9/Cmax;
        if (*dt_<1.e-9) *dt_=1.e-9;
        printf("Using ... dt=%e\n",*dt_);
        return MacC2D(dt_,dX,dY,gam);
      }
      for (n=0;n<NN;++n)                          /* perform corrector step */
          for (m=0;m<MM;++m) {
              C= -m*dY*(n== 0 ? w[n+1]-w[n] : w[n]-w[n-1])/dX/w[n];
             B=C*Pu[m][n]+Pv[m][n]/w[n];
             Pden=exp(Plnr[m][n]);
             Clnr[m][n]=lnr[m][n]-dt*( Pu[m][n]*_delXc(Plnr) + _delXc(Pu)
                 +B*_delYc(Plnr)+C*_delYc(Pu)+_delYc(Pv)/w[n] );
             Cu[m][n]=u[m][n]-dt*( Pu[m][n]*_delXc(Pu) + B*_delYc(Pu)
                 + (_delXc(Ppi) + C*_delYc(Ppi))/Pden );
             Cv[m][n]=v[m][n]-dt*( Pu[m][n]*_delXc(Pv) + B*_delYc(Pv)
                 + _delYc(Ppi)/w[n]/Pden );
      }
for (n=1;n<NN;++n) {        /* average predictor and corrector results */
         for (m=0;m<MM;++m) {
         lnr[m][n]=0.5*(Plnr[m][n]+Clnr[m][n]);
         u[m][n]  =0.5*(Pu[m][n]  +Cu[m][n]);
            v[m][n]  =0.5*(Pv[m][n]  +Cv[m][n]);
            pi[m][n] =exp(gam*lnr[m][n]-log(gam));
         }
       v[0][n]  =0.;
       v[MM-1][n]=u[MM-1][n]*(w[n]-w[n-1])/dX;
     }
     return pi[0][NN-1];
}

int main(void)
{
    FILE *fp;
    int n,m,cnt=0;
    extern double u[MM][NN],v[MM][NN],pi[MM][NN],w[NN];
    double M,u2,piExit,piExitLast=0.,piExitDel;
    double gam=1.4,dX=10./(NN-1),dY=1.0/(MM-1),dt=1.0e-2;
    double Me=2.8,ue_Me=sqrt((gam+1.)/(2.+Me*Me*(gam-1.)));   /* guess Me */
  for (n=0;n<NN;++n) w[n]=1.+0.025*(n*dX)*(n*dX);    /* nozzle profile  */
   lnr[0][0]=v[0][0]=0.;                         /* initialize the domain */
    u[0][0]=1.;
    pi[0][0]=1./gam;
    u[0][NN-1]=ue_Me*Me;
    pi[0][NN-1]=ue_Me/gam/w[NN-1]/Me;
    for (n=0;n<NN;++n) {      /* initial guess at profiles through nozzle */
         u[0][n]=u[0][0]+(u[0][NN-1]-u[0][0])*n/(NN-1);
         pi[0][n]=pi[0][0]+(pi[0][NN-1]-pi[0][0])*n/(NN-1);
         lnr[0][n]=(log(pi[0][n])+log(gam))/gam;
         for (m=0;m<MM;++m) {
            lnr[m][n]=lnr[0][n];
            u[m][n]=u[0][n];
            v[m][n]=u[0][n]*m*dY*(n==0 ? 0.0 : w[n]-w[n-1])/dX;
            pi[m][n]=pi[0][n];
         }
  }
  do {                                       /* integrate forward in time */
       piExit=MacC2D(&dt,dX,dY,gam);
       if (!(cnt%100)) {
            piExitDel=fabs((piExit-piExitLast)/pi[0][NN-1]);
            piExitLast=piExit;
            printf("cnt=%d piExitDel=%e piExit=%e\n",cnt,piExitDel,piExit);
     }
   } while ((++cnt<50000) && (piExitDel>1.0e-3));
   fp=fopen("soln2D.dat","w");                   /* output solution */
   for (n=0;n<NN;++n)
     for (m=0;m<MM;++m) {
        u2=u[m][n]*u[m][n]+v[m][n]*v[m][n];
        M=sqrt(u2*exp(lnr[m][n])/gam/pi[m][n]);
        fprintf(fp,"%e %e %e %e %e %e %e\n",
            n*dX,m*dY*w[n],M,pi[m][n],u[m][n],v[m][n],exp(lnr[m][n]));
     }
   fclose(fp);
   return 1;
}
```

22.1.3 Nozzle with a Transient Inlet Temperature

Reconsider the sonic flow through a diverging section of a nozzle. Now suppose the sonic inlet conditions change with time as the result of heat added to the stagnation state of the gas, as illustrated in Figure 22-4. Heat added at constant volume raises the stagnation temperature by $\delta T_o = \delta Q/C_{\bar{v}}$ and entropy by $\delta s_o = \delta Q/T_o$. Although the flow remains sonic (choked), the inlet throat conditions change with the degree of heat added. The inlet gas temperature can be calculated from Eq. (21-20) for a calorically perfect ideal gas:

$$T^* = \frac{2(T_o + \delta T_o)}{1+\gamma}. \tag{22-51}$$

With this result, the velocity of the sonic flow $a^* = \sqrt{\gamma R T^*}$ can be found. Since flow from the stagnation state to the nozzle is isentropic, the inlet density is found by integrating the isentropic relation $ds = C_{\bar{v}} dT/T - R d\rho/\rho = 0$ for the result

$$\frac{\rho^*}{\rho_o} = \left(\frac{T^*}{T_o + \delta T_o}\right)^{\frac{1}{\gamma-1}}. \tag{22-52}$$

Or with Eq. (22-51),

$$\frac{\rho^*}{\rho_o} = \left(\frac{2}{1+\gamma}\right)^{\frac{1}{\gamma-1}}. \tag{22-53}$$

Since the stagnation density ρ_o is unchanged by heat added at a constant volume, ρ^* is not a function of the added heat. With the ideal gas law $P^* = \rho^* R T^*$, the inlet pressure is found from Eq. (22-51) and (22-53) to be

$$P^* = \left(\frac{2}{1+\gamma}\right)^{\frac{\gamma}{\gamma-1}} \rho_o R(T_o + \delta T_o). \tag{22-54}$$

In the previous section, the dependent flow variables were made dimensionless by the inlet state. However, in the current situation the inlet state changes with time. Therefore, it is appropriate to redefine the dimensionless dependent variables as

$$u = v_x/a^{ref}, \quad v = v_y/a^{ref}, \quad r = \rho/\rho^{ref}, \quad \pi = P/(\rho^{ref}(a^{ref})^2),$$

$$\Sigma = (s - s^{ref})/C_{\bar{v}} - \ln(\gamma), \quad \text{and} \quad \tau = t\, a^{ref}/W^*, \tag{22-55}$$

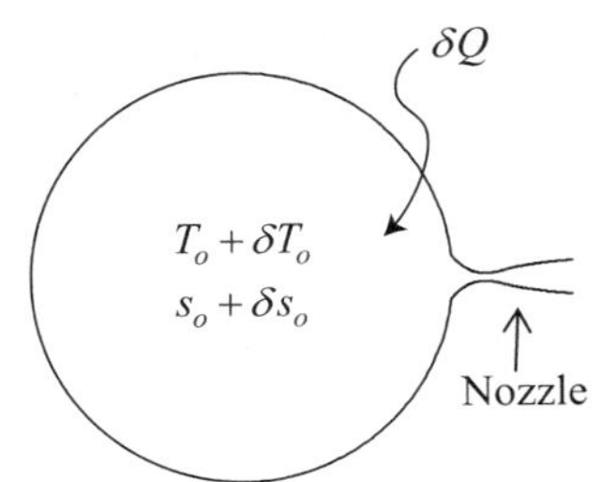

Figure 22-4 Heat addition to stagnation state of nozzle.

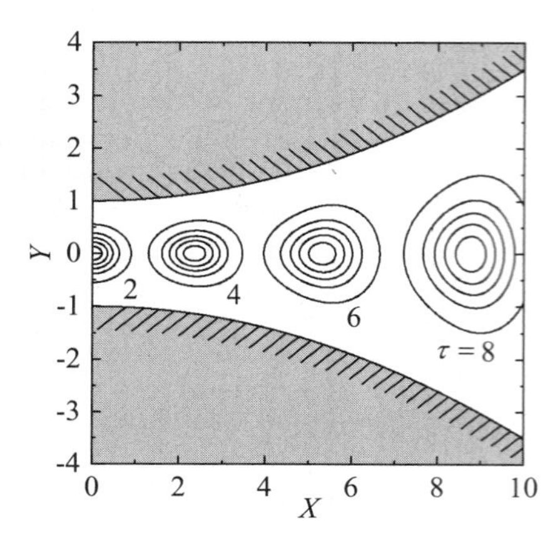

Figure 22-5 Transport of added heat through a diverging nozzle.

where $a^{ref} = a^*(\delta Q = 0)$, $\rho^{ref} = \rho^*(\delta Q = 0)$, and $s^{ref} = s^*(\delta Q = 0)$ correspond to sonic inlet conditions with no added heat. With these new definitions, the inlet boundary conditions for the flow through the nozzle become

$$\text{at } X = 0: \quad \pi = 1/\gamma, \ \ln r = 0, \ u = \sqrt{1 + \delta T_o/T_o}, \ v = 0, \ \Sigma = \delta T_o/T_o - \ln(\gamma). \quad (22\text{-}56)$$

As an example, consider a nonuniform release of heat into the flow that results in the nozzle inlet profile:

$$\delta T_o/T_o = 0.1 \ \exp(-9Y^2)\exp(-4(\tau - 2)^2). \quad (22\text{-}57)$$

The motion of this added heat can be tracked down the length of the nozzle. The added heat is best viewed by the entropy rise, since other gas variables change with the motion of a material element downstream. However, entropy remains constant as it is transported, since $Ds/Dt = 0$. Outside of the region of added heat the entropy remains $\Sigma = -\ln(\gamma)$. Figure 22-5 shows four superimposed snapshots of the region of added heat at times $\tau = 2, 4, 6,$ and 8. The center of the added heat region has a stagnation temperature elevated by $\delta T_o/T_o = 0.1$ and an entropy of $\Sigma = 0.1 - \ln(\gamma)$. The downstream expansion of the heat-added region is a consequence of the drop in density of the gas traveling with the supersonic flow.

22.2 PROBLEMS

22-1 Reconsider the solution to the supersonic flow in a diverging nozzle, discussed in this chapter. Instead of calculating pressure from the state of constant entropy in the flow, develop an expression to determine pressure from a state of constant stagnation enthalpy. Apply this condition to solving the continuity and momentum equations. Contrast the calculated exit Mach numbers between these two methods. When $h_o(x,t) = const.$ is applied as a constraint to the time integration of the continuity and momentum equations, is the description of the flow for times leading up to the steady-state condition physically correct? Does the constraint $s(x,t) = const.$ violate the true transient nature of the solution to this problem?

22-2 Reconsider the flow in the diverging nozzle discussed in this chapter. Suppose a reaction occurs in the flow that releases heat at a rate of $\dot{q}$(W/kg). Reformulate the entropy transport equation to include this heating source. Defining the variables:

$$Y = y/W(x), \ X = x/W^*, \ \tau = t\,a^*/W^*, \ \beta = \chi\,u + v/w, \ w = W(x)/W^*, \ \chi = -(Y/w)(\partial w/\partial X)$$

$$\Sigma = \frac{s - s^*}{C_{\bar{v}}} - \ln(\gamma), \quad u = v_x/a^*, \quad v = v_y/a^*, \quad r = \rho/\rho^*, \quad \pi = P/(\rho^*(a^*)^2), \quad \text{and} \quad Q = \dot{q}W^*/(a^*)^3,$$

show that the entropy transport equation becomes

$$\frac{\partial \Sigma}{\partial \tau} = -\left[u\frac{\partial \Sigma}{\partial X} + \beta\frac{\partial \Sigma}{\partial Y}\right] + \frac{(\gamma - 1)Q}{\exp(\Sigma + (\gamma - 1)\ln r)}.$$

Assume the flow in the diverging nozzle is choked. Solve the entropy, continuity, and momentum equations in the diverging nozzle for $Q = 1$ to determine the steady-state shock-free solution. With heating, the transition from subsonic to supersonic conditions occurs downstream of the inlet. Given the pressure and density of the flow at the inlet, $\pi(X = 0) = 1/\gamma$ and $\ln r(X = 0) = 0$, determine the Mach number at the inlet. At what downstream position does the flow become sonic along the centerline of the nozzle? Plot the Mach lines in the nozzle.

Chapter 23

Runge-Kutta Integration

23.1 Fourth-Order Runge-Kutta Integration of First-Order Equations

23.2 Runge-Kutta Integration of Higher Order Equations

23.3 Numerical Integration of Bubble Dynamics

23.4 Numerical Integration with Shooting

23.5 Problems

Many problems describing advection transport discussed in Chapters 17 through 22 benefited from numerical solutions. Because of the role of advection, it would be reasonable to suspect further utility of numerical techniques for solving convection transport equations in the coming chapters. Although the transient partial differential equation describing one-dimensional convection was solved by numerical MacCormack integration in Section 17.3, the nature of most convection problems yet to be addressed is significantly different. In Chapters 24 through 26, convection equations arising in boundary layer problems will reduce to ordinary differential equations using a similarity variable. (This technique was developed in Chapter 10.) Additionally, in Chapters 27 and 28, the convection equations arising in internal flows will also reduce to ordinary differential equations when transport is "fully developed." When these ordinary differential equations are linear, analytic solution techniques, such as the power series expansion discussed in Section 10.4, will find utility. However, when the ordinary differential equations are nonlinear, numerical integration will be required. Convection transport in nonlinear turbulent flows, as discussed in Chapters 30 through 32, will rely heavily on numerical solutions. Therefore, to address coming topics in convection, it is desired to have a numerical recipe for integrating ordinary differential equations. Although there are many methods for numerically solving ordinary differential equations, in this chapter only the popular fourth-order Runge-Kutta integration is discussed. For a broader perspective on this and related numerical techniques, the interested reader may refer to references [1] and [2].

23.1 FOURTH-ORDER RUNGE-KUTTA INTEGRATION OF FIRST-ORDER EQUATIONS

To illustrate numerical integration of an ordinary differential equation, consider solving the first-order initial value problem:

$$\frac{df}{dx} + 2xf - x = 0, \tag{23-1}$$

with

$$f(0) = 3/2. \tag{23-2}$$

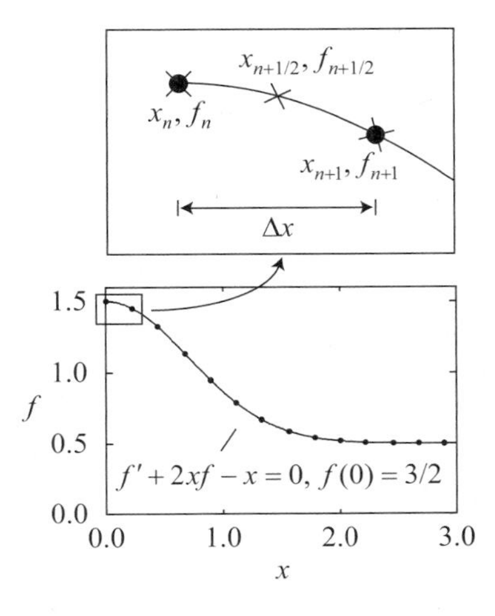

Figure 23-1 Illustration of discretized points used in numerical integration.

For this problem an analytic solution $f = 1/2 + e^{-x^2}$ exists.

Forward integration of (23-1) requires that an initial value of f is known. A numerical solution can be obtained by marching f forward in discrete steps from the initial condition. Consider using the algorithm

$$f_{n+1} = f_n + K, \tag{23-3}$$

where K represents the estimated change in f over the interval x_n to x_{n+1}. Denoting $\Delta x = x_{n+1} - x_n$ and df/dx as a representative value of the slope in f over the interval x_n to x_{n+1}, one can estimate that $K \approx \Delta x \, df/dx$. Different numerical schemes use different approaches for evaluating K. For most functions, the slope df/dx is not constant over the step interval. Therefore, a good estimate of K should be weighted by more than one value of the slope between x_n and x_{n+1}. The fourth-order Runge-Kutta method* uses a weighted average of estimates at x_n, $x_{n+1/2}$, and x_{n+1}, as illustrated in Figure 23-1. $x_{n+1/2} = (x_{n+1} + x_n)/2$ represents the midpoint between x_n and x_{n+1}.

Since f_n is known at node n, determining the slope of the function at x_n is easily obtained from the governing equation (23-1):

$$\left.\frac{df}{dx}\right|_{x_n} = x_n - 2x_n f_n. \tag{23-4}$$

Therefore, a first estimate for the change in f over the interval x_n to x_{n+1} can be made with

$$K_1 = \Delta x \left.\frac{df}{dx}\right|_{x_n} = \Delta x (x_n - 2x_n f_n). \tag{23-5}$$

To estimate the slope of f from the governing equation at the midpoint $x_{n+1/2}$ requires knowledge of $f_{n+1/2}$. However, since this is unknown, an estimate for $f_{n+1/2}$ can be made with

$$f_{n+1/2}^{(1)} = f_n + \frac{\Delta x}{2} \left.\frac{df}{dx}\right|_{x_n} = f_n + \frac{K_1}{2}. \tag{23-6}$$

---

*Named in honor of the German mathematicians Carl David Tolmé Runge (1856–1927) and Martin Wilhelm Kutta (1867–1944).

Here, the superscript (1) is used to denote this result as the first estimate of $f_{n+1/2}$. With this result, an estimate for the slope of the function at $x_{n+1/2}$ can be made from the governing equation (23-1):

$$\left.\frac{df}{dx}\right|_{n+1/2}^{(1)} = x_{n+1/2} - 2x_{n+1/2}f_{n+1/2}^{(1)}. \tag{23-7}$$

Therefore, a second estimate for the change in f over the interval x_n to x_{n+1} can be made by substituting (23-6) into (23-7) for the result

$$K_2 = \Delta x\left.\frac{df}{dx}\right|_{n+1/2}^{(1)} = \Delta x\left(x_{n+1/2} - 2x_{n+1/2}\left(f_n + \frac{K_1}{2}\right)\right). \tag{23-8}$$

Notice that $f_{n+1/2}^{(1)} = f_n + K_1/2$ is an estimate based on $df/dx|_{x_n}$. A second estimate of $f_{n+1/2}$ can be obtained using K_2. Specifically,

$$f_{n+1/2}^{(2)} = f_n + \frac{K_2}{2} \tag{23-9}$$

is an estimate based on $df/dx|_{n+1/2}^{(1)}$. With this second estimate of $f_{n+1/2}$, a second estimate for the function slope at $x_{n+1/2}$ can be made from the governing equation (23-1):

$$\left.\frac{df}{dx}\right|_{n+1/2}^{(2)} = x_{n+1/2} - 2x_{n+1/2}f_{n+1/2}^{(2)}. \tag{23-10}$$

Therefore, a third estimate for the change in f over the interval x_n to x_{n+1} can be made by substituting (23-9) into (23-10) for the result

$$K_3 = \Delta x\left.\frac{df}{dx}\right|_{n+1/2}^{(2)} = \Delta x\left(x_{n+1/2} - 2x_{n+1/2}\left(f_n + \frac{K_2}{2}\right)\right). \tag{23-11}$$

Finally, to estimate the slope of f at x_{n+1} from the governing equation requires knowledge of f_{n+1}. An estimate of f_{n+1} can be made with

$$f_{n+1} = f_n + K_3, \tag{23-12}$$

which is based on $df/dx|_{n+1/2}^{(2)}$. With this estimate of f_{n+1}, an the estimate for the function slope at x_{n+1} can be made from the governing equation (23-1):

$$\left.\frac{df}{dx}\right|_{n+1} = x_{n+1} - 2x_{n+1}f_{n+1}. \tag{23-13}$$

Therefore, a fourth estimate for the change in f over the interval x_n to x_{n+1} can be made by substituting (23-12) into (23-13), for the result

$$K_4 = \Delta x\left.\frac{df}{dx}\right|_{n+1} = \Delta x(x_{n+1} - 2x_{n+1}(f_n + K_3)). \tag{23-14}$$

The fourth-order Runge-Kutta method uses a weighted mean of the four estimates (23-5), (23-8), (23-11), and (23-14) to predict the change in f over the interval x_n to x_{n+1}. Therefore, the final prediction for f_{n+1} given by this weighted average is

$$f_{n+1} = f_n + \frac{K_1 + 2K_2 + 2K_3 + K_4}{6}. \tag{23-15}$$

It can be shown that Eq. (23-15) satisfies the Taylor series expansion of f carried out to five terms [1]. This dictates that the integration method is fourth-order, meaning that the error per step is on the order of Δx^5, while the total accumulated error has an order of Δx^4.

To generalize the fourth-order Runge-Kutta method, notice that the governing equation is being utilized in the form

$$\frac{df}{dx} = F(x,f). \tag{23-16}$$

For the problem at hand, $F(x,f) = x - 2xf$. The numerical solution for f is obtained by marching Eq. (23-15) forward with discrete steps of Δx. The change in f over each step is predicted from Eqs. (23-5), (23-8), (23-11), and (23-14), as summarized by

$$K_1 = \Delta x\, F(x_n, f_n) \tag{23-17a}$$

$$K_2 = \Delta x\, F(x_{n+1/2}, f_n + K_1/2) \tag{23-17b}$$

$$K_3 = \Delta x\, F(x_{n+1/2}, f_n + K_2/2) \tag{23-17c}$$

$$K_4 = \Delta x\, F(x_{n+1}, f_n + K_3) \tag{23-17d}$$

Each prediction ($K_1 \cdots K_4$) requires evaluating $\Delta x\, F(x,f)$ using the governing equation. However, for each prediction, appearances of x and f in the governing equation are handled differently. For K_1, x and f are evaluated simply by x_n and f_n. For K_2, appearances of x are evaluated at $x_{n+1/2}$ and appearances of f are evaluated with $f_n + K_1/2$. For K_3, appearances of x are again evaluated at $x_{n+1/2}$, but appearances of f are now evaluated with $f_n + K_2/2$. Finally, for K_4, appearances of x are evaluated at x_{n+1} and appearances of f are evaluated with $f_n + K_3$.

23.2 RUNGE-KUTTA INTEGRATION OF HIGHER ORDER EQUATIONS

Often the ordinary differential equation to be solved has an order higher than a first-order equation. Fortunately, the methodology for Runge-Kutta integration can easily be extended to a higher order equation by representing it as a coupled system of first-order equations. Consider a second-order equation for $f(x)$ that is generalized into the form

$$\frac{d^2f}{dx^2} = F(x,f,f'). \tag{23-18}$$

For notational simplicity, $df/dx = f'$ is used to represent the first derivative of f. The governing equation (23-18) can be written as two first-order equations for the dependent variables f' and f as follows:

$$\frac{df'}{dx} = F(x,f,f') \tag{23-19}$$

and

$$\frac{df}{dx} = f'. \tag{23-20}$$

The first equation follows from the governing equation (23-18), and the second equation simply conveys a definition. Solutions for f' and f can be marched forward with fourth-order Runge-Kutta integration, using

$$f'_{n+1} = f'_n + \frac{K_1^{f'} + 2K_2^{f'} + 2K_3^{f'} + K_4^{f'}}{6} \tag{23-21a}$$

and

$$f_{n+1} = f_n + \frac{K_1^{f} + 2K_2^{f} + 2K_3^{f} + K_4^{f}}{6}, \tag{23-21b}$$

where f' and f appear as superscripts for the K values to distinguish between the two dependent variables.

For the f' problem, each predicted change in f' ($K_1^{f'} \cdots K_4^{f'}$) over an interval Δx requires evaluating $\Delta x\, F(x, f, f')$ with the governing equation in a procedure similar to that used in the preceding section. However, now the governing equation may be dependent on both f' and f. Using Eq. (23-17) as a guide, the $K_1^{f'} \cdots K_4^{f'}$ values are expressed as

$$K_1^{f'} = \Delta x\, F(x_n, f_n, f'_n) \tag{23-22a}$$

$$K_2^{f'} = \Delta x\, F(x_{n+1/2}, f_n + K_1^{f}/2, f'_n + K_1^{f'}/2) \tag{23-22b}$$

$$K_3^{f'} = \Delta x\, F(x_{n+1/2}, f_n + K_2^{f}/2, f'_n + K_2^{f'}/2) \tag{23-22c}$$

$$K_4^{f'} = \Delta x\, F(x_{n+1}, f_n + K_3^{f}, f'_n + K_3^{f'}). \tag{23-22d}$$

For the f problem, each predicted change in f ($K_1^{f} \cdots K_4^{f}$) over an interval Δx requires evaluating $\Delta x\, f'$. However, changes in f' over the interval can be evaluated with predictions from $K_1^{f'}$, $K_2^{f'}$, and $K_3^{f'}$ in a straightforward way:

$$K_1^{f} = \Delta x\, f'_n \tag{23-23a}$$

$$K_2^{f} = \Delta x\, (f'_n + K_1^{f'}/2) \tag{23-23b}$$

$$K_3^{f} = \Delta x\, (f'_n + K_2^{f'}/2) \tag{23-23c}$$

$$K_4^{f} = \Delta x\, (f'_n + K_3^{f'}). \tag{23-23d}$$

With this recipe for calculating $K_1^{f'} \cdots K_4^{f'}$ and $K_1^{f} \cdots K_4^{f}$, the dependent variables f'_{n+1} and f_{n+1} can be marched forward with Eqs. (23-21a, b) from the initial conditions of $f'(0)$ and $f(0)$, respectively. When all the required initial conditions for a problem are given, forward integration is straightforward. This will be the case for the problem discussed in the next section. However, other problems may have spatially separated constraints. For example, if $f(0) = 1$ and $f(1) = 0$ were the boundary conditions on a problem, the value of $f'(0)$ needed for forward integration of a second-order equation would be part of the unknown solution. In such a case, it is impossible to know the full set of "initial" conditions required for Runge-Kutta integration, and a method for dealing with this is needed. This will be the situation for the problem treated in Section 23.4.

23.3 NUMERICAL INTEGRATION OF BUBBLE DYNAMICS

Momentum transport associated with the growth and collapse of a spherical bubble is often dominated by advection. Since the governing equation is nonlinear, numerical integration is generally required to describe bubble dynamics. Application of fourth-order Runge-Kutta integration is illustrated in this section by solving the second-order bubble dynamics equation developed in Section 14.6.

Consider a spherical bubble of radius R, surrounded by an infinite domain of fluid, as illustrated in Figure 23-2. Far from the bubble, the fluid is at a temperature T_∞ and pressure P_∞. Suppose that initially the fluid surrounding the bubble is static, the bubble radius is $R = R_o$, the ambient pressure is $P_\infty(t < 0) = P_o$, and the bubble pressure is in equilibrium with its surroundings:

$$P_B(t < 0) = p^v(t < 0) + p^g(t < 0) = P_o. \tag{23-24}$$

The pressures p^v and p^g are the partial pressures of the liquid vapor and noncondensable gas in the bubble. For $t \geq 0$, the fluid pressure surrounding the bubble changes with time, such that $P_\infty = P_\infty(t)$. As was developed in Section 14.6, the bubble dynamics is described by the equation

$$R\ddot{R} + \frac{3}{2}\dot{R}^2 + 4\nu\frac{\dot{R}}{R} + \frac{2\sigma}{\rho R} = \frac{p^v - P_\infty(t)}{\rho} + \frac{p_o^g}{\rho}\left(\frac{R_o}{R}\right)^{3k}, \tag{23-25}$$

where

$$\ddot{R} = d^2R/dt^2 \quad \text{and} \quad \dot{R} = dR/dt. \tag{23-26}$$

In Eq. (23-25), $p_o^g = p^g(t < 0)$ is the initial partial pressure of noncondensable gas in the bubble. With $k = 1$, the governing equation describes an isothermal process, while with $k = \gamma = C_p/C_v$ an isentropic process is described.

Suppose that for the present problem, the effect of viscosity and surface tension can be omitted. Nondimensionalization of the remaining terms in the governing equation can be accomplished by letting

$$\eta = \frac{R}{R_o}, \quad \tau = \frac{t}{R_o\sqrt{\rho/P_o}}, \quad \pi^v = \frac{p^v}{P_o}, \quad \pi_o^g = \frac{p_o^g}{P_o}, \quad \text{and} \quad \Pi_\infty(t) = \frac{P_\infty(t)}{P_o}. \tag{23-27}$$

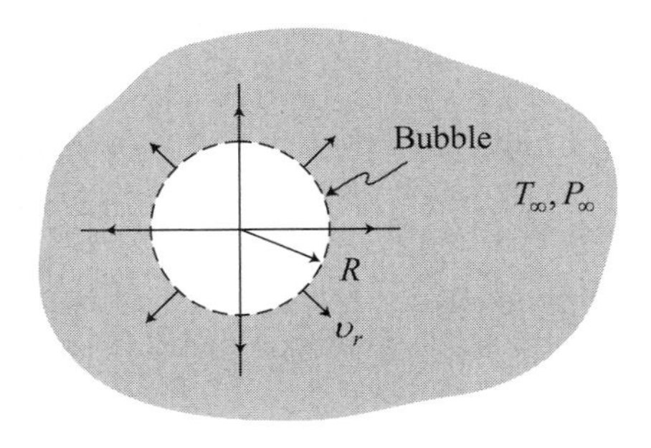

Figure 23-2 Illustration for spherical bubble dynamics.

Without the viscosity and surface tension terms, Eq. (23-25) expressed in terms of the dimensionless variables becomes

$$\eta\ddot{\eta} + \frac{3}{2}\dot{\eta}^2 = \pi^v - \Pi_\infty(\tau) + \pi_o^g \eta^{-3k}. \tag{23-28}$$

This equation may be integrated forward in time for a specified function $\Pi_\infty(\tau)$ describing an ambient pressure fluctuation. Defining

$$F(\tau, \eta, \dot{\eta}) = \eta^{-1}\left[\pi^v - \Pi_\infty(\tau) + \pi_o^g \eta^{-3k} - \frac{3}{2}\dot{\eta}^2\right], \tag{23-29}$$

the first-order equations for $\dot{\eta}$ and η are

$$\frac{\partial \dot{\eta}}{\partial \tau} = F(\tau, \eta, \dot{\eta}), \qquad \dot{\eta}(0) = 0, \tag{23-30}$$

and

$$\frac{\partial \eta}{\partial \tau} = \dot{\eta}, \qquad \eta(0) = 1. \tag{23-31}$$

For this problem, all initial conditions are known and Runge-Kutta integration is straightforward. The equations for $\dot{\eta}$ and η are marched forward in time using

$$\eta_{n+1} = \eta_n + \frac{K_1^\eta + 2K_2^\eta + 2K_3^\eta + K_4^\eta}{6} \quad \text{and} \quad \dot{\eta}_{n+1} = \dot{\eta}_n + \frac{K_1^{\dot{\eta}} + 2K_2^{\dot{\eta}} + 2K_3^{\dot{\eta}} + K_4^{\dot{\eta}}}{6}. \tag{23-32}$$

The K values for the two dependent variables $\dot{\eta}$ and η are

$$\begin{aligned}
K_1^\eta &= \Delta\tau\, \dot{\eta}_n & K_1^{\dot{\eta}} &= \Delta\tau F(\tau_n,\ \eta_n,\ \dot{\eta}_n) \\
K_2^\eta &= \Delta\tau\, (\dot{\eta}_n + K_1^{\dot{\eta}}/2) & K_2^{\dot{\eta}} &= \Delta\tau\, F(\tau_n + \Delta\tau/2,\ \eta_n + K_1^\eta/2,\ \dot{\eta}_n + K_1^{\dot{\eta}}/2) \\
K_3^\eta &= \Delta\tau\, (\dot{\eta}_n + K_2^{\dot{\eta}}/2) & K_3^{\dot{\eta}} &= \Delta\tau\, F(\tau_n + \Delta\tau/2,\ \eta_n + K_2^\eta/2,\ \dot{\eta}_n + K_2^{\dot{\eta}}/2) \\
K_4^\eta &= \Delta\tau\, (\dot{\eta}_n + K_3^{\dot{\eta}}) & K_4^{\dot{\eta}} &= \Delta\tau\, F(\tau_n + \Delta\tau,\ \eta_n + K_3^\eta,\ \dot{\eta}_n + K_3^{\dot{\eta}})
\end{aligned}$$

where the function F is defined by Eq. (23-29).

To consider a particular example, suppose $\pi_v = 0$, $\pi_{g,o} = 1$, and the thermodynamic changes to the gas in the bubble are adiabatic and isentropic (such that $k = \gamma$ in the governing equation). Furthermore, suppose the ambient pressure experiences a dip described by the function

$$\Pi_\infty(\tau) = \begin{cases} (1 + \cos(\pi\,\tau/5))/2 & 0 \le \tau < 10 \\ 1 & \textit{otherwise}. \end{cases} \tag{23-33}$$

Runge-Kutta integration is implemented for the present problem in Code 23-1. The time step for numerical integration should be small compared to significant time scales of the bubble dynamics. Although the time scale for the ambient pressure fluctuation is of order 10, the time step for numerical integration was selected to be much smaller ($d\tau = 0.01$) because of the rapid rebounding events seen in the solution.

Code 23-1 Runge-Kutta solution to bubble dynamics

```
#include <stdio.h>
#include <math.h>

inline double Pinf(double t)
{
   return (t>10 ? 1 : (1.+cos(M_PI*t/5))/2.);
}

inline double F(double t,double R,double dR)
{
   double Pv=0.0,Pg=1.0,k=1.4;
   return (Pv-Pinf(t)+Pg*pow(R,-3.*k)-3.*dR*dR/2.)/R;
}

int main()
{
   int n,N=2000;
   double R[N],dR[N];
   double del_t=0.01;
   double t,K1R,K1dR,K2R,K2dR,K3R,K3dR,K4R,K4dR;
   FILE *fp=fopen("out.dat","w");
   R[0]=        1.;
   dR[0]=       0.;
   for (n=0;n<N-1;++n) {
      t=del_t*n;
      K1R=del_t*dR[n];
      K1dR=del_t*F(t,R[n],dR[n]);
      K2R=del_t*(dR[n]+.5*K1dR);
     K2dR=del_t*F(t+.5*del_t,R[n]+.5*K1R,dR[n]+.5*K1dR);
      K3R=del_t*(dR[n]+.5*K2dR);
     K3dR=del_t*F(t+.5*del_t,R[n]+.5*K2R,dR[n]+.5*K2dR);
      K4R=del_t*(dR[n]+K3dR);
      K4dR=del_t*F(t+del_t,R[n]+K3R,dR[n]+K3dR);
      R[n+1]=R[n]+(K1R+2*K2R+2*K3R+K4R)/6;
      dR[n+1]=dR[n]+(K1dR+2*K2dR+2*K3dR+K4dR)/6;
      fprintf(fp,"%e %e %e\n",t,R[n],Pinf(t));
   }
   fclose(fp);
   return 0;
}
```

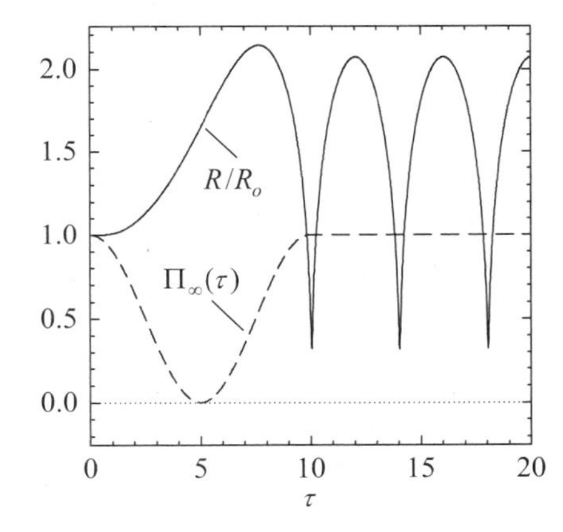

Figure 23-3 Bubble radius responding to an ambient pressure dip.

The numerical result for $\eta = R/R_o$ is plotted in Figure 23-3. The dip in ambient pressure causes an initial expansion of the bubble that collapses after the ambient pressure is restored to the initial value. Bubble collapse causes a compression of the gas that is explored in Problem 23-4. The rise in bubble pressure during collapse eventually reverses the radial flow of the surrounding fluid. This causes the bubble to grow again, even in the presence of the now static ambient pressure. As shown in Figure 23-3, the bubble growth and collapse is repeated, and would continue indefinitely in the absence of any viscous damping.

23.4 NUMERICAL INTEGRATION WITH SHOOTING

As a second illustration of Runge-Kutta integration, consider the mass transfer problem solved analytically in Section 10.5. It was determined that a species concentration through a semi-infinite solid over time can be expressed as $c(x,t) = b\,t^{\gamma}\phi$, where b and γ are constants and ϕ satisfies the governing equation

$$\phi'' + 2\eta\phi' - 4\gamma\phi = 0, \tag{23-34}$$

in which

$$\phi'' = d^2\phi/d\eta^2 \quad \text{and} \quad \phi' = d\phi/d\eta. \tag{23-35}$$

The governing equation for ϕ is a function of the similarity variable,

$$\eta = x/\sqrt{4\,Đ_A t}, \tag{23-36}$$

and is subject to the boundary conditions

$$\phi(0) = 1 \quad \text{and} \quad \phi(\infty) = 0. \tag{23-37}$$

The governing equation may be expressed as two first-order equations for ϕ' and ϕ:

$$\frac{\partial\phi'}{\partial\eta} = 4\gamma\phi - 2\eta\phi' = F(\eta,\phi,\phi'), \qquad \phi'(0) = \underline{?}\,, \text{ leading to } \phi(\infty) = 0 \tag{23-38}$$

and

$$\frac{\partial\phi}{\partial\eta} = \phi', \qquad\qquad \phi(0) = 1. \tag{23-39}$$

To integrate these equations forward with Runge-Kutta requires initial values of $\phi'(0)$ and $\phi(0)$. However, the problem description dictates $\phi(0) = 1$ and $\phi(\infty) = 0$, which makes $\phi'(0)$ part of the unknown solution. Therefore, to apply Runge-Kutta, $\phi'(0)$ must be guessed. The guess is evaluated by checking to see whether integration of the governing equation for ϕ yields $\phi(\infty) = 0$. If it does not, $\phi'(0)$ is guessed again until it does. This approach to a solution is known as the *shooting method*.

Runge-Kutta integration is implemented for the present problem in Code 23-2. The equations for ϕ' and ϕ are marched forward using

$$\phi_{n+1} = \phi_n + \frac{K_1^{\phi} + 2K_2^{\phi} + 2K_3^{\phi} + K_4^{\phi}}{6} \quad \text{and} \quad \phi'_{n+1} = \phi'_n + \frac{K_1^{\phi'} + 2K_2^{\phi'} + 2K_3^{\phi'} + K_4^{\phi'}}{6}. \tag{23-40}$$

The K values for the two dependent variables ϕ' and ϕ are

$$\begin{aligned}
K_1^{\phi} &= \Delta\eta\,\phi'_n & K_1^{\phi'} &= \Delta\eta\,[4\gamma\,\phi_n - 2\eta\phi'_n]\\
K_2^{\phi} &= \Delta\eta\,(\phi'_n + K_1^{\phi'}/2) & K_2^{\phi'} &= \Delta\eta\,[4\gamma(\phi_n + K_1^{\phi}/2) - 2(\eta_n + \Delta\eta/2)(\phi'_n + K_1^{\phi'}/2)]\\
K_3^{\phi} &= \Delta\eta\,(\phi'_n + K_2^{\phi'}/2) & K_3^{\phi'} &= \Delta\eta\,[4\gamma(\phi_n + K_2^{\phi}/2) - 2(\eta_n + \Delta\eta/2)(\phi'_n + K_2^{\phi'}/2)_n]\\
K_4^{\phi} &= \Delta\eta\,(\phi'_n + K_3^{\phi'}) & K_4^{\phi'} &= \Delta\eta\,[4\gamma(\phi_n + K_3^{\phi}) - 2(\eta_n + \Delta\eta)(\phi'_n + K_3^{\phi'})].
\end{aligned}$$

The results for ϕ' and ϕ are plotted in Figure 23-4. The unknown initial value of ϕ' is determined from the numerical solution to be $\phi'(0) = -2.2568$, when $\gamma = 1$. This numerical result is in agreement with the analytic result of Section 10.5.

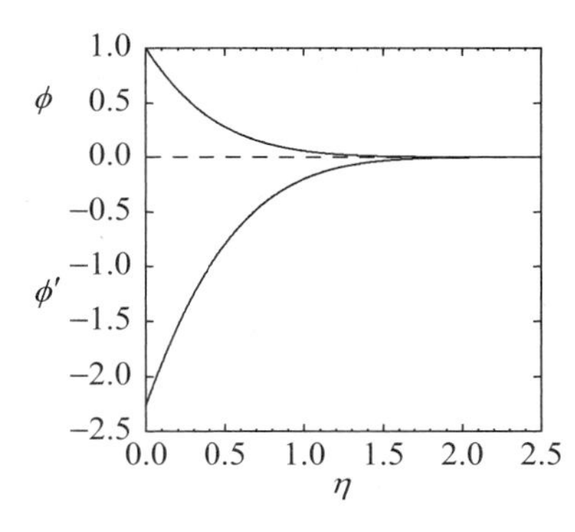

Figure 23-4 Numerical solution to Eq. (23-34) for example mass transfer problem.

Several points about Code 23-2 deserve additional comment. The first point regards the meaning of $\eta \rightarrow \infty$. The mathematical problem stipulates that $\phi(\eta \rightarrow \infty) = 0$. However, "infinity" must be replaced with a suitably large, but finite, number for numerical integration. Consulting Figure 23-4, it is seen that $\phi(\eta > 2) \approx 0$. Therefore, $\eta_\infty = 2.5$ is a good representation of infinity for this problem. However, initially one does not have the solution to consult for this advice. Therefore, one must guess an appropriately large value of η_∞. If some number η_∞ is picked to represent infinity that is too small, then the numerical solution will force $\phi(\eta = \eta_\infty) = 0$ to be satisfied. However, the solution will be wrong because $\phi'(\eta = \eta_\infty) \neq 0$. In other words, the numerical solution for ϕ will be forced to cross the value of zero at η_∞, as opposed to asymptotically approaching zero.

One may be tempted to pick η_∞ much larger than is necessary just to be "safe." However, for a fixed number of numerical steps from $\eta = 0$ to $\eta = \eta_\infty$, the larger η_∞ becomes the coarser $\Delta\eta$ becomes. If $\Delta\eta$ is too coarse over the interval in which important changes in ϕ occur, the quality of the solution will suffer. One should check that a numerical solution is not sensitive to the chosen value of $\Delta\eta$. In other words, if $\Delta\eta$ is appropriately small for a good solution, then replacing $\Delta\eta$ by $\Delta\eta/2$ should yield the same result.

23.4.1 Bisection Method

Finally, attention should be given to the process of guessing $\phi'(0)$ to satisfy $\phi(\eta \rightarrow \infty) = 0$. In Code 23-2, a bisection method is used to make consecutive guesses of $\phi'(0)$. As illustrated in Figure 23-5, numerical integration will result in $\phi(\eta \rightarrow \infty) > 0$ if $\phi'(0)$ is guessed too high and $\phi(\eta \rightarrow \infty) < 0$ if $\phi'(0)$ is guessed too low. To use the bisection method, the true value of $\phi'(0)$ must be bounded by a high ϕ'_{max} and low ϕ'_{min} value. Each iteration of the bisection method makes the guess that $\phi'(0) = (\phi'_{\mathrm{max}} + \phi'_{\mathrm{min}})/2$. Integration of the governing equation then establishes $\phi(\eta \rightarrow \infty)$ corresponding to the guess. For the current example, if $\phi(\eta \rightarrow \infty) > 0$ then the guessed $\phi'(0)$ is too high and one can update $\phi'_{\mathrm{max}} = \phi'(0)$. If $\phi(\eta \rightarrow \infty) < 0$ then the guessed $\phi'(0)$ is too low and one can update $\phi'_{\mathrm{min}} = \phi'(0)$. In this way the difference between ϕ'_{max} and ϕ'_{min} is reduced, and the subsequent iteration proceeds with the next guess that $\phi'(0) = (\phi'_{\mathrm{max}} + \phi'_{\mathrm{min}})/2$.

The bisection method is not the fastest converging method, but it is stable and easy to code. Shooting methods are generally susceptible to the difficulty that if the $\phi'(0)$ guess is too far from the truth, ϕ may become very large (in the positive or negative sense) before numerical integration reaches η_∞. This can cause numerical overflow and bring the calculation to a halt. This event is avoided in the code by inspecting ϕ as integration proceeds, and breaking off the integration before reaching η_∞ if ϕ becomes too large (as either a positive or negative number).

Code 23-2 Runge-Kutta solution to example mass transfer problem

```
#include <stdio.h>
#include <math.h>

inline double F(double eta,double C,double dC)
{
   double gam=1.;
   return 4.*gam*C-2.*eta*dC;
}

int main()
{
   int N=1000;
   int n,iter=0;
   double C[N],dC[N];
   double eta_inf=2.5;
   double del_eta=eta_inf/(double)(N-1);
   double eta,K1C,K1dC,K2C,K2dC,K3C,K3dC,K4C,K4dC;
   FILE *fp;

   // put bounds on the possibilities for dC0
   double dC0_high=0.;
   double dC0_low=-10.;

   double C_inf=0.;  // end BC
   C[0]=        1.;  // init BC

   do {
      dC[0]=(dC0_low+dC0_high)/2.0;
      for (n=0;n<N-1;++n) {
         eta=del_eta*n;

         K1C=del_eta*dC[n];
         K1dC=del_eta*F(eta,C[n],dC[n]);
         K2C=del_eta*(dC[n]+.5*K1dC);
         K2dC=d_eta*
           F(eta+.5*del_eta,C[n]+.5*K1C,dC[n]+.5*K1dC);
         K3C=del_eta*(dC[n]+.5*K2dC);
         K3dC=del_eta*
           F(eta+.5*del_eta,C[n]+.5*K2C,dC[n]+.5*K2dC);
         K4C=del_eta*(dC[n]+K3dC);
       K4dC=del_eta*F(eta+del_eta,C[n]+K3C,dC[n]+K3dC);
         C[n+1]=C[n]+(K1C+2*K2C+2*K3C+K4C)/6;
         dC[n+1]=dC[n]+(K1dC+2*K2dC+2*K3dC+K4dC)/6;

         if (fabs(C[n+1])>10.) { // soln. is blowing up
            C[N-1]=C[n+1];         // jump to end
            break;
         }
      }
      // Bi-section method (slow but stable)
      if (C[N-1]>C_inf) dC0_high=dC[0]; // shoot lower
      else dC0_low=dC[0];               // shoot higher
   } while (++iter<200 && fabs(C[N-1]-C_inf)>1.e-6);
   printf("\niter=%d",iter);
   if (iter==200) printf("\nSoln. Failed!");
   else {
      printf("\ndC0=%e\n",dC[0]);
      fp=fopen("soln.dat","w");
      for (n=0;n<N;++n)
         fprintf(fp,"%e %e %e\n",n*del_eta,C[n],dC[n]);
      fclose(fp);
   }

   return 0;
}
```

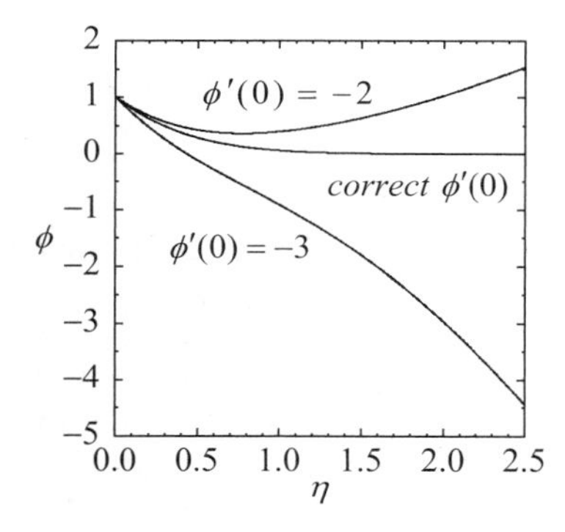

Figure 23-5 Illustration of shooting to determine $\phi'(0)$.

23.4.2 Newton-Raphson Method

The Newton-Raphson method† is a more efficient guessing algorithm for shooting than the bisection method, but requires a few more steps to implement. Let $\phi_i'(0)$ be the ith

†This method is named after the English physicist Sir Isaac Newton (1643–1727) and the English mathematician Joseph Raphson (1648–1715).

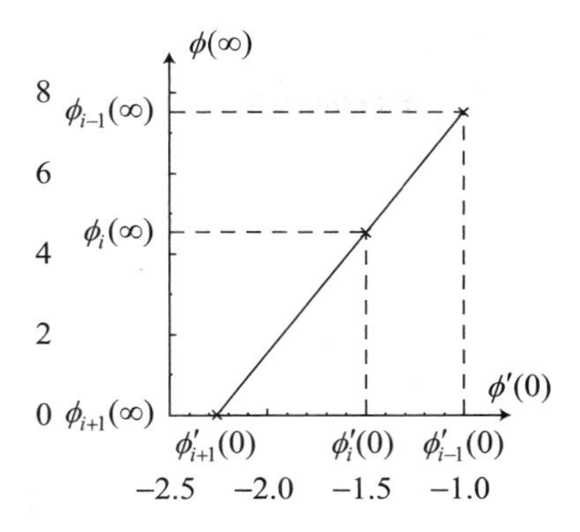

Figure 23-6 Using Newton-Raphson to find $\phi'(0)$ satisfying $\phi(\infty) = 0$.

guess and $\phi'_{i-1}(0)$ be the previous guess for the value of $\phi'(0)$. Suppose that both $\phi'_i(0)$ and $\phi'_{i-1}(0)$ have yielded values of $\phi_i(\infty)$ and $\phi_{i-1}(\infty)$, respectively, that do not satisfy the problem requirement that $\phi(\eta \to \infty) = 0$. With the aid of Figure 23-6, it is seen that a good next guess for $\phi'_{i+1}(0)$ is given by

$$\overbrace{\phi_{i+1}(\infty)}^{want\,=\,0} - \overbrace{\phi_i(\infty)}^{current} = \overbrace{\frac{\partial \phi(\infty)}{\partial \phi'(0)}}^{slope} \left(\overbrace{\phi'_{i+1}(0)}^{next} - \overbrace{\phi'_i(0)}^{current} \right), \tag{23-41}$$

where

$$\frac{\partial \phi(\infty)}{\partial \phi'(0)} \approx \frac{\phi_i(\infty) - \phi_{i-1}(\infty)}{\phi'_i(0) - \phi'_{i-1}(0)}. \tag{23-42}$$

Equation 23-41 is rearranged for

$$\phi'_{i+1}(0) = \phi'_i(0) - \frac{\phi'_i(0) - \phi'_{i-1}(0)}{\phi_i(\infty) - \phi_{i-1}(\infty)} \phi_i(\infty). \tag{23-43}$$

It is apparent that the Newton-Raphson algorithm for guessing $\phi'_{i+1}(0)$ with Eq. (23-43) requires the governing equation to have been integrated twice previously using $\phi'_i(0)$ and $\phi'_{i-1}(0)$ in order to establish corresponding values of $\phi_i(\infty)$ and $\phi_{i-1}(\infty)$. Additionally, for highly nonlinear problems, some care is required to start the Newton-Raphson procedure with a reasonable guess for $\phi'(0)$. In contrast, the bisection method only requires the ability to bound the correct value of $\phi'(0)$ with high and low limiting values.

23.5 PROBLEMS

23-1 For the ordinary differential equation given below, set up the governing equations and K equations needed for Runge-Kutta integration of each first-order equation:

$$x^2 \frac{d^2y}{dx^2} + x\frac{dy}{dx} + (x^2 - n^2)y = 0, \quad \text{where } n \text{ is a constant (Bessel equation)}$$

Integrate the $n=0$ Bessel equation over the interval $0 \le x \le 10$, using the initial conditions $y(0)=1$ and $y'(0)=0$. Note: The $n=0$ Bessel equation observes the limiting behavior that $\lim_{x\to 0} d^2y/dx^2 \to -1/2$. Contrast y and y' with the Bessel functions $J_0(x)$ and $J_1(x)$, as evaluated with the math library.

23-2 Consider a semi-infinite solid initially at a low temperature whose surface is brought to a high temperature at time zero. The similarity solution to this problem is governed by $\theta'' + 2\eta\theta' = 0$, subject to the boundary conditions $\theta(0)=1$ and $\theta(\infty)=0$. Numerically solve this ordinary differential equation using fourth-order Runge-Kutta integration with the shooting method, and compare your results with the exact solution $\theta = erfc(\eta)$.

23-3 Using the Newton-Raphson method for shooting, solve the equation $\phi'' + 2\eta\phi' - 4\gamma\phi = 0$, with $\phi(0)=1$ and $\phi(\infty)=0$, as discussed in Section 23.4. Contrast the number of guesses required for the Newton-Raphson method to determine $\phi'(0)$ versus the bisection method.

23-4 Consider the motion of an adiabatic and isentropic gas bubble as it experiences a dip in the ambient pressure, as described in Section 23.3. At what bubble radius does the effect of viscosity and surface tension start to become important for a fluid like water? Investigate the gas pressure in the bubble as a function of time when the ambient pressure change is described by Eq. (23-33). Assume the initial conditions are $P_o = 101$ kPa, $T_o = 293$ K, $R_o = 2\ \mu$m, and include the effects of viscosity and surface tension. The liquid is water, with $\rho = 998$ kg/m^3, $\sigma = 0.071$ N/m, and $\nu = 1. \times 10^{-6}$ m^2/s.

23-5 The governing equation for the liquid pressure distribution in a journal bearing was derived in Problem 13-3 to be

$$\frac{d\Pi}{d\theta} = 6\frac{1 + (a/\varepsilon)\cos\theta + A}{[1 + (a/\varepsilon)\cos\theta]^3}, \quad \text{where} \quad \Pi = \frac{P(\theta) - P(\theta = 0)}{\mu\omega(R/\varepsilon)^2}.$$

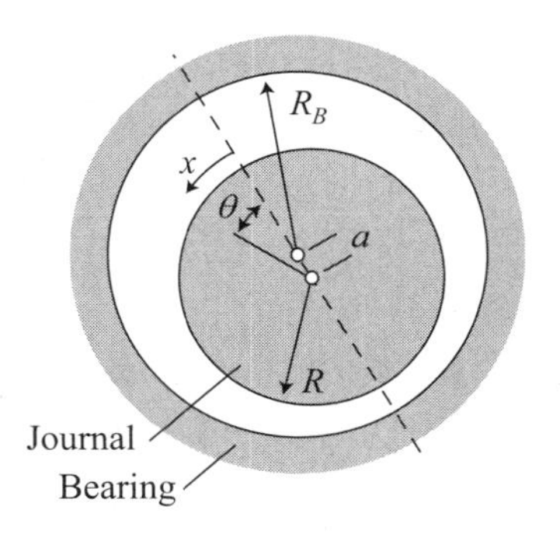

a is the displacement of the journal center from the bearing center, and $\varepsilon = R_B - R$ is the difference in radii between the bearing and the journal. The integration constant A has yet to be determined. Using $a/\varepsilon = 0.5$, numerically integrate the pressure equation, and evaluate the integration constant A such that $P(2\pi) = P(0)$. What is the value of A? Contrast the numerical solution with Sommerfold's analytic solution [3]:

$$\frac{P(\theta) - P(0)}{\mu\omega(R/\varepsilon)^2} = \frac{6(a/\varepsilon)\sin\theta\,[2 + (a/\varepsilon)\cos\theta]}{[2 + (a/\varepsilon)^2][1 + (a/\varepsilon)\cos\theta]^2}.$$

23-6 Sommerfeld's solution to the journal-bearing pressure distribution (see Problem 23-5) can predict unrealistically large negative pressures where the bearing gap is diverging. In this region the fluid can cavitate and exhibit an approximately constant pressure. The Swift-Stieber model [4, 5] attempts to capture this effect by determining the integration constant A in the journal-bearing pressure equation that allows the pressure distribution to smoothly approach $P(0)$ at an unknown angle θ_{cav} where fluid cavitation begins. The pressure distribution is then assumed to be constant until the beginning of the converging section of the journal-bearing gap: $P(\theta_{cav} \leq \theta \leq 2\pi) = P(0)$. Employ a shooting method to determine A for the Swift-Stieber model, such that at $\theta = \theta_{cav}$, $P = P(0)$, and $dP/d\theta = 0$. Plot the pressure distribution in the journal bearing for $a/\varepsilon = 0.5$. At what angle does cavitation begin?

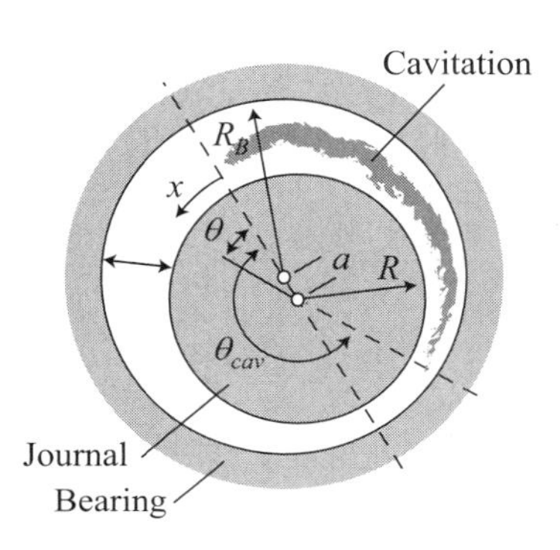

23-7 A reaction is conducted in a liquid film, $A + B \rightarrow 2C$. The volumetric rate of formation of species C is given by $R_C = \kappa c_A c_B$ (moles/m^3/s). Consider a problem in which the concentrations of species at the boundaries of the liquid film are given as in the illustration. Assume dilute concentrations of all species, and that the liquid diffusivities of all species are the same. Show that the governing equation for the concentration of species A is

$$\frac{d^2\theta}{d\eta^2} = Da(2\eta - 1 + \theta)\theta,$$

where $\theta = c_A/c_o$, $\eta = x/L$, and $Da = \kappa c_o L^2/\mathcal{D}$. Numerically solve the governing equation for $\theta(\eta)$ when $Da = 10$. Plot c_A/c_o and c_B/c_o.

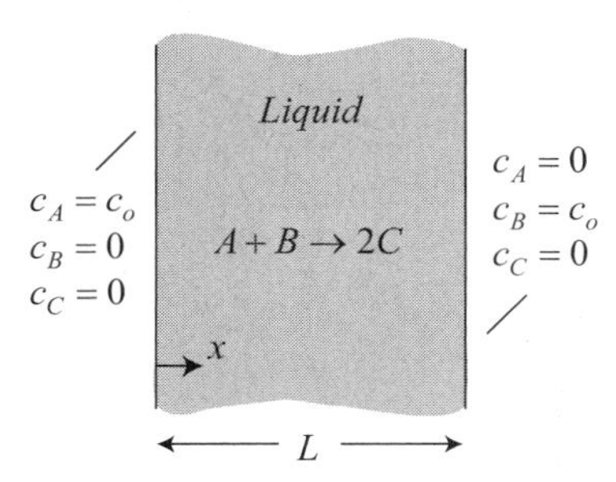

REFERENCES

[1] J. C. Butcher, *Numerical Methods for Ordinary Differential Equations*, Second Revised Edition. New York, NY: Wiley, 2003.

[2] J. D. Lambert, *Numerical Methods for Ordinary Differential Systems*. Chichester, UK: John Wiley & Sons, 1991.

[3] A. Sommerfeld, "Zur hydrodynamischen Theorie der Schmiermittelreibung." *Zeitschrift für Mathematik und Physik*, **50**, 97 (1904).
[4] H. W. Swift, "Stability of Lubricating Films in Journal Bearings." *Proceedings Institute Civil Engineers (London)*, **233**, 267 (1932).
[5] W. Stieber, *Das Schwimmlager: Hydrodynamische Theorie des Gleitlagers*. Berlin, Germany: VDI-Verlag, 1933.

Chapter 24

Boundary Layer Convection

24.1 Scanning Laser Heat Treatment
24.2 Convection to an Inviscid Flow
24.3 Species Transfer to a Vertically Conveyed Liquid Film
24.4 Problems

Convection describes transport in which both advection and diffusion are important. Convection transport generally requires a solution to the nonlinear momentum equation and is most commonly associated with boundary layers that form when a flow interacts with a surface. Since surfaces are normally impenetrable to the flow, only diffusion transport can directly interact with the surface; advection cannot provide transport across the surface boundary. However, as diffusion transports momentum, heat, and species between the surface and the flow, further from the surface advection takes over sweeping properties of the flow downstream with the fluid. Ludwig Prandtl is credited for developing the notion that the effects of fluid viscosity are only experienced in a small region of the flow near a bounding wall [1]. This notion led to a whole branch of fluid mechanics devoted to boundary layer theory [2].

A distinguishing characteristic of boundary layer problems is that the streamwise length scale for advection L is much greater than the transverse length scale for diffusion δ (the boundary layer thickness), as shown in Figure 24-1. As a result of this disparity in scales, $\delta/L \ll 1$, diffusion across the boundary layer is generally large compared with streamwise diffusion. The *boundary layer approximation* is to retain only the dominant diffusion term and thereby simplify the mathematical description of convection transport. An extension of this notion to fully developed internal flows through ducts and pipes will prove useful in Chapter 27.

In the present chapter, analysis of boundary layers of heat and mass transfer is introduced with flows that are inviscid or hydrodynamically fully developed. Solving the nonlinear form of the momentum equation is thereby avoided. The nonlinear problems arising from viscous external boundary layers are treated in Chapters 25 and 26.

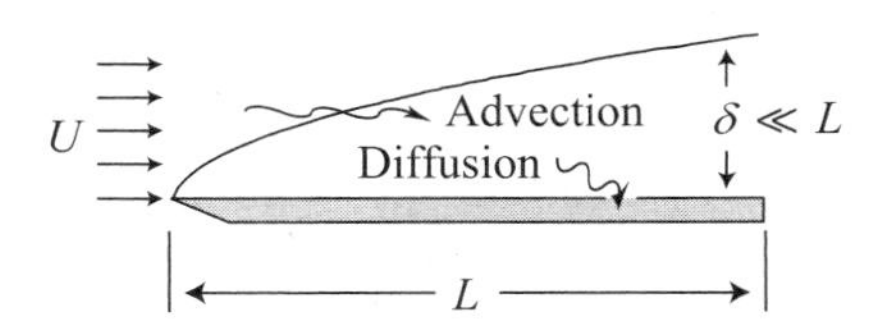

Figure 24-1 Boundary layer forming over a flat plate.

24.1 SCANNING LASER HEAT TREATMENT

To illustrate boundary layer analysis, consider a bar being heat treated with a laser as shown in Figure 24-2. Heating occurs over a patch of length w on the surface of the bar

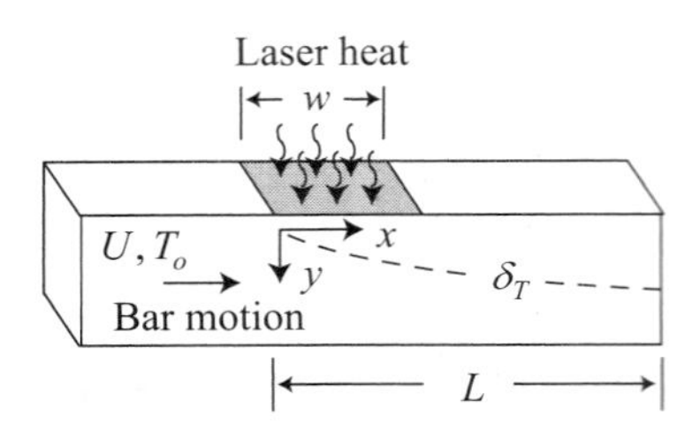

Figure 24-2 Scanning laser heat treatment.

that moves with a speed U past the laser. For successful heat treatment, the maximum temperature rise in the bar must be determined.

Due to the ridged body motion of the bar, the velocity field description is trivial: $v_x = U$ and $v_y = 0$. Furthermore, with the boundary layer approximation that $\partial^2 T/\partial y^2 \gg \partial^2 T/\partial x^2$, the steady heat equation simplifies greatly,

$$\overbrace{v_x}^{U} \frac{\partial T}{\partial x} + \overbrace{v_y}^{0} \frac{\partial T}{\partial y} = \alpha \left(\overbrace{\frac{\partial^2 T}{\partial x^2}}^{\approx 0} + \frac{\partial^2 T}{\partial y^2} \right), \tag{24-1}$$

such that the governing equation becomes

$$U \frac{\partial T}{\partial x} = \alpha \frac{\partial^2 T}{\partial y^2}. \tag{24-2}$$

A consequence of the boundary layer approximation is that the governing equation is only first-order differential with respect to x, requiring only an initial temperature condition for the bar moving beneath the laser. Notice that without diffusion in the x-direction, heat cannot propagate upstream into the material advancing toward the laser. This is a reasonable description only if transport in the x-direction by advection is large compared with diffusion. This justification is built into the boundary layer scaling argument that $\delta_T/w \ll 1$. When the characteristic scales,

$$x \sim w, \quad y \sim \delta_T, \quad v_x \sim U, \quad \text{and} \quad T \sim \Delta T, \tag{24-3}$$

are applied to the boundary layer heat equation (24-2), one finds

$$U \frac{\Delta T}{w} \sim \alpha \frac{\Delta T}{\delta_T^2} \quad \text{or} \quad \frac{Uw}{\alpha} \left(\frac{\delta_T}{w} \right)^2 \sim 1. \tag{24-4}$$

As long as $\delta_T/w \ll 1$, it must follow that $Uw/\alpha \gg 1$, which is simply a statement that transport in the x-direction by advection is large compared with diffusion. This dimensionless quantity contrasting advection transport with diffusion is the heat transfer *Péclet number*:

$$\text{Pe}_w = \frac{Uw}{\alpha}. \tag{24-5}$$

Notice that the Péclet number has a form similar to the Reynolds number, but references the thermal diffusivity of the fluid α instead of the momentum diffusivity ν.

The governing equation (24-2) is second-order differential with respect to y, requiring two boundary conditions in that direction. At the top surface of the bar, the heat diffusion flux must equal the laser energy flux. As long as the thermal penetration depth is small compared with the thickness of the bar, the second condition imposed in y could be that the temperature field approaches the initial value T_o with distances "far" from the surface. With the additional statement that the initial temperature of the bar upstream of the laser is T_o, a solution to Eq. (24-2) for the temperature field beneath the laser ($0 \leq x \leq w$) is prescribed by the set of boundary conditions:

$$T(x = 0) = T_o \tag{24-6}$$

$$I_o = -k \cdot dT/dy|_{y=0}, \quad \text{and} \quad T(y \to \infty) = T_o. \tag{24-7}$$

As was seen in Section 10.2, a similarity solution does not exist to this problem, as stated. The difficulty is associated with having a nonhomogeneous boundary condition of the second kind (related to $\partial T/\partial y$), which cannot be expressed in terms of the similarity variable alone. However, the original problem can be made amenable to a similarity solution by transforming the dependent variable into one that describes the y-direction component of the heat flux:

$$q_y = -k\frac{\partial T}{\partial y}, \quad \text{such that } T - T_o = \int_y^{\infty} (q_y/k)dy. \tag{24-8}$$

The governing equation is transformed by differentiating all terms with respect to y, multiplying all terms by the constant $-k$, and changing the order of differentiation of the term with mixed derivatives. The transformed problem is summarized in Figure 24-3. A similarity solution is sought for this problem by letting

$$\eta = \frac{y}{2\sqrt{\alpha x/U}}. \tag{24-9}$$

This similarity variable resembles that used in Section 10.2, when the dependent variable t in the former definition is replaced by x/U for the current problem. The transformed governing equation becomes

$$\frac{\partial^2 q_y}{\partial^2 \eta} + 2\eta\frac{\partial q_y}{\partial \eta} = 0. \tag{24-10}$$

The transformed initial and boundary conditions become

$$\underbrace{\begin{aligned} q_y(x = 0) &= 0 \\ q_y(y \to \infty) &= 0 \\ q_y(y = 0) &= I_o \end{aligned}}_{3\ \textit{original conditions}} \qquad \underbrace{\begin{aligned} &\left.\begin{aligned} q_y(\eta \to \infty) &= 0 \\ q_y(\eta \to \infty) &= 0 \end{aligned}\right\} \textit{same} \\ &q_y(\eta = 0) = I_o \end{aligned}}_{2\ \textit{final conditions}} \tag{24-11}$$

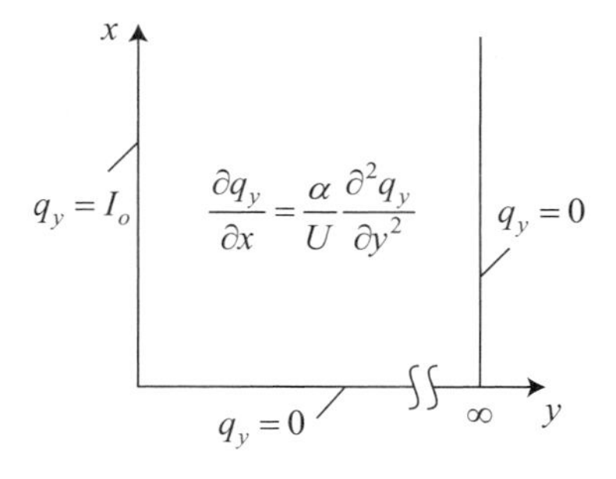

Figure 24-3 Heat flux sub-problem for the laser heat treatement of a moving bar.

A collapse of two of the conditions imposed on the problem occurs, as required for the similarity solution to be successful. The solution for q_y following from the ordinary differential equation (24-10) was determined in Section 10.2 to be

$$q_y = I_o\, erfc\left(\frac{y}{2\sqrt{\alpha x/U}}\right) \quad \text{where} \quad erfc(\eta) = 1 - erf(\eta). \tag{24-12}$$

Returning to the original variables of the problem for $T(x, y)$ results in the solution

$$\begin{aligned}\frac{T - T_o}{I_o/k} &= \int_y^\infty erfc\left(\frac{y}{2\sqrt{\alpha x/U}}\right) dy \\ &= \left[\sqrt{\frac{4\alpha x/U}{\pi}}\exp\left(-\frac{y^2}{4\alpha x/U}\right) - y \cdot erfc\left(\frac{y}{\sqrt{4\alpha x/U}}\right)\right].\end{aligned} \tag{24-13}$$

This solution is valid only for $x \le w$, since beyond this distance the surface boundary condition changes.

A solution for $x > w$ can be found by superposition of two related problems, as shown in Figure 24-4. It is assumed that the top surface of the bar is adiabatic for distances beyond the region where laser heat is delivered. Consequently, the T_2 problem constructed for $x > w$ has a surface heat flux boundary condition that is opposite in sign to the original problem, T_1. Consequently, when the two subproblems are superimposed for $x > w$, the surface heat flux sums to zero. Notice that the solution for T_1 is already given by Eq. (24-13). The solution for $T_2(x > w)$ can be obtained by slight modifications to the T_1 solution, which require that $T_o = 0$, the sign of I_o to be reversed, and the distance x is offset by w. Making the spatial variables dimensionless with $x^* = x/w$ and $y^* = y/w$, the solutions for the two subproblems are given by

$$\frac{T_1 - T_o}{I_o w/k} = \left[\frac{2}{\sqrt{\pi}}\sqrt{\frac{x^*}{\text{Pe}_w}}\exp\left(-\left(\frac{y^*}{2}\right)^2\frac{\text{Pe}_w}{x^*}\right) - y^*\, erfc\left(\frac{y^*}{2}\sqrt{\frac{\text{Pe}_w}{x^*}}\right)\right] \tag{24-14}$$

$$\frac{T_2}{I_o w/k} = -\left[\frac{2}{\sqrt{\pi}}\sqrt{\frac{x^* - 1}{\text{Pe}_w}}\exp\left(-\left(\frac{y^*}{2}\right)^2\frac{\text{Pe}_w}{x^* - 1}\right) - y^*\, erfc\left(\frac{y^*}{2}\sqrt{\frac{\text{Pe}_w}{x^* - 1}}\right)\right]. \tag{24-15}$$

Therefore, the solution to the original problem is constructed from

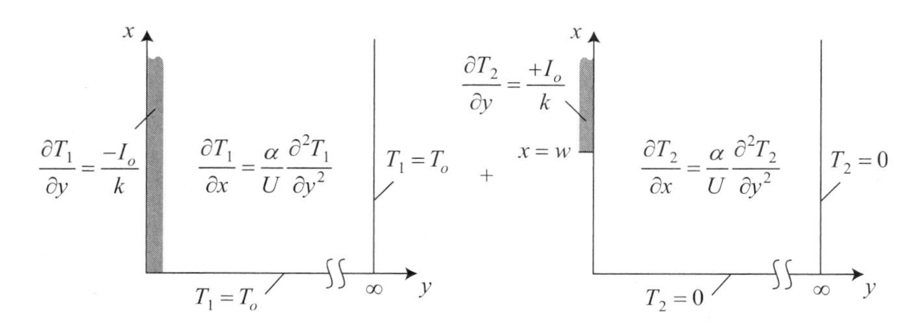

Figure 24-4 Superposition of two sub-problems for the laser heat treatment solution.

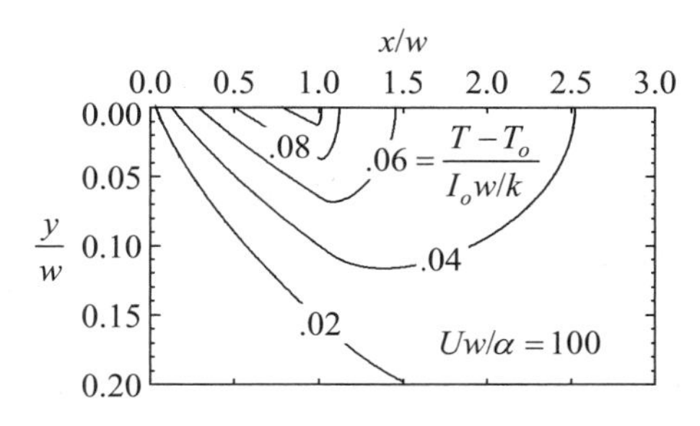

Figure 24-5 Isotherms in the moving bar during laser heat treatment.

$$T(x,y) = \begin{cases} T_1 & x \le w \\ T_1 + T_2 & x > w \end{cases}. \tag{24-16}$$

Observe that the solution depends on the Péclet number $\mathrm{Pe}_w = Uw/\alpha$. For the boundary layer assumption $\delta_T/w \ll 1$ to be valid, it must be true that $\mathrm{Pe}_w \gg 1$, as was seen from scaling of the heat equation. The solution for the temperature field in the bar is presented in Figure 24-5 for the case where $\mathrm{Pe}_w = 100$. Notice that depth into the bar is plotted with an expanded scale.

The maximum temperature occurs at the surface of the bar ($y = 0$) and at the end of the irradiated region ($x = w$). The maximum temperature at this location is given by

$$T_{\max} = T_o + \frac{2}{\sqrt{\pi}} \frac{I_o w}{k} \sqrt{\frac{\alpha}{Uw}}, \tag{24-17}$$

which can be used to design a laser treatment to achieve a required maximum temperature.

24.2 CONVECTION TO AN INVISCID FLOW

To evaluate the limitations of the boundary layer approximation, it is useful to contrast solutions to a particular problem that can be solved with and without streamwise diffusion. To this end, consider an inviscid flow approaching a flat plate oriented parallel to the flow. The flow is heated over the length of the plate by a constant surface heat flux q_s, and forms a thermal boundary layer of thickness δ_T, as illustrated in Figure 24-6. Since the flow is inviscid, no momentum boundary layer develops, and v_x has the constant value U. Because the fluid motion is entirely in the x-direction, the steady-state heat equation retains only one term related to advection:

$$\text{Exact heat equation:} \quad U\frac{\partial T}{\partial x} = \alpha\left(\frac{\partial^2 T}{\partial x^2} + \frac{\partial^2 T}{\partial y^2}\right). \tag{24-18}$$

When the boundary layer approximation is made, streamwise diffusion is neglected and the heat equation becomes

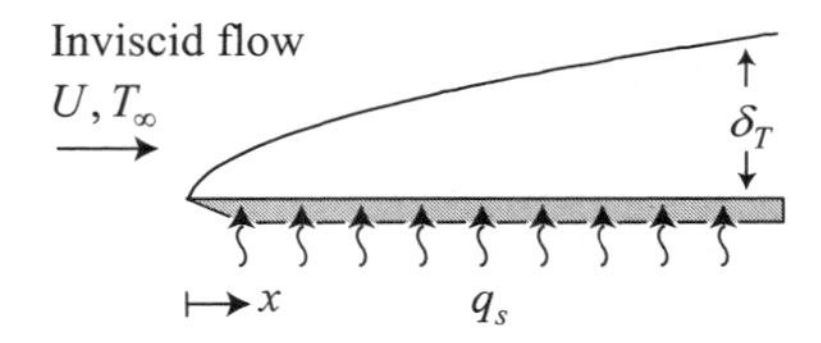

Figure 24-6 A constant heat flux plate in an inviscid flow.

$$\text{Boundary layer heat equation:} \quad U\frac{\partial T}{\partial x} = \alpha\frac{\partial^2 T}{\partial y^2}. \tag{24-19}$$

The boundary layer heat equation (24-19) is solved in Problem 24-1 (through a slight modification of the analysis performed in Section 24-1) for the resulting description of the temperature field in the inviscid flow over the heated plate:

$$\frac{T - T_o}{q_s/k} = \left[2L\sqrt{\frac{x/L}{\pi\,\mathrm{Pe}_L}}\exp\left(-\left(\frac{y}{2L}\right)^2\frac{\mathrm{Pe}_L}{x/L}\right) - y\cdot\mathrm{erfc}\left(\frac{y}{2L}\sqrt{\frac{\mathrm{Pe}_L}{x/L}}\right)\right]. \tag{24-20}$$

In this solution, the downstream distance x is measured from the leading edge of the plate, and the cross-stream distance y is measured from the surface of the plate. The approximate boundary layer solution (24-20) is expressed in terms of the dimensionless Péclet number $\mathrm{Pe}_L = UL/\alpha$ and an arbitrary downstream length scale L.

Next, the exact solution to Eq. (24-18) is addressed. To facilitate an analytic approach to this problem, a second identical plate is positioned over the original, as shown in Figure 24-7. Notice that the top plate is a heat sink while the lower plate is a heat source. As long as thermal boundary layer thicknesses are less than half the distance between the plates $\delta_T < a/2$, heat transfer to the second plate has minimal impact on heat transfer from the first. The top and bottom surfaces defined by $y = \pm\, a/2$ are adiabatic upstream of the constant heat flux plates. Therefore, along the top and bottom surfaces

$$-k\frac{\partial T}{\partial y}\bigg|_{y=-a/2} = -k\frac{\partial T}{\partial y}\bigg|_{y=+a/2} = \begin{cases} 0 & x < 0 \\ q_s & x \geq 0 \end{cases}. \tag{24-21}$$

Far upstream and far downstream of the leading edges of the plates, the conditions for heat transfer across the flow become independent of y:

$$-k\frac{\partial T}{\partial y}\bigg|_{x\to-\infty} = 0 \quad \text{and} \quad -k\frac{\partial T}{\partial y}\bigg|_{x\to+\infty} = q_s. \tag{24-22}$$

The boundary conditions for this problem suggest the transformation:

$$q_y = -k\frac{\partial T}{\partial y} \quad \text{such that } T - T(y=0) = \int_y^0 (q_y/k)dy, \tag{24-23}$$

as was employed in Section 24-1. The variable q_y is the vertical heat flux. Letting $\theta = q_y/q_s$, Eq. (24-18) can be transformed into a governing equation for the normalized vertical heat flux:

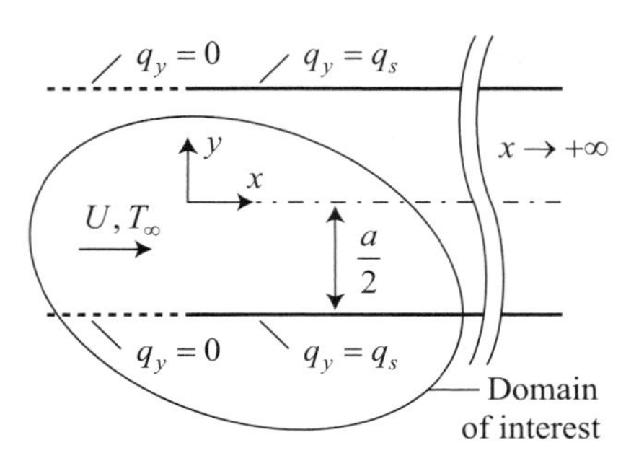

Figure 24-7 Boundary conditions required for the exact analytic solution.

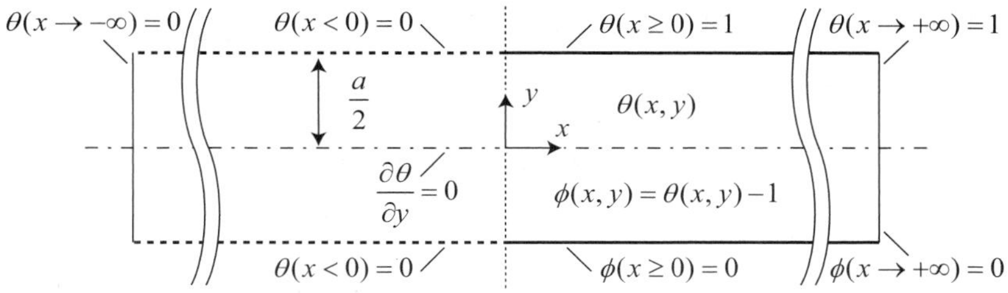

Figure 24-8 Boundary conditions imposed on the vertical heat flux.

$$U\frac{\partial\theta}{\partial x}=\alpha\left(\frac{\partial^2\theta}{\partial x^2}+\frac{\partial^2\theta}{\partial y^2}\right). \tag{24-24}$$

The boundary conditions are also transformed, and the problem for θ is summarized in Figure 24-8.

The problem for the vertical heat flux θ may be solved for $-a/2\leq y\leq 0$ using symmetry, and by splitting the solution into two parts: one for $x<0$ and the other for $x\geq 0$. As can be seen in Figure 24-8, the boundary conditions for $x<0$ are all homogeneous with the exception of the vertical surface cutting through the flow at $x=0$. In contrast, none of the boundary conditions for θ are homogeneous when $x\geq 0$. However, the same governing equation can be solved for an auxiliary problem described by $\phi=\theta-1$. In that problem, ϕ has all homogeneous boundary conditions, with the exception of when $x=0$. Therefore, to obtain a solution for $\theta(x\geq 0)$, the auxiliary problem for $\phi(x\geq 0)$ is solved first. Solutions for θ and ϕ are sought for the whole domain by separation of variables, as discussed in Chapter 8. Solutions of the form

$$(\theta\ or\ \phi)=X(x)Y(y) \tag{24-25}$$

are substituted into the governing equation to separate variables:

$$\frac{1}{X}\left(\frac{U}{\alpha}\frac{\partial X}{\partial x}-\frac{\partial^2 X}{\partial x^2}\right)=\frac{1}{Y}\frac{\partial^2 Y}{\partial y^2}=-\lambda_n^2. \tag{24-26}$$

The separation constant is $-\lambda_n^2$. After functions of the independent variables are separated, the governing equation yields two ordinary differential equations that are easily integrated:

$$\frac{\partial^2 X}{\partial x^2}-\frac{U}{\alpha}\frac{\partial X}{\partial x}-\lambda_n^2 X=0 \qquad \frac{\partial^2 Y}{\partial y^2}+\lambda_n^2 Y=0$$

$$\downarrow \qquad \downarrow$$

$$X=C_1\exp\left\{\frac{x}{2L}[\mathrm{Pe}_L+\beta_n]\right\}+C_2\exp\left\{\frac{x}{2L}[\mathrm{Pe}_L-\beta_n]\right\} \qquad Y=C_3\cos(\lambda_n y)+C_4\sin(\lambda_n y) \tag{24-27}$$

where

$$\beta_n=\sqrt{\mathrm{Pe}_L^2+(2\lambda_n L)^2}.$$

To satisfy $\theta(x\rightarrow -\infty)=0$ requires that $C_2=0$ in the equation for $X(x<0)$. Likewise, to satisfy $\phi(x\rightarrow +\infty)=0$ requires that $C_1=0$ in the equation for $X(x\geq 0)$. Additionally, to satisfy the centerline ($y=0$) boundary conditions for all x requires that $C_4=0$, such that $Y'(y=0)=0$. Furthermore, the eigenvalues $\lambda_n=(2n+1)\pi/a$ (for $n=0, 1, 2, \ldots$) forces $Y(y=-a/2)=0$, satisfying the homogeneous boundary conditions for both θ and ϕ along the lower surface. Therefore, the solutions for θ and ϕ take the forms

$$\theta(x<0,y)=\sum_{n=0}^{\infty}C_n\exp\left\{\frac{x}{2L}[\mathrm{Pe}_L+\beta_n]\right\}\cos(\lambda_n y) \tag{24-28}$$

$$\phi(x\geq 0,y)=\sum_{n=0}^{\infty}D_n\exp\left\{\frac{x}{2L}[\mathrm{Pe}_L-\beta_n]\right\}\cos(\lambda_n y). \tag{24-29}$$

The remaining integration constants, renamed C_n and D_n, must be determined to satisfy the final nonhomogeneous conditions along the boundary between the two solutions at $x=0$. For the final solution to be continuous, the problems for θ and ϕ must satisfy

$$\theta(x=0)=1+\phi(x=0) \tag{24-30}$$

$$\left.\frac{\partial\theta}{\partial x}\right|_{x=0}=\left.\frac{\partial\phi}{\partial x}\right|_{x=0}. \tag{24-31}$$

The first relation yields the condition that

$$\sum_{n=0}^{\infty}C_n\cos(\lambda_n y)=1+\sum_{n=0}^{\infty}D_n\cos(\lambda_n y) \tag{24-32}$$

or

$$\int_{-a/2}^{0}\sum_{n=0}^{\infty}C_n\cos(\lambda_n y)\cos(\lambda_m y)dy=\int_{-a/2}^{0}\cos(\lambda_m y)dy$$
$$+\int_{-a/2}^{0}\sum_{n=0}^{\infty}D_n\cos(\lambda_n y)\cos(\lambda_m y)dy. \tag{24-33}$$

Making use of orthogonality, and the fact that $\sin(\lambda_n y)=(-1)^n$, the first condition simplifies to

$$C_n=4\frac{(-1)^n}{\lambda_n a}+D_n. \tag{24-34}$$

The second condition (24-31) requires

$$\sum_{n=0}^{\infty}\frac{C_n}{2L}[\mathrm{Pe}_L+\beta_n]\cos(\lambda_n y)=\sum_{n=0}^{\infty}\frac{D_n}{2L}[\mathrm{Pe}_L-\beta_n]\cos(\lambda_n y), \tag{24-35}$$

or

$$\int_{-a/2}^{0}\sum_{n=0}^{\infty}\frac{C_n}{2L}[\mathrm{Pe}_L+\beta_n]\cos(\lambda_n y)\cos(\lambda_m y)dy$$
$$=\int_{-a/2}^{0}\sum_{n=0}^{\infty}\frac{D_n}{2L}[\mathrm{Pe}_L-\beta_n]\cos(\lambda_n y)\cos(\lambda_m y)dy. \tag{24-36}$$

Again making use of orthogonality, the second condition simplifies to

$$C_n[\mathrm{Pe}_L + \beta_n] = D_n[\mathrm{Pe}_L - \beta_n]. \tag{24-37}$$

Combining Eqns. (24-34) and (24-37) yields

$$C_n = +2\frac{(-1)^n}{\lambda_n a}\left[1 - \frac{\mathrm{Pe}_L}{\beta_n}\right] \quad \text{and} \quad D_n = -2\frac{(-1)^n}{\lambda_n a}\left[1 + \frac{\mathrm{Pe}_L}{\beta_n}\right]. \tag{24-38}$$

Therefore, with Eqs. (24-28) and (24-29), the solutions for θ and ϕ become

$$\theta(x<0, y) = +2\sum_{n=0}^{\infty}\frac{(-1)^n}{\lambda_n a}\left[1 - \frac{\mathrm{Pe}_L}{\beta_n}\right]\exp\left\{\frac{x}{2L}[\mathrm{Pe}_L + \beta_n]\right\}\cos(\lambda_n y) \tag{24-39}$$

$$\phi(x\geq 0, y) = -2\sum_{n=0}^{\infty}\frac{(-1)^n}{\lambda_n a}\left[1 + \frac{\mathrm{Pe}_L}{\beta_n}\right]\exp\left\{\frac{x}{2L}[\mathrm{Pe}_L - \beta_n]\right\}\cos(\lambda_n y). \tag{24-40}$$

The temperature field is recovered from the solution for the vertical heat flux θ using

$$\Theta(x, y) = \frac{T(x, y) - T(y=0)}{q_s a/k} = \frac{1}{a}\int_y^0 \theta(x, y)\, dy. \tag{24-41}$$

As long as the centerline temperature remains at the free stream value $T(y=0) = T_\infty$, or, equivalently, as long as $\theta(x, y=0) \approx 0$, the temperature fields can easily be evaluated from the solution for the vertical heat flux:

$$\Theta(x, y) = \begin{cases} \dfrac{1}{a}\displaystyle\int_y^0 \theta\, dy & x<0 \\ \dfrac{-y}{a} + \dfrac{1}{a}\displaystyle\int_y^0 \phi\, dy & x\geq 0 \end{cases} \quad \left(\text{as long as} \quad \theta(x, y=0) \approx 0\right). \tag{24-42}$$

The integrals in Eq. (24-42) evaluate to

$$\frac{1}{a}\int_y^0 \theta(x<0, y)dy = -2\sum_{n=0}^{\infty}\frac{(-1)^n}{(\lambda_n a)^2}\left[1 - \frac{\mathrm{Pe}_L}{\beta_n}\right]\exp\left\{\frac{x}{2L}(\mathrm{Pe}_L + \beta_n)\right\}\sin(\lambda_n y) \tag{24-43}$$

$$\frac{1}{a}\int_y^0 \phi(x\geq 0, y)dy = +2\sum_{n=0}^{\infty}\frac{(-1)^n}{(\lambda_n a)^2}\left[1 + \frac{\mathrm{Pe}_L}{\beta_n}\right]\exp\left\{\frac{x}{2L}(\mathrm{Pe}_L - \beta_n)\right\}\sin(\lambda_n y). \tag{24-44}$$

Equation (24-42) is the "exact" solution for heat transfer to the inviscid flow. The temperature field overlying the lower boundary is plotted in Figure 24-9 for $\mathrm{Pe}_L = 10$ and $L/a = 1/4$. Near the surface of the plate, the fluid temperature rise is seen to begin in advance of the leading edge. This is possible because of the streamwise diffusion that is accounted for in the exact solution.

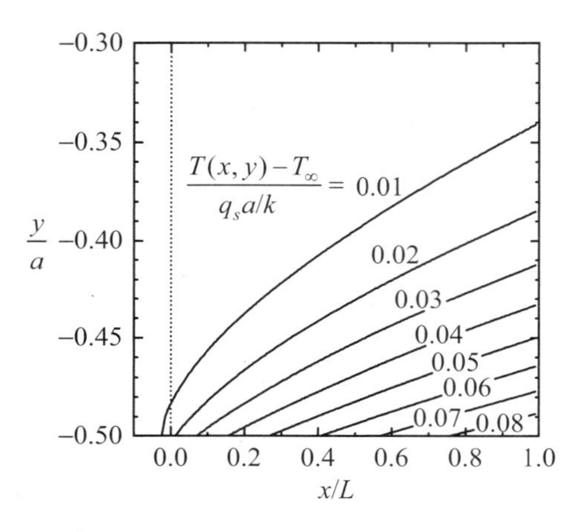

Figure 24-9 Isotherms derived from Eq. (24–42) for the exact solution.

The boundary layer thickness can serve as a point of comparison between the approximate boundary layer solution given by Eq. (24-20) and the exact solution given by Eq. (24-42). For the current problem, it is difficult to define the thermal boundary thickness in terms of the temperature field, since the surface temperature of the plate is not a constant. In problems having a heat flux boundary condition q_s, the edge of the boundary layer can be defined to occur where the magnitude of the (total) diffusion heat flux ($\sqrt{q_x^2 + q_y^2}$) falls to 1% of q_s. To evaluate this condition, the cross-stream diffusion flux can be expressed with Eqs. (24-39) and (24-40), since $q_y(x < 0)/q_s = \theta$ and $q_y(x \geq 0)/q_s = 1 + \phi$. The streamwise diffusion flux is obtained from

$$\frac{q_x}{q_s} = -\frac{k}{q_s}\frac{\partial T}{\partial x} = -a\frac{\partial \Theta}{\partial x} = \begin{cases} -\dfrac{\partial}{\partial x}\displaystyle\int_y^0 \theta\, dy & x < 0 \\ -\dfrac{\partial}{\partial x}\displaystyle\int_y^0 \phi\, dy & x \geq 0 \end{cases} \tag{24-45}$$

or,

$$\frac{q_x}{q_s} = \begin{cases} +\displaystyle\sum_{n=0}^{\infty} \frac{(-1)^n}{(a\lambda_n)^2}\left[1 - \frac{\mathrm{Pe}_L}{\beta_n}\right]\left[\frac{\mathrm{Pe}_L + \beta_n}{L/a}\right]\exp\left\{\frac{x}{2L}(\mathrm{Pe}_L + \beta_n)\right\}\sin(\lambda_n y) & x < 0 \\ -\displaystyle\sum_{n=0}^{\infty} \frac{(-1)^n}{(a\lambda_n)^2}\left[1 + \frac{\mathrm{Pe}_L}{\beta_n}\right]\left[\frac{\mathrm{Pe}_L - \beta_n}{L/a}\right]\exp\left\{\frac{x}{2L}(\mathrm{Pe}_L - \beta_n)\right\}\sin(\lambda_n y) & x \geq 0 \end{cases}. \tag{24-46}$$

The edge of the thermal boundary layer can now be determined from the exact solution to compare with the approximate boundary layer solution. As determined in Problem 24-1, the approximate boundary layer solution (with origin moved to the position shown in Figure 24-7) yields the diffusion flux components

$$\left(\frac{q_x}{q_s}\right)_{BL\ approx.} = -\frac{\exp\left(\dfrac{-\mathrm{Pe}_L\,(1 + 2y/a)^2}{16(L/a)^2(x/L)}\right)}{\sqrt{\pi\,\mathrm{Pe}_L(x/L)}} \tag{24-47}$$

and

$$\left(\frac{q_y}{q_s}\right)_{BL\ approx.} = erfc\left(\frac{1+2y/a}{4L/a}\sqrt{\frac{\mathrm{Pe}_L}{x/L}}\right). \tag{24-48}$$

As noted previously, heat can diffuse upstream of the leading edge of the heated plate in the exact solution when the Péclet number is low. In contrast, the approximate boundary layer solution does not account for streamwise diffusion. When streamwise diffusion is significant, the solution obtained with the boundary layer approximation is poor, as shown in Figure 24-10. However, as the Péclet number increases, the importance of streamwise diffusion, in comparison to advection, decreases. For $\mathrm{Pe}_L = 100$, only small differences in the two solutions are seen in Figure 24-10 near the leading edge of the heated plate. When $\mathrm{Pe}_L = 500$, the two solutions are virtually identical.

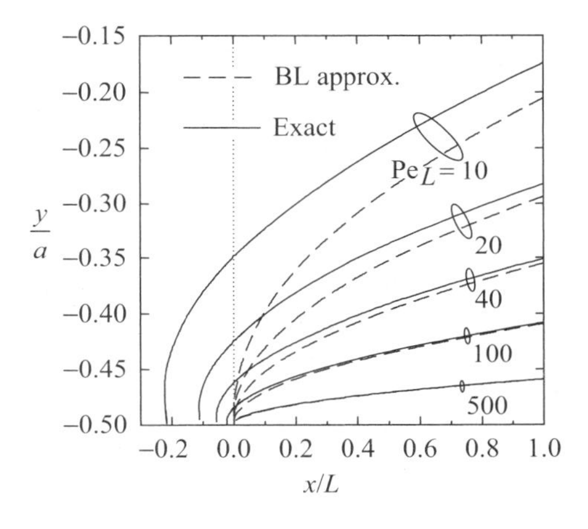

Figure 24-10 Comparison between boundary layer thickness of exact and approximate solutions, with $L/a=1/4$.

24.3 SPECIES TRANSFER TO A VERTICALLY CONVEYED LIQUID FILM

In the examples of the last two sections, the flow was trivially described by inviscid or rigid body motion. However, most convection problems must consider a real viscous fluid flow, as illustrated in this section.

Consider a conveyor belt lifting a liquid film vertically in front of a vapor box, as illustrated in Figure 24-11. The vapor box contains a chemical species that is introduced in low concentrations to the passing film. Mass diffusion into the liquid is sufficiently slow that (1) the penetration depth is small compared to the film thickness, and (2) the liquid film surface concentration (c_s) inside the vapor box is in thermodynamic equilibrium with the vapor. The latter condition implies that c_s is a constant. The conveyor speed is established to cause the free surface of the liquid film to be stationary: $v_x(y=0)=0$.

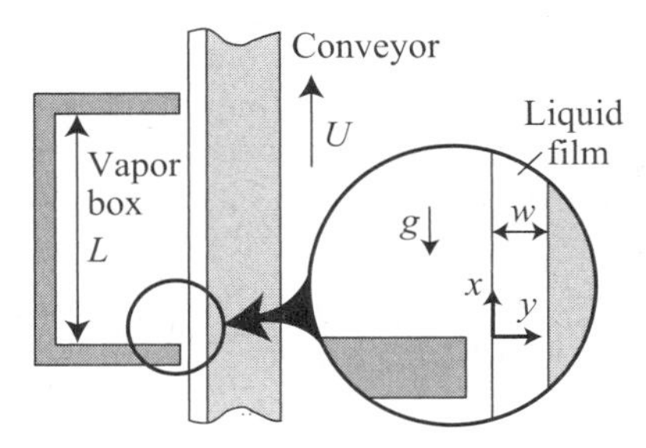

Figure 24-11 Species transfer to liquid film on vertical conveyor belt.

The velocity profile in the liquid film is fully developed, meaning that $v_x(y)$ does not change as a function of the vertical distance along the film (in the x-direction). The velocity distribution in the film is prescribed by the steady-state momentum diffusion equation:

$$0 = \nu \frac{\partial^2 v_x}{\partial y^2} - g \tag{24-49}$$

subject to the boundary conditions

$$v_x(y=0) = 0 \quad \text{and} \quad \partial v_x/\partial y\big|_{y=0} = 0, \tag{24-50}$$

and has the solution

$$v_x = \frac{g}{2\nu} y^2. \tag{24-51}$$

The concentration of chemical species "A" in the film is not fully developed, and increases with distance along the direction of conveyance. The dilute species concentration is governed by the steady-state convection equation

$$v_x \frac{\partial c_A}{\partial x} + v_y \frac{\partial c_A}{\partial y} = Đ_A \frac{\partial^2 c_A}{\partial y^2}, \tag{24-52}$$

where

$$c_A(y=0) = c_s, \; c_A(x=0) = 0, \quad \text{and} \quad c_A(y \to \infty) = 0. \tag{24-53}$$

Notice that the boundary layer approximation ($\partial^2 c_A/\partial y^2 \gg \partial^2 c_A/\partial x^2$) has been used to simplify the governing equation. Furthermore, as long as the boundary layer thickness for species diffusion is much smaller than the film thickness $\delta_A < w$, the film can be treated as though semi-infinite, as implied by the final boundary condition for $y \to \infty$. Since v_x is known, and $v_y = 0$, the governing equation simplifies to

$$\frac{g}{2\nu} y^2 \frac{\partial c_A}{\partial x} = Đ_A \frac{\partial^2 c_A}{\partial y^2}. \tag{24-54}$$

It is reasonable to look for a similarity solution to this problem. Scaling of the governing equation suggests

$$\eta = \left(\frac{g}{8\nu Đ_A} \right)^{1/4} \frac{y}{x^{1/4}}. \tag{24-55}$$

The coefficient of $8^{-1/4}$ was introduced to yield the simplest form of the final transformed equation (although this is not revealed by the scaling arguments). In general, this factor could appear as an unknown constant, to be decided later. Transforming the derivatives in the governing equation requires

$$\frac{\partial(\,)}{\partial x} = \frac{\partial \eta}{\partial x}\frac{\partial(\,)}{\partial \eta} = -\frac{\eta}{4x}\frac{\partial(\,)}{\partial \eta}, \; \frac{\partial(\,)}{\partial y} = \frac{\partial \eta}{\partial y}\frac{\partial(\,)}{\partial \eta} = \frac{\eta}{y}\frac{\partial(\,)}{\partial \eta}, \quad \text{and} \quad \frac{\partial^2(\,)}{\partial y^2} = \left(\frac{\eta}{y}\right)^2 \frac{\partial^2(\,)}{\partial \eta^2}. \tag{24-56}$$

The transformed governing equation becomes

$$\frac{\partial^2\theta}{\partial\eta^2}+\eta^3\frac{\partial\theta}{\partial\eta}=0 \quad \text{with} \quad \theta(0)=1 \quad \text{and} \quad \theta(\infty)=0, \tag{24-57}$$

where

$$\theta=c_A/c_s. \tag{24-58}$$

Notice that all three original boundary conditions are satisfied by the two remaining conditions imposed on the transformed governing equation.

The problem defined by Eq. (24-57) can be solved numerically (see Problem 24-2), using the methods developed in Chapter 23. However, since the problem is linear, it is also solvable by analytic methods, as is pursued here. Using the method of power series, discussed in Section 10.4, it is assumed that a solution for θ can be found with the form

$$\theta=\sum_{n=0}^{\infty}a_n\eta^n. \tag{24-59}$$

Substituting the assumed solution into the governing equation yields

$$\sum_{n=2}^{\infty}n(n-1)a_n\eta^{n-2}+\sum_{n=1}^{\infty}na_n\eta^{n+2}=0. \tag{24-60}$$

Or, writing the equation in a form where the coefficients to η^n are easily determined:

$$\sum_{n=0}^{\infty}(n+2)(n+1)a_{n+2}\eta^n+\sum_{n=3}^{\infty}(n-2)a_{n-2}\eta^n=0. \tag{24-61}$$

Expanding the power series form of the governing equation yields

$$2a_2\eta^0+6a_3\eta^1+12a_4\eta^2+[20a_5+a_1]\eta^3+\cdots+[(n+2)(n+1)a_{n+2}+(n-2)a_{n-2}]\eta^n+\cdots=0. \tag{24-62}$$

By inspection of the coefficients, it can be seen that the governing equation is satisfied when $a_2=0$, $a_3=0$, $a_4=0$, $a_5=(-1/20)a_1$, and $(n+2)(n+1)a_{n+2}+(n-2)a_{n-2}=0$ for $n\geq 3$. The last condition expresses a recursion relation between values of a_n's. Specifically,

$$a_{n+2}=\frac{-(n-2)}{(n+2)(n+1)}a_{n-2} \quad \text{for} \quad n\geq 3. \tag{24-63}$$

Or, equivalently,

$$a_n=\frac{-(n-4)}{n(n-1)}a_{n-4} \quad \text{for} \quad n\geq 5. \tag{24-64}$$

Therefore, the series solution to the governing equation for θ becomes

$$\theta = a_0 + \left(a_1\eta^1 + \overbrace{a_1\left(\frac{-1}{20}\right)}^{a_5}\eta^5 + \cdots + \overbrace{\frac{-(n-4)}{n(n-1)}a_{n-4}}^{a_n}\eta^n + \cdots \right). \tag{24-65}$$

Notice that most of the terms in this series are zero by virtue of the recursion relation and the fact that $a_2 = 0, a_3 = 0$, and $a_4 = 0$. Factoring out a_1 and changing the indexing scheme to $i = 0, 1, 2, \ldots$ (such that $n = 4i + 1$ may be used to evaluate the nonzero terms $n = 1, 5, 9, 13, \cdots$ in the old series), yields

$$\theta = a_0 + a_1\left(\eta^1 + \overbrace{\left(\frac{-1}{20}\right)}^{b_1}\eta^5 + \cdots + \overbrace{\frac{-(4i-3)b_{i-1}}{4i(4i+1)}}^{b_i}\eta^{4i+1} + \cdots \right). \tag{24-66}$$

The solution (24-66) is now expressed in terms of a series function:

$$\phi(\eta) = \left(b_0\eta^1 + \sum_{i=1}^{\infty} b_i\eta^{4i+1} \right) \tag{24-67}$$

where

$$b_0 = 1 \quad \text{and} \quad b_i = \frac{-(4i-3)b_{i-1}}{4i(4i+1)}. \tag{24-68}$$

Figure 24-12 plots the series function (24-67). Notice that as $\eta \rightarrow \infty$ this series approaches a constant value, which evaluates to $\phi(\infty) = 1.28185$.

At $\eta = 0$, the solution for θ is required to evaluate to $\theta(0) = 1$. Therefore, the solution (24-66) requires $a_0 = 1$. The second boundary condition $\theta(\infty) = 0$ requires $0 = 1 + a_1\phi(\infty)$. Therefore, with $a_1 = -1/\phi(\infty)$, the final solution for the species concentration field is given by

$$c_A/c_s = \theta = 1 - \phi(\eta)/\phi(\infty) \tag{24-69}$$

where

$$\phi(\eta) = \left(b_0\eta^1 + \sum_{i=1}^{\infty} b_i\eta^{4i+1} \right) \tag{24-70}$$

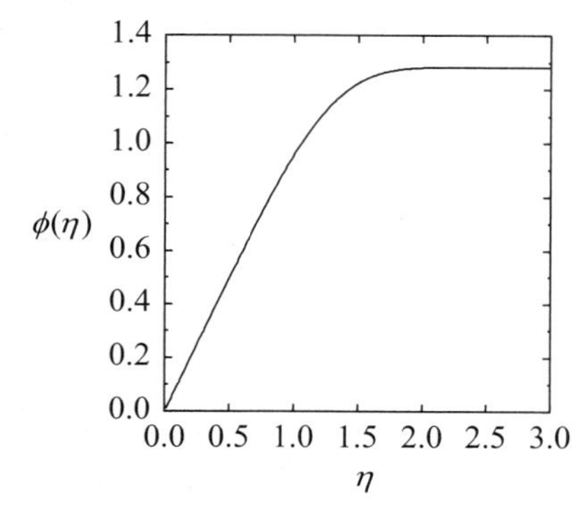

Figure 24-12 Plot of the series function given by Eq. (24-67).

with

$$b_0 = 1,\ b_i = \frac{-(4i-3)b_{i-1}}{4i(4i+1)} \quad \text{and} \quad \phi(\infty) = 1.28185 \tag{24-71}$$

and

$$\eta = \left(\frac{g}{8\nu\, Đ_A}\right)^{1/4} \frac{y}{x^{1/4}}. \tag{24-72}$$

The species flux into the liquid film can be evaluated from Fick's law:

$$\begin{aligned} J_A^* = -Đ_A \frac{\partial c_A}{\partial y}\bigg|_{y=0} &= -c_s\left(\frac{g\, Đ_A^3}{8\nu x}\right)^{1/4} \frac{\partial \theta}{\partial \eta}\bigg|_{\eta=0} \\ &= \frac{c_s}{\phi(\infty)}\left(\frac{g\, Đ_A^3}{8\nu x}\right)^{1/4} = 0.780\, c_s \left(\frac{g\, Đ_A^3}{8\nu x}\right)^{1/4}. \end{aligned} \tag{24-73}$$

The species flux is seen to scale as $1/x^{1/4}$. The result for species transfer can also be expressed in terms of the dimensionless local *Sherwood number*:*

$$Sh_x = \frac{h_A x}{Đ_A} = \frac{J_A^* x}{c_s Đ_A} = 0.464\left(\frac{g\, x^3}{\nu Đ_A}\right)^{1/4}. \tag{24-74}$$

The local Sherwood number is a dimensionless presentation of the convection coefficient h_A for species transfer, which is defined such that

$$J_A^* = h_A(c_s - c_\infty). \tag{24-75}$$

In the present problem, $c_\infty = c_A(y \to \infty) = 0$.

The result for Sh_x can also be estimated by scaling arguments

$$Sh_x = \frac{J_A^* x}{c_s Đ_A} = -\frac{x}{c_s}\frac{\partial c}{\partial y}\bigg|_{y=0} \sim \frac{x}{\delta_A}, \tag{24-76}$$

where δ_A, which is a function of x, is the liquid penetration depth of the chemical species delivered from the vapor box. Scaling the governing equation with $y \sim \delta_A$ allows the development of δ_A to be estimated:

$$\frac{g}{2\nu} y^2 \frac{\partial c_A}{\partial x} = Đ_A \frac{\partial^2 c_A}{\partial y^2} \quad \rightarrow \quad \frac{g}{2\nu}\frac{\delta_A^2}{x} \sim \frac{Đ_A}{\delta_A^2} \quad \rightarrow \quad \frac{\delta_A}{x} \sim \left(\frac{2\nu\, Đ_A}{g\, x^3}\right)^{1/4}. \tag{24-77}$$

Therefore, the scaling expectation for the Sherwood number is that

$$Sh_x \sim \frac{x}{\delta_A} \sim \left(\frac{g\, x^3}{2\nu\, Đ_A}\right)^{1/4}, \tag{24-78}$$

which is in agreement with the exact result to within a constant coefficient.

---

*Named in honor of the American chemical engineer Thomas Kilgore Sherwood (1903–1976).

24.4 PROBLEMS

24-1 Consider inviscid flow over a heated plate. The plate has a constant wall heat flux q_s. Making use of the boundary layer approximation, derive the temperature field overlying the plate and show that it is the same as given by Eq. (24-20). Derive an expression for the diffusion heat flux (vector) everywhere in the fluid.

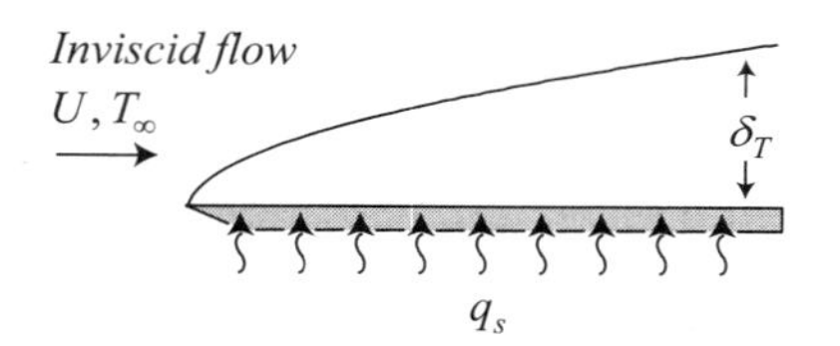

24-2 Solve the problem described by Eq. (24-57) numerically, using the methods discussed in Chapter 23. Compare the numerical solution with the analytic solution derived in Section 24-3.

24-3 Consider the problem of heat transfer to a falling film from a constant temperature heat patch on a vertical wall. The film has a fully developed velocity profile, and, over the length of the patch, the thermal penetration depth δ_T is small compared to the film thickness δ. For the condition that $\delta_T \ll \delta$, show that the velocity distribution in the thermal boundary layer is linear, $v_z \approx Ay$. Find A. Simplify the heat equation for the boundary layer problem, making use of the linear expression for velocity. Then propose the existence of a similarity solution by finding a similarity variable that transforms the governing partial differential equation into an ordinary differential equation having the form

$$\frac{d^2\theta}{d\eta^2} + 3\eta^2 \frac{d\theta}{d\eta} = 0$$

where $\theta = (T - T_o)/(T_s - T_o)$, and T_s is the surface temperature of the patch and T_o is the initial temperature of the film. Transform the boundary conditions and show that a similarity solution exists. Solve the ordinary differential equation using a power series solution and determine the local Nusselt number from the definition $Nu_x = h\,z/k$, where the convection coefficient h is defined through the wall heat flux $q_s = h(T_s - T_o)$.

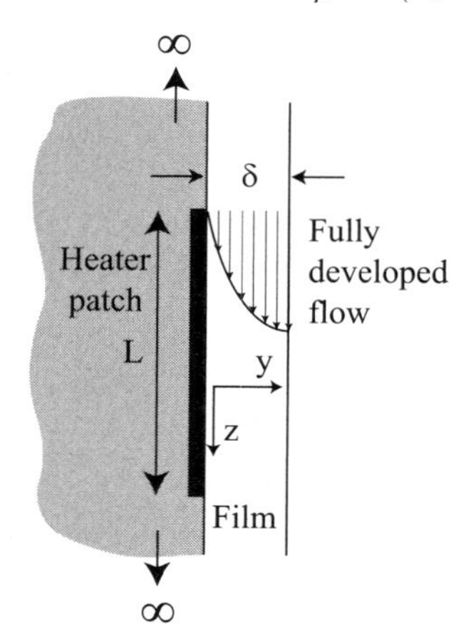

24-4 A fluid is bounded by two large plates separated by a distance δ. The top plate moves with speed U relative to the lower plate, as shown. Two possible states in pressure gradient exist: $dP/dx = +2\mu U/\delta^2$ or $dP/dx = 0$. For each of these states, determine the velocity profile between the two plates. Now suppose the lower wall has a contaminated region with a surface concentration of c_w. Upstream of the contaminated region, the fluid has a contaminate concentration of $c_\infty = 0$. Which of the two pressure gradient states discussed will cause a higher contaminant flux into the fluid? Why?

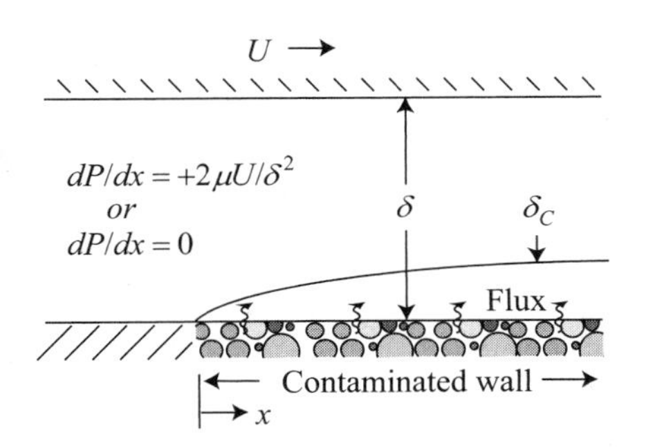

For the two pressure gradient states, estimate the contaminant transport flux from the wall $J_A^*(y=0) = J_w$, as a function of downstream distance x. If the contaminant flux from the wall is expressed in terms of the Sherwood number, show through scaling arguments that

$$Sh = \frac{J_w\, x}{(c_w - c_\infty)\, Đ_c} \sim \left(\frac{U\delta}{\nu}\right)^m \left(\frac{x}{\delta}\right)^n \left(\frac{\nu}{Đ_c}\right)^o.$$

What are the constants m, n, and o for the two possible pressure gradient states discussed?

24-5 Consider Problem 24-4 for the case when $dP/dx = +2\mu U/\delta^2$ and solve for the concentration distribution in the fluid adjacent to the contaminated wall. Determine the exact expression for the Sherwood number.

REFERENCES

[1] L. Prandtl, "Über Flüssigkeitsbewegung bei sehr kleiner Reibung." *Proc. 3rd International Congress of Mathematicians, Kong, Heidelberg*, 484 (1904).

[2] H. Schlichting, *Boundary-Layer Theory.* New York, NY: McGraw-Hill, 1979.

Chapter 25

Convection into Developing Laminar Flows

25.1 Boundary Layer Flow over a Flat Plate (Blasius Flow)

25.2 Species Transfer across the Boundary Layer

25.3 Heat Transfer across the Boundary Layer

25.4 A Correlation for Forced Heat Convection from a Flat Plate

25.5 Transport Analogies

25.6 Boundary Layers Developing on a Wedge (Falkner-Skan Flow)

25.7 Viscous Heating in the Boundary Layer

25.8 Problems

In this chapter, *external boundary layers* that form on surfaces interacting with open flows are analyzed. This subject is simplified by restricting attention to flows that are bounded by a flat surface. Unlike the problems discussed in Chapter 24, hydrodynamic development of the flow is central to the boundary layers investigated in this chapter. Problems in which boundary layers of momentum, heat, and species transport develop in a similar and simultaneous way are considered. Additional topics in boundary layer theory can be found in reference [1].

For the external flows addressed in this chapter, it is assumed that no opposing surfaces will ever impede the thickening of boundary layers. In contrast, boundary layers that form in the entrance region of internal flows eventually envelop the cross-sectional dimension of the flow and become *fully developed*, as will be discussed in Chapters 27 and 28.

25.1 BOUNDARY LAYER FLOW OVER A FLAT PLATE (BLASIUS FLOW)

A boundary layer grows in a thin region adjacent to a surface bounding the flow, as shown in Figure 25-1. The boundary layer defines a region of the flow adjusting between the wall conditions and the free stream conditions, far from the wall. The momentum boundary layer thickness δ is defined as the distance from the surface where the velocity reaches 99% of the external (free steam) flow velocity. In forced convection, the imposed external flow is minimally affected by the thickness of the boundary layer. However, inclined or curved surfaces can accelerate the flow such that the fluid speed external to the boundary layer changes with distance along the surface. The special case in which the external flow has a constant velocity U in a direction parallel to the surface is considered first. This problem is known as Blasius flow.*

*Named in honor of the German fluid dynamicist Paul Richard Heinrich Blasius (1883–1970).

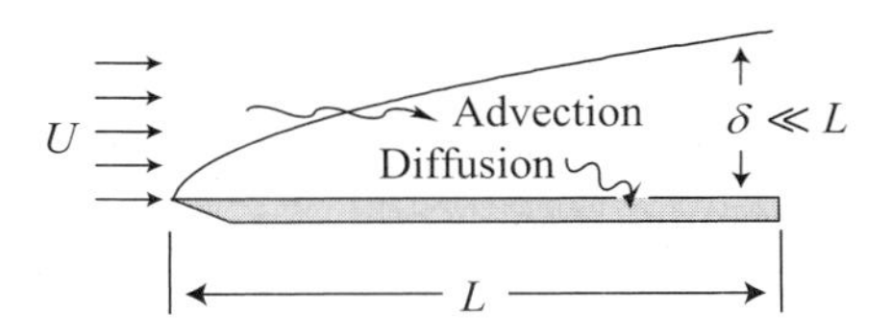

Figure 25-1 Momentum boundary layer forming over a flat plate.

Analysis of the boundary layer region starts with the momentum equation, in which the effect of gravity is usually negligible. For a steady-state incompressible flow, the momentum equation is written as

$$\underbrace{\partial_o v_i}_{=0} + v_j \partial_j v_i = \nu \partial_j \partial_j v_i - (1/\rho)\partial_i P + \underbrace{g_i}_{=0}. \tag{25-1}$$

For consideration of the two-dimensional Blasius flow, the momentum equation is expanded into Cartesian coordinates as

$$v_x \frac{\partial v_x}{\partial x} + v_y \frac{\partial v_x}{\partial y} = \nu \left(\frac{\partial^2 v_x}{\partial x^2} + \frac{\partial^2 v_x}{\partial y^2} \right) - \frac{1}{\rho} \underbrace{\frac{\partial P}{\partial x}}_{=0}. \tag{25-2}$$

Since the external flow speed is constant, pressure outside the boundary layer is known to be constant by consideration of Bernoulli's equation. Consequently, it is expected that the pressure inside the thin region of the boundary layer is unchanged from the value just outside the boundary layer, such that $\partial P/\partial x = 0$. (In contrast, flows that are accelerated by the surface would have a nonzero pressure gradient, as discussed in Section 25.6.)

Further simplification of the flow description can be justified by simple scaling arguments, which require the use of the incompressible continuity equation:

$$\frac{\partial v_x}{\partial x} + \frac{\partial v_y}{\partial y} = 0. \tag{25-3}$$

Letting $v_x \sim U$ (the external flow speed), $x \sim L$ (the downstream length of the boundary layer), and $y \sim \delta$ (the boundary layer thickness), scaling the continuity equation (25-3) for an incompressible ($\rho = const.$) flow yields

$$\text{Continuity:} \quad \frac{U}{L} \sim \frac{v_y}{\delta} \quad \text{or} \quad v_y \sim \frac{U\delta}{L}. \tag{25-4}$$

Using this result for v_y in the scaling of the momentum equation (25-2) yields

$$U\frac{U}{L} \text{ "+" } \frac{U\delta}{L}\frac{U}{\delta} \sim \nu\frac{U}{L^2} \text{ "+" } \nu\frac{U}{\delta^2}. \tag{25-5}$$

The scaled expression is cosmetically left in the form of an equation. Notice that both advection terms have the same order of magnitude: U^2/L. Inspecting both diffusion terms reveals that $U/L^2 \ll U/\delta^2$ (since $\delta/L \ll 1$), and the boundary layer approximation can be made where $\partial^2 v_x/\partial x^2$ is ignored in comparison to $\partial^2 v_x/\partial y^2$. Therefore, the scaled momentum equation reveals that

$$\frac{U^2}{L} \sim \nu\frac{U}{\delta^2}, \tag{25-6}$$

which describes a balance between advection and diffusion transport. This balance is an important feature of boundary layers. For the boundary layer assumption ($\delta/L \ll 1$) to be valid, the Reynolds number for the flow must be large ($UL/\nu \gg 1$), as is seen from rearranging the scaled momentum equation:

$$\text{Boundary layer problems:} \quad \overbrace{\frac{UL}{\nu}}^{large}\overbrace{\left(\frac{\delta}{L}\right)^2}^{small} \sim 1. \tag{25-7}$$

Retaining only the significant terms in Eq. (25-2), the momentum equation becomes

$$v_x\frac{\partial v_x}{\partial x} + v_y\frac{\partial v_x}{\partial y} = \nu\frac{\partial^2 v_x}{\partial y^2}. \tag{25-8}$$

Equations (25-3) and (25-8) for continuity and momentum govern the boundary layer problem for a flat plate oriented parallel to the flow. These equations must be solved in a way that satisfies the boundary conditions

$$\begin{aligned} &v_x(x=0) = U \qquad v_y(y=0) = 0 \\ &v_x(y=0) = 0 \\ &v_x(y\to\infty) = U \end{aligned} \tag{25-9}$$

A similarity solution to this problem can be sought. After returning the scales L and δ back to the variables x and y, the scaled momentum equation (25-6) suggests the relation

$$\frac{U^2}{x} \sim \nu\frac{U}{y^2}. \tag{25-10}$$

Therefore, a suitable choice for the similarity variable is

$$\eta = \frac{y}{\sqrt{\nu x/U}}. \tag{25-11}$$

With this similarity variable, the momentum equation can (hopefully) be transformed into an ordinary differential equation. However, the momentum equation still governs two dependent variables, v_x and v_y. Therefore, the continuity equation must also be addressed in the solution. One approach is to formulate the problem in terms of the stream function, as was done in the treatment of ideal plane flows in Section 14.3. The stream function definition automatically satisfies the continuity equation. As before, the stream function is defined such that

$$v_x = \frac{\partial\psi}{\partial y} \quad \text{and} \quad v_y = -\frac{\partial\psi}{\partial x}. \tag{25-12}$$

Introducing the similarity variable to the relation between v_x and the stream function yields

$$v_x = \frac{\partial\psi}{\partial y} = \frac{\partial\eta}{\partial y}\frac{\partial\psi}{\partial\eta} = \frac{U}{\sqrt{\nu x U}}\frac{\partial\psi}{\partial\eta}. \tag{25-13}$$

Inspection of this expression for v_x suggests defining a dimensionless stream function with the variable

$$f = \frac{\psi}{\sqrt{\nu x U}}, \tag{25-14}$$

such that

$$v_x = \frac{\partial \psi}{\partial y} = \frac{\partial}{\partial y}\left(f\sqrt{\nu x U}\right) = \sqrt{\nu x U}\,\frac{\partial \eta}{\partial y}\,\frac{\partial f}{\partial \eta} = Uf'. \tag{25-15}$$

Derivatives with respect to the similarity variable will be denoted with primes for conciseness—that is, $f' = df/d\eta$, $f'' = d^2f/d\eta^2$, and so on.

Scaling the dimensional stream function by $\sqrt{\nu x U}$ accomplishes more than merely making the stream function dimensionless. One could use a fixed length scale (say L) instead of x for the same dimensional effect. However, one can start with the assumption that

$$\psi = \sqrt{\nu L U}\,g(x)f(\eta), \tag{25-16}$$

and show that

$$g(x) \sim \sqrt{x/L} \tag{25-17}$$

is required for the momentum equation to fully transform into an ordinary differential equation. This is left as an exercise (see Problem 25-1), and the wisdom of defining the dimensionless stream function with Eq. (25-14) is accepted here without further justification.

In terms of f, v_y becomes

$$v_y = -\frac{\partial}{\partial x}(\psi) = -\frac{\partial}{\partial x}\left(\sqrt{\nu x/U f}\right) = -\frac{1}{2}\sqrt{\frac{\nu U}{x}}f - \sqrt{\nu U x}\,\frac{\partial f}{\partial x}. \tag{25-18}$$

Furthermore, using

$$\frac{\partial f}{\partial x} = \frac{\partial \eta}{\partial x}\frac{\partial f}{\partial \eta} = -\frac{\eta}{2x}\frac{\partial f}{\partial \eta}, \tag{25-19}$$

one obtains

$$v_y = \frac{1}{2}\sqrt{\frac{\nu U}{x}}[\eta f' - f]. \tag{25-20}$$

The derivatives in the momentum equation (25-8) are transformed with

$$\frac{\partial(\ )}{\partial x} = \frac{\partial \eta}{\partial x}\frac{\partial(\ )}{\partial \eta} = \frac{-\eta}{2x}\frac{\partial(\ )}{\partial \eta}, \tag{25-21a}$$

$$\frac{\partial(\ )}{\partial y} = \frac{\partial \eta}{\partial y}\frac{\partial(\ )}{\partial \eta} = \frac{\eta}{y}\frac{\partial(\ )}{\partial \eta}, \tag{25-21b}$$

and

$$\frac{\partial^2(\ )}{\partial y^2} = \left(\frac{\partial \eta}{\partial y}\right)^2 \frac{\partial^2(\ )}{\partial \eta^2} = \left(\frac{\eta}{y}\right)^2 \frac{\partial^2(\ )}{\partial \eta^2}. \tag{25-21c}$$

Therefore, transforming the momentum equation into the new variables yields

$$\overbrace{(Uf')\frac{-\eta}{2x}\frac{\partial}{\partial \eta}(Uf')}^{v_x \frac{\partial v_x}{\partial x}} + \overbrace{\frac{1}{2}\sqrt{\frac{vU}{x}}[\eta f' - f]\frac{\eta}{y}\frac{\partial}{\partial \eta}(Uf')}^{v_y \frac{\partial v_x}{\partial y}} = \overbrace{v\left(\frac{\eta}{y}\right)^2 \frac{\partial}{\partial \eta^2}(Uf')}^{v \frac{\partial^2 v_x}{\partial y^2}} \tag{25-22}$$

or

$$f''' + \frac{1}{2} f f'' = 0. \tag{25-23}$$

The final result given by Eq. (25-23) is known as the *Blasius equation* [2]. This equation embodies two important transformations from the original mathematical statement given by Eqs. (25-3) and (25-8). The first transformation reduced the two coupled equations for v_x and v_y to a single equation for the stream function. This came at the expense of increasing the momentum equation from a second-order to a third-order differential equation. The second transformation changed the momentum equation from a partial differential equation dependent on x and y to an ordinary differential equation dependent on the similarity variable η.

It has yet to be established whether all of the original boundary conditions (25-9) can be satisfied by the problem expressed in terms of the similarity variable. To transform the boundary conditions, recall the relations

$$\eta = \frac{y}{\sqrt{\nu x/U}}, \quad v_x = Uf', \quad \text{and} \quad v_y = \frac{1}{2}\sqrt{\frac{\nu U}{x}}[\eta f' - f]. \tag{25-24}$$

Therefore, the boundary conditions can be restated in terms of $f(\eta)$ as

$$\underbrace{\begin{matrix} v_x(x=0)=U \\ v_x(y\to\infty)=0 \\ v_x(y=0)=0 \\ v_y(y=0)=0 \end{matrix}}_{\text{4 original conditions}} \qquad \underbrace{\begin{matrix} \left.\begin{matrix} f'(\eta\to\infty)=1 \\ f'(\eta\to\infty)=1 \end{matrix}\right\} \textit{same} \\ f'(\eta=0)=0 \\ f(\eta=0)=0 \end{matrix}}_{\text{3 final conditions}} \tag{25-25}$$

Notice that the four original conditions collapse to three conditions imposed on the problem expressed in terms of the similarity variable. Since four independent conditions cannot be satisfied by the third-order momentum equation (25-23), this reduction in the number of imposed boundary conditions is required for a similarity solution to exist. The nonlinear ordinary differential equation can now be solved by numerical integration. The fourth-order Runge-Kutta method discussed in Chapter 23 is used by expressing the third-order equation for f as three first-order equations for f, f', and f'':

$$\frac{df''}{d\eta} = -\frac{1}{2} f \cdot f'' = F(f, f'') \quad \text{with} \quad f''(0) = \underline{?}, \quad \text{leading to} \quad f'(\infty) = 1 \tag{25-26}$$

$$\frac{df'}{d\eta} = f'' \quad \text{with} \quad f'(0) = 0 \tag{25-27}$$

$$\frac{df}{d\eta} = f' \quad \text{with} \quad f(0) = 0. \tag{25-28}$$

The equations for f, f', and f'' are marched forward with Runge-Kutta integration using

$$f_{n+1} = f_n + \frac{K_1^f + 2K_2^f + 2K_3^f + K_4^f}{6} \tag{25-29}$$

$$f'_{n+1} = f'_n + \frac{K_1^{f'} + 2K_2^{f'} + 2K_3^{f'} + K_4^{f'}}{6} \tag{25-30}$$

$$f''_{n+1} = f''_n + \frac{K_1^{f''} + 2K_2^{f''} + 2K_3^{f''} + K_4^{f''}}{6} \tag{25-31}$$

where

$$\begin{array}{lll}
K_1^f = \Delta\eta f'_n & K_1^{f'} = \Delta\eta f''_n & K_1^{f''} = \Delta\eta(-0.5 f_n f''_n) \\
K_2^f = \Delta\eta\left(f'_n + 0.5K_1^{f'}\right) & K_2^{f'} = \Delta\eta\left(f''_n + 0.5K_1^{f''}\right) & K_2^{f''} = \Delta\eta\left[-0.5\left(f_n + 0.5\cdot K_1^f\right)\left(f''_n + 0.5K_1^{f''}\right)\right] \\
K_3^f = \Delta\eta\left(f'_n + 0.5K_2^{f'}\right) & K_3^{f'} = \Delta\eta\left(f''_n + 0.5K_2^{f''}\right) & K_3^{f''} = \Delta\eta\left[-0.5\left(f_n + 0.5\cdot K_2^f\right)\left(f''_n + 0.5K_2^{f''}\right)\right] \\
K_4^f = \Delta\eta\left(f'_n + K_3^{f'}\right) & K_4^{f'} = \Delta\eta\left(f''_n + K_3^{f''}\right) & K_4^{f''} = \Delta\eta\left[-0.5\left(f_n + K_3^f\right)\left(f''_n + K_3^{f''}\right)\right]
\end{array}.$$

The initial condition required to integrate f'' is unknown, and the shooting method discussed in Section 23.4 is used to guess the correct initial condition $f''(0)$ that satisfies the final condition $f'(\infty) = 1$.

The subroutine given in Code 25-1 can be used to solve the Blasius equation (25-23). Figure 25-2 plots the solution for f, f', and f''. There are several useful results that can be quantified from the velocity solution, such as the extent of the boundary layer thickness and the fluid drag on the plate.

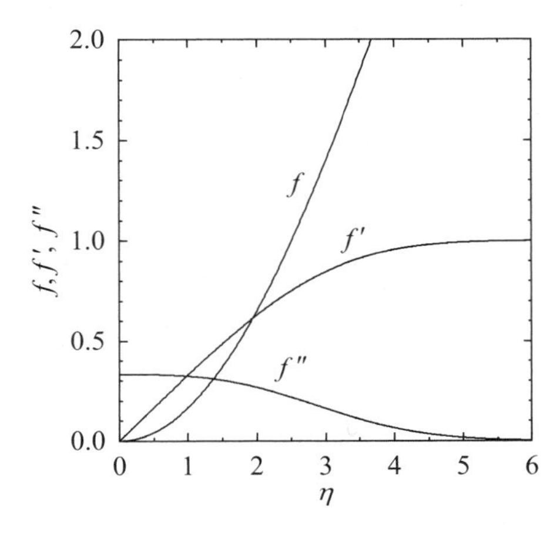

Figure 25-2 Solution to the Blasius equation (25-23) for flow over a flat plate.

Code 25-1 Blasius equation for flat plate

```
#include <stdio.h>
#include <stdlib.h>
#include <math.h>

void blasius(double InfEta,int N,double *f,double *df,double *ddf)
{
   int n,iter=0;
   double del_eta=InfEta/(N-1);
   double K1f,K1df,K1ddf,K2f,K2df,K2ddf,K3f,K3df,K3ddf,K4f,K4df,K4ddf;

   double ddf0_low=0.1;
   double ddf0_high=5.0;
   double df_inf=1.0;     /* end BC */
   df[N-1]=0.0;           /* make sure not 1.0 yet */
   f[0]=0.0;              /* initial conditions */
   df[0]=0.0;
   do {
      ddf[0]=(ddf0_low+ddf0_high)/2.0;
      for (n=0;n<N-1;++n) {
            K1f=del_eta*df[n];
            K1df=del_eta*ddf[n];
            K1ddf=del_eta*(-0.5*f[n]*ddf[n]);
            K2f=del_eta*(df[n]+0.5*K1df);
            K2df=del_eta*(ddf[n]+0.5*K1ddf);
            K2ddf=del_eta*(-0.5*(f[n]+0.5*K1f)*(ddf[n]+0.5*K1ddf));
            K3f=del_eta*(df[n]+0.5*K2df);
            K3df=del_eta*(ddf[n]+0.5*K2ddf);
            K3ddf=del_eta*(-0.5*(f[n]+0.5*K2f)*(ddf[n]+0.5*K2ddf));
            K4f=del_eta*(df[n]+K3df);
            K4df=del_eta*(ddf[n]+K3ddf);
            K4ddf=del_eta*(-0.5*(f[n]+K3f)*(ddf[n]+K3ddf));
            f[n+1]=f[n]+(K1f+2*K2f+2*K3f+K4f)/6;
            df[n+1]=df[n]+(K1df+2*K2df+2*K3df+K4df)/6;
            ddf[n+1]=ddf[n]+(K1ddf+2*K2ddf+2*K3ddf+K4ddf)/6;

            if (fabs(df[n+1]) > df_inf) {
               df[N-1]=df[n+1];
               break;
            }
         }
      /* continue Bi-section method (slow but stable) */
      if (df[N-1]>df_inf) ddf0_high=ddf[0];
      else                  ddf0_low=ddf[0];
   } while (++iter < 200 && (ddf0_high-ddf0_low)>1.0e-6);
   if (iter==200) {
      printf("\nBlasius Soln. Failed!");
      exit(1);
   }
}
```

Code 25-2 Constant temperature plate

```
#include <stdio.h>
#include <stdlib.h>
#include <math.h>
void blasiusT(double InfEta,double Pr,int N,double *f,double *T,double *dT)
{
   int n,iter=0;
   double del_eta=InfEta/(N-1);
   double K1T,K1dT,K2T,K2dT,K3T,K3dT, K4T,K4dT;
   double dT0_high=5.,dT0_low=0.;
   double T_inf=1.;     /* end BC */
   T[0]=0.;             /* initial conditions */
   do {
      dT[0]=(dT0_low+dT0_high)/2.0;
      for (n=0;n<N-1;++n) {
         K1T=del_eta*dT[n];
         K1dT=del_eta*(-0.5*Pr*f[n]*dT[n]);
         K2T=del_eta*(dT[n]+0.5*K1dT);
         K2dT=del_eta*(-0.25*Pr*(f[n]+f[n+1])*(dT[n]+0.5*K1dT));
         K3T=del_eta*(dT[n]+0.5*K2dT);
         K3dT=del_eta*(-0.25*Pr*(f[n]+f[n+1])*(dT[n]+0.5*K2dT));
         K4T=del_eta*(dT[n]+K3dT);
         K4dT=del_eta*(-0.5*Pr*f[n+1]*(dT[n]+0.5*K3dT));
         T[n+1]=T[n]+(K1T+2*K2T+2*K3T+K4T)/6;
         dT[n+1]=dT[n]+(K1dT+2*K2dT+2*K3dT+K4dT)/6;
         if (fabs(T[n+1])>T_inf) {
            T[N-1]=T[n+1];
            break;
         }
      }
      if (T[N-1]>T_inf) dT0_high=dT[0];   /* shoot lower */
      else              dT0_low= dT[0];   /* shoot higher */
   } while (++iter<200 && (dT0_high-dT0_low)>1.0e-6);
   if (iter==200) { printf("\nBlasiusT Soln. Failed!");exit(1); }
}
void blasius(double InfEta,int N,double *f,double *df,double *ddf);
int main()
{
   int n,N=1000;
   double Pr=1.,InfEta=7.;
   double del_eta=InfEta/(N-1);
   double f[N],df[N],ddf[N],T[N],dT[N];
   FILE *fp=fopen("out.dat","w");
   blasius(InfEta,N,f,df,ddf);
   blasiusT(InfEta,Pr,N,f,T,dT);
   for (n=0;n<N;n++)
      fprintf(fp,"%e %e %e %e %e %e\n",
         n*del_eta,f[n],df[n],ddf[n],T[n],dT[n]);
   fclose(fp);
   return 1;
}
```

The momentum boundary layer thickness δ is defined by the distance from the wall at which $v_x/U = f' = 0.99$. From the solution it is found that $f' = 0.99$ occurs when $\eta = 4.95$. Therefore, the momentum boundary layer thickness $y = \delta$ can be determined using the definition of the similarity variable:

$$4.95 = \frac{\delta}{\sqrt{\nu x/U}} \quad \text{or} \quad \delta = 4.95\sqrt{\nu x/U}. \tag{25-32}$$

Therefore, for parallel flow over a flat plate, the boundary layer is expected to grow as the square root of the distance from the leading edge.

The local value of the coefficient of friction (skin-friction coefficient) can also be determined from the solution. By definition:

$$c_f = \frac{\tau(x)\big|_{y=0}}{\rho U^2/2}, \quad \text{where} \quad \tau(x)\big|_{y=0} = \mu\frac{\partial v_x}{\partial y}\bigg|_{y=0}. \tag{25-33}$$

This expression is transformed into the variables of the solution. Using

$$\frac{\partial v_x}{\partial y} = \frac{\partial^2 \psi}{\partial y^2} = \frac{1}{\sqrt{\nu x/U}}\frac{\partial^2 f}{\partial y^2} = \frac{1}{\sqrt{\nu x/U}}\left(\frac{\partial \eta}{\partial y}\right)^2\frac{\partial^2 f}{\partial \eta^2} = U\sqrt{\frac{U}{\nu x}}f'', \tag{25-34}$$

one obtains

$$c_f = 2\sqrt{\frac{\nu}{Ux}}f''(0) = 0.664\text{Re}_x^{-1/2}, \tag{25-35}$$

where $f''(0) = 0.332$ is determined from the numerical solution. The total drag coefficient, which is defined in relation to the total drag on the plate, is given by

$$C_D = \frac{\int_0^L \tau(x)\big|_{y=0}dx}{L\rho U^2/2} = 4\sqrt{\frac{\nu}{UL}}f''(0) = 1.328\text{Re}_L^{-1/2}. \tag{25-36}$$

25.2 SPECIES TRANSFER ACROSS THE BOUNDARY LAYER

Consider the transport of species across a boundary layer. In the absence of chemical reactions, the source of species is on one side of the boundary layer while the sink is on the other. For example, in chemical vapor deposition, a species being deposited on the surface is introduced with the external flow. However, the reverse is also possible when a chemical diffuses into the boundary layer from the surface and is swept away by the moving fluid.

Mass transfer across a boundary layer involves solving coupled equations of species and momentum transport. Two boundary layers develop, one for momentum and one for species concentration, as shown in Figure 25-3. The relative size of the boundary layer thicknesses depends on whether species transport or momentum transport has the higher value of diffusivity. The case illustrated in Figure 25-3 has larger species diffusivity. Although species transport depends on the velocity field established by momentum transfer, the momentum equation does not depend on the concentration field for the problem at hand. Therefore, the

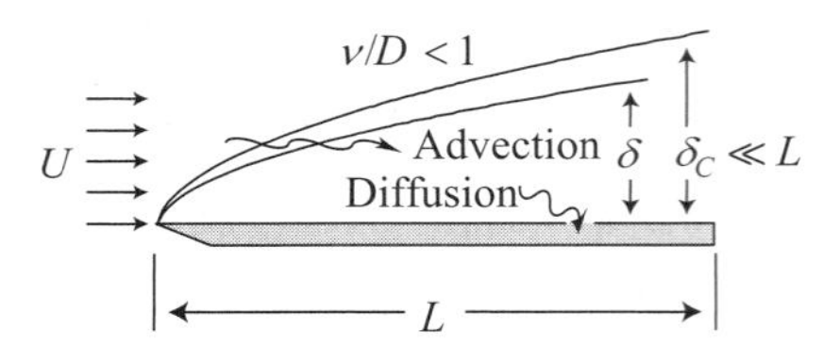

Figure 25-3 Species boundary layer forming with momentum boundary layer over a flat plate.

momentum equation can be solved independently as described in the preceding section. The remaining task is to solve the species transport equation.

Using the boundary layer approximation, diffusion of species down the length of the boundary layer may be neglected in comparison with diffusion across the boundary layer. Therefore, the dilute species transport equation for a steady-state incompressible flow becomes

$$v_x \frac{\partial C}{\partial x} + v_y \frac{\partial C}{\partial y} = Đ_A \frac{\partial^2 C}{\partial y^2} \tag{25-37}$$

where

$$C = c_A / c_\infty. \tag{25-38}$$

In this equation, the concentration of some species "A" has been normalized by the free stream value (c_∞) for convenience. Suppose the species transport problem is described by the boundary conditions

$$C(y=0) = 0 \quad C(y \to \infty) = 1 \quad C(x=0) = 1. \tag{25-39}$$

These boundary conditions might be used to describe a situation in which "A" is a precursor gas that reacts upon contact with the plate surface. The reaction results in chemical deposition, and maintains a zero concentration of the precursor gas at the surface.

It is logical to solve the species transport equation using the variables with which the momentum equation was solved. Therefore, instead of expressing the species transport equation in terms of the velocity field, the dimensionless stream function is introduced and, instead of using the independent variables x and y, a solution in terms of the similarity variable is sought. Using the similarity variable defined for solving the momentum equation,

$$\eta = y / \sqrt{\nu x / U}, \tag{25-40}$$

and making use of the previous results:

$$v_x = Uf' \quad \text{and} \quad v_y = \sqrt{\nu U / x}\,[\eta f' - f]/2, \quad \text{where} \quad f = \psi / \sqrt{\nu x U}, \tag{25-41}$$

the species transport equation can be transformed with the aid of the derivative expressions given by Eqs. (25-21a–c). Transformation of the species boundary layer equation (25-37) into the new variables yields

$$\overbrace{(Uf')\frac{-\eta}{2x}C'}^{v_x \frac{\partial C}{\partial x}} + \overbrace{\frac{1}{2}\sqrt{\frac{\nu U}{x}}[\eta f' - f]\frac{\eta}{y}C'}^{v_y \frac{\partial C}{\partial y}} = \overbrace{Đ_A \left(\frac{\eta}{y}\right)^2 C''}^{Đ_A \frac{\partial^2 C}{\partial y^2}} \tag{25-42}$$

or

$$C'' + \frac{1}{2} Sc\ f C' = 0. \tag{25-43}$$

The *Schmidt number*[†] has been introduced into the species transport equation, and is defined by the ratio of momentum diffusivity to species diffusivity:

$$Sc = \frac{\nu}{Đ_A}. \tag{25-44}$$

The boundary conditions (25-39) for the problem can be expressed in terms of the similarity variable:

$$\underbrace{\begin{matrix} C(x=0)=1 \\ C(y\to\infty)=1 \\ C(y=0)=0 \end{matrix}}_{3\ original\ conditions} \qquad \underbrace{\begin{matrix} \left.\begin{matrix} C(\eta\to\infty)=1 \\ C(\eta\to\infty)=1 \end{matrix}\right\} same \\ C(\eta=0)=0 \end{matrix}}_{2\ final\ conditions} \quad . \tag{25-45}$$

Again, notice that the number of conditions imposed on the problem has collapsed, as is required for the existence of a similarity solution.

Since f is expressed as a numerical solution, the species equation (25-43) is also numerically solved. The fourth-order Runge-Kutta method is used to solve the species transport equation by expressing the second-order equation for C as two first-order equations for C and C':

$$\frac{dC'}{d\eta} = -\frac{1}{2} Sc\, f\, C' = F(Sc, f, C') \qquad \text{with } C'(0) = \underline{?}, \quad \text{leading to } C(\infty) = 1 \tag{25-46}$$

$$\frac{dC}{d\eta} = C' \quad \text{with } C(0) = 0. \tag{25-47}$$

The initial condition required to integrate the first equation for C' is unknown, and the shooting method discussed in Section 23.4 is used to guess the correct initial condition $C'(0)$ that satisfies the final condition $C(\infty) = 1$.

The equations for C and C' are marched forward with Runge-Kutta integration using

$$C_{n+1} = C_n + \frac{K_1^C + 2K_2^C + 2K_3^C + K_4^C}{6} \text{ and } C'_{n+1} = C'_n + \frac{K_1^{C'} + 2K_2^{C'} + 2K_3^{C'} + K_4^{C'}}{6} \tag{25-48}$$

where the K values for the two dependent variables C and C' are given by

$$\begin{aligned} K_1^C &= \Delta\eta\, C'_n & K_1^{C'} &= \Delta\eta(-0.5\, Sc\, f_n\, C'_n) \\ K_2^C &= \Delta\eta\, (C'_n + 0.5\, K_1^{C'}) & K_2^{C'} &= \Delta\eta[-0.25\, Sc\, (f_n + f_{n+1})\, (C'_n + 0.5\, K_1^{C'})] \\ K_3^C &= \Delta\eta\, (C'_n + 0.5\, K_2^{C'}) & K_3^{C'} &= \Delta\eta[-0.25\, Sc\, (f_n + f_{n+1})\, (C'_n + 0.5\, K_2^{C'})] \\ K_4^C &= \Delta\eta\, (C'_n + K_3^{C'}) & K_4^{C'} &= \Delta\eta[-0.5\, Sc\, f_{n+1}\, (C'_n + K_3^{C'})] \end{aligned} \quad .$$

[†]The Schmidt number is named after the German thermodynamics Ernst Heinrich Wilhelm Schmidt (1892–1975).

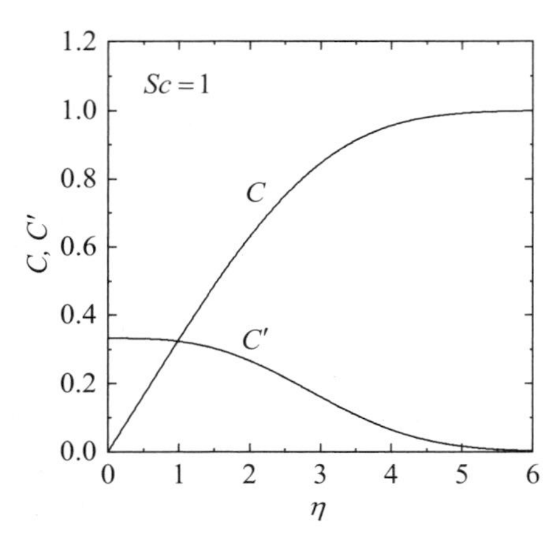

Figure 25-4 Solution to Eq. (25-43) for species transport in Blasius flow over a flat plate.

Figure 25-4 plots the solution for C and C' for the case when $Sc = 1$. The flux of the precursor gas to the surface can be calculated from

$$J_A^* = -Đ_A \left.\frac{\partial c_A}{\partial y}\right|_{y=0} = -Đ_A c_\infty \left.\frac{\partial \eta}{\partial y}\frac{\partial C}{\partial \eta}\right|_{\eta=0} = \frac{-Đ_A c_\infty C'(0)}{\sqrt{\nu x/U}} \tag{25-49}$$

where $C'(0)$ is determined from the numerical solution, and depends on the value of the Schmidt number Sc. For $Sc = 1$, $C'(0) = 0.332$. Notice that when $Sc = 1$ (i.e., $\nu = Đ_A$), $C'(0) = f''(0)$, as determined in the last section, because the problems for C and f' are mathematically identical.

A relation between the species flux to the surface and a mass convection coefficient h_A for transport can be defined:

$$-J_A^* = h_A(c_\infty - c_s), \tag{25-50}$$

where $c_s = c(y = 0) = 0$ in the present problem. The local Sherwood number is a dimensionless presentation of the convection coefficient for species transfer:

$$Sh_x = \frac{h_A x}{Đ_A} = \frac{J_A^*}{(c_s - c_\infty)}\frac{x}{Đ_A} = \frac{Đ_A C'(0)}{\sqrt{\nu x/U}}\frac{x}{Đ_A} = C'(0)\mathrm{Re}_x^{1/2}. \tag{25-51}$$

When evaluating the Sherwood number, it is understood that $C'(0)$ is a function of the Schmidt number Sc, as determined from the numerical solution. Figure 25-5 plots

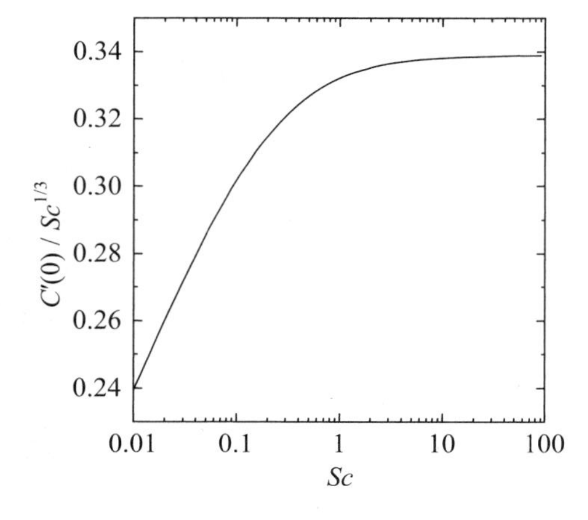

Figure 25-5 Numerical results for $C'(0)$ as a function of Sc.

$C'(0)/Sc^{1/3}$ as a function of Sc to demonstrate the dependency. It is seen that $C'(0) \sim Sc^{1/3}$ in the limit of large-Schmidt-number fluids ($Sc \gg 1$).

25.3 HEAT TRANSFER ACROSS THE BOUNDARY LAYER

Heat transfer across the boundary layer involves solving coupled equations of heat and momentum transport. In this section, it is assumed that the fluid density is temperature independent. The effect of temperature dependent fluid properties will be considered later in Chapter 29.

In the presence of heat transfer, two boundary layers develop over the plate, as shown in Figure 25-6, one for momentum and one for temperature. The relative size of the boundary layer thicknesses depends on whether diffusion is larger for heat or momentum transport. The case illustrated in Figure 25-6 shows a fluid with larger momentum diffusivity than heat diffusivity.

The heat transfer problem is solved in a way that is completely analogous to the species transfer problem of the previous section. The incompressible heat transfer equation (without viscous heat generation) can be written for the boundary layer as

$$v_x \frac{\partial \theta}{\partial x} + v_y \frac{\partial \theta}{\partial y} = \alpha \frac{\partial^2 \theta}{\partial y^2} \qquad (25\text{-}52)$$

where

$$\theta = \frac{T - T_s}{T_\infty - T_s}. \qquad (25\text{-}53)$$

For a constant-temperature surface, the boundary conditions for the thermal boundary layer are

$$\theta(y=0) = 0 \quad \theta(y \to \infty) = 1 \quad \theta(x=0) = 1. \qquad (25\text{-}54)$$

The heat equation is transformed with the similarity variable used to solve the momentum equation,

$$\eta = y/\sqrt{\nu x/U}. \qquad (25\text{-}55)$$

Making use of the previous results:

$$v_x = Uf' \quad \text{and} \quad v_y = \sqrt{\nu U/x}\,[\eta f' - f]/2, \quad \text{where } f = \psi/\sqrt{\nu x U}, \qquad (25\text{-}56)$$

the heat transport equation for the boundary layer becomes

$$\theta'' + \frac{1}{2}\mathrm{Pr}\, f\, \theta' = 0 \qquad (25\text{-}57)$$

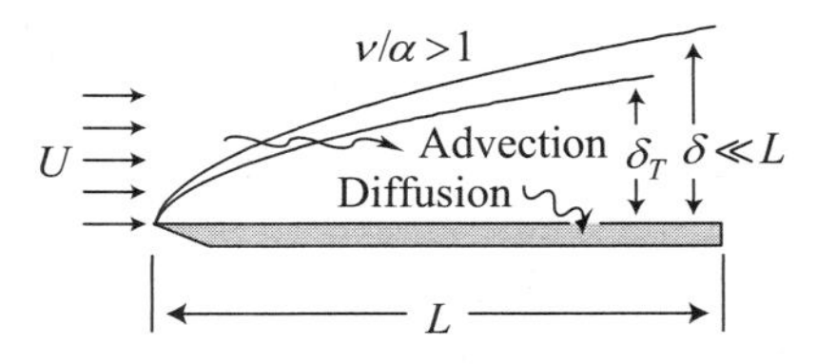

Figure 25-6 Heat and momentum boundary layers over a flat plate.

subject to

$$\theta(0) = 0 \quad \text{and} \quad \theta(\infty) = 1. \tag{25-58}$$

The *Prandtl number*‡ has been introduced into the heat equation, and is defined by the ratio of momentum diffusivity to heat diffusivity:

$$\mathrm{Pr} = \frac{\nu}{\alpha}. \tag{25-59}$$

The same process by which the species equation was solved in the previous section can be used to find a numerical solution for the fluid temperature distribution θ. Figure 25-7 illustrates the isotherms in the boundary layer.

From the definition of the similarity variable (25-55), it can be seen that the thermal boundary layer thickness is given by

$$\frac{\delta_T}{x} = \eta_{99}\mathrm{Re}_x^{-1/2}, \tag{25-60}$$

where η_{99} corresponds to the value of the similarity variable at which the fluid temperature is within 1% of the free stream value (i.e., where $\theta = 0.99$). However, the numerical value for η_{99} will be dependent on the Prandtl number of the fluid.

The local heat transfer from the surface can also be evaluated from the solution:

$$q_s = -k\frac{\partial T}{\partial y}\bigg|_{y=0} = k(T_s - T_\infty)\frac{\partial \eta}{\partial y}\frac{\partial \theta}{\partial \eta}\bigg|_{\eta=0} = k(T_s - T_\infty)\frac{\theta'(0)}{\sqrt{\nu x/U}}. \tag{25-61}$$

The heat transfer depends on the value of $\theta'(0)$ obtained from the solution, which in turn depends on the Prandtl number of the fluid. The local *Nusselt number*§ for heat transfer from the plate can be determined:

$$Nu_x = \frac{hx}{k} = \frac{q_s}{(T_s - T_\infty)}\frac{x}{k} = \frac{\theta'(0)}{\sqrt{\nu x/U}}x = \theta'(0)\mathrm{Re}_x^{1/2}. \tag{25-62}$$

The Nusselt number is a dimensionless presentation of the convection coefficient h for heat transfer. The convection coefficient is defined such that the wall heat flux is given by Newton's convection law:

$$q_s = h(T_s - T_\infty). \tag{25-63}$$

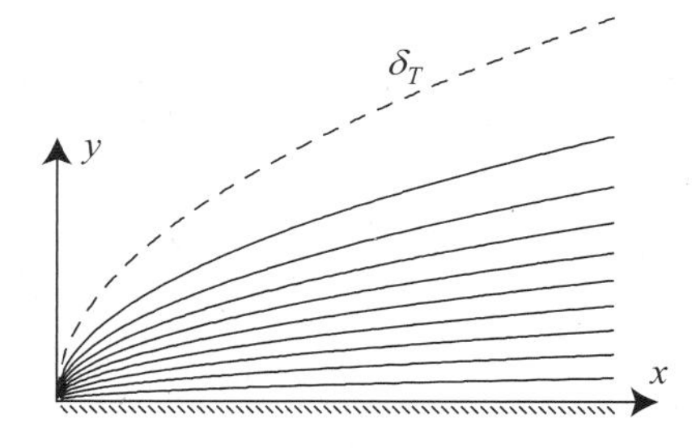

Figure 25-7 Isotherms above a constant temperature plate at 0.1 intervals between 0 and 1.

‡The Prandtl number is named after the German physicist Ludwig Prandtl (1875–1953).
§The Nusselt number is named after the German engineer Ernst Kraft Wilhelm Nusselt (1882–1957).

Code 25-2 on page 382 solves the constant temperature flat plate problem described. For $\mathrm{Pr} = 1$, it is found that $\theta'(0) = 0.332$, which is the same as $f''(0)$. Similar to the observation made with respect to species transport, discussed in Section 25.2, this is no coincidence since the problems for θ and f' are mathematically identical when $\nu = \alpha$. More generally, it will be shown in Section 25.4 by scaling arguments that $\theta'(0) \sim \mathrm{Pr}^{1/3}$ (for $\mathrm{Pr} \geq 1$), as is verified with the exact solution in Problem 25-5.

25.4 A CORRELATION FOR FORCED HEAT CONVECTION FROM A FLAT PLATE

It is of significant practical importance to be able to calculate heat transfer from a plate for the type of problem considered in Section 25.3. However, since the heat equation was solved numerically, the nature of the heat transfer dependence on Pr was not explicitly revealed. In this section, the effect of the Pr number is deduced through scaling arguments.

Two of the characteristic scales for the thermal boundary layer problem are

$$y \sim \delta_T \quad \text{and} \quad \theta \sim \Delta T = T_s - T_\infty. \tag{25-64}$$

When these scales are applied to the heat equation, Eq. (25-52), one finds

$$v_x^T \frac{\Delta T}{x} \text{ ``}+\text{''}\ v_y^T \frac{\Delta T}{\delta_T} \sim \alpha \frac{\Delta T}{\delta_T^2}. \tag{25-65}$$

Thus far the characteristic scales of v_x^T and v_y^T have not been identified. The "T" superscript is added as a reminder that these velocities are characteristic of the thermal boundary layer (to be distinguished from the velocity scales in the momentum equation). The characteristic scales of v_x^T and v_y^T are related to each other through the scaling of the continuity equation:

$$\frac{v_x^T}{x} \sim \frac{v_y^T}{\delta_T}, \quad \text{such that } v_y^T \sim \frac{v_x^T \delta_T}{x}. \tag{25-66}$$

When this relation is applied to the heat equation, it becomes apparent that both advection terms in the scaled equation have the same order of magnitude of $v_x^T \Delta T/x$. Therefore, the scaled boundary layer heat equation becomes

$$v_x^T \frac{\Delta T}{x} \sim \alpha \frac{\Delta T}{\delta_T^2}. \tag{25-67}$$

One might be tempted to scale v_x^T with the free stream velocity U. However, this could be a very poor scaling argument if the thermal boundary layer was much thinner than the momentum boundary layer $\delta_T \ll \delta$. With the restriction that $\delta_T \lesssim \delta$, a better scaling argument for v_x^T would be

$$v_x^T \sim U \frac{\delta_T}{\delta}, \quad \text{when } \mathrm{Pr} = \frac{\nu}{\alpha} \sim \frac{\delta}{\delta_T} \gtrsim 1. \tag{25-68}$$

Since the momentum and thermal boundary layer thicknesses scale with their respective diffusivities, requiring $\delta_T \lesssim \delta$ is equivalent (in scaling terms) to saying $\mathrm{Pr} \gtrsim 1$.

The final form of the scaled boundary layer heat equation becomes

$$U \frac{\delta_T}{\alpha} \frac{\Delta T}{x} \sim \alpha \frac{\Delta T}{\delta_T^2} \quad \text{for } \mathrm{Pr} \gtrsim 1. \tag{25-69}$$

It is left as an exercise (see Problem 25-2) to show that scaling arguments applied to the momentum equation yield $\delta/x \sim \mathrm{Re}_x^{-1/2}$. Since δ_T is the only remaining unknown scale, the scaled heat equation can be used to establish how δ_T depends on the other known scales of the problem:

$$\frac{\delta_T^3}{x^3} \sim \frac{\alpha}{\nu}\frac{\delta}{x}\frac{\nu}{Ux} \sim \mathrm{Pr}^{-1}\mathrm{Re}_x^{-1/2}\mathrm{Re}_x^{-1} = \mathrm{Pr}^{-1}\mathrm{Re}_x^{-3/2} \tag{25-70}$$

or

$$\frac{\delta_T}{x} \sim \mathrm{Pr}^{-1/3}\mathrm{Re}_x^{-1/2} \quad \text{for } \mathrm{Pr} \gtrsim 1. \tag{25-71}$$

Fourier's law for the heat conduction from the surface can also be scaled:

$$q_s \sim k\frac{\Delta T}{\delta_T}. \tag{25-72}$$

Therefore, since $q_s \sim h\ \Delta T$, the heat transfer convection coefficient scales as

$$h \sim \frac{k}{\delta_T} = \frac{k}{x}\frac{x}{\delta_T} \sim \frac{k}{x}\mathrm{Pr}^{1/3}\mathrm{Re}_x^{1/2} \quad \text{for } \mathrm{Pr} \gtrsim 1. \tag{25-73}$$

The Nusselt number can present the convection coefficient in dimensionless form, such that

$$Nu_x = \frac{hx}{k} \sim \mathrm{Pr}^{1/3}\mathrm{Re}_x^{1/2} \quad \text{for } \mathrm{Pr} \gtrsim 1. \tag{25-74}$$

Therefore, the scaling of the heat and momentum equations suggest the functional dependency of Nu_x on Pr and Re_x. The exact analysis of the preceding section established that $Nu_x = \theta'(0)\mathrm{Re}_x^{1/2}$, where $\theta'(0)$ is a function of Pr. Equating this exact analysis with the present scaling arguments requires that $\theta'(0) \sim \mathrm{Pr}^{1/3}$. Since for $\mathrm{Pr} = 1$ it was determined that $\theta'(0) = 0.332$, one can tentatively suggest that $\theta'(0) \approx 0.332\mathrm{Pr}^{1/3}$, which results in the correlation

$$Nu_x = 0.332\ \mathrm{Pr}^{1/3}\mathrm{Re}_x^{1/2} \quad \text{for } \mathrm{Pr} \gtrsim 1. \tag{25-75}$$

It will be left as an exercise (see Problem 25-5) to test this hypothesis.

25.5 TRANSPORT ANALOGIES

In the previous sections of this chapter, the momentum, heat, and mass transfer equations for Blasius flow over a flat plate were solved. The boundary layer equations for these problems can be written in fully dimensionless form with the following definitions:

$$v_x^* = \frac{v_x}{U},\ v_y^* = \frac{v_y}{U},\ T^* = \frac{T - T_s}{T_\infty - T_s},\ C^* = \frac{c_A - c_s}{c_\infty - c_s},\ x^* = \frac{x}{L},\ \text{and}\ \ y^* = \frac{y}{L}. \tag{25-76}$$

With these definitions, the boundary layer equations become

$$x\text{-dir. Mom:}\quad v_x^*\frac{\partial v_x^*}{\partial x^*} + v_y^*\frac{\partial v_x^*}{\partial y^*} = \frac{1}{\mathrm{Re}_L}\frac{\partial^2 v_x^*}{\partial y^{*\,2}} \tag{25-77}$$

Heat: $$v_x^* \frac{\partial T^*}{\partial x^*} + v_y^* \frac{\partial T^*}{\partial y^*} = \frac{1}{\mathrm{Re}_L \mathrm{Pr}} \frac{\partial^2 T^*}{\partial y^{*\,2}} \tag{25-78}$$

Species: $$v_x^* \frac{\partial C^*}{\partial x^*} + v_y^* \frac{\partial C^*}{\partial y^*} = \frac{1}{\mathrm{Re}_L Sc} \frac{\partial^2 C^*}{\partial y^{*\,2}} \tag{25-79}$$

where

$$\mathrm{Re}_L = \frac{UL}{\nu} \quad \mathrm{Pr} = \frac{\nu}{\alpha} \quad Sc = \frac{\nu}{Đ_A}. \tag{25-80}$$

For the momentum, heat, and mass transfer problems discussed in the previous sections, the boundary conditions for the dimensionless variables are

$$v_x^*(y = 0) = 0 \quad v_x^*(y \to \infty) = 1 \quad v_x^*(x = 0) = 1 \tag{25-81}$$

$$T^*(y = 0) = 0 \quad T^*(y \to \infty) = 1 \quad T^*(x = 0) = 1 \tag{25-82}$$

$$C^*(y = 0) = 0 \quad C^*(y \to \infty) = 1 \quad C^*(x = 0) = 1\,. \tag{25-83}$$

One sees that if $\mathrm{Pr} = 1$ and $Sc = 1$, the mathematical problems for momentum, heat, and mass transfer become identical, and therefore the solutions to these problems are equivalent. This observation is useful in experimental situations, since transport phenomena of one kind can be inferred from measurements of another. For example, by definition

$$c_f = \frac{\tau(x)\big|_{y=0}}{\rho U^2/2} = \frac{2}{\mathrm{Re}_L} \frac{\partial v_x^*}{\partial y^*}\bigg|_{y^*=0} \tag{25-84}$$

and

$$Nu_L = \frac{hL}{k} = \frac{(\partial T/\partial y)\big|_{y=0} L}{(T_s - T_\infty)} = \frac{\partial T^*}{\partial y^*}\bigg|_{y^*=0}. \tag{25-85}$$

If $\mathrm{Pr} = 1$, the solution for v_x^* and T^* are identical, and therefore

$$\frac{c_f \mathrm{Re}_L}{2} = \frac{\partial v_x^*}{\partial y^*}\bigg|_{y^*=0} = \frac{\partial T^*}{\partial y^*}\bigg|_{y^*=0} = Nu_L. \tag{25-86}$$

The result, suggesting that

$$\frac{c_f}{2} \mathrm{Re}_L = Nu_L \quad (\mathrm{Pr} = 1), \tag{25-87}$$

is one of the best known analogies between heat and momentum transport, which is attributed to Reynolds [3]. If $\mathrm{Pr} \neq 1$, it can be shown that, to a good approximation,

$$Nu_L = \frac{c_f}{2} \mathrm{Pr}^{1/3} \mathrm{Re}_L \quad \text{for } 0.6 < \mathrm{Pr} < 60, \tag{25-88}$$

which is known as the Colburn analogy [4]. Since it was determined in Section 25.1 that $c_f = 0.664\mathrm{Re}_x^{-1/2}$, one can infer that Eq. (25-88) is equivalent to Eq. (25-75) for Blasius flow, the validity of which is demonstrated in Problem 25-5.

It is interesting to note that, although both the Reynolds analogy (25-87) and Colburn analogy (25-88) were developed here in the context of laminar flow, both analogies are rooted in the studies of turbulent flows. Therefore, it is apparent that the strength of these analogies transcends many details of the flow condition, and attests to the underlying commonality of transport phenomena.

25.6 BOUNDARY LAYERS DEVELOPING ON A WEDGE (FALKNER-SKAN FLOW)

In the preceding sections, boundary layers that develop on flat surfaces oriented parallel to the flow have been analyzed. An extension to this problem considers boundary layers that are developing on a flat surface oriented at some angle to the flow, as illustrated in Figure 25-8. The velocity distribution in the boundary layer is governed by the momentum and continuity equations:

$$\text{Momentum}: \quad v_x \frac{\partial v_x}{\partial x} + v_y \frac{\partial v_x}{\partial y} = \nu \frac{\partial^2 v_x}{\partial y^2} - \frac{1}{\rho}\frac{\partial P}{\partial x} \tag{25-89}$$

$$\text{Continuity}: \quad \frac{\partial v_x}{\partial x} + \frac{\partial v_y}{\partial y} = 0 \tag{25-90}$$

Notice that the momentum equation is written in boundary layer form. The pressure gradient cannot be dropped from the momentum equation, since the external flow is now accelerated over the wedge. It was found in Section 15.5.1 that the velocity external to the boundary layer increases with distance from the tip of the wedge as

$$v_e = dx^m, \quad \text{where } m = \frac{\beta/\pi}{2 - \beta/\pi} \tag{25-91}$$

and β is the wedge angle, as illustrated in Figure 25-8. As discussed in Section 15.5.1, the pressure distribution of the potential flow outside the boundary layer is governed by Bernoulli's equation, such that

$$-\frac{1}{\rho}\frac{dP}{dx} = v_e \frac{dv_e}{dx}. \tag{25-92}$$

Therefore, the momentum boundary layer equation may be written as

$$v_x \frac{\partial v_x}{\partial x} + v_y \frac{\partial v_x}{\partial y} = \nu \frac{\partial^2 v_x}{\partial y^2} + v_e \frac{dv_e}{dx} \tag{25-93}$$

where v_e is the flow velocity external to the boundary layer. In the form of Eq. (25-93), the boundary layer momentum equation can be applied to surfaces of arbitrary shape and

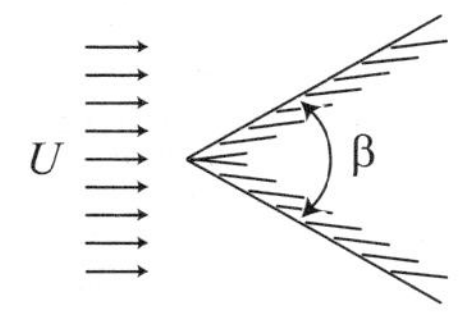

Figure 25-8 Flow over a wedge.

orientation to the flow. To solve the momentum equation for flow over a wedge, a solution is sought using the similarity variable

$$\eta = \frac{y}{\sqrt{\nu x / v_e}} = y\left(\frac{d\,x^{m-1}}{\nu}\right)^{1/2} \tag{25-94}$$

and a dimensionless stream function given by

$$f = \frac{\psi}{\sqrt{\nu x v_e}} = \frac{\psi}{\sqrt{\nu d\,x^{m+1}}}. \tag{25-95}$$

Notice that the forms of the similarity variable and dimensionless stream function are the same as those used for Blasius flow, given by Eqs. (25-11) and (25-14), except that the constant U appearing in those previous definitions is replaced with Eq. (25-91) for $v_e(x)$.

The components of the velocity field in the boundary layer are expressed in terms of the dimensionless stream function as

$$v_x = v_e f' = d\,x^m f' \tag{25-96}$$

$$v_y = -\sqrt{\frac{\nu v_e}{x}}\left[\frac{m+1}{2}f + \frac{m-1}{2}\eta f'\right]. \tag{25-97}$$

Fully transforming the boundary layer momentum equation (25-93) into the new variables yields

$$f''' + \frac{m+1}{2}f\,f'' + m\left[1 - (f')^2\right] = 0 \tag{25-98}$$

with boundary conditions

$$f(0) = 0, \quad f'(0) = 0, \quad \text{and} \quad f'(\infty) = 1. \tag{25-99}$$

Equation (25-98) is known as the Falkner-Skan equation [5]. Notice that when $\beta = 0$, such that $m = 0$, the momentum equation becomes the same as the Blasius equation (25-23). Flow over the wedge is solved numerically in Problem 25-7.

25.6.1 Heat and Mass Transfer for Flows over a Wedge

The boundary layer equations for heat and species transport can be transformed into ordinary differential equations using the same variables that describe the transformed momentum equation (25-98). The result for heat transfer is

$$\theta'' + \frac{m+1}{2}\mathrm{Pr}\,f\,\theta' = 0, \quad \text{where } \theta = \frac{T - T_s}{T_\infty - T_s} \tag{25-100}$$

and is solved subject to the boundary conditions

$$\theta(0) = 0 \quad \text{and} \quad \theta(\infty) = 1. \tag{25-101}$$

The result for species transport is

$$\phi'' + \frac{m+1}{2}\mathrm{Sc}\,f\,\phi' = 0, \quad \text{where } \phi = \frac{c - c_s}{c_\infty - c_s} \tag{25-102}$$

and is solved subject to the boundary conditions

$$\phi(0) = 0 \quad \text{and} \quad \phi(\infty) = 1. \tag{25-103}$$

The heat and mass transfer between a flow and the bounding surface of a wedge is solved numerically in Problem 25-8.

25.7 VISCOUS HEATING IN THE BOUNDARY LAYER

Consider a flow over a flat plate that is steady, isobaric, and incompressible (Figure 25-9). The only source of thermal energy is viscous heat generation. Although the loss of kinetic energy to viscous dissipation is accounted for in the momentum equation, the reappearance of this thermal energy in the heat equation is most often neglected. In this section, viscous heat generation will be included. It is left as an exercise (see Problem 25-9) to show that the heat equation in boundary layer form is given by

$$v_x \frac{\partial \theta}{\partial x} + v_y \frac{\partial \theta}{\partial y} = \alpha \frac{\partial^2 \theta}{\partial y^2} + \frac{\nu}{C_p T_\infty} \left(\frac{\partial v_x}{\partial y} \right)^2, \tag{25-104}$$

where the dimensionless temperature is defined by

$$\theta = T/T_\infty. \tag{25-105}$$

For an adiabatic plate, the boundary conditions for the thermal boundary layer are

$$\theta(y \to \infty) = 1 \quad \theta(x = 0) = 1 \quad \partial\theta/\partial y|_{y=0} = 0. \tag{25-106}$$

Using the similarity variable defined for Blasius flow in Section 25.1,

$$\eta = y/\sqrt{\nu x/U}, \tag{25-107}$$

and making use of the previous results,

$$v_x = Uf' \quad \text{and} \quad v_y = \sqrt{\nu U/x}\,[\eta f' - f]/2, \quad \text{where } f = \psi/\sqrt{\nu x U}, \tag{25-108}$$

the heat equation can be transformed with the aid of the derivative expressions given by Eqs. (25-21a–c). Transformation of the boundary layer heat equation (25-104) into the new variables yields

$$\overbrace{(Uf')\frac{-\eta}{2x}\theta'}^{v_x \frac{\partial \theta}{\partial x}} + \overbrace{\frac{1}{2}\sqrt{\frac{\nu U}{x}}[\eta f' - f]\frac{\eta}{y}\theta'}^{v_y \frac{\partial \theta}{\partial y}} = \overbrace{\alpha\left(\frac{\eta}{y}\right)^2 \theta''}^{\alpha \frac{\partial^2 \theta}{\partial y^2}} + \overbrace{\frac{\nu}{C_p T_\infty}\left(\frac{\eta}{y}\frac{\partial}{\partial \eta}(Uf')\right)^2}^{\frac{\nu}{C_p T_\infty}\left(\frac{\partial v_x}{\partial y}\right)^2}, \tag{25-109}$$

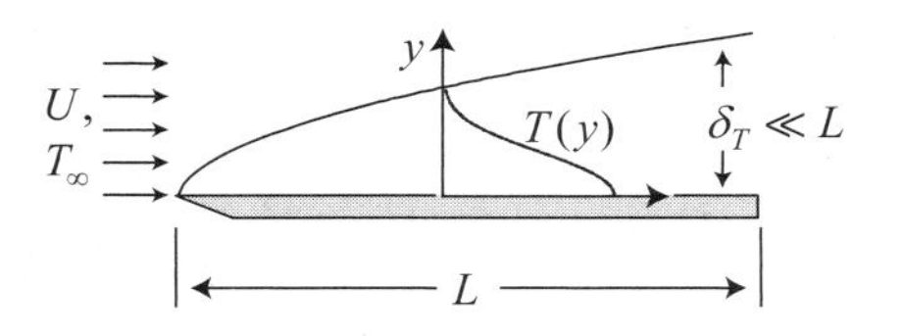

Figure 25-9 Viscous heating in flow over an adiabatic flat plate.

after which the heat equation can be simplified to

$$\theta'' + \Pr\left(\frac{f\,\theta'}{2} + Ec(f'')^2\right) = 0. \tag{25-110}$$

The final form of the heat equation employs the *Eckert number*,** as defined by

$$Ec = \frac{U^2}{C_p T_\infty}, \tag{25-111}$$

which expresses the ratio of kinetic energy to the enthalpy content in the free stream. The transformed heat equation (25-110) is subject to the boundary conditions

$$\theta'(0) = 0 \quad \text{and} \quad \theta(\infty) = 1. \tag{25-112}$$

A numerical solution for θ can be found by the same process used in Section 25.2. The shooting method, as discussed in Section 23.4, is used to guess the correct initial value of $\theta(0)$ that satisfies the condition $\theta(\infty) = 1$.

It is interesting to note that the adiabatic wall temperature $\theta(0)$ is not a function of x. Although there is an accumulation of heat in the downstream flow, the heat content spreads by diffusion over the thickening dimension of the thermal boundary layer. Furthermore, the volumetric strength of the generation term in the heat equation diminishes downstream as shear in the momentum boundary layer is reduced. In this way, the fluid temperature at the adiabatic wall is able to remain constant.

The adiabatic wall temperature $\theta(0)$ as a function of the Prandtl number and the Eckert number is shown in Figure 25-10. To achieve a significant Eckert number—say, $Ec/2 \sim 0.1$—typical fluids must move at hundreds of meters per second. Consequently, many flows do not require inclusion of viscous heating for an accurate description of the heat equation.

Figure 25-10 demonstrates that the adiabatic wall temperature obeys the relation $\theta(0) = 1 + Ec/2$ when $\Pr = 1$. The simplicity of this result suggests the existence of an analytic solution for this special case. For $\Pr = 1$, the temperature field obeys the solution

$$\theta = 1 + Ec(1 - f'^2)/2, \tag{25-113}$$

as is easily demonstrated by direct substitution into both the heat equation:

$$\theta'' + \frac{f\,\theta'}{2} + Ec(f'')^2 = 0 \;\rightarrow\; -Ec\,f'\overbrace{\left(f''' + \frac{1}{2}ff''\right)}^{=0} = 0 \tag{25-114}$$

and the boundary conditions:

$$\theta(\infty) = 1 + Ec(1 - [f'(\infty)]^2)/2 = 1 \tag{25-115}$$

$$\theta'(0) = -Ec\,f'(0)f''(0) = 0. \tag{25-116}$$

Evaluating Eq. (25-113) at the wall yields the adiabatic wall temperature $\theta(0) = 1 + Ec/2$ for the case when $\Pr = 1$.

**The Eckert number is named after the aeronautical engineer Ernst R. G. Eckert (1904–2004).

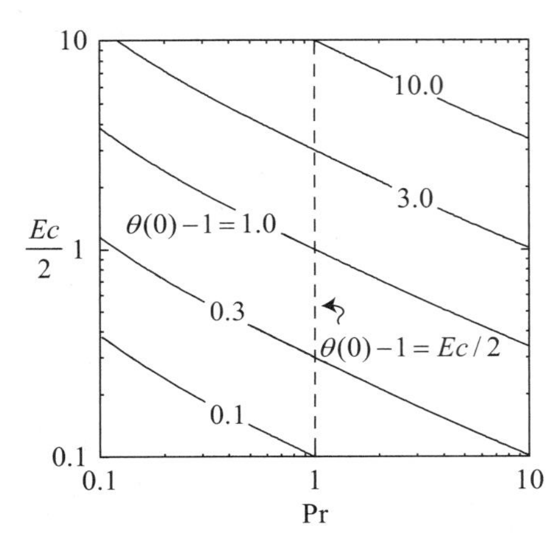

Figure 25-10 Adiabatic wall temperature mapped as a function of Pr and *Ec*.

25.8 PROBLEMS

25-1 Start with the assumption that

$$f = \frac{\psi(x,y)}{g(x)\sqrt{\nu LU}},$$

and show that $g(x) = \sqrt{x/L}$ permits the momentum equation to be expressed as an ordinary differential equation of the nondimensional stream function f, with $\eta = y/\sqrt{\nu x/U}$ as the similarity variable.

25-2 Scale the momentum equation to show that $\delta/x \sim \mathrm{Re}_x^{-1/2}$.

25-3 Using a similarity solution approach, derive the thermal energy boundary layer equation for Blasius flow and show that it has the form $\theta'' + \mathrm{Pr}\, f\, \theta'/2 = 0$, where $\theta = (T - T_s)/(T_\infty - T_s)$.

25-4 Find an exact expression for the growth of the thermal boundary layer thickness $\delta_T(x)$ when water flows over a heated flat plate. The Prandtl number for water at 300 K is Pr = 5.8.

25-5 Show that $\mathrm{Nu}_x/\mathrm{Re}_x^{1/2} = \theta'(0)$, where $\theta(\eta)$ is the solution to the heat equation for flow over a flat plate (Blasius' problem). Determine $\theta'(0)$ for four fluids: Pr = 0.07, 0.7, 7.0, and 70, and plot $\mathrm{Nu}_x/\mathrm{Re}_x^{1/2}$ versus Pr on log-log axes. Superimpose on your results a plot of Eq. (25-75). What can you conclude?

25-6 Scale the heat equation for the case when $\mathrm{Pr} \ll 1$ to estimate the dependence of the Nusselt number on the Reynolds and Prandtl numbers. Your result should be of the form $\mathrm{Nu_x} \sim \mathrm{Pr}^n \mathrm{Re}^m$.

25-7 Using the transformations suggested in Section 25.6, demonstrate that the momentum boundary layer equation for flow over a wedge is given by

$$f''' + \frac{m+1}{2} f\, f'' + m(1 - f'^2) = 0.$$

Derive the boundary conditions for this problem. Solve the momentum equation using Runge-Kutta integration when m= 0.2. Plot f, f', and f''.

25-8 You are designing a chemical vapor deposition (CVD) system, based on the flow configuration shown. A precursor gas carries a molecular element "A" that reacts with the heated surface of a wafer. The wafer temperature is maintained with a resistively heated plate. This plate is located below a thermally conductive substrate on which the wafer is placed. (The wafer is not shown in the illustration.) The system parameters and thermophysical properties of the fluid are given in the tables. At the tail edge of the wafer, the speed of the flow external to the boundary layer is measured to be 10 cm/s.

CVD System Parameters

	Temperature (T)	*Flow speed* (v)	*Mole fraction* (c_A)
Free stream:	25 °C	$d\,x^m$	0.2
Wafer surface:	600 °C	0 m/s	0.0

Thermophysical Properties of the Fluid

Property	*Value*
Density (ρ)	$0.60\ \mathrm{kg/m^3}$
Viscosity (μ)	$30 \cdot 10^{-6}\ \mathrm{Pa \cdot s}$
Mass diffusion coef. ($Ð_A$)	$14 \cdot 10^{-6}\ \mathrm{Pa \cdot s}$
Heat capacity (C_p)	$1050\ \mathrm{J/(kg \cdot K)}$
Thermal conductivity (k)	$0.045\ \mathrm{W/(m \cdot K)}$

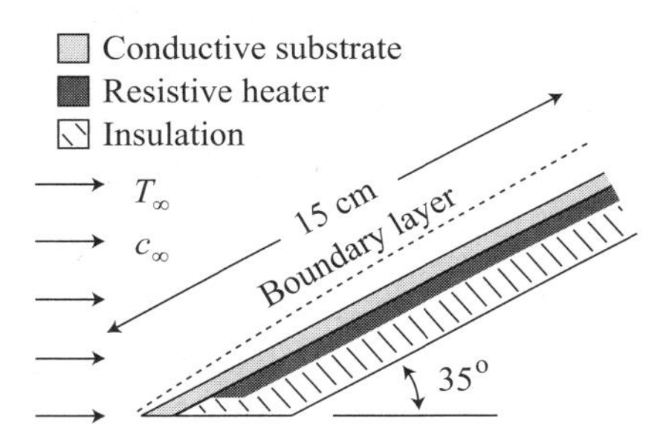

Derive the boundary layer equations for thermal energy and species concentration using the same similarity parameter and velocity field transformation as for the momentum equation. Solve both problems using Runge-Kutta numerical integration and plot θ, θ', C_A, and C'_A, where

$$\theta = \frac{T - T_s}{T_\infty - T_s} \quad \text{and} \quad C = \frac{c_A}{c_\infty}.$$

Calculate the momentum, heat, and concentration boundary layer thicknesses at the middle of the wafer. Explain the relative scales of the three boundary layers.

25-9 Show that the heat equation for a boundary layer with viscous heating can be simplified to

$$v_x \frac{\partial \theta}{\partial x} + v_y \frac{\partial \theta}{\partial y} = \alpha \frac{\partial^2 \theta}{\partial y^2} + \frac{\nu}{C_p T_\infty} \left(\frac{\partial v_x}{\partial y} \right)^2,$$

where the dimensionless temperature is defined by $\theta = T/T_\infty$. Note all assumptions used to arrive at this form.

REFERENCES

[1] H. Schlichting, *Boundary-Layer Theory.* New York, NY: McGraw-Hill, 1979.
[2] H. Blasius, "Grenzschichten in Flüssigkeiten mit Kleiner Reibung." *Zeitschrift für Mathematik und Physik*, **56**, 1 (1908).
[3] O. Reynolds, "On the Extent and Action of the Heating Surface for Steam Boilers." *Proc. Manchester Literary and Philosophical Society*, **14**, 7 (1874).
[4] A. P. Colburn, "A Method of Correlating Forced Convection Heat-Transfer Data and a Comparison with Fluid Friction." *American Institute of Chemical Engineers*, **29**, 174 (1933).
[5] V. M. Falkner and S. W. Skan, "Some Approximate Solutions of the Boundary-Layer Equations." *Philosophical Magazine*, **12**, 865 (1930).

Chapter 26

Natural Convection

26.1 Buoyancy

26.2 Natural Convection from a Vertical Plate

26.3 Scaling Natural Convection from a Vertical Plate

26.4 Exact Solution to Natural Convection Boundary Layer Equations

26.5 Problems

The momentum equation has been independent of the heat equation in the convection problems considered so far. However, in some problems the velocity field will depend on the temperature field. In *natural convection*, the buoyancy of a heated fluid drives fluid motion, and serves as a good example in which the momentum equation is coupled to the heat equation.

26.1 BUOYANCY

The origin of buoyancy is related to the hydrostatic pressure in a fluid. With the body force of gravity expressed as a potential $g_i = -g\partial_i h$ (see Section 14.2), where h is a measure of vertical distance (along the direction of $-\vec{g}$) the momentum equation for a constant-density hydrostatic fluid is

$$\text{Hydrostatic momentum:}\quad \overbrace{\rho(\partial_o v_i + v_j\partial_j v_i) - \mu\partial_j\partial_j v_i}^{=0} = -\partial_i P_\infty - \rho_\infty g\partial_i h. \tag{26-1}$$

Therefore, the hydrostatic pressure P_∞ is prescribed by the relation

$$P_\infty + \rho_\infty gh = const. \tag{26-2}$$

The hydrostatic pressure can often be ignored in solving hydrodynamic problems (without interfaces between fluids or free surfaces). However, if there is a local change in density of an otherwise homogeneous fluid, as illustrated in Figure 26-1, a force imbalance is created that drives motion. The magnitude of this buoyancy force is given by the density difference relative to the surrounding fluid: $-g\Delta\rho/\rho_\infty$ (see Problem 5-4). If the density difference is due to a relative change in temperature, one can write for the buoyancy force:

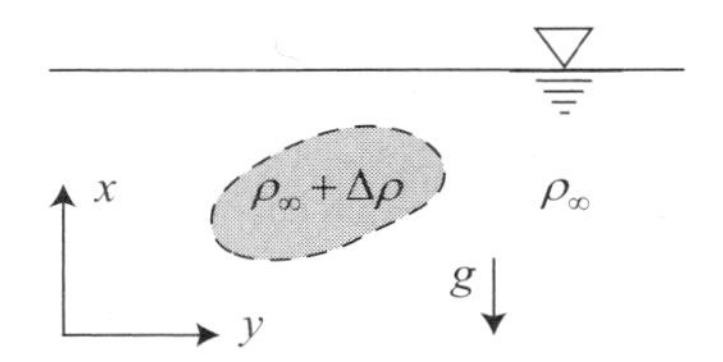

Figure 26-1 Illustration of a local change in density.

$$-g\Delta\rho/\rho_\infty = g\beta\,\Delta T, \tag{26-3}$$

where the coefficient of thermal expansion is defined by

$$\beta = -\frac{1}{\rho}\left(\frac{\partial\rho}{\partial T}\right)_p, \tag{26-4}$$

and $\Delta T = T - T_\infty$ is the amount by which the local fluid temperature is elevated above the surrounding fluid.

26.2 NATURAL CONVECTION FROM A VERTICAL PLATE

Consider the problem of a hot vertical wall submerged in a cool fluid, as illustrated in Figure 26-2. The fluid far from the wall is stationary. However, due to the elevated temperature of the wall, the fluid near the surface rises by natural convection. Therefore, two boundary layers develop in front of the vertical wall, one associated with the temperature field and one associated with the motion of the fluid induced by buoyancy.

The density change $\rho = \rho_\infty + \Delta\rho$ caused by a temperature rise $\Delta T = T - T_\infty$ near the wall is assumed to be small, and the pressure on the fluid in the vicinity of the wall remains P_∞. For small changes in density, it is satisfactory to describe the flow with the incompressible form of the continuity equation:

$$\partial_j v_j = 0. \tag{26-5}$$

Furthermore, with $\rho = \rho_\infty + \Delta\rho$ and $P = P_\infty$, the x-direction momentum equation can be written as

$$(\rho_\infty + \Delta\rho)(\partial_o v_x + v_j\partial_j v_x) - \overbrace{\mu\partial_j\partial_j v_x}^{incompressible} = -\partial_x P_\infty - (\rho_\infty + \Delta\rho)g\partial_x h. \tag{26-6}$$

Notice that momentum diffusion is expressed in a form appropriate for incompressible flow. The right-hand side of the momentum equation (26-6) simplifies to

$$-\partial_x P_\infty - (\rho_\infty + \Delta\rho)g\partial_x h = -\underbrace{\frac{\partial}{\partial x}(P_\infty + \rho_\infty g h)}_{=const.} - \Delta\rho g\underbrace{\frac{\partial h}{\partial x}}_{=1} = -\Delta\rho g. \tag{26-7}$$

Since $P_\infty + \rho_\infty g h$ is a constant, its spatial derivative is zero. Furthermore, if $\Delta\rho \ll \rho_\infty$, the change in density $\Delta\rho$ can be neglected on the left-hand side of the momentum equation (26-6), where it appears with the inertial term. This simplification to the momentum equation is known as the Boussinesq approximation [1]. With these simplifications, the momentum equation for a steady-state flow becomes

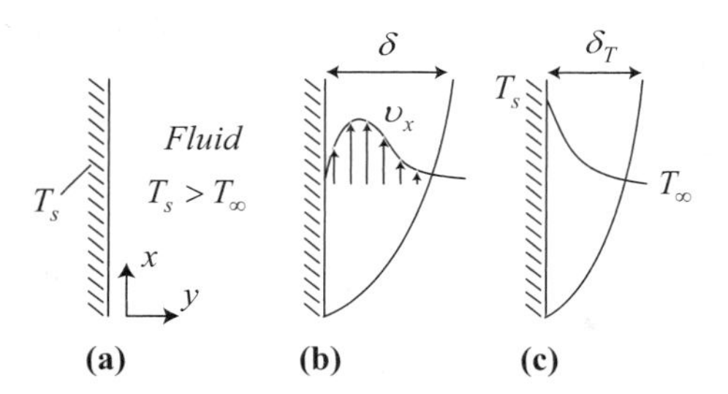

Figure 26-2 Illustration of natural convection from a vertical heated plate.

$$v_j \partial_j v_x - \nu \partial_j \partial_j v_x = -\frac{\Delta \rho}{\rho_\infty} g = g\beta \Delta T \tag{26-8}$$

where the density change is expressed in terms of a temperature change, as related to the coefficient of thermal expansion (26-4). Finally, as long as changes in density are small, it is appropriate to use the incompressible form of the heat equation (without viscous heating) to describe the temperature field in the fluid:

$$\partial_o T + v_j \partial_j T = \alpha\, \partial_j \partial_j T. \tag{26-9}$$

The solution to the flow field near the heated wall requires coupling solutions to the continuity (26-5), momentum (26-8), and heat (26-9) equations describing the three dependent variables v_y, v_x, and T. These equations can be written for two-dimensional Cartesian coordinates, in a form appropriate for boundary layer flow (which neglects streamwise diffusion):

$$\text{Continuity:} \quad \frac{\partial v_x}{\partial x} + \frac{\partial v_y}{\partial y} = 0 \tag{26-10}$$

$$\text{Momentum:} \quad v_x \frac{\partial v_x}{\partial x} + v_y \frac{\partial v_x}{\partial y} = \nu \frac{\partial^2 v_x}{\partial y^2} + g\beta(T - T_\infty) \tag{26-11}$$

$$\text{Heat:} \quad v_x \frac{\partial T}{\partial x} + v_y \frac{\partial T}{\partial y} = \alpha \frac{\partial^2 T}{\partial y^2}. \tag{26-12}$$

Suppose the wall temperature is uniform. The boundary conditions imposed on the flow are expressed by

$$\begin{aligned} y = 0: &\quad v_x, v_y = 0, \quad T = T_s \\ x = 0: &\quad v_x = 0, \quad T = T_\infty \\ y \to \infty: &\quad v_x \to 0, \quad T \to T_\infty. \end{aligned} \tag{26-13}$$

For these boundary conditions, the governing equations (26-10) through (26-12) can be solved with a similarity transformation (as will be done in Section 26.4) [2, 3]. However, it is informative to first investigate what can be learned about the solution from scaling arguments.

26.3 SCALING NATURAL CONVECTION FROM A VERTICAL PLATE

The thermal boundary layer thickness δ_T is an important scale for estimating heat transfer from the wall, since the heat transfer coefficient scales as

$$h = \frac{q}{\Delta T} \sim \frac{k}{\Delta T}\frac{\Delta T}{\delta_T} = \frac{k}{\delta_T}. \tag{26-14}$$

The other important scale is the momentum boundary layer thickness δ. These two scales are physically related in natural convection such that $\delta_T \leq \delta$. It is impossible for $\delta_T > \delta$ to occur, since one cannot have a region of the fluid away from the wall that is heated (and therefore buoyant) but not moving.

The governing equations that control the thermal and momentum boundary layer thicknesses can be scaled to provide the desired estimates:

$$\text{Continuity:} \quad \frac{v_x}{x} \sim \frac{v_y}{\delta} \quad \text{and} \quad \frac{v_x^T}{x} \sim \frac{v_y^T}{\delta_T} \tag{26-15}$$

$$\text{Momentum:} \quad v_x \frac{v_x}{x} \text{ ``+'' } v_y \frac{v_x}{\delta} \sim \nu \frac{v_x}{\delta^2} + g\beta\Delta T \tag{26-16}$$

$$\text{Heat:} \quad v_x^T \frac{\Delta T}{x} \text{ ``+'' } v_y^T \frac{\Delta T}{\delta_T} \sim \alpha \frac{\Delta T}{\delta_T^2}, \tag{26-17}$$

where the change in the temperature field scales as $\Delta T \sim T_s - T_\infty$. Two scalings of the continuity equation result from alternately considering the boundary layer scales $y \sim \delta$ and $y \sim \delta_T$. The velocities in the momentum equation are denoted by v_x and v_y, while v_x^T and v_y^T are the velocities in the heat equation. However, with continuity,

$$v_y \sim v_x \delta / x \quad \text{and} \quad v_y^T \sim v_x^T \delta_T / x. \tag{26-18}$$

Combing this result with the momentum and heat equations yields

$$\text{Momentum}: \quad \frac{v_x^2}{x} \sim \nu \frac{v_x}{\delta^2} \text{ ``+'' } g\beta\Delta T \tag{26-19}$$

$$\text{Heat}: \quad \frac{v_x^T}{x} \sim \frac{\alpha}{\delta_T^2}. \tag{26-20}$$

The physics of natural convection dictate a common vertical velocity scale for both heat and momentum transfer:

$$v_x^T / x \sim v_x / x. \tag{26-21}$$

This is unlike forced convection, discussed in Section 25.4, where heat transfer can experience a velocity scale much smaller than momentum transfer (when $\Pr \gg 1$). Using the common velocity scale for heat and momentum transfer in natural convection, a relation between the two boundary layer thicknesses can be found. Equating the scales of advection and diffusion in the equations for momentum transfer (26-19) and heat transfer (26-20) suggests

$$\underbrace{\frac{\nu}{\delta^2} \sim \frac{v_x}{x}}_{momentum} \sim \underbrace{\frac{v_x^T}{x} \sim \frac{\alpha}{\delta_T^2}}_{heat} \tag{26-22}$$

from which it follows that

$$\frac{\delta}{\delta_T} \sim \sqrt{\frac{\nu}{\alpha}} \quad (\Pr \geq 1). \tag{26-23}$$

However, since natural convection requires that $\delta_T \leq \delta$, the scaling result $\delta/\delta_T \sim \Pr^{1/2}$ must break down for fluids where $\Pr \ll 1$. In such a case, the diffusion term in the momentum equation can no longer be compared to the advection term, as done in Eq. (26-22). Instead, the advection term must scale with the buoyancy term in Eq. (26-19), while the diffusion term is relatively small. With this insight, it is realized that the momentum equation has two limiting forms: one for the limit when $\delta_T \to \delta$ (where $\Pr \ll 1$)

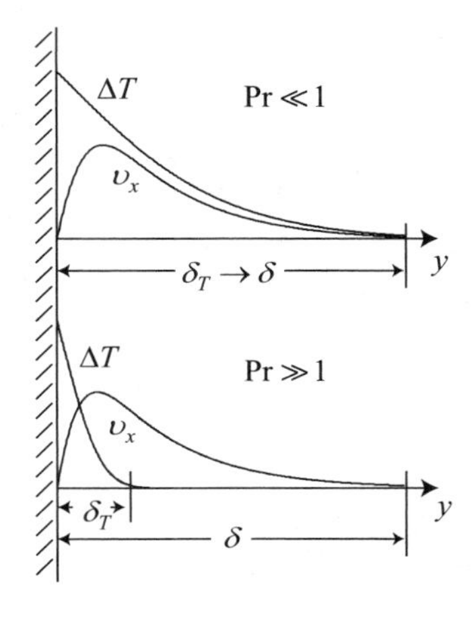

Figure 26-3 Natural convection boundary layers for large and small Prandtl number.

and the other for the limit when $\delta_T \ll \delta$ (where $\mathrm{Pr} \gg 1$). The temperature and velocity profiles are illustrated in Figure 26-3 for both limiting cases.

The thermal boundary layer thickness δ_T can be investigated for the two limiting cases. When $\delta_T \approx \delta$, scaling the momentum equation yields

$$\begin{array}{c}\text{Momentum:}\\ \text{(scaled over } \delta)\end{array} \left\{ \delta_T \approx \delta: \quad \underbrace{\frac{v_x^2}{x}}_{advection} \sim \underbrace{g\beta\Delta T}_{buoyancy} \gg \underbrace{\nu \frac{v_x}{\delta^2}}_{diffusion} \quad (\mathrm{Pr} \ll 1). \right. \tag{26-24}$$

When the heat and momentum transfer are coupled through the common velocity scale, using Eqs (26-20) and (26-24), the result is

$$\delta_T \approx \delta: \quad \underbrace{\frac{\alpha}{\delta_T^2} \sim \frac{v_x^T}{x}}_{heat} \sim \underbrace{\frac{v_x}{x} \sim \sqrt{\frac{g\beta\Delta T}{x}}}_{mom.\ (over\ \delta)}, \quad (\mathrm{Pr} \ll 1). \tag{26-25}$$

However, when $\delta_T \ll \delta$, the scaling given by Eq. (26-25) is no longer valid. To determine v_x/x in this case requires rescaling the momentum equation for the region in front of the wall defined by δ_T. In this subregion, the momentum equation scales as

$$\begin{array}{c}\text{Momentum:}\\ \text{(scaled over } \delta_T)\end{array} \left\{ \delta_T \ll \delta: \quad \underbrace{\frac{v_x^2}{x}}_{advection} \ll \underbrace{\nu \frac{v_x}{\delta_T^2}}_{diffusion} \sim \underbrace{g\beta\Delta T}_{buoyancy} \quad (\mathrm{Pr} \gg 1). \right. \tag{26-26}$$

The advection term is now smaller than the other terms by a factor of $(\delta_T/\delta)^2$. Using Eqs (26-20) and (26-26), heat and momentum transfer are again compared through the common velocity scale:

$$\delta_T \ll \delta: \quad \underbrace{\frac{\alpha}{\delta_T^2} \sim \frac{v_x^T}{x}}_{heat} \sim \underbrace{\frac{v_x}{x} \sim \frac{g\beta\Delta T}{x}\frac{\delta_T^2}{\nu}}_{mom.\ (over\ \delta_T)} \quad (\mathrm{Pr} \gg 1). \tag{26-27}$$

The scaling arguments given by Eqs. (26-25) and (26-27) reveal the dependence of δ_T on the physical parameters of the problem for the two limiting cases of high and low Prandtl number. These scales for δ_T can be used to estimate the heat transfer coefficient, via Eq. (26-14), leading to

$$Nu_x = \frac{hx}{k} \sim \frac{x}{\delta_T} \sim \begin{cases} \delta_T \approx \delta: \quad \dfrac{x}{\delta_T} \sim \left(\dfrac{g\beta x^3 \Delta T}{\nu^2}\dfrac{\nu^2}{\alpha^2}\right)^{1/4} \sim Gr_x^{1/4}\mathrm{Pr}^{1/2} \quad (\mathrm{Pr} \ll 1) & \text{(26-28a)} \\ \delta_T \ll \delta: \quad \dfrac{x}{\delta_T} \sim \left(\dfrac{g\beta x^3 \Delta T}{\nu^2}\dfrac{\nu}{\alpha}\right)^{1/4} \sim Gr_x^{1/4}\mathrm{Pr}^{1/4} \quad (\mathrm{Pr} \gg 1) & \text{(26-28b)} \end{cases}$$

in which the *Grashof number** has been introduced:

$$Gr_x = \frac{g\beta x^3 \Delta T}{\nu^2}. \tag{26-29}$$

This number is often said to express the ratio of buoyancy forces to viscous forces. However, in natural convection the Grashof number must be a large number in order to satisfy the boundary layer expectation that $\delta_T \ll x$, as can be seen from Eqs. (26-28a, b). This is similar to having a large Reynolds number in forced convection boundary layer problems.

It is a common alternative for problems in natural convection to be described in terms of the *Rayleigh number*, which is related to the Grashof number through the definition

$$Ra_x = \frac{g\beta x^3 \Delta T}{\alpha\nu} = Gr_x \mathrm{Pr}. \tag{26-30}$$

26.4 EXACT SOLUTION TO NATURAL CONVECTION BOUNDARY LAYER EQUATIONS

Natural convection from a vertical plate admits a similarity solution to the boundary layer equations for momentum (26-11) and heat (26-12) transfer. Experience with forced convection in Chapter 25 suggests that a suitable similarity variable will have the form $\eta \sim y/\delta$, where δ is the momentum boundary layer thickness. The dependence of δ on x can be established from the scaling arguments of the preceding section, with the result

$$\delta \sim \mathrm{Pr}^n \left(\frac{\nu^2 x}{g\beta\Delta T}\right)^{1/4}. \tag{26-31}$$

The momentum boundary layer thickness depends on the Prandtl number raised to some power, n. The exponent depends on the fluid, with $n = -1/2$ for $\mathrm{Pr} \ll 1$ and $n = 1/4$ for $\mathrm{Pr} \gg 1$ (see Problem 26-1). With Eq. (26-31), a similarity variable based on $\eta \sim y/\delta$ can now be expressed as

$$\eta = C\frac{y}{x^{1/4}}, \quad \text{where } C = \left(\frac{g\beta(T_s - T_\infty)}{4\nu^2}\right)^{1/4} = \left(\frac{Gr_x}{4x^3}\right)^{1/4} \tag{26-32}$$

$$\text{and} \quad Gr_x = \frac{g\beta x^3 (T_s - T_\infty)}{\nu^2}. \tag{26-33}$$

The boundary layer thickness dependence on Pr^n has no functional impact on the similarity variable, and therefore is omitted. The appearance of the constant $(1/4)^{1/4}$ is a cosmetic change used to tidy up the form of the final ordinary differential equation. Recall

*The Grashof number is named after the German engineer Franz Grashof (1826–1893).

that the similarity variable can be scaled by any dimensionless constant (e.g., Pr^n or $(1/4)^{1/4}$) with the only effect being a change in the constant coefficients appearing in the final ordinary differential equation.

In Section 25.1, for forced convection, a dimensionless stream function was defined by $f = \psi/\sqrt{\nu x U}$. However, in natural convection, the velocity scale $v_x \sim U$ is undefined and part of the solution. From scaling arguments (for the case where $\mathrm{Pr} \ll 1$), it was found that $v_x \sim \nu x/\delta^2$. Therefore, replacing U with this expression for v_x in the dimensionless stream function, it is anticipated that

$$f = \psi/(\nu x/\delta). \tag{26-34}$$

Inserting the scaling expression for δ, a dimensionless stream function may be defined as

$$f = \frac{\psi}{4\nu C x^{3/4}}, \tag{26-35}$$

where C is as defined in Eq. (26-32). The appearance of a constant factor of $1/4$ is again only a cosmetic change. Although scaling the stream function by $x^{3/4}$ in Eq. (26-35) and the y spatial variable by $x^{1/4}$ in Eq. (26-32) have both been deduced from previous experience, the required powers of x for a similarity solution to exist can also be determined by a systematic approach, as suggested in Problem 26-2.

The dimensional velocity appearing in the transport equations can be expressed in terms of the dimensionless stream function as

$$v_x = 4\nu C^2 x^{1/2} f' \quad \text{and} \quad v_y = \nu C x^{-1/4}[\eta f' - 3f]. \tag{26-36}$$

To normalize the boundary conditions associated with the heat equation, a dimensionless temperature is defined by

$$\theta = \frac{T - T_s}{T_\infty - T_s}. \tag{26-37}$$

The governing boundary layer equations (26-11) and (26-12) can now be transformed with the similarity variable, Eq. (26-32), and the dimensionless dependent variables for the stream function (26-35) and temperature (26-37) to yield

$$\text{Momentum:} \quad f''' + 3f f'' - 2(f')^2 + (1 - \theta) = 0 \tag{26-38}$$

$$\text{Heat:} \quad \theta'' + 3\mathrm{Pr}\, f\, \theta' = 0. \tag{26-39}$$

The continuity equation no longer appears as a governing equation because it is explicitly satisfied by the stream function definition. The boundary conditions on the original problem are transformed:

$$\begin{array}{lll}
T(x=0) = T_\infty & \eta \to \infty & \left.\begin{array}{l}\theta(\infty) = 1 \\ \theta(\infty) = 1\end{array}\right\} \textit{same} \\
T(y \to \infty) = T_\infty & \eta \to \infty & \\
T(y=0) = T_s & \eta = 0 & \theta(0) = 0 \\
v_x(x=0) = 0 & \eta \to \infty & \left.\begin{array}{l}f'(\infty) = 0 \\ f'(\infty) = 0\end{array}\right\} \textit{same} \\
v_x(y \to \infty) = 0 & \eta \to \infty & \\
v_x(y=0) = 0 & \eta = 0 & f'(0) = 0 \\
v_y(y=0) = 0 & \eta = 0 & f(0) = 0
\end{array}$$

7 original conditions — *5 final conditions*

Notice that the conditions at $x = 0$ and $y \to \infty$ collapse to the same condition for $\eta \to \infty$ in the transformed problem.

The fourth-order Runge-Kutta method can be used to integrate the coupled governing equations for f and θ. However, there are two unknown initial conditions imposed on integration: $f''(0)$ and $\theta'(0)$. While shooting methods can be employed to solve for both unknowns, simple bisection methods are impaired by the coupled (and highly nonlinear) nature of the problem. Specifically, it is difficult to evaluate a particular guess of $\theta'(0)$, based on the result for $\theta(\infty)$, when this result is also influenced by the guessed value of $f''(0)$. To circumvent this issue, a related problem for θ is sought that can be solved without shooting. Let $\Theta(\zeta)$ be related to the problem for $\theta(\eta)$ through the scaling

$$\theta(\eta) = a\,\Theta(\zeta) \quad \text{where} \quad \zeta = a\,\eta, \tag{26-40}$$

and "a" is a constant. The motivation for this transformation is understood by looking at the boundary conditions for the $\Theta(\zeta)$ problem. Transforming the original boundary conditions yields

$$\eta = 0: \quad \theta(0) = 0 \;\rightarrow\; \theta(0) = a\Theta(0) \;\rightarrow\; \Theta(0) = 0 \tag{26-41a}$$

$$\eta \to \infty: \quad \theta(\infty) = 1 \;\rightarrow\; \theta(\infty) = a\Theta(\infty) \;\rightarrow\; \Theta(\infty) = 1/a. \tag{26-41b}$$

Since "a" is not prescribed, it does not have to be imposed on the problem for $\Theta(\zeta)$. The trick, therefore, is to impose another condition on $\Theta(\zeta)$, and determine the corresponding value of "a" that follows. For example, suppose that $\Theta'(0) = 1$. With $\Theta(0) = 0$ also known, determining $\Theta(\zeta)$ by numerical integration is straightforward. However, the transformed boundary condition (26-41b) requires that $a = 1/\Theta(\infty)$. Therefore, once a is known from the solution to $\Theta(\zeta)$, $\theta'(0)$ for the original problem can be determined without shooting. Since

$$\frac{d\theta}{d\eta} = a\,\frac{d\zeta}{d\eta}\frac{d\Theta}{d\zeta} = a^2\frac{d\Theta}{d\zeta} \quad \text{or} \quad \theta' = a^2\Theta', \tag{26-42}$$

the initial condition for θ' may be evaluated from

$$\theta'(0) = a^2\Theta'(0) \rightarrow \theta'(0) = 1/\Theta^2(\infty), \tag{26-43}$$

where the last step makes use of the earlier choice that $\Theta'(0) = 1$.

Therefore, the solution for $\theta(\eta)$ can be found without any guessing for $\theta'(0)$, by solving the related problem for $\Theta(\zeta)$. The equation for $\Theta(\zeta)$ is obtained by transforming the governing equation for $\theta(\eta)$. However, $\theta(\eta)$ also depends on $f(\eta)$. This suggests that the dimensionless stream function be scaled with the same definitions:

$$f(\eta) = a\,F(\zeta) \quad \text{where} \quad \zeta = a\eta. \tag{26-44}$$

Therefore, with $f = a\,F$, $\theta' = a^2\Theta'$, and $\theta'' = a^3\Theta''$, the heat equation (26-39) can be expressed in terms of the new variables as

$$\text{Heat:} \quad \Theta'' + 3\text{Pr}\; F\; \Theta' = 0 \tag{26-45}$$

with

$$\Theta(0) = 0 \quad \text{and} \quad \Theta'(0) = 1. \tag{26-46}$$

The equation for $F(\zeta)$ is obtained by transforming the governing equation for $f(\eta)$. Using the relations: $f = aF$, $f' = a^2F'$, $f'' = a^3F''$, $f''' = a^4F'''$, and $\theta = a\,\Theta$, where $a = 1/\Theta(\infty)$, the momentum equation (26-38) can be expressed in terms of the new variables as

$$\text{Momentum:} \quad F''' + 3F\ F'' - 2(F')^2 + \Theta^3(\infty)[\Theta(\infty) - \Theta] = 0 \tag{26-47}$$

with

$$F(0) = 0, \ F'(0) = 0, \quad \text{and} \quad F'(\infty) = 0. \tag{26-48}$$

The fourth-order Runge-Kutta method is used to integrate the governing equations, by expressing the third-order equation (26-47) for F as three first-order equations, and expressing the second-order equation (26-45) for Θ as two first-order equations:

$$\frac{d}{d\zeta}(F) = F' \qquad \text{with } F(0) = 0 \tag{26-49}$$

$$\frac{d}{d\zeta}(F') = F'' \qquad \text{with } F'(0) = 0 \tag{26-50}$$

$$\frac{d}{d\zeta}(F'') = -3F\ F'' + 2(F')^2 - \Theta^3(\infty)[\Theta(\infty) - \Theta] \qquad \text{with } F''(0) = ?, \quad \text{leading to } F'(\infty) = 0 \tag{26-51}$$

and

$$\frac{d}{d\zeta}(\Theta) = \Theta' \qquad \text{with } \Theta(0) = 0 \tag{26-52}$$

$$\frac{d}{d\zeta}(\Theta') = -3\text{Pr}\ F\,\Theta \qquad \text{with } \Theta'(0) = 1. \tag{26-53}$$

One required initial condition is still unknown, and the shooting method is used to guess the correct value of $F''(0)$ that satisfies the final condition $F'(\infty) = 0$. Additionally, the solution for $F(\zeta)$ depends on knowledge of $\Theta(\infty)$, such that the two problems cannot be integrated simultaneously.

Code 26-1 solves for $F(\zeta)$ and $\Theta(\zeta)$ by alternately integrating the two governing equations. The F equation is solved using a previous solution of $\Theta(\zeta)$, and the Θ equation is solved using a previous solution of $F(\zeta)$. The process is repeated until neither solution changes. Because of the nonlinear nature of the problem, some tricks are employed to slow the iterative change in F and Θ, to help with the stability of convergence. Specifically, an averaging of previous iteration results for $\Theta(\infty)$ is used in the F equation rather than the most recent result, and the solution for $\Theta(\zeta)$ is not allowed to change too much in one iteration, as is likely to happen initially in the calculation.

Figure 26-4 plots the solutions for $f = F/\Theta(\infty)$, $f' = F'/\Theta^2(\infty)$, $f'' = F''/\Theta^3(\infty)$, $(1-\theta) = 1 - \Theta/\Theta(\infty)$, and $\theta' = \Theta'/\Theta^2(\infty)$. Note that the quantity $(1-\theta) = (T - T_\infty)/(T_s - T_\infty)$ is descriptive of the temperature rise relative to the ambient. For Pr = 0.1, the region of moving fluid is contained within the region of heated fluid; the elevated temperature extends far from the heated surface, as does the velocity field. In contrast, for Pr = 10.0, the thermal boundary layer is confined close to the wall surface, but the velocity field extends well beyond the heated region. In this case, the relatively high momentum diffusivity is responsible for the large momentum boundary layer.

From the solution, the heat transfer rate from the wall can be determined:

$$q'' = -k\frac{\partial T}{\partial y}\bigg|_{y=0} = k(T_s - T_\infty)\frac{\partial \eta}{\partial y}\frac{\partial \theta}{\partial \eta}\bigg|_{\eta=0} = k(T_s - T_\infty)Cx^{-1/4}\theta'(0) = \frac{k(T_s - T_\infty)C}{\Theta^2(\infty)}x^{-1/4}, \tag{26-54}$$

Code 26-1 Natural convection from vertical plate

```
#include <stdio.h>
#include <math.h>

int main()
{
   int n,N=5000,iter=0,twice=0,MAX_iter=500;
   double Pr=1.;
   double zeta_inf=50.,del_zeta=zeta_inf/(N-1);
   double dF_inf=0.0;                          /* end BC */
   double F[N],dF[N],ddF[N],T[N],dT[N];
   double K1F,K1dF,K1ddF,K2F,K2dF,K2ddF,K3F,K3dF,K3ddF,K4F,K4dF,K4ddF;
   double K1T,K1dT,K2T,K2dT,K3T,K3dT,K4T,K4dT;
   double ddFhi0,ddFhi,ddFlo0,ddFlo,Tinf,Tinf3,Tinf3Last;
   FILE *fp;
   Tinf=Tinf3=Tinf3Last=1.;                    /* guess  */
   ddFhi=ddFhi0=10.;
   ddFlo=ddFlo0=0.;


   n=0;
   while ((T[n]=n*del_zeta)<1.) ++n;           /* initial T[n] */
   while (n<N) T[n++]=1.;
   F[0]=dF[0]=T[0]=0.0;                        /* initial conditions */
   dT[0]=1.0;


   do {
      ddFhi=ddFhi0;
      ddFlo=ddFlo0;
      ddF[0]=(ddFhi+ddFlo)/2.;
      do {
         for (n=0;n<N-1;++n) {
            K1F=del_zeta*dF[n];
            K1dF=del_zeta*ddF[n];
            K1ddF=del_zeta*(2.0*dF[n]*dF[n]
                  -3.0*F[n]*ddF[n]-Tinf3*(Tinf-T[n]));
            K2F=del_zeta*(dF[n]+0.5*K1dF);
            K2dF=del_zeta*(ddF[n]+0.5*K1ddF);
            K2ddF=del_zeta*(2.0*(dF[n]+0.5*K1dF)*(dF[n]+0.5*K1dF)
                  -3.0*(F[n]+0.5*K1F)*(ddF[n]+0.5*K1ddF)
                  -Tinf3*(Tinf-0.5*(T[n]+T[n+1])));
            K3F=del_zeta*(dF[n]+0.5*K2dF);
            K3dF=del_zeta*(ddF[n]+0.5*K2ddF);
            K3ddF=del_zeta*(2.0*(dF[n]+0.5*K2dF)*(dF[n]+0.5*K2dF)
                  -3.0*(F[n]+0.5*K2F)*(ddF[n]+0.5*K2ddF)
                  -Tinf3*(Tinf-0.5*(T[n]+T[n+1])));
            K4F=del_zeta*(dF[n]+K3dF);
            K4dF=del_zeta*(ddF[n]+K3ddF);
            K4ddF=del_zeta*(2.0*(dF[n]+K3dF)*(dF[n]+K3dF)
                  -3.0*(F[n]+K3F)*(ddF[n]+K3ddF)-Tinf3*(Tinf-T[n+1]));
            F[n+1]=F[n]+(K1F+2*K2F+2*K3F+K4F)/6;
            dF[n+1]=dF[n]+(K1dF+2*K2dF+2*K3dF+K4dF)/6;
            ddF[n+1]=ddF[n]+(K1ddF+2*K2ddF+2*K3ddF+K4ddF)/6;
            if (dF[n+1] > 100. || dF[n+1] < 0. || ddF[n+1] > ddF[0]) {
               dF[N-1]=dF[n+1];
               break;
            }
         }
         if (dF[N-1] < dF_inf) ddFlo=ddF[0];
         if (dF[N-1] > dF_inf) ddFhi=ddF[0];
         ddF[0]=(ddFhi+ddFlo)/2.;
         if (ddFhi0-ddF[0] < 1.e-3) ddFhi=(ddFhi0*=2.0); /* if needed */
      } while (ddFhi-ddFlo > 1.e-6 || n!=N-1);
      printf("\niter=%d ddF0=%e dFinf=%e Tinf=%e",iter,ddF[0],dF[n],T[n]);

      for (n=0;n<N-1;++n) {
         K1T=del_zeta*dT[n];
         K1dT=del_zeta*(-3.0*Pr*F[n]*dT[n]);
         K2T=del_zeta*(dT[n]+0.5*K1dT);
         K2dT=del_zeta*(-3.0*Pr*0.5*(F[n]+F[n+1])*(dT[n]+0.5*K1dT));
         K3T=del_zeta*(dT[n]+0.5*K2dT);
         K3dT=del_zeta*(-3.0*Pr*0.5*(F[n]+F[n+1])*(dT[n]+0.5*K2dT));
         K4T=del_zeta*(dT[n]+K3dT);
         K4dT=del_zeta*(-3.0*Pr*F[n+1]*(dT[n]+K3dT));
         T[n+1]=T[n]+(K1T+2*K2T+2*K3T+K4T)/6;
         dT[n+1]=dT[n]+(K1dT+2*K2dT+2*K3dT+K4dT)/6;
         if (dT[n+1]<1.e-6) break;
      }
      for ( ;n<N-1;++n) {T[n+1]=T[n]; dT[n+1]=0.;}
      if (fabs(T[n]-Tinf)/Tinf > .2) { /* limit change for stability */
         if (T[n] > Tinf) for (n=0;n<N-1;++n) T[n+1]*=1.2*Tinf/T[N-1];
         else             for (n=0;n<N-1;++n) T[n+1]*=0.8*Tinf/T[N-1];
      }
      if (fabs(T[n]*T[n]*T[n]-Tinf3Last) < .0001) { /* Tinf converged ? */
         if (!twice) twice=1;
         else break;
      } else twice=0;
      Tinf3=(Tinf*Tinf*Tinf+Tinf3Last)/2.;
      Tinf3Last=Tinf3;
      Tinf=T[n];
   } while (++iter < MAX_iter);
   if (iter == MAX_iter) printf("\nincrease MAX_iter=%d",MAX_iter);

   Tinf=T[N-1];
   printf("\n\nPr=%f  f''(0)=%f  t'(0)=%f  Nu/(Gr^1/4)/(Pr^1/4)=%f\n",
      Pr,ddF[0]/Tinf/Tinf/Tinf,1./Tinf/Tinf,
      1./Tinf/Tinf/sqrt(2.)/sqrt(sqrt(Pr)));
   fp=fopen("out.dat","w");
   for (n=0;n<N;++n)
      fprintf(fp,"%e %e %e %e %e %e\n",n*del_zeta*Tinf,F[n]/Tinf,
         dF[n]/Tinf/Tinf,ddF[n]/Tinf/Tinf/Tinf,
         1.-T[n]/Tinf,dT[n]/Tinf/Tinf);
   fclose(fp);
   return 1;
}
```

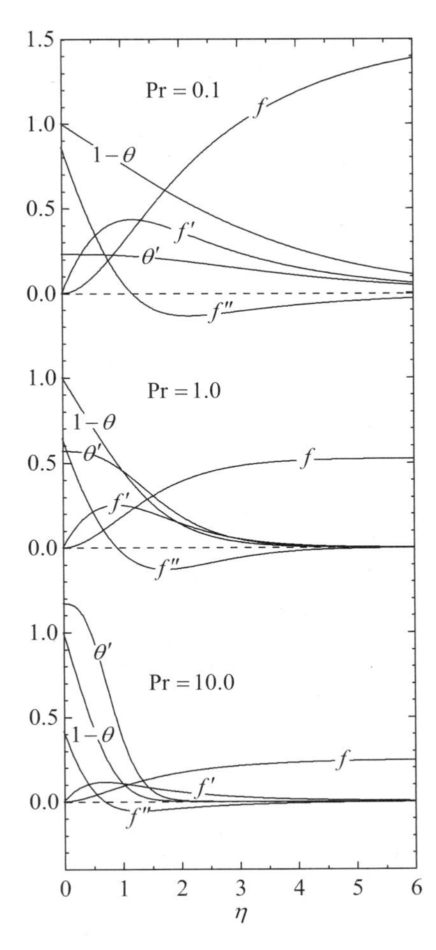

Figure 26-4 Numerical solution to natural convection for Pr = 0.1, 1.0, 10.0.

with

$$C = \left(\frac{g\beta(T_s - T_\infty)}{4\nu^2}\right)^{1/4}. \qquad (26\text{-}55)$$

The convection coefficient for heat transfer can be found from

$$h = \frac{q''}{(T_s - T_\infty)} = \frac{kC}{\Theta^2(\infty)} x^{-1/4}. \qquad (26\text{-}56)$$

Using this result for the convection coefficient, the dimensionless Nusselt number can be expressed as

$$Nu_x = \frac{hx}{k} = \frac{C\,x}{\Theta^2(\infty)} x^{-1/4} = \frac{1}{\sqrt{2}\Theta^2(\infty)} \left(\frac{g\beta x^3(T_s - T_\infty)}{\nu^2}\right)^{1/4} = \frac{Gr_x^{1/4}}{\sqrt{2}\Theta^2(\infty)}. \qquad (26\text{-}57)$$

Notice that this exact result for the Nusselt number fulfills the scaling expectations given by Eq. (26-28a, b) as long as $1/\Theta^2(\infty) \sim \mathrm{Pr}^{1/2}$ for $\mathrm{Pr} \ll 1$ and $1/\Theta^2(\infty) \sim \mathrm{Pr}^{1/4}$ for $\mathrm{Pr} \gg 1$. Figure 26-5 plots $Nu_x/Gr_x^{1/4}$ as a function of Pr to demonstrate that the exact solution exhibits these limiting behaviors.

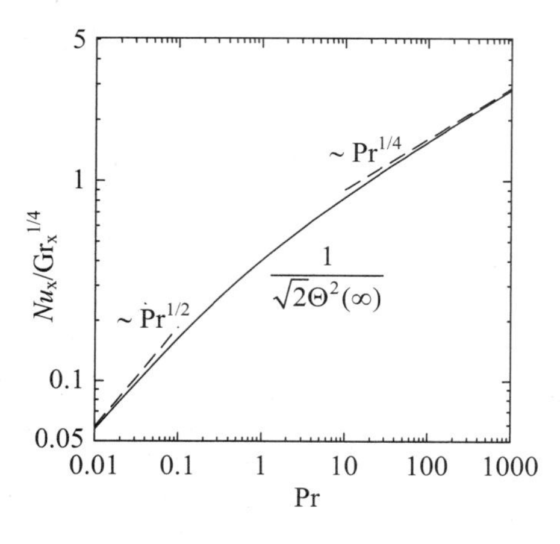

Figure 26-5 Numerical solution for Nu_x as a function of Pr.

A second result obtained from the solution is related to the viscous forces on the wall. The wall shear stress can be evaluated from

$$\tau(x)\big|_{y=0} = \mu \frac{\partial v_x}{\partial y}\bigg|_{y=0} = 4\mu\nu C^3 x^{1/4} f''(0). \tag{26-58}$$

Defining a coefficient of friction c_f based on a characteristic stress scale $(\rho g \beta x \Delta T)$, one can establish that

$$c_f = \frac{\tau(x)\big|_{y=0}}{\rho g \beta x \Delta T} = \frac{f''(0)}{C x^{3/4}} = \sqrt{2}\, f''(0) Gr_x^{-1/4} = \frac{\sqrt{2} F''(0)}{\Theta^3(\infty)} Gr_x^{-1/4}. \tag{26-59}$$

Figure 26-6 plots $c_f/Gr_x^{-1/4}$ as a function of Prandtl number. Notice in the limit that $\mathrm{Pr} \gg 1$ the exact solution has the scaling $F''(0)/\Theta^3(\infty) \sim \mathrm{Pr}^{-1/4}$, and for $\mathrm{Pr} \ll 1$ the scaling $F''(0)/\Theta^3(\infty) \sim \mathrm{Pr}^0$ is independent of Prandtl number. It is left as an exercise to demonstrate these results through scaling arguments (see Problems 26-3 and 26-4).

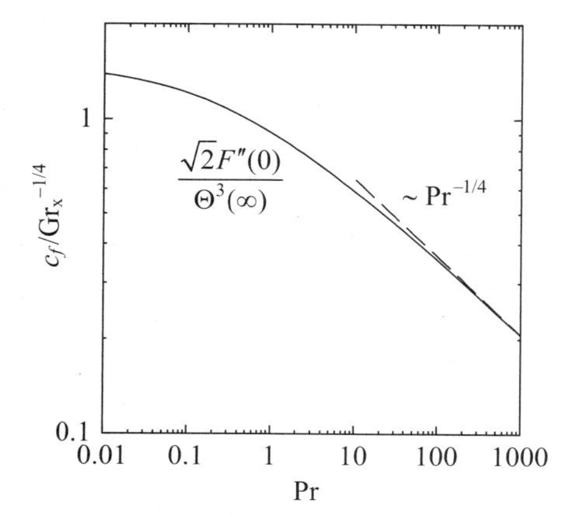

Figure 26-6 Numerical solution for c_f as a function of Pr.

26.5 PROBLEMS

26-1 Consider natural convection from a vertical plate. Using scaling arguments, demonstrate that

$$\frac{\delta}{x} \sim \mathrm{Pr}^n Gr_x^{-1/4}$$

where $n = -1/2$ for $\mathrm{Pr} \ll 1$ and $n = 1/4$ for $\mathrm{Pr} \gg 1$.

26-2 Consider natural convection from a vertical plate. Starting with the assumption that

$$\eta = C_1 \frac{y}{x^n} \quad \text{and} \quad f = \frac{\psi}{C_2 x^m},$$

transform the boundary layer equations (26-11) and (26-12) to determine the required values of n and m. What requirements do the transformed equations impose on C_1 and C_2?

26-3 Consider natural convection from a vertical plate for a fluid having $\mathrm{Pr} \gg 1$. Show by careful scaling arguments that the coefficient of friction goes as

$$c_f = \frac{\tau(x)|_{y=0}}{\rho g \beta x \Delta T} \sim \frac{\delta_T}{x} \sim Gr_x^{-1/4}\mathrm{Pr}^{-1/4} \quad (\mathrm{Pr} \gg 1).$$

26-4 Consider natural convection from a vertical plate for a fluid having $\mathrm{Pr} \ll 1$. Show by scaling arguments that the length scale from the wall δ^*, over which viscous effects are significant, is given by

$$\frac{\delta^*}{x} \sim Gr_x^{-1/4}.$$

Show by scaling arguments that the coefficient of friction goes as

$$c_f = \frac{\tau(x)|_{y=0}}{\rho g \beta x \Delta T} \sim Gr_x^{-1/4} \quad (\mathrm{Pr} \ll 1),$$

independent of Prandtl number.

REFERENCES

[1] J. V. Boussinesq, *Théorie Analytique de la Chaleur Mise en Harmonie avec la Thermodynamique et avec la Théorie Mécanique de la Lumière. Tome II : Refroidissement et Echauffement par Rayonnement. Conductibilité des Tiges, Lames et Masses Cristallines. Courants de Convection. Théorie Mécanique de la Lumière* (1903).

[2] E. Schmidt and W. Beekmann, "Das Temperatur- und Geschwindigkeitsfeld vor einer Wärmeabgebenden Senkrechten Platte bei Natürlicher Konvektion." *Technische Mechanik und Thermodynamik*, 1, 341 and 391 (1930).

[3] S. Ostrach, "An Analysis of Laminar Free-Convection Flow of Laminar Free-Convection Boundary Layers on a Vertical and Heat Transfer about a Flat Plate Parallel to the Direction of the Generating Body Forces." *NACA Report*, 1111, (1953).

Chapter 27

Internal Flow

27.1 Entrance Region
27.2 Heat Transport in an Internal Flow
27.3 Entrance Region of Plug Flow between Plates of Constant Heat Flux
27.4 Plug Flow between Plates of Constant Temperature
27.5 Fully Developed Transport Profiles
27.6 Fully Developed Heat Transport in Plug Flow between Plates of Constant Heat Flux
27.7 Fully Developed Species Transport in Plug Flow Between Surfaces of Constant Concentration
27.8 Problems

Convection transports momentum, heat, and species between a fluid and the bounding surfaces of a flow by advection and diffusion. For the boundary layer problems treated in Chapters 24 through 26, there is no length scale normal to the surface imposed on transport; the only physically significant length scale in this direction is the boundary layer thickness itself. In contrast, (see Figure 27-1) internal flows have a geometric length scale imposed on the cross-stream transport that introduces important physical and mathematical consequences.

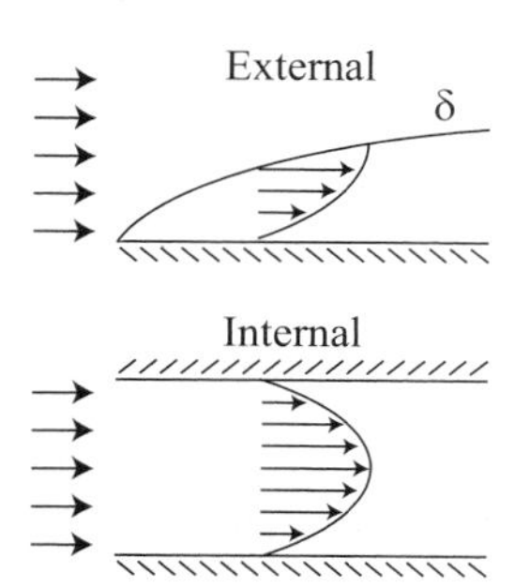

Figure 27-1 Contrasting external and internal flows.

27.1 ENTRANCE REGION

Figure 27-2 illustrates the development of a velocity profile in an internal flow. Boundary layers form on the surfaces downstream of the channel inlet. These boundary layers grow in thickness over the *entrance region* of the channel. Eventually the thicknesses of the boundary layers envelop the cross-stream dimension of the channel. Downstream of this point the flow is *fully developed*, and the channel diameter becomes the pertinent length scale for transport.

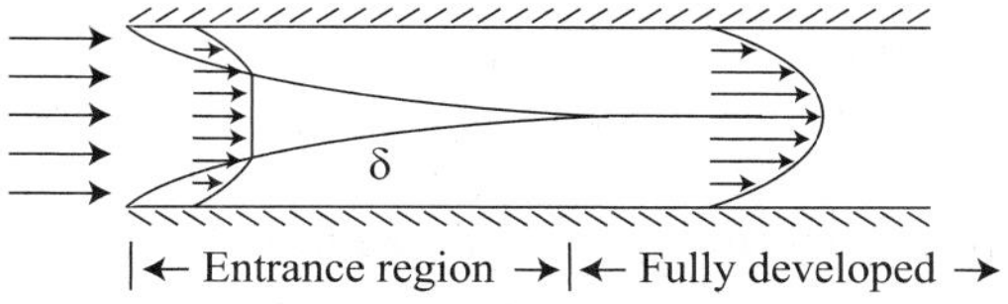

Figure 27-2 Illustration of entrance region.

In the entrance region, transport fluxes between the flow and the surface scale with the boundary layer thickness. For example, the coefficient of friction, associated with momentum transport, scales as

$$c_f = \frac{\tau_w}{\rho v_m^2/2} \sim \frac{\mu v_m/\delta}{\rho v_m^2/2} \sim 2\frac{\nu}{v_m D}\frac{D}{\delta}. \tag{27-1}$$

It is useful to contrast the boundary layer thickness with the geometric length scale, or diameter D, defining the distance between the bounding surfaces. Continuity prohibits the average velocity v_m from changing downstream when D is constant. Therefore, the boundary layer thickness is the only variable scale associated with the coefficient of friction. Approaching the end of the entrance region, the boundary layer thickness approaches the same magnitude as the geometric length scale ($\delta \approx D$) and the coefficient of friction (27-1) becomes a constant that depends only on the Reynolds number. At this point, the flow is hydrodynamically fully developed.

As a second example, consider heat transfer between a flow and the walls bounding the flow. The Nusselt number scales as

$$Nu_D = \frac{hD}{k} = \frac{q_s/k}{(T_s - T_i)} D \sim \frac{\Delta T/\delta_T}{\Delta T} D \sim \frac{D}{\delta_T}. \tag{27-2}$$

When the flow becomes fully developed, δ_T is fixed by the geometric scale of D, and Nu_D becomes a constant, of order 1. The downstream invariance of the Nusselt number is central to the meaning of a thermally fully developed flow.

Likewise, the Sherwood number for mass transfer of a chemical species "A" scales as

$$Sh_D = \frac{h_A D}{Ð_A} = \frac{J_A^*/Ð_A}{(c_s - c_i)} D \sim \frac{\Delta c/\delta_A}{\Delta c} D \sim \frac{D}{\delta_A}. \tag{27-3}$$

Mass transfer will be fully developed when $\delta_A \approx D$ and the Sherwood number becomes a constant, of order 1.

It is clear that the geometric scale D has significant physical importance to internal flows. It is convention to use the *hydraulic diameter* for this length scale, as defined by

$$D = \frac{4 \times (\textit{cross-sectional area of flow})}{(\textit{perimeter wetted by flow})}. \tag{27-4}$$

Notice that the hydraulic diameter of a pipe is the same as the geometric diameter. Even when multiple scales describe the cross-sectional geometry of the flow, the hydraulic diameter permits the definition of a single length scale.

After transport is fully developed, diffusion fluxes between the walls and the flow are established over the scale of the hydraulic diameter. Thereafter, in some respects, transport ceases to change with downstream distance (such as having a constant convection coefficient). The nature of this transition to fully developed conditions is investigated for heat transfer in the sections that follow.

27.2 HEAT TRANSPORT IN AN INTERNAL FLOW

Heat can be introduced to, or removed from, a flow through the surrounding walls. The heat transfer equation is simplified for internal flows through the following considerations. In many instances hydrodynamic transport is fully developed, providing a unidirectional flow downstream (e.g., $v_x(y)$ and $v_y = 0$). This simplifies the advection terms in the heat equation. Furthermore, when the downstream length scale is large compared with the distance between bounding surfaces, diffusion transport to the walls tends to be much greater than downstream diffusion. This situation is identical to that exploited in the boundary layer approximation (Chapter 24), and allows one component of diffusion to be neglected. Under these conditions, the heat equation simplifies to

$$v_x \frac{\partial T}{\partial x} + \overbrace{v_y}^{=0} \frac{\partial T}{\partial y} = \alpha \left(\overbrace{\frac{\partial^2 T}{\partial x^2}}^{\approx 0} + \frac{\partial^2 T}{\partial y^2} \right), \tag{27-5}$$

as long as the compressibility and viscous heating of the flow can be neglected. The remaining terms in the heat equation still retain the form of a partial differential equation:

$$v_x \frac{\partial T}{\partial x} = \alpha \frac{\partial^2 T}{\partial y^2}. \tag{27-6}$$

For fully developed heat transfer, it will be shown in later analysis that $\partial T/\partial x$ can be related in a simple way to the downstream change in mean temperature $\partial T_m/\partial x$. In turn, $\partial T_m/\partial x$ can be related to the heat transfer to (or from) the walls in the form of Newton's convection law:

$$q_s = h(T_s - T_m). \tag{27-7}$$

It is important to notice that for internal flows, the temperature difference in the convection law (27-7) is defined with the nonconstant mean temperature T_m, in contrast to a constant free-stream temperature arising in boundary layer problems. While q_s depends on the local value of T_m, the coefficient h no longer depends on downstream position when heat transfer is fully developed (since $h \sim k/D = const.$). To facilitate use of Newton's convection law, the *mean temperature* of an internal flow is defined by

$$T_m = \int_A v_x T dA \bigg/ \int_A v_x dA, \tag{27-8}$$

in which

$$v_m = \int_A v_x dA \tag{27-9}$$

defines the mean velocity of the flow for the cross-sectional area A. Notice that T_m is a transport average (dependent on v_x). In contrast, v_m is simply a geometric average.

Before jumping into the analysis of heat transfer in fully developed transport, the entrance region will be investigated for the special case of an inviscid flow, where the velocity profile may be approximated as a spatial constant $v_x = U$. Under this situation Eq. (27-6) may be solved by separation of variables (as developed in Chapter 8). The analytic description of heat transfer in the entrance region will help shed light on the physical meaning and mathematical requirements of fully developed heat transfer.

27.3 ENTRANCE REGION OF PLUG FLOW BETWEEN PLATES OF CONSTANT HEAT FLUX

Consider the problem illustrated in Figure 27-3, in which a fluid of initial temperature T_i flows between parallel plates that (each) deliver a constant heat flux q_s to the fluid. The fluid is sufficiently inviscid that the flow is essentially uniform between the plates (plug flow), with $v_x = U$. Using the dimensionless temperature and spatial variables

$$\vartheta = \frac{T - T_i}{aq_s/k}, \quad x^* = \frac{x}{a}, \quad \text{and} \quad y^* = \frac{y}{a}, \tag{27-10}$$

the heat equation for internal flows (27-6) may be expressed in dimensionless form as

$$\frac{Ua}{\alpha}\frac{\partial \vartheta}{\partial x^*} = \frac{\partial^2 \vartheta}{\partial y^{*2}}. \tag{27-11}$$

The heat equation is subject to the boundary conditions

$$\left.\frac{\partial \vartheta}{\partial y^*}\right|_{y^*=0} = 0, \quad \left.\frac{\partial \vartheta}{\partial y^*}\right|_{y^*=1/2} = 1, \quad \text{and} \quad \vartheta(x^*=0) = 0. \tag{27-12}$$

It is left as an exercise (see Problem 27-1) to demonstrate that the solution for $\vartheta(x^*, y^*)$ is

$$\vartheta = \frac{4\alpha}{Ua}x^* + \sum_{n=1}^{\infty} \frac{4(-1)^n}{\lambda_n^{*2}}\left[1 - \exp\left(-\frac{\alpha}{Ua}\lambda_m^{*2}x^*\right)\right]\cos(\lambda_n^* y^*), \quad \text{with} \quad \lambda_n^* = n(2\pi). \tag{27-13}$$

From this solution, the local Nusselt number can be found (see Problem 27-2):

$$Nu_D = \frac{h(2a)}{k} = \frac{q_s(2a)}{(T_s - T_m)k} = \left\{\sum_{n=1}^{\infty} \frac{2}{\lambda_n^{*2}}\left[1 - \exp\left(-\frac{\alpha}{Ua}\lambda_m^{*2}x^*\right)\right]\right\}^{-1}. \tag{27-14}$$

Notice that the convection coefficient h is defined by Eq. (27-7), as is appropriate for internal flows, and the mean temperature between the plates is calculated from

$$T_m = \int_{-a/2}^{+a/2} U\,T dy \Big/ \int_{-a/2}^{+a/2} U dy = 2\int_0^{1/2} T dy^*. \tag{27-15}$$

Notice that for the special case of plug flow, T_m becomes a geometric average.

Solutions for the temperature field, Eq. (27-13), and the Nusselt number, Eq. (27-14), are illustrated in Figure 27-4. The rising temperature field is shown at five downstream locations. The exact solution demonstrates that the fluid entering the channel remains at the initial temperature T_i while outside of the boundary layer regions. As the boundary layers grow, the Nusselt number for heat transfer decreases, as expected from the scaling

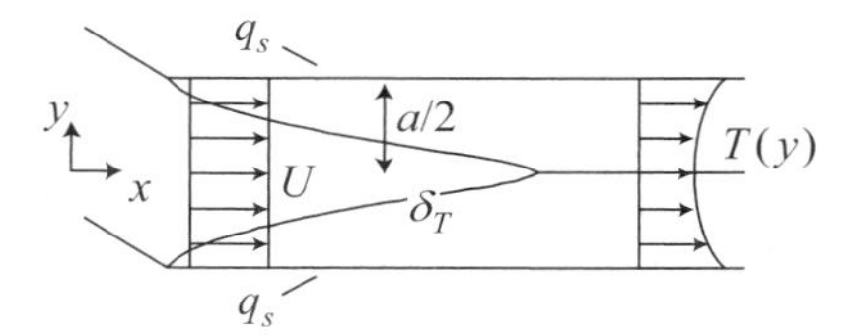

Figure 27-3 Plug flow between plates of constant heat flux.

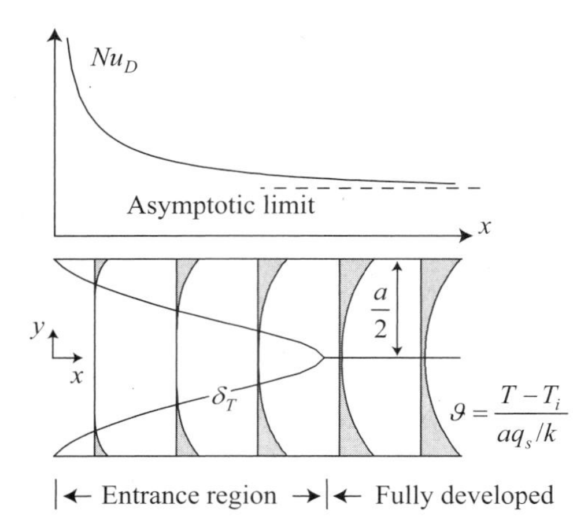

Figure 27-4 The developing Nusselt number and temperature field in the entrance region.

result $Nu_D \sim D/\delta_T$. The Nusselt number asymptotically approaches a constant in the fully developed region of the flow, which occurs after the boundary layers join in the center of the channel. Equation (27-14) can be used to evaluate this limiting value of the Nusselt number, for the result that $Nu_D(x^* \to \infty) = 12$. This result will also be derived from analysis of fully developed heat transfer in Section 27-6.

It is significant that the existence of opposing surfaces is unfelt by the heat transfer process until the thermal boundary layers join at the centerline of the flow. Prior to this event, heat transfer can be solved using boundary layer theory, as was done for an analogous problem in Section 24.1 (see Eq. 24-13). This analysis yields the temperature profile in the boundary layer:

$$\frac{T - T_i}{aq_s/k} = \sqrt{\frac{4}{\pi}\frac{\alpha x^*}{Ua}}\exp\left(-\frac{Ua}{\alpha x^*}\frac{(1-2|y^*|)^2}{16}\right) - \frac{1-2|y^*|}{2}\,erfc\left(\sqrt{\frac{Ua}{\alpha x^*}}\frac{1-2|y^*|}{4}\right). \tag{27-16}$$

From the boundary layer result, the Nusselt number in the entrance region can be evaluated (see Problem 27-3) as a function of downstream distance:

$$Nu_D = \frac{h(2a)}{k} = \frac{q_s(2a)}{(T_s - T_m)k}$$
$$= \left\{\sqrt{\frac{1}{\pi}\frac{\alpha x^*}{Ua}}\left[1 - \frac{1}{2}\exp\left(-\frac{1}{16}\frac{Ua}{\alpha x^*}\right)\right] + \frac{1}{8}\left[1 - \left(1 + 8\frac{\alpha x^*}{Ua}\right)erf\left(\frac{1}{4}\sqrt{\frac{Ua}{\alpha x^*}}\right)\right]\right\}^{-1}. \tag{27-17}$$

It should be noted that Eq. (27-17) reports the Nusselt number in a manner appropriate for internal flows; the convection coefficient is defined by the relation $q_s = h(T_s - T_m)$ rather than by following the boundary layer convention of $q_s = h(T_s - T_i)$.

It is instructive to contrast the Nusselt numbers derived by the two different methods for heat transfer in the entrance region of inviscid plug flow. Figure 27-5 compares the exact result, given by Eq. (27-14), with the approximate result, Eq. (27-17), just derived from boundary layer (BL) theory, and the asymptotic result for fully developed heat transfer ($Nu_D = 12$). The crossover between the approximate solutions occurs when $\alpha x/(Ua^2) = 0.067$. For $\alpha x/(Ua^2) > 0.1$, the exact Nusselt number is within 1% of the value

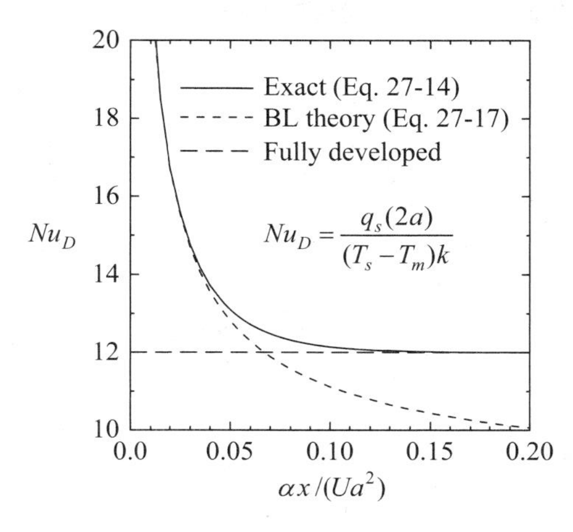

Figure 27-5 Nusselt number for plug flow between plates of constant heat flux.

for fully developed heat transfer. This provides some guidance for estimating the downstream distance at which heat transfer is fully developed. Generalizing this finding, the thermal entry length is approximately

$$0.1 \approx 4\frac{\alpha}{\nu}\frac{\nu}{U(2a)}\frac{x}{(2a)} = 4\frac{x/D}{\text{Re}_D\text{Pr}} \quad \text{or} \quad \left(\frac{x}{D}\right)_{\substack{\text{thermal}\\ \text{entry-length}}} \approx 0.025\ \text{Re}_D\ \text{Pr}, \tag{27-18}$$

where D is the hydraulic diameter as defined by Eq. (27-4).

27.4 PLUG FLOW BETWEEN PLATES OF CONSTANT TEMPERATURE

Consider the problem of heat transfer in plug flow between isothermal plates, as shown in Figure 27-6. For this problem, the fluid at the inlet to the channel has a specified temperature profile, $T(x=0) = f(y)$, elevated above the wall temperature T_s. Neglecting downstream diffusion, compressibility, and viscous heating, the heat equation governing this problem is

$$\frac{Ua}{\alpha}\frac{\partial \theta}{\partial x^*} = \frac{\partial^2 \theta}{\partial y^{*2}} \tag{27-19}$$

where

$$x^* = x/a \quad \text{and} \quad y^* = y/a \tag{27-20}$$

as in the last section, but the dimensionless fluid temperature is defined by

$$\theta = \frac{T - T_s}{T_s}. \tag{27-21}$$

Consider a specific inlet temperature profile that leads to the dimensionless initial condition

$$\theta(x^* = 0) = C_1 \cos(\pi y^*), \tag{27-22}$$

where C_1 is a constant. Using separation of variables, Eq. (27-19) can be solved with the boundary conditions

$$\partial\theta/\partial y^*|_{y^*=0} = 0 \quad \text{and} \quad \theta(y^* = \pm 1/2) = 0. \tag{27-23}$$

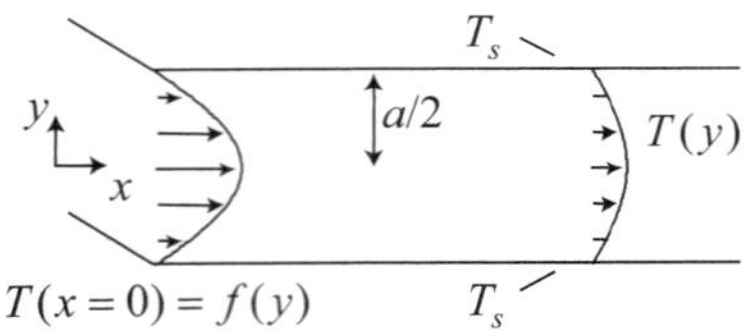

Figure 27-6 Plug flow between plates of constant temperature.

The solution is

$$\theta = C_1 \cos(\pi y^*) \exp\left(-\frac{\alpha \pi^2}{Ua} x^*\right). \tag{27-24}$$

From this solution, the mean fluid temperature, as defined by Eq. (27-8), can be determined as a function of downstream distance:

$$T_m = 2\int_0^{1/2} T_s(\theta + 1)dy^* \quad \text{or} \quad \frac{T_m - T_s}{T_s} = 2\int_0^{1/2} \theta \, dy^* = \frac{2C_1}{\pi} \exp\left(-\frac{\alpha \pi^2}{Ua} x^*\right). \tag{27-25}$$

Using the results for the temperature distribution given by Eq. (27-24), and mean temperature given by Eq. (27-25), the Nusselt number for heat transfer between the fluid and the plates can be determined:

$$\begin{aligned} Nu_D = \frac{h(2a)}{k} &= 2\left[\frac{T_s}{(T_s - T_m)}\right]\left[-\frac{\partial \theta}{\partial y^*}\bigg|_{y^*=-1/2}\right] \\ &= 2\left[\frac{\pi}{2C_1}\exp\left(\frac{\alpha\pi^2}{Ua}x^*\right)\right]\left[\pi C_1 \exp\left(-\frac{\alpha \pi^2}{Ua}x^*\right)\right] = \pi^2. \end{aligned} \tag{27-26}$$

It is noteworthy that the Nusselt number is a constant, independent of downstream position. This result implies that the temperature profile imposed as an inlet condition on the channel is already fully developed. However, if the initial condition were changed to $\theta(x^* = 0) = C_2 \cos(3\pi y^*)$, the solution to this problem would yield a different value for the Nusselt number, $Nu_D = 9\pi^2$, also a constant. This indicates that the Nusselt number for heat transfer is dependent on the initial condition, and implies that fully developed conditions between isothermal plates may not be unique. This quandary is resolved by extending the previous analysis to an arbitrary initial temperature distribution, which yields a solution with the form

$$\theta(x^*, y^*) = \sum_{n=1}^{\infty} C_n \cos(\lambda_n y^*) \exp\left(-\frac{\alpha \lambda_n^2}{Ua} x^*\right) \quad \text{with } \lambda_n = (2n - 1)\pi. \tag{27-27}$$

The magnitude of each C_n in the series is determined in the usual way to satisfy the initial condition for the temperature distribution at the inlet of the channel. However, the temperature field described by Eq. (27-27) is not fully developed; each term in the series solution decays as $\exp(-\lambda_n^2 \alpha x^*/(Ua))$. For example, the first term decays as $\exp(-\pi^2 \alpha x^*/(Ua))$ and the second term decays as $\exp(-9\pi^2 \alpha x^*/(Ua))$, as illustrated in Figure 27-7. During the decay of competing terms, the Nusselt number for heat transfer changes. However, notice that the second term decays by almost an order of magnitude faster than the first, as illustrated in Figure 27-7. Consequently, irrespective of what the actual initial condition is, the temperature solution rapidly approaches Eq. (27-24).

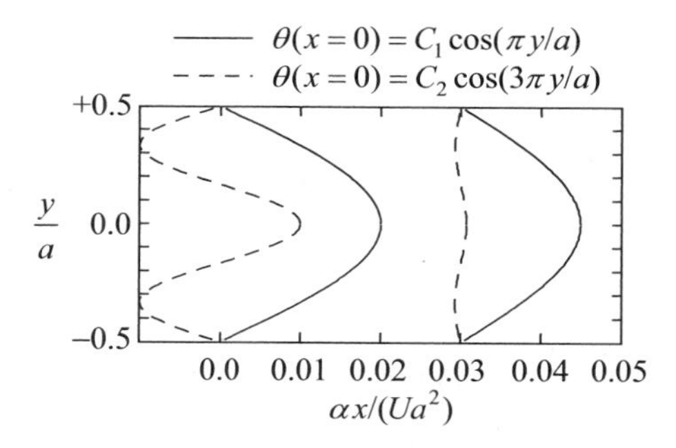

Figure 27-7 Decay of temperature field from two different initial conditions.

This temperature profile reflects the thermally fully developed condition, and corresponds to a Nusselt number of $Nu_D = \pi^2$. In the unlikely event that the initial temperature distribution at the inlet of the channel is described by a series (27-27) in which $C_1 \approx 0$, the temperature field will rapidly approach a profile corresponding to the smallest eigenvalue λ_n with non-negligible C_n in the solution. In this case, the Nusselt number approaches the value $Nu_D = \lambda_n^2$ related to the smallest eigenvalue. Fortunately, in most problems of practical importance, the first term (n = 1) in the temperature series solution corresponds to the persisting fully developed profile.

27.5 FULLY DEVELOPED TRANSPORT PROFILES

For fully developed transport, profiles of the dependent variable must be scaled appropriately to reveal its "self-similar condition." Scaled profiles satisfying this condition do not change with downstream distance. To explore this scaling, the temperature profile associated with heat transfer is considered initially. The conclusions of this investigation are extended by analogy to other forms of transport.

27.5.1 Scaling of the Fully Developed Temperature Field

To explore the scaling of the fully developed temperature profile, consider again plug flow between plates of constant heat flux. The effect of heat added over a differential distance dx is shown schematically in Figure 27-8. Notice that the nonconstant surface temperature provides an appropriate reference point to view the self-similar profile. Additionally, the magnitude of the temperature profile (relative to the surface temperature) can be scaled by the magnitude of the surface heat flux. Therefore, one may consider scaling the temperature field as

$$\frac{T_s(x) - T(x,y)}{aq_s/k}. \tag{27-28}$$

For the constant heat flux boundary condition illustrated in Figure 27-8, the fully developed temperature field is elevated in a uniform way with downstream distance. Therefore, for the heat flux boundary condition, neither $(T_s - T)$ nor aq_s/k will change with downstream position, and one concludes that

$$\frac{\partial}{\partial x}\left(\frac{T_s - T}{aq_s/k}\right) = 0. \tag{27-29}$$

For other boundary conditions, both $(T_s - T)$ and aq_s/k may change with downstream position. Nevertheless, if the proposed scaling reveals the constant self-similar condition

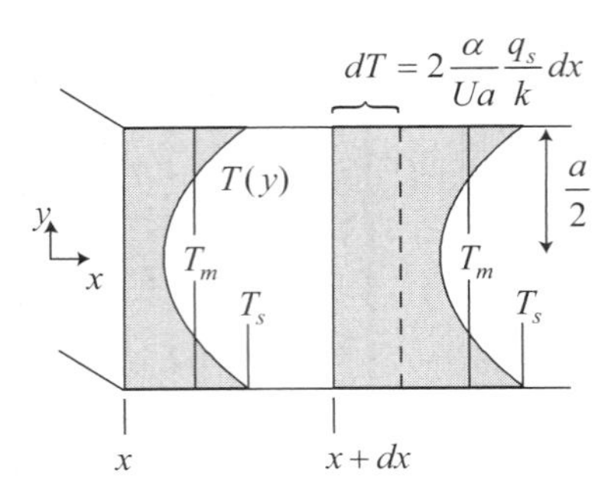

Figure 27-8 Fully developed temperature profile between plates of constant heat flux.

of the temperature profile, then Eq. (27-29) should remain valid. For boundary conditions other than a constant heat flux, normalization of the temperature field is more transparent when using $q_s = h(T_s - T_m)$. Substituting this expression for q_s into Eq. (27-29) yields

$$\frac{\partial}{\partial x}\left(\frac{T_s - T}{aq_s/k}\right) = \frac{\partial}{\partial x}\left(\frac{k}{ah}\frac{T_s - T}{T_s - T_m}\right) = 0. \tag{27-30}$$

As implied by scaling arguments in Section 27.1 and demonstrated with a specific example in Section 27.3, the Nusselt number $(h2a/k)$ approaches a constant value for fully developed heat transfer. Therefore, the last statement suggests an equivalent scaling of the temperature profile that reveals the constant self-similar condition

$$\frac{\partial}{\partial x}\left(\frac{T_s - T(x,y)}{T_s - T_m(x)}\right) = 0. \tag{27-31}$$

The validity of Eq. (27-31) for a constant temperature boundary condition is suggested by Figure 27-9. In this case, the nonconstant $T_s - T_m$ is a "stretching" factor in the normalization of the temperature field that reveals the self-similar condition.

For plug flow, the fully developed temperature profile between isothermal plates was determined to be $T = T_s(1 + \theta)$, where θ is given by Eq. (27-24). The mean temperature for this profile is given by Eq. (27-25). Substituting these expressions into Eq. (27-31) shows that the self-similar temperature profile is indeed invariant to changes in downstream distance:

$$\frac{\partial}{\partial x}\left(\frac{T_s - T}{T_s - T_m}\right) = \frac{\partial}{\partial x}\left(\frac{-T_s\phi}{T_s - T_m}\right) = \frac{\partial}{\partial x}\left[\frac{\pi}{2}\cos(\pi y^*)\right] = 0. \tag{27-32}$$

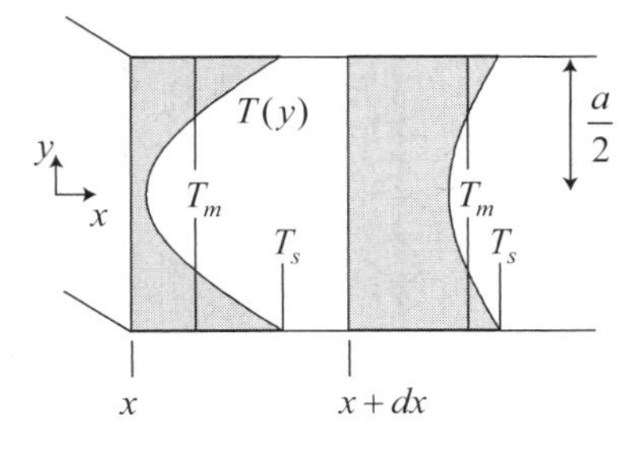

Figure 27-9 Fully developed temperature profile between isothermal plates.

27.5.2 Constant Self-Similar Transport Profiles

The notion of fully developed transport leads to an expectation that the dependent variable exhibits a constant self-similar profile, irrespective of whether one is describing the heat, species, or momentum content of the fluid. Therefore, for fully developed heat, species, and momentum transport:

$$\frac{\partial}{\partial x}\left(\frac{T_s - T(x,y)}{T_s - T_m(x)}\right) = 0 \quad (or \approx 0) \quad \text{(heat)} \tag{27-33a}$$

$$\frac{\partial}{\partial x}\left(\frac{c_s - c_A(x,y)}{c_s - c_m(x)}\right) = 0 \quad (or \approx 0) \quad \text{(species)} \tag{27-33b}$$

$$\frac{\partial}{\partial x}\left(\frac{v_s - v_x(x,y)}{v_s - v_m(x)}\right) = 0 \quad (or \approx 0) \quad \text{(momentum)} \tag{27-33c}$$

It is noteworthy that the mean values for temperature T_m and species c_m are transport averages. The definition of a transport average was illustrated for T_m with Eq. (27-8). In contrast, v_m is the geometric average defined by Eq. (27-9).

For internal flows, the no-slip condition requires that $v_s = 0$ and the incompressible continuity equation stipulates that $dv_m/dx = 0$. Therefore, the self-similar profile (27-33c) for momentum transport reduces to the familiar expectation that $\partial v_x(y)/\partial x = 0$ for fully developed conditions. The downstream momentum content in the flow remains unchanged and invariance of the self-similar velocity profile $v(y)$ is exact. In such a case, advection plays no net role in transport, and the remaining terms in the governing equation describe a balance between the momentum source (pressure gradient) and diffusion transport to the wall.

In contrast, many heat and species transport problems do not have source terms in the governing equations. In such a case, changes in the downstream advection are required to balance diffusion transport to the wall. When the dependent variable (T or c) is held constant at the boundary of the flow, transport causes differences between the dependent variable in the flow and at the boundary to disappear with downstream distance. Now the invariance of the self-similar profile describing the dependent variable is only exactly true in the limit of $x \to \infty$ (at which point the dependent variable is identical to the boundary value). However, for fully developed transport, it is practical to approximate invariance of the self-similar profile well in advance of this limit. Once the transport boundary layers have spanned the width between the walls of an internal flow, the convection coefficients for heat and species transfer will be within a few percent of the fully developed limit.

The utility of the constant self-similar profiles are that when the conditions (27-33a, b) are imposed on the governing internal flow equations for heat and species transfer, the partial differential equations can be transformed into ordinary differential equations, as will be shown for transport in inviscid flows in Sections 27.6 and 27.7. This fact will be exploited in the treatment of viscous laminar flows in Chapter 28, as well as viscous turbulent flows in Chapter 33.

27.6 FULLY DEVELOPED HEAT TRANSPORT IN PLUG FLOW BETWEEN PLATES OF CONSTANT HEAT FLUX

Reconsider the problem of heat transfer in plug flow between plates of constant heat flux, as discussed in Section 27.3. In the present analysis, assume that heat transfer is fully developed everywhere in the flow, as illustrated in Figure 27-10. The convection

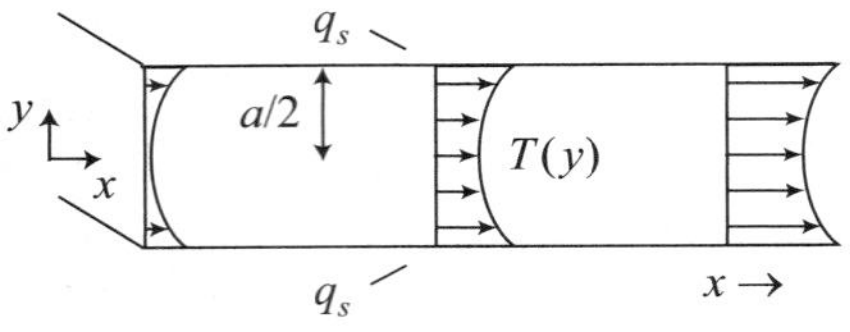

Figure 27-10 Temperature field in fully developed region between plates of constant heat flux.

coefficient h is a constant. Consequently, since the flow is subject to a constant heat flux from the bounding plates, Newton's cooling law (27-7) dictates that the rising temperature field obeys the relation

$$\frac{dT_s}{dx} = \frac{dT_m}{dx} \quad \left(= \frac{\partial T}{\partial x} \text{ fully developed}\right) \quad \text{(for const. heat flux BC).} \tag{27-34}$$

The last equality is a consequence of Eq. (27-33a), which asserts that for fully developed conditions

$$(T_s - T_m)\frac{\partial}{\partial x}(T_s - T) - (T_s - T)\overbrace{\frac{\partial}{\partial x}(T_s - T_m)}^{=0} = 0 \;\rightarrow\; \frac{\partial T_s}{\partial x} = \frac{\partial T}{\partial x}. \tag{27-35}$$

The relation between the changing mean fluid temperature and the applied heat flux q_s can be found by integrating the heat Equation (27-6) across the flow:

$$\int_{-a/2}^{+a/2} v_x \frac{\partial T}{\partial x} dy = \int_{-a/2}^{+a/2} \alpha \frac{\partial^2 T}{\partial y^2} dy. \tag{27-36}$$

For a hydrodynamically fully developed flow, $\partial v_x/\partial x = 0$. Therefore,

$$\int_{-a/2}^{+a/2} v_x \frac{\partial T}{\partial x} dy + \int_{-a/2}^{+a/2} \overbrace{\frac{\partial v_x}{\partial x}}^{=0} T dy = \int_{-a/2}^{+a/2} \frac{\partial}{\partial x}(v_x T) dy = \frac{\partial}{\partial x} \int_{-a/2}^{+a/2} v_x T dy, \tag{27-37}$$

and the integrals in the heat Equation (27-36) can be evaluated such that

$$\rho C_p \frac{\partial}{\partial x} \int_{-a/2}^{+a/2} v_x T dy = k \left.\frac{\partial T}{\partial y}\right|_{-a/2}^{+a/2} = 2q_s. \tag{27-38}$$

With $v_x = U$, the definition (27-8) for T_m applied to Eq. (27-38) yields

$$\frac{dT_m}{dx} = 2\frac{\alpha}{Ua}\frac{q_s}{k} = 2\frac{\alpha}{Ua}\frac{h}{k}(T_s - T_m), \tag{27-39}$$

where Newton's cooling law (27-7) is used to express the wall heat flux q_s in terms of the convection coefficient h. Equating $\partial T/\partial x = dT_m/dx$, as required by (27-34) for fully developed conditions in the flow, and using Eq. (27-39) to express $\partial T_m/\partial x$ in terms of h, the heat Equation (27-6) may be written as

$$U\frac{dT_m}{dx} = \alpha \frac{\partial^2 T}{\partial y^2} \;\rightarrow\; 2\frac{\alpha}{a}\frac{h}{k}(T_s - T_m) = \alpha \frac{\partial^2 T}{\partial y^2}. \tag{27-40}$$

Notice that the heat equation becomes an ordinary differential equation in its final form. Using dimensionless temperature and spatial variables defined by

$$\theta = \frac{T - T_s}{T_s} \tag{27-41}$$

and

$$y^* = \frac{y}{a}, \tag{27-42}$$

the heat Equation (27-40) can be expressed as

$$\frac{\partial^2 \theta}{\partial y^{*2}} + Nu_D \theta_m = 0 \tag{27-43}$$

where

$$Nu_D = h(2a)/k \tag{27-44}$$

is the Nusselt number based on the hydraulic diameter of the flow ($2a$). The heat Equation (27-43) is subject to the boundary conditions

$$\partial\theta/\partial y^*\big|_{y^*=0} = 0 \quad \text{and} \quad \theta(y^* = \pm 1/2) = 0 \tag{27-45}$$

and may be directly integrated for the result

$$\theta = \left(\frac{1}{4} - y^{*2}\right)\frac{Nu_D \theta_m}{2}. \tag{27-46}$$

However, the solution is not complete because the value for Nu_D is unspecified. Applying the definition (27-41) for θ to the definition of mean temperature (27-8) demonstrates that

$$T_m \int_A U dA = \int_A U\, T dA \rightarrow \frac{T_m - T_s}{T_s}\int_A dA = \int_A \theta dA. \tag{27-47}$$

Therefore, this last result stipulates that

$$\theta_m = 2\int_0^{1/2} \theta\, dy^*, \tag{27-48}$$

to satisfy the definition of θ. This final requirement is used to determine Nu_D. Substituting Eq. (27-46) into Eq. (27-48) and integrating leads to the result that

$$Nu_D = 12. \tag{27-49}$$

This is the same result for the Nusselt number as found in Section 27.3, where the exact solution was evaluated in the limit of $x \rightarrow \infty$. Although inviscid plug flow was considered here for the sake of simplicity, application of the constant self-similar profiles (27-33a, b) to the governing equations for transport in viscid flows, as considered in Chapter 28, can be solved with a similar approach.

27.7 FULLY DEVELOPED SPECIES TRANSPORT IN PLUG FLOW BETWEEN SURFACES OF CONSTANT CONCENTRATION

Suppose a two-dimensional description of fully developed species transport is sought for an inviscid incompressible flow between planar plates. As shown in Figure 27-11, the plates in this problem impose constant concentration of a species "A" on the boundaries of the flow. In analogy to Eq. (27-6), the dilute species transport equation for an internal flow is

$$v_x \frac{\partial c_A}{\partial x} = Đ_A \frac{\partial^2 c_A}{\partial y^2}. \tag{27-50}$$

Solving Eq. (27-50) for fully developed transport is the subject of Problem 27-5. However, key steps of this analysis are highlighted here.

For the boundary conditions of this problem, the constant self-similar profile of fully developed species transport (27-33b) requires that

$$-(c_s - c_m)\frac{\partial c_A}{\partial x} + (c_s - c_A)\frac{\partial c_m}{\partial x} = 0, \tag{27-51}$$

or

$$\frac{\partial c_A}{\partial x} = \frac{c_s - c_A}{c_s - c_m}\frac{dc_m}{dx} \quad \text{(for const. concentration BC)}, \tag{27-52}$$

where

$$c_m = \int_A v_x c_A \, dA \Big/ \int_A v_x dA. \tag{27-53}$$

Additionally, the change in mean concentration of the flow with respect to downstream position can be expressed in terms of the convection coefficient h_A by

$$\frac{dc_m}{dx} = 2\frac{h_A}{Ua}(c_s - c_m). \tag{27-54}$$

Using the relations (27-52) and (27-54), the species transport Equation (27-50) can be cast into an ordinary differential equation:

$$\frac{\partial^2 \phi}{\partial y^{*2}} + Sh_D \phi = 0, \tag{27-55}$$

where the dimensionless variables are defined by

$$\phi = \frac{c_s - c_A}{c_s - c_m}, \tag{27-56}$$

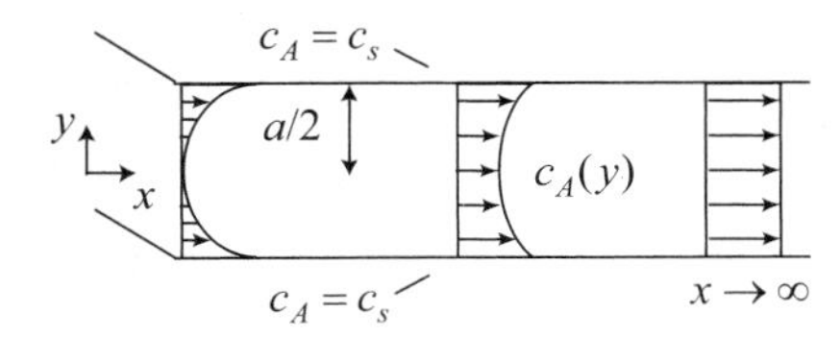

Figure 27-11 Fully developed species transport between surfaces of constant concentration.

$$y^* = \frac{y}{a}, \tag{27-57}$$

and

$$Sh_D = h_A(2a)/Đ_A \quad \text{(the Sherwood number).} \tag{27-58}$$

The solution to the fully developed species transport Equation (27-55) is subject to the boundary conditions

$$\partial\phi/\partial y^*\big|_{y^*=0} = 0 \quad \text{and} \quad \phi(y^* = \pm 1/2) = 0. \tag{27-59}$$

However, applying the definition (27-56) for ϕ to the definition of mean concentration (27-53) demonstrates that ϕ must satisfy the normalization constraint

$$2\int_0^{1/2} \phi \, dy^* = 1. \tag{27-60}$$

The solution to the transport Equation (27-55), which satisfies the boundary conditions (27-59) and the normalization constraint (27-60), is readily seen to be

$$\phi(y^*) = \frac{\pi}{2}\cos\left(\sqrt{Sh_D}y^*\right), \quad \text{where } Sh_D = \pi^2. \tag{27-61}$$

Since ϕ describes the concentration distribution relative to the mean value, to complete the solution with respect to the downstream position x requires a description of the changing mean concentration $c_m(x)$. Integrating Eq. (27-54) for the downstream change in c_m yields

$$c_s - c_m(x) = [c_s - c_m(0)]\exp\left(-Sh_D\frac{Đ_A x}{Ua^2}\right), \tag{27-62}$$

where $c_m(0)$ is the mean concentration of the flow at $x = 0$. Combining this result for $c_m(x)$ with the solution (27-61) for $\phi(y^*)$ yields the complete two-dimensional solution for fully developed species transport between the plates:

$$\frac{c_s - c_A(x,y)}{c_s - c_m(0)} = \frac{\pi}{2}\cos\left(\sqrt{Sh_D}\frac{y}{a}\right)\exp\left(-Sh_D\frac{Đ_A x}{Ua^2}\right), \quad \text{where } Sh_D = \pi^2. \tag{27-63}$$

Figure 27-12 plots the downstream development of c_A/c_s between the two plates, using Eq. (27-63) with $c_m(0) = c_s(\pi - 2)/\pi$. By a downstream distance of $Đ_A x/Ua^2 = 0.3$, the average fluid concentration is within 5% of the surface concentration c_s.

It is interesting to note that for the solution to Eq. (27-55), the Sherwood number $Sh_D = \pi^2$ was selected to satisfy the boundary conditions of the problem. However, it is apparent that other discrete values—$Sh_D = (2n-1)^2\pi^2$, $n = 1, 2, \ldots$—will also satisfy the boundary conditions. Justifying the choice of $Sh_D = \pi^2$ requires recollection of the

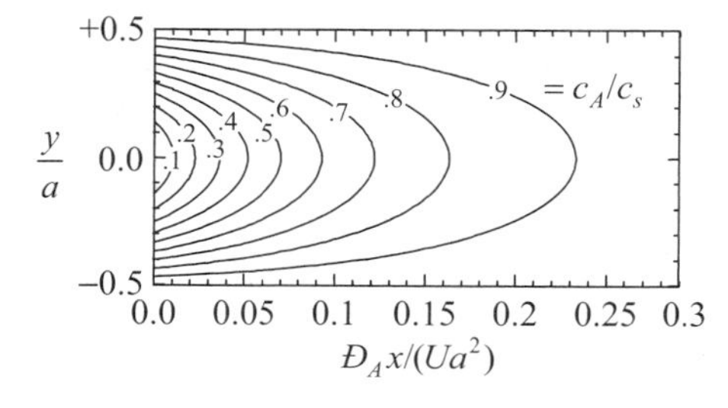

Figure 27-12 Fully developed iso-concentration lines in plug flow between constant concentration surfaces.

observations made in Section 27-4. Each possible choice of the Sherwood number corresponds to a different eigenvalue for the eigenfunction describing the concentration solution. However, only the smallest eigenvalue describes the fully developed condition, since contributions from larger eigenvalues decay from the solution before fully developed conditions occur.

In Problem 27-6, the exact solution to the species transport Equation (27-6) is solved for a uniform inlet concentration c_i. This solution is evaluated to determine the Sherwood number as a function of downstream distance in Problem 27-7. This final result can be evaluated in the limit that $x \to \infty$, to show that $Sh_D = \pi^2$ for fully developed species transfer in plug flow between boundaries of constant concentration.

27.8 PROBLEMS

27-1 Consider the problem illustrated in Figure 27-3, in which an inviscid fluid of initial temperature T_i is directed between parallel plates that (each) deliver a constant heat flux q_s to the fluid. Using the dimensionless temperature and spatial variables $\vartheta = (T - T_i)/(aq_s/k)$, $x^* = x/a$, and $y^* = y/a$, show that the heat equation may be expressed in dimensionless form as

$$\frac{Ua}{\alpha}\frac{\partial \vartheta}{\partial x^*} = \frac{\partial^2 \vartheta}{\partial y^{*2}},$$

with the boundary conditions $\partial\vartheta/\partial y^*|_{y^*=0} = 0$, $\partial\vartheta/\partial y^*|_{y^*=1/2} = 1$, and $\vartheta(x^* = 0) = 0$. Solve the heat equation to demonstrate the solution given by Eq. (27-13).

27-2 Solve the problem described in Problem 27-1 and show that the local Nusselt number for heat transfer to the flow is given by Eq. (27-14).

27-3 Evaluate the Nusselt number from the boundary layer temperature field in the entrance region of plug flow, with the heat flux boundary conditions illustrated in Figure 27-3. Demonstrate that the result given by Eq. (27-17) is obtained when the convection coefficient is defined such that $q_s = h(T_s - T_m)$, which is appropriate for an internal flow.

27-4 Beyond the thermal entry region of a pipe, does one expect the Nusselt number to be a function of the Reynolds number? Demonstrate why or why not.

27-5 Consider the problem discussed in Section 27-7 of fully developed species transport in an inviscid flow between plates at $y = \pm a/2$. The plates impose surfaces of constant concentration c_s as illustrated by Figure 27-11. Because of the boundary conditions, show that the constant self-similar concentration profile requires

$$\frac{\partial c_A}{\partial x} = \frac{c_s - c_A}{c_s - c_m}\frac{dc_m}{dx}.$$

Show that the rise in mean concentration with downstream position between the plates is given by

$$\frac{dc_m}{dx} = 2\frac{h_A}{Ua}(c_s - c_m).$$

Show that when these conditions are imposed on the species transport Equation (27-50) for internal flows, the governing equation becomes

$$\frac{\partial^2 \phi}{\partial y^{*2}} + Sh_D\phi = 0$$

where

$$\phi = \frac{c_s - c_A}{c_s - c_m}, \quad y^* = \frac{y}{a} \quad \text{and} \quad Sh_D = \frac{h_A(2a)}{\mathcal{D}_A}.$$

In addition to the boundary conditions $\partial\phi/\partial y^*|_{y^*=0} = 0$ and $\phi(y^* = 1/2) = 0$, show that the solution to ϕ is also required to satisfy the constraint

$$2\int_0^{1/2} \phi \, dy^* = 1.$$

Show that the solution to the dimensionless concentration profile is

$$\phi = \frac{\pi}{2}\cos\left(\sqrt{Sh_D}y^*\right), \quad \text{where } Sh_D = \pi^2.$$

27-6 Consider the problem of an inviscid flow between parallel plates. Suppose the fluid inlet has a dilute species concentration of c_i and the plates impose a constant surface concentration of c_s on the flow boundaries. Using the dimensionless concentration and spatial variables $\vartheta = (c_s - c_A)/(c_s - c_i)$, $x^* = x/a$, and $y^* = y/a$, derive the species transport equation

$$\frac{Ua}{Đ_A}\frac{\partial\vartheta}{\partial x^*} = \frac{\partial^2\vartheta}{\partial y^{*2}}$$

noting any simplifying assumptions that are made. What are the boundary conditions for ϑ? At what dimensionless distance $(Đ_A/Ua)(x/a)$ do the boundary layers join at the centerline?

27-7 Using the results of Problem 27-6, show that the local Sherwood number for species transfer to an inviscid flow between constant concentration boundaries is

$$Sh_D = \frac{h_A(2a)}{Đ_A} = \frac{J_A^*(2a)}{(c_s - c_m)Đ_A} = \frac{\sum_{n=0}^{\infty} \exp[-(\lambda_n^*)^2(Đ_A/Ua)(x/a)]}{\sum_{n=0}^{\infty} \frac{\exp[-(\lambda_n^*)^2(Đ_A/Ua)(x/a)]}{[(2n+1)\pi]^2}}, \quad \text{where } \lambda_n^* = (2n+1)\pi.$$

What value does Sh_D approach in the fully developed limit? Is this in agreement with the result of Section 27-7 for a fully developed flow? At what dimensionless distance $(Đ_A/Ua)(x/a)$ does the exact solution agree to within 1% of the result $Sh_D = \pi^2$ for fully developed species transport?

27-8 Consider the problem of inviscid pipe flow with a fully developed temperature profile and constant wall temperature T_s. Using the dimensionless variables

$$\theta = \frac{T_s - T}{T_s - T_m} \quad \text{and} \quad \eta = \frac{r}{R},$$

where T_m is the mean flow temperature defined by Eq. (27-8), derive the fully developed heat transfer equation for inviscid pipe flow. Solve the heat equation to determine the Nusselt number for heat transfer Nu_D. Determine the solution for $T(r, z)$, assuming that the initial mean temperature of the pipe flow is known, $T_m(z = 0) = T_m(0)$.

27-9 Consider an inviscid fluid flow between parallel surfaces. The bottom surface is adiabatic and the top surface has a convective heat flux from a hot overlying fluid at T_∞. The top surface temperature T_s changes with downstream position, and $T_m(x) < T_s(x) < T_\infty$.

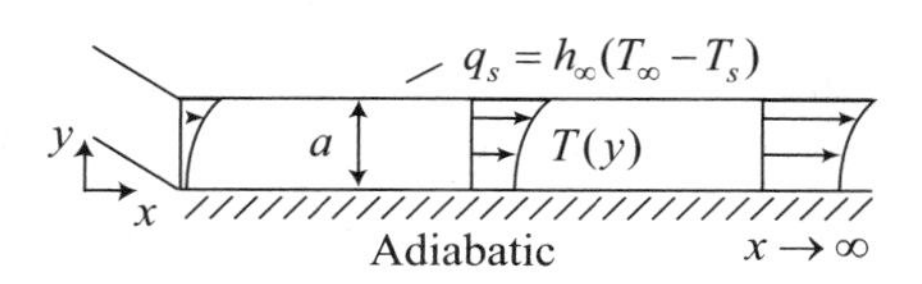

Show that for conditions of fully developed heat transfer,

$$\frac{\partial T}{\partial x} = \frac{\dfrac{T_s - T}{T_s - T_m} + \dfrac{h}{h_\infty}}{1 + h/h_\infty} \frac{\partial T_m}{\partial x},$$

where h is the unknown heat transfer coefficient between the nonadiabatic wall and the inviscid flow, and T_m is the mean temperature of the flow. Using the dimensionless variables

$$\vartheta = \frac{T_s - T}{T_s - T_m}, \quad y^* = \frac{y}{a}, \quad Nu_D = \frac{h(2a)}{k}, \quad \text{and} \quad Bi_\infty = \frac{h_\infty a}{k},$$

derive the heat equation for fully developed heat transfer:

$$\frac{\partial^2 \vartheta}{\partial y^{*2}} + \frac{Bi_\infty \vartheta + Nu_D/2}{Bi_\infty + Nu_D/2} \frac{Nu_D}{2} = 0.$$

Show that the Nusselt number for fully developed heat transfer to the inviscid flow is given by

$$Nu_D = \frac{2Bi_\infty \lambda^2}{Bi_\infty - \lambda^2}, \quad \text{where } \lambda \text{ is the root of } \lambda \tan(\lambda) = Bi_\infty, \quad \text{and } \lambda < \pi/2.$$

Plot Nu_D as a function of Bi_∞, for $0.001 < Bi_\infty < 1000$. What is the significance of the limiting values of Nu_D for high and low Bi_∞?

Chapter 28

Fully Developed Transport in Internal Flows

28.1 Momentum Transport in a Fully Developed Flow

28.2 Heat Transport in a Fully Developed Flow

28.3 Species Transport in a Fully Developed Flow

28.4 Problems

The mathematical description of fully developed internal flows, developed in Chapter 27, is applied to incompressible viscous flows in this chapter. Although solutions to partial differential equations are generally sought, the fully developed state of the flow allows transport equations to be formulated as ordinary differential equations. In addition to its application to laminar flows, as treated in this chapter, the mathematical simplification of fully developed transport is exploited again in Chapter 33 for turbulent flows.

28.1 MOMENTUM TRANSPORT IN A FULLY DEVELOPED FLOW

Consider pressure-driven flow, known as Poiseuille flow,* of a viscous incompressible liquid between parallel plates separated by a distance "a", as shown in Figure 28-1. The flow is governed by the continuity and momentum equations, written for fully developed conditions. Continuity for a constant density flow dictates

$$\underbrace{\frac{\partial v_x}{\partial x}}_{=\,0} + \frac{\partial v_y}{\partial y} = 0. \tag{28-1}$$

Since the streamwise velocity profile is not changing with downstream distance ($\partial v_x/\partial x = 0$), v_y must be a constant to satisfy continuity. Furthermore, v_y is zero since any other constant would imply flow penetration of the walls.

The steady-state momentum equation for a constant density fluid simplifies considerably when the flow is fully developed:

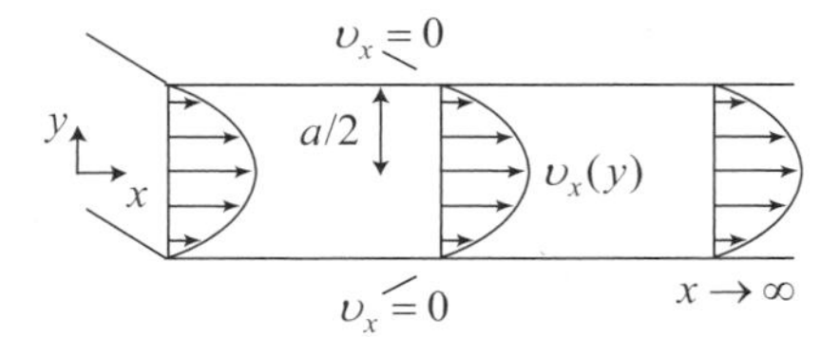

Figure 28-1 Fully developed Poiseuille flow between infinite parallel plates.

*Pressure-driven flow is named after the French physician Jean Louis Marie Poiseuille (1797–1869).

$$v_x \overbrace{\frac{\partial v_x}{\partial x}}^{=0} + \overbrace{v_y}^{=0} \frac{\partial v_x}{\partial y} = \nu \left(\overbrace{\frac{\partial^2 v_x}{\partial x^2}}^{=0} + \frac{\partial^2 v_x}{\partial y^2} \right) - \frac{1}{\rho}\frac{\partial P}{\partial x}. \tag{28-2}$$

Both advection terms are zero since $\partial v_x/\partial x = 0$ and $v_y = 0$. Furthermore, diffusion cannot contribute to transport in the x-direction because $\partial v_x/\partial x = 0$. In Poiseuille flows, the momentum lost by diffusion to the walls is balanced by the pressure gradient $\partial P/\partial x$. Therefore, the momentum equation becomes

$$\frac{\partial^2 v_x}{\partial y^2} = \frac{1}{\mu}\frac{\partial P}{\partial x} \tag{28-3}$$

and is subject to the boundary conditions $v_x(y = \pm\, a/2) = 0$ at the surfaces bounding the flow. It is a straightforward matter to integrate the momentum equation and apply the boundary conditions to determine the fully developed velocity distribution between the two plates:

$$v_x = \frac{1}{2\mu}\left(\frac{-\partial P}{\partial x}\right)\left[(a/2)^2 - y^2\right]. \tag{28-4}$$

This result can also be expressed in terms of the mean flow velocity:

$$v_m = \int_{-a/2}^{+a/2} v_x dy/a. \tag{28-5}$$

Using this definition to eliminate the pressure gradient from the velocity description yields

$$\frac{v_x}{v_m} = \frac{3}{2}\left[1 - 4(y/a)^2\right]. \tag{28-6}$$

One useful result of this analysis is the ability to quantify the fluid shear stress on the walls, τ_w. This can be reported with the dimensionless coefficient of friction, c_f:

$$c_f = \frac{\tau_w}{\rho v_m^2/2} = \frac{-\mu(\partial v_x/\partial y)\big|_{y=a}}{\rho v_m^2/2} = \frac{24\nu}{v_m(2a)} = \frac{24}{\mathrm{Re}_D}. \tag{28-7}$$

The length scale $D = 2a$ used in the Reynolds number is the hydraulic diameter of the flow as defined in Section 27.1 by

$$D = \frac{4 \times (\textit{cross-sectional area of flow})}{(\textit{perimeter wetted by flow})}. \tag{28-8}$$

28.2 HEAT TRANSPORT IN A FULLY DEVELOPED FLOW

As was discussed in the previous section for momentum transport, when the downstream length scale is large compared with the distance between bounding surfaces, diffusion transport to the walls will dominate over streamwise diffusion. Additionally,

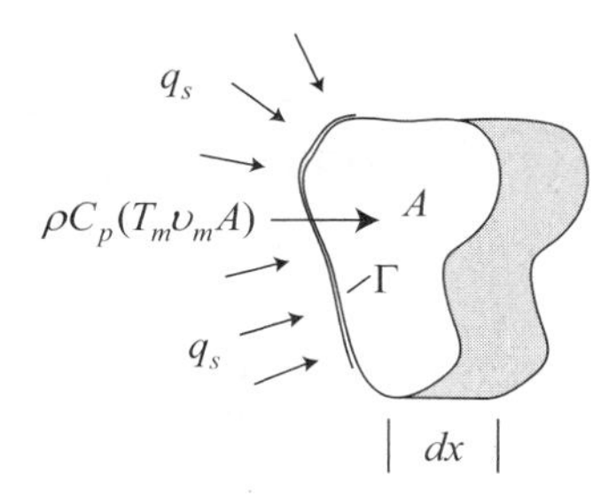

Figure 28-2 Heat balance over a unit streamwise distance.

with fully developed hydrodynamic conditions, there is no transverse advection since $v_y = 0$. Therefore, in this situation, the heat equation for internal flows simplifies to

$$v_x \frac{\partial T}{\partial x} = \alpha \frac{\partial^2 T}{\partial y^2}. \tag{28-9}$$

Next, the simplification that arises when heat transfer is fully developed is considered. As was discussed in Chapter 27, the governing partial differential equation may be transformed into an ordinary differential equation when heat transfer is fully developed. As illustrated in Section 27.5, fully developed conditions reveal a self-similarity in the temperature profile that results in

$$\frac{\partial}{\partial x}\left[\frac{T_s - T(x,y)}{T_s - T_m(x)}\right] = 0 \quad (or \quad \approx 0). \tag{28-10}$$

This condition describes a *thermally fully developed* flow, where the mean flow temperature is defined by

$$T_m = \int_A v_x T dA \Big/ \int_A v_x dA, \tag{28-11}$$

and A is the cross-sectional area of the flow. The mean temperature T_m is a transport average, not a geometric mean like the average velocity. To illustrate the utility of mean flow properties, the advected energy crossing the area A in Figure 28-2 is

$$\rho C_p \int_A v_x T dA = \rho C_p T_m v_m A. \tag{28-12}$$

Since $\rho v_m A$ is the total mass flow rate, $C_p T_m$ must be the average thermal energy per unit mass carried by the flow. Heat transfer to or from the fluid can be expressed in terms of the change in mean temperature $\partial T_m/\partial x$. For example, a heat balance over a differential downstream distance dx of the flow illustrated in Figure 28-2 yields

$$\frac{d}{dx}\left(\rho C_p T_m v_m A\right) = \Gamma q_s, \tag{28-13}$$

where Γ is the peripheral length along the circumference of A through which heat transfer q_s occurs. Since $d(\rho v_m A)/dx = 0$ by continuity, the heat balance can be rearranged to yield

$$v_m \frac{dT_m}{dx} = \frac{\Gamma \alpha}{Ak} q_s = \frac{\Gamma \alpha}{Ak} h(T_s - T_m) \tag{28-14}$$

as long as the specific heat C_p is constant. Notice that the convection coefficient h for heat transfer in Eq. (28-14) is defined such that the heat flux from the wall to the fluid is $q_s = h(T_s - T_m)$.

28.2.1 Heat Transport with Isothermal Boundaries

To illustrate the analysis of fully developed heat transfer, consider the problem illustrated in Figure 28-3 of a viscous Poiseuille flow between two hot isothermal plates. The flow is assumed to be fully developed in both the hydrodynamic and thermal senses. As heat is transported to the fluid, the temperature distribution rises until eventually becoming uniform across the gap between the two plates, as shown.

Since the temperature field is fully developed, and T_s is a constant, Eq. (28-10) requires that

$$-\frac{\partial T}{\partial x}(T_s - T_m) + \frac{dT_m}{dx}(T_s - T) \approx 0. \tag{28-15}$$

Therefore, $\partial T/\partial x$ can be related to changes in the mean temperature through the relation

$$\frac{\partial T}{\partial x} \approx \frac{T_s - T}{T_s - T_m}\frac{dT_m}{dx}. \tag{28-16}$$

The change in mean temperature $\partial T_m/\partial x$ is an expression of the overall rate of heat transfer to the fluid given by Eq. (28-14). Therefore, combining Eqs. (28-14) and (28-16) for application to the advection term in the heat equation (28-9) results in

$$\frac{v_x}{v_m}(T_s - T)\frac{\Gamma h}{Ak} = \frac{\partial^2 T}{\partial y^2} \quad \text{(thermally and hydrodynamically fully developed flow with isothermal boundaries).} \tag{28-17}$$

In its present form, the heat equation is an ordinary differential equation with respect to y. It is noteworthy that the velocity profile v_x/v_m for the laminar flow solution has no dependence on the Reynolds number. Therefore, laminar flow solutions to the fully developed heat equation will be Reynolds number independent. However, this will not be the case for turbulent flows addressed in Chapter 31, since the shape of the turbulent velocity profile v_x/v_m will change with Reynolds number.

Using $\Gamma/A = 2/a$ for the geometry shown in Figure 28-3, and Eq. (28-6) for v_x/v_m, the heat equation for the Poiseuille flow can be rewritten as

$$3\left[1 - 4(y/a)^2\right](T_s - T)\frac{h}{a\,k} = \frac{\partial^2 T}{\partial y^2}. \tag{28-18}$$

To simplify the mathematical presentation of the heat equation, one can define

$$\phi = \frac{T - T_i}{T_s - T_i}, \quad \eta = \frac{y}{a} \quad \text{and} \quad Nu_{D,T} = \frac{h(2a)}{k}, \tag{28-19}$$

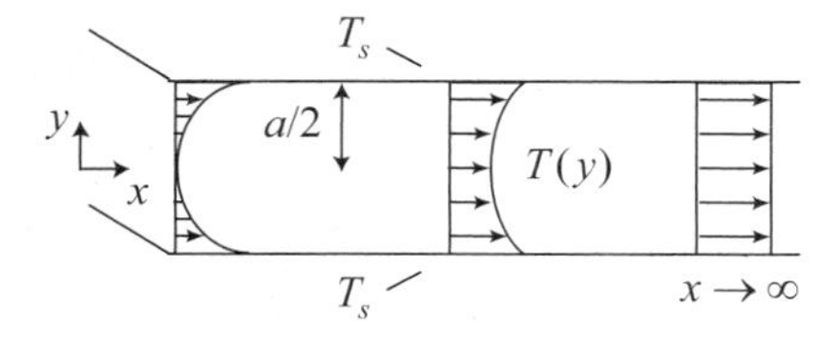

Figure 28-3 Fully developed temperature field between isothermal parallel plates.

where the length scale used in the Nusselt number is the hydraulic diameter for the flow. The "T" subscript serves as a reminder that, in this problem, the Nusselt number reflects heat transfer from a constant wall temperature. A reference temperature T_i has been introduced to identify the lowest possible temperature in the flow. With these definitions, the dimensionless heat equation becomes

$$\frac{3}{2}\left[1-4\eta^2\right](1-\phi)Nu_{D,T}=\frac{\partial^2\phi}{\partial\eta^2}, \tag{28-20}$$

and is subject to the boundary conditions

$$\phi'(0)=0 \quad \text{and} \quad \phi(1/2)=1. \tag{28-21}$$

Since the heat equation must describe the temperature distribution at any downstream distance x, there are an infinite number of possible solutions to Eq. (28-20) that satisfy the boundary conditions (28-21). These solutions correspond to different values of $\phi(0)$ between 0 and 1. This is illustrated in Figure 28-3 for a few centerline temperatures that increase as a function of x. Since the governing equation (28-20) is linear, an analytic solution can be sought. An effective approach to solving linear homogeneous ordinary differential equations with variable coefficients is the method of power series solutions, as discussed in Section 10.4. To make the governing equation homogeneous, let $\theta=1-\phi$, and the problem statement becomes

$$\theta''+b(1-4\eta^2)\theta=0 \quad \text{with} \quad b=3Nu_{D,T}/2 \tag{28-22}$$

subject to the boundary conditions

$$\theta'(0)=0 \quad \text{and} \quad \theta(1/2)=0. \tag{28-23}$$

The solution can be expressed as a power series of the form

$$\theta=\sum_{n=0}^{\infty}a_n\eta^n. \tag{28-24}$$

Substituting the assumed solution into the governing equation yields

$$\sum_{n=2}^{\infty}n(n-1)a_n\eta^{n-2}+b\sum_{n=0}^{\infty}a_n\eta^n-4b\sum_{n=0}^{\infty}a_n\eta^{n+2}=0. \tag{28-25}$$

The solution for θ will be realized if the a_n's can be determined to satisfy the governing equation. By expanding Eq. (28-25) fully, it can be reorganized to have the form

$$(\cdots)\eta^0+(\cdots)\eta^1+(\cdots)\eta^2+\cdots=0. \tag{28-26}$$

For this power series representation of the governing equation to be satisfied for all values of η, each coefficient of η^n in the series must be identically zero. Therefore, Eq. (28-25) is written in a form where the coefficients to η^n are easily determined:

$$\sum_{n=0}^{\infty}(n+2)(n+1)a_{n+2}\eta^n+b\sum_{n=0}^{\infty}a_n\eta^n-4b\sum_{n=2}^{\infty}a_{n-2}\eta^n=0. \tag{28-27}$$

Notice that the last summation does not start until $n = 2$. Therefore, the last summation does not contribute to the first two terms in the expanded form of this equation. However, when $n \geq 2$, all three summations contribute to each coefficient of η^n. Therefore, fully expanded, the power series expression for the governing equation becomes

$$(2a_2 + ba_0)\eta^0 + (6a_3 + ba_1)\eta^1 + \cdots + ((n+2)(n+1)a_{n+2} + ba_n - 4ba_{n-2})\eta^n + \cdots = 0. \quad (28\text{-}28)$$

By inspection of the coefficients, it is seen that the governing equation is satisfied when $a_2 = (-b/2)a_0$, $a_3 = (-b/6)a_1$, and $(n+2)(n+1)a_{n+2} + ba_n - 4ba_{n-2} = 0$ for $n \geq 2$. The last condition expresses a recursion relation between values of a_n's. Specifically,

$$a_{n+2} = b\frac{4a_{n-2} - a_n}{(n+2)(n+1)} \quad \text{for } n \geq 2. \quad (28\text{-}29)$$

Or, equivalently,

$$a_n = b\frac{4a_{n-4} - a_{n-2}}{n(n-1)} \quad \text{for } n \geq 4. \quad (28\text{-}30)$$

Notice that a_n will be a multiple of a_0 for even values of n and a multiple of a_1 for odd values of n. However, since $\theta'(0) = 0$, it is required that $a_1 = 0$, and therefore all odd values of a_n are zero in this solution. Consequently, only even values of n are needed to evaluate a_n, and the series solution for θ becomes

$$\theta = \left(a_0 + \overbrace{a_0\left(\frac{-b}{2}\right)}^{a_2}\eta^2 + \overbrace{a_0\frac{b}{12}\left(4 - \frac{-b}{2}\right)}^{a_4}\eta^4 + \cdots + \overbrace{b\frac{4a_{n-4} - a_{n-2}}{n(n-1)}}^{a_n}\eta^n + \cdots \right). \quad (28\text{-}31)$$

Factoring out a_0, and changing the indexing scheme to $i = 0, 1, 2, \ldots$ yields

$$\theta = a_0\left(1 + \overbrace{\left(\frac{-b}{2}\right)}^{c_1}\eta^2 + \overbrace{\frac{b}{12}\left(4 - \frac{-b}{2}\right)}^{c_2}\eta^4 + \cdots + \overbrace{\frac{b(4c_{i-2} - c_{i-1})}{2i(2i-1)}}^{c_i}\eta^{2i} + \cdots \right). \quad (28\text{-}32)$$

When $\eta = 0$, the solution evaluates to $\theta(0) = a_0$. But, by definition, $\theta(0) = 1 - \phi(0)$. Therefore $a_0 = 1 - \phi(0)$. The second boundary condition $\theta(1/2) = 0$ requires finding b to satisfy

$$0 = 1 + \overbrace{\left(\frac{-b}{2}\right)}^{c_1}\left(\frac{1}{2}\right)^2 + \overbrace{\frac{b}{12}\left(4 - \frac{-b}{2}\right)}^{c_2}\left(\frac{1}{2}\right)^4 + \cdots + \overbrace{\frac{b(4c_{i-2} - c_{i-1})}{2i(2i-1)}}^{c_i}\left(\frac{1}{2}\right)^{2i} + \cdots. \quad (28\text{-}33)$$

The first root of this polynomial series occurs at $b = 11.311$. Other roots correspond to temperature profiles that may be disregarded on physical grounds, as discussed in Section 27.4. Recalling that $b = 3Nu_{D,T}/2$, it can be established that $Nu_{D,T} = 7.541$.

The solution for $T(y)$ is obtained from $\phi = (T - T_i)/(T_s - T_i)$, where $\phi = 1 - \theta$, and can be written as

$$\frac{T_s - T(y)}{T_s - T(0)} = \left(c_0 + c_1(y/a)^2 + \sum_{i=2}^{\infty} c_i(y/a)^{2i} \right) \quad (28\text{-}34)$$

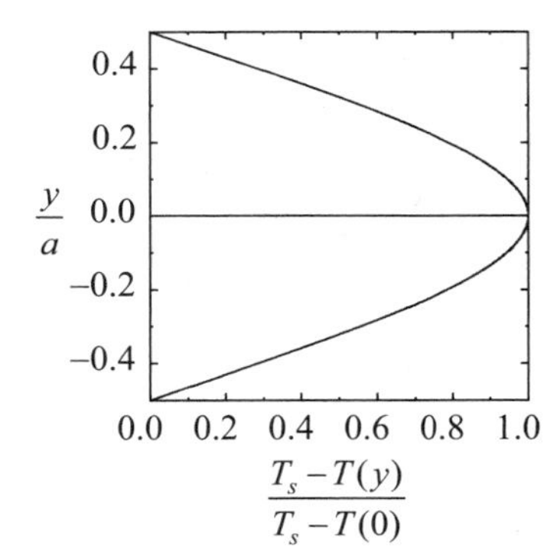

Figure 28-4 Temperature distribution between constant temperature plates.

with

$$c_0 = 1, \; c_1 = -3Nu_{D,T}/4, \tag{28-35}$$

$$c_i = \frac{3Nu_{D,T}(4c_{i-2} - c_{i-1})}{4i(2i-1)}, \tag{28-36}$$

where

$$Nu_{D,T} = 7.541. \tag{28-37}$$

Equation (28-34) is plotted in Figure 28-4. The centerline temperature $T(0)$ depends on the downstream distance x, as will be addressed further in Section 28.2.3.

28.2.2 Heat Transport with Constant Heat Flux Boundaries

Consider the problem illustrated in Figure 28-5 of a Poiseuille flow between two plates that deliver a constant heat flux to the fluid. The flow is assumed to be fully developed both hydrodynamically and thermally. As heat is transported to the fluid, the temperature distribution rises indefinitely in the downstream direction.

When the temperature field is fully developed, Eq. (28-10) dictates that

$$(T_s - T_m)\left(\frac{\partial T_s}{\partial x} - \frac{\partial T}{\partial x}\right) - (T_s - T)\left(\frac{\partial T_s}{\partial x} - \frac{dT_m}{dx}\right) = 0. \tag{28-38}$$

However, with a constant heat flux $q_s = h(T_s - T_m)$, it is determined from $dq_s/dx = 0$ that

$$\frac{\partial T_s}{\partial x} = \frac{\partial T_m}{\partial x}. \tag{28-39}$$

Consequently, Eq. (28-38) becomes

$$(T_s - T_m)\left(\frac{\partial T_s}{\partial x} - \frac{\partial T}{\partial x}\right) = 0, \tag{28-40}$$

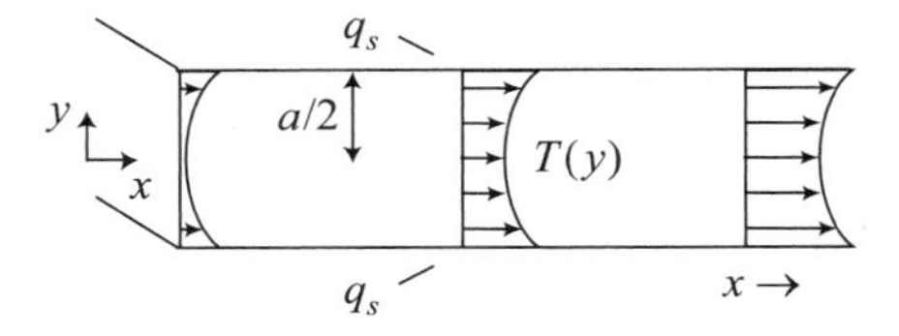

Figure 28-5 Fully developed temperature field between plates with constant heat flux.

which yields for a constant heat flux boundary condition

$$\frac{\partial T}{\partial x} = \frac{\partial T_s}{\partial x} \quad \left(= \frac{\partial T_m}{\partial x} \right). \tag{28-41}$$

Using the fact that $\partial T/\partial x = \partial T_m/\partial x$, and recalling through Eq. (28-14) that $\partial T_m/\partial x$ is an expression of the overall rate of heat transfer to the fluid, the heat equation Eq. (28-9) can be written as

$$\frac{v_x}{v_m}\frac{\Gamma q_s}{Ak} = \frac{\partial^2 T}{\partial y^2} \qquad \text{(thermally and hydrodynamically fully developed flow with constant heat flux boundaries).}$$

Equation (28-6) is used to evaluate v_x/v_m for Poiseuille flow, and for the present geometry of flow, $\Gamma/A = 2/a$. Again defining $\phi = (T - T_i)/(T_s - T_i)$ and $\eta = y/a$, the heat equation can be written in dimensionless form

$$\frac{3}{2}[1 - 4\eta^2] Nu_{D,H}(1 - \phi_m) = \frac{\partial^2 \phi}{\partial \eta^2}, \tag{28-42}$$

where

$$Nu_{D,H}(1 - \phi_m) = 2q_s a/(kT_s) \quad \text{and} \quad \phi_m = \int_0^1 (v_x/v_m)\phi \, d\eta = 2\int_0^{1/2} (v_x/v_m)\phi \, d\eta. \tag{28-43}$$

The H subscript on the Nusselt number serves as a reminder that heat transfer is from a constant heat flux boundary condition. The boundary conditions are $\phi'(0) = 0$ and $\phi(1/2) = 1$. The solution to the heat equation is readily found:

$$\phi = 1 + \frac{1}{4}\left(3\eta^2 - 2\eta^4 - \frac{5}{8}\right) Nu_{D,H}(1 - \phi_m). \tag{28-44}$$

The Nusselt number $Nu_{D,H}$ can be determined from the mean temperature of the solution. Using Eqs. (28-6) and (28-44),

$$\phi_m = 2\int_0^{1/2} (v_x/v_m)\,\phi \, d\eta = 3\int_0^{1/2} (1 - 4\eta^2)\left[1 + \frac{1}{4}\left(3\eta^2 - 2\eta^4 - \frac{5}{8}\right) Nu_{D,H}(1 - \phi_m)\right] d\eta \tag{28-45}$$

or

$$\phi_m = 1 - \frac{17}{140} Nu_{D,H}(1 - \phi_m). \tag{28-46}$$

Therefore,

$$Nu_{D,H} = \frac{140}{17} = 8.24. \tag{28-47}$$

It is interesting that the Nusselt number for the constant heat flux boundary condition is higher by 9% than the constant temperature boundary condition, found in Section 28.2.1.

28.2.3 Downstream Development of Temperature in a Heat Exchanger

The preceding sections have illustrated how the fully developed temperature distribution of an internal flow may be determined. However, solutions so far are somewhat incomplete. For the constant surface temperature boundary condition discussed in Section 28.2.1, the centerline temperature must be specified before a unique temperature profile can be calculated. For the constant heat flux boundary condition discussed in Section 28.2.2, the surface temperature must be specified. In either case, the missing information is related to the heat added upstream of the position at which the temperature distribution is evaluated. To provide closure, a look at the streamwise accumulation of heat is required.

To demonstrate the methodology, consider the specific problem of a parallel-flow heat exchanger illustrated in Figure 28-6. Hot and cold fluids flow in two parallel channels that permit heat transfer between the two fluids. It is desired to determine the total heat transfer between the two fluids and the temperature distributions of both hot and cold fluids at any position down the length of the heat exchanger. It is assumed that the flow and temperature fields are fully developed everywhere in the heat exchanger. If the hot and cold fluids have equal mass flow rates and the same thermal properties, each degree that the hot fluid cools causes a degree rise in temperature of the cold fluid. Although the mean temperatures of the hot and cold fluids are changing in the streamwise direction, the wall temperature between the two fluids has a constant value equal to the average of the hot and cold mean temperatures.

A differential heat transfer rate between the two fluids can be expressed as

$$dQ = C^h(-dT^h_m) = h^h\, \Gamma\, dx(T^h_m - T_s) \tag{28-48}$$

and

$$dQ = C^c(+dT^c_m) = h^c\, \Gamma\, dx(T_s - T^c_m). \tag{28-49}$$

Here $\Gamma\, dx$ is the differential area of the wall between the hot and cold fluids through which heat transfer occurs. For notational simplicity, $C^h = \dot{m}^h C^h_p$ is the product of the mass flow rate and specific heat of the hot fluid, and T^h_m is mean temperature of the hot fluid. Likewise, for the cold fluid $C^c = \dot{m}^c C^c_p$, and T^c_m is the mean temperature of the cold fluid. The wall convection coefficients for the hot and cold sides of the heat exchanger are h^h and h^c, respectively. Equations (28-48) and (28-49) for the differential heat transfer rate can be

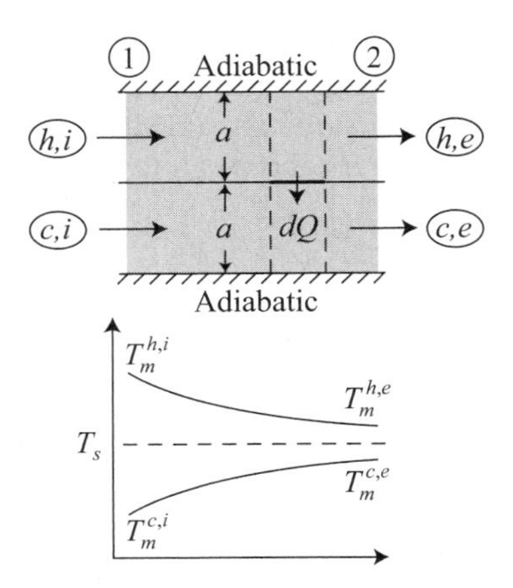

Figure 28-6 A parallel flow heat exchanger.

used to derive expressions for the temperature difference and change in the temperature difference between the mean values of the hot and cold fluids:

$$\Delta T_m = T_m^h - T_m^c = \frac{dQ}{\Gamma\, dx}\left(\frac{1}{h^h} + \frac{1}{h^c}\right) \tag{28-50}$$

$$d(\Delta T_m) = dT_m^h - dT_m^c = -dQ\left(\frac{1}{C^h} + \frac{1}{C^c}\right). \tag{28-51}$$

Eliminating dQ from these two equations yields

$$\frac{d(\Delta T_m)}{\Delta T_m} = -\frac{1/C^h + 1/C^c}{1/h^h + 1/h^c}\,\Gamma\, dx, \tag{28-52}$$

which integrated over the limits of the heat exchanger yields

$$\ln\left(\frac{\Delta T_m^e}{\Delta T_m^i}\right) = -\,A_H\frac{1/C^h + 1/C^c}{1/h^h + 1/h^c}. \tag{28-53}$$

In this result, $\Delta T_m^i = T_m^{h,i} - T_m^{c,i}$, $\Delta T_m^e = T_m^{h,e} - T_m^{c,e}$, and $A_H = \Gamma L$ (the product of the width and length) is the total area through which heat transfer occurs. With the total heat transfer rate expressed as

$$Q = C^h(T_m^{h,i} - T_m^{h,e}) = C^c(T_m^{c,e} - T_m^{c,i}), \tag{28-54}$$

the total heat transfer rate can be found in relation to the differences in mean temperature at the inlet and exit of the heat exchanger:

$$Q = \frac{\Delta T_m^i - \Delta T_m^e}{1/C^h + 1/C^c}. \tag{28-55}$$

Combined with Eq. (28-53), the total heat transfer rate can be calculated from differences in inlet and outlet temperatures:

$$Q = \frac{A_H}{1/h^h + 1/h^c}\,\frac{\Delta T_m^e - \Delta T_m^i}{\ln[\Delta T_m^e/\Delta T_m^i]}. \tag{28-56}$$

The mean temperature as a function of distance from the inlet can be determined from Eqs. (28-48) and (28-49), and may be subsequently used to determine the adiabatic wall temperature $T(x, y = 0)$ as a function of downstream distance. For the cold fluid side, Eq. (28-49) yields a differential equation for the mean temperature change:

$$\frac{dT_m^c}{dx} = \frac{h^c(T_s - T_m^c)}{v_m^c a \rho^c C_p^c} = \frac{h^c(2a)}{k^c}\,\frac{\alpha^c}{\nu^c}\,\frac{\nu^c}{v_m^c(2a)}\,\frac{T_s - T_m^c}{a} = \frac{Nu_D^c}{\Pr^c \mathrm{Re}_D^c}\,\frac{T_s - T_m^c}{a}, \tag{28-57}$$

in which ν^c and α^c are the cold fluid momentum and thermal diffusivities, respectively, and the following definitions have been used:

$$\text{Nusselt number:}\quad Nu_D^c = \frac{h^c(2a)}{k^c} \tag{28-58}$$

$$\text{Prandtl number:} \qquad \mathrm{Pr}^c = \frac{\nu^c}{\alpha^c} \tag{28-59}$$

$$\text{Reynolds number:} \qquad \mathrm{Re}_D^c = \frac{v_m^c(2a)}{\nu^c}. \tag{28-60}$$

The differential equation for the mean temperature on the cold side can be integrated for the result

$$\frac{T_s - T_m^c(x)}{T_s - T_m^c(x=0)} = \exp\left(\frac{-Nu_D^c}{\mathrm{Pr}^c \mathrm{Re}_D^c}\frac{x}{a}\right). \tag{28-61}$$

A similar result can be found for the hot fluid side.

Heat transfer in the current problem differs from the isothermal plates considered in Section 28.2.1. In the heat exchanger, as illustrated in Figure 28-7, only one wall is isothermal, while the opposing wall is adiabatic. Therefore, the Nusselt number for heat transfer in the heat exchanger is expected to be different from the earlier result. As discussed in Section 27.1, the scaling estimate for the heat transfer Nusselt number is

$$Nu_D \sim \frac{D}{\delta_T}, \tag{28-62}$$

in which D is, by convention, the hydraulic diameter. The hydraulic diameter is a common scale to the problems addressed here and in Section 28.2.1. However, the length scale for heat transfer δ_T in the current problem, illustrated by Figure 28-7, is roughly twice that expected for the previous situation illustrated in Figure 28-3. Therefore, it is expected that the Nusselt number in the current situation will be roughly one-half that found for the previous situation ($Nu_{D,T} = 7.541$).

After moving the origin to a location on the adiabatic wall, Eq. (28-6) for Poiseuille flow becomes

$$\frac{v_x}{v_m} = 6\eta(1-\eta), \quad \text{where } \eta = \frac{y}{a}. \tag{28-63}$$

For the configuration of heat transfer shown in Figure 28-7, $\Gamma/A = 1/a$ where A is the cross-sectional area to the flow and Γ is the peripheral length around A that accommodates heat transfer. Therefore, the heat equation, Eq. (28-17), for Poiseuille flow becomes

$$6\frac{y}{a}\left(1-\frac{y}{a}\right)(T_s - T)\frac{h}{a\,k} = \frac{\partial^2 T}{\partial y^2}. \tag{28-64}$$

The boundary conditions for this problem are $dT/dy|_{y=0} = 0$ and $T(y=a) = T_s$. It is left as an exercise (see Problem 28-1) to show that the solution for the temperature field is given by

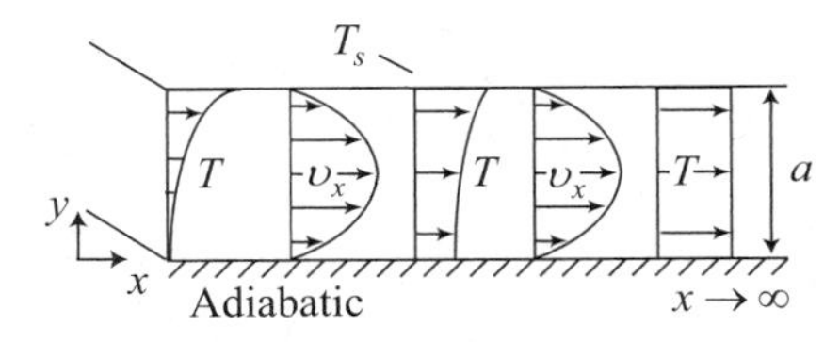

Figure 28-7 Cold side of heat exchanger.

$$\frac{T_s - T(x,y)}{T_s - T(x,y=0)} = c_0 + c_3\left(\frac{y}{a}\right)^3 + \sum_{n=4}^{\infty} c_n \left(\frac{y}{a}\right)^n \tag{28-65}$$

where

$$c_0 = 1,\ \ c_1 = c_2 = 0,\ \ c_3 = -Nu_D^c/2, \tag{28-66}$$

$$c_n = 3Nu_D^c \frac{c_{n-4} - c_{n-3}}{n(n-1)}, \quad \text{and} \quad Nu_D^c = 4.861. \tag{28-67}$$

The Nusselt number is defined as

$$Nu_D^c = \frac{h(2a)}{k}. \tag{28-68}$$

As expected from scaling arguments, the numeric value of Nu_D^c is roughly one-half of the value found for the situation discussed in Section 28.2.1.

Notice that Eq. (28-65) requires knowledge of the adiabatic wall temperature $T(x, y = 0)$ to provide a description for the temperature profile spanning the flow. To provide a fully two-dimensional description of the fluid temperature requires combining Eq. (28-65) for the temperature profile with Eq. (28-61) for the streamwise development of the mean temperature. These two descriptions are interrelated through the condition of fully developed heat transfer, given by Eq. (28-10); this condition requires that the adiabatic wall temperature $T(x, y = 0)$ obeys the relation

$$\frac{T_s - T(x=0, y=0)}{T_s - T_m^c(x=0)} = \frac{T_s - T(x, y=0)}{T_s - T_m^c(x)}, \tag{28-69}$$

or

$$\frac{T_s - T(x, y=0)}{T_s - T(x=0, y=0)} = \frac{T_s - T_m^c(x)}{T_s - T_m^c(x=0)}. \tag{28-70}$$

Combining this result with Eq. (28-61) yields an expression for the rise in the adiabatic wall temperature:

$$\frac{T_s - T(x, y=0)}{T_s - T(x=0, y=0)} = \exp\left(\frac{-Nu_D^c}{\mathrm{Pr}^c \mathrm{Re}_D^c} \frac{x}{a}\right). \tag{28-71}$$

Combining this result with Eq. (28-65) for the temperature profile spanning the flow yields

$$\frac{T_s - T(x,y)}{T_s - T(x=0, y=0)} = \left(c_0 + c_3\left(\frac{y}{a}\right)^3 + \sum_{n=4}^{\infty} c_n \left(\frac{y}{a}\right)^n\right) \exp\left(\frac{-Nu_D^c}{\mathrm{Pr}^c \mathrm{Re}_D^c} \frac{x}{a}\right). \tag{28-72}$$

Defining the reference temperature $T^r = T(x = 0, y = 0)$, the two-dimensional description of the fluid temperature field becomes

$$\frac{T(x,y) - T^r}{T_s - T^r} = 1 - \left(c_0 + c_3(y/a)^3 + \sum_{n=4}^{\infty} c_n (y/a)^n\right) \exp\left(\frac{-Nu_D^c}{\mathrm{Pr}^c\ \mathrm{Re}_D^c} \frac{x}{a}\right), \tag{28-73}$$

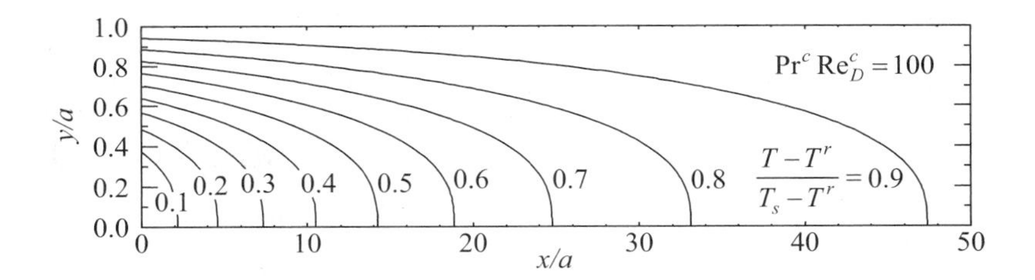

Figure 28-8 Isotherms on cold side of the heat exchanger shown in Figure 28-6.

where the constants are the same as those defined for Eq. (28-65). To illustrate this result, Eq. (28-73) is plotted in Figure 28-8 for $\mathrm{Pr}^c\mathrm{Re}_D^c = 100$. An expression similar to (28-73) can be found for the hot side of the heat exchanger.

28.3 SPECIES TRANSPORT IN A FULLY DEVELOPED FLOW

The treatment of fully developed mass transport is in close analogy to the subject of heat transfer discussed in the previous sections. As an illustration, the leaching of a contaminant from the walls of a circular pipe is investigated in this section.

28.3.1 Contaminant Leaching from a Constant Concentration Pipe Wall

Consider the problem illustrated in Figure 28-9 of a contaminant leaching into a fully developed pipe flow. The radius of the pipe is R, and the contaminant concentration at the wall is a constant c_w. It is desired to determine the total rate of contaminant leaching for a given length of pipe.

The contamination rate through a streamwise differential length of the pipe wall can be expressed as

$$\pi R^2 v_m(+dc_m) = 2\pi R\, h_C(c_w - c_m)dz, \tag{28-74}$$

where h_C is an unknown convection coefficient for the contamination flux, and v_m is the mean velocity of the flow through the pipe. For a fully developed Poiseuille flow, the pipe velocity profile is given by

$$\frac{v_z(r)}{v_m} = 2\left[1 - (r/R)^2\right]. \tag{28-75}$$

In analogy to Eq. (28-11), the local mean concentration c_m in the pipe is a flow average given by

$$c_m = 2\int_0^R c(r)\frac{v_z(r)}{v_m}\frac{rdr}{R^2}. \tag{28-76}$$

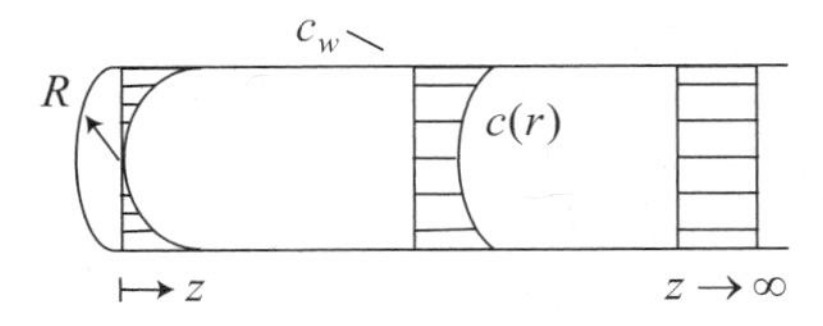

Figure 28-9 Contamination leaching into fully developed pipe flow.

From Eq. (28-74), the mean concentration in the pipe is governed by the ordinary differential equation

$$\frac{dc_m}{dz} = \frac{2\,h_C}{R v_m}(c_w - c_m) = 2\frac{h_C(2R)}{Đ_C}\frac{Đ_C}{\nu}\frac{\nu}{v_m(2R)}\frac{c_w - c_m}{R} = \frac{2\,Sh_D}{ScRe_D}\frac{c_w - c_m}{R}, \tag{28-77}$$

in which ν and $Đ$ are momentum and species diffusivities, respectively, and the following definitions have been used:

$$\text{Sherwood number:} \quad Sh_D = \frac{h_C(2R)}{Đ_C} \tag{28-78}$$

$$\text{Schmidt number:} \quad Sc = \frac{\nu}{Đ_C} \tag{28-79}$$

$$\text{Reynolds number:} \quad Re_D = \frac{v_m(2R)}{\nu}. \tag{28-80}$$

Integrating the governing equation down the length of the pipe yields the mean concentration profile

$$c_m(z) = c_w - (c_w - c^i_m)\exp\left(-\frac{2\,Sh_D}{ScRe_D}\frac{z}{R}\right), \tag{28-81}$$

where $c_m(z=0) = c^i_m$ is the initial mean concentration. The rate at which contaminates are leached from the pipe can be estimated from the volume flow rate $\pi R^2 v_m$ and the change in mean concentration between the start and the end of the pipe $c_m(L) - c^i_m$, where $c_m(L)$ is evaluated from Eq. (28-81). The key to quantifying this rate is determining the unknown Sherwood number Sh_D for this problem.

It is left as an exercise (see Problem 28-2) to show that, for fully developed conditions, the concentration distribution in the pipe is governed by

$$\frac{v_z}{v_m}(c_w - c)\frac{h_C(2R)}{Đ_C} = \frac{R^2}{r}\frac{\partial}{\partial r}\left(r\frac{\partial c}{\partial r}\right). \tag{28-82}$$

Introducing the fully developed velocity profile, Eq. (28-75), and letting $\chi = 1 - c/c_w$ and $\eta = r/R$, the dimensionless species transport equation becomes

$$2\left[1-\eta^2\right]\chi\,Sh_D = -\frac{1}{\eta}\frac{\partial}{\partial\eta}\left(\eta\frac{\partial\chi}{\partial\eta}\right), \tag{28-83}$$

or

$$\eta\chi'' + \chi' + s(\eta - \eta^3)\chi = 0 \quad \text{with } s = 2\,Sh_D, \tag{28-84}$$

and subject to

$$\chi'(0) = 0 \quad \text{and} \quad \chi(1) = 0. \tag{28-85}$$

This linear transport problem can be solved with a power series solution of the form

$$\chi = \sum_{n=0}^{\infty} a_n\eta^n, \tag{28-86}$$

which can satisfy the governing equation when

$$
\begin{aligned}
(a_1)\eta^0 + (4a_2 + sa_0)\eta^1 + (9a_3 + sa_1)\eta^2 + \cdots \\
+ \Big((n+1)^2 a_{n+1} + sa_{n-1} - sa_{n-3}\Big)\eta^n + \cdots = 0,
\end{aligned}
\tag{28-87}
$$

as demonstrated by direct substitution. By inspection it is seen that the coefficients must evaluate to $a_1 = 0$, $a_2 = (-s/4)a_0$, $a_3 = (-s/9)a_1 = 0$, and $(n+1)^2 a_{n+1} + sa_{n-1} - sa_{n-3} = 0$ for $n \geq 3$. The last condition expresses a recursion relation between values of a_n's that is equivalent to

$$
a_n = s\frac{a_{n-4} - a_{n-2}}{n^2} \quad \text{for } n \geq 4. \tag{28-88}
$$

Notice that a_n is a multiple of a_0 when the index is even and a multiple of a_1 when the index is odd. However, since $a_1 = 0$, all odd index values of a_n are zero. Consequently, only a_n for even values of n are needed. Factoring out a_0 from the series, changing the indexing scheme to $i = 0, 1, 2, \ldots$, and renaming the coefficients such that $d_i = a_{2i}/a_0$ yields a solution for χ in the form

$$
\chi = a_0\left(1 + \overbrace{\left(\frac{-s}{4}\right)}^{d_1}\eta^2 + \overbrace{\frac{s}{16}\left(1 - \frac{-s}{4}\right)}^{d_2}\eta^4 + \cdots + \overbrace{s\frac{d_{i-2} - d_{i-1}}{4i^2}}^{d_i}\eta^{2i} + \cdots\right). \tag{28-89}
$$

When $\eta = 0$, the solution evaluates to $\chi(0) = a_0$. The second boundary condition $\chi(1) = 0$ requires finding s to satisfy

$$
0 = 1 + \overbrace{\left(\frac{-s}{4}\right)}^{d_1} + \overbrace{\frac{s}{16}\left(1 - \frac{-s}{4}\right)}^{d_2} + \cdots + \overbrace{s\frac{d_{i-2} - d_{i-1}}{4i^2}}^{d_i} + \cdots. \tag{28-90}
$$

The smallest root of this polynomial series occurs at $s = 7.3136$. Higher roots are disregarded on physical grounds, as discussed in Section 27.4. Recalling $s = 2\,Sh_D$, it is determined that the Sherwood number for mass transfer from the pipe wall is

$$
Sh_D = 3.6568. \tag{28-91}
$$

The rate at which contaminants are leached into the flow over a length L of the pipe is given by $(\pi R^2 v_m)[c_m(L) - c_m^i]$. With a Sherwood number of $Sh_D = 3.6568$, Eq. (28-81) can be evaluated for $c_m(L)$. It would be convenient to suppose that $c_m^i = 0$. However, if this were the case, the inlet condition of the pipe could not be fully developed. In such a case, one expects the Sherwood number to be initially higher for the entrance region, and this would be a source of error for any estimates based on the assumption of fully developed conditions everywhere in the pipe.

Using the fact that $a_0 = \chi(0) = 1 - c(0)/c_w$, the local solution for $c(r)$ can be expressed from $\chi = 1 - c/c_w$, and is written as

$$
\frac{c_w - c(r)}{c_w - c(r=0)} = \left(d_0 + d_1(r/R)^2 + \sum_{i=2}^{\infty} d_i(r/R)^{2i}\right) \tag{28-92}
$$

with

$$
d_0 = 1, \;\; d_1 = -Sh_D/2, \tag{28-93}
$$

$$d_i = Sh_D \frac{d_{i-2} - d_{i-1}}{2i^2}, \quad \text{and} \quad Sh_D = 3.6568. \tag{28-94}$$

To reconstruct the full concentration profile in the pipe as it changes with downstream position requires relating the local centerline concentration $c(r=0)$ to the local mean concentration value c_m. Once this relation is established, the full concentration profile in the pipe $c(r,z)$ can be constructed from Eqs. (28-81) and (28-92), as has been left as an exercise in Problem 28-2.

28.4 PROBLEMS

28-1 Demonstrate that Eq. (28-65) is the cold side temperature field in the heat exchanger illustrated in Figure 28-7. Show that the Nusselt number is $Nu_D^c = 4.86$ for this configuration of heat transfer.

28-2 Consider the problem illustrated in Figure 28-9 of a contaminant leaching into a fully developed pipe flow. Show that, for fully developed conditions, the concentration distribution in the pipe is governed by Eq. (28-82). Determine the full concentration profile in the pipe $c(r,z)$, assuming that $c(r=0, z=0) = 0$.

28-3 Consider fully developed Poiseuille flow through a pipe having an isothermal wall temperature T_s. Solve the heat equation by the fourth-order Runge-Kutta method to determine the Nusselt number Nu_D for fully developed heat transfer. Make the problem dimensionless by letting

$$\theta = \frac{T - T_i}{T_s - T_i} \quad \text{and} \quad \eta = \frac{r}{R},$$

where T_i is an arbitrary "inlet" temperature. Derive an expression for the downstream centerline temperature of the flow in terms of Nu_D, Re_D, and Pr. Plot the dimensionless temperature distribution θ through the pipe at the distances $z/R = 0$, 75, 150, and 300, for the case where $\text{Re}_D = 200$ and $\text{Pr} = 7$.

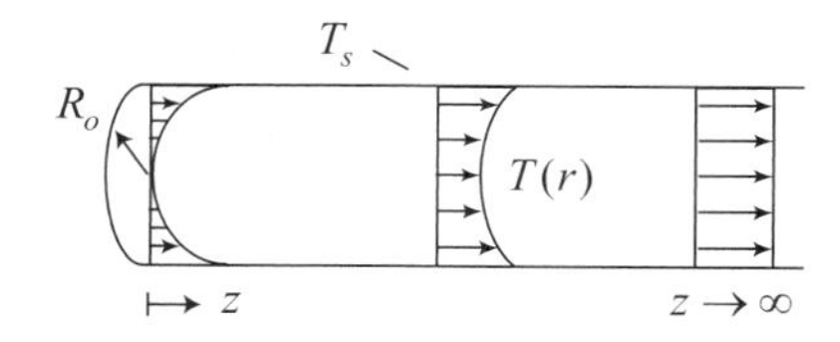

28-4 Surface water flows over a contaminated bed. The average velocity v_m and depth H of the flow are known. It is desired to determine the local contaminate flux to the water using $J^* = h_C(c_w - c_m)$. The surface concentration of the contaminated bed c_w is constant; however, the mean concentration of contaminate in the flow c_m increases downstream. The mean concentration is defined by

$$c_m = \int_0^H v_x c dy / (v_m H) \quad \text{and} \quad v_m = \int_0^H v_x dy / H.$$

Fully developed
Flow
H
Flux
x Contaminated bed

Determine the convection coefficient h_C for fully developed mass transfer conditions. Generalize your analysis by determining the Sherwood number for mass transfer, defined by $Sh = h_C H/Đ_C$, where H is depth of the flow and $Đ_C$ is the contaminate diffusivity in water. Plot the concentration c/c_w as a function of y at the beginning of the fully developed region. Derive an expression for the mean local concentration c_m as a function of distance downstream from the beginning of the fully developed region. Derive an expression for the distance downstream at which the water is "fully contaminated." Define fully contaminated to mean $c_m/c_w = 0.99$.

28-5 Consider the problem of a counterflow heat exchanger, as illustrated. Equal mass flow rates of hot and cold fluids flow in two parallel channels that permit heat transfer between the two fluids. Assume that the flow and temperature fields are fully developed everywhere in the heat exchanger.

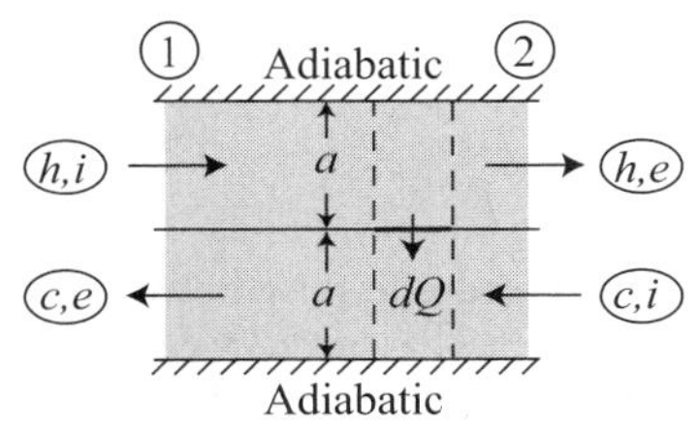

Sketch the mean hot fluid temperature and mean cold fluid temperature as a function of distance along the heat exchanger. Why is the heat flux between the two fluids constant with respect to distance along the heat exchanger? Solve the fully developed heat equation to demonstrate that the Nusselt number for heat transfer between the dividing wall and either fluid is $Nu_{D,H} = 70/13$. Show that the temperature distributions in the heat exchanger are given by

$$\frac{T^c(x,y) - [T^h_m(0) + T^c_m(0)]/2}{[T^h_m(0) - T^c_m(0)]/2} = Nu_{D,H}\left[\frac{x/a}{\text{PrRe}_D} - \frac{1}{4}\left(1 - 2\left(\frac{y}{a}\right)^3 + \left(\frac{y}{a}\right)^4\right)\right] \quad \text{(cold side)}$$

and

$$\frac{T^h(x,y) - [T^h_m(0) + T^c_m(0)]/2}{[T^h_m(0) - T^c_m(0)]/2} = Nu_{D,H}\left[\frac{x/a}{\text{PrRe}_D} + \frac{1}{4}\left(1 - 2\left(\frac{2a-y}{a}\right)^3 + \left(\frac{2a-y}{a}\right)^4\right)\right] \quad \text{(hot side)}$$

where y is a measure of distance from the lower adiabatic wall, and $T^h_m(0)$ and $T^c_m(0)$ are the mean fluid temperatures at $x = 0$ on the hot and cold sides of the heat exchanger.

28-6 Demonstrate that the equation for fully developed heat transfer in an annulus is

$$\frac{v_z}{v_m}\frac{(1-\theta)Nu_D}{(1+1/\eta_i)(1-\eta_i)^2} = \frac{1}{\eta}\frac{\partial}{\partial \eta}\left(\eta \frac{\partial \theta}{\partial \eta}\right)$$

where

$$\theta = \frac{T}{T_s} \quad \text{and} \quad \eta = \frac{r}{R_o}.$$

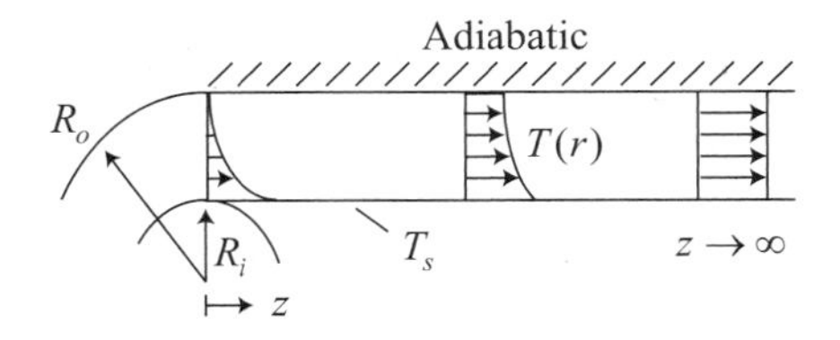

Derive an expression for the velocity profile v_z/v_m for Poiseuille flow. Numerically solve the heat equation for the annulus flow, and determine the Nu_D for heat transfer with a constant inner wall temperature. Find Nu_D for $R_i/R_o = 0.25$, 0.5, and 0.75. For $R_i/R_o = 0.5$, plot the dimensionless temperature profile θ.

28-7 Solve the equation for fully developed heat transfer in an annulus Poiseuille flow when the inner surface is adiabatic. Determine the Nu_D for heat transfer with a constant outer wall temperature for $R_i/R_o = 0.25$, 0.5, and 0.75. For $R_i/R_o = 0.5$, plot the dimensionless temperature profile θ.

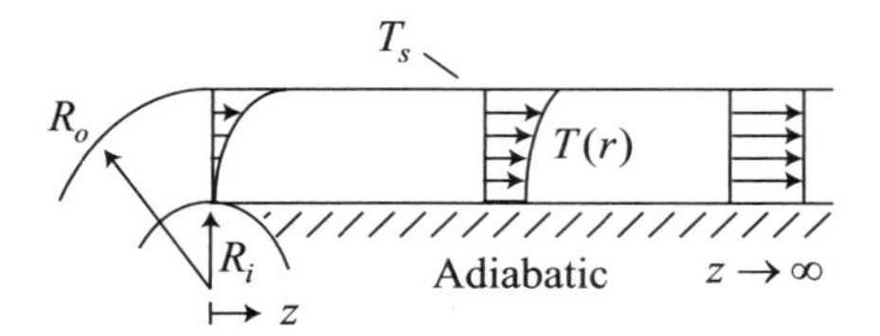

28-8 Consider Poiseuille flow between parallel surfaces. The bottom surface is adiabatic and the top surface has a convective heat flux from a hot overlying fluid at T_∞. The top surface temperature T_s changes with downstream position, and $T_m(x) < T_s(x) < T_\infty$.

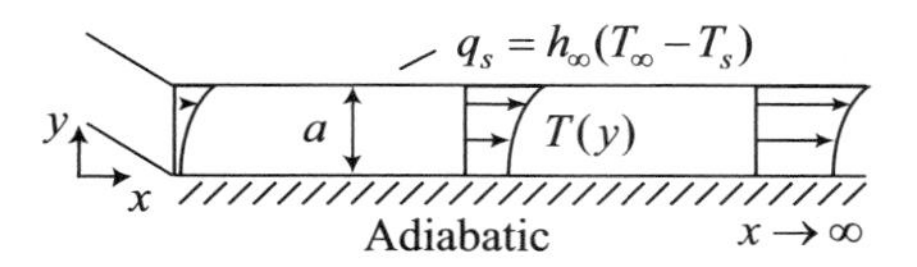

Show that for conditions of fully developed heat transfer,

$$\frac{\partial T}{\partial x} = \frac{\dfrac{T_s - T}{T_s - T_m} + \dfrac{h}{h_\infty}}{1 + h/h_\infty} \frac{\partial T_m}{\partial x}$$

where h is the unknown heat transfer coefficient between the nonadiabatic wall and the flow, and T_m is the mean temperature of the flow. Using the dimensionless variables

$$\theta = \frac{T_s - T}{T_s - T_m}, \quad \eta = \frac{y}{a}, \quad Nu_D = \frac{h(2a)}{k}, \quad \text{and} \quad Bi_\infty = \frac{h_\infty a}{k},$$

derive the heat equation for fully developed heat transfer in the flow:

$$\left(\frac{2Bi_\infty}{Nu_D} + 1\right) \frac{\partial^2 \theta}{\partial \eta^2} + 6Bi_\infty \eta(1 - \eta)\theta + 3Nu_D \eta(1 - \eta) = 0.$$

What conditions are imposed on the solution to this heat equation? Solve the heat equation and plot Nu_D as a function of Bi_∞ for $0.001 \leq Bi_\infty \leq 1000$. What is the significance of the limiting values of Nu_D for high and low Bi_∞?

Chapter 29

Influence of Temperature-Dependent Properties

29.1 Temperature-Dependent Conductivity in a Solid

29.2 Temperature-Dependent Diffusivity in Internal Convection

29.3 Temperature-Dependent Gas Properties in Boundary Layer Flow

29.4 Problems

This chapter addresses a departure from the linear diffusion laws treated thus far. Specifically, temperature-dependent diffusivities will be considered. Fourier's law becomes nonlinear when the heat flux depends on the product of a temperature-dependent conductivity and the temperature gradient: $q_j = k(T)\partial_j T$. Problems illustrating additional temperature dependencies of the fluid, such as viscosity and density in viscous flows, will also be considered. In addition to being an interesting and important topic in its own right, the following treatment provides a useful transition into the topic of highly nonlinear transport that follows in chapters on turbulent flow.

29.1 TEMPERATURE-DEPENDENT CONDUCTIVITY IN A SOLID

Consider the problem of uniform heat generation in a solid slab of thickness L, as illustrated in Figure 29-1. One surface of the slab is insulated, while the opposing surface experiences convective heat transfer to a fluid at temperature T_∞. The solid has a constant density ρ and specific heat C_p, but a linear temperature-dependent thermal conductivity, such that

$$k = k_o[1 + a(T - T_o)]. \tag{29-1}$$

Before the onset of heat generation in the slab, the solid is initially at a temperature $T_o = T_\infty$. It is desired to find the steady-state temperature distribution in the slab. Heat transfer in the solid is governed by the steady-state diffusion equation

$$\underbrace{\partial_o T}_{=0} + \underbrace{v_j \partial_j T}_{=0} = \partial_j(\alpha\, \partial_j T) + q_{gen} \tag{29-2}$$

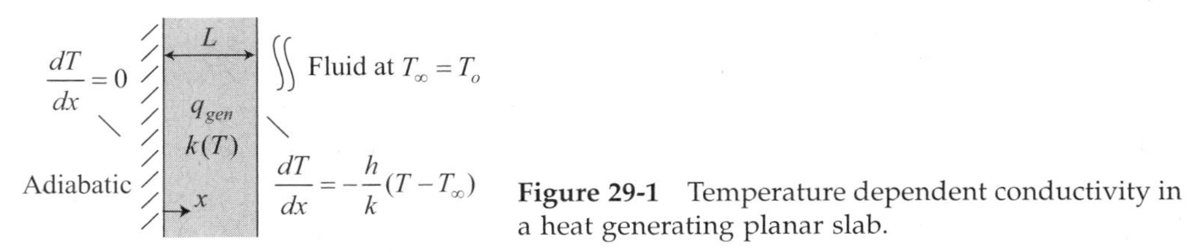

Figure 29-1 Temperature dependent conductivity in a heat generating planar slab.

where q_{gen} is the constant volumetric rate of heat generation. For one-dimensional heat transfer, the governing equation becomes

$$\frac{\partial}{\partial x}\left\{k_o[1+a(T-T_o)]\frac{\partial T}{\partial x}\right\}+q_{gen}=0, \tag{29-3}$$

which is subject to the adiabatic and convective boundary conditions

$$dT/dx|_{x=0}=0 \quad \text{and} \quad -kdT/dx|_{x=L}=h[T(L)-T_o]. \tag{29-4}$$

Defining the dimensionless variables

$$\eta=\frac{x}{L},\ \theta=\frac{T-T_o}{q_{gen}/k_o}, \quad \text{and} \quad \varepsilon=\frac{aq_{gen}L^2}{k_o}, \tag{29-5}$$

the heat equation becomes

$$\frac{\partial}{\partial \eta}\left[(1+\varepsilon\theta)\frac{\partial \theta}{\partial \eta}\right]+1=0 \tag{29-6}$$

and the boundary conditions become

$$d\theta/d\eta\big|_{\eta=0}=0 \quad \text{and} \quad d\theta/d\eta\big|_{\eta=1}=-Bi\theta(1). \tag{29-7}$$

The dimensionless *Biot number* $Bi = hL/k$, appearing in the convective boundary condition, is a measure of the relative ease of heat convection into the fluid, as measured by h, compared with the thermal conductance of the slab k/L.

29.1.1 Solution by Regular Perturbation

Equation (29-6) is nonlinear, as a consequence of the temperature dependent conductivity $k(T)$. Nonlinear equations are generally difficult to solve analytically. Although this problem can be readily solved by numerical integration, before proceeding with this approach, an analytic solution suitable for small values of ε is pursued. The method to be used is called *regular perturbation* analysis [1, 2]. The method proceeds with the assumption that the solution can be written in the form

$$\theta=\theta_0\varepsilon^0+\theta_1\varepsilon^1+\theta_2\varepsilon^2+\cdots=\sum_{n=0}^{\infty}\theta_n\varepsilon^n. \tag{29-8}$$

Notice that for $\varepsilon = 0$, the governing equation (29-6) is linear and has a solution equal to the first term θ_0 of the series solution for the nonlinear problem. Unfortunately, the nonlinear nature of the problem does not lend itself to formulating a recursion relation between terms in the series. Therefore, each term in the solution $\theta_0, \theta_1, \theta_2, \ldots$ must be laboriously determined, making evaluation of a large number of terms impractical. However, if $\varepsilon \ll 1$, the series should converge rapidly, and a reasonably good approximate solution can be had by truncating the series after a few terms:

$$\theta=\theta_0\varepsilon^0+\theta_1\varepsilon^1+\theta_2\varepsilon^2+O(\varepsilon^3). \tag{29-9}$$

The expression $O(\varepsilon^3)$ represents the order of magnitude of the truncated terms. The truncated series may be substituted into the governing equation:

$$\frac{\partial}{\partial\eta}\left[\left(1+\varepsilon\{\theta_0\varepsilon^0+\theta_1\varepsilon^1+\theta_2\varepsilon^2+O(\varepsilon^3)\}\right)\frac{\partial}{\partial\eta}\{\theta_0\varepsilon^0+\theta_1\varepsilon^1+\theta_2\varepsilon^2+O(\varepsilon^3)\}\right]+1=0, \quad (29\text{-}10)$$

which can be reorganized with terms in ascending powers of ε:

$$\frac{\partial}{\partial\eta}\left[\frac{\partial\theta_0}{\partial\eta}\varepsilon^0+\left(\frac{\partial\theta_1}{\partial\eta}+\theta_0\frac{\partial\theta_0}{\partial\eta}\right)\varepsilon^1+\left(\frac{\partial\theta_2}{\partial\eta}+\theta_0\frac{\partial\theta_1}{\partial\eta}+\theta_1\frac{\partial\theta_0}{\partial\eta}\right)\varepsilon^2+O(\varepsilon^3)\right]+1=0. \quad (29\text{-}11)$$

Substituting the truncated series into the boundary conditions as well yields

$$\eta=0: \quad \frac{d}{d\eta}\left[\theta_0\varepsilon^0+\theta_1\varepsilon^1+\theta_2\varepsilon^2+O(\varepsilon^3)\right]=0 \quad (29\text{-}12)$$

$$\eta=1: \quad \frac{d}{d\eta}\left[\theta_0\varepsilon^0+\theta_1\varepsilon^1+\theta_2\varepsilon^2+O(\varepsilon^3)\right]=-Bi\left[\theta_0\varepsilon^0+\theta_1\varepsilon^1+\theta_2\varepsilon^2+O(\varepsilon^3)\right]. \quad (29\text{-}13)$$

The lowest order solution θ_0 is determined by neglecting all terms of $O(\varepsilon^1)$ and higher in the governing equation and boundary conditions. For determining θ_0, the governing equation (29-11) becomes

$$\frac{d}{d\eta}\left[\frac{\partial\theta_0}{\partial\eta}\right]+1=0, \quad (29\text{-}14)$$

while the boundary conditions given by Eqs. (29-12) and (29-13) become

$$\eta=0: \quad \frac{d\theta_0}{d\eta}=0 \quad (29\text{-}15)$$

$$\eta=1: \quad \frac{d\theta_0}{d\eta}=-Bi\,\theta_0. \quad (29\text{-}16)$$

Notice that the governing equation and boundary conditions for the θ_0 problem are linear, such that the solution can readily be determined:

$$\theta_0=\frac{1}{Bi}+\frac{1-\eta^2}{2}. \quad (29\text{-}17)$$

Next, θ_1 is found by neglecting all terms of $O(\varepsilon^2)$ and higher in the governing equation (29-11) and boundary conditions (29-12) and (29-13). The governing equation becomes

$$\frac{\partial}{\partial\eta}\left[\frac{\partial\theta_0}{\partial\eta}+\left(\theta_0\frac{\partial\theta_0}{\partial\eta}+\frac{\partial\theta_1}{\partial\eta}\right)\varepsilon\right]+1=0, \quad (29\text{-}18)$$

into which the solution for θ_0 can be substituted:

$$\frac{\partial}{\partial\eta}\left[-\eta+\left(\frac{-\eta}{Bi}-\eta\frac{1-\eta^2}{2}+\frac{\partial\theta_1}{\partial\eta}\right)\varepsilon\right]+1=0. \quad (29\text{-}19)$$

The boundary conditions become

$$\eta=0: \quad \underbrace{d\theta_0/d\eta}_{=0}+\frac{d\theta_1}{d\eta}\varepsilon=0 \quad (29\text{-}20)$$

$$\eta = 1: \quad \underbrace{d\theta_0/d\eta + Bi\theta_0}_{=0} + \frac{d\theta_1}{d\eta}\varepsilon = -Bi\theta_1\varepsilon \tag{29-21}$$

from which the θ_0 boundary conditions are removed. Notice that the governing equation and boundary conditions for the θ_1 problem are all linear. Integrating the governing equation for θ_1 once yields

$$\left(\frac{-\eta}{Bi} - \eta\frac{1-\eta^2}{2} + \frac{\partial\theta_1}{\partial\eta}\right)\varepsilon = C_1. \tag{29-22}$$

To satisfy the boundary condition at $\eta = 0$, $d\theta_1/d\eta\big|_{\eta=0} = 0$, requires that $C_1 = 0$. Integrating the governing equation for θ_1 a second time yields

$$\theta_1 = \frac{C_2}{\varepsilon} + \left(\frac{1}{Bi} + \frac{1}{2}\right)\frac{\eta^2}{2} - \frac{\eta^4}{8}. \tag{29-23}$$

To satisfy the boundary condition at $\eta = 1$, $d\theta_1/d\eta\big|_{\eta=1} = -Bi\,\theta_1(1)$, requires

$$\frac{C_2}{\varepsilon} = -\frac{1}{Bi^2} - \frac{1}{2Bi} - \frac{1}{8}. \tag{29-24}$$

Therefore, the solution for θ_1 becomes

$$\theta_1 = -\frac{1}{Bi^2} - \frac{1}{2Bi} - \frac{1}{8} + \left(\frac{1}{Bi} + \frac{1}{2}\right)\frac{\eta^2}{2} - \frac{\eta^4}{8}. \tag{29-25}$$

This process can be continued to find θ_2. However, if a solution of order $\theta \cong \theta_0 + \theta_1\varepsilon + O(\varepsilon^2)$ is satisfactory, the results for θ_0 and θ_1 may be combined for

$$\theta = \theta_0 + \theta_1\varepsilon + O(\varepsilon^2) \cong Bi^{-1} + \frac{1}{2}(1-\eta^2)\frac{1}{Bi^2} - \frac{1}{2Bi} - \frac{1}{8} + \frac{1}{2}\eta^2\left(\frac{1}{Bi} + \frac{1}{2}\right) - \frac{1}{8}\eta^4. \tag{29-26}$$

29.1.2 Numerical Solution by Runge-Kutta Integration

The governing equation (29-6) can also be solved numerically using the fourth-order Runge-Kutta method, as discussed in Chapter 23. The second-order governing equation (29-6) can be expressed as two first-order equations for θ and θ', which are integrated forward in space from the initial conditions:

$$\frac{d\theta'}{d\eta} = -\frac{1+\varepsilon(\theta')^2}{1+\varepsilon\,\theta} \quad \text{with } \theta'(0) = 0 \tag{29-27}$$

$$\frac{d\theta}{d\eta} = \theta' \quad \text{with } \theta(0) = \underline{?}, \quad \text{leading to } \theta'(1)/\theta(1) = -Bi. \tag{29-28}$$

Since the initial condition for θ is unknown, the shooting method is used to guess the value of $\theta(0)$ that satisfies the final condition $\theta'(1)/\theta(1) = -Bi$.

Figure 29-2 plots the exact numerical result against the approximate analytic solution given by Eq. (29-26), for $Bi = 1$ and $\varepsilon = 0.05$. Also shown is the θ_0 solution, which

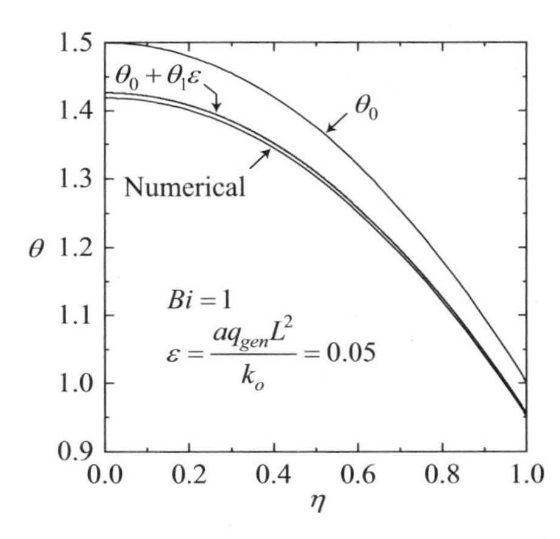

Figure 29-2 Comparison of analytic and exact temperature profiles in heat generating slab.

corresponds to the case when thermal conductivity has no temperature dependence. The analytic solution could be improved further by seeking the next higher-order solution corresponding to θ_2.

Any analytic result will be limited to problems for which $\varepsilon = aq_{gen}L^2/k_o \ll 1$, if the series solution is to converge rapidly enough to be practical. Although the analytic solution is only approximate, once established it may be evaluated readily for any value of *Bi*. In contrast, a graphically reported numerical solution can be presented for only a limited number of cases.

29.2 TEMPERATURE-DEPENDENT DIFFUSIVITY IN INTERNAL CONVECTION

In fully developed flows with constant fluid properties, the advection terms may vanish with exact statements, such as $\partial v_x/\partial x = 0$ for the case of momentum transport (see Section 28.1). When transport is influenced by variable fluids properties, fully developed conditions may no longer be characterized as precisely. However, it is often satisfactory to represent fully developed conditions in a more approximate sense, such as $\partial v_x/\partial x \approx 0$ for momentum transport. This is similar to the boundary layer approximation discussed in Chapter 24, where streamwise changes are gradual in contrast to changes that occur across the flow.

In this section, temperature-dependent diffusivities are considered for a fully developed Poiseuille flow between isothermal parallel plates, as illustrated in Figure 29-3. This geometry of flow was discussed for constant fluid properties in Chapter 28. Now, because of temperature effects, it is anticipated that changes in the downstream viscosity will cause gradual changes in the v_x velocity profile. Nevertheless, fully developed conditions will imply that

$$\partial v_x/\partial x \approx 0 \quad \text{and} \quad v_y \approx 0. \tag{29-29}$$

For this reason, advection terms in the momentum equation are neglected, and the resulting equation becomes

$$0 = \frac{\partial}{\partial y}\left[\frac{\mu(T)}{\rho}\frac{\partial v_x}{\partial y}\right] - \frac{1}{\rho}\frac{\partial P}{\partial x}. \tag{29-30}$$

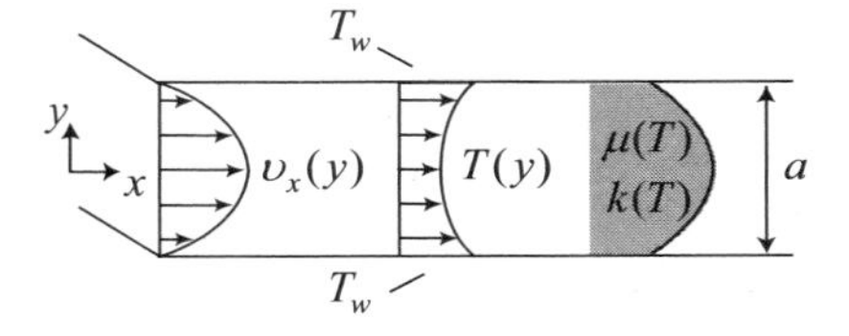

Figure 29-3 Temperature-dependent diffusion in Poiseuille flow between isothermal plates.

The pressure gradient dP/dx acting on a volume of flow between the plates is balanced by the viscous stresses transmitted to the walls. Consequently,

$$\frac{1}{\rho}\frac{dP}{dx} = \frac{2}{a}\frac{\mu_w}{\rho}\frac{\partial v_x}{\partial y}\bigg|_{y=a/2} \tag{29-31}$$

where μ_w is the fluid viscosity at the wall temperature. The momentum equation (29-30) can be integrated once and written in terms of the dimensionless variables $\eta = y/a$ and $u = v_x/v_m$, yielding

$$\frac{\mu(T)}{\mu_w}u' = 2\eta u'(1/2). \tag{29-32}$$

The dimensionless velocity gradient at the wall $u'(1/2)$ in Eq. (29-32) is not arbitrary. The value of $u'(1/2)$ is dictated by the normalization requirement that

$$\int_{-1/2}^{1/2} u \, d\eta = 1, \tag{29-33}$$

which is a consequence of the nondimensionalization of v_x by the mean flow velocity

$$v_m = \int_{-a/2}^{a/2} v_x dy/a. \tag{29-34}$$

For example, if the fluid viscosity were constant, $\mu(T)/\mu_w = 1$, then normalization of the velocity profile dictates that $u'(1/2) = -6$.

Equation (29-32) can be integrated for the velocity profile using the boundary condition $u(1/2) = 0$, as long as the viscosity function $\mu(T)/\mu_w$ is known. To establish this function requires determining the temperature profile across the flow. For fully developed hydrodynamic conditions, where $v_y \approx 0$, the heat equation for an internal flow is given by

$$v_x\frac{\partial T}{\partial x} = \frac{\partial}{\partial y}\left[\frac{k(T)}{\rho C_p}\frac{\partial T}{\partial y}\right]. \tag{29-35}$$

Notice that the heat equation is nonlinear because of the temperature dependence of thermal conductivity. Fully developed thermal conditions imply that

$$\frac{\partial}{\partial x}\left[\frac{T_w - T(x,y)}{T_w - T_m(x)}\right] \approx 0. \tag{29-36}$$

Therefore, for a constant wall temperature T_w,

$$\frac{\partial T}{\partial x} \approx \frac{T_w - T}{T_w - T_m}\frac{dT_m}{dx}. \tag{29-37}$$

Furthermore, for the parallel plate geometry, an energy balance over a differential downstream distance dx requires that

$$v_m \frac{dT_m}{dx} = \frac{2\alpha_w}{a\,k_w} h(T_w - T_m). \tag{29-38}$$

Equations (29-37) and (29-38) combined with Eq. (29-35) permit the heat equation to be written in the form

$$\frac{v_x}{v_m}(T_w - T)\frac{2\alpha_w}{a\,k_w} h = \frac{\partial}{\partial y}\left[\frac{k(T)}{\rho C_p}\frac{\partial T}{\partial y}\right]. \tag{29-39}$$

A dimensionless temperature variable is defined as

$$\theta = \frac{T - T_i}{T_w - T_i}, \tag{29-40}$$

where T_i is the inlet temperature of the fluid before heat is added. With this temperature definition, the heat equation may be nondimensionalized using $\eta = y/a$ and $u = v_x/v_m$ to yield

$$u(1-\theta)\, Nu_D = \frac{\partial}{\partial \eta}\left[\frac{k(\theta)}{k_w}\frac{\partial \theta}{\partial \eta}\right], \tag{29-41}$$

in which $Nu_D = h(2a)/k_w$ is the Nusselt number for heat transfer between the flow and plates. Although $\theta'(0) = 0$ is required, $\theta(0)$ could be any number between 0 and 1 in the solution to Eq. (29-41). However, the correct value of Nu_D in Eq. (29-41) is identified by satisfying the boundary condition $\theta(1/2) = 1$.

To explore a particular situation, assume that the diffusion coefficients for this flow can be described by a linear temperature dependence over the range T_i to T_w, such that

$$\frac{\mu(\theta)}{\mu_w} = \frac{1 + \mathrm{M}\theta}{1 + \mathrm{M}} \quad \text{where } \mathrm{M} = \frac{\mu(T_w) - \mu(T_i)}{\mu(T_i)} \tag{29-42}$$

and

$$\frac{k(\theta)}{k_w} = \frac{1 + \mathrm{K}\theta}{1 + \mathrm{K}} \quad \text{where } \mathrm{K} = \frac{k(T_w) - k(T_i)}{k(T_i)}. \tag{29-43}$$

With these expressions for the temperature dependent diffusion coefficients, the governing equations (29-32) and (29-41) become

$$\text{Momentum:} \quad u' = \frac{1 + \mathrm{M}}{1 + \mathrm{M}\theta} 2\eta u'(1/2) \tag{29-44}$$

$$\text{with} \quad u(1/2) = 0 \quad \text{and} \quad u'(1/2) \quad \text{leading to} \quad \int_{-1/2}^{1/2} u\, d\eta = 1 \tag{29-45}$$

and

$$\text{Heat:} \quad \theta'' = \frac{u(1-\theta)(1+\mathrm{K})Nu_D - \mathrm{K}(\theta')^2}{1 + \mathrm{K}\theta} \tag{29-46}$$

$$\text{given} \quad \theta(0), \quad \text{with} \quad \theta'(0) = 0 \quad \text{and} \quad Nu_D \quad \text{leading to} \quad \theta(1/2) = 1. \tag{29-47}$$

Solutions can be found by solving the coupled equations for any centerline temperature $\theta(0)$. However, because the velocity profile depends on the temperature profile, each choice of centerline temperature $\theta(0)$ will yield a different solution to the coupled momentum and heat equations. As $\theta(0) \to 1$, the solution for the Nusselt number approaches that describing a fluid with constant diffusion coefficients, where $Nu_D = 7.541$, as determined in Section 28.2.1.

Code 29-1 implements a solution for $u(\eta)$ and $\theta(\eta)$ by alternately integrating the two governing equations using the fourth-order Runge-Kutta method, as discussed in Chapter 23. The u equation is solved using a previous solution of $\theta(\eta)$, and the θ equation is solved using a previous solution of $u(\eta)$. The process is repeated until neither solution changes, which is assessed by looking at the Nu_D result after each solution of the heat equation.

Figure 29-4 plots the solution above the centerline of the flow $0 \le \eta \le 1/2$, when the centerline temperature is $\theta(0) = 0$. Contrasting cases of constant diffusivities ($M = K = 0$) and variable diffusivities, with $M = K = -1/2$, are illustrated. The latter case corresponds to a factor of two decrease in both the viscosity and conductivity over the temperature range from T_i to T_w. Figure 29-4 shows that lowering the viscosity near the heated walls causes a slight flattening and broadening of the velocity profile, while lowering the thermal conductivity near the heated walls inhibits penetration of heat to the center of the flow.

Description of the downstream development of the flow is complicated by the fact that the Nusselt number Nu_D is not constant. The Nusselt number changes as a result of downstream developments in the velocity and temperature fields brought on by the temperature-dependent viscosity and thermal conductivity. In terms of the local value of the Nusselt number, a differential relation between the mean temperature and the downstream distance can be found from the overall energy balance given by Eq. (29-38), or

$$\frac{dT_m}{dx} = \frac{h(T_w - T_m)}{v_m a \rho C_p} = \frac{h(2a)}{k_w} \frac{\alpha_w}{\nu_w} \frac{\nu_w}{v_m(2a)} \frac{T_w - T_m}{a} = \frac{Nu_D}{\mathrm{PrRe}_D} \frac{T_w - T_m}{a}. \tag{29-48}$$

In dimensionless terms, this differential relation can be expressed as

$$\frac{d\theta_m}{dx} = \frac{Nu_D}{\mathrm{PrRe}_D} \frac{1 - \theta_m}{a} \quad \text{or} \quad \frac{dx}{a} = \frac{\mathrm{PrRe}_D}{Nu_D} \frac{d\theta_m}{1 - \theta_m}. \tag{29-49}$$

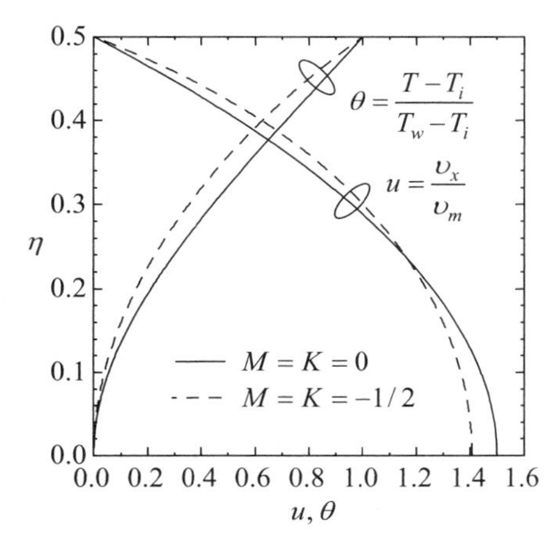

Figure 29-4 Effect of temperature dependent diffusion on flow illustrated in Figure 29-3.

Code 29-1 Temperature-dependent diffusivities in Poiseuille flow

```
#include <stdio.h>
#include <stdlib.h>
#include <math.h>


double ddT(double u,double NuD,double eta,double T,double dT) {
   double K= -.50;
   return  ( u*(1.-T)*(1.+K)*NuD - K*dT*dT ) / ( 1.+K*T );
}


double solve_T(double T0,int N,double *u,double *T,double *dT) {
   int n,iter=0;
   double del_eta=0.5/(N-1);
   double NuD,K1T,K2T,K3T,K4T,K1dT,K2dT,K3dT,K4dT;


   double NuD_lo=0.0;
   double NuD_hi=100.0;
   T[0]= T0;     /* initial conditions */
   dT[0]=.0;     /* initial conditions */
   do {
      NuD=(NuD_lo+NuD_hi)/2.0;
      for (n=0;n<N-1;++n) {
         K1T=   del_eta*dT[n];
         K1dT=  del_eta*ddT(u[n],NuD,n*del_eta,T[n],dT[n]);
         K2T=   del_eta*(dT[n]+0.5*K1dT);
         K2dT=  del_eta*ddT((u[n]+u[n+1])/2.,NuD,
                        (n+.5)*del_eta,T[n]+0.5*K1T,dT[n]+0.5*K1dT);
         K3T=   del_eta*(dT[n]+0.5*K2dT);
         K3dT=  del_eta*ddT((u[n]+u[n+1])/2.,NuD,
                        (n+.5)*del_eta,T[n]+0.5*K2T,dT[n]+0.5*K2dT);
         K4T=   del_eta*(dT[n]+K3dT);
         K4dT=  del_eta*ddT(u[n+1],NuD,(n+1.)*del_eta,
                        T[n]+K3T,dT[n]+K3dT);
         T[n+1]=  T[n]+(K1T+ 2.*K2T+ 2.*K3T+ K4T )/6.;
         dT[n+1]=dT[n]+(K1dT+2.*K2dT+2.*K3dT+K4dT)/6.;

         if (T[n+1]>1. || T[n+1]<0.) {
            T[N-1]=T[n+1];
            break;
         }
      }
      if (T[N-1]<=1.) NuD_lo=NuD;
      else if (T[N-1]>1.) NuD_hi=NuD;
   } while (++iter < 500 && (NuD_hi-NuD_lo)/NuD>1.0e-6);

   if (fabs(T[N-1]-1.) > 1.0e-5) {
      printf("\nSoln. for NuD failed T[N-1]=%f\n",T[N-1]);
      exit(1);
   }
   return (NuD_lo+NuD_hi)/2.0;
}
```

```
double findTm(int N,double *u,double *T) {
   int n;
   double del_eta=0.5/(N-1);
   double sum=(del_eta/2.)*(u[0]*T[0]+u[N-1]*T[N-1]);
   for (n=1;n<N-1;++n) sum+=del_eta*u[n]*T[n];
   return 2.*sum;
}
inline double du(double eta,double T) {
   double M= -.50;
   return 2.*eta*(1.+M)/(1.+M*T);
}
double solve_u(int N,double *u,double *T) {
   int n;
   double del_eta=0.5/(N-1),K1u,K2u,K3u,K4u,iu,K1iu,K2iu,K3iu,K4iu;
   u[N-1]=iu=.0;                                /*initial condition*/
   for (n=N-1;n>0;--n) {                        /*reverse integration*/
      K1u=-del_eta*du(n*del_eta,T[n]);
      K2u=-del_eta*du((n-.5)*del_eta,(T[n]+T[n-1])/2.);
      K3u=-del_eta*du((n-.5)*del_eta,(T[n]+T[n-1])/2.);
      K4u=-del_eta*du((n- 1)*del_eta,T[n-1]);
      K1iu=del_eta*u[n];
      K2iu=del_eta*(u[n]+.5*K1u);
      K3iu=del_eta*(u[n]+.5*K2u);
      K4iu=del_eta*(u[n]+K3u);
      u[n-1]=u[n]+(K1u+2.*K2u+2.*K3u+K4u)/6.;
      iu+=(K1iu+2.*K2iu+2.*K3iu+K4iu)/6.;
   }
   iu*=2.;
   for (n=0;n<N;++n) u[n]/=iu;                 /*enforce normalization*/
   return -4./iu;                              /*cfReD*/
}

int main() {
{
   FILE *fp;
   int n,N=201;
   double cfReD,Tm,last_NuD,u[N],T[N],dT[N];
   double NuD=.0,T0=.0;
   for (n=0;n<N;++n) u[n]=T[n]=0.;
   do {
      last_NuD=NuD;
      cfReD=solve_u(N,u,T);    // solve for u & du0
      NuD=solve_T(T0,N,u,T,dT);
   } while (fabs(last_NuD-NuD)/last_NuD > 1.e-6);
   Tm=findTm(N,u,T);
   printf("\nT0=%f Tm=%f cfReD=%f NuD=%f",T0,Tm,cfReD,NuD);
   fp=fopen("out.dat","w");
   for (n=0;n<N;++n)
      fprintf(fp,"%e %e %e\n",.5*n/(N-1),u[n],T[n]);
   fclose(fp);
   return 0;
}
```

The fluid properties required in the definitions of Nu_D, Pr, and Re_D are all evaluated at the wall temperature T_w. The second form of Eq. (29-49) can be numerically integrated when the downstream position is discretized with respect to specific values of the centerline temperature $\theta(0)$. Values of $\theta(0)$ can range downstream from 0 to 1. For each discretized value of $\theta(0)$, the local Nu_D value can be determined from the solutions of $u(\eta)$ and $\theta(\eta)$, and the local mean temperature calculated from

$$\theta_m = 2\int_0^{1/2} u\theta d\eta. \tag{29-50}$$

The downstream locations of these discretized points can be established numerically by integrating Eq. (29-49) using the known values of Nu_D and θ_m. Once these locations are known, the temperature and velocity profiles can be reported as a function of downstream distance. The two-dimensional solution for $u(x, y)$ is constructed by this procedure and plotted in Figure 29-5. Without the effects of variable viscosity, the solution would yield the result $\partial u/\partial x = 0$, which would exhibit iso-u lines all parallel to the x-direction. However, the effect of the downstream temperature development on viscosity causes the observed topography in iso-u lines shown in Figure 29-5. The initial velocity profile $u(x = 0, y)$ is as shown in Figure 29-4 corresponding to the case where $M = K = -1/2$. Since the downstream temperature field becomes uniform (with uniform fluid properties), the downstream velocity profile $u(x \to \infty, y)$ approaches the limit shown in Figure 29-4 for $M = K = 0$.

Because the velocity profile is evolving downstream, viscous stresses transmitted to the walls are not constant. This can be characterized with the coefficient of friction:

$$c_f = \frac{\tau_w}{\rho v_m^2/2} = \frac{-\mu_w(\partial v_x/\partial y)\big|_{y=a/2}}{\rho v_m^2/2} = -4\frac{\nu_w}{v_m(2a)}u'(1/2) = -4\frac{u'(1/2)}{Re_D}. \tag{29-51}$$

The downstream evolution of the coefficient of friction is plotted in Figure 29-6, along with the variable Nusselt number. As the downstream temperature becomes uniform, it is

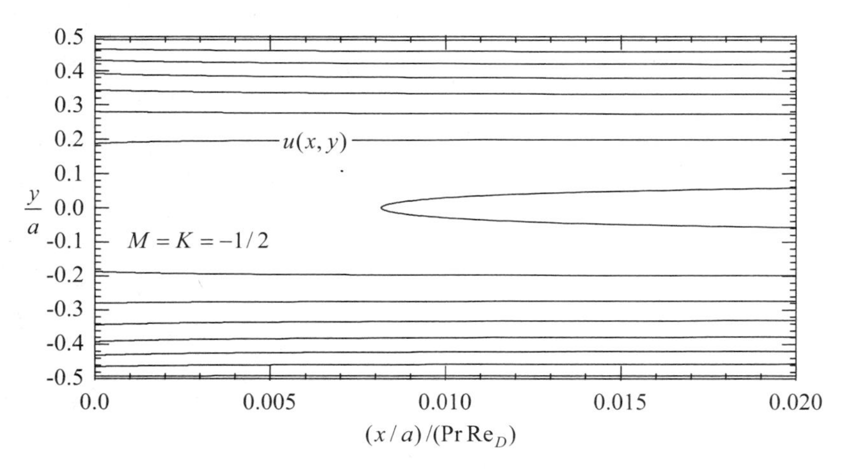

Figure 29-5 Downstream development of iso-velocity lines caused by temperature dependent diffusion. Contour intervals of 0.2 start from the wall, with an additional contour at 1.46.

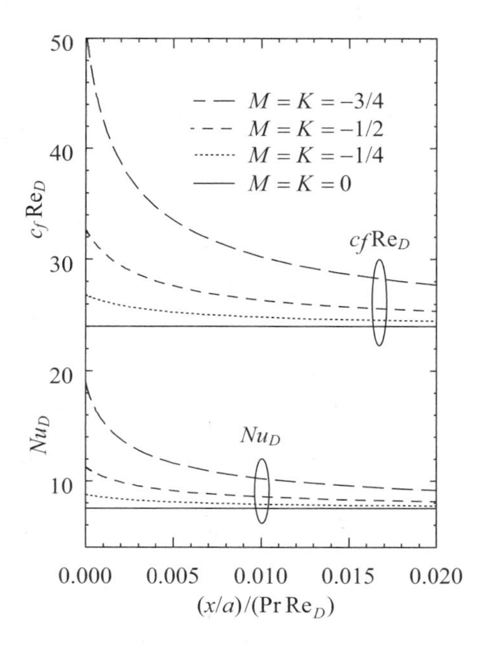

Figure 29-6 Nonconstant transfer coefficients caused by temperature-dependent diffusion.

observed that the coefficients for heat and momentum transfer approach the values corresponding to constant fluid diffusivities: $Nu_D = 7.54$ and $c_f \mathrm{Re}_D = 24.0$.

29.3 TEMPERATURE-DEPENDENT GAS PROPERTIES IN BOUNDARY LAYER FLOW

Consider a steady two-dimensional boundary layer flow over a heated flat plate oriented parallel to the flow. Heat and momentum transport across the boundary layer was considered for constant fluid properties in Chapter 25. In the present section, the effect of temperature-dependent fluid properties is considered. The governing equations can be written for a boundary layer (see Sections 25.1 and 25.3):

$$\text{Momentum:} \quad \rho\left(v_x \frac{\partial v_x}{\partial x} + v_y \frac{\partial v_x}{\partial y}\right) = \frac{\partial}{\partial y}\left[\mu \frac{\partial v_x}{\partial y}\right] \tag{29-52}$$

$$\text{Heat:} \quad \rho C_p\left(v_x \frac{\partial T}{\partial x} + v_y \frac{\partial T}{\partial y}\right) = \frac{\partial}{\partial y}\left[k \frac{\partial T}{\partial y}\right]. \tag{29-53}$$

To put the momentum equation into the present form requires that the plate is oriented parallel to the flow, such that an isobaric (constant P) solution can be sought. Additionally, the diffusion fluxes exhibited within the flow are such that $(-\partial_y M_{yx}) \gg (-\partial_x M_{xx})$ and $M_{yx} \approx -\mu(\partial_y v_x)$. This situation is suggested by the usual boundary layer approximations as long as the flow divergence, caused by the temperature-dependent density change, is small compared with the across-flow gradients ($\partial_y v_x \gg \partial_k v_k$). Notice, however, that μ and ρ cannot be combine as ν in the momentum equation without assuming that either μ or ρ is constant. The present form of the heat equation (29-53) also prohibits combining k and ρC_p into the thermal diffusivity α for the same reason. The heat equation is valid for nonconstant gas properties, including ρ, but it has been assumed that the flow is isobaric and that viscous generation of heat is negligible. The effect of viscous generation of heat is included in Problem 29-3.

A similarity solution to the momentum and heat equations can be formulated for a gas having temperature-dependent properties: ρ, μ, C_p, and k [3]. The variables required to seek this solution are defined by

$$\rho v_x = \rho_\infty \frac{\partial \psi}{\partial y}, \quad \rho v_y = -\rho_\infty \frac{\partial \psi}{\partial x}, \quad f(\eta) = \frac{\psi}{\sqrt{\nu_\infty x U}}, \quad \text{and} \quad \eta = \sqrt{\frac{U}{\nu_\infty x}} \int_0^y \frac{\rho}{\rho_\infty} dy. \tag{29-54}$$

It is easy to verify that the stream function definitions satisfy the steady continuity equation $\partial_i(\rho v_i) = 0$; expanding in x and y, the continuity equation requires

$$\frac{\partial(\rho v_x)}{\partial x} + \frac{\partial(\rho v_y)}{\partial y} = \frac{\partial}{\partial x}\left(\rho_\infty \frac{\partial \psi}{\partial y}\right) + \frac{\partial}{\partial y}\left(-\rho_\infty \frac{\partial \psi}{\partial x}\right) = \rho_\infty \left(\frac{\partial}{\partial x}\frac{\partial \psi}{\partial y} - \frac{\partial}{\partial y}\frac{\partial \psi}{\partial x}\right) = 0, \tag{29-55}$$

which is always satisfied.

To transform the momentum and heat equations, it is observed that

$$\rho v_x = \rho_\infty \frac{\partial \psi}{\partial y} = \rho_\infty \frac{\partial}{\partial y}(\sqrt{\nu_\infty x U}\, f) = \rho_\infty \sqrt{\nu_\infty x U} \frac{\partial f}{\partial y} \quad \text{and} \tag{29-56}$$

$$\rho v_y = -\rho_\infty \frac{\partial \psi}{\partial x} = -\rho_\infty \frac{\partial}{\partial x}(\sqrt{\nu_\infty x U}\, f) = -\frac{\rho_\infty}{2}\sqrt{\frac{\nu_\infty U}{x}} f - \rho_\infty \sqrt{\nu_\infty U x} \frac{\partial f}{\partial x}. \tag{29-57}$$

Using

$$\frac{\partial(\;)}{\partial x} = \frac{\partial \eta}{\partial x}\frac{\partial(\;)}{\partial \eta} = -\frac{\eta}{2x}\frac{\partial(\;)}{\partial \eta} \quad \text{and} \quad \frac{\partial(\;)}{\partial y} = \frac{\partial \eta}{\partial y}\frac{\partial(\;)}{\partial \eta} = \sqrt{\frac{U}{\nu_\infty x}}\frac{\rho}{\rho_\infty}\frac{\partial(\;)}{\partial \eta}, \tag{29-58}$$

one obtains the results

$$v_x = Uf' \tag{29-59}$$

$$v_y = \frac{\rho_\infty}{2\rho}\sqrt{\frac{\nu_\infty U}{x}}[\eta f' - f]. \tag{29-60}$$

Therefore, transforming the boundary layer momentum equation (29-52) gives

$$\underbrace{(\rho U f')\frac{-\eta}{2x}\frac{\partial}{\partial \eta}(Uf')}_{\rho v_x \frac{\partial v_x}{\partial x}} + \underbrace{\frac{\rho_\infty}{2}\sqrt{\frac{\nu_\infty U}{x}}[\eta f' - f]\sqrt{\frac{U}{\nu_\infty x}}\frac{\rho}{\rho_\infty}\frac{\partial}{\partial \eta}(Uf')}_{\rho v_y \frac{\partial v_x}{\partial y}} = \underbrace{\frac{U}{\nu_\infty x}\frac{\rho}{\rho_\infty}\frac{\partial}{\partial \eta}\left(\mu \frac{\rho}{\rho_\infty}\frac{\partial (Uf')}{\partial \eta}\right)}_{\frac{\partial}{\partial y}\left(\mu \frac{\partial v_x}{\partial y}\right)} \tag{29-61}$$

which can be simplified to

$$\frac{\partial}{\partial \eta}\left(\frac{\mu}{\mu_\infty}\frac{\rho}{\rho_\infty} f''\right) + \frac{1}{2} f f'' = 0. \tag{29-62}$$

Applying the transformations to the boundary layer heat equation (29-53) yields

$$\overbrace{\rho C_p(Uf')\frac{-\eta}{2x}\frac{\partial T}{\partial \eta}}^{\rho C_p v_x \frac{\partial T}{\partial x}}+\overbrace{\frac{\rho_\infty C_p}{2}\sqrt{\frac{\nu_\infty U}{x}}[\eta f'-f]\sqrt{\frac{U}{\nu_\infty x}}\frac{\rho}{\rho_\infty}\frac{\partial T}{\partial \eta}}^{\rho C_p v_y \frac{\partial T}{\partial y}}=\overbrace{\frac{U}{\nu_\infty x}\frac{\rho}{\rho_\infty}\frac{\partial}{\partial \eta}\left(k\frac{\rho}{\rho_\infty}\frac{\partial T}{\partial \eta}\right)}^{\frac{\partial}{\partial y}\left(k\frac{\partial T}{\partial y}\right)} \tag{29-63}$$

or

$$\frac{\partial}{\partial \eta}\left(\frac{k}{k_\infty}\frac{\rho}{\rho_\infty}T'\right)+\frac{\Pr}{2}\frac{C_p}{C_{p,\infty}}f\,T'=0, \tag{29-64}$$

where the Prandtl number is based on the gas properties at the temperature far from the heated plate $\Pr = \nu_\infty/\alpha_\infty$. To solve the coupled momentum and heat equations requires that the gas properties dependence on temperature be established. For an isobaric flow, an ideal gas obeys the relation

$$\frac{\rho}{\rho_\infty}=\frac{T_\infty}{T}. \quad (P = const.) \tag{29-65}$$

Simple kinetic theory suggests that ideal gas properties have temperature dependences given by [4]:

$$\frac{C_p}{C_{p,\infty}}=1,\ \frac{k}{k_\infty}=\left(\frac{T}{T_\infty}\right)^{1/2}, \quad \text{and} \quad \frac{\mu}{\mu_\infty}=\left(\frac{T}{T_\infty}\right)^{1/2}. \tag{29-66}$$

Real gases may deviate from this behavior. Adopting as a dimensionless temperature definition

$$\theta=\frac{T}{T_\infty}, \tag{29-67}$$

Eqs. (29-65) through (29-67) may be applied to the momentum (29-62) and heat (29-64) equations for the results

$$\text{Momentum:}\quad f'''+\frac{1}{2}\left(\theta^{1/2}f-\frac{\theta'}{\theta}\right)f''=0 \tag{29-68}$$

$$\text{subject to}\quad f(0)=0,\ f'(0)=0, \quad \text{and} \quad f'(\infty)=1. \tag{29-69}$$

$$\text{Heat:}\quad \theta''+\frac{1}{2}\left(\Pr\,\theta^{1/2}f-\frac{\theta'}{\theta}\right)\theta'=0 \tag{29-70}$$

$$\text{subject to}\quad \theta(0)=T_s/T_\infty \quad \text{and} \quad \theta(\infty)=1. \tag{29-71}$$

Both equations are differential with respect to the similarity variable:

$$\eta = \sqrt{\frac{U}{\nu_\infty x}} \int_0^y \frac{\rho}{\rho_\infty} dy = \sqrt{\frac{U}{\nu_\infty x}} \int_0^y \frac{dy}{\theta}. \tag{29-72}$$

Notice that the solution will depend on two dimensionless parameters: the Prandtl number $\mathrm{Pr} = \nu_\infty/\alpha_\infty$ and the relative plate temperature $\theta(0) = T_s/T_\infty$.

The fourth-order Runge-Kutta method can be used to integrate the coupled governing equations. There are two unknown initial conditions imposed on integration: $f''(0)$ and $\theta'(0)$, for which shooting methods can be employed in the usual manner (see Section 23.4). Code 29-2 solves for $f(\eta)$ and $\theta(\eta)$ by alternately integrating the two governing equations. The f equation is solved using a previous solution of $\theta(\eta)$, and the θ equation is solved using a previous solution of $f(\eta)$. The process is repeated until neither solution changes, as can be assessed by looking at the values of $f''(0)$ and $\theta'(0)$.

Two interesting results that are obtained from the numerical solution are the coefficient of friction c_f for the flow and the Nusselt number Nu for the heat transfer. For the coefficient of friction,

$$c_f = \frac{2}{\rho_\infty U^2} \left(\mu \frac{\partial v_x}{\partial y} \right) \bigg|_{y=0} = 2 \frac{\mu_s}{\mu_\infty} \frac{\rho_s}{\rho_\infty} f''(0) \sqrt{\frac{\nu_\infty}{Ux}} = \frac{2f''(0)}{\sqrt{\theta(0)}} \mathrm{Re}_x^{-1/2}, \tag{29-73}$$

in which $\mu_s/\mu_\infty = (T_s/T_\infty)^{1/2} = \sqrt{\theta(0)}$ and $\rho_s/\rho_\infty = T_\infty/T_s = 1/\theta(0)$ have been used. The local coefficient of drag is seen to scale with distance from the leading edge of the heated plate as $c_f \sim \mathrm{Re}_x^{-1/2}$. The coefficient $2f''(0)/\sqrt{\theta(0)}$ appearing in Eq. (29-73) may be evaluated from the numerical solution and will depend on both the Prandtl number $\mathrm{Pr} = \nu_\infty/\alpha_\infty$ and the relative plate temperature $\theta(0) = T_s/T_\infty$. Figure 29-7 plots $2f''(0)/\sqrt{\theta(0)}$ as a function of Pr for the plate temperatures $T_s/T_\infty = 8$, 4, 2, and the limiting case $T_s/T_\infty \to 1$. Since typical gases have $\mathrm{Pr} \approx 1$, the high and low Prandtl ranges shown in Figure 29-7 are mostly of academic interest. As the surface temperature approaches the gas temperature far from the plate ($T_s/T_\infty \to 1$), the coefficient of drag approaches the relation found in Chapter 25 for flows without temperature-dependent fluid properties: $c_f = 0.664\mathrm{Re}_x^{-1/2}$. Figure 29-7 also shows that the coefficient of drag approaches the constant fluid property value as $\mathrm{Pr} \to \infty$; when the Prandtl number is large enough, most of the momentum boundary layer is at T_∞, and momentum transfer to the surface is less influenced by the temperature-dependent gas properties.

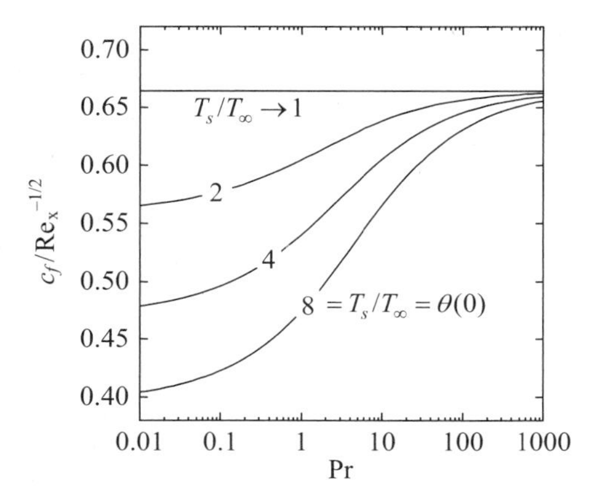

Figure 29-7 Influence of plate temperature and Pr on the coefficient of drag.

Code 29-2 Temperature-dependent gas boundary layer flow

```c
#include <stdio.h>
#include <math.h>
#define _N_ 5000

inline double dddf(double f,double ddf,double T,double dT) {
   return -(sqrt(T)*f-dT/T)*ddf/2.;
}
inline double ddT(double T,double dT,double Pr,double f) {
   return -(Pr*sqrt(T)*f-dT/T)*dT/2.;
}

int main()
{
   int n,iter=0,MAX_iter=20;
   double f[_N_],df[_N_],ddf[_N_],T[_N_],dT[_N_];
   double K1f,K1df,K1ddf,K2f,K2df,K2ddf,K3f,K3df,K3ddf,K4f,K4df,K4ddf;
   double K1T,K1dT,K2T,K2dT,K3T,K3dT,K4T,K4dT;
   double ddf0_hi0,ddf0_hi,ddf0_lo0,ddf0_lo;
   double dT0_hi0,dT0_hi,dT0_lo0,dT0_lo;
   double Pr,eta_inf,del_eta,ddf0_last,dT0_last;
   FILE *fp;

   f[0]=df[0]=0.0;  /* initial conditions */
   for (n=0;n<_N_;++n) {T[n]=1.; dT[n]=0.;}
   ddf0_hi=ddf0_hi0=5.;
   ddf0_lo=ddf0_lo0=0.;
   dT0_hi=dT0_hi0=0.;
   dT0_lo=dT0_lo0=-100.;
   ddf[0]=dT[0]=0.;

   T[0]=2.0;
   Pr=0.7;
   eta_inf=(Pr<.7 ? 70. : 7.);
   del_eta=eta_inf/(_N_-1);
   do {
      ddf0_hi=ddf0_hi0;
      ddf0_lo=ddf0_lo0;
      ddf0_last=ddf[0];
      do {
         ddf[0]=(ddf0_hi+ddf0_lo)/2.;
         for (n=0;n<_N_-1;++n) {
            K1f=del_eta*df[n];
            K1df=del_eta*ddf[n];
            K1ddf=del_eta*dddf(f[n],ddf[n],T[n],dT[n]);
            K2f=del_eta*(df[n]+0.5*K1df);
            K2df=del_eta*(ddf[n]+0.5*K1ddf);
            K2ddf=del_eta*dddf(f[n]+0.5*K1f,ddf[n]+0.5*K1ddf,
                           0.5*(T[n]+T[n+1]),0.5*(dT[n]+dT[n+1]));
            K3f=del_eta*(df[n]+0.5*K2df);
            K3df=del_eta*(ddf[n]+0.5*K2ddf);
            K3ddf=del_eta*dddf(f[n]+0.5*K2f,ddf[n]+0.5*K2ddf,
                           0.5*(T[n]+T[n+1]),0.5*(dT[n]+dT[n+1]));

            K4f=del_eta*(df[n]+K3df);
            K4df=del_eta*(ddf[n]+K3ddf);
            K4ddf=del_eta*dddf(f[n]+K3f,ddf[n]+K3ddf,T[n+1],dT[n+1]);
            f[n+1]=f[n]+(K1f+2*K2f+2*K3f+K4f)/6;
            df[n+1]=df[n]+(K1df+2*K2df+2*K3df+K4df)/6;
            ddf[n+1]=ddf[n]+(K1ddf+2*K2ddf+2*K3ddf+K4ddf)/6;
            if (df[n+1]>1.1 || df[n+1]<0.) {
               df[_N_-1]=df[n+1];
               for ( ;n<_N_-1;++n) f[n+1]=f[n];
            }
         }
         if (df[_N_-1]<1.) ddf0_lo=ddf[0];
         else              ddf0_hi=ddf[0];
      } while (ddf0_hi-ddf0_lo>1.e-6);

      dT0_hi=dT0_hi0;
      dT0_lo=dT0_lo0;
      dT0_last=dT[0];
      do {
         for (n=0;n<_N_-1;++n) {
            dT[0]=(dT0_hi+dT0_lo)/2.;
            K1T=del_eta*dT[n];
            K1dT=del_eta*ddT(T[n],dT[n],Pr,f[n]);
            K2T=del_eta*(dT[n]+0.5*K1dT);
            K2dT=del_eta*ddT(T[n]+0.5*K1T,dT[n]+0.5*K1dT,
               Pr,0.5*(f[n]+f[n+1]));
            K3T=del_eta*(dT[n]+0.5*K2dT);
            K3dT=del_eta*ddT(T[n]+0.5*K2T,dT[n]+0.5*K2dT,
               Pr,0.5*(f[n]+f[n+1]));
            K4T=del_eta*(dT[n]+K3dT);
            K4dT=del_eta*ddT(T[n]+K3T,dT[n]+K3dT,Pr,f[n+1]);
            T[n+1]=T[n]+(K1T+2*K2T+2*K3T+K4T)/6;
            dT[n+1]=dT[n]+(K1dT+2*K2dT+2*K3dT+K4dT)/6;
            if (T[n+1]>T[0] || T[n+1]<0.9)
               for (++n;n<_N_-1;++n) {T[n+1]=T[n]; dT[n+1]=dT[n];}
         }
         if (T[_N_-1]<1.) dT0_lo=dT[0];
         else             dT0_hi=dT[0];
      } while (dT0_hi-dT0_lo>1.e-6);

      if (fabs(ddf[0]-ddf0_last)<1.e-6
         && fabs(dT[0]-dT0_last)<1.e-6) break;
   } while (++iter<MAX_iter);
   if (iter == MAX_iter) printf("\nincrease MAX_iter=%d\n\n",MAX_iter);

   printf("\ncf=(%f)Rex^-1/2",2.*ddf[0]/sqrt(T[0]));
   printf("\nNu=(%f)Rex^+1/2\n",-dT[0]/sqrt(T[0])/(T[0]-1));
   fp=fopen("out.dat","w");
   for (n=0;n<_N_;++n)
      fprintf(fp,"%e %e %e %e %e %e\n"
            ,n*del_eta,f[n],df[n],ddf[n],T[n],dT[n]);
   fclose(fp);
   return 1;
}
```

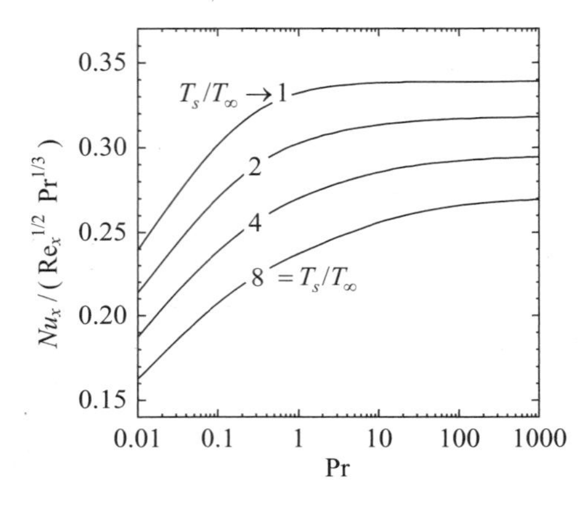

Figure 29-8 Influence of plate temperature and Pr on the Nusselt number.

The Nusselt number for heat transfer between the flow and the plate can also be obtained from the numerical solution:

$$Nu_x = \frac{hx}{k_\infty} = \frac{-x(k_s/k_\infty)}{(T_s - T_\infty)}\frac{\partial T}{\partial y}\bigg|_{y=0} = \frac{-k_s/k_\infty}{\theta(0)-1}\frac{\rho_s}{\rho_\infty}\theta'(0)\sqrt{\frac{Ux}{\nu_\infty}} = \frac{-\theta'(0)}{\theta(0)-1}\frac{\mathrm{Re}_x^{1/2}}{\sqrt{\theta(0)}}. \tag{29-74}$$

In this result, $k_s/k_\infty = (T_s/T_\infty)^{1/2} = \sqrt{\theta(0)}$ and $\rho_s/\rho_\infty = T_\infty/T_s = 1/\theta(0)$ have been used. Since it was shown in Section 25.4 (without temperature-dependent fluid properties) that $\theta'(0) \sim \mathrm{Pr}^{1/3}$ (for $\mathrm{Pr} \geq 1$), the result given by Eq. (29-74) is rearranged into the form

$$Nu_x = \left(\frac{-\theta'(0)/\mathrm{Pr}^{1/3}}{(\theta(0)-1)\sqrt{\theta(0)}}\right)\mathrm{Pr}^{1/3}\mathrm{Re}_x^{1/2}. \tag{29-75}$$

The coefficient to Eq. (29-75) may be evaluated from the numerical solution and is shown in Figure 29-8 as a function of Pr and plate temperature, with $T_s/T_\infty = 8$, 4, 2, and the limiting case $T_s/T_\infty \to 1$. Again, the high and low Prandtl number limits are atypical for normal gases. As the surface temperature approaches the gas temperature far from the plate ($T_s/T_\infty \to 1$), the Nusselt number approaches the relation found in Chapter 25 for flows without temperature-dependent fluid properties (see Problem 25-5).

29.4 PROBLEMS

29-1 Consider a two-dimensional fully developed steady pressure driven flow of a non-Newtonian incompressible fluid between two flat plates. Suppose the constitutive equation for the non-Newtonian fluid is $\tau_{ji} = \mu(\tau)(\partial_j v_i + \partial_i v_j)$, where the fluid viscosity $\mu(\tau)$ is a function of the magnitude of the local shear stress in the flow.

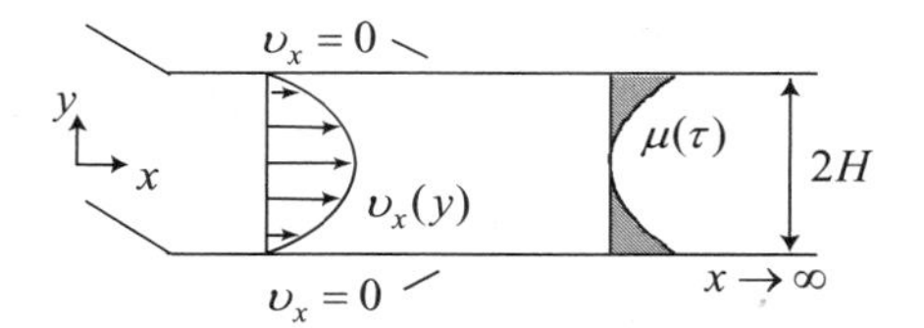

Simplify both the x-direction and the y-direction momentum equations assuming a fully developed flow. Show how the local pressure gradient $\partial P/\partial x$ is related to the local wall shear stress τ_w. Using the definitions $u = v_x\mu_w/(H\tau_w)$ and $\eta = y/H$, show that the x-direction momentum equation becomes

$$\frac{\partial}{\partial \eta}\left[\frac{\mu(\tau)}{\mu_w}\frac{\partial u}{\partial \eta}\right] = -1.$$

Assume that the viscosity shear stress relation is given by $\mu(\tau) = A_o \tau^{1/2}$, where τ is the magnitude of the local shear stress τ_{xy} in the flow. Solve this problem and determine how the mass flow rate of this non-Newtonian fluid compares with that of a Newtonian fluid.

29-2 Assuming that $\varepsilon \to 0$, solve the following ordinary differential equation by regular perturbation, retaining terms through $O(\varepsilon^2)$.

$$\frac{d^2\theta}{d\eta^2} + \frac{\sin(\varepsilon\theta)}{\varepsilon} = 0, \quad \theta(0) = 1, \quad \text{and} \quad d\theta/d\eta\big|_{\eta=0} = 0$$

Hint: The Maclaurin series for the sine function is $\sin(X) = X - \frac{X^3}{3!} + \frac{X^5}{5!} - \cdots$.

29-3 Consider a steady isobaric gas flow over a flat plate, as shown. The gas has temperature-dependent properties: ρ, μ, and k. The only source of thermal energy is viscous heating, and the surface of the plate is adiabatic.

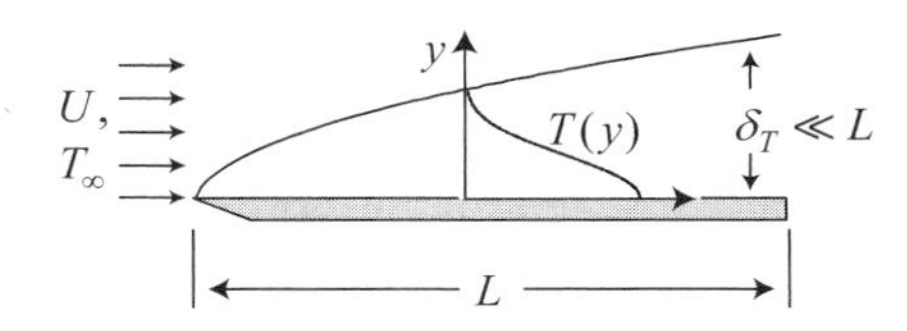

State the assumptions that are needed to put the momentum and heat equations into the following forms:

$$\text{Momentum:} \quad \rho\left(v_x\frac{\partial v_x}{\partial x} + v_y\frac{\partial v_x}{\partial y}\right) = \frac{\partial}{\partial y}\left[\mu\frac{\partial v_x}{\partial y}\right]$$

$$\text{Heat:} \quad \rho C_p\left(v_x\frac{\partial T}{\partial x} + v_y\frac{\partial T}{\partial y}\right) = \frac{\partial}{\partial y}\left[k\frac{\partial T}{\partial y}\right] + \mu\left(\frac{\partial v_x}{\partial y}\right)^2.$$

Making use of the relations

$$\frac{C_p}{C_{p,\infty}} = 1, \quad \frac{k}{k_\infty} = \left(\frac{T}{T_\infty}\right)^{1/2}, \quad \text{and} \quad \frac{\mu}{\mu_\infty} = \left(\frac{T}{T_\infty}\right)^{1/2}$$

from the kinetic theory of gases, transform the governing boundary layer equations and appropriate boundary conditions using the variables defined by

$$\rho v_x = \rho_\infty\frac{\partial \psi}{\partial y}, \quad \rho v_y = -\rho_\infty\frac{\partial \psi}{\partial x}, \quad f(\eta) = \frac{\psi}{\sqrt{\nu_\infty x U}}, \quad \theta = \frac{T}{T_\infty}, \quad \text{and} \quad \eta = \sqrt{\frac{U}{\nu_\infty x}}\int_0^y \frac{\rho}{\rho_\infty}dy.$$

Show that the boundary layer equations become governed by

$$\text{Momentum:} \quad f''' + \frac{f''}{2}\left(f\sqrt{\theta} - \frac{\theta'}{\theta}\right) = 0, \quad f(0) = 0, \ f'(0) = 0, \ f'(\infty) = 1$$

$$\text{Heat:} \quad \theta'' + \frac{\theta'}{2}\left(\Pr f\sqrt{\theta} - \frac{\theta'}{\theta}\right) + \Pr Ec(f'')^2 = 0, \quad \theta'(0) = 0, \ \theta(\infty) = 1$$

where $\Pr = \nu_\infty/\alpha_\infty$ and $Ec = U^2/(C_p T_\infty)$. Numerically solve this problem for the case where $\Pr = 1$ and $Ec = 1$ to determine the numerical value for $c_f \mathrm{Re}_x^{1/2}$. How does this result compare with the case when viscous heating is ignored?

REFERENCES

[1] A. Aziz and T. Y. Na, *Perturbation Methods in Heat Transfer.* Washington, DC: Hemisphere Publishing, 1984.
[2] B. A. Finlayson, *Nonlinear Analysis in Chemical Engineering*. New York, NY: McGraw-Hill, 1980.
[3] W. Kays and M. Crawford, *Convective Heat and Mass Transfer*, Second Edition. New York, NY: McGraw-Hill, 1987.
[4] T. I. Gombosi, *Gaskinetic Theory*, Cambridge Atmospheric and Space Science Series. Cambridge, UK: Cambridge University Press, 1994.

Chapter 30

Turbulence

30.1 The Transition to Turbulence

30.2 Reynolds Decomposition

30.3 Decomposition of the Continuity Equation

30.4 Decomposition of the Momentum Equation

30.5 The Mixing Length Model of Prandtl

30.6 Regions in a Wall Boundary Layer

30.7 Parameters of the Mixing Length Model

30.8 Problems

Turbulence describes flows exhibiting irregular and unpredictable fluctuations in the velocity field. The unpredictable nature of turbulence does not reflect a failure of the inescapable principles used in deriving the transport equations solved for laminar flows. Rather, it reflects the multiplicative nature of solutions that obey these nonlinear equations. The existence of multiple solutions to the equations of motion has been illustrated with the hydraulic jump discussed in Chapter 18 and the shock wave discussed in Chapter 20. For the most part, in turbulence, effects that are too subtle to correctly discern distinguish multiple solutions. It is like trying to answer the question "which side of a spinning coin will land facing up?" The equations of motion are known, but the outcome remains unpredictable because the deciding factors appear random. However, one can predict how many times the coin lands face up, on average, even if one cannot predict the outcome of every toss. Similarly, although the instantaneous details of a turbulent flow may be unpredictable, over some longer time scales, the averaged effects of turbulence on transport are predictable. Therefore, the goal of turbulence modeling is to correctly describe the averaged transport quantities.

The same transport equations used for laminar flows could, in principle, be integrated to gain sufficient information about a turbulent flow to predict averaged transport quantities. However, this requires integration over the small spatial scales and the short time scales important to describing fluctuations in the flow. Additionally, the integration period must be sufficiently long to permit meaningful averaging. With fast computational capabilities, this approach may eventually become more practical for engineering purposes. However, currently most turbulent transport calculations involve integrating equations that are expressed in terms of time-averaged variables of the flow. These averaged equations require an empirical model that describes the influence of turbulent fluctuations on the transport of averaged variables. One such model, *the mixing length model*, will be introduced in Section 30.5, and will be applied to turbulent transport in Chapters 31 through 35. A more advanced model, *the k-epsilon model*, will be introduced in Chapter 36. The k-epsilon model is one of the most common turbulence models used in computational fluid dynamics.

30.1 THE TRANSITION TO TURBULENCE

To gain a better physical sense of how laminar flows transition to turbulence, consider pressure-driven (Poiseuille) flow between parallel plates, illustrated in Figure 30-1. It is helpful to change one's mental picture of this flow from being governed by transport of linear momentum to the transport of vorticity. Vorticity is a measure of the rotation rate in the fluid. The vorticity equation embodies conservation of angular momentum, and is most simply obtained by taking the curl, term by term, of the linear momentum equation. For the special case of a two-dimensional plane flow of a constant density fluid, the vorticity equation becomes

$$\frac{\partial \omega_z}{\partial t} + v_x \frac{\partial \omega_z}{\partial x} + v_y \frac{\partial \omega_z}{\partial y} = \nu \left(\frac{\partial^2 \omega_z}{\partial x^2} + \frac{\partial^2 \omega_z}{\partial y^2} \right), \tag{30-1}$$

where $\omega_z = (\partial v_y/\partial x - \partial v_x/\partial y)$. The vorticity equation (30-1) has the prototypical form of a transport equation, describing both advection and diffusion. For a steady-state laminar and fully developed flow between parallel plates, the vorticity equation simplifies to

$$0 = \nu \frac{\partial^2 \omega_z}{\partial y^2}, \tag{30-2}$$

which describes the cross-stream transport of vorticity by diffusion. The boundary conditions on ω_z are understood from the conditions imposed on $\partial v_x/\partial y$ at the walls by the pressure gradient in the flow. Alternatively, one can solve the momentum equation for v_x (see Section 28.1) and obtain ω_z directly from $\omega_z = -\partial v_x/\partial y$. (This exercise is not important for grasping the main point of this section.) The vorticity solution between the plates for a laminar flow is given by

$$\omega_z = \frac{-\partial v_x}{\partial y} = \frac{y}{\mu} \frac{\partial P}{\partial x}. \tag{30-3}$$

Some structure is given to the solution by thinking of streamwise spans of equal-strength vorticity or *"vortex sheets."* Vortex sheets are subject to streamwise advection and simultaneous diffusion across the flow, as illustrated in Figure 30-2.

Notice that a planar vortex sheet does not experience any self-induced velocity. As illustrated in Figure 30-3(a), each vortex in a sheet feels equal and opposing influence

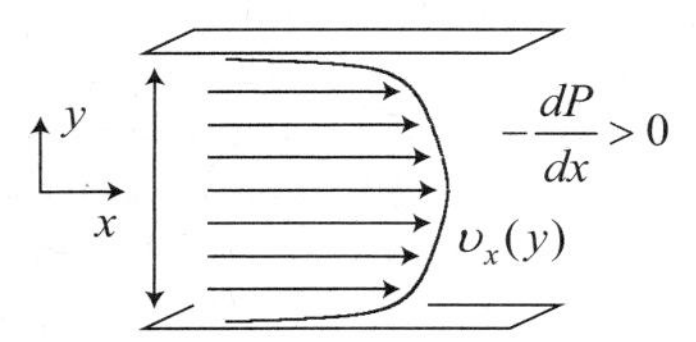

Figure 30-1 Pressure-driven flow between infinite parallel plates.

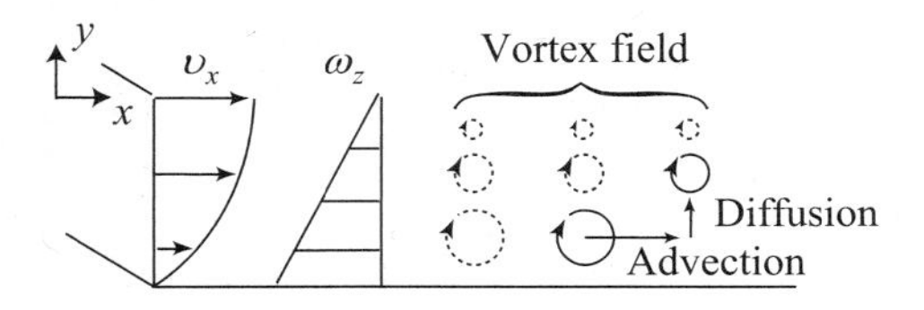

Figure 30-2 Vortex field in pressure-driven flow between infinite parallel plates.

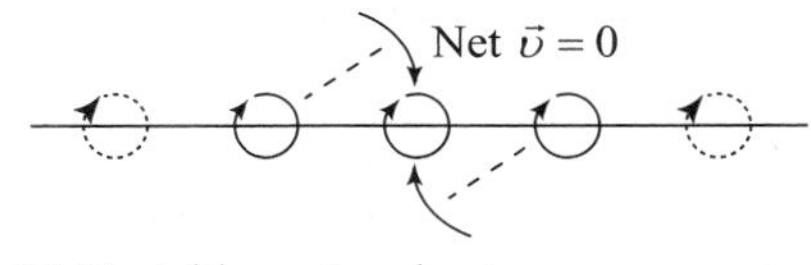

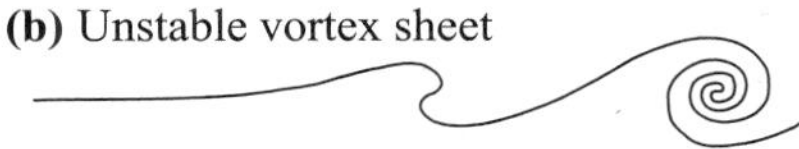

Figure 30-3 Vortex sheets in a laminar (a) and a turbulent (b) flow.

of vortices from in front and behind. However, imagine that a disturbance occurs in the flow. This disturbance could be viewed as a wrinkle in the vortex sheet. Once the sheet is no longer planar, any vortex in that sheet may become subject to a nonzero velocity caused by the integrated effect from the rest of the vortex sheet. This disturbance can cause the vortex sheet to roll up upon itself, as illustrated in Figure 30-3(b). This is symptomatic of a transition to turbulence.

The initial disturbance can also be viewed as a departure from the parabolic shape of the laminar velocity profile for Poiseuille flow. The viscous forces, associated with momentum diffusion, act to restore the parabolic profile. Equivalently, viscous forces act to restore the vortex sheet to a planar state. Therefore, viscous forces are a proponent to laminar flow.

The time scale associated with diffusion over a length scale ℓ of a disturbance is ℓ^2/ν. The time scale for advection over a length scale ℓ is ℓ/U, where U is the characteristic velocity of the flow. The competing effects of advection that drives roll up of the vortex sheet and diffusion that flattens the vortex sheet decides whether a flow remains laminar or transitions to turbulence. The ratio of these two time scales is

$$\frac{\ell^2/\nu}{\ell/U} = \frac{U\ell}{\nu} = \text{Re}_\ell, \tag{30-4}$$

which describes a Reynolds number based on the length scale of a disturbance. Flows having a high Reynolds number offer insufficient time for diffusion to smooth out disturbances, and tend to transition into turbulence. The Reynolds number can also be interpreted as the ratio of inertial to viscous forces. In that light, high Reynolds number flows are driven by large inertial effects, and are prone to turbulence because of the weak dampening effect of viscous forces.

Reynolds numbers are generally reported using a geometric length scale of the flow, such as the diameter of a pipe D or length of a plate L. Since these length scales are typically much larger than ℓ, the *critical Reynolds number* that delineates the transition from a laminar to a turbulent flow tends to be a large number. Some "rule of thumb" values for the critical Reynolds number are as follows:

$$\text{Re}_D \approx 2300 \quad \text{critical Reynolds number for flows through pipes} \tag{30-5}$$

and

$$\text{Re}_L \approx 5 \times 10^5 \quad \text{critical Reynolds number for flows over a flat plate.} \tag{30-6}$$

However, the transition to turbulence can occur at significantly lower Reynolds numbers if the flow is intentionally "tripped" (destabilized) and at significantly higher Reynolds numbers if care is taken not to perturb the flow.

30.2 REYNOLDS DECOMPOSITION

Determining the mean (time-averaged) quantities and the effect of fluctuations on the transport of mean quantities are usually central to the goals of an engineering calculation. The physical quantities characterizing a turbulent flow can be written as a sum of mean and fluctuating values, such as $A = \overline{A} + A'$ where $\overline{A}$ is the mean value and A' is the fluctuating value. Figure 30-4 illustrates this decomposition principle, which was introduced by Osborne Reynolds [1].

To transform governing equations, some rules of time averaging are required:

$$\text{(a)}\ \overline{A'} = 0 \tag{30-7}$$

$$\text{(b)}\ \overline{A+B} = \overline{A} + \overline{B} \tag{30-8}$$

$$\text{(c)}\ \overline{\overline{A}B'} = 0 \tag{30-9}$$

$$\text{(d)}\ \overline{AB} = \overline{(\overline{A}+A')(\overline{B}+B')} = \overline{\overline{A}\,\overline{B} + \overline{A}B' + A'\overline{B} + A'B'} = \overline{\overline{A}\,\overline{B}} + \overline{A'B'} = \overline{A}\,\overline{B} + \overline{A'B'} \tag{30-10}$$

$$\text{(e)}\ \overline{\partial_i A} = \partial_i \overline{A},\quad \overline{\partial_i \partial_j A} = \partial_i \partial_j \overline{A},\quad \overline{\partial_i \partial_i A} = \partial_i \partial_i \overline{A},\quad \text{etc.} \tag{30-11}$$

$$\text{(f)}\ \overline{\partial_o A} = \partial_o \overline{A} \tag{30-12}$$

Rule (a) states that the fluctuating component of a quantity must time average to zero. Rule (b) states that the time averaging of a sum of two quantities is equivalent to time averaging the two quantities individually and then performing the sum. Rule (c) states that the product between an already time-averaged quantity and the fluctuating component of a second quantity will time average to zero. Rule (d) carries out the time averaging of a product between two quantities; the result is demonstrated using rules (a) and (b). Rule (e) states that time averaging of various spatial derivatives of a quantity is equivalent to performing the spatial derivatives on the time-averaged quantity. Rule (f) states that time averaging of the time derivative of a quantity is equivalent to performing the time derivative on the time-averaged quantity. Implicit to this rule is that the time scale required for time averaging out fluctuations in a turbulent flow is much shorter than the time scale for which transient change of time-averaged variables in the flow is being evaluated.

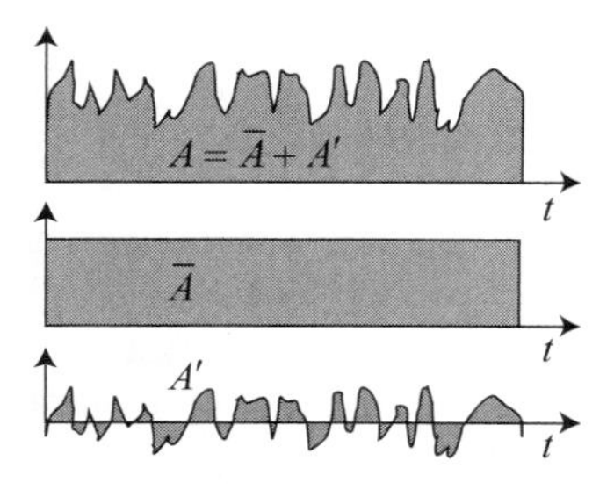

Figure 30-4 Variable decomposition into mean and fluctuating components.

Rule (d) illustrates an important result that will arise in the time averaging of transport equations. Only nonlinear terms result in a persisting effect of fluctuations on the transport of mean quantities. For example, the linear diffusion of momentum retains the same form in the time-averaged equation:

$$\overline{\mu \partial v_x / \partial y} = \mu \partial \overline{v}_x / \partial y. \tag{30-13}$$

However, the nonlinear advection of momentum,

$$\overline{v_y v_x} = \overline{(\overline{v}_y + v'_y)(\overline{v}_x + v'_x)} = \overline{v}_y \overline{v}_x + \overline{v'_y v'_x}, \tag{30-14}$$

reveals two contributions to transport in the time-averaged form: the gross effect of advection in the mean flow, $\overline{v}_y \overline{v}_x$, plus transport resulting from the *correlation* in fluctuating velocities, $\overline{v'_y v'_x}$. If v'_x and v'_y were uncorrelated, then the time average $\overline{v'_y v'_x}$ would be zero. However, the physics of fluid flow dictates that all variables describing the local condition of the flow tend to be correlated. For example, continuity suggests that it is improbable for v'_x to change without a reactionary change in v'_y. Therefore, the correlation between fluctuating quantities tends to contribute to the transport of the mean quantities described by the time-averaged equation.

30.3 DECOMPOSITION OF THE CONTINUITY EQUATION

Consider the steady-state continuity equation

$$\partial_j(\rho v_j) = 0. \tag{30-15}$$

Substituting the decomposed expressions for density and velocity gives

$$\partial_j[(\overline{\rho} + \rho')(\overline{v}_j + v'_j)] = 0 \quad \text{or} \quad \partial_j(\overline{\rho}\,\overline{v}_j + \overline{\rho} v'_j + \rho' \overline{v}_j + \rho' v'_j) = 0. \tag{30-16}$$

Time averaging the resulting equation gives

$$\overline{\partial_j(\overline{\rho}\,\overline{v}_j + \overline{\rho} v'_j + \rho' \overline{v}_j + \rho' v'_j)} = 0 \quad \text{or} \quad \partial_j(\overline{\overline{\rho}\,\overline{v}_j} + \overline{\overline{\rho} v'_j} + \overline{\rho' \overline{v}_j} + \overline{\rho' v'_j}) = 0. \tag{30-17}$$

Since time averaging the product of mean and fluctuating terms yields zero, the continuity equation becomes

$$\partial_j(\overline{\rho}\,\overline{v}_j + \overline{\rho' v'_j}) = 0. \tag{30-18}$$

If the discussion of turbulence is restricted to constant-density fluids, $\rho = const.$, then $\rho' = 0$ and the continuity equation becomes

$$\partial_j \overline{v}_j = 0. \tag{30-19}$$

Therefore, the divergence of the time-averaged velocity field in a turbulent incompressible flow is zero (which is identical in form to the non-time-averaged version of the continuity equation when $\rho = const.$). Furthermore, when $\rho = const.$, then it must also be true that

$$\partial_j v'_j = 0 \tag{30-20}$$

since continuity of the total (instantaneous) velocity requires that $\partial_j(\overline{v}_j + v'_j) = 0$.

30.4 DECOMPOSITION OF THE MOMENTUM EQUATION

Starting with the incompressible ($\rho = const.$) form of the Navier-Stokes equations and substituting the decomposed expressions for the flow variables gives

$$(\overline{v}_j + v'_j)\partial_j(\overline{v}_i + v'_i) = \partial_j\left[\nu\partial_j(\overline{v}_i + v'_i)\right] - \frac{1}{\rho}\partial_i(\overline{P} + P'). \tag{30-21}$$

Time averaging the right-hand side of the equation requires

$$\overline{\partial_j[\nu\partial_j(\overline{v}_i + v'_i)]} = \partial_j[\nu\partial_j\overline{v}_i] \tag{30-22}$$

and

$$\overline{\partial_i(\overline{P} + P')} = \partial_i\overline{P}, \tag{30-23}$$

and for the advection term on the left-hand side,

$$\overline{(\overline{v}_j + v'_j)\partial_j(\overline{v}_i + v'_i)} = \overline{\overline{v}_j\partial_j\overline{v}_i} + \overline{v'_j\partial_j\overline{v}_i} + \overline{\overline{v}_j\partial_j v'_i} + \overline{v'_j\partial_j v'_i} = \overline{v}_j\partial_j\overline{v}_i + \overline{v'_j\partial_j v'_i}. \tag{30-24}$$

Since $\partial_j v'_j = 0$ for an incompressible flow, the substitution $\overline{v'_j\partial_j v'_i} = \partial_j\overline{v'_j v'_i}$ can be made in the final expression of the advection term. Therefore, the time-averaged steady-state momentum equation becomes

$$\underbrace{\overline{v}_j\partial_j\overline{v}_i}_{\substack{\text{advection in}\\ \text{mean flow}}} + \underbrace{\partial_j\overline{v'_j v'_i}}_{\substack{\text{turbulent scale}\\ \text{advection}}} = \underbrace{\partial_j\left[\nu\partial_j\overline{v}_i\right]}_{\substack{\text{diffusion in}\\ \text{mean flow}}} - \underbrace{\frac{1}{\rho}\partial_i\overline{P}}_{\text{pressure source}}. \tag{30-25}$$

The result of using Reynolds decomposition on the Navier-Stokes equations is called the Reynolds-Averaged Navier-Stokes Equations (RANS).

Several important realizations follow from the Reynolds decomposed Navier-Stokes equations. Although much of the time-averaged momentum equation appears identical in form to the original equation, a new term has emerged from the contribution of non-linear advection. The new term, describing the gradient in a correlation between fluctuating velocity components $\partial_j\overline{v'_j v'_i}$, can be interpreted in several equivalent ways. Physically it is allied with advection, but does not describe advection on the same scale as that of the mean flow. Instead, this term describes transport on the scale of turbulent fluctuations. It is associated with the transport of momentum, but equivalently describes a state of time-averaged stress that is commonly referred to as the "*Reynolds stress*." In that light, this term is likened to the viscous stress cause by diffusion, although it can be orders of magnitude larger. The central problem in turbulence modeling is how to relate the Reynolds stress back to the mean variables of the flow (i.e., to the dependent variables of the transport equations). This demonstrates the closure problem faced by all turbulent modeling of the time-averaged equations.

For further development of turbulent transport, it is useful to look at the somewhat simplified boundary layer problem. The Reynolds decomposed momentum equation (30-25) can be written in Cartesian coordinates for a two-dimensional flow as

$$\overline{v}_x\frac{\partial\overline{v}_x}{\partial x} + \overline{v}_y\frac{\partial\overline{v}_x}{\partial y} = \frac{\partial}{\partial x}\left(\nu\frac{\partial\overline{v}_x}{\partial x} - \overline{v'_x v'_x}\right) + \frac{\partial}{\partial y}\left(\nu\frac{\partial\overline{v}_x}{\partial y} - \overline{v'_y v'_x}\right) - \frac{1}{\rho}\frac{\partial\overline{P}}{\partial x}. \tag{30-26}$$

For the treatment of boundary layers, spatial scales of the problem require that when gradients of comparable magnitude terms are considered, streamwise gradients are small compared with cross-stream gradients: $\partial(\ )/\partial x \ll \partial(\ )/\partial y$. Since $\overline{v'_x v'_x} \sim \overline{v'_y v'_x}$, as will be demonstrated in the next section, the momentum equation simplifies to

$$\overline{v}_x \frac{\partial \overline{v}_x}{\partial x} + \overline{v}_y \frac{\partial \overline{v}_x}{\partial y} = \frac{\partial}{\partial y}\left[\nu \frac{\partial \overline{v}_x}{\partial y} - \overline{v'_y v'_x}\right] - \frac{1}{\rho}\frac{\partial \overline{P}}{\partial x} \quad \text{(for boundary layer)}. \tag{30-27}$$

To solve the turbulent boundary layer momentum equation requires a model that allows the Reynolds stress $\overline{v'_y v'_x}$ to be expressed in terms of properties of the mean flow. In the next section, a model is introduced that draws attention to the mathematical similarity of the Reynolds stress to a diffusion term, in the time-averaged turbulent momentum equation.

30.5 THE MIXING LENGTH MODEL OF PRANDTL

The mixing length model offers a very simple picture of turbulence. Imagine that the irregular fluctuations in the velocity field of a turbulent flow can be characterized by a rolling motion of *eddies* over some length scale ℓ. This length describes advection over the turbulent scale associated with the Reynolds stress $\overline{v'_y v'_x}$, appearing in the momentum boundary layer equation (30-27). The physical idea behind the mixing length model is that fluid "elements" can interact with the surrounding flow over the turbulent length scale ℓ, as illustrated in Figure 30-5. Associated with this length scale is a magnitude of velocity fluctuations v'_x. Prandtl proposed that a fluid element should experience a velocity fluctuation v'_x that is associated with the distance traveled ℓ and the change in mean flow velocity over that distance [2]. As illustrated in Figure 30-5, this physically suggests that

$$v'_x \sim \ell \left|\frac{\partial \overline{v}_x}{\partial y}\right|. \tag{30-28}$$

In the absence of turbulence, the local flow condition can "sample" the surrounding conditions only over a length scale associated with molecular diffusion (the mean free path length of molecular collisions). In contrast, the mixing length scales of turbulence offer a much larger length scale for interactions. For this reason, turbulence can exhibit significantly enhanced transport over a corresponding laminar flow.

The continuity equation for turbulent fluctuations (30-20) provides the scaling argument needed to relate the components of the fluctuating velocity:

$$v'_x/\ell \text{ ``}+\text{'' } v'_y/\ell \sim 0. \tag{30-29}$$

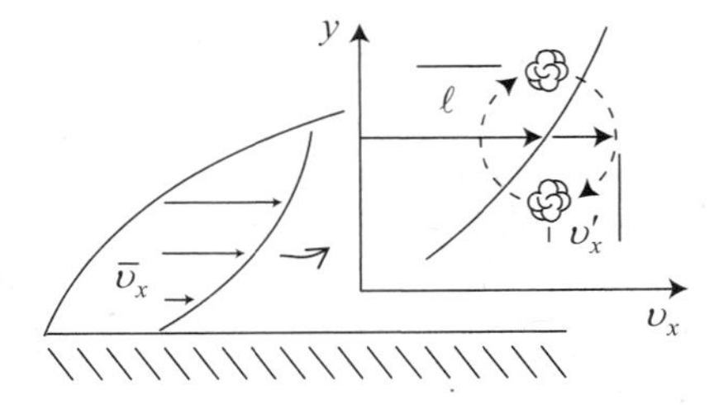

Figure 30-5 Illustration of Prandtl's mixing length model.

In other words: $v'_x \sim v'_y$. Combining this result with Eq. (30-28) yields a scaling expectation for the Reynolds stress:

$$\overline{v'_y v'_x} = -\ell \left| \frac{\partial \bar{v}_x}{\partial y} \right| \ell \frac{\partial \bar{v}_x}{\partial y}. \tag{30-30}$$

The Reynolds stress describes a flux that carries momentum from a high content region to a low content region in the flow. To fulfill this expectation, a negative sign was added to Eq. (30-30) to contrast with the sign of the velocity gradient in the mean flow. Equation (30-30) can be considered as a formal definition for ℓ, since through experimental measurements of $\overline{v'_y v'_x}$ and $\partial \bar{v}_x / \partial y$, the mixing length scale ℓ can be calculated.

To achieve a mathematical likeness of small turbulent scale advection to molecular scale diffusion, a turbulent momentum diffusivity ε_M is defined such that

$$\overline{v'_y v'_x} = -\varepsilon_M \frac{\partial \bar{v}_x}{\partial y}. \tag{30-31}$$

Equation (30-31), for the turbulent momentum flux $\overline{v'_y v'_x}$, has the prototypical form of a diffusion flux. The turbulent diffusivity, or *eddy diffusivity*, ε_M depends upon the mixing length scale and the gradient in mean flow velocity. Combining Eqs. (30-30) and (30-31) gives

$$\varepsilon_M = \ell^2 \left| \frac{\partial \bar{v}_x}{\partial y} \right|. \tag{30-32}$$

Therefore, with Eq. (30-31), the momentum equation (30-27) can be written in terms of the eddy diffusivity as

$$\overbrace{\bar{v}_x \frac{\partial \bar{v}_x}{\partial x} + \bar{v}_y \frac{\partial \bar{v}_x}{\partial y}}^{\text{advection}} = \overbrace{\frac{\partial}{\partial y}\left[(\overbrace{\nu}^{\text{molecular}} + \overbrace{\varepsilon_M}^{\text{turbulent}}) \frac{\partial \bar{v}_x}{\partial y} \right]}^{\text{diffusion}} - \frac{1}{\rho}\frac{\partial \bar{P}}{\partial x} \tag{30-33}$$

for a two-dimensional boundary layer. In this form, the turbulent equation has a similar mathematical form to the laminar boundary layer equation. However, the current equation has two distinct contributions to the momentum diffusion term, one molecular and one turbulent.

Although the problem formulation has undergone significant transformations, the underlying closure problem still remains. The Reynolds stress, written in any form possible to this point, retains an unknown variable:

$$\underset{?}{\overline{v'_y v'_x}} \rightarrow -\underset{?}{\varepsilon_M} \frac{\partial \bar{v}_x}{\partial y} \rightarrow -\underset{?}{\ell^2} \left(\frac{\partial \bar{v}_x}{\partial y} \right)^2. \tag{30-34}$$

However, since the mixing length ℓ is a variable on which one can impose some expectations, some progress can be claimed. It makes physical sense that the mixing length must go to zero approaching the wall, as illustrated in Figure 30-6. In the immediate vicinity of the wall, the mixing length is negligibly small. At its largest, the mixing length should be proportional to the boundary layer thickness ($\approx \gamma\, \delta$). Between these two extremes, one might expect a linear dependence of the mixing length with distance from the wall,

$$\ell = \kappa y \quad (\text{for} \quad 0 \leq \ell \leq \gamma\, \delta). \tag{30-35}$$

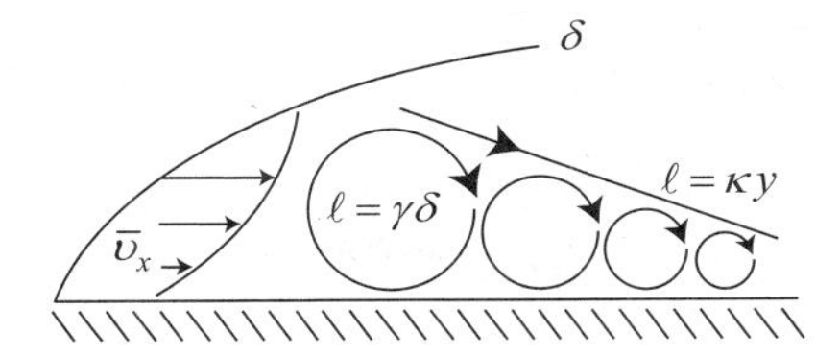

Figure 30-6 Influence of boundary layer thickness and wall proximity on mixing length scale.

The coefficient for this linear growth in the mixing length with distance from the wall is known as von Kármán constant* κ. Expectations asserted to this point arise from purely geometric constraints imposed on the scale of a turbulent eddy, and are supported reasonably well by experimental investigations, as will be shown in the next section.

30.6 REGIONS IN A WALL BOUNDARY LAYER

Experimental investigations have revealed that the turbulent boundary has regions distinguished by the relative importance of various transport phenomena. The structure of a boundary layer is illustrated schematically in Figure 30-7. The vertical scale of this sketch is logarithmic to reveal features of the boundary layer that are very close to the wall. The momentum equation will be consulted together with Prandtl's mixing length model to help classify distinct regions in the boundary layer. Flow over a flat plate oriented parallel to the flow is considered, for which the pressure gradient is zero. Therefore, the momentum equation for a turbulent boundary layer becomes

$$\overline{v}_x \frac{\partial \overline{v}_x}{\partial x} + \overline{v}_y \frac{\partial \overline{v}_x}{\partial y} = \frac{\partial}{\partial y}\left[(\nu + \varepsilon_M)\frac{\partial \overline{v}_x}{\partial y}\right] - \overbrace{\frac{1}{\rho}\frac{\partial \overline{P}}{\partial x}}^{=0}. \tag{30-36}$$

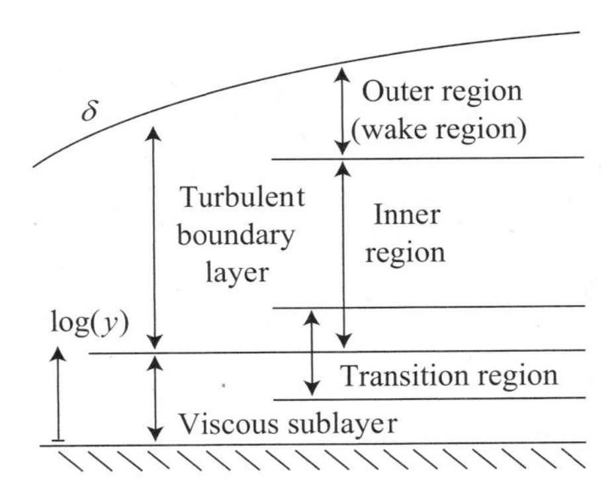

Figure 30-7 Regions of a turbulent boundary layer.

30.6.1 The Viscous Sublayer: (Advection) ≪ (Diffusion) and $\varepsilon_M \ll \nu$

The region of the boundary layer closest to the wall shown in Figure 30-7 is the *viscous sublayer*. Because of the no-slip condition at the wall, advection is negligible compared to diffusion. Additionally, the mixing length model has argued that $\ell \to 0$ approaching the wall, with the consequence that the turbulent diffusivity, given by Eq. (30-32), becomes

*Named after the Hungarian-American aerospace engineer Theodore von Kármán (1881–1963).

negligible. Consequently, the viscous sublayer is a region of the boundary layer in which the flow remains laminar. In this region, the momentum equation (30-36) becomes

$$0 = \frac{\partial}{\partial y}\left[\frac{\mu}{\rho}\frac{\partial \overline{v}_x}{\partial y}\right], \tag{30-37}$$

which dictates that the momentum flux $-\mu(\partial \overline{v}_x/\partial y)$ is constant through the viscous sublayer and equal to the stress imparted on the wall $\tau_w = \mu(\partial \overline{v}_x/\partial y)|_{y=0}$. The solution to the momentum equation in the viscous sublayer can be expressed in terms of τ_w as

$$\overline{v}_x = \frac{\tau_w/\rho}{\nu} y. \tag{30-38}$$

Analysis of turbulent flow near the wall can be generalized by introducing dimensionless *wall variables* that are based on the wall shear stress:

$$u^+ = \frac{\overline{v}_x}{\sqrt{\tau_w/\rho}} \quad \text{and} \quad y^+ = \frac{y\sqrt{\tau_w/\rho}}{\nu}. \tag{30-39}$$

The quantity $\sqrt{\tau_w/\rho}$ is often referred to as the *friction velocity*. In terms of the wall variables, the solution (30-38) to the momentum equation in the viscous sublayer becomes

$$u^+ = y^+. \tag{30-40}$$

This velocity profile remains valid only for a finite distance $0 \leq y^+ \leq E_v^+$ from the wall, the distance over which the effects of turbulence are negligible.

30.6.2 Inner Region: (Advection) ≪ (Diffusion) and $\varepsilon_M \gg \nu$

Moving out from the viscous sublayer, the mixing length is assumed to grow linearly with distance from the wall, such that $\ell = \kappa y$. Since the eddy diffusivity grows as ℓ^2, at some distance from the wall the eddy diffusivity becomes the significant contributor to the diffusion flux. Experimental investigations show that this can initially happen close to the wall where transport by advection remains small compared with diffusion. Therefore, the next region in the boundary layer shown in Figure 30-7, called the *inner region*, is distinguished by large turbulent diffusivity ($\varepsilon_M \gg \nu$) and negligible advection. In this region, the momentum equation (30-36) becomes

$$0 = \frac{\partial}{\partial y}\left[\varepsilon_M \frac{\partial \overline{v}_x}{\partial y}\right]. \tag{30-41}$$

Although the nature of momentum diffusion has changed, the magnitude of the momentum flux remains the same, such that, after integrating once, the momentum equation requires

$$\tau_w/\rho = \varepsilon_M\, \partial \overline{v}_x/\partial y. \tag{30-42}$$

Expressed in terms of the wall variables, the momentum equation for the inner region of the boundary layer becomes

$$\frac{\varepsilon_M}{\nu}\frac{\partial u^+}{\partial y^+} = 1. \tag{30-43}$$

Since $\ell = \kappa y$ in the inner region, the turbulent diffusivity (30-32) can be expressed in terms of wall variables as

$$\frac{\varepsilon_M}{\nu} = (\kappa y^+)^2 \frac{\partial u^+}{\partial y^+}. \tag{30-44}$$

Combining Eq. (30-43) with Eq. (30-44) to eliminate $\partial u^+/\partial y^+$ yields

$$\frac{\varepsilon_M}{\nu} = \kappa y^+. \tag{30-45}$$

This result indicates that, in the inner region of the boundary layer, the ratio between turbulent and molecular diffusivities is proportional to the distance from the wall y^+. Substituting Eq. (30-45) into Eq. (30-44) yields a form for the momentum equation that may be integrated through the inner region of the boundary layer:

$$\frac{du^+}{dy^+} = \frac{1}{\kappa y^+}. \tag{30-46}$$

Since this form of the momentum equation is not applicable to the viscous sublayer, integration must start from the edge of the viscous sublayer, where $u^+ = y^+ = E_\nu^+$ (by Eq. 30-40). Integration yields

$$\int_{E_\nu^+}^{u^+} du^+ = \frac{1}{\kappa} \int_{E_\nu^+}^{y^+} \frac{dy^+}{y^+} \tag{30-47}$$

or

$$u^+ - E_\nu^+ = \frac{1}{\kappa} \ln(y^+/E_\nu^+). \tag{30-48}$$

Equation (30-48) is a prediction of the velocity profile through the inner region of the turbulent boundary layer published by von Kármán [3]. It is a prediction that will be borne out by experimental investigations and is known as the "*law of the wall*" to reflect the generality of this result to turbulent boundary layers. There are two unknown empirical constants associated with the law of the wall: the von Kármán constant κ and the viscous sublayer thickness E_ν^+. It is important to realize that the "solution" offered by the law of the wall is incomplete without knowledge of the wall shear stress τ_w utilized in the normalized wall variables.

30.6.3 Outer Region: (Advection) $\sim$ (Diffusion) and $\varepsilon_M \gg \nu$

Moving past the inner region of the boundary layer, even further from the wall, one comes to the *outer region*, as illustrated in Figure 30-7. In the outer region, the influence of advection on transport can no longer be ignored. The momentum equation becomes

$$\overline{v}_x \frac{\partial \overline{v}_x}{\partial x} + \overline{v}_y \frac{\partial \overline{v}_x}{\partial y} = \frac{\partial}{\partial y}\left[\varepsilon_M \frac{\partial \overline{v}_x}{\partial y}\right], \tag{30-49}$$

which neglects molecular diffusivity in comparison with turbulent diffusivity. With the nonlinear advection term present, the momentum equation requires numerical integration to determine the velocity profile through the outer region.

Integration of the momentum equation through the viscous sublayer and inner region of the boundary layer was facilitated by specifying two boundary conditions at the wall: first, the no-slip condition and, second, the wall shear stress. However, in reality, the second boundary condition is prescribed on the other side of the boundary layer, where $\overline{v}_x$ approaches the free-stream value. Therefore, the wall shear stress is part of the unknown solution. Consequently, numerical integration through the outer region of the boundary layer is required to complete the solution. In the outer region, it is easiest to assume that the mixing length is a constant that scales with the boundary layer thickness: $\ell = \gamma\,\delta$ [4].

30.7 PARAMETERS OF THE MIXING LENGTH MODEL

Prandtl's mixing length model, applied to the various regions of the boundary layer, helps solidify a number of constants that must be specified in order to utilize the model. They are the extent of the viscous sublayer E_ν^+, the proportionality constant κ between the mixing length size and distance from the wall in the inner region of the boundary layer, and the proportionality constant γ between the mixing length size and the boundary layer thickness in the outer region. Table 30-1 summarizes the mixing length model as broken down by region. The law of the wall behavior (30-48) suggests an approach for analyzing experimental measurements to reveal two of these constants (E_ν^+ and κ).

Figure 30-8 contrasts the turbulent velocity profile for a flow over a smooth wall with the model behavior outlined in the preceding analysis. The velocity profile is plotted across the boundary layer utilizing wall variables. The law of the wall (30-48) is fitted to the experimental data to determine $E_\nu^+ = 10.8$ and $\kappa = 0.41$ [5]. Using these values, the law of the wall becomes

$$u^+ = 2.44 \ln y^+ + 5.0. \tag{30-50}$$

Table 30-1 Summary of mixing length model for turbulent momentum diffusivity

$\varepsilon_M = \ell^2 \left\lvert \dfrac{\partial \overline{v}_x}{\partial y} \right\rvert$	Viscous sublayer $y^+ < E_\nu^+$	Inner region $(E_\nu^+ < y^+)$ *and* $(\kappa y < \gamma\delta)$	Outer region $\kappa y > \gamma\delta$
Velocity profile:	$u^+ = y^+$	$u^+ = E_\nu^+ + (1/\kappa)\ln(y^+/E_\nu^+)$	?
Mixing length:	$\ell = 0$	$\ell = \kappa y$	$\ell = \gamma\delta$
Van Driest:	$\ell = \kappa y\,[1 - \exp(-y^+/A_\nu^+)]$		$\ell = \gamma\delta$

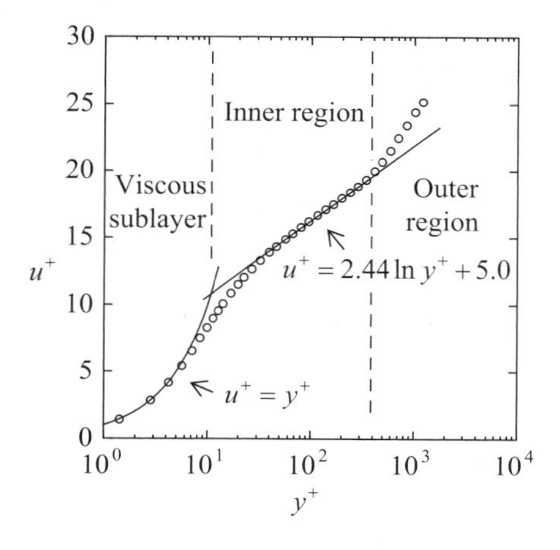

Figure 30-8 Comparison of simplified model versus experimental trends.

The mixing length constants have been investigated extensively in the literature, with agreement generally falling within 20–30%. A review of some of this literature can be found in reference [6].

Figure 30-8 illustrates two features of the experimental data that are not well described by the analysis of the preceding sections. First, the velocity profile exhibits a smooth transition from the viscous sublayer to the inner region in the experimental data that is not captured correctly by the piecewise joining of analytic functions used to describe the two regions. To correct this, Van Driest [7] proposed an empirical function that provides a smooth transition from the viscous sublayer to the inner region:

$$\ell = \kappa y\left[1 - \exp(-y^+/A_\nu^+)\right]. \tag{30-51}$$

It is seen that the Van Driest damping function has the correct limiting behavior of $\ell \to 0$ as $y \to 0$ and $\ell \to \kappa y$ as $y^+ \to \infty$. A_ν^+ has a similar function to E_ν^+ in specifying the thickness of the viscous sublayer, and is found by fitting (30-51) to the transition region data. For the smooth wall data shown in Figure 30-8, a value of $A_\nu^+ = 25.0$ reproduces well the transition region.

The second experimental trend seen in Figure 30-8 that is not described by the preceding analysis is the velocity profile in the outer region of the boundary layer. The law of the wall is not valid in the outer region because advection has a significant role in the momentum equation and the turbulent mixing length scale becomes proportional to the boundary layer thickness $\ell = \gamma\delta$ (and not the distance from the wall). As observed earlier, the momentum equation must be numerically integrated to determine the velocity profile in the outer region. Experimental investigations of ℓ/δ in the outer region indicate that the largest mixing length scale is typically much less than the boundary layer thickness, with $\gamma = 0.085$ being a representative value [8].

In Chapter 31, the mixing length model is applied to the turbulent hydrodynamic description of internal flows. The model is extended to transport of heat and species in Chapter 32, and is applied to fully developed transport in Chapter 33. The mixing length model is extended again in Chapter 34 to flows over rough surfaces. Finally, in Chapter 35 the mixing length model is applied to transport across developing boundary layers.

30.8 PROBLEMS

30-1 Can the "regular" Navier-Stokes equations be used to describe turbulence? (Why or why not?) What is the purpose of a "turbulence-model," and why is it desirable for describing turbulent flows?

30-2 How many fitted parameters are there in the mixing length model of turbulence? What is the physical significance of each?

30-3 The following is a time-averaged equation of turbulence, not in its final form. What is this equation describing? Organize this equation into the form of a traditional transport equation and provide a physical interpretation for each part.

$$\overline{v}_j\partial_j\left(\frac{\overline{v_i'v_i'}}{2}\right) + \overline{v_i'v_j'}\partial_j\overline{v}_i + \partial_j\left(\overline{v_j'\frac{v_i'v_i'}{2}}\right) - \nu\partial_j\partial_j\left(\frac{\overline{v_i'v_i'}}{2}\right) + \nu\overline{\partial_j v'\partial_j v_i'} + \partial_j\left(\overline{v_j'P'}/\rho\right) = 0$$

30-4 Consider Poiseuille flow between parallel plates separated by a distance W. Try to evaluate the appropriateness of the critical Reynolds number $\mathrm{Re}_D \approx 2300$ (rule of thumb) based on our understanding of the transition between the laminar viscous sublayer and turbulent inner region described by the law of the wall. To this end, find an expression for the wall coordinate at mid-distance between the plates $y^+(W/2)$ in terms of the coefficient of friction c_f and the Reynolds number Re_D (based on hydraulic diameter). Assume that the viscous sublayer mostly envelops the flow between the plates, and determine c_f as a function of Re_D for a *laminar* flow. Combine this result with the critical Reynolds to evaluate $y^+(W/2)$. Does $y^+(W/2)$ fall within the transition region between the viscous sublayer and the inner region of the wall turbulence? Comment of the significance of this result.

30-5 Suppose that turbulent pipe flow has been solved using the variables $u = \overline{v}_z/\overline{v}_m$ and $\eta = 1 - r/R$. Using the notation that $u'(0) = \partial u/\partial\eta|_{\eta=0}$, establish relations for y^+, u^+, and c_f in terms of η, u, $u'(0)$, and Re_D.

30-6 Consider a steady turbulent flow of a constant-property fluid in a long duct formed by two parallel plates. Consider a point sufficiently far removed from the duct entrance that the y-component of mean velocity is zero and the mean flow is entirely in the x direction. Perform Reynolds decomposition of the Navier-Stokes equations in the x and y directions. What can you deduce about the pressure gradients? Obtain an exact expression for computing $\overline{v'_x v'_y}$.

REFERENCES

[1] O. Reynolds, "On the Dynamical Theory of Incompressible Viscous Fluids and the Determination of the Criterion." *Philosophical Transactions of the Royal Society of London*, **186**, 123 (1895).

[2] L. Prandtl, "Bericht über Untersuchungen zur Ausgebildeten Turbulenz." *Zeitschrift für Angewandte Mathematik und. Mechanik*, **5**, 136 (1925).

[3] T. von Kármán, "Mechanische Ähnlichkeit und Turbulenz." *Proceedings of the 3rd International Congress on Applied Mechanics* (Stockholm, 1930), **1**, 85 (1931).

[4] M. P. Escudier, *"The Distribution of Mixing-Length in Turbulent Flows Near Walls."* Heat Transfer Section Report TWF/TN/1, Imperial College, London, UK, 1966.

[5] L. P. Purtell, P. S. Klebanoff, and F. T. Buckley, "Turbulent Boundary Layer at Low Reynolds Number." *Physics of Fluids*, **24**, 802 (1981).

[6] E.-S. Zanoun and F. Dursta, "Evaluating the Law of the Wall in Two-Dimensional Fully Developed Turbulent Channel Flows." *Physics of Fluids*, **15**, 3079 (2003).

[7] E. R. Van Driest, "On Turbulent Flow Near a Wall." *Journal of Aeronautical Science*, **23**, 1007 (1956).

[8] P. S. Anderson, W. M. Kays, and R. J. Moffat, "Experimental Results for the Transpired Turbulent Boundary Layer in an Adverse Pressure Gradient." *Journal of Fluid Mechanics*, **69**, 353 (1975).

Chapter 31

Fully Developed Turbulent Flow

31.1 Turbulent Poiseuille Flow between Smooth Parallel Plates

31.2 Turbulent Couette Flow between Smooth Parallel Plates

31.3 Turbulent Poiseuille Flow in a Smooth-Wall Pipe

31.4 Utility of the Hydraulic Diameter

31.5 Turbulent Poiseuille Flow in a Smooth Annular Pipe

31.6 Reichardt's Formula for Turbulent Diffusivity

31.7 Poiseuille Flow with Blowing between Walls

31.8 Problems

This chapter treats turbulent momentum transport in fully developed internal flows using the mixing length model developed in the context of boundary layers in Chapter 30. There would be little hope of quantitative success in applying the same model if similar turbulent behavior were not exhibited for these two types of flows. Fortunately, experimental measurements of turbulent flows through smooth pipes have demonstrated the same near wall velocity profile as for turbulent boundary layers. Nikuradse [1] found that the velocity profile near the wall is given by

$$u^+ = 2.5 \ln y^+ + 5.5 \tag{31-1}$$

where the wall coordinates are defined the same as in Chapter 30,

$$u^+ = \frac{\overline{v}_x}{\sqrt{\tau_w/\rho}} \qquad y^+ = \frac{y\sqrt{\tau_w/\rho}}{\nu}, \tag{31-2}$$

and $y = R - r$ is the distance from the wall of the pipe of radius R. Nikuradse's equation implies mixing length constants of $\kappa = 0.40$ and $E_\nu^+ = 11.6$. In contrast, the values used for boundary layers in Chapter 30 were $\kappa = 0.41$ and $E_\nu^+ = 10.8$. Other experimental investigations into κ and E_ν^+ show some scatter around the values given by the Nikuradse equation. (A review of some of this literature can be found in reference [2].) A value of $E_\nu^+ = 11.6$ corresponds to $A_\nu^+ = 27$ in the Van Driest damping function (although a value of $A_\nu^+ = 26$ is used more often in the literature).

For boundary layers, the dimension of the momentum boundary layer thickness limits the largest scale of the mixing length. In contrast, the dimension of the pipe or channel for internal flows limits the largest scale of the mixing length. It seems reasonable to assume that for a pipe of radius R, the mixing length is limited in scale to $\ell \leq \gamma R$, where, for lack of better information, the scaling factor can be assumed the same as that used in Chapter 30 for a boundary layer $\gamma = 0.085$.

A conceptual failure of the mixing length model becomes apparent with the consideration of internal flows. Because the model assumes that turbulent diffusivity is proportional to the velocity gradient of the time-averaged flow, it predicts that turbulent diffusivity goes to zero along the centerline of internal flows:

$$\varepsilon_M = \ell^2 \left| \frac{\partial \overline{v}_x}{\partial y} \right| \to 0 \quad \text{(along lines of flow symmetry)}. \tag{31-3}$$

This consequence of the mixing length model is not reconciled with a physically correct picture of turbulence. However, since the momentum flux goes to zero along lines of flow symmetry anyway, it will be shown that this deficiency in the mixing length model will not have a great quantitative impact on many calculations. To patch shortcomings of the mixing length model, empirical relations for ε_M can be employed for the outer region of turbulence, such as Reichardt's formula discussed in Section 31.6. However, some of the main issues of the mixing length model will be overcome in a conceptually more satisfying way when the k-epsilon model of turbulence is discussed in Chapter 36.

31.1 TURBULENT POISEUILLE FLOW BETWEEN SMOOTH PARALLEL PLATES

To illustrate application of the mixing length model to an internal flow, consider Poiseuille (pressure-driven) flow between smooth parallel plates, as illustrated in Figure 31-1. The flow is assumed to be fully developed, such that $\partial \overline{v}_x/\partial x = 0$ and $\overline{v}_y = 0$ everywhere in the flow. The momentum equation, with appropriate terms set to zero, becomes

$$0 = \frac{\partial}{\partial y}\left[(\nu + \varepsilon_M)\frac{\partial \overline{v}_x}{\partial y}\right] - \frac{1}{\rho}\frac{\partial \overline{P}}{\partial x}. \tag{31-4}$$

Upon integrating with respect to y, the momentum equation becomes

$$c_1 = (\nu + \varepsilon_M)\frac{\partial \overline{v}_x}{\partial y} - \frac{1}{\rho}\frac{\partial \overline{P}}{\partial x}y. \tag{31-5}$$

This result can be evaluated at the wall ($y = 0$), where $\varepsilon_M = 0$, to determine the integration constant:

$$c_1 = \nu \left.\frac{\partial \overline{v}_x}{\partial y}\right|_{y=0}. \tag{31-6}$$

Furthermore, the pressure gradient, acting on the streamwise area between the plates ($a \cdot 1$), must be balanced by the viscous stresses transmitted to the walls ($\tau_w = \mu(\partial \overline{v}_x/\partial y)|_{y=0}$). Consequently,

$$2\tau_w = a\left(-\frac{d\overline{P}}{dx}\right) \quad \text{or} \quad -\frac{1}{\rho}\frac{d\overline{P}}{dx} = \frac{2}{a}\nu\left.\frac{\partial \overline{v}_x}{\partial y}\right|_{y=0}. \tag{31-7}$$

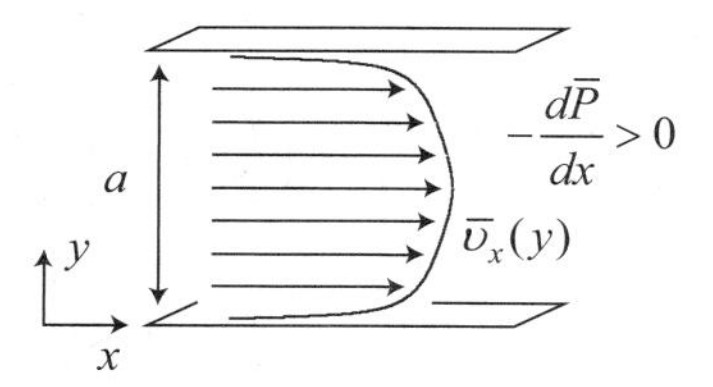

Figure 31-1 Turbulent Poiseuille flow between smooth parallel plates.

Therefore, the momentum equation, Eq. (31-5), can be written as

$$\nu \frac{d\overline{v}_x}{dy}\bigg|_{y=0} = (\nu + \varepsilon_M)\frac{d\overline{v}_x}{dy} + \frac{2\nu}{a}\frac{d\overline{v}_x}{dy}\bigg|_{y=0} y, \tag{31-8}$$

where, recognizing that the time averaged velocity in the x-direction is only a function of y, the partial derivatives have been replaced with total derivatives. Using the dimensionless variables

$$\eta = y/a \quad \text{and} \quad u = \overline{v}_x/\overline{v}_m, \quad \text{where } \overline{v}_m = \int_0^a \overline{v}_x dy/a, \tag{31-9}$$

the momentum equation can be expressed as

$$(1 + \varepsilon_M/\nu)u' = (1 - 2\eta)u'(0), \tag{31-10}$$

where primes are used to denote derivatives with respect to η. The solution for u must satisfy the no-slip condition at the wall, $u(0) = 0$. Notice that the second boundary condition, $u'(1/2) = 0$, is enforced automatically by the governing equation. However, the solution for u must also satisfy

$$\int_0^1 u d\eta = 1, \tag{31-11}$$

which follows from the definitions of u and $\overline{v}_m$. This condition replaces the second (centerline) boundary condition.

The mixing length model is employed to evaluate turbulent diffusivity in the momentum equation (31-10). Defining a dimensionless mixing length Λ, the turbulent diffusivity becomes

$$\frac{\varepsilon_M}{\nu} = \frac{\ell^2}{\nu}\left|\frac{\partial \overline{v}_x}{\partial y}\right| = \frac{\overline{v}_m a}{\nu}\Lambda^2|u'| \quad \text{where} \quad \Lambda = \frac{\ell}{a}. \tag{31-12}$$

Notice that a Reynolds number is required to evaluate the turbulent diffusivity from Eq. (31-12). This is how the expected dependency of the solution on the Reynolds number enters into the problem. It is customary to define the Reynolds number based on the hydraulic diameter of the flow. For flow between parallel plates, the hydraulic diameter is given by

$$D = \frac{4A}{\Gamma} = 2a, \tag{31-13}$$

where A is the flow cross-sectional area and Γ is the wetted perimeter. Therefore, defining the Reynolds number to be

$$\mathrm{Re}_D = \frac{\overline{v}_m(2a)}{\nu}, \tag{31-14}$$

Eq. (31-12) for the turbulent diffusivity becomes

$$\frac{\varepsilon_M}{\nu} = (\mathrm{Re}_D/2)\Lambda^2 u'. \tag{31-15}$$

As previously mentioned, the mixing length model predicts that turbulent diffusivity goes to zero at the centerline of the flow. This is a significant conceptual error, since it implies that the flow returns to a laminar state along the centerline! However, the resulting quantitative error is less significant since symmetry requires that the momentum flux be zero along the centerline of the flow.

The mixing length scale can be evaluated with the Van Driest damping function,

$$\ell = \kappa y \left[1 - \exp\left(-\frac{y}{\nu A_\nu^+}\sqrt{\frac{\tau_w}{\rho}}\right)\right] \quad \text{(from the wall through inner region)}, \tag{31-16}$$

which was discussed in Section 30.7. Assuming that the largest eddies in the flow must scale with the half-distance between the plates (because of the flow symmetry), the largest mixing length is evaluated as

$$\ell = \gamma\, a/2 \quad \text{(in the outer region)}. \tag{31-17}$$

Therefore, the dimensionless mixing length $\Lambda = \ell/a$ is selected to satisfy the rule

$$\Lambda = \begin{cases} \Lambda^* & \Lambda^* < \gamma/2 \\ \gamma/2 & \textit{otherwise} \end{cases} \tag{31-18}$$

where (31-16) becomes

$$\Lambda^* = \kappa\eta\left[1 - \exp\left(\frac{-\eta}{A_\nu^+}\sqrt{(\mathrm{Re}_D/2)u'(0)}\right)\right]. \tag{31-19}$$

Substituting Eq. (31-15) for the turbulent diffusivity into Eq. (31-10) for the momentum equation yields

$$\left[1 + (\mathrm{Re}_D/2)\Lambda^2 u'\right]u' = (1 - 2\eta)u'(0) \tag{31-20}$$

or

$$\frac{du}{d\eta} = \begin{cases} u'(0) & , \quad \eta = 0 \\ \dfrac{\sqrt{1 + 2\Lambda^2 \mathrm{Re}_D u'(0)(1 - 2\eta)} - 1}{\Lambda^2 \mathrm{Re}_D}, & 0 < \eta \le 1. \end{cases} \tag{31-21}$$

A numerical solution for u can be sought from Eq. (31-21) subject to the conditions that

$$u(0) = 0 \tag{31-22}$$

and

$$\int_0^{1/2} u\, d\eta = 1/2, \tag{31-23}$$

where the solution is sought for $0 \le \eta \le 1/2$ using the symmetry of the problem.

Equation (31-21) can be numerically integrated using the fourth-order Runge-Kutta method described in Chapter 23. Since $u'(0)$ is required for integration but is unknown, a shooting method is employed where consecutive guesses of $u'(0)$ are evaluated by checking the requirement given by Eq. (31-23). The bisection method (see Section 23.4) requires establishing upper and lower bounds on possible guesses for $u'(0)$. For a laminar flow with normalized velocity and spatial scales, one might expect $u'(0)$ to be of order 1. However, this expectation fails for turbulent flows because the spatial scale over which large changes in velocity occur is much smaller than the hydraulic diameter. Therefore, additional care must be taken to ensure that $u'(0)$ is correctly bounded in the implementation of the shooting method.

The large change in transport that occurs over the viscous sublayer and the inner region, as discussed in Section 30.6, is best resolved on a log scale of distance from the wall. Numerical integration of the momentum equation with uniform spatial discretization would require an excessively large number of nodes to resolve the near wall region. Therefore, it is desirable to integrate the turbulent momentum equation with a nonuniform spatial discretization that brings a high density of nodes to the near wall region. One approach is to uniformly discretize the log-of-distance from the wall. This approach is used in Code 31-1 to solve the turbulent Poiseuille flow problem between the wall and the centerline of the flow: $0 \le \eta \le 1/2$. With $0 \le n < N$, $\eta_{n=0} = 0$, and $\eta_{N-1} = 1/2$, distances away from the wall ($n > 0$) are assigned by

$$\eta_n = \eta_{N-1} \times 10^{(\text{log\_min} + (n-1) \times \text{log\_del})} \tag{31-24}$$

where log_min is the log-distance to the first node away from the wall, and log_del = (−log_min)/(N − 2) is the log-space increment between nodes.

Figure 31-2 shows velocity profiles for $\mathrm{Re}_D = 10^4$, 10^5, and 10^6, calculated with Code 31-1 using the mixing length model. The model constants used in this calculation are $\kappa = 0.40$, $\gamma = 0.085$, and $A_\nu^+ = 26$. The wall variables for position and velocity are calculated from the solution with

$$y^+ = \eta\sqrt{u'(0)\mathrm{Re}_D/2} \quad \text{and} \quad u^+ = u/\sqrt{2u'(0)/\mathrm{Re}_D}. \tag{31-25}$$

The solution in the inner region obeys the law of the wall, as it should since the mixing length model defines this expectation. However, at sufficiently large distances from the wall, the velocity profile departs from the law of the wall as the solution enters the outer

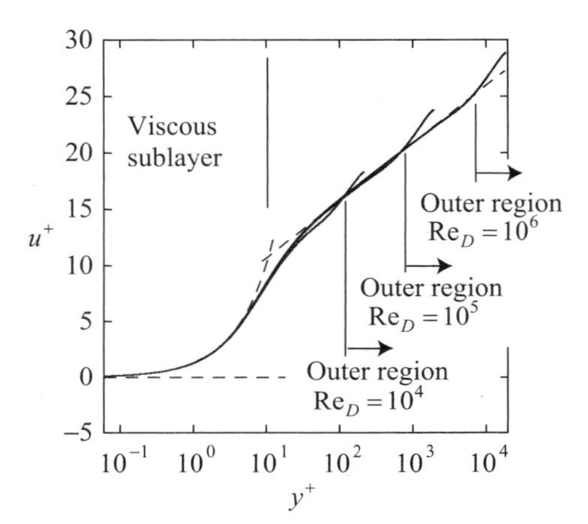

Figure 31-2 Velocity profile for Poiseuille flow in wall variables.

Code 31-1 Turbulent Poiseuille flow between smooth parallel plates

```
#include <stdio.h>
#include <stdlib.h>
#include <math.h>
inline double VanDriest(double eta,double Ap,double Kappa,
                        double Gamma,double ReD_du0)
{
  double yp=eta*sqrt(ReD_du0/2.);
  double Lam=Kappa*eta*(1.-exp(-yp/Ap));
  if (Lam>Gamma/2.) return Gamma/2.;
  return Lam;
}
inline double du(double ReD,double du0,double eta)
{
  double Kappa=0.4,Gamma=0.085,Ap=26.;
  if (eta==0.) return du0;
  double Lam=VanDriest(eta,Ap,Kappa,Gamma,ReD*du0);
  if (Lam==0.) return (1.-2.*eta)*du0;
  return (sqrt(1.+2.*ReD*Lam*Lam*du0*(1.-2.*eta))-1.)/(ReD*Lam*Lam);
}
double solve_u(int N,double *eta,double *u,double ReD)
{
  int n,iter=0;
  double del_eta,K1u,K2u,K3u,K4u;
  double du0,um,K1um,K2um,K3um,K4um;
  double du0_low=1.; /* lower bound on du0 */
  double du0_high=1000000.0; /* upper bound on du0 */
  u[0]=0.0; /* initial conditions */
  do   /* RK integration */
    {
      um=0.;
      du0=(du0_low+du0_high)/2.0; /* use bisection method */
      for (n=0; n<N-1; ++n)
        {
           del_eta=eta[n+1]-eta[n];
           K1u=del_eta*du(ReD,du0,eta[n]);
           K2u=K3u=del_eta*du(ReD,du0,(eta[n]+eta[n+1])/2.);
           K4u=del_eta*du(ReD,du0,eta[n+1]);
           K1um= del_eta*u[n];
           K2um= del_eta*(u[n]+0.5*K1um);
           K3um= del_eta*(u[n]+0.5*K2um);
           K4um= del_eta*(u[n]+K3um);
           K1um= del_eta*u[n];
           K2um= del_eta*(u[n]+0.5*K1u);
           K3um= del_eta*(u[n]+0.5*K2u);
           K4um= del_eta*(u[n]+K3u);
           u[n+1]=u[n]+(K1u+2.*K2u+2.*K3u+K4u)/6.;
           um+=(K1um+2.*K2um+2.*K3um+K4um)/6.;
           if (um>0.5)
           {
              du0_high=du0; /* shoot lower */
              break;
            }
        }
      if (n==N-1) du0_low=du0; /* shoot higher */
    }
  while (++iter<200 && (du0_high-du0_low)>1.0e-7);
  if (fabs(um-0.5)>.00001)
    {
      printf("\nSoln. failed, du0=%e %e\n",du0,um);
      exit(1);
    }
  return du0;
}
int main()
{
   int n,N=2000;
   double yp,up,du0,eta[N],u[N];
   eta[0]=0.;
   eta[N-1]=0.5;
   double log_min=-6.; // use log steps
   double log_del=(-log_min)/(N-2);
   for (n=1; n<N-1; ++n)
     eta[n]=eta[N-1]*pow(10.,log_min+(n-1)*log_del);
   FILE *fp_up=fopen("up.dat","w");
   double ReD=1.0e5;
   du0=solve_u(N,eta,u,ReD); /* solve for u & du0 */
   double cf=4.*du0/ReD;
   printf("%e %e\n",ReD,cf);
   for (n=0; n<N; n++) /* output velocity */
    {
       yp=eta[n]*sqrt(du0*ReD/2.);
       up=u[n]/sqrt(2.*du0/ReD);
       fprintf(fp_up,"%e %e\n",yp,up);
    }
  fclose(fp_up);
  return 0;
}
```

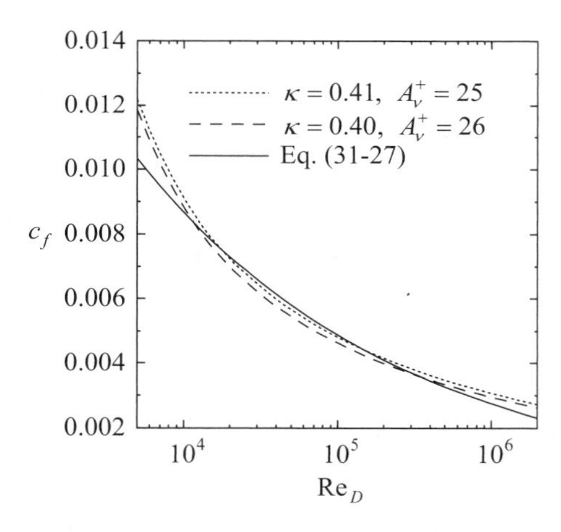

Figure 31-3 Coefficient of friction for Poiseuille flow between smooth parallel plates.

region of turbulence. As the Reynolds number increases, this departure occurs at greater distances (in wall coordinates) from the wall.

One important result that can be obtained from the solution is the friction coefficient:

$$c_f = \frac{\tau_w}{\rho \overline{v}_m^2/2} = 2\frac{\nu}{\overline{v}_m a} u'(0) = \frac{4u'(0)}{\mathrm{Re}_D}. \tag{31-26}$$

Based on a comprehensive study of available data, Dean [3] suggested that the following correlation be used for fully developed turbulent Poiseuille flow between parallel plates:

$$c_f = \frac{0.0868}{(\mathrm{Re}_D)^{1/4}}. \tag{31-27}$$

This correlation is plotted alongside the results of the mixing length model in Figure 31-3. The model results are shown for two different sets of mixing length constants, one used for boundary layer flow ($\kappa = 0.41$, $A_\nu^+ = 25$) in Chapter 30, and the other thought to be more appropriate for pipe flow ($\kappa = 0.40$, $A_\nu^+ = 26$). It is seen that small changes in the constants yield little quantitative difference in the results of the mixing length model.

In many cases of wall turbulence, only modest departures from the law of the wall are observed in the outer region of the boundary layer (as illustrated in Figure 31-2 for the present problem). As far as calculating the coefficient of friction is concerned, much of the success of the mixing length model is guaranteed by conforming to the law of the wall. For this reason, the coefficient of friction determined from the mixing length model is relatively insensitive to the choice of γ. For example, changing γ by an order of magnitude (from 0.085 to 1.0) only results in a 10% increase in the value of c_f for $\mathrm{Re}_D = 10^5$ Poiseuille flow between parallel plates.

31.2 TURBULENT COUETTE FLOW BETWEEN SMOOTH PARALLEL PLATES

Consider fully developed Couette flow between smooth parallel plates, as illustrated in Figure 31-4. With $\partial P/\partial x = 0$, $\partial \overline{v}_x/\partial x = 0$, and $\overline{v}_y = 0$, the momentum equation governing this flow becomes

$$0 = \frac{\partial}{\partial y}\left[(\nu + \varepsilon_M)\frac{\partial \overline{v}_x}{\partial y}\right] \tag{31-28}$$

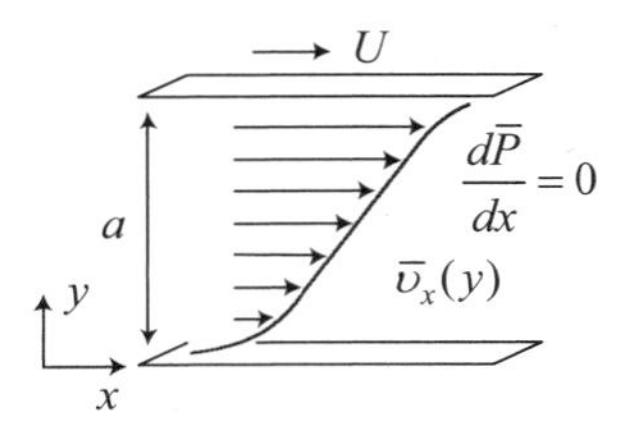

Figure 31-4 Turbulent Couette flow between smooth parallel plates.

and is subject to the boundary conditions

$$\bar{v}_x(0) = 0 \quad \text{and} \quad \bar{v}_x(a/2) = U/2. \tag{31-29}$$

Specifying the velocity at the mid-position, $y = a/2$, is made possible by exploiting the point symmetry in the solution. Integrating the momentum equation with respect to y yields

$$\nu \frac{\partial \bar{v}_x}{\partial y}\bigg|_{y=0} = (\nu + \varepsilon_M)\frac{\partial \bar{v}_x}{\partial y} \tag{31-30}$$

where the integration constant is evaluated at $y = 0$ (same as for Poiseuille flow in the previous section). Introducing the dimensionless variables $\eta = y/a$ and $u = \bar{v}_x/U$, the momentum equation becomes

$$u'(0) = \left(1 + \frac{\varepsilon_M}{\nu}\right)u', \tag{31-31}$$

where primes are used to denote derivatives with respect to η. For Couette flow, the turbulent diffusivity is well described by the mixing length model everywhere in the flow, since the velocity gradient never goes to zero. Defining a dimensionless mixing length as $\Lambda = \ell/a$, Eq. (31-12) for the turbulent diffusivity becomes

$$\frac{\varepsilon_M}{\nu} = \text{Re}_a\, \Lambda^2 u', \tag{31-32}$$

where

$$\text{Re}_a = \frac{\bar{v}_m a}{\nu}. \tag{31-33}$$

The Reynolds number is now defined based on the distance between the two plates (and not the hydraulic diameter). The Van Driest damping function for mixing length is employed again, such that

$$\Lambda^* = \kappa\eta\left[1 - \exp\left(\frac{-\eta}{A_\nu^+}\sqrt{\text{Re}_a u'(0)}\right)\right], \tag{31-34}$$

and the dimensionless mixing length is selected to satisfy the rule

$$\Lambda = \begin{cases} \Lambda^* & \Lambda^* < \gamma \\ \gamma & \textit{otherwise} \end{cases}. \tag{31-35}$$

Notice that for Couette flow, it is supposed that $\ell = \gamma\, a$ in the outer region of turbulence, while for Poiseuille flow $\ell = \gamma\, a/2$ was adopted. The logic behind this difference is based on the observation that Poiseuille flow has a sign reversal in the velocity gradient at $a/2$, while Couette flow does not. Therefore, it seems reasonable that turbulent eddies could fill the entire gap between the plates in Couette flow, while for Poiseuille flow they should not. Substituting Eq. (31-32) for the turbulent diffusivity into Eq. (31-31) for the momentum equation yields

$$\frac{du}{d\eta} = \begin{cases} u'(0) & , \quad \eta = 0 \\ \dfrac{\sqrt{1 + 4\mathrm{Re}_a\,\Lambda^2 u'(0)} - 1}{2\mathrm{Re}_a \Lambda^2}, & \quad \eta \neq 0 \end{cases} \tag{31-36}$$

and is subject the boundary conditions

$$u(0) = 0 \tag{31-37}$$

and

$$u(1/2) = 1/2. \tag{31-38}$$

A numerical solution to Eq. (31-36) for u can be sought using the fourth-order Runge-Kutta method. Since $u'(0)$ is required for integration but is unknown, a shooting method is employed where consecutive guesses at $u'(0)$ are evaluated using the second boundary condition given by Eq. (31-38). Figure 31-5 shows the velocity profile calculated from the mixing length model for the Reynolds numbers $\mathrm{Re}_D = 10^4,\ 10^5$, and 10^6. The mixing length constants are evaluated using $\kappa = 0.41$, $A_\nu^+ = 25$, and $\gamma = 0.085$.

Unlike laminar Couette flow, the turbulent velocity profile shown in Figure 31-5 is highly nonlinear despite the fact that the momentum flux (and the shear stress) remains constant across the flow. Equation (31-30) illustrates that as the turbulent eddy diffusivity ε_M increases, moving away from the wall, the velocity field gradient must decrease to maintain a constant momentum flux. This trend is exhibited clearly through the inner region of turbulence.

The friction coefficient for Couette flow can be expressed in terms of the Reynolds number by

$$c_f = \frac{\tau_w}{\rho \bar{v}_m^2/2} = 2\frac{\nu}{\bar{v}_m a} u'(0) = \frac{2u'(0)}{\mathrm{Re}_a}. \tag{31-39}$$

In Figure 31-6, the friction coefficient is contrasted between Couette flow and Poiseuille flow, as calculated using the mixing length model. It is observed that Couette flow incurs

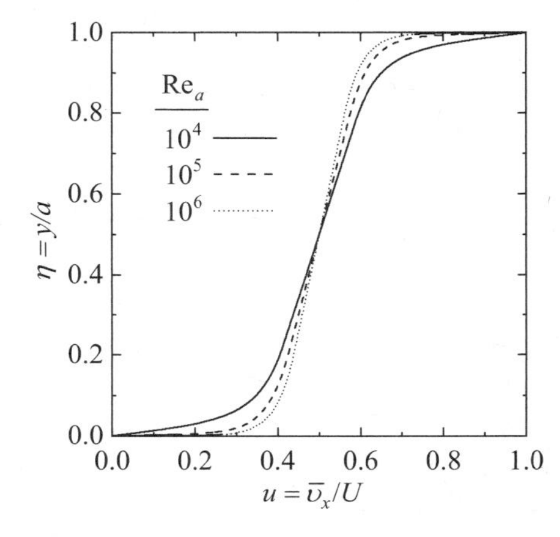

Figure 31-5 Turbulent Couette velocity profile between smooth parallel plates.

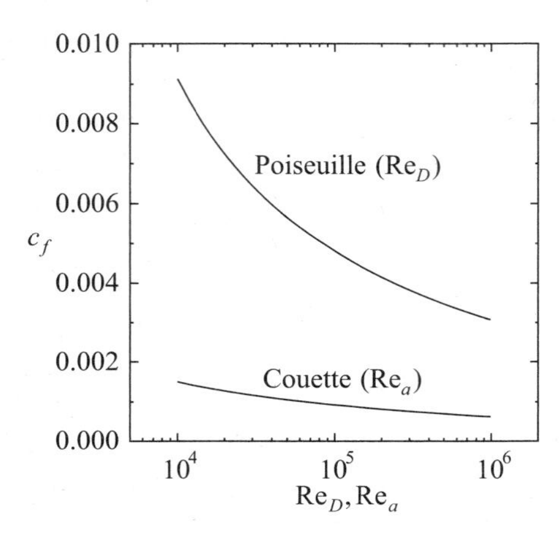

Figure 31-6 Friction coefficients for turbulent Couette and Poiseuille flows between smooth parallel plates.

significantly less friction (shear stress) between the flow and plates for the same Reynolds number. This difference can be attributed to the difference in velocity profile across the flow, since Poiseuille flow creates steeper velocity gradients near the walls than Couette flow for the same Reynolds number.

31.3 TURBULENT POISEUILLE FLOW IN A SMOOTH-WALL PIPE

To illustrate the application of the mixing length model to a significantly different flow geometry, consider Poiseuille flow in a smooth pipe of radius R, as illustrated in Figure 31-7. The flow is assumed to be fully developed, such that $\partial \overline{v}_z / \partial z = 0$ and $\overline{v}_r = 0$ everywhere in the flow. The time-averaged turbulent momentum equation in cylindrical form, with appropriate terms set to zero, is

$$0 = \frac{1}{r} \frac{\partial}{\partial r} \left[r(\nu + \varepsilon_M) \frac{\partial \overline{v}_z}{\partial r} \right] - \frac{1}{\rho} \frac{\partial \overline{P}}{\partial z}. \tag{31-40}$$

Implementing the mixing length model, it is left as an exercise (see Problem 31-1) to show that the momentum equation takes the form

$$\frac{du}{d\eta} = \begin{cases} u'(0) & , \quad \eta = 0 \\ \dfrac{\sqrt{1 + 2\mathrm{Re}_D \Lambda^2 u'(0)(1-\eta)} - 1}{\mathrm{Re}_D \Lambda^2}, & \eta \neq 0 \end{cases} \tag{31-41}$$

subject to

$$u(0) = 0 \quad \text{and} \quad 2 \int_0^1 u(1-\eta) d\eta = 1 \tag{31-42}$$

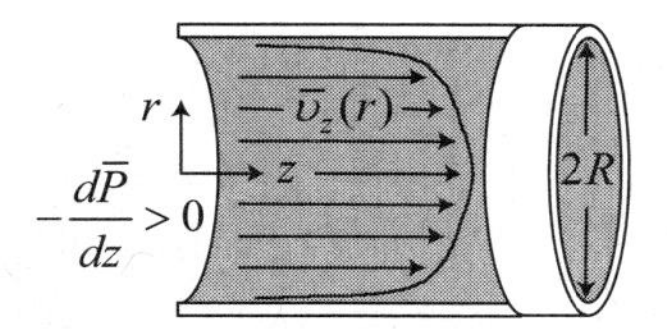

Figure 31-7 Turbulent Poiseuille flow in a smooth-wall pipe.

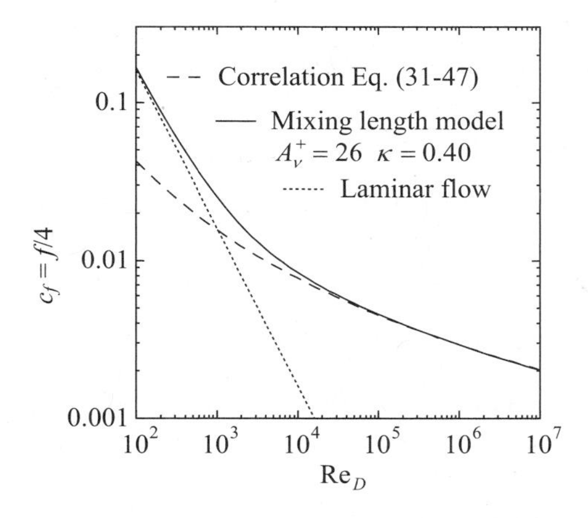

Figure 31-8 Coefficient of friction for turbulent Poiseuille flow in a smooth-wall pipe.

where

$$u = \overline{v}_z/\overline{v}_m, \quad \eta = 1 - r/R, \quad \Lambda = \ell/R, \quad \text{and} \quad \text{Re}_D = \overline{v}_m(2R)/\nu. \tag{31-43}$$

The Van Driest damping function is used to evaluate the mixing length from the wall through the inner region of the wall turbulence. In the outer region of turbulence, $\ell = \gamma R$ is used. Therefore, the dimensionless mixing length is selected to satisfy the rule

$$\Lambda = \begin{cases} \Lambda^* & \Lambda^* < \gamma \\ \gamma & \textit{otherwise} \end{cases} \tag{31-44}$$

where

$$\Lambda^* = \kappa\eta\left[1 - \exp\left(\frac{-\eta}{A_\nu^+}\sqrt{u'(0)\text{Re}_D/2}\right)\right]. \tag{31-45}$$

Equation (31-41) can be solved by the fourth-order Runge-Kutta method with shooting to determine $u'(0)$. From the solution, the friction coefficient can be determined:

$$c_f = \frac{\tau_w}{\rho\overline{v}_m^2/2} = 2\frac{\nu}{\overline{v}_m R}u'(0) = \frac{4u'(0)}{\text{Re}_D}. \tag{31-46}$$

As shown in Figure 31-8, the mixing length model results are in good agreement with the predictions made using Prandtl's correlation [4] for turbulent smooth pipe flow:

$$1/f^{1/2} = 2.0\log\left(\text{Re}_D f^{1/2}\right) - 0.8. \tag{31-47}$$

Notice that Prandtl's correlation makes use of Darcy's friction factor*:

$$f = \frac{(-dP/dx)D}{\rho\overline{v}_m^2/2}. \tag{31-48}$$

For the current geometry of Poiseuille flow, $f = 4c_f$. For the comparison with Prandtl's correlation, the mixing length model uses the constants $A_\nu^+ = 26$, $\kappa = 0.40$, and $\gamma = 0.085$.

*Named after the French engineer Henry Philibert Gaspard Darcy (1803–1858).

Although Prandtl's correlation is not valid for laminar flows, in the limit of low Reynolds numbers the mixing length model yields the laminar flow solution.

31.4 UTILITY OF THE HYDRAULIC DIAMETER

The reason the hydraulic diameter was introduced in Section 27.1 was to "normalize" the effect of the cross-sectional geometry of the flow. Having solved the turbulent Poiseuille flow problem for two different geometries, now is a good opportunity to revisit this rationale.

Figure 31-9 contrasts Poiseuille flow results for c_f between the parallel plate and the pipe flow geometries. Results for the parallel plate geometry are presented as a function of the Reynolds number based on two different length scales: Re_a uses the gap between the plates and Re_D uses the hydraulic diameter. When the Reynolds number is defined with the hydraulic diameter, results for the parallel plate geometry lie close to results for the pipe flow geometry. This suggests that correlations derived for one turbulent flow geometry may be used to a good approximation for different flow geometries so long as the hydraulic diameter is used as the characteristic length scale in the definition of the Reynolds number.

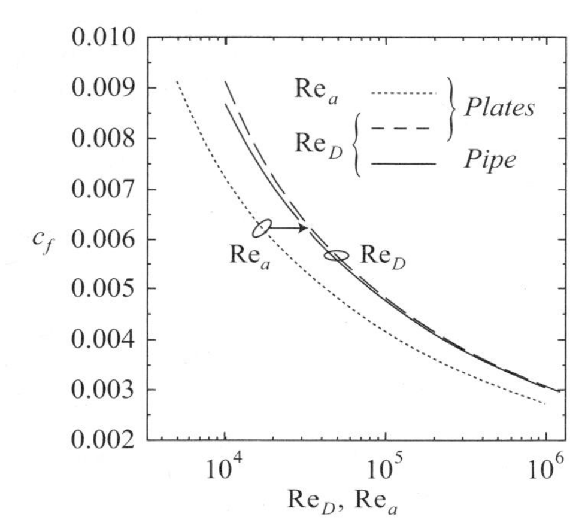

Figure 31-9 Influence of the hydraulic diameter in the Reynolds number on reporting Poiseuille flow c_f results.

31.5 TURBULENT POISEUILLE FLOW IN A SMOOTH ANNULAR PIPE

It is advantageous when an internal turbulent flow solution can make use of a known plane of symmetry (or point of symmetry), as was the case in Sections 31.1 through 31.3. However, when such symmetry does not exist, a numerical solution must be integrated across the entire flow, as is illustrated next.

Consider Poiseuille flow in a smooth annulus, as illustrated in Figure 31-10. The turbulent flow is assumed to be fully developed, such that $\partial \overline{v}_z/\partial z = 0$ and $\overline{v}_r = 0$. Therefore, the appropriate momentum equation to be solved in cylindrical coordinates is

$$0 = \frac{1}{r}\frac{\partial}{\partial r}\left[r(\nu + \varepsilon_M)\frac{\partial \overline{v}_z}{\partial r}\right] - \frac{1}{\rho}\frac{\partial \overline{P}}{\partial z}. \tag{31-49}$$

After integrating once, the momentum equation becomes

$$c_1 = r(\nu + \varepsilon_M)\frac{\partial \overline{v}_z}{\partial r} - \frac{1}{\rho}\frac{\partial \overline{P}}{\partial z}\frac{r^2}{2}. \tag{31-50}$$

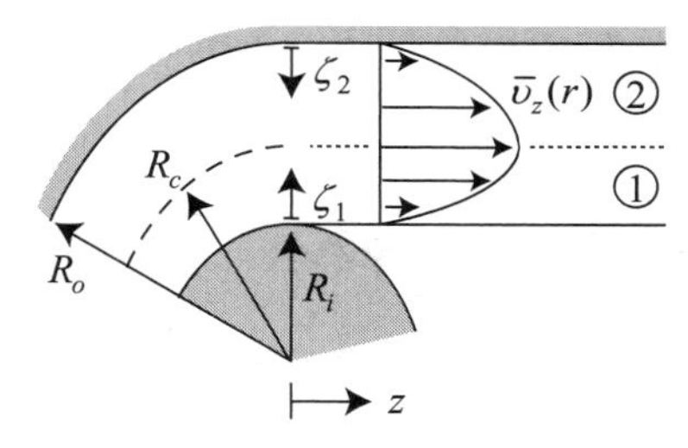

Figure 31-10 Turbulent Poiseuille flow in a smooth-wall annular pipe.

Further integration of the momentum equation for annular flow is complicated by the fact that the velocity field lacks symmetry. However, the flow may be subdivided into the two regions shown in Figure 31-10. Region 1 covers the range $R_i \le r \le R_c$ and region 2 covers the range $R_c \le r \le R_o$, where R_i and R_o are the inner and outer radii of the annulus, respectively. The two regions share a common boundary at $r = R_c$, which is defined as the position where

$$\text{at } r = R_c: \quad \partial \overline{v}_z/\partial r\big|_{R_c} = 0. \tag{31-51}$$

Because of the lack of symmetry, R_c is an unknown part of the solution. The integration constant in the momentum equation (31-50) can be evaluated three ways:

$$\text{at } r = R_i: \quad c_1 = R_i \nu \frac{\partial \overline{v}_z}{\partial r}\bigg|_{R_i} - \frac{1}{\rho}\frac{\partial \overline{P}}{\partial z}\frac{R_i^2}{2}, \tag{31-52}$$

$$\text{at } r = R_c: \quad c_1 = -\frac{1}{\rho}\frac{\partial \overline{P}}{\partial z}\frac{R_c^2}{2}, \tag{31-53}$$

$$\text{at } r = R_o: \quad c_1 = R_o \nu \frac{\partial \overline{v}_z}{\partial r}\bigg|_{R_o} - \frac{1}{\rho}\frac{\partial \overline{P}}{\partial z}\frac{R_o^2}{2}. \tag{31-54}$$

The pressure gradient may be expressed in two alternate forms by combining Eqs. (31-52) and (31-53) in the first case, and combining Eqs. (31-53) and (31-54) in the second. These two alternate forms are

$$-\frac{1}{\rho}\frac{\partial \overline{P}}{\partial z} = \frac{2R_i \nu}{R_c^2 - R_i^2}\frac{\partial \overline{v}_z}{\partial r}\bigg|_{R_i} \tag{31-55}$$

and

$$-\frac{1}{\rho}\frac{\partial \overline{P}}{\partial z} = \frac{2R_o \nu}{R_c^2 - R_o^2}\frac{\partial \overline{v}_z}{\partial r}\bigg|_{R_o}. \tag{31-56}$$

The axial pressure gradient may be eliminated between Eqs. (31-55) and (31-56) to reveal the relation between the velocity gradients at the inner and outer walls:

$$\frac{\partial \overline{v}_z}{\partial r}\bigg|_{R_o} = \frac{R_i}{R_o}\frac{R_c^2 - R_o^2}{R_c^2 - R_i^2}\frac{\partial \overline{v}_z}{\partial r}\bigg|_{R_i}. \tag{31-57}$$

Additionally, the integration constant may be eliminated from the momentum equation (31-50) using Eq. (31-53) for the result

$$r(\nu+\varepsilon_M)\frac{\partial\overline{v}_z}{\partial r}=-\frac{1}{\rho}\frac{\partial\overline{P}}{\partial z}\left(\frac{R_c^2-r^2}{2}\right). \tag{31-58}$$

Then the momentum equation may be written in two forms by substituting the alternate expressions for the pressure gradient, Eqs. (31-55) and (31-56), into Eq. (31-58) for the results

$$r\left(1+\frac{\varepsilon_M}{\nu}\right)\frac{\partial\overline{v}_z}{\partial r}=R_i\frac{R_c^2-r^2}{R_c^2-R_i^2}\frac{\partial\overline{v}_z}{\partial r}\bigg|_{R_i} \tag{31-59}$$

and

$$r\left(1+\frac{\varepsilon_M}{\nu}\right)\frac{\partial\overline{v}_z}{\partial r}=R_o\frac{R_c^2-r^2}{R_c^2-R_o^2}\frac{\partial\overline{v}_z}{\partial r}\bigg|_{R_o}. \tag{31-60}$$

Notice that when $r=R_c$, both momentum equations enforce the condition $\partial\overline{v}_z/\partial r|_{R_c}=0$. Equations (31-59) and (31-60) can be solved as coupled equations for $\overline{v}_z$, subject to the conditions that both equations evaluate to the same velocity $\overline{v}_z(r=R_c)$ at the common boundary, and that Eq. (31-57) relating the velocity gradients at the inner and outer walls is satisfied.

The problem is made dimensionless using the variables:

$$\eta=r/R_o \quad \text{and} \quad u=\overline{v}_z/\overline{v}_m, \tag{31-61}$$

where

$$\overline{v}_m\pi(R_o^2-R_i^2)=\int_{R_i}^{R_o}2\pi r\overline{v}_z dr. \tag{31-62}$$

The relation between u and $\overline{v}_m$ requires that

$$\int_{\eta_i}^{1}u(\eta)\,\eta\,d\eta=\frac{1-\eta_i^2}{2}. \tag{31-63}$$

The momentum equations (31-59) and (31-60) written in dimensionless form become

$$\eta\left(1+\frac{\varepsilon_M}{\nu}\right)\frac{\partial u}{\partial\eta}=\eta_i\frac{\eta_c^2-\eta^2}{\eta_c^2-\eta_i^2}\frac{\partial u}{\partial\eta}\bigg|_{\eta_i} \tag{31-64}$$

and

$$\eta\left(1+\frac{\varepsilon_M}{\nu}\right)\frac{\partial u}{\partial\eta}=\frac{\eta_c^2-\eta^2}{\eta_c^2-1}\frac{\partial u}{\partial\eta}\bigg|_{1}. \tag{31-65}$$

The two regions in the annulus are integrated for u separately. For region 1, where $\eta_i\le\eta\le\eta_c$, Eq. (31-64) will be integrated with respect to a new dependent variable

$\zeta_1 = \eta - \eta_i$. For region 2, where $\eta_c \le \eta \le 1$, Eq. (31-65) will be integrated with respect to a new dependent variable $\zeta_2 = 1 - \eta$. Both new variables measure distances from the walls, as illustrated in Figure 31-10. In terms of the new variables, the momentum equations (31-64) and (31-65) may be written as

$$(\eta_i + \zeta_1)\left(1 + \left(\frac{\varepsilon_M}{\nu}\right)_1\right)\frac{\partial u_1}{\partial \zeta_1} = \eta_i \frac{\eta_c^2 - (\eta_i + \zeta_1)^2}{\eta_c^2 - \eta_i^2}\left.\frac{\partial u_1}{\partial \zeta_1}\right|_0 \quad (0 \le \zeta_1 < \eta_c - \eta_i) \tag{31-66}$$

and

$$(1 - \zeta_2)\left(1 + \left(\frac{\varepsilon_M}{\nu}\right)_2\right)\frac{\partial u_2}{\partial \zeta_2} = \frac{\eta_c^2 - (1 - \zeta_2)^2}{\eta_c^2 - 1}\left.\frac{\partial u_2}{\partial \zeta_2}\right|_0 \quad (0 \le \zeta_2 < 1 - \eta_c)\,. \tag{31-67}$$

Turbulent diffusivity is also expressed in terms of the new dimensionless variables:

$$\left(\frac{\varepsilon_M}{\nu}\right)_{j=1,2} = \frac{\ell^2}{\nu}\left|\frac{\partial \overline{v}_z}{\partial r}\right| = \frac{\Lambda_j^2 \mathrm{Re}_D}{2(1 - \eta_i)}\left.\frac{\partial u}{\partial \zeta}\right|_{j=1,2}, \tag{31-68}$$

where

$$\Lambda_j = \ell_j / R_o \quad \text{and} \quad \mathrm{Re}_D = \frac{\overline{v}_m 2R_o(1 - \eta_i)}{\nu}. \tag{31-69}$$

Notice that the Reynolds number is defined using the hydraulic diameter $D = 2R_o(1 - \eta_i)$. The mixing length Λ_j is evaluated with the Van Driest damping function (31-16) through the inner region of the wall turbulence, and becomes a constant in the outer region. In dimensionless form, the mixing lengths in the two regions of the annulus flow are evaluated from

$$\Lambda_{j=1,2} = \begin{cases} \Lambda_j^*(\zeta_j) & \Lambda_j^*(\zeta_j) < \gamma\, \zeta_{c,j} \\ \gamma\, \zeta_{c,j} & \textit{otherwise} \end{cases}, \tag{31-70}$$

where

$$\zeta_{c,j} = \begin{cases} \eta_c - \eta_i & (j = 1) \\ 1 - \eta_c & (j = 2) \end{cases} \tag{31-71}$$

and

$$\Lambda_j^*(\zeta_j) = \kappa \zeta_j \left[1 - \exp\left(\frac{-\zeta_j}{A_\nu^+}\sqrt{\frac{\mathrm{Re}_D u_j'(0)}{2(1 - \eta_i)}}\right)\right]. \tag{31-72}$$

With Eq. (31-68) for the turbulent diffusivity, the momentum equations (31-66) and (31-67) can be expressed as

$$\frac{du_j}{d\zeta_j} = \begin{cases} u_j'(0), & \zeta_j = 0 \\ \dfrac{2(1 - \eta_i)}{\Lambda_j^2 \mathrm{Re}_D}\left(\dfrac{\varepsilon_M}{\nu}\right)_j, & 0 < \zeta_j < \zeta_{c,j} \end{cases} \tag{31-73}$$

where

$$\left(\frac{\varepsilon_M}{\nu}\right)_j = \frac{1}{2}\left[\sqrt{1+\frac{2\Lambda_j^2 \mathrm{Re}_D}{1-\eta_i} g_j(\zeta_j)} - 1\right], \tag{31-74}$$

$$g_j(\zeta_j) = \begin{cases} \dfrac{\eta_i u_1'(0)}{\eta_i + \zeta_1}\left(1 - \dfrac{2\eta_i + \zeta_1}{\eta_c^2 - \eta_i^2}\zeta_1\right) & (j=1) \\ \dfrac{u_2'(0)}{1-\zeta_2}\left(1 - \dfrac{2-\zeta_2}{1-\eta_c^2}\zeta_2\right) & (j=2) \end{cases}, \tag{31-75}$$

and in which $u_1'(0) = du_1/d\zeta_1\big|_{\zeta_1=0}$ and $u_2'(0) = du_2/d\zeta_2\big|_{\zeta_2=0}$.

The boundary conditions at the walls require that $u_1(0) = 0$ and $u_2(0) = 0$. The solutions to the momentum equation (31-73) in both regions of the flow are coupled by three requirements. The first is given by Eq. (31-57), relating the velocity gradients at the outer and inner walls; the second is given by Eq. (31-63), which enforces the normalization of the velocity scale with respect to $\overline{v}_m$; the last is that the velocity solution be continuous at $r = R_c$. These requirements in dimensionless form are given as follows: from Eq. (31-57)

$$u_2'(0) = u_1'(0)\frac{\eta_i(1-\eta_c^2)}{\eta_c^2 - \eta_i^2}; \tag{31-76}$$

from Eq. (31-63):

$$\int_0^{\eta_c - \eta_i} u_1(\zeta_1)(\eta_i + \zeta_1)d\zeta_1 + \int_0^{1-\eta_c} u_2(\zeta_2)(1-\zeta_2)d\zeta_2 = \frac{1-\eta_i^2}{2}; \tag{31-77}$$

finally,

$$u_1(\eta_c - \eta_i) = u_2(1-\eta_c). \tag{31-78}$$

The solution to the momentum equation (31-73) involves determining the unknown values of η_c, $u_1'(0)$, and $u_2'(0)$ that satisfy the constraints given by Eqs. (31-76) through (31-78). The solution may be iteratively approached with the following steps:

1. Guess η_c.
2. Guess $u_1'(0)$ and calculate $u_2'(0)$ from Eq. (31-76).
3. Solve for $u_1(\zeta_1)$ and $u_2(\zeta_2)$ from Eq. (31-73).
4. Iterate back to step (2) if Eq. (31-77) is not satisfied with current $u_1'(0)$.
5. Iterate back to step (1) if Eq. (31-78) is not satisfied with current η_c.
6. When Eqs. (31-76) through (31-78) are satisfied with $u_1'(0)$, $u_2'(0)$, and η_c, the solution has converged.

The bisection method is used to improve each guess of $u_1'(0)$ and η_c, based on the results of the previous iteration. The solution procedure described above can be implemented using the fourth-order Runge-Kutta method to solve Eq. (31-73).

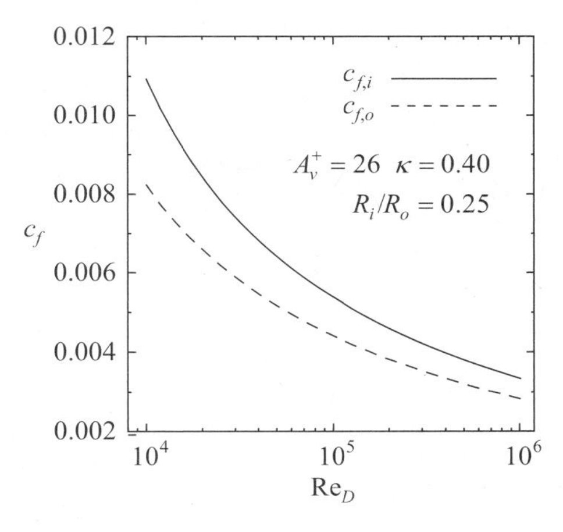

Figure 31-11 Coefficients of friction for turbulent Poiseuille flow in a smooth-wall annulus.

From the solution for the turbulent annulus flow, the friction coefficient $c_f = \tau_w/(\rho\overline{v}_m^2/2)$ can be determined for the inner and outer walls:

$$c_{f,i} = \frac{4(1-\eta_i)}{\mathrm{Re}_D} u_1'(0) \quad \text{and} \quad c_{f,o} = \frac{4(1-\eta_i)}{\mathrm{Re}_D} u_2'(0). \tag{31-79}$$

Figure 31-11 plots the coefficient of friction for both walls of the annulus as a function of the Reynolds number for an annulus having an inner wall radius that is one-quarter the outer wall radius, $R_i/R_o = 0.25$. The mixing length constants $\kappa = 0.40$, $\gamma = 0.085$, and $A_\nu^+ = 26$ have been used for this calculation. The numerical results reveal that the maximum time-averaged velocity in the annulus occurs at a radius that is slightly closer to the inner radius than the outer radius. As a result, the time-averaged velocity gradients in the inner region of the annulus are steeper than in the outer region. This causes the coefficient of friction to be higher for the inner wall than for the outer wall, as shown in Figure 31-11.

31.6 REICHARDT'S FORMULA FOR TURBULENT DIFFUSIVITY

One shortcoming of the mixing length model is the prediction that turbulent diffusivity goes to zero when the gradient in the time-averaged velocity goes to zero, such as at the centerline of internal flows. To address this, Reichardt proposed an empirical formula for the turbulent diffusivity in pipe flow outside of the viscous sublayer [5]:

$$\frac{\varepsilon_M}{\nu} = \frac{\kappa y^+}{6}(1 + r/R)\left(1 + 2(r/R)^2\right). \tag{31-80}$$

Reichardt's formula extrapolates the turbulent diffusivity from the inner region of turbulence to the centerline of the flow, and is independent of the velocity gradient. Approaching the wall ($r \to R$) Reichardt's formula approaches the inner region behavior where $\varepsilon_M(r \to R)/\nu = \kappa y^+$, as realized in Section 30.6.2 for the mixing length model. The functional form of Reichardt's extrapolation formula provides for a finite diffusivity at the centerline of the flow, where $\varepsilon_M(r \to 0)/\nu = \kappa R\sqrt{\tau_w/\rho}/(6\nu)$ that is 8/9ths (about 89%) of the maximum value predicted by the formula. This mimics experimental data that suggests the centerline turbulent diffusivity should be about 85% of the maximum value [6]. However, Reichardt's formula does not reproduce the correct behavior approaching the viscous sublayer. Therefore, this formula must be used in conjunction with another model, such as the mixing length model, to describe the transition region and the viscous sublayer.

31.6.1 Turbulent Poiseuille Flow Between Smooth Parallel Plates

To illustrate the application of Reichardt's formula, reconsider the turbulent Poiseuille flow between smooth parallel plates that was treated in Section 31.1. Since Reichardt's formula cannot be applied to the viscous sublayer, diffusivities over the distance from the wall to the inner region of turbulence are calculated from the mixing length model. Combining Eq. (31-15) and (31-21), the resulting mixing length model for turbulent diffusivity is

$$\left(\frac{\varepsilon_M}{\nu}\right)_{mix} = \frac{\sqrt{1 + 2\Lambda^2 \mathrm{Re}_D u'(0)(1-2\eta)} - 1}{2}, \tag{31-81}$$

where the mixing length Λ is evaluated with Van Driest damping function (31-19). Since the turbulent diffusivity evaluated with Eq. (31-81) approaches inner region behavior $\varepsilon_M/\nu = \kappa y^+$ moving away from the wall and Reichardt's formula approaches the same limit moving toward the wall, the two models intersect somewhere in the inner region of turbulence, very close to the wall as shown in Figure 31-12. When the turbulent diffusivities calculated from the two approaches are identical (at $\eta = \eta_x$), a transition from the mixing length model to Reichardt's formula can be made.

For application of Reichardt's formula to Poiseuille flow between parallel plates, distances measured from the centerline of the flow are expressed as $r/R \to 1 - 2\eta$, and distances from the wall are expressed by $y^+ \to \eta\sqrt{\mathrm{Re}_D u'(0)/2}$. With these modifications, Reichardt's formula (31-80) for turbulent diffusivity becomes

$$\left(\frac{\varepsilon_M}{\nu}\right)_{Reich} = \kappa\eta\sqrt{\frac{\mathrm{Re}_D u'(0)}{2}}\,(1-\eta)\left(1 - \frac{8}{3}\eta(1-\eta)\right). \tag{31-82}$$

The turbulent Poiseuille flow between smooth parallel plates is solved by integrating the momentum equation as evaluated with

$$\frac{du}{d\eta} = \begin{cases} u'(0) \quad , & \eta = 0 \\ \dfrac{2(\varepsilon_M/\nu)_{mix}}{\Lambda^2 \mathrm{Re}_D} \quad , & 0 < \eta < \eta_x \\ \dfrac{(1-2\eta)u'(0)}{1 + (\varepsilon_M/\nu)_{Reich}}, & \eta_x \le \eta \le 1 \ . \end{cases} \tag{31-83}$$

The form of the momentum equation (31-83) changes at the crossover position η_x, where the mixing length model and Reichardt's formula yield the same diffusivity:

$$\left[\left(\frac{\varepsilon_M}{\nu}\right)_{mix} = \left(\frac{\varepsilon_M}{\nu}\right)_{Reich}\right]_{\eta=\eta_x}. \tag{31-84}$$

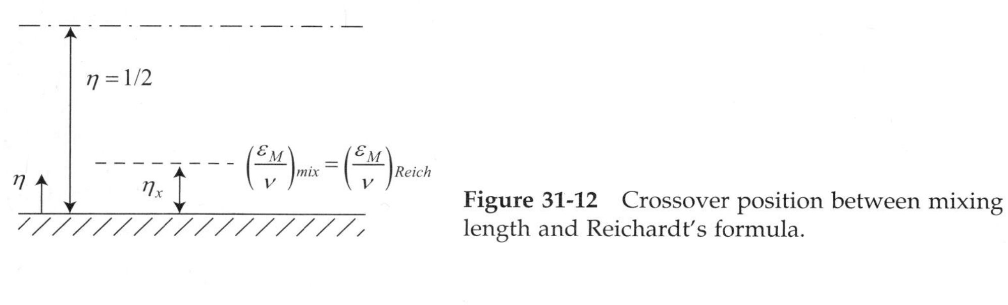

Figure 31-12 Crossover position between mixing length and Reichardt's formula.

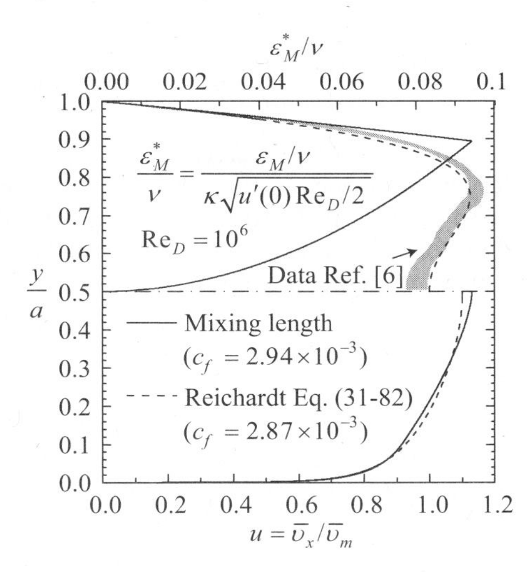

Figure 31-13 Comparison of turbulent diffusivity results based on mixing length model and Reichardt's formula.

The crossover position η_x can be found numerically by equating Eq. (31-81) and Eq. (31-82). For $0 < \eta < \eta_x$, the momentum equation is equivalent to that used in Section 31.1, where the mixing length Λ is evaluated from Eq. (31-19). For $\eta_x \leq \eta \leq 1$, the turbulent diffusivity is evaluated from Reichardt's formula (31-82).

As in Section 31.1, the momentum equation (31-83) is solved subject to the constraints

$$u(0) = 0 \quad \text{and} \quad \overline{u}_m = \int_0^{1/2} u d\eta = 1/2, \tag{31-85}$$

and can be integrated using the fourth-order Runge-Kutta method.

Results from the mixing length model solution, as described in Section 31.1, and the solution derived with Reichardt's formula are compared in Figure 31-13 for turbulent flow with $\mathrm{Re}_D = 10^6$. Shown above the centerline are the turbulent diffusivities for the two approaches. Near the wall, both methods converge to the same behavior. However, the mixing length model exhibits an abrupt change in turbulent diffusivity where the model transitions from the behavior where $\ell = \kappa y$ in the inner region to where the mixing length becomes a constant $\ell = \gamma a/2$ in the outer region. From this transition point forward to the centerline, the turbulent diffusivity rapidly drops to zero in the mixing length model. In contrast, Reichardt's formula exhibits a smooth trend for the turbulent diffusivity that mimics experimental data and does not go to zero at the centerline.

Despite the disparity between turbulent diffusivities, the difference in the velocity profiles produced by the two models is less severe, as compared below the centerline of Figure 31-13. The mixing length model yields a velocity profile that is slightly more pointed than that produced with Reichardt's formula for the turbulent diffusivity. Additionally, the two approaches yield virtually indistinguishable results for the coefficient of friction, with a difference of only about 2%.

31.7 POISEUILLE FLOW WITH BLOWING BETWEEN WALLS

Consider a steady-state, fully developed turbulent pressure-driven flow between parallel walls. Fluid is injected into the flow through the lower wall and is removed at an equal rate from the upper wall. The blowing velocity between the walls causes the velocity distribution $\overline{v}_x(y)$ to be asymmetric. For fully developed conditions, the steady-state momentum equation for turbulent flow between the walls simplifies to

$$\underbrace{\overline{v}_x \frac{\partial \overline{v}_x}{\partial x}}_{=0} + \overline{v}_y \frac{\partial \overline{v}_x}{\partial y} = \underbrace{\frac{\partial}{\partial x}\left((\nu + \varepsilon_M)\frac{\partial \overline{v}_x}{\partial x}\right)}_{=0} + \frac{\partial}{\partial y}\left((\nu + \varepsilon_M)\frac{\partial \overline{v}_x}{\partial y}\right) - \frac{1}{\rho}\frac{\partial \overline{P}}{\partial x}. \tag{31-86}$$

The momentum equation retains one of the advection terms because the blowing between the walls imposes a $\overline{v}_y$ velocity. In addition, turbulent diffusion only contributes to transport across the flow for fully developed conditions. The remaining terms in the momentum equation are integrated once, yielding

$$\overline{v}_y\,\overline{v}_x = (\nu + \varepsilon_M)\frac{\partial \overline{v}_x}{\partial y} - \frac{1}{\rho}\frac{\partial \overline{P}}{\partial x}y + c_1. \tag{31-87}$$

Analysis of the flow between the walls can be subdivided into two regions: region 1 covers the range $0 \le y \le y_c$ and region 2 covers the range $y_c \le y \le a$, where y_c coincides with the point in the flow where $\partial \overline{v}_x/\partial y = 0$. The integration constant in the momentum equation (31-87) can be evaluated three ways:

$$\text{at } y = 0: \quad c_1 = -\nu\frac{\partial \overline{v}_x}{\partial y}\bigg|_{y=0}, \tag{31-88}$$

$$\text{at } y = y_c: \quad c_1 = \overline{v}_y\,\overline{v}_c + \frac{1}{\rho}\frac{\partial \overline{P}}{\partial x}y_c, \tag{31-89}$$

$$\text{at } y = a: \quad c_1 = -\nu\frac{\partial \overline{v}_x}{\partial y}\bigg|_{y=a} + \frac{1}{\rho}\frac{\partial \overline{P}}{\partial x}a, \tag{31-90}$$

where $\overline{v}_c = \overline{v}_x(y = y_c)$. Combining Eq. (31-89) with Eq. (31-88) and Eq. (31-90), the pressure gradient may be expressed in two alternate forms:

$$-\frac{1}{\rho}\frac{\partial \overline{P}}{\partial x} = \frac{1}{y_c}\left(\nu\frac{\partial \overline{v}_x}{\partial y}\bigg|_{y=0} + \overline{v}_y\,\overline{v}_c\right) \tag{31-91}$$

and

$$-\frac{1}{\rho}\frac{\partial \overline{P}}{\partial x} = \frac{-1}{a - y_c}\left(\nu\frac{\partial \overline{v}_x}{\partial y}\bigg|_{y=a} + \overline{v}_y\,\overline{v}_c\right). \tag{31-92}$$

The pressure gradient may be eliminated between equations (31-91) and (31-92) to reveal a relation between the velocity gradients at the top and bottom walls:

$$\nu\frac{\partial \overline{v}_x}{\partial y}\bigg|_{y=a} = -\frac{a - y_c}{y_c}\nu\frac{\partial \overline{v}_x}{\partial y}\bigg|_{y=0} - \frac{a}{y_c}\overline{v}_y\,\overline{v}_c. \tag{31-93}$$

Using Eq. (31-89) for the integration constant, the momentum equation (31-87) becomes

$$(\nu + \varepsilon_M)\frac{\partial \overline{v}_x}{\partial y} = -(y_c - y)\frac{1}{\rho}\frac{\partial \overline{P}}{\partial x} - \overline{v}_y(\overline{v}_c - \overline{v}_x). \tag{31-94}$$

The momentum equation (31-94) may be written in two forms by substituting the alternate expressions for the pressure gradient, Eq. (31-91) and Eq. (31-92). These two forms of the momentum equation are

$$(\nu + \varepsilon_M)\frac{\partial \overline{v}_x}{\partial y} = \frac{y_c - y}{y_c}\nu \frac{\partial \overline{v}_x}{\partial y}\bigg|_{y=0} - \overline{v}_y\left(\frac{y}{y_c}\overline{v}_c - \overline{v}_x\right) \tag{31-95}$$

and

$$(\nu + \varepsilon_M)\frac{\partial \overline{v}_x}{\partial y} = -\frac{y_c - y}{a - y_c}\nu \frac{\partial \overline{v}_x}{\partial y}\bigg|_{y=a} - \overline{v}_y\left(\frac{a - y}{a - y_c}\overline{v}_c - \overline{v}_x\right). \tag{31-96}$$

Notice that when $y = y_c$, both momentum equations enforce the condition $\partial \overline{v}_x/\partial y|_{y=y_c} = 0$. The problem is made dimensionless using the variables

$$\eta = y/a \quad \text{and} \quad u = \overline{v}_x/\overline{v}_m, \tag{31-97}$$

where

$$\overline{v}_m = \int_0^a \overline{v}_x dy/a. \tag{31-98}$$

The relation between u and $\overline{v}_m$ requires that

$$\int_0^1 u d\eta = 1. \tag{31-99}$$

The momentum equations (31-95) and (31-96) in dimensionless form become

$$(1 + \frac{\varepsilon_M}{\nu})\frac{\partial u}{\partial \eta} = \frac{\eta_c - \eta}{\eta_c}\frac{\partial u}{\partial \eta}\bigg|_{\eta=0} - \frac{\text{Re}_B}{2}\left(\frac{\eta}{\eta_c}u_c - u(\eta)\right) \tag{31-100}$$

and

$$(1 + \frac{\varepsilon_M}{\nu})\frac{\partial u}{\partial \eta} = -\frac{\eta_c - \eta}{1 - \eta_c}\frac{\partial u}{\partial \eta}\bigg|_{\eta=1} - \frac{\text{Re}_B}{2}\left(\left(\frac{1 - \eta}{1 - \eta_c}\right)u_c - u(\eta)\right), \tag{31-101}$$

where

$$\text{Re}_B = \frac{\overline{v}_y(2a)}{\nu}. \tag{31-102}$$

For region 1, where $0 \le \eta \le \eta_c$, the first momentum equation (31-100) will be integrated with respect to the dependent variable $\zeta_1 = \eta$. For region 2, where $\eta_c \le \eta \le 1$, the second momentum equation (31-101) will be integrated with respect to a new dependent variable $\zeta_2 = 1 - \eta$. These new spatial variables are illustrated in Figure 31-14. The momentum equations (31-100) and (31-101) may be written in terms of the new variables as

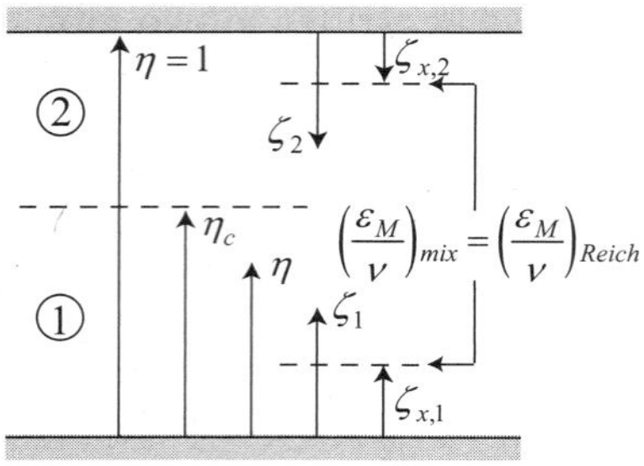

Figure 31-14 Relationship between distance variables.

$$\left(1+\frac{\varepsilon_M}{\nu}\right)\frac{\partial u}{\partial \zeta}\bigg|_{j=1,2} = g_j(\zeta_j), \quad (0 \le \zeta_j \le \zeta_{c,j}) \tag{31-103}$$

where

$$g_j(\zeta_j) = \left(1-\frac{\zeta_j}{\zeta_{c,j}}\right)u_j'(0) + \frac{(-1)^j \mathrm{Re}_B}{2}\left(\frac{\zeta_j}{\zeta_{c,j}}u_c - u_j(\zeta_j)\right), \tag{31-104}$$

$$u_j'(0) = \frac{\partial u_j}{\partial \zeta_j}\bigg|_{\zeta_j=0}, \tag{31-105}$$

$$\zeta_{c,j} = \begin{cases} \eta_c & (j=1) \\ 1-\eta_c & (j=2) \end{cases}, \tag{31-106}$$

and the index $j = 1$ or 2 is used to specify the flow region.

The momentum equation can be solved using Reichardt's formula for the turbulent diffusivity outside of the inner region of turbulence. To this end, the two main regions 1 and 2 of the flow are further subdivided into regions where the turbulent diffusivity is evaluated by either the mixing length model or Reichardt's formula, depending on the distance from the wall. The mixing length model will be used for distances $0 < \zeta_j < \zeta_{x,j}$ from the wall and Reichardt's formula for distances $\zeta_{x,j} \le \zeta_j \le \zeta_{c,j}$. At the distance $\zeta_{x,1}$ from the lower wall and the distance $\zeta_{x,2}$ from the upper wall, the turbulent diffusivity must be the same for both the mixing length model and Reichardt's formula, as illustrated in Figure 31-14.

For the mixing length model, turbulent diffusivity is expressed through the inner region of turbulence by

$$\left(\frac{\varepsilon_M}{\nu}\right)_{\substack{j=1,2\\ mix}} = \frac{\ell^2}{\nu}\left|\frac{\partial \overline{v}_x}{\partial y}\right| = \frac{\Lambda_j^2\,\mathrm{Re}_D}{2}\frac{\partial u}{\partial \zeta}\bigg|_{j=1,2}, \tag{31-107}$$

where

$$\Lambda_j = \frac{\ell_j}{a} \quad \text{and} \quad \mathrm{Re}_D = \frac{\overline{v}_m 2a}{\nu}. \tag{31-108}$$

With the mixing length model expression for turbulent diffusivity (31-107), the momentum equation (31-103) can be expressed as

$$\left.\frac{\partial u}{\partial \zeta}\right|_{j=1,2} = \frac{2(\varepsilon_M/\nu)_{j,mix}}{\Lambda_j^2 \mathrm{Re}_D} \quad (0 < \zeta_j < \zeta_{x,j}), \tag{31-109}$$

where

$$\left(\frac{\varepsilon_M}{\nu}\right)_{\substack{j=1,2\\ mix}} = \frac{1}{2}\left[\sqrt{1 + 2\Lambda_j^2 \mathrm{Re}_D g_j(\zeta_j)} - 1\right] \tag{31-110}$$

and $g_j(\zeta_j)$ is given by Eq. (31-104). The mixing length Λ_j is evaluated with the Van Driest damping function, given in dimensionless form by

$$\Lambda_j(\zeta_j) = \kappa\zeta_j\left[1 - \exp\left(\frac{-\zeta_j}{A_\nu^+}\sqrt{\frac{\mathrm{Re}_D u_j'(0)}{2}}\right)\right]. \tag{31-111}$$

For the regions $\zeta_{x,j} \leq \zeta_j \leq \zeta_{c,j}$, the momentum equation (31-103) is evaluated with Reichardt's formula for turbulent diffusivity. For the parallel wall geometry, with blowing, Reichardt's formula (31-80), can be expressed as

$$\left(\frac{\varepsilon_M}{\nu}\right)_{j,Reich} = \frac{\kappa\zeta_j}{6}\sqrt{\frac{\mathrm{Re}_D u_j'(0)}{2}}\, s_j(\rho_j)(1+\rho_j)(1+2\rho_j^2), \tag{31-112}$$

where

$$\rho_j = 1 - \frac{\zeta_j}{\zeta_{c,j}} \tag{31-113}$$

and

$$s_j(\rho_j) = \begin{cases} \rho_1 + (1-\rho_1)\dfrac{\zeta_{c,2}}{\zeta_{c,1}}\sqrt{\dfrac{u_2'(0)}{u_1'(0)}} & (j=1) \\ 1 & (j=2) \end{cases}. \tag{31-114}$$

To create a continuous turbulent diffusivity function between region 1 and 2, Reichardt's formula is stretched in region 1 by the function $s_1(\rho_1)$. This is required by the asymmetry of the problem.

In summary, the final form of the momentum equation to be integrated is

$$\left.\frac{\partial u}{\partial \zeta}\right|_{j=1,2} = \begin{cases} u_j'(0) \quad , & \zeta_j = 0 \\ \dfrac{2(\varepsilon_M/\nu)_{j,mix}}{\Lambda_j^2 \mathrm{Re}_D} \quad , & 0 < \zeta_j < \zeta_{x,j} \\ \dfrac{g_j(\zeta_j)}{1 + (\varepsilon_M/\nu)_{j,Reich}}, & \zeta_{x,j} \leq \zeta_j \leq \zeta_{c,j} \end{cases}, \tag{31-115}$$

where $(\varepsilon_M/\nu)_{j,mix}$ is evaluated with Eq. (31-110), Λ_j is evaluated with Eq. (31-111), $g_j(\zeta_j)$ is evaluated with Eq. (31-104), and $(\varepsilon_M/\nu)_{j,Reich}$ is evaluated with Eq. (31-112). The boundary conditions at the walls require

$$u_1(0) = 0 \quad \text{and} \quad u_2(0) = 0. \tag{31-116}$$

The solution to the momentum equation (31-115) must additionally satisfy three requirements: the first relates the velocity gradients at the upper and lower walls (31-93); the second enforces the normalization of the velocity scale (31-99); the last is that the velocity solution be continuous at $y = y_c$. These requirements in dimensionless form are given as follows:

$$u_2'(0) = \frac{1-\eta_c}{\eta_c} u_1'(0) + \frac{\mathrm{Re}_B}{2}\frac{u_c}{\eta_c}; \tag{31-117}$$

$$\int_0^{\eta_c} u_1(\zeta_1)d\zeta_1 + \int_0^{1-\eta_c} u_2(\zeta_2)d\zeta_2 = 1; \tag{31-118}$$

$$u_1(\eta_c) = u_2(1-\eta_c). \tag{31-119}$$

The momentum equation is solved by guessing $u_1'(0)$ to satisfy the normalization requirement (31-118) and guessing η_c to satisfy the continuity requirement (31-119). For each pair of $u_1'(0)$and η_c, the momentum equation must be solved iteratively to establish the correct value of $u_c = \overline{v}_x(y = y_c)/\overline{v}_m$ appearing in the equation.

The blowing velocity $\overline{v}_y$ leads to differences in the coefficient of friction for the top and bottom walls. In terms of the variables of the solution, the friction coefficients for the top and bottom wall are given by

$$c_{f,1} = \frac{\tau_w\big|_{y=0}}{\rho\overline{v}_m^2/2} = \frac{4u_1'(0)}{\mathrm{Re}_D} \quad \text{and} \quad c_{f,2} = \frac{\tau_w\big|_{y=a}}{\rho\overline{v}_m^2/2} = \frac{4u_2'(0)}{\mathrm{Re}_D}, \tag{31-120}$$

respectively.

The momentum equation (31-115) is solved with Code 31-2, where the model parameters $A_\nu^+ = 26$ and $\kappa = 0.4$ have been used. Because it is difficult to iterate to a solution by guessing for more than one unknown with the bisection method, the approach to the correct solution for η_c is made by using a Newton-Raphson method, while the approach to the correct solution for $u_1'(0)$ is made by the bisection method. Both these methods are discussed in Section 23.4.

It is known that wall blowing or suction influences the extent of the viscous sublayer; that is, it alters the value of A_ν^+. For turbulent boundary layers, Kays and Crawford [7] proposed a condition for determining A_ν^+ from the flow solution whereby

$$A_\nu^+ = \frac{25.0}{\sqrt{(\tau/\tau_w)_{y^+=3A_\nu^+}}}, \tag{31-121}$$

and

$$\frac{\tau}{\tau_w} = \left(1 + \frac{\varepsilon_M}{\nu}\right)\frac{\partial\overline{v}_x/\partial y}{\partial\overline{v}_x/\partial y\big|_{y=0}}. \tag{31-122}$$

This condition is used by solving the turbulent flow with assumed values of A_ν^+ for each wall. From this proposed solution, improved estimates of A_ν^+ for each wall can be determined using the transcendental equation (31-121). This process of solving the turbulent flow is repeated until the flow solution yields the same values of A_ν^+ from

Code 31-2 Turbulent Poiseuille flow with blowing

```
#include <stdio.h>
#include <math.h>
double g(int id,double zet,double u,double etc,double uc,
        double ReB,double du0) {
   double zetc=(id==1 ? etc : 1.-etc);
   return (1.-zet/zetc)*du0 + pow(-1,id)*ReB*(zet*uc/zetc-u)/2.;
}
double Em_mix(int id,double Ap,double Kap,double zet,double u,
              double etc,double uc,double ReB,double ReD,
              double du0,double *Lam) {
   *Lam=Kap*zet*(1.-exp(-zet*sqrt(ReD*du0/2.)/Ap));
   double _g=g(id,zet,u,etc,uc,ReB,du0);
   return (sqrt(1.+2.*(*Lam)*(*Lam)*ReD*_g)-1.)/2.;
}
double Em_Reich(int id,double Kap,double zet,
                double etc,double ReDdu10,double ReDdu20) {
   double ReDdu0=(id==1 ? ReDdu10 : ReDdu20);
   double s,r=(id==1 ? (etc-zet)/etc : (1.-etc-zet)/(1.-etc) );
   s=(id==1 ? r+(1.-r)*(1.-etc)*sqrt(ReDdu20/ReDdu10)/etc : 1. );
   return Kap*zet*sqrt(ReDdu0/2.)*s*(1.+r)*(1.+2.*r*r)/6.;
}
double Em(int id,double Ap,double Kap,double zet,double u,
          double etc,double uc,double ReB,double ReD,
          double du10,double du20,double *Lam) {
   double Em_R=Em_Reich(id,Kap,zet,etc,ReD*du10,ReD*du20);
   if (zet>0.2)    {*Lam=0.; return Em_R;}
   double Em_M=Em_mix(id,Ap,Kap,zet,u,etc,uc,ReB,ReD,
                  (id==1 ? du10 : du20),Lam);
   if (Em_M<Em_R) return Em_M;
   *Lam=0.;
   return Em_R;
}
double du(int id,double Ap,double Kap,double zet,double u,double etc,
          double uc,double ReB,double ReD,double du10,double du20) {
   double _du,du0=(id==1 ? du10 : du20);
   if (zet==0.) return du0;
   double Lam,_Em=Em(id,Ap,Kap,zet,u,etc,uc,ReB,ReD,du10,du20,&Lam);
   if (Lam) _du=2.*_Em/Lam/Lam/ReD;
   else _du=g(id,zet,u,etc,uc,ReB,du0)/(1.+_Em);
   return (_du>0 ? _du : 0.);
}
double intgrl_u(int id,double Ap,double Kap,int N,double *zet,double *u,
   double etc,double uc,double ReB,double ReD,double du10,double du20) {
   int n;
   double delz,half,um,K1u,K2u,K3u,K4u,K1um,K2um,K3um,K4um;
   for (u[0]=um=n=0;n<N-1;++n) {
      delz=zet[n+1]-zet[n];
      half=(zet[n]+zet[n+1])/2.;
      K1u= delz*du(id,Ap,Kap,zet[n],u[n],etc,uc,ReB,ReD,du10,du20);
      K1um=delz*u[n];
      K2u= delz*du(id,Ap,Kap,half,u[n]+0.5*K1u,etc,uc,ReB,ReD,du10,du20);
      K2um=delz*(u[n]+0.5*K1u);
      K3u= delz*du(id,Ap,Kap,half,u[n]+0.5*K2u,etc,uc,ReB,ReD,du10,du20);
      K3um=delz*(u[n]+0.5*K2u);
      K4u= delz*du(id,Ap,Kap,zet[n+1],u[n+1]+K3u,etc,uc,ReB,ReD,du10,du20);
      K4um=delz*(u[n]+K3u);
      u[n+1]=u[n]+(K1u+2.*K2u+2.*K3u+K4u)/6.;
      um+=(K1um+2.*K2um+2.*K3um+K4um)/6.;
   }
   return um;
}
void solve_u(double Ap,double Kap,int N,double *zet1,double *zet2,
             double *u1,double *u2,double ReB,double ReD,
             double *etc,double *du10,double *du20) {
   int n,iter;
   double log_min=-6.,log_del=(-log_min)/(N-2),um1,um2;
   double slope=0.,etc_last=0.,last_uc,uc=1.1,dif_last;
   zet1[0]=zet2[0]=0.;
   *etc=0.5;
   do {
      if (slope && fabs(dif_last/slope)<*etc/5.) *etc-=dif_last/slope;
      else *etc *=1.1; // arb step before using Newton-Raphoson
      zet1[N-1]= *etc;
      zet2[N-1]=1.- *etc;
      for (n=1;n<N-1;++n) {
         zet1[n]=(*etc)*pow(10.,log_min+(n-1)*log_del);
         zet2[n]=(1.- *etc)*pow(10.,log_min+(n-1)*log_del);
      }
      double du10_hi=40000.0,du10_lo=0.0;
      do {
         *du10=(du10_hi+du10_lo)/2.;
         iter=0;
         do {
            *du20= (*du10)*(1.-(*etc))/(*etc)+ReB*uc/(*etc)/2.;
            um1=intgrl_u(1,Ap,Kap,N,zet1,u1,*etc,uc,ReB,ReD,*du10,*du20);
            um2=intgrl_u(2,Ap,Kap,N,zet2,u2,*etc,uc,ReB,ReD,*du10,*du20);
            last_uc=uc;
            uc=(uc+u1[N-1]+u2[N-1])/3.; // next guess
            if (uc<1.) uc=1.;
         } while ( fabs(uc-last_uc)/uc>0.01);
         if (um1+um2>1.) du10_hi= *du10;
         else du10_lo= *du10;
      } while ((du10_hi-du10_lo)/(*du10)>.0001);
      if (etc_last) slope=(fabs(u1[N-1]-u2[N-1])-dif_last)/(*etc-etc_last);
      etc_last=*etc;
   } while ( (dif_last=fabs(u1[N-1]-u2[N-1])) /u1[N-1]>0.01);
}
int main() {
   int N=1000;
   double Kap=0.4,Ap=26.0,ReD=1.e6,ReB=1500.0;
   double etc,du10,du20,zet1[N],u1[N],zet2[N],u2[N];
   for (ReB=0.;ReB<=1600.;ReB+=100.) {
      solve_u(Ap,Kap,N,zet1,zet2,u1,u2,ReB,ReD,&etc,&du10,&du20);
      printf("ReB=%e cf1=%e cf2=%e\n",ReB,4.*du10/ReD,4.*du20/ReD);
   }
   return 1;
}
```

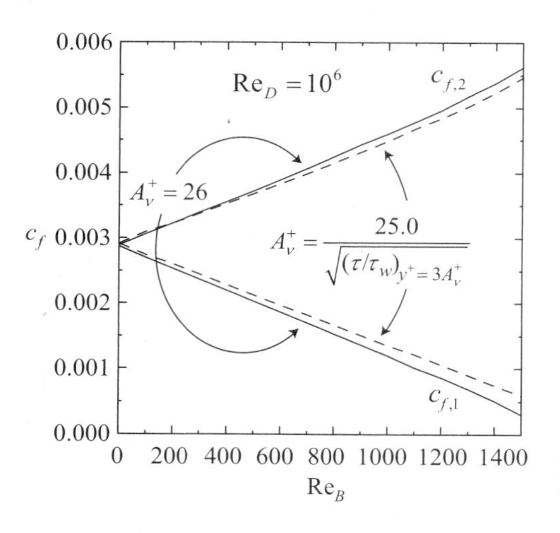

Figure 31-15 Friction coefficients for turbulent Poiseuille flow with blowing.

Eq. (31-121) as were assumed. It is reassuring that when this procedure is followed for the present Poiseuille flow problem without blowing, the Kays and Crawford condition (31-121) yields a value of $A_\nu^+ = 25.1$, which is consistent with expectation.

For a streamwise Reynolds number of $\mathrm{Re}_D = 10^6$, the friction coefficients calculated for both walls are shown in Figure 31-15 as a function of the blowing Reynolds number Re_B. The blowing between the walls causes a reduction of the lower wall friction coefficient $c_{f,1}$ and an increase in the upper wall friction coefficient $c_{f,2}$. Solid line results are for a constant viscous sublayer thickness $A_\nu^+ = 26$, as determined using Code 31-2. It is left as an exercise (see Problem 31-5) to establish the results using the A_ν^+ condition (31-121) proposed by Kays and Crawford. Results for the variable viscous sublayer thickness are shown in Figure 31-15 with dashed lines.

31.8 PROBLEMS

31-1 Derive the final form of the turbulent momentum equation for pipe flow, given by Eq. (31-41). Integrate the momentum equation using the mixing length model with the parameters $A_\nu^+ = 27.0$, $\kappa = 0.40$, and $\gamma = 0.085$. Plot the velocity profile between the wall and centerline of the flow using the wall variables u^+ and y^+. Overlay on these plots the viscous sublayer behavior $u^+ = y^+$ and law of the wall behavior $u^+ = 2.5 \ln y^+ + 5.5$.

31-2 Numerically implement a solution to the fully developed turbulent Couette flow between smooth parallel plates, as discussed in Section 31.2. Plot the velocity profile using wall variables for $\mathrm{Re}_D = 10^6$. Indicate on the plot the extent of the viscous sublayer, as well as the inner and outer regions of turbulence.

31-3 Use Reichardt's formula to solve the turbulent pipe flow problem, and determine the coefficient of friction as a function of Reynolds number $10^4 \le \mathrm{Re}_D \le 10^6$. For $\mathrm{Re}_D = 10^5$, plot $u = \overline{v}_x/\overline{v}_m$ between the wall and the centerline of the pipe.

31-4 Consider fully developed turbulent flow between smooth parallel plates as shown. The top plate moves with a speed U, while the bottom plate is held stationary. Divide the flow field into two parts that are separated by a plane at $y = y_c$, defined to coincide with $\partial\overline{v}_x/\partial y = 0$. Derive coupled momentum equations for the flow field above and below $y = y_c$. Express the momentum equations in terms of the dimensionless velocity $u = \overline{v}_z/\overline{v}_m$, and using the dimensionless spatial variables $\eta_1 = y/a$ (for $y \le y_c$) and $\eta_2 = 1 - y/a$ (for $y \ge y_c$). Express the momentum equations for $u_1(\eta_1)$ and $u_2(\eta_2)$ exclusively in terms of the three unknowns:

$u_1'(0) = du_1/d\eta_1|_{\eta_1=0}$, $u_2'(0) = du_2/d\eta_2|_{\eta_2=0}$, and $\eta_c = y_c/a$. Implement a numerical solution to this problem using the mixing length model. For $\mathrm{Re}_D = 10^6$, evaluate coefficient of friction $c_f = \tau_w/(\rho\bar{v}_m^2/2)$ for the top and bottom plates as a function of the top plate speed $U = b\,\bar{v}_m$. Plot c_f results for $0 < b < 1$. Plot $u(\eta)$when $b = 0.5$.

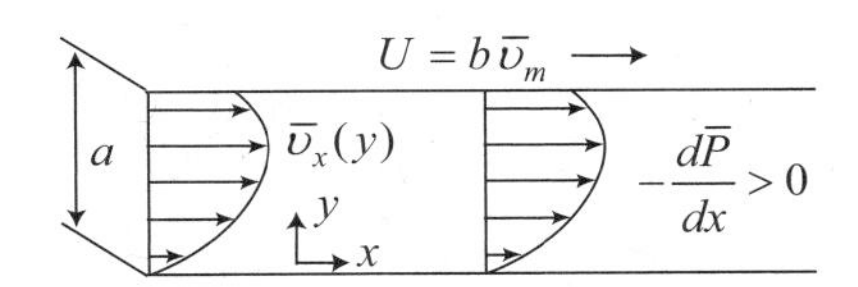

31-5 Solve the Poiseuille flow problem with blowing, as discussed in Section 31.7. Use the Kays and Crawford condition (31-121) to determine the viscous sublayer thickness A_ν^+ for the top and bottom walls as a function of the blowing Reynolds number $0 \le \mathrm{Re}_B \le 1500$, where $\mathrm{Re}_B = \bar{v}_y(2a)/\nu$. What is the effect of changing the constant in the numerator of Eq. (31-121) to 26.0?

31-6 Solve the momentum equation in the annulus shown using Reichardt's formula through the central region of turbulence. In terms of both the mixing length model and Reichardt's formula for turbulent diffusivity, the momentum equation can be expressed as

$$\frac{du_j}{d\zeta_j} = \begin{cases} u_j'(0) \quad , & \zeta_j = 0 \\ \dfrac{2(1-\eta_i)}{\Lambda_j^2 \mathrm{Re}_D}\left(\dfrac{\varepsilon_M}{\nu}\right)_{j,mix}, & 0 < \zeta_j < \zeta_{x,j} \\ \dfrac{h_j(\zeta_j)u_j'(0)}{1 + (\varepsilon_M/\nu)_{j,Reich}}, & \zeta_{x,j} \le \zeta_j \le \zeta_{c,j} \end{cases}$$

where

$$\zeta_{c,j} = \begin{cases} \eta_c - \eta_i & (j=1) \\ 1 - \eta_c & (j=2) \end{cases}$$

and

$$h_1(\zeta_1) = \eta_i \frac{\eta_c^2 - (\eta_i + \zeta_1)^2}{(\eta_i + \zeta_1)(\eta_c^2 - \eta_i^2)}$$

$$h_2(\zeta_2) = \frac{\eta_c^2 - (1-\zeta_2)^2}{(1-\zeta_2)(\eta_c^2 - 1)}$$

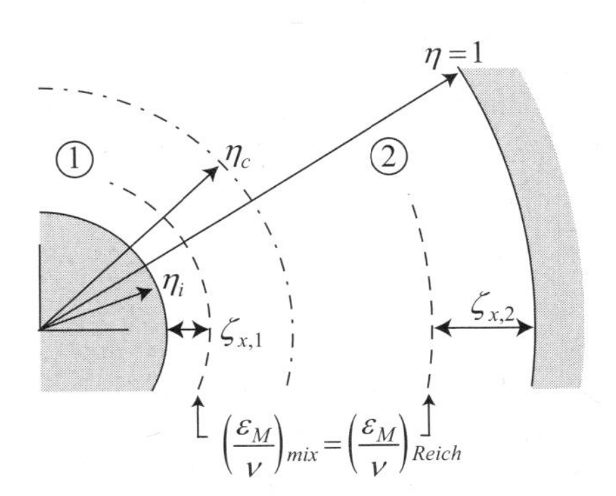

For the regions closest to the walls, $0 < \zeta_j < \zeta_{x,j}$, the momentum equation is evaluated in the same manner as in Section 31.5, where $(\varepsilon_M/\nu)_{j,mix}$ is given by Eq. (31-74). For the regions $\zeta_{x,j} \leq \zeta_j \leq \zeta_{c,j}$, the momentum equation is evaluated with Reichardt's formula (31-80) for turbulent diffusivity, which for the annular geometry can be expressed as

$$\left(\frac{\varepsilon_M}{\nu}\right)_{j,Reich} = \frac{\kappa\zeta_j}{6}\sqrt{\frac{\mathrm{Re}_D u_j'(0)}{2(1-\eta_i)}}\, s_j(\rho_j)(1+\rho_j)(1+2\rho_j^2),$$

where

$$\rho_j = \begin{cases} \dfrac{\eta_c - \eta_i - \zeta_1}{\eta_c - \eta_i} & (j=1) \\ \dfrac{1 - \eta_c - \zeta_2}{1 - \eta_c} & (j=2) \end{cases}$$

and

$$s_j(\rho_j) = \begin{cases} \rho_1 + (1-\rho_1)\dfrac{1-\eta_c}{\eta_c - \eta_i}\sqrt{\dfrac{u_2'(0)}{u_1'(0)}} & (j=1) \\ 1 & (j=2) \end{cases}.$$

The stretching function $s_j(\rho_j)$ is used to allow Reichardt's formula to be continuous between the two main regions. The crossover position $\zeta_{x,j}$ between use of the mixing-length model and Reichardt's formula can be found numerically by equating $(\varepsilon_M/\nu)_{j,mix}$ and $(\varepsilon_M/\nu)_{j,Reich}$. The solution to the momentum equation may be obtained iteratively with the same approach outlined for the mixing length model solution in Section 31.5. Write a fourth-order Runge-Kutta integration code to solve the momentum equation, using the model parameters $A_\nu^+ = 26$ and $\kappa = 0.4$. For $R_i/R_o = \eta_i = 0.25$ and $\mathrm{Re}_D = 10^6$, plot u and ε_M/ν across the entire flow. Determine the coefficients of friction for the inner and outer wall.

REFERENCES

[1] J. Nikuradse, *Gesetzmässigkeiten der Turbulenten Strömung in Glatten Röhren*, Forschungscheft no. 356, Berlin, Germany: VDI Verlag, 1932.

[2] E.-S. Zanoun, F. Durst, and H. Nagib, "Evaluating the Law of the Wall in Two-Dimensional Fully Developed Turbulent Channel Flows." *Physics of Fluids*, **15**, 3079 (2003).

[3] R. B. Dean, "Reynolds Number Dependence of Skin Friction and Other Bulk Flow Variables in Two-Dimensional Rectangular Duct Flow." *Journal of Fluid Engineering*, **100**, 215 (1978).

[4] L. Prandtl, *Führer, durch die Strömungslehre*. Braunschweig, Germany: Vieweg, 1944.

[5] H. Reichardt, "Die Grundlagen des Turbulenten Wärmeüberganges." *Archiv gesamte Wärmetechnik*, **6/7**, 129 (1951).

[6] A. K. M. F Hussain, and W.C. Reynolds, "Measurements in Fully Developed Turbulent Channel Flow." *Journal of Fluids Engineering, Transactions of the ASME*, **97**, 568 (1975).

[7] W. Kays and M. Crawford, *Convective Heat and Mass Transfer*, Second Edition. New York, NY: McGraw-Hill, 1987.

Chapter 32

Turbulent Heat and Species Transfer

32.1 Reynolds Decomposition of the Heat Equation

32.2 The Reynolds Analogy

32.3 Thermal Profile Near the Wall

32.4 Mixing Length Model for Heat Transfer

32.5 Mixing Length Model for Species Transfer

32.6 Problems

In this chapter, Reynolds decomposition is performed on the heat and species transport equations for turbulent flows. Using an extension to Prandtl's mixing length model, the turbulent fluxes of heat $\overline{v_i'T'}$ and species $\overline{v_i'c'}$ that appear in the time-averaged equations can be evaluated in terms of the mean flow characteristics. Using the mixing length model, the heat equation can be integrated analytically for regions close to the wall. When expressed in terms of the wall coordinates, the result is a thermal equivalence to the law of the wall for momentum transfer. By analogy, a similar investigation into the species transport equation can be made.

32.1 REYNOLDS DECOMPOSITION OF THE HEAT EQUATION

Ignoring viscous heating and other sources of thermal energy, the steady-state heat equation for a constant ρ and C_p fluid is

$$v_i \partial_i T = \partial_i(\alpha\, \partial_i T). \tag{32-1}$$

To describe a turbulent flow, variables in the heat equation are decomposed into mean and fluctuating components. When the resulting decomposed heat equation is time averaged,

$$\overline{(\overline{v}_i + v_i')\partial_i(\overline{T} + T')} = \overline{\partial_i\left[\alpha\, \partial_i(\overline{T} + T')\right]}, \tag{32-2}$$

the advection term on the left-hand side of the equation becomes

$$\overline{\overline{v}_i\partial_i\overline{T} + v_i'\partial_i\overline{T} + \overline{v}_i\partial_i T' + v_i'\partial_i T'} = \overline{v}_i\partial_i\overline{T} + \overline{v_i'\partial_i T'} = \overline{v}_i\partial_i\overline{T} + \partial_i\overline{v_i'T'} - \overline{T'\underbrace{\partial_i v_i'}_{=0}}. \tag{32-3}$$

Since continuity of an incompressible flow requires that $\partial_i v_i' = 0$, the only remaining advection term involving a correlation between fluctuating quantities is $\partial_i\overline{v_i'T'}$. On the right-hand side of the heat equation (32-2), the fluctuating temperature term will time average to zero, leaving only the contribution from the mean temperature. Therefore, the turbulent heat equation becomes

$$\overline{v}_i \partial_i \overline{T} = \partial_i \left[\alpha \, \partial_i \overline{T} - \overline{v_i' T'} \right], \tag{32-4}$$

where the advection term involving $\overline{v_i' T'}$ has been moved to the right-hand side of the equation. This term describes transport of heat on the scale of turbulent fluctuations, and is analogous to the Reynolds stress for momentum transport.

The heat equation can be written for a steady two-dimensional flow in Cartesian coordinates as

$$\overline{v}_x \frac{\partial \overline{T}}{\partial x} + \overline{v}_y \frac{\partial \overline{T}}{\partial y} = \frac{\partial}{\partial x}\left(\alpha \frac{\partial \overline{T}}{\partial x} - \overline{v_x' T'} \right) + \frac{\partial}{\partial y}\left(\alpha \frac{\partial \overline{T}}{\partial y} - \overline{v_y' T'} \right). \tag{32-5}$$

For the treatment of boundary layers, streamwise gradients are small compared with cross-stream gradients. Therefore, the heat equation for the turbulent boundary layer is

$$\overline{v}_x \frac{\partial \overline{T}}{\partial x} + \overline{v}_y \frac{\partial \overline{T}}{\partial y} = \frac{\partial}{\partial y}\left(\alpha \frac{\partial \overline{T}}{\partial y} - \overline{v_y' T'} \right) \quad \text{(boundary layer)}. \tag{32-6}$$

To solve the boundary layer heat equation requires a model to evaluate $\overline{v_y' T'}$ in terms of the mean conditions of the turbulent flow. The mixing length model, introduced in Section 30.5, is extended for this purpose by making an analogy between heat and momentum transport.

32.2 THE REYNOLDS ANALOGY

The very simple picture of turbulent transport offered by the mixing length model can be extended to all quantities carried by the fluid. Recall that turbulent eddies are characterized by a rolling motion of the fluid over some length scale ℓ. Any quantity that can be correlated with turbulent velocity fluctuations will be transported over this length scale by the motion of eddies. For example, the Reynolds stress $\overline{v_y' v_x'}$ describes the transport of x-direction momentum (v_x') by velocity fluctuations in the y-direction (v_y'). By the same mechanism, velocity fluctuations in the y-direction can also carry heat $\overline{v_y' T'}$, species $\overline{v_y' c'}$, and so forth.

Figure 32-1 illustrates the turbulent transport of heat and momentum in a boundary layer at some vertical distance from the wall. As was argued in Section 30.5, a velocity fluctuation v_x' should scale as the product of the gradient in the mean flow $\partial \overline{v}_x / \partial y$ and the distance traveled ℓ. Exactly the same argument can be made for heat, species, or any other quantity carried by the fluid, such that

$$v_x' \sim \ell \frac{\partial \overline{v}_x}{\partial y}, \quad T' \sim \ell_H \frac{\partial \overline{T}}{\partial y}, \quad c' \sim \ell_C \frac{\partial \overline{c}}{\partial y}, \cdots \tag{32-7}$$

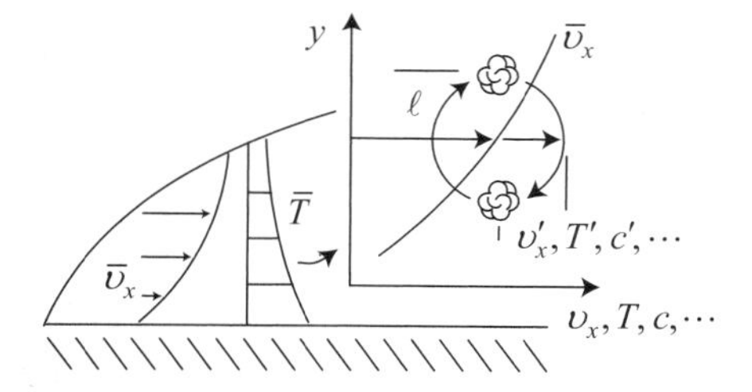

Figure 32-1 Mixing length model applied to other fluid quantities.

Notice that the mixing length scale for heat ℓ_H and species ℓ_C transport is not assumed to be identical to momentum ℓ transport (at this point). These estimates of the fluctuating quantities v'_x, T', and c' can be used to establish their correlations with velocity fluctuations v'_y. In this manner, the relations

$$\overline{v'_y v'_x} = -\ell\left|\frac{\partial \overline{v}_x}{\partial y}\right|\ell\frac{\partial \overline{v}_x}{\partial y}, \quad \overline{v'_y T'} = -\ell\left|\frac{\partial \overline{v}_x}{\partial y}\right|\ell_H\frac{\partial \overline{T}}{\partial y}, \quad \overline{v'_y c'} = -\ell\left|\frac{\partial \overline{v}_x}{\partial y}\right|\ell_C\frac{\partial \overline{c}}{\partial y}, \cdots \tag{32-8}$$

can be defined for small turbulent scale advection. The first expression in (32-8) is the Reynolds stress, which was obtained in Section 30.5 using the argument that $v'_y \sim v'_x$. Since the equations in (32-8) will be used as formal definitions for the mixing length scales, ℓ, ℓ_H, and ℓ_C, the scaling signs ($\sim$) have been replaced with equality signs. To achieve a mathematical likeness to diffusion, eddy diffusivities can be defined for turbulent scale advection, as was done previously for momentum transport. Drawing upon the equations in Eq. (32-8), turbulent diffusivities for momentum, heat, and species transport can be formally defined:

$$\varepsilon_M = \ell\,\ell\left|\frac{\partial \overline{v}_x}{\partial y}\right|, \quad \varepsilon_H = \ell_H\,\ell\left|\frac{\partial \overline{v}_x}{\partial y}\right|, \quad \varepsilon_C = \ell_C\,\ell\left|\frac{\partial \overline{v}_x}{\partial y}\right|, \ldots. \tag{32-9}$$

The physical picture proposed for turbulent transport over the scale of an eddy suggests equivalent mixing length scales: $\ell \approx \ell_H \approx \ell_C$. It follows from this expectation that the associated eddy diffusivities in (32.9) should be the same. This is the basis for the *Reynolds analogy* for turbulent diffusion, suggesting that $\varepsilon_M \approx \varepsilon_H \approx \varepsilon_C$. However, careful studies have revealed some quantitative differences between turbulent diffusivities for momentum, heat, and species transport. Consequently, there must be another factor influencing the effective values of the turbulent length scales ℓ, ℓ_H, and ℓ_C that is different for the different flow properties being carried by the motion of an eddy. To comprehend this difference, it is observed that there must be a molecular-level process to exchange flow properties between fluid elements for the eddy motion to have any net effect on transport. Differences in this molecular-level process required for interelement transfer may explain departures from the expectation set forward by the Reynolds analogy. This observation is revisited in Section 32.4 during the discussion of turbulent heat transfer.

One particular situation to be avoided in application of the mixing length model is illustrated in Figure 32-2. Because of symmetry in the flow, the turbulent diffusivity goes to zero along the centerline. For the hydrodynamic problem, this is not a critical failure because momentum transport is also required to go to zero at the centerline. However, coupled to a second asymmetric transport problem, such as that shown in Figure 32-2, a difficulty does arise. For the problem illustrated, the mixing length model predicts that the turbulent contribution to heat transfer through the centerline vanishes because

$$\varepsilon_H = \ell_H\,\ell\left|\frac{\partial \overline{v}_x}{\partial y}\right| \to 0 \quad \text{(along lines of flow symmetry)}. \tag{32-10}$$

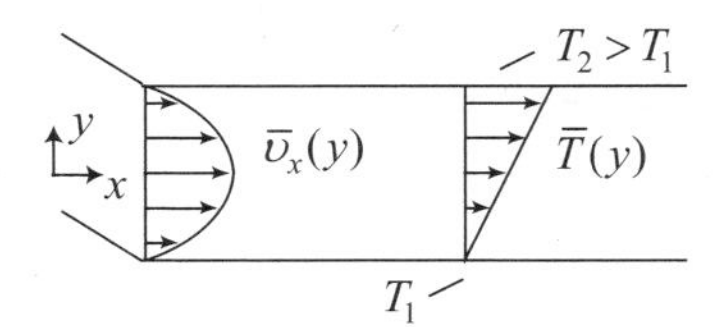

Figure 32-2 A problematic situation for the mixing length model.

Of course, this does not happen physically. Reichardt's empirical formula for turbulent diffusivity, discussed in Section 31.6, can be used to address this difficulty. Additionally, the k-epsilon model for turbulence, developed in Chapter 36, does not suffer from this shortcoming.

32.3 THERMAL PROFILE NEAR THE WALL

One can establish a simple expression for the temperature profile near the wall. This exercise is valuable both for evaluating the parameters of the mixing length model from experimental data and for verifying correct implementation of the model in numerical code. In light of the arguments of the preceding section, the boundary layer heat equation may be written as

$$\overline{v}_x \frac{\partial \overline{T}}{\partial x} + \overline{v}_y \frac{\partial \overline{T}}{\partial y} = \frac{\partial}{\partial y}\left[(\alpha + \varepsilon_H)\frac{\partial \overline{T}}{\partial y}\right], \tag{32-11}$$

where the turbulent heat flux is expressed as

$$\overline{v_y' T'} = -\varepsilon_H(\partial \overline{T}/\partial y). \tag{32-12}$$

32.3.1 The Diffusion Sublayer: (Advection) ≪ (Diffusion) and $\varepsilon_H \ll \alpha$

Since the velocity field approaches zero at the wall, advection becomes negligible in this region. Additionally, since the mixing lengths go to zero, $\ell, \ell_H \to 0$ approaching the wall, turbulent diffusion becomes negligible compared to molecular diffusion. This region closest to the wall is the diffusion sublayer, for which the heat equation becomes

$$0 = \frac{\partial}{\partial y}\left[\alpha \frac{\partial \overline{T}}{\partial y}\right]. \tag{32-13}$$

In the diffusion sublayer, the heat equation describes a constant diffusion heat flux equal to that at the wall q_s. Therefore, with $\alpha(\partial \overline{T}/\partial y) = -q_s/\rho C_p$, the heat equation may be solved in the diffusion sublayer for

$$T_s - \overline{T} = \frac{q_s/\rho C_p}{\alpha} y = \frac{q_s/\rho C_p}{\nu} \mathrm{Pr}\, y, \tag{32-14}$$

where T_s is the wall surface temperature. Recalling the dimensionless wall variable for distance y^+ (see Section 30.6.1), and defining a new wall variable for temperature

$$y^+ = \frac{y\sqrt{\tau_w/\rho}}{\nu} \quad \text{and} \quad T^+ = \frac{T_s - \overline{T}}{q_s/\rho C_p}\sqrt{\tau_w/\rho}, \tag{32-15}$$

the solution to the heat equation in the diffusion sublayer becomes

$$T^+ = \mathrm{Pr}\, y^+. \tag{32-16}$$

32.3.2 Inner Region: (Advection) ≪ (Diffusion) and $\varepsilon_H \gg \alpha$

Beyond the diffusion sublayer, the next region in the boundary layer is the *inner region*, where advection is still negligible compared to diffusion but transport by turbulent diffusion is large compared with molecular diffusion ($\varepsilon_H \gg \alpha$). Fluids of high thermal diffusivity will not necessarily satisfy this second requirement. The governing equation for the temperature field in the inner region is

$$0 = \frac{\partial}{\partial y}\left[\varepsilon_H \frac{\partial \overline{T}}{\partial y}\right]. \tag{32-17}$$

Integrating the heat equation once yields

$$\frac{-q_s}{\rho C_p} = \varepsilon_H \frac{\partial \overline{T}}{\partial y}, \tag{32-18}$$

which can be expressed in terms of wall variables as

$$\frac{dT^+}{dy^+} = \frac{\nu}{\varepsilon_H}. \tag{32-19}$$

Recall that in Section 30.6.2 the turbulent momentum diffusivity for the inner region was found to be

$$\frac{\varepsilon_M}{\nu} = \kappa y^+. \tag{32-20}$$

Equation (32-20) can be combined with Eq. (32-19) to eliminate the viscosity if the thickness of the viscous sublayer is smaller than or equal to the diffusion sublayer for heat transfer. The result allows the heat equation to be expressed in the form

$$\frac{dT^+}{dy^+} = \frac{\varepsilon_M/\varepsilon_H}{\kappa y^+}. \tag{32-21}$$

The heat equation can be integrated from the edge of the diffusion sublayer, where $y^+ = E_\alpha^+$ and $T^+ = \Pr y^+$, into the inner region of the boundary layer:

$$\int_{\Pr y^+}^{T^+} dT^+ = \int_{E_\alpha^+}^{y^+} \frac{\varepsilon_M/\varepsilon_H}{\kappa} \frac{dy^+}{y^+}. \tag{32-22}$$

Adopting Reynolds analogy, one could equate $\varepsilon_H = \varepsilon_M$. However, one can also make a less restrictive assumption that $\varepsilon_M/\varepsilon_H$ is constant over the inner region (not necessarily equal to 1). In this case,

$$\varepsilon_M/\varepsilon_H = \frac{\ell^2|\partial \overline{v}_x/\partial y|}{\ell_H\, \ell|\partial \overline{v}_x/\partial y|} = \ell/\ell_H = \kappa/\kappa_H. \tag{32-23}$$

As discussed in Section 30.5, the mixing length is expected to scale with distance from the wall. Therefore, let $\ell_H = \kappa_H y$ be written in an analogous fashion to the mixing length for momentum transport $\ell = \kappa y$. Using Eq. (32-23), the heat equation for the inner region (32-22) is integrated for the result

$$T^+ - \Pr E_\alpha^+ = \frac{\kappa/\kappa_H}{\kappa} \ln\left(\frac{y^+}{E_\alpha^+}\right). \tag{32-24}$$

Equation (32-24) is a prediction of the temperature profile through the inner region of the turbulent boundary layer, and has a functional form that agrees well with experimental data [1]. Experimental data for a variety of conditions suggests $\kappa/\kappa_H \approx 0.85$ [2]. However, without requiring $E_\alpha^+ = E_\nu^+$, it is impossible for κ/κ_H to remain constant approaching the viscous sublayer. Therefore, Eq. (32-24) has little utility for investigating the extent of the diffusion sublayer E_α^+.

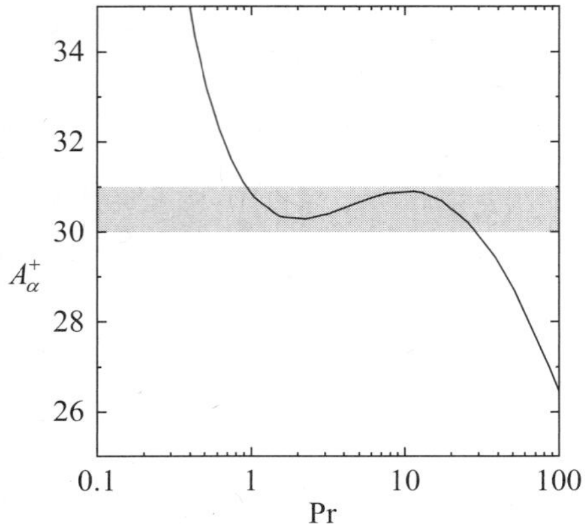

Figure 32-3 Heat transfer damping function constant dependency on Prandtl number.

Kander [3] found through a survey of experimental data that the temperature profiles through the inner region of turbulence can be described by the correlation

$$T^{+} = \frac{\kappa/\kappa_H}{\kappa}\ln(y^{+}) + B(\mathrm{Pr}) \tag{32-25}$$

where

$$B(\mathrm{Pr}) = (3.85\mathrm{Pr}^{1/3} - 1.3)^2 + 2.12\ln(\mathrm{Pr}). \tag{32-26}$$

The experimental data used in formulating this result are comprised of fluids having molecular Prandtl numbers in the range $0.7 \leq \mathrm{Pr} \leq 60.0$. Kander's correlation has the same functional form as Eq. (32-24), but alludes to the fact that the heat diffusion sublayer thickness E_{α}^{+} is not fluid independent, as was the case for the momentum viscous sublayer. The Van Driest damping function constant A_{α}^{+} can be determined by fitting Kander's correlation for the dependency on Prandtl number. The result is shown in Figure 32-3. For air ($\mathrm{Pr} = 0.7$), the best choice of the damping constant appears to be $A_{\alpha}^{+} = 32$. For smaller Prandtl numbers, A_{α}^{+} increases steeply. For $\mathrm{Pr} > 30$, A_{α}^{+} decreases steeply. However, for Prandtl numbers $0.9 \leq \mathrm{Pr} \leq 30$, the damping constant falls in the range $30 \leq A_{\alpha}^{+} \leq 31$.

32.3.3 Outer Region: (Advection) ~ (Diffusion) and $\varepsilon_H \gg \alpha$

The outer region of the turbulent boundary layer begins when advection becomes important to transport. The heat equation for the outer region becomes

$$\overline{v}_x \frac{\partial \overline{T}}{\partial x} + \overline{v}_y \frac{\partial \overline{T}}{\partial y} = \frac{\partial}{\partial y}\left[\varepsilon_H \frac{\partial \overline{T}}{\partial y}\right]. \tag{32-27}$$

With the advection term present, the heat equation requires numerical integration to determine the temperature profile through the outer region. One can relate the turbulent diffusivities of heat and momentum transfer to each other through the definition of a *turbulent Prandtl number*:

$$\mathrm{Pr}_t = \frac{\varepsilon_M}{\varepsilon_H}. \tag{32-28}$$

To assess the turbulent heat diffusivity in the outer region of the boundary layer, it is reasonable to assume that the turbulent Prandtl number can be evaluated in the same manner as in the inner region where $\mathrm{Pr}_t = \kappa/\kappa_H \approx 0.85$. With this assumption, ε_M may be calculated in the outer region with a mixing length that scales with the momentum boundary layer thickness $\ell = \gamma\,\delta$. Subsequently, turbulent heat diffusivity is evaluated from $\varepsilon_H = \varepsilon_M/\mathrm{Pr}_t$.

32.4 MIXING LENGTH MODEL FOR HEAT TRANSFER

When the extent of the momentum viscous sublayer and the heat diffusion sublayer are the same ($E_\alpha^+ = E_\nu^+$), there is only one additional parameter to incorporate into a mixing length model for heat transfer. This parameter is the turbulent Prandtl number. In terms of the turbulent Prandtl number, the boundary layer equation for turbulent heat transfer can be written as

$$\overline{v}_x \frac{\partial \overline{T}}{\partial x} + \overline{v}_y \frac{\partial \overline{T}}{\partial y} = \frac{\partial}{\partial y}\left[\left(\alpha + \frac{\varepsilon_M}{\mathrm{Pr}_t}\right)\frac{\partial \overline{T}}{\partial y}\right]. \tag{32-29}$$

Since $\overline{v}_x$, $\overline{v}_y$, and ε_M are all determined in the process of solving the turbulent momentum equation, the additional task of solving the heat Equation (32-29) is not too arduous.

When $E_\alpha^+ \neq E_\nu^+$, there will be a region near the viscous sublayer for which the Prandtl number $\mathrm{Pr}_t = \varepsilon_M/\varepsilon_H$ cannot equal a constant value. In such a case, ε_H in the heat equation can be evaluated from its definition:

$$\varepsilon_H = \ell_H\,\ell\left|\frac{\partial \overline{v}_x}{\partial y}\right|. \tag{32-30}$$

One can evaluate the mixing length scale for heat transfer ℓ_H in an analogous way to its momentum counterpart. The procedure is illustrated in Table 32-1, where E_α^+ is the extent of the diffusion sublayer for heat transfer, and A_α^+ is its counterpart expression used in the Van Driest function. The use of the Van Driest damping function is recommended for numerical work. The constant $\kappa/\kappa_H \approx 0.85$ is the ratio between momentum and heat transfer mixing lengths in the inner region of turbulence.

When $E_\alpha^+ \neq E_\nu^+$, the turbulent Prandtl number cannot be constant approaching the sublayers. Using the Van Driest damping function, the turbulent Prandtl number can be described by the constants A_ν^+, A_α^+, and κ/κ_H, through the relation [4]:

$$\mathrm{Pr}_t = \frac{\varepsilon_M}{\varepsilon_H} = \frac{\ell}{\ell_H} = \frac{\kappa}{\kappa_H}\frac{1-\exp(-y^+/A_\nu^+)}{1-\exp(-y^+/A_\alpha^+)}. \tag{32-31}$$

Equation (32-31) is plotted in Figure 32-4 for a fluid with a heat diffusion sublayer that extends significantly beyond the viscous sublayer ($A_\alpha^+ = 50 > A_\nu^+ = 25$). Consequently, the turbulent Prandtl number Pr_t is seen to rise approaching the wall as ε_H becomes small before ε_M.

Table 32-1 Mixing length model for turbulent heat diffusivity

$\varepsilon_H = \ell_H \ell\left\|\frac{\partial \overline{v}_x}{\partial y}\right\|$	Diffusion sublayer $y^+ < E_\alpha^+$	Inner region $(E_\alpha^+ < y^+)$ *and* $(\kappa y < \gamma\delta)$	Outer region $\kappa y > \gamma\delta$
Mixing length:	$\ell_H = 0$	$\ell_H = \kappa_H y$	$\ell_H = \gamma\delta/(\kappa/\kappa_H)$
Van Driest:	$\ell_H = \kappa_H y[1-\exp(-y^+/A_\alpha^+)]$		$\ell_H = \gamma\delta/(\kappa/\kappa_H)$

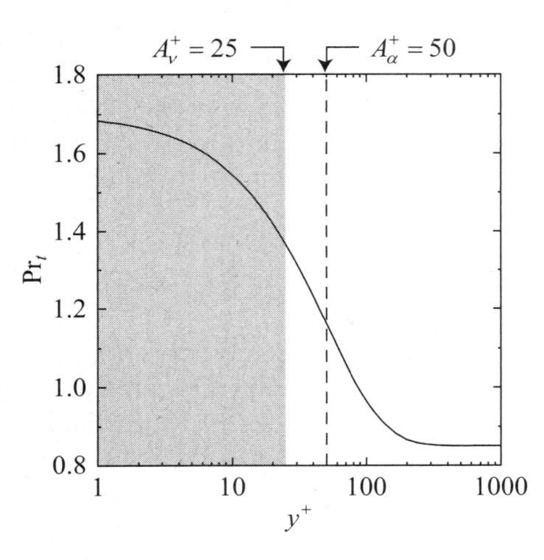

Figure 32-4 Turbulent Prandtl number as described by Eq. (32-31).

The premise of the mixing length model is that turbulent diffusion results from fluid elements interacting over the length scale $\ell \approx \ell_H$ of an eddy. So long as the time scale for interelement transfer is small compared to the time scale for the rolling motion of an eddy, one could expect turbulence to be indiscriminate with respect to the properties being transported. However, as the eddy length scale and time scale are reduced approaching the sublayer, the time allowed for interelement transfer can be restricted. Therefore, one interpretation of the rise in turbulent Prandtl number shown in Figure 32-4 is that the time scale required for interelement heat transfer is significantly longer than for momentum transfer. When this is the case, the reduction in eddy time scale caused by $\ell \to 0$ will affect first the effectiveness with which turbulence transports heat. This leads to an increase in $\mathrm{Pr}_t = \varepsilon_M/\varepsilon_H$ because of the diminished value of ε_H relative to ε_M.

32.5 MIXING LENGTH MODEL FOR SPECIES TRANSFER

By analogy to the heat transfer equation, one can show (see Problem 32-1) that the incompressible boundary layer equation for dilute turbulent species transfer is

$$\bar{v}_x \frac{\partial \bar{c}}{\partial x} + \bar{v}_y \frac{\partial \bar{c}}{\partial y} = \frac{\partial}{\partial y}\left[(D + \varepsilon_C)\frac{\partial \bar{c}}{\partial y}\right]. \tag{32-32}$$

The turbulent species diffusivity ε_C can be evaluated from its definition:

$$\varepsilon_C = \ell_C\, \ell \left|\frac{\partial \bar{v}_x}{\partial y}\right|, \tag{32-33}$$

where ℓ_C is the effective mixing length for species transfer. The mixing length can be evaluated as indicated in Table 32-2, where E_D^+ is the extent of the diffusion sublayer for species transfer, and A_D^+ is its counterpart expression used in the Van Driest function. In the inner region of the turbulent boundary layer, the effective species mixing length scales as $\ell_C = \kappa_C y$. The quantity κ/κ_C, appearing in Table 32-2, is the ratio of the momentum and species mixing lengths in the inner region of turbulence, and describes a turbulent Schmidt number,

$$Sc_t = \varepsilon_M/\varepsilon_C, \tag{32-34}$$

that is analogous to the turbulent Prandtl number.

Table 32-2 Mixing length model for turbulent species diffusivity

$\varepsilon_C = \ell_C \ell \left\lvert \frac{\partial \overline{v}_x}{\partial y} \right\rvert$	Diffusion sublayer $y^+ < E_D^+$	Inner region $(E_D^+ < y^+)$ *and* $(\kappa y < \gamma\delta)$	Outer region $\kappa y > \gamma\delta$
Mixing length:	$\ell_C = 0$	$\ell_C = \kappa_C y$	$\ell_C = \gamma\,\delta/(\kappa/\kappa_C)$
Van Driest:	$\ell_C = \kappa_C y[1 - \exp(-y^+/A_D^+)]$		$\ell_C = \gamma\,\delta/(\kappa/\kappa_C)$

32.6 PROBLEMS

32-1 Derive the turbulent species transport equation for a boundary layer, making note of all assumptions required to obtain the final form given by Eq. (32-32).

32-2 Consider flow through a pipe of radius R that experiences a constant heat flux q''_w through the wall. For fully developed heat transfer through the pipe, the turbulent flow heat equation is

$$\frac{2q''_w\overline{v}_z}{\rho C R \overline{v}_m} = \frac{1}{r}\frac{\partial}{\partial r}\left[r(\alpha + \varepsilon_H)\frac{\partial \overline{T}}{\partial r}\right] \quad \text{where} \quad \overline{v}_m = \int_0^R \overline{v}_z 2r dr/R^2.$$

As a simplifying approximation, let $\overline{v}_z = \overline{v}_m$ (plug flow). Using $y = R - r$, and the wall coordinates:

$$T^+ = \frac{T_w - \overline{T}}{q''_w/(\rho C)}\sqrt{\frac{\tau_w}{\rho}} \quad v^+ = \frac{\overline{v}_z}{\sqrt{\tau_w/\rho}} \quad y^+ = \frac{\sqrt{\tau_w/\rho}}{\nu}y,$$

integrate the heat equation to show that

$$T^+ = \int_0^{y^+} \frac{1 - y/R}{1/\mathrm{Pr} + \varepsilon_H/\nu} dy^+.$$

Split the task of integration into two regions: first over the diffusion sublayer which extends to $y^+ = 13.2$, then over the core region of turbulence. Make appropriate simplifications for each region. Use Reichardt's equation to evaluate the turbulent momentum diffusivity:

$$\frac{\varepsilon_M}{\nu} = \frac{\kappa y^+}{6}(1 + r/R)\left[1 + 2(r/R)^2\right]$$

and show that the temperature distribution is

$$T^+ = \begin{cases} \mathrm{Pr}y^+ & y^+ < 13.2 \\ 13.2\mathrm{Pr} + \frac{\mathrm{Pr}_t}{\kappa}\ln\left[y^+\frac{1.5(1 + r/R)}{1 + 2(r/R)^2}\right] - \frac{\mathrm{Pr}_t}{\kappa}\ln(13.2) & y^+ \geq 13.2 \end{cases}.$$

REFERENCES

[1] L. Fulachier, E. Verollet and I. Dekeyser, "Résultats Expérimentaux Concernant une Couche Limite Turbulente avec Aspiration et Chauage á la Paroi." *International Journal of Heat and Mass Transfer*, **20**, 731 (1977).

[2] W. M. Kays, "Turbulent Prandtl Number—Where Are We?" *Journal of Heat Transfer*, **116**, 284 (1994).

[3] B. A. Kander, "Temperature and Concentration Profiles in Fully Turbulent Boundary Layers." *International Journal of Heat and Mass Transfer*, **24**, 1541 (1981).

[4] T. Cebeci, "A Model for Eddy Conductivity and Turbulent Prandtl Number." *Journal of Heat Transfer*, **95**, 227 (1973).

Chapter 33

Fully Developed Transport in Turbulent Flows

33.1 Chemical Vapor Deposition in Turbulent Tube Flow with Generation

33.2 Heat Transfer in a Fully Developed Internal Turbulent Flow

33.3 Heat Transfer in a Turbulent Poiseuille Flow between Smooth Parallel Plates

33.4 Fully Developed Transport in a Turbulent Flow of a Binary Mixture

33.5 Problems

The mixing length model developed in Chapters 30 and 32 is applied in this chapter to turbulent heat and species transport in fully developed internal flows. For this purpose, the analysis developed in Chapter 28 for fully developed transport in laminar flows is extended to turbulence.

33.1 CHEMICAL VAPOR DEPOSITION IN TURBULENT TUBE FLOW WITH GENERATION

Consider a chemical vapor deposition process occurring in a smooth tube of radius R (Figure 33-1). The flow is a fully developed incompressible turbulent Poiseuille flow. The interior surface of the tube is being coated by a species that is carried by the flow. There is a constant chemical reaction in the flow to sustain a dilute concentration of the deposition species. A surface reaction at the tube wall drives deposition and causes the species concentration in the flow to go to zero at the wall. It is desired to determine the mean concentration $\bar{c}_m$ of the deposition species in the tube from a knowledge of the sustaining chemical reaction rate R_C (moles/m^3/s). The constant reaction rate is balanced by the deposition rate, as described by the convection coefficient h_C for species transport to the tube wall:

$$\underbrace{\pi R^2 R_C}_{production} = \underbrace{(2\pi R)h_C(\bar{c}_m - \overbrace{c_s}^{=0})}_{deposition} \quad \text{or} \quad (R^2/\DJ)R_C = Sh_D \bar{c}_m. \tag{33-1}$$

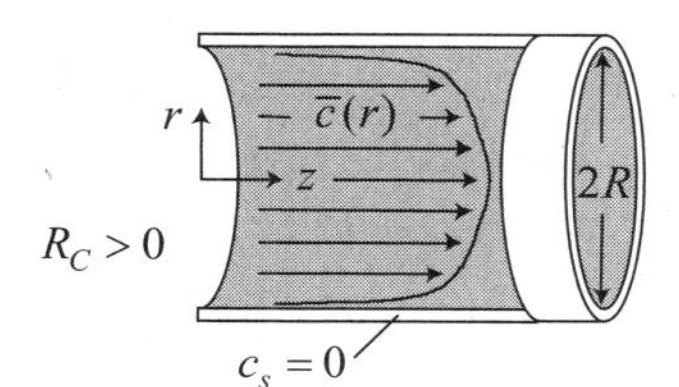

Figure 33-1 Chemical vapor deposition in turbulent tube flow.

Therefore, to establish the relation between the mean concentration $\bar{c}_m$ and the sustaining reaction rate R_C, it is necessary to determine the Sherwood number Sh_D for the turbulent flow as defined by

$$Sh_D = \frac{h_C(2R)}{Ð} = \frac{-(2R)\partial\bar{c}/\partial r\big|_{r=R}}{(\bar{c}_m - c_s)Ð}. \tag{33-2}$$

Notice that the Sherwood number is a dimensionless presentation of the convection coefficient h_C, which is defined in relation to the mean concentration $\bar{c}_m$ in the flow. The mean concentration is a transport average that is dependent on the flow field:

$$\bar{c}_m = \frac{\int_0^R \bar{c}\,\bar{v}_z\, rdr}{\int_0^R \bar{v}_z\, rdr}. \tag{33-3}$$

The steady-state dilute concentration of deposition species in the tube is governed by the transport equation

$$\overbrace{v_y}^{=0}\frac{\partial\bar{c}}{\partial r} + \bar{v}_z\overbrace{\frac{\partial\bar{c}}{\partial z}}^{=0} = \frac{1}{r}\frac{\partial}{\partial r}\left[r(Ð+\varepsilon_C)\frac{\partial\bar{c}}{\partial r}\right] + \frac{\partial}{\partial z}\left[(Ð+\varepsilon_C)\overbrace{\frac{\partial\bar{c}}{\partial z}}^{=0}\right] + R_C. \tag{33-4}$$

For fully developed conditions, the advection terms drop out of the transport equation, as well as the streamwise diffusion. The remaining two terms in the equation describe a balance between the source of deposition species and diffusion to the wall of the tube. This transport equation is analogous to the fully developed momentum equation for pipe flow, where the pressure gradient is the source term balancing momentum loss to the wall by diffusion.

The species equation (33-4) can be integrated once for the result

$$C_1 = r(Ð+\varepsilon_C)\frac{\partial\bar{c}}{\partial r} + R_C\frac{r^2}{2}. \tag{33-5}$$

The integration constant C_1 can be evaluated at the tube wall $r = R$, where $\varepsilon_C = 0$. Substituting this result back into the transport equation (33-5) yields

$$r(Ð+\varepsilon_C)\frac{\partial\bar{c}}{\partial r} = R\left[Ð\frac{\partial\bar{c}}{\partial r}\right]_{r=R} + R_C\frac{(R^2-r^2)}{2}. \tag{33-6}$$

However, the species flux at the wall must be balanced by the chemical reaction rate R_C in the flow, such that

$$\pi R^2 R_C = (2\pi R)\left[-Ð\frac{\partial\bar{c}}{\partial r}\right]_{r=R}. \tag{33-7}$$

Using this result to eliminate R_C from the species transport equation (33-6) yields

$$\left(1+\frac{\varepsilon_C}{Ð}\right)\frac{\partial\bar{c}}{\partial r} = \frac{\partial\bar{c}}{\partial r}\bigg|_{r=R}\frac{r}{R} \quad \text{or} \quad \left(1+\frac{\varepsilon_C}{Ð}\right)\frac{\partial\bar{c}}{\partial r} = -\frac{h_c\bar{c}_m}{Ð}\frac{r}{R}. \tag{33-8}$$

The second form of this last result makes use of the relation between the species gradient at the wall and the definition of the convection coefficient h_C.

The turbulent species diffusivity can be related to the momentum diffusivity through a turbulent Schmidt number:

$$Sc_t = \frac{\varepsilon_M}{\varepsilon_C} = \frac{\ell}{\ell_C}, \tag{33-9}$$

where ℓ and ℓ_C are the mixing lengths for momentum and species transport. Therefore,

$$\frac{\varepsilon_C}{Đ} = \frac{Sc}{Sc_t}\frac{\varepsilon_M}{\nu} = Sc\frac{\ell_C}{\ell}\frac{\varepsilon_M}{\nu}, \tag{33-10}$$

where $Sc = \nu/Đ$ is the molecular Schmidt number. The turbulent momentum diffusivity is determined from the mixing length model by

$$\frac{\varepsilon_M}{\nu} = \frac{\ell^2}{\nu}\left|\frac{\partial \overline{v}_z}{\partial r}\right| = \frac{\mathrm{Re}_D}{2}\Lambda^2\left|\frac{\partial u}{\partial \eta}\right|, \tag{33-11}$$

where

$$\eta = 1 - r/R, \quad u = \overline{v}_z/\overline{v}_m, \quad \Lambda = \ell/R, \quad \text{and} \quad \mathrm{Re}_D = \overline{v}_m(2R)/\nu. \tag{33-12}$$

The dimensionless velocity profile for turbulent Poiseuille flow through a pipe was discussed in Section 31.3, and must satisfy the equation for $u' = \partial u/\partial\eta$ as given by

$$\begin{aligned} u' &= \frac{\sqrt{1 + 2\mathrm{Re}_D\,\Lambda^2 u'(0)(1-\eta)} - 1}{\mathrm{Re}_D\Lambda^2}, \quad \eta \neq 0 \\ u' &= u'(0) \qquad\qquad\qquad\qquad\qquad\quad, \quad \eta = 0 \end{aligned} \tag{33-13}$$

subject to

$$u(0) = 0 \quad \text{and} \quad \int_0^1 u(1-\eta)d\eta = 1/2. \tag{33-14}$$

The Van Driest damping function is used to evaluate the mixing length from the wall through the inner region of turbulence, and $\ell = \gamma R$ is used to assign a constant mixing length in the outer region of turbulence. Therefore, the dimensionless mixing length $\Lambda = \ell/R$ is selected to satisfy the rule

$$\Lambda = \begin{cases} \Lambda^* & \Lambda^* < \gamma \\ \gamma & \textit{otherwise} \end{cases}, \quad \text{where} \quad \Lambda^* = \kappa\eta\left[1 - \exp(-y^+/A_v^+)\right], \tag{33-15}$$

and

$$y^+ = \eta\sqrt{(\mathrm{Re}_D/2)u'(0)}. \tag{33-16}$$

The constants appearing in the mixing length model are: the extent of the viscous sublayer, as reflected by A_ν^+; the proportionality coefficient κ between the mixing length and distance from the wall (used in the inner region of turbulence); and the proportionality coefficient γ between the mixing length size and the geometric constraint of the pipe radius (used in the outer region of turbulence).

The species concentration can be nondimensionalized with the definition

$$\phi = \bar{c}/\bar{c}_m, \tag{33-17}$$

where

$$\bar{c}_m = \frac{\int_0^R \bar{c}\,\bar{v}_z r dr}{\int_0^R \bar{v}_z r dr} = \frac{\int_0^1 \bar{c}u(1-\eta)d\eta}{\int_0^1 u(1-\eta)d\eta} = 2\int_0^1 \bar{c}u(1-\eta)d\eta. \tag{33-18}$$

Or,

$$2\int_0^1 \phi u(1-\eta)d\eta = 1. \tag{33-19}$$

In terms of dimensionless variables $\eta = 1 - r/R$ and $\phi = \bar{c}/\bar{c}_m$, the species transport equation (33-8) becomes

$$\frac{\partial \phi}{\partial \eta} = \frac{\phi'(0)(1-\eta)}{1+\varepsilon_C/Ɖ} \quad \text{or} \quad \frac{\partial \phi}{\partial \eta} = \frac{(Sh_D/2)(1-\eta)}{1+\varepsilon_C/Ɖ}. \tag{33-20}$$

The latter form makes use of the definition of the Sherwood number, Eq. (33-2), resulting in

$$Sh_D = \frac{-(2R)\partial\bar{c}/\partial r\big|_{r=R}}{(\bar{c}_m - c_s)Ɖ} = 2\phi'(0), \tag{33-21}$$

since $c_s = 0$. Because the heat equation (33-20) is first order, only one initial condition is needed for integration, which is given by the wall condition $\phi(0) = 0$.

Defining a new variable Φ, such that $\phi = (Sh_D/2)\Phi$, the governing equation (33-20) becomes

$$\frac{\partial \Phi}{\partial \eta} = \frac{(1-\eta)}{1+\varepsilon_C/Ɖ} \tag{33-22}$$

with

$$\Phi(0) = 0. \tag{33-23}$$

Combining Eqs. (33-10) and (33-11), the turbulent species diffusivity can be evaluated from

$$\frac{\varepsilon_C}{Ɖ} = \frac{Sc}{Sc_t}\frac{\mathrm{Re}_D}{2}\Lambda^2 u', \tag{33-24}$$

where Λ is evaluated from Eq. (33-15). Using this result for the turbulent species diffusivity, the governing equation (33-22) for Φ becomes

$$\frac{\partial \Phi}{d\eta} = \frac{(1-\eta)}{1+(Sc/Sc_t)(\mathrm{Re}_D/2)\Lambda^2 u'} \tag{33-25}$$

with

$$\Phi(0) = 0. \tag{33-26}$$

In general, the turbulent Schmidt number Sc_t is not constant. However, for the present problem, it is supposed that the species diffusion sublayer has the same thickness as the viscous sublayer, such that $A_D^+ = A_\nu^+$. In this case, the turbulent Schmidt number $Sc_t = \ell/\ell_C = \kappa/\kappa_C$ can be assumed constant for the inner and outer regions of turbulence.

Once the solution for Φ is found, then Eq. (33-19) with $\phi = (Sh_D/2)\Phi$ permits the Sherwood number to be evaluated from

$$Sh_D = \frac{1}{\int_0^1 \Phi u(1-\eta)d\eta}. \tag{33-27}$$

By solving the governing equation (33-25) for Φ, the Sherwood number Sh_D for the turbulent flow through the tube can be determined as a function of the molecular Schmidt number Sc and the Reynolds number Re_D. Figure 33-2 plots the result of this analysis, using $A_D^+ = A_\nu^+ = 26$, $\kappa = 0.40$, $\gamma = 0.085$, and $Sc_t = 0.9$ in the mixing length model.

When the Sherwood number is normalized by the product of the coefficient of friction and Reynolds number, it is interesting that the result, $Sh_D/(c_f Re_D)$, becomes almost independent of the Reynolds number at a Schmidt number just below one (see Figure 33-2). The explanation for this lies in the similarity between momentum and species transport in the tube, similar to the Reynolds analogy discussed in Section 25.5 for the boundary layer. In terms of the dimensional dependent variables, the equations governing momentum and species transport in the tube are nearly identical:

$$\text{Momentum:} \quad \frac{\partial \overline{v}_z}{d\eta} = \frac{\overline{v}_z'(0)(1-\eta)}{1+\varepsilon_M/\nu}, \quad \text{with} \quad \overline{v}_z(0) = 0 \tag{33-28}$$

$$\text{Species:} \quad \frac{\partial \overline{c}}{\partial \eta} = \frac{\overline{c}'(0)(1-\eta)}{1+\varepsilon_C/Đ}, \quad \text{with} \quad \overline{c}(0) = 0. \tag{33-29}$$

The transport equations would be identical in form if $\varepsilon_C/Đ = \varepsilon_M/\nu$ (or equivalently if $Sc = Sc_t$). When the dependent variables are normalized, such that $u = \overline{v}_z/\overline{v}_m$ and $\phi = \overline{c}/\overline{c}_m$, it is easily demonstrated that

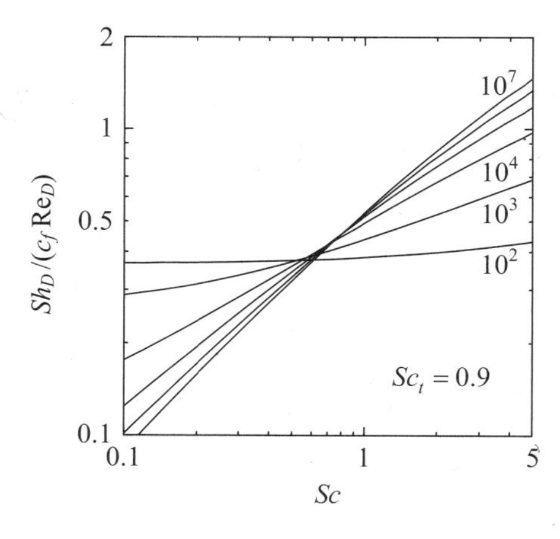

Figure 33-2 Sherwood number for turbulent Poiseuille flow in a smooth-wall tube.

$$u'(0) = \frac{c_f \mathrm{Re}_D}{4} \quad \text{and} \quad \phi'(0) = \frac{Sh_D}{2}. \tag{33-30}$$

If the solutions to u and ϕ were identical (such that $u'(0) = \phi'(0)$), then the scaling relation $Sh_D \sim 2c_f \mathrm{Re}_D$ should follow. However, from the results in Figure 33-2, it is apparent that the solutions for u and ϕ must not be identical. The reason for this becomes apparent when considering the normalization requirements for $u = \overline{v}_z/\overline{v}_m$ and $\phi = \overline{c}/\overline{c}_m$:

$$\text{Momentum:} \quad 2\int_0^1 u(1-\eta)d\eta = 1 \tag{33-31}$$

$$\text{Species:} \quad 2\int_0^1 \phi u(1-\eta)d\eta = 1. \tag{33-32}$$

Notice that the normalization of u is independent of ϕ, but the normalization of ϕ is dependent on u. For this reason, solutions to u and ϕ are not identical even though their governing equations are the same. In contrast, definitions of mean values do not enter into the problem formulation for boundary layers. Consequently, simple analogies between identical transport solutions can be made for boundary layer problems, as famously done by Osborne Reynolds for momentum and heat transport (see Section 25.5).

33.2 HEAT TRANSFER IN A FULLY DEVELOPED INTERNAL TURBULENT FLOW

Consider heat transfer between a fluid and its bounding walls when there is no source of heating in the flow. In such a case, consideration of advection is required to balance the heat transport to the walls by diffusion. Consequently, the turbulent heat equation for a fully developed internal flow ($\overline{v}_y = 0$) is

$$\overline{v}_x \frac{\partial \overline{T}}{\partial x} = \frac{\partial}{\partial y}\left[(\alpha + \varepsilon_H)\frac{\partial \overline{T}}{\partial y}\right]. \tag{33-33}$$

This equation differs from its laminar counterpart in Chapter 28 only in that it is time averaged and contains the additional term of turbulent diffusivity. As discussed in Section 27.5, a fully developed temperature field exhibits the behavior

$$\frac{\partial}{\partial x}\left[\frac{T_s - \overline{T}(x,y)}{T_s - \overline{T}_m(x)}\right] = 0 \quad (or \approx 0). \tag{33-34}$$

In other words, a fully developed temperature profile ceases to change with downstream distance when it is evaluated relative to the surface temperature T_s and normalized relative to the mean temperature $\overline{T}_m$. The mean temperature of the flow is defined by

$$\overline{T}_m = \int_A \overline{v}_x \overline{T} dA/(\overline{v}_m A), \tag{33-35}$$

where

$$\overline{v}_m = \int_A \overline{v}_x dA/A \tag{33-36}$$

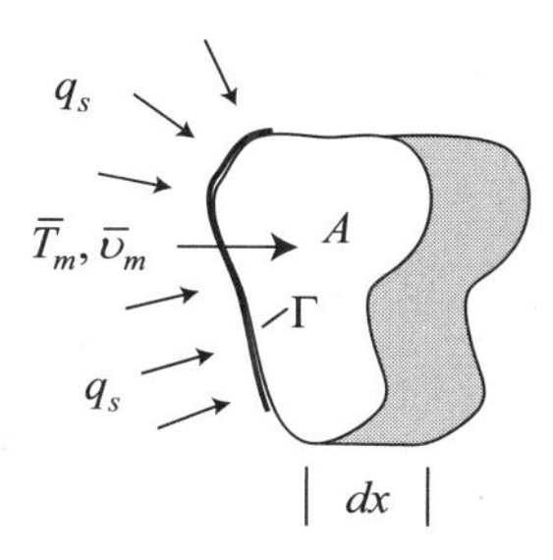

Figure 33-3 Heat balance over a unit distance streamwise.

is the mean velocity and A is the cross-sectional area of the flow. These definitions are unchanged from the treatment of laminar flows in Chapter 28.

An energy balance over a differential distance dx in the streamwise direction of the flow, as illustrated in Figure 33-3, yields

$$\overline{v}_m \frac{d\overline{T}_m}{dx} = \frac{\Gamma\alpha}{Ak} q_s = \frac{\Gamma\alpha}{Ak} h(T_s - \overline{T}_m), \tag{33-37}$$

where Γ is the peripheral length around the cross-sectional area A through which the heat transfer flux q_s occurs, and h is the associated heat transfer coefficient. Using Eq. (33-37), the mean temperature gradient $\partial\overline{T}_m/\partial x$ can be expressed in terms of heat transfer from the walls of the flow. In turn, Eq. (33-34) relates $\partial\overline{T}_m/\partial x$ to the temperature gradient at any point in the flow $\partial\overline{T}/\partial x$. Through these relations, the turbulent heat equation (33-33) can be transformed into an ordinary differential equation, as will be demonstrated in the following section.

33.3 HEAT TRANSFER IN A TURBULENT POISEUILLE FLOW BETWEEN SMOOTH PARALLEL PLATES

The turbulent Poiseuille flow between smooth parallel plates, as discussed in Section 31.1, will be considered again to demonstrate the calculation of turbulent heat transfer. Highlights of the momentum equation solution are repeated here, since they will be used in subsequent analysis of heat transfer.

33.3.1 Summary of Turbulent Momentum Transport

The solution to the momentum equation for fully developed turbulent Poiseuille flow between parallel plates is expressed in terms of the dimensionless variables

$$\eta = y/a \quad \text{and} \quad u = \overline{v}_x/\overline{v}_m, \tag{33-38}$$

where a is the gap between the plates and $\overline{v}_m = \int_0^a \overline{v}_x dy/a$ is the mean flow velocity. As demonstrated in Section 31.1, the turbulent momentum equation can be expressed as

$$(1 + \varepsilon_M/\nu)u' = (1 - 2\eta)u'(0), \tag{33-39}$$

and is subject to the conditions $u(0) = 0$ and $\int_0^{1/2} u d\eta = 1/2$. The turbulent diffusivity, appearing in the momentum equation, is evaluated from the mixing length model using

$$\frac{\varepsilon_M}{\nu} = (\mathrm{Re}_D/2)\Lambda^2 u', \tag{33-40}$$

where $\mathrm{Re}_D = \overline{v}_m(2a)/\nu$ is the Reynolds number based on the hydraulic diameter. Using the Van Driest damping function, the dimensionless mixing length for momentum transport $\Lambda = \ell/a$ is evaluated by

$$\Lambda = \begin{cases} \Lambda^* & \Lambda^* < \gamma/2 \\ \gamma/2 & \textit{otherwise} \end{cases} \quad \text{where} \quad \Lambda^* = \kappa\eta\left[1 - \exp\left(-y^+/A_\nu^+\right)\right] \tag{33-41}$$

and

$$y^+ = \eta\sqrt{(\mathrm{Re}_D/2)u'(0)}. \tag{33-42}$$

Equation (34-20) can be numerically integrated with the fourth-order Runge-Kutta method using Code 31-1. Subroutines of that code are used again for solving turbulent heat transfer, as discussed next.

33.3.2 Turbulent Heat Transfer with Constant Temperature Boundary

Consider the problem illustrated in Figure 33-4 of a flow between hot isothermal smooth plates. The turbulent flow is fully developed both hydrodynamically and thermally. As heat transfer to the fluid occurs, the temperature of the flow rises until eventually becoming uniform between the plates, as illustrated.

Since the temperature field is fully developed, and T_s is a constant, Eq. (33-34) dictates that the temperature gradient $\partial\overline{T}/\partial x$ everywhere in the flow can be related to changes in the mean temperature:

$$\frac{\partial\overline{T}}{\partial x} \approx \frac{T_s - \overline{T}}{T_s - \overline{T}_m}\frac{d\overline{T}_m}{dx}. \tag{33-43}$$

This result is combined with Eq. (33-37) to eliminate $\partial\overline{T}_m/\partial x$, allowing the heat equation (33-33) to be written in the form

$$\frac{\overline{v}_x}{\overline{v}_m}(T_s - \overline{T})\frac{\Gamma h}{Ak} = \frac{\partial}{\partial y}\left[\left(1 + \frac{\varepsilon_H}{\alpha}\right)\frac{\partial\overline{T}}{\partial y}\right] \quad \text{(thermally-hydrodynamically fully developed turbulent flow with isothermal boundaries).} \tag{33-44}$$

Notice that the heat equation is now an ordinary differential equation. With $\Gamma/A = 2/a$ for the parallel plate geometry, $\eta = y/a$ and $u = \overline{v}_x/\overline{v}_m$, as defined previously, the heat equation becomes

$$u(1 - \theta)Nu_{D,T} = \frac{\partial}{\partial\eta}\left[\left(1 + \frac{\varepsilon_H}{\alpha}\right)\frac{\partial\theta}{\partial\eta}\right], \tag{33-45}$$

where the dimensionless temperature variable is defined by

$$\theta = \frac{\overline{T} - \overline{T}_m}{T_s - \overline{T}_m}. \tag{33-46}$$

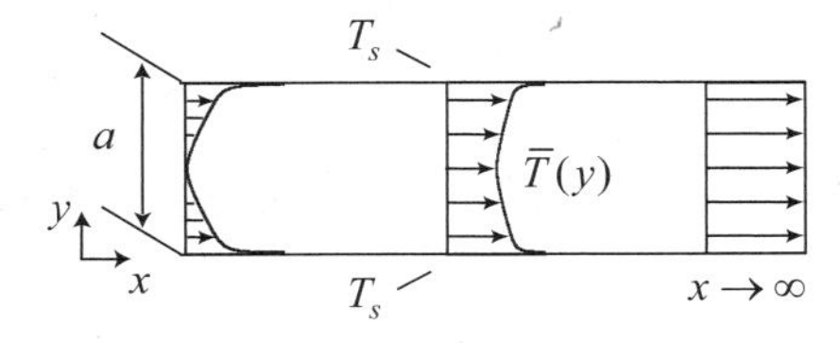

Figure 33-4 Fully developed turbulent temperature field between isothermal plates.

With the mean temperature definition

$$\overline{T}_m = \frac{\int_0^{a/2} \overline{v}_x \overline{T} dy}{\overline{v}_m a/2} = \frac{\int_0^{a/2} \overline{v}_x (\theta(T_s - \overline{T}_m) + \overline{T}_m) dy}{\overline{v}_m a/2}, \tag{33-47}$$

the dimensionless temperature variable requires that

$$\int_0^{1/2} u\theta \frac{T_s - \overline{T}_m}{\overline{T}_m} d\eta + \int_0^{1/2} u d\eta = \frac{1}{2} \tag{33-48}$$

or

$$\int_0^{1/2} u\theta d\eta = 0. \tag{33-49}$$

The Nusselt number in the governing equation (33-45) is defined with the hydraulic diameter ($2a$) and is related to the temperature solution by

$$Nu_{D,T} = \frac{h(2a)}{k} = -\frac{2a}{T_s - \overline{T}_m} \frac{\partial \overline{T}}{\partial y}\bigg|_{y=0} = -2\theta'(0). \tag{33-50}$$

The additional "T" subscript serves as a reminder that, in this problem, the Nusselt number reflects heat transfer from a constant wall temperature. With $Nu_{D,T} = -2\theta'(0)$, the heat equation (33-45) can be integrated once for the result

$$-2\theta'(0) \int_0^{\eta} u(1-\theta) = \left(1 + \frac{\varepsilon_H}{\alpha}\right)\theta' - \theta'(0). \tag{33-51}$$

The result can be rearranged to express the ordinary differential equation for θ in a form suitable for Runge-Kutta integration:

$$\theta' = \frac{1 - g(\eta)}{1 + \varepsilon_H/\alpha} \theta'(0), \tag{33-52}$$

where

$$g(\eta) = 2\int_0^{\eta} u(1-\theta) d\eta. \tag{33-53}$$

The constraint given by Eq. (33-49) on the solution requires that $g(1/2) = 1$. Furthermore, the symmetry in the solution, stipulated by $\theta'(1/2) = 0$, is already enforced in the governing equation (33-52) through the requirement that $g(1/2) = 1$. The second boundary condition is specified by the isothermal wall temperature. Therefore, the two conditions imposed on the solution are

$$\theta(0) = 1 \quad \text{and} \quad g(1/2) = 1. \tag{33-54}$$

The turbulent heat diffusivity in the heat equation (33-52) is given by

$$\frac{\varepsilon_H}{\alpha} = \frac{\ell_H \ell}{\alpha}\left|\frac{\partial \overline{v}_x}{\partial y}\right| = \text{Pr}\frac{\text{Re}_D}{2}\Lambda_H \Lambda u'. \tag{33-55}$$

The momentum mixing length Λ is evaluated with Eq. (33-41). As discussed in Section 32.4, the mixing length for heat transport Λ_H is evaluated to satisfy the rule

$$\Lambda_H = \begin{cases} \Lambda_H^* & \Lambda_H^* < \gamma/[2(\kappa/\kappa_H)] \\ \gamma/[2(\kappa/\kappa_H)] & \textit{otherwise} \end{cases}, \tag{33-56}$$

where Λ_H^* is calculated from the Van Driest damping function

$$\Lambda_H^* = \frac{\kappa\eta}{\kappa/\kappa_H}\left[1 - \exp\left(-y^+/A_\alpha^+\right)\right], \tag{33-57}$$

using the wall variable

$$y^+ = \eta\sqrt{(\text{Re}_D/2)u'(0)}. \tag{33-58}$$

With Eq. (33-56), the dimensionless mixing length for heat transfer is evaluated with the Van Driest damping function in the viscous sublayer and inner region of turbulence, and is held constant in the outer region of turbulence. The extent of the heat diffusion sublayer is reflected by the damping function constant A_α^+, and κ_H is the proportionality coefficient between the heat transfer mixing length and distance from the wall (used in the inner region of turbulence). The damping function constant A_α^+ can be estimated from the Prandtl number of the fluid using Figure 32-3. In the outer region, the mixing length for heat transfer is evaluated with $\gamma/[2(\kappa/\kappa_H)]$ to reflect both the largest scale allowed for the momentum mixing length between the plates ($\gamma/2$) and the relative size of the momentum and heat transfer mixing length scales (κ/κ_H), as was determined from the inner region.

The heat equation (33-52) can be numerically integrated using the fourth-order Runge-Kutta method described in Chapter 23. The governing equations for momentum (33-39), heat (33-52) transport, and the constraint (33-53) may be expressed as three coupled first-order equations for u, θ, and g:

$$\frac{du}{d\eta} = \frac{\sqrt{1 + 2\Lambda^2 \text{Re}_D\, u'(0)(1 - 2\eta)} - 1}{\Lambda^2 \text{Re}_D}, \quad u(0) = 0 \tag{33-59}$$

$$\frac{\partial\theta}{\partial\eta} = \frac{(1 - g)\theta'(0)}{1 + \text{Pr}(\text{Re}_D/2)\Lambda_H \Lambda u'}, \quad \theta(0) = 1 \tag{33-60}$$

$$\frac{\partial g}{\partial\eta} = 2u(1 - \theta), \quad g(0) = 0 \text{ and } g(1/2) = 1. \tag{33-61}$$

Code 33-1 is used to solve for the heat transfer between isothermal smooth parallel plates and a fully developed turbulent Poiseuille flow. Transport is calculated with the mixing length model using $\kappa = 0.40$, $A_\nu^+ = 26.0$, $\gamma = 0.085$, $A_\alpha^+ = 32.0$, and $\kappa/\kappa_H = 0.9$. The spatial variable η is uniformly discretized with respect to the log-of-distance from the wall for the

Code 33-1 Solves turbulent Poiseuille heat convection between parallel isothermal plates (uses solve_u( ), du( ), and VanDriest () from Code 31-1)

```
#include <stdio.h>
#include <stdlib.h>
#include <math.h>

double VanDriest(double eta,double Ap,double Kappa,
                 double Gamma,double ReD_du0);
double du(double ReD,double du0,double eta);
double solve_u(int N,double *eta,double *u,double ReD);

double NuT_dT(double ReD,double Pr,double ApH,double du0,double dT0,
              double eta,double g)
{
   double Kappa=0.4,Gamma=0.085,Ap=26.0,Prt=.85;
   double _du=du(ReD,du0,eta);
   double Lam= VanDriest(eta,Ap,Kappa,Gamma,ReD*du0);
   double LamH=VanDriest(eta,ApH,Kappa/Prt,Gamma/Prt,ReD*du0);
   double EH=Pr*(ReD/2.)*LamH*Lam*_du;
   return (1.-g)*dT0/(1.+EH);
}

double solve_NuT(int N,double *eta,double *T,
                 double du0,double ReD,double Pr,double ApH)
{
   int n,iter=0;
   double del_eta,dT0,u,g,K1u,K2u,K3u,K4u,K1T,K2T,K3T,K4T,K1g,K2g,K3g,K4g;
   double dT0_lo=-1000000.0;     // lower bound on dT0
   double dT0_hi=0.0;            // upper bound on dT0

   T[0]=1.;                      // initial condition
   do {                          // RK integration
      u=g=0.;
      dT0=(dT0_lo+dT0_hi)/2.0;   // use bisection method
      for (n=0;n<N-1;++n) {
         del_eta=eta[n+1]-eta[n];
         K1u=del_eta*du(ReD,du0,eta[n]);
         K1T= del_eta*NuT_dT(ReD,Pr,ApH,du0,dT0,eta[n],g);
         K1g= del_eta*2.*u*(1.-T[n]);
         K2u=del_eta*du(ReD,du0,(eta[n]+eta[n+1])/2.);
         K2T= del_eta*NuT_dT(ReD,Pr,ApH,du0,dT0,
                             (eta[n+1]+eta[n])/2.,g+0.5*K1g);
         K2g= del_eta*2.*(u+0.5*K1u)*(1.-T[n]-0.5*K1T);
         K3u=del_eta*du(ReD,du0,(eta[n]+eta[n+1])/2.);
         K3T= del_eta*NuT_dT(ReD,Pr,ApH,du0,dT0,
                             (eta[n+1]+eta[n])/2.,g+0.5*K2g);
         K3g= del_eta*2.*(u+0.5*K2u)*(1.-T[n]-0.5*K2T);
         K4u=del_eta*du(ReD,du0,eta[n+1]);
         K4T= del_eta*NuT_dT(ReD,Pr,ApH,du0,dT0,eta[n+1],g+K3g);
         K4g= del_eta*2.*(u+K3u)*(1.-T[n]-K3T);
         u+=            (K1u+2.*K2u+2.*K3u+K4u)/6.;
         T[n+1]=  T[n]+(K1T+ 2.*K2T+ 2.*K3T+ K4T )/6.;
         g+=            (K1g+ 2.*K2g+ 2.*K3g+ K4g )/6.;
         if (g>1.) {
            dT0_lo=dT0;      // shoot higher
            break;
         }
      }
      if (n==N-1) dT0_hi=dT0; // shoot lower

   } while (++iter<200 && (dT0_hi-dT0_lo)>1.0e-7);

   if (fabs(g-1.0)>.00001) {
      printf("\nSoln. failed, dT0=%e g=%e\n",dT0,g);
      exit(1);
   }
   return dT0;
}

int main()
{
   int n,N=2001;
   double yp,Tp,du0,dT0;
   double eta[N],u[N],T[N];

   eta[0]=0.;
   eta[N-1]=0.5;
   double log_min=-6.;
   double log_del=(log10(eta[N-1])-log_min)/(N-2);
   for (n=1;n<N-1;++n) eta[n]=pow(10.,log_min+(n-1)*log_del);

   double ReD=1.e6,Pr=0.7,ApH=32.;
   double B=pow(3.85*pow(Pr,1./3.)-1.3,2.)+2.12*log(Pr);

   du0=solve_u(N,eta,u,ReD);    // solve for u & du0
   dT0=solve_NuT(N,eta,T,du0,ReD,Pr,ApH);
   printf("\nNuD=%f\n",-2.*dT0);

   FILE *fp=fopen("Tp.dat","w");
   for (n=0;n<N;n++) {          // output the results
      yp=eta[n]*sqrt(du0*ReD/2.);
      Tp=Pr*(T[n]-1.)*sqrt(ReD*du0/2.)/dT0;
      fprintf(fp,"%e %e %e %e\n",yp,Tp,(.85/.40)*log(yp)+B,Pr*yp);
   }
   fclose(fp);
   return 0;
}
```

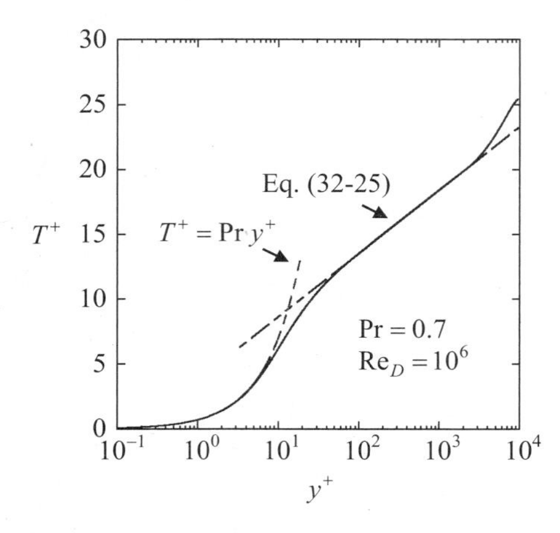

Figure 33-5 Temperature profile in wall variables.

numerical solution. This process of discretization was discussed in Section 31.1 for the solution to the momentum equation. The value of $u'(0)$ is known from the previous solution of the momentum equation, and the shooting method is used to determine $\theta'(0)$ satisfying the constraint $g(1/2) = 1$ for the heat equation.

Figure 33-5 illustrates the temperature profile in wall variables for a flow rate of $\mathrm{Re}_D = 10^6$ and a Prandtl number of $\mathrm{Pr} = 0.7$ (corresponding to air, with $A_\alpha^+ = 32.$). The wall-variable temperature

$$T^+ = \frac{T_s - \overline{T}}{q_s/\rho C_p} \sqrt{\tau_w/\rho} \tag{33-62}$$

is related to the other variables of the solution by

$$T^+ = \mathrm{Pr}\frac{\theta - 1}{\theta'(0)} \sqrt{(\mathrm{Re}_D/2)u'(0)}. \tag{33-63}$$

As shown in Figure 33-5, the temperature solution follows the $T^+ = \mathrm{Pr}\, y^+$ behavior through the diffusion sublayer, and follows the logarithmic behavior through the inner region of turbulence. Kander's correlation (32-15) for the logarithmic temperature profile is plotted in Figure 33-5 to demonstrate agreement with the mixing length solution through the inner region of turbulence.

Nusselt number results from the mixing length model are investigated in Figure 33-6 for a matrix of flow rates: $\mathrm{Re}_D = 10^4$, 10^5, and 10^6, and fluid Prandtl numbers: $\mathrm{Pr} = 0.7$, 7.0, and 70.0. The Van Driest damping function constant A_α^+ is determined from the Prandtl number using Figure 32-3. The mixing length model results are contrasted in Figure 33-6 with Gnielinski's correlation [1]:

$$Nu_D = \frac{(c_f/2)(\mathrm{Re}_D - 10^3)\mathrm{Pr}}{1 + 12.7(c_f/2)^{1/2}(\mathrm{Pr}^{2/3} - 1)} \quad (0.5 \leq \mathrm{Pr} \leq 2000, 2300 \leq \mathrm{Re}_D \leq 5 \times 10^6). \tag{33-64}$$

The coefficient of friction c_f used in the correlation is determined from the mixing length model for Poiseuille flow between smooth parallel plates, as discussed in Section 31.1. A good agreement between the correlation and mixing length model is found for all cases considered.

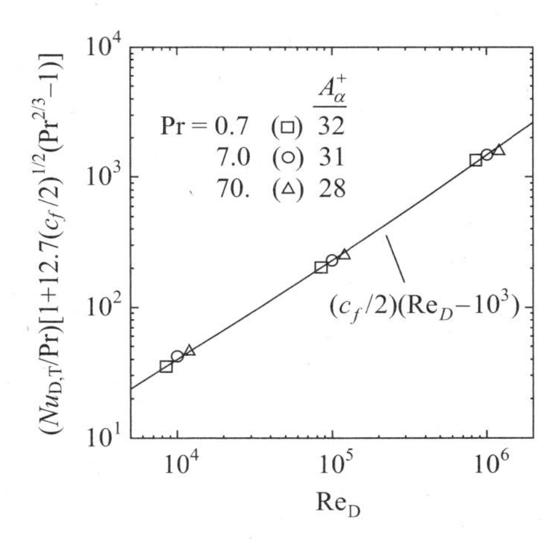

Figure 33-6 Correlation for turbulent flow Nusselt number.

33.3.3 Heat Transfer with Constant Heat Flux Boundary

Consider the problem illustrated in Figure 33-7 of a flow between plates that deliver a constant flux of heat to the fluid. The flow is fully developed both hydrodynamically and thermally. As heat is transferred to the fluid, the downstream temperature distribution between the two plates rises indefinitely, as shown.

For fully developed heat transfer, it was found in Section 28.2.2 that a constant heat flux boundary condition results in the temperature field rising uniformly with downstream distance:

$$\frac{\partial \overline{T}}{\partial x} = \frac{\partial T_s}{\partial x}\left(= \frac{\partial \overline{T}_m}{\partial x}\right). \tag{33-65}$$

The rise in mean temperature is related to the overall heat transfer rate to the fluid through Eq. (33-37), which yields

$$\frac{d\overline{T}_m}{dx} = \frac{1}{\overline{v}_m}\frac{\Gamma\alpha}{Ak}h(T_s - \overline{T}_m). \tag{33-66}$$

This result, combined with the fact that $\partial\overline{T}/\partial x = \partial\overline{T}_m/\partial x$, allows the heat equation (33-33) to be written as

$$\frac{\overline{v}_x}{\overline{v}_m}\frac{\Gamma}{Ak}h(T_s - \overline{T}_m) = \frac{\partial}{\partial y}\left[\left(1 + \frac{\varepsilon_H}{\alpha}\right)\frac{\partial \overline{T}}{\partial y}\right] \quad \begin{pmatrix}\text{thermally and hydrodynamically} \\ \text{developed turbulent flow with} \\ \text{constant heat flux boundary}\end{pmatrix}. \tag{33-67}$$

For the parallel plate geometry, $\Gamma/A = 2/a$. Letting

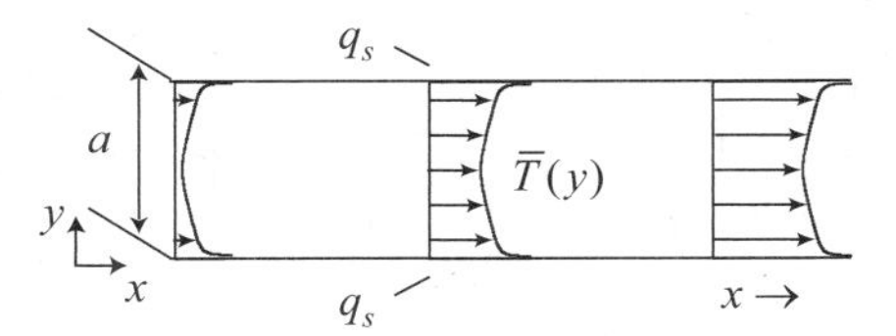

Figure 33-7 Fully developed temperature field between plates with constant heat flux.

$$\eta = \frac{y}{a}, u = \frac{\overline{v}_x}{\overline{v}_m}, \quad \text{and} \quad \theta = \frac{\overline{T} - T_m}{T_s - T_m}, \tag{33-68}$$

the heat equation becomes

$$u\, Nu_{D,H} = \frac{\partial}{\partial \eta}\left[\left(1 + \frac{\varepsilon_H}{\alpha}\right)\frac{\partial \theta}{\partial \eta}\right], \tag{33-69}$$

where

$$Nu_{D,H} = \frac{h(2a)}{k} = \frac{(2a)}{k}\frac{q_s}{(T_s - T_m)} = \frac{(2a)}{k}\frac{k\partial T/\partial y|_{y=0}}{(T_s - T_m)} = -2\theta'(0). \tag{33-70}$$

The additional "H" subscript serves as a reminder that, in this problem, the Nusselt number reflects heat transfer from a wall of specified heat flux. The heat equation (33-69) may be integrated once for

$$Nu_{D,H}\int_0^\eta u d\eta = \left(1 + \frac{\varepsilon_H}{\alpha}\right)\theta' - \theta'(0) = \left(1 + \frac{\varepsilon_H}{\alpha}\right)\theta' + Nu_{D,H}/2 \tag{33-71}$$

or

$$\theta' = -Nu_{D,H}\frac{1/2 - g(\eta)}{1 + \varepsilon_H/\alpha}, \quad \text{where} \quad g(\eta) = \int_0^\eta u d\eta. \tag{33-72}$$

Notice that symmetry at the centerline, as stipulated by $\theta'(1/2) = 0$, is guaranteed with the heat equation (33-72) since $g(1/2) = 1/2$. Integrating the heat equation a second time yields

$$\theta = 1 - Nu_{D,H}\, t(\eta), \qquad \text{where} \quad t(\eta) = \int_0^\eta \frac{1/2 - g(\eta)}{1 + \varepsilon_H/\alpha}, \tag{33-73}$$

where the turbulent diffusivity for heat transfer ε_H/α is evaluated from Eq. (33-55). However, since

$$T_m = \frac{\int_0^1 \overline{T}u(1-\eta)d\eta}{\int_0^1 u(1-\eta)d\eta} = \frac{\int_0^1 [\theta(T_s - T_m) + T_m]u(1-\eta)d\eta}{\int_0^1 u(1-\eta)d\eta}$$

$$= \frac{\int_0^1 [\theta(T_s - T_m)]u(1-\eta)d\eta}{\int_0^1 u(1-\eta)d\eta} + T_m, \tag{33-74}$$

the temperature solution must satisfy the relation

$$\int_0^1 \theta u(1-\eta)d\eta = 0. \tag{33-75}$$

Upon substituting the temperature solution (33-73) into (33-75), one finds

$$Nu_{D,H} \int_0^1 t\,u(1-\eta)d\eta = 1/2. \tag{33-76}$$

This may be solved for the heat transfer Nusselt number

$$Nu_{D,H} = \frac{1/2}{h(1/2)}, \quad \text{where} \quad h(\eta) = \int_0^{\eta} t\,u\,d\eta. \tag{33-77}$$

The numerical task (see Problem 33-3) of seeking a solution for $Nu_{D,H}$ can be tackled by the fourth-order Runge-Kutta integration method discussed in Chapter 23. The Nusselt number is evaluated from Eq. (33-77) after solving the system of coupled equations:

$$\frac{\partial u}{\partial \eta} = \frac{\sqrt{1 + 2\Lambda^2 \mathrm{Re}_D u'(0)(1-2\eta)} - 1}{\Lambda^2 \mathrm{Re}_D}, \quad u(0) = 0 \tag{33-78}$$

$$\frac{\partial g}{\partial \eta} = u, \quad g(0) = 0 \tag{33-79}$$

$$\frac{\partial t}{\partial \eta} = \frac{1/2 - g}{1 + \mathrm{Pr}(\mathrm{Re}_D/2)\Lambda_H \Lambda u'}, \quad t(0) = 0 \tag{33-80}$$

$$\frac{\partial h}{\partial \eta} = ut, \quad h(0) = 0. \tag{33-81}$$

Since $u'(0)$ is known from the previous solution of the turbulent momentum equation discussed in Section 31.1, no further shooting is required to solve the equations governing heat transfer.

Figure 33-8 contrasts the calculated values of the Nusselt number for the constant heat flux boundary condition with the constant temperature boundary condition of the preceding section. The difference between $Nu_{D,T}$ and $Nu_{D,H}$ has the same order as observed for laminar flows. However, for most turbulent flows with $\mathrm{Pr} > 0.1$, this difference is negligible because of the generally high magnitude of the Nusselt number. When the Prandtl number is very small (as can be the case with liquid metals), turbulent diffusion no longer dominates heat transport, and the resulting Nusselt number can be small enough that the difference between $Nu_{D,T}$ and $Nu_{D,H}$ is noticeable, as illustrated in Figure 33-8.

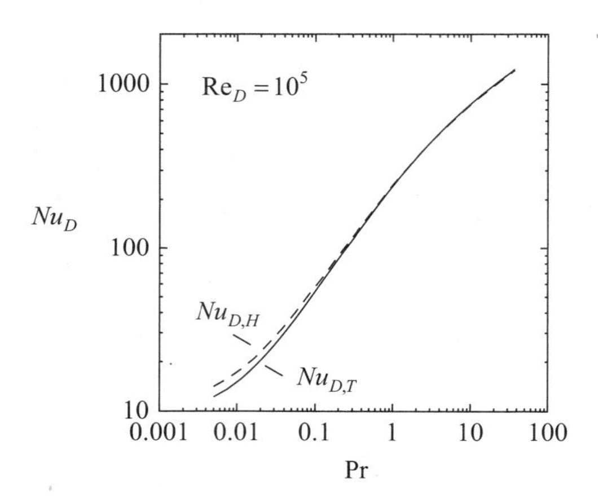

Figure 33-8 Turbulent flow Nusselt number for constant temperature and heat flux wall conditions.

33.4 FULLY DEVELOPED TRANSPORT IN A TURBULENT FLOW OF A BINARY MIXTURE

Consider Poiseuille flow of a mixture between parallel walls, as shown in Figure 33-9. The fluid consists of air (A) and a second species (B). Both walls are impermeable to air. However, species B is introduced to the flow at the lower wall at a mole fraction χ_{B0}. The top wall is a sink material that forces the species B mole fraction to zero. Between the walls the total molar concentration of air and species B is constant (corresponding to isothermal and isobaric conditions across the flow). The concentrations and gradients of species B are sufficient to introduce Stefan flow between the two walls, as discussed in Section 4.2.1. However, because the molecular weight of B is sufficiently similar to that of air, there is no gradient in density and the mass-averaged velocity across the flow is the same as the molar-averaged velocity $\overline{v}_y = \overline{v}_y^*$.

The turbulent transport of species across the flow is described by the steady-state equation:

$$\underbrace{\partial_o(\overline{c}\,\overline{\chi}_i)}_{=0} + \partial_j(\overline{c}\,\overline{\chi}_i\overline{v}_j^*) = \partial_j(\overline{c}(Đ_{BA} + \varepsilon_C)\partial_j\overline{\chi}_i) \qquad i = \text{A or B.} \tag{33-82}$$

The free index in this equation is used to represent either air (species A) or species B. The governing equation is expanded into Cartesian coordinates:

$$\underbrace{\frac{\partial}{\partial x}(\overline{c}\,\overline{\chi}_i\,\overline{v}_x^*)}_{=0} + \frac{\partial}{\partial y}(\overline{c}\,\overline{\chi}_i\overline{v}_y^*) = \underbrace{\frac{\partial}{\partial x}\left(\overline{c}(Đ_{BA} + \varepsilon_C)\frac{\partial\overline{\chi}_i}{\partial x}\right)}_{=0} + \frac{\partial}{\partial y}\left(\overline{c}(Đ_{BA} + \varepsilon_C)\frac{\partial\overline{\chi}_i}{\partial y}\right), \tag{33-83}$$

and simplified for fully developed conditions that exhibit no downstream gradients. The species transport equation can be written separately for air (A) and species B in the mixture:

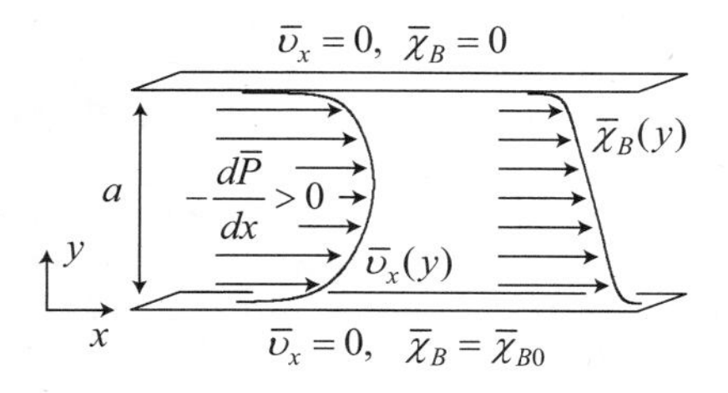

Figure 33-9 Turbulent Poiseuille flow of a binary mixture between plates with mass transfer.

$$\text{transport of A:} \qquad 0 = \frac{\partial}{\partial y}\left(\overline{c}_A \overline{v}_y^* - \overline{c}(Ð_{BA} + \varepsilon_C)\frac{\partial \overline{\chi}_A}{\partial y}\right) \tag{33-84}$$

$$\text{transport of B:} \qquad 0 = \frac{\partial}{\partial y}\left(\overline{c}_B \overline{v}_y^* - \overline{c}(Ð_{BA} + \varepsilon_C)\frac{\partial \overline{\chi}_B}{\partial y}\right). \tag{33-85}$$

Integrating the transport equation for air once yields

$$\overline{c}\left(\overline{\chi}_A \overline{v}_y^* - (Ð_{BA} + \varepsilon_C)\frac{\partial \overline{\chi}_A}{\partial y}\right) = \overline{J}_A^* \; (= 0). \tag{33-86}$$

However, since the air flux is zero at the walls, the integration constant is $\overline{J}_A^* = 0$. Therefore, the air transport equation (33-86) with $\overline{\chi}_A = (1 - \overline{\chi}_B)$ can be used to determine the Stefan flow velocity:

$$\overline{v}_y^* = \frac{-(Ð_{BA} + \varepsilon_C)}{1 - \overline{\chi}_B}\frac{\partial \overline{\chi}_B}{\partial y}. \tag{33-87}$$

The flow velocity is substituted back into the transport equation (33-85) for the species B, yielding

$$\frac{\partial}{\partial y}\left(\overline{c}_B \frac{(Ð_{BA} + \varepsilon_C)}{1 - \overline{\chi}_B}\frac{\partial \overline{\chi}_B}{\partial y} + \overline{c}(Ð_{BA} + \varepsilon_C)\frac{\partial \overline{\chi}_B}{\partial y}\right) = 0 \tag{33-88}$$

or

$$\frac{\partial}{\partial \eta}\left(\frac{\overline{c}(Ð_{BA} + \varepsilon_C)}{1 - \overline{\chi}_B}\frac{\partial \overline{\chi}_B}{\partial \eta}\right) = 0 \quad \text{with } \eta = \frac{y}{a}. \tag{33-89}$$

When the total concentration $\overline{c}$ and species diffusivity $Ð_{BA}$ are constant, the governing equation for species B becomes

$$\frac{\partial}{\partial \eta}\left(\frac{(1 + \varepsilon_C/Ð_{BA})}{1 - \overline{\chi}_B}\frac{\partial \overline{\chi}_B}{\partial \eta}\right) = 0. \tag{33-90}$$

The solution to the transport equation for species B is subject to the boundary conditions

$$\overline{\chi}_B(0) = \chi_{B0} \quad \text{and} \quad \overline{\chi}_B(1) = 0. \tag{33-91}$$

Integrating the species B transport equation once yields

$$\frac{(1 + \varepsilon_C/Ð_{BA})}{1 - \overline{\chi}_B}\frac{\partial \overline{\chi}_B}{\partial \eta} = d_1 \left(= \frac{\overline{\chi}_B'(0)}{1 - \overline{\chi}_{B0}}\right), \tag{33-92}$$

where the integration constant d_1 can be determined from conditions at the ($\eta = 0$) wall. Therefore, the species transport equation for $\overline{\chi}_B$ becomes

$$\frac{\partial \overline{\chi}_B}{\partial \eta} = \frac{1 - \overline{\chi}_B}{1 - \overline{\chi}_{B0}}\frac{\overline{\chi}_B'(0)}{1 + \varepsilon_C/Ð_{BA}}, \quad \text{with} \quad \overline{\chi}_B(0) = \chi_{B0}. \tag{33-93}$$

The constant $\overline{\chi}'_B(0)$ in the governing equation is dictated by the second boundary condition $\overline{\chi}_B(1) = 0$. However, the species transport equation cannot be integrated further without quantifying the species turbulent diffusivity ε_C. This requires the turbulent flow solution to the momentum equation to determine the turbulent momentum diffusivity ε_M, from which

$$\frac{\varepsilon_C}{Ð_{BA}} = \frac{Sc}{Sc_t}\frac{\varepsilon_M}{\nu} = \frac{Sc\Lambda_C}{\Lambda}\frac{\varepsilon_M}{\nu}, \tag{33-94}$$

where the turbulent Schmidt number Sc_t is defined by Eq. (33-9). The steady-state momentum equation for turbulent Poiseuille flow with blowing ($\overline{v}_y \neq 0$) between the walls simplifies for fully developed conditions:

$$\underbrace{\overline{v}_x\frac{\partial \overline{v}_x}{\partial x}}_{=0} + \overline{v}_y\frac{\partial \overline{v}_x}{\partial y} = \underbrace{\frac{\partial}{\partial x}\left((\nu + \varepsilon_M)\frac{\partial \overline{v}_x}{\partial x}\right)}_{=0} + \frac{\partial}{\partial y}\left((\nu + \varepsilon_M)\frac{\partial \overline{v}_x}{\partial y}\right) - \frac{1}{\rho}\frac{\partial \overline{P}}{\partial x}. \tag{33-95}$$

The fully developed momentum equation retains one of the advection terms, because of the blowing velocity between the walls, and the term describing diffusion between the walls. The terms in the momentum equation are integrated once, yielding

$$\int_0^y \overline{v}_y\frac{\partial \overline{v}_x}{\partial y}dy = (\nu + \varepsilon_M)\frac{\partial \overline{v}_x}{\partial y} - \frac{1}{\rho}\frac{\partial \overline{P}}{\partial x}y + c_1. \tag{33-96}$$

The integration constant in the momentum equation (33-96) can be evaluated three ways:

$$\text{at } y = 0\text{: } c_1 = -\nu\left.\frac{\partial \overline{v}_x}{\partial y}\right|_{y=0}, \tag{33-97}$$

$$\text{at } y = y_c\text{: } c_1 = \int_0^{y_c} \overline{v}_y\frac{\partial \overline{v}_x}{\partial y}dy + \frac{1}{\rho}\frac{\partial \overline{P}}{\partial x}y_c, \tag{33-98}$$

$$\text{at } y = a\text{: } c_1 = -\nu\left.\frac{\partial \overline{v}_x}{\partial y}\right|_{y=a} + \frac{1}{\rho}\frac{\partial \overline{P}}{\partial x}a + \int_0^a \overline{v}_y\frac{\partial \overline{v}_x}{\partial y}dy. \tag{33-99}$$

The distance y_c coincides with the point in the flow where $\partial\overline{v}_x/\partial y = 0$. Combining Eq. (33-98) alternately with Eq. (33-97) and Eq. (33-99), the pressure gradient may be expressed in two forms:

$$-\frac{1}{\rho}\frac{\partial \overline{P}}{\partial x} = \frac{1}{y_c}\left(\nu\left.\frac{\partial \overline{v}_x}{\partial y}\right|_{y=0} + \int_0^{y_c} \overline{v}_y\frac{\partial \overline{v}_x}{\partial y}dy\right) \tag{33-100}$$

and

$$-\frac{1}{\rho}\frac{\partial \overline{P}}{\partial x} = \frac{1}{a - y_c}\left(-\nu\left.\frac{\partial \overline{v}_x}{\partial y}\right|_{y=a} + \int_0^{a} \overline{v}_y\frac{\partial \overline{v}_x}{\partial y}dy - \int_0^{y_c} \overline{v}_y\frac{\partial \overline{v}_x}{\partial y}dy\right). \tag{33-101}$$

The pressure gradient may be eliminated between equations (33-100) and (33-101) to reveal a relation between the velocity gradients at the top and bottom walls:

$$\nu \frac{\partial \overline{v}_x}{\partial y}\bigg|_{y=a} = -\frac{a-y_c}{y_c}\nu \frac{\partial \overline{v}_x}{\partial y}\bigg|_{y=0} + \int_0^a \overline{v}_y \frac{\partial \overline{v}_x}{\partial y} dy - \frac{a}{y_c}\int_0^{y_c} \overline{v}_y \frac{\partial \overline{v}_x}{\partial y} dy. \tag{33-102}$$

Using Eq. (33-98) for the integration constant, the momentum equation (33-96) becomes

$$(\nu + \varepsilon_M)\frac{\partial \overline{v}_x}{\partial y} = -(y_c - y)\frac{1}{\rho}\frac{\partial \overline{P}}{\partial x} - \int_0^{y_c} \overline{v}_y \frac{\partial \overline{v}_x}{\partial y} dy + \int_0^{y} \overline{v}_y \frac{\partial \overline{v}_x}{\partial y} dy. \tag{33-103}$$

The momentum equation (33-103) may be written in two forms by substituting the alternate expressions for the pressure gradient, Eq. (33-100) and Eq. (33-101). These resulting forms of the momentum equation are

$$(\nu + \varepsilon_M)\frac{\partial \overline{v}_x}{\partial y} = \frac{y_c - y}{y_c}\nu \frac{\partial \overline{v}_x}{\partial y}\bigg|_{y=0} - \frac{y}{y_c}\int_0^{y_c} \overline{v}_y \frac{\partial \overline{v}_x}{\partial y} dy + \int_0^{y} \overline{v}_y \frac{\partial \overline{v}_x}{\partial y} dy \tag{33-104}$$

and

$$(\nu + \varepsilon_M)\frac{\partial \overline{v}_x}{\partial y} = -\frac{y_c - y}{a - y_c}\nu \frac{\partial \overline{v}_x}{\partial y}\bigg|_{y=a} - \frac{y - y_c}{a - y_c}\int_0^{a} \overline{v}_y \frac{\partial \overline{v}_x}{\partial y} dy - \frac{a - y}{a - y_c}\int_0^{y_c} \overline{v}_y \frac{\partial \overline{v}_x}{\partial y} dy + \int_0^{y} \overline{v}_y \frac{\partial \overline{v}_x}{\partial y} dy. \tag{33-105}$$

Notice that when $y = y_c$, both momentum equations enforce the condition $\partial \overline{v}_x/\partial y|_{y=y_c} = 0$. The problem is made dimensionless using the variables

$$\eta = y/a \quad \text{and} \quad u = \overline{v}_z/\overline{v}_m, \tag{33-106}$$

where

$$\overline{v}_m = \int_0^a \overline{v}_x dy/a. \tag{33-107}$$

The relation between u and $\overline{v}_m$ requires that

$$\int_0^1 u d\eta = 1. \tag{33-108}$$

Additionally, defining

$$\phi = \overline{\chi}_B/\chi_{B0} \tag{33-109}$$

and

$$Sc = \nu/Ð_{BA}, \tag{33-110}$$

the Stefan flow velocity given by Eq. (33-87) is employed (with $\overline{v}_y = \overline{v}_y^*$) to write

$$\overline{v}_y = -\frac{Ð_{BA}}{a}\frac{1 + \varepsilon_C/Ð_{BA}}{1/\chi_{B0} - \phi}\frac{\partial \phi}{\partial \eta}. \tag{33-111}$$

Then, the momentum equations (33-104) and (33-105) in dimensionless form become

$$\left(1+\frac{\varepsilon_M}{\nu}\right)\frac{\partial u}{\partial \eta}=\frac{\eta_c-\eta}{\eta_c}\frac{\partial u}{\partial \eta}\bigg|_{\eta=0}+\frac{1}{Sc}\left(\frac{\eta}{\eta_c}\mathrm{v}(\eta_c)-\mathrm{v}(\eta)\right) \tag{33-112}$$

and

$$\left(1+\frac{\varepsilon_M}{\nu}\right)\frac{\partial u}{\partial \eta}=-\frac{\eta_c-\eta}{1-\eta_c}\frac{\partial u}{\partial \eta}\bigg|_{\eta=1}+\frac{1}{Sc}\left(\frac{\eta-\eta_c}{1-\eta_c}\mathrm{v}(1)+\left(\frac{1-\eta}{1-\eta_c}\right)\mathrm{v}(\eta_c)-\mathrm{v}(\eta)\right), \tag{33-113}$$

where

$$\mathrm{v}(\eta)=\int_0^\eta \frac{1+\varepsilon_C/Ð_{BA}}{1/\chi_{B0}-\phi}\frac{\partial \phi}{\partial \eta}\frac{\partial u}{\partial \eta}d\eta. \tag{33-114}$$

Analysis of the flow between the walls can be subdivided into two regions. For region 1, where $0\le\eta\le\eta_c$, the first momentum equation (33-112) will be integrated with respect to the dependent variable $\zeta_1=\eta$. For region 2, where $\eta_c\le\eta\le 1$, the second momentum equation (33-113) will be integrated with respect to a new dependent variable, $\zeta_2=1-\eta$. These spatial variables are illustrated in Figure 33-10. The momentum equations (33-112) and (33-113) may be written in terms of the new variables as

$$\left(1+\frac{\varepsilon_M}{\nu}\right)\frac{\partial u}{\partial \zeta}\bigg|_{j=1,2}=g_j(\zeta_j),\quad (0\le\zeta_j\le\zeta_{c,j}) \tag{33-115}$$

where

$$g_j(\zeta_j)=\left(1-\frac{\zeta_j}{\zeta_{c,j}}\right)u_j'(0)+\frac{1}{Sc}\left(\frac{\zeta_j}{\zeta_{c,j}}\mathrm{V}_j(\zeta_{c,j})-\mathrm{V}_j(\zeta_j)\right), \tag{33-116}$$

and

$$\zeta_{c,j}=\begin{cases}\eta_c & (j=1)\\ 1-\eta_c & (j=2)\end{cases}. \tag{33-117}$$

The index $j=1$ or 2 is used to specify the flow region. Presentation of the momentum equation has been simplified with the definitions

$$u_j'(0)=\frac{\partial u_j}{\partial \zeta_j}\bigg|_{\zeta_j=0} \tag{33-118}$$

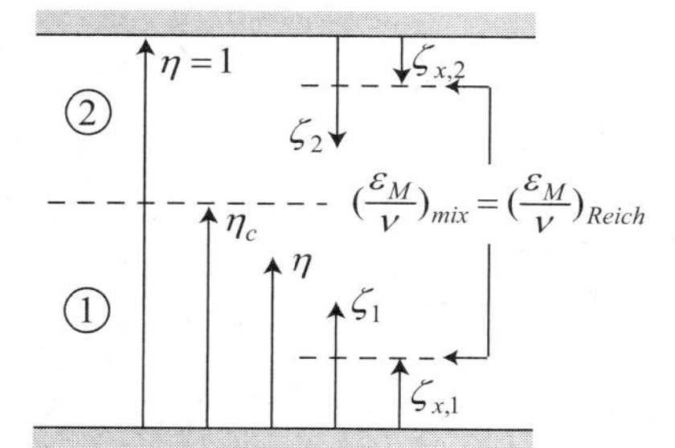

Figure 33-10 Relationship between distance variables.

and

$$V_j(\zeta_j) = \int_0^{\zeta_j} \frac{1 + \varepsilon_C/Ð_{BA}}{1/\chi_{B0} - \phi_j} \frac{\partial \phi_j}{\partial \zeta_j} \frac{\partial u_j}{\partial \zeta_j} d\zeta_j, \tag{33-119}$$

where

$$v_1(\zeta_1) = V_1(\zeta_1) \quad \text{and} \quad v_2(\zeta_2) = V_1(\eta_c) + V_2(1 - \eta_c) - V_2(\zeta_2). \tag{33-120}$$

Both main regions 1 and 2 are further subdivided into regions where the turbulent diffusivity is evaluated by either the mixing length model or Reichardt's formula, depending on the distance from the wall. This is needed because the mixing length model alone will predict a laminar central region for the flow (when $\partial \overline{v}_x/\partial y = 0$). Since turbulent species diffusivity flux through this central region is expected to be significant, this failure of the mixing length model is unacceptable. Therefore, the mixing length model will only be used for distances $0 < \zeta_j < \zeta_{x,j}$ from the wall and Reichardt's formula, discussed in Section 31.6, will be used for distances $\zeta_{x,j} \leq \zeta_j \leq \zeta_{c,j}$. At the distance $\zeta_{x,1}$ from the lower wall and the distance $\zeta_{x,2}$ from the upper wall, the turbulent diffusivity calculated by both the mixing length model and Reichardt's formula should be identical, as shown in Figure 33-10.

For the mixing length model, turbulent diffusivity is expressed for the region from the wall through the inner region by

$$\left(\frac{\varepsilon_M}{\nu}\right)_{\substack{j=1,2 \\ mix}} = \frac{\ell^2}{\nu} \left| \frac{\partial \overline{v}_x}{\partial y} \right| = \frac{\Lambda_j^2 \mathrm{Re}_D}{2} \left. \frac{\partial u}{\partial \zeta} \right|_{j=1,2}, \tag{33-121}$$

where

$$\Lambda_j = \frac{\ell_j}{a} \quad \text{and} \quad \mathrm{Re}_D = \frac{\overline{v}_m 2a}{\nu}. \tag{33-122}$$

With the mixing length model expression for turbulent diffusivity (33-121), the momentum equation (33-115) can be expressed as

$$\left. \frac{\partial u}{\partial \zeta} \right|_{j=1,2} = \frac{2(\varepsilon_M/\nu)_{j,mix}}{\Lambda_j^2 \mathrm{Re}_D} \quad (0 < \zeta_j < \zeta_{x,j}), \tag{33-123}$$

where

$$\left(\frac{\varepsilon_M}{\nu}\right)_{\substack{j=1,2 \\ mix}} = \frac{1}{2}\left[\sqrt{1 + 2\Lambda_j^2 \mathrm{Re}_D \, g_j(\zeta_j)} - 1\right] \tag{33-124}$$

and $g_j(\zeta_j)$ is given by Eq. (33-116). The dimensionless mixing length Λ_j is evaluated with the Van Driest damping function. In terms of the variables of the solution,

$$\Lambda_j(\zeta_j) = \kappa \zeta_j \left[1 - \exp\left(\frac{-\zeta_j}{A_\nu^+} \sqrt{\frac{\mathrm{Re}_D \, u_j'(0)}{2}}\right)\right]. \tag{33-125}$$

For the regions $\zeta_{x,j} \leq \zeta_j \leq \zeta_{c,j}$, the momentum equation (33-115) is evaluated with Reichardt's formula for turbulent diffusivity, as discussed in Section 31.6. For the parallel plate geometry, Reichardt's formula becomes

$$\left(\frac{\varepsilon_M}{\nu}\right)_{j,\,Reich} = \frac{\kappa\zeta_j}{6}\sqrt{\frac{\mathrm{Re}_D\, u_j'(0)}{2}}\, s_j(\rho_j)(1+\rho_j)(1+2\rho_j^2), \tag{33-126}$$

where

$$\rho_j = 1 - \frac{\zeta_j}{\zeta_{c,j}} \tag{33-127}$$

and

$$s_j(\rho_j) = \begin{cases} \rho_1 + (1-\rho_1)\dfrac{\zeta_{c,2}}{\zeta_{c,1}}\sqrt{\dfrac{u_2'(0)}{u_1'(0)}} & (j=1) \\ 1 & (j=2) \end{cases}. \tag{33-128}$$

To create a continuous turbulent diffusivity function between region 1 and 2, Reichardt's formula is stretched in region 1 by the function $s_1(\rho_1)$.

In summary, the final form of the momentum equation to be integrated is

$$\left.\frac{\partial u}{\partial \zeta}\right|_{j=1,2} = \begin{cases} u_j'(0), & \zeta_j = 0 \\ \dfrac{2(\varepsilon_M/\nu)_{j,\,mix}}{\Lambda_j^2 \mathrm{Re}_D}, & 0 < \zeta_j < \zeta_{x,j} \\ \dfrac{g_j(\zeta_j)}{1+(\varepsilon_M/\nu)_{j,\,Reich}}, & \zeta_{x,j} \leq \zeta_j \leq \zeta_{c,j} \end{cases}, \tag{33-129}$$

where $(\varepsilon_M/\nu)_{j,\,mix}$ is evaluated with Eq. (33-124), Λ_j is evaluated with Eq. (33-125), $g_j(\zeta_j)$ is evaluated with Eq. (33-116), and $(\varepsilon_M/\nu)_{j,\,Reich}$ is evaluated with Eq. (33-126). The boundary conditions at the walls require

$$u_j(\zeta_j = 0) = 0. \tag{33-130}$$

The solution to the momentum equation (33-129) must additionally satisfy three requirements: the first relates the velocity gradients at the upper and lower walls (33-102); the second enforces the normalization of the velocity scale (33-108); and the last is that the velocity solution be continuous at $y = y_c$. These requirements in dimensionless form are given as follows:

$$u_2'(0) = \frac{1-\eta_c}{\eta_c} u_1'(0) + \frac{1}{Sc}\left(\mathrm{V}_2(1-\eta_c) - \frac{1-\eta_c}{\eta_c}\mathrm{V}_1(\eta_c)\right), \tag{33-131}$$

$$\int_0^{\eta_c} u_1(\zeta_1)d\zeta_1 + \int_0^{1-\eta_c} u_2(\zeta_2)d\zeta_2 = 1, \tag{33-132}$$

$$u_1(\eta_c) = u_2(1-\eta_c). \tag{33-133}$$

The momentum equation is solved by guessing $u_1'(0)$ to satisfy the normalization requirement (33-132) and guessing η_c to satisfy the continuity requirement (33-133). The solution to the momentum equation is coupled to the species equation (33-93). Expressed in terms of

$$\phi = \overline{\chi}_B/\chi_{B0}, \tag{33-134}$$

the species equation becomes

$$\frac{\partial \phi}{\partial \eta} = \frac{1/\overline{\chi}_{B0} - \phi}{1/\overline{\chi}_{B0} - 1} \frac{\phi'(0)}{1 + \varepsilon_C/Đ_{BA}} \tag{33-135}$$

with boundary conditions

$$\phi(0) = 1 \quad \text{and} \quad \phi(1) = 0. \tag{33-136}$$

In terms of the spatial variables ζ_j, the solution for ϕ_j is given by

$$\phi_j(\zeta_j) = \begin{cases} 1 + \Phi_1(\zeta_1) & (j = 1) \\ 1 + \Phi_1(\zeta_{c,1}) + \Phi_2(\zeta_{c,2}) - \Phi_2(\zeta_2) & (j = 2) \end{cases}, \tag{33-137}$$

where

$$\Phi_j = \int_0^{\zeta_j} \frac{1/\overline{\chi}_{B0} - \phi}{1/\overline{\chi}_{B0} - 1} \frac{\phi'(0)}{1 + \varepsilon_C/Đ_{BA}} d\zeta_j. \tag{33-138}$$

The species equation is solved by guessing $\phi'(0) = \partial\phi/\partial\eta|_{\eta=0}$ to satisfy the condition $\phi(1) = 0$, or equivalently,

$$1 + \Phi_1(\zeta_{c,1}) + \Phi_2(\zeta_{c,2}) = 0. \tag{33-139}$$

To determine all of the unknowns of the current problem, the following iterative procedure can be used:

1. Solve Eq. (33-129) for $u(\eta)$ with $\mathrm{v}(\eta) = 0$ and $\phi'(0) = 0$ (basic Poiseuille flow solution).
2. Solve Eq. (33-137) for $\phi(\eta)$ using the current $u(\eta)$ to evaluate $\varepsilon_C/Đ_{BA}$.
3. Solve Eq. (33-119/120) for $\mathrm{v}(\eta)$ using the current $u(\eta)$ and $\phi(\eta)$distributions.
4. Solve Eq. (33-129) for $u(\eta)$ using the current $\mathrm{v}(\eta)$ distribution.
5. Iterate back to step (2) if $\phi'(0)$ has changed from the last iteration. Otherwise, the solution has converged.

The transport of species B between the walls is solved with Code 33-2, where the model parameters $A_D^+ = A_\nu^+ = 26$, $\kappa = 0.4$, and $Sc_t = 0.9$ have been used. Although it is not done so here, it would be appropriate to consider modifying A_D^+ and A_ν^+ for change in the diffusion sublayer thicknesses brought about by the blowing flux between the walls. One approach for doing this would be to apply the Kays and Crawford condition discussed in Section 31.7.

The flux of species B between the walls is expressed from the contributions of advection and diffusion by

$$\overline{J}_B^* = c_B\overline{v}_y^* - c(Đ_{BA} + \varepsilon_C)\frac{\partial \overline{\chi}_B}{\partial y}. \tag{33-140}$$

Using Eq. (33-87) for the Stefan flow velocity, the flux can be evaluated at the lower wall:

$$\overline{J}_B^* = \left[\frac{-cĐ_{BA}}{1 - \overline{\chi}_B}\frac{\partial \overline{\chi}_B}{\partial y}\right]_{y=0} = -\phi'(0)\frac{\chi_{B0}}{1 - \chi_{B0}}\frac{cĐ_{BA}}{a}. \tag{33-141}$$

Defining a convection coefficient h_B for species transport between the walls (such that $\overline{J}_B^* = ch_B\chi_{B0}$), the Sherwood number for mass transfer can be determined from

$$Sh_a = \frac{h_B a}{Đ_{BA}} = \frac{-\phi'(0)}{1-\chi_{B0}}. \tag{33-142}$$

The Sherwood number is shown in Figure 33-11 as a function of the lower wall concentration $\overline{\chi}_{B0}$ for $\mathrm{Re}_D = 10^6$ and $Sc = 1$.

The Stefan flow causes the velocity distribution between the walls to be asymmetric. This leads to differences between the coefficient of friction for the top and bottom walls. In terms of the variables of the solution, the skin-friction coefficients for the top and bottom wall are given by:

$$c_{f,1} = \frac{\tau_w\big|_{y=0}}{\rho\overline{v}_m^2/2} = \frac{4u_1'(0)}{\mathrm{Re}_D} \quad \text{and} \quad c_{f,2} = \frac{\tau_w\big|_{y=a}}{\rho\overline{v}_m^2/2} = \frac{4u_2'(0)}{\mathrm{Re}_D}, \tag{33-143}$$

respectively. The skin-friction coefficients are shown in Figure 33-12 as a function of the lower wall concentration $\overline{\chi}_{B0}$, again for the flow conditions $\mathrm{Re}_D = 10^6$ and $Sc = 1$.

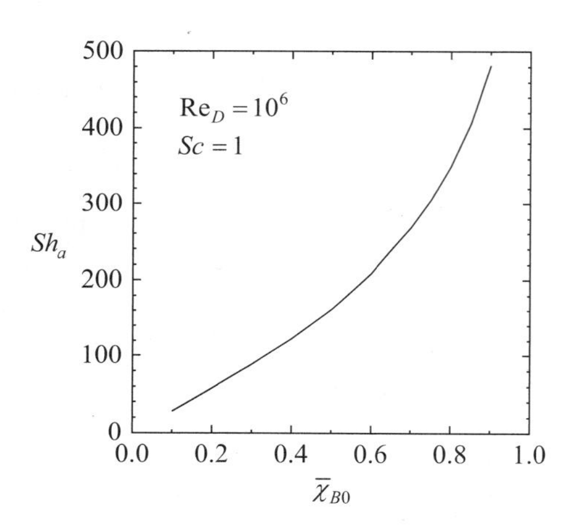

Figure 33-11 Sherwood number for a turbulent Poiseuille flow of a binary mixture between plates with mass transfer.

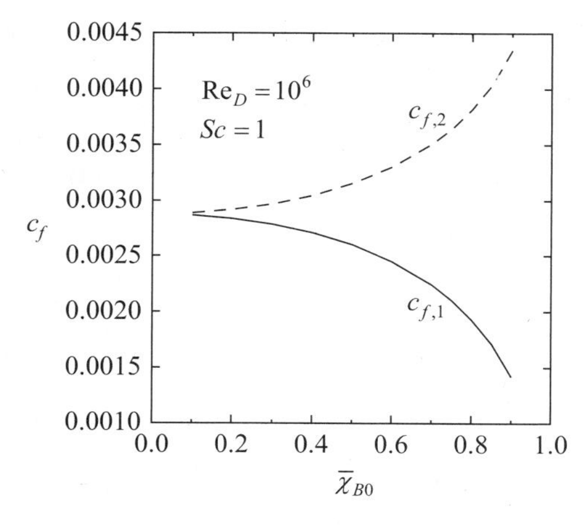

Figure 33-12 Skin-friction coefficients for a turbulent Poiseuille flow of a binary mixture between plates with mass transfer.

Code 33-2 Transport of a binary mixture in turbulent Poiseuille flow.

```
#include <stdio.h>
#include <stdlib.h>
#include <math.h>
double g(int id,double zet,double V,double etc,double Vc,
         double Sc,double du0) {
   double zetc=(id==1 ? etc : 1.-etc);
   return (1.-zet/zetc)*du0 + (zet*Vc/zetc-V)/Sc;
}
double Em_mix(int id,double Ap,double Kap,double zet,double V,
              double etc,double Vc,double Sc,double ReD,
              double du0,double *Lam) {
   *Lam=Kap*zet*(1.-exp(-zet*sqrt(ReD*du0/2.)/Ap));
   double _g=g(id,zet,V,etc,Vc,Sc,du0);
   return (sqrt(1.+2.*(*Lam)*(*Lam)*ReD*_g)-1.)/2.;
}
double Em_Reich(int id,double Kap,double zet,
                double etc,double ReDdu10,double ReDdu20) {
   double ReDdu0=(id==1 ? ReDdu10 : ReDdu20);
   double s,r=(id==1 ? (etc-zet)/etc : (1.-etc-zet)/(1.-etc) );
   s=(id==1 ? r+(1.-r)*(1.-etc)*sqrt(ReDdu20/ReDdu10)/etc : 1. );
   return Kap*zet*sqrt(ReDdu0/2.)*s*(1.+r)*(1.+2.*r*r)/6.;
}
double Em(int id,double Ap,double Kap,double zetX,double zet,double V,
          double etc,double Vc,double Sc,double ReD,
          double du10,double du20,double *Lam) {
   if (zet>zetX) return Em_Reich(id,Kap,zet,etc,ReD*du10,ReD*du20);
   return Em_mix(id,Ap,Kap,zet,V,etc,Vc,Sc,ReD,(id==1 ? du10 : du20),Lam);
}
double dx(int id,double Ap,double Kap,double zetX,double zet,double V,
          double x,double etc,double Vc,double B0,double Sc,double Sct,
          double ReD,double du10,double du20,double dx0) {
   if (id==2) dx0*=-1.;
   else if (zet==0.) return dx0;
   double Lam,_Em=Em(id,Ap,Kap,zetX,zet,V,etc,Vc,Sc,ReD,du10,du20,&Lam);
   return (1./B0-x)*dx0/(1./B0-1.)/(1.+Sc*_Em/Sct);
}
double du(int id,double Ap,double Kap,double zetX,double zet,double V,
       double etc,double Vc,double Sc,double ReD,double du10,double du20) {
   double du0=(id==1 ? du10 : du20);
   if (zet==0.) return du0;
   double Lam,_Em=Em(id,Ap,Kap,zetX,zet,V,etc,Vc,Sc,ReD,du10,du20,&Lam);
   if (zet<zetX) return 2.*_Em/Lam/Lam/ReD;
   return g(id,zet,V,etc,Vc,Sc,du0)/(1.+_Em);
}
double dV(int id,double Ap,double Kap,double zetX,double zet,double V,
          double x,double etc,double Vc,double B0,double Sc,double Sct,
          double ReD,double du10,double du20,double dx0) {
   if (x<0.) return 0.;
   double _dx=dx(id,Ap,Kap,zetX,zet,V,x,etc,Vc,B0,Sc,Sct,ReD,du10,du20,dx0);
   double _du=du(id,Ap,Kap,zetX,zet,V,etc,Vc,Sc,ReD,du10,du20);
   double Lam,_Em=Em(id,Ap,Kap,zetX,zet,V,etc,Vc,Sc,ReD,du10,du20,&Lam);
   return (1.+Sc*_Em/Sct)*_dx*_du/(1./B0-x);
}

double find_zetX(int id,double Ap,double Kap,int N,double *z,double *V,
                 double etc,double Vc,
                 double Sc,double ReD,double du10,double du20) {
   int n=0;
   double Lam,Em_m,Em_R,Vx,zet,zetx_lo=0.0;
   double zetx_hi=(id==1 ? etc : 1.-etc)/5.;
   do {
      zet=(zetx_hi+zetx_lo)/2.;
      int n_hi=N-2,n_lo=0;
      do {
         n=(n_hi+n_lo)/2;
         if (z[n]>zet) n_hi=n;
         else n_lo=n;
      } while (n_hi-n_lo>1);
      if (zet>z[n]) Vx=V[n]+(V[n+1]-V[n])*(zet-z[n])/(z[n+1]-z[n]);
      else Vx=V[n-1]+(V[n]-V[n-1])*(zet-z[n-1])/(z[n]-z[n-1]);
      Em_m=Em_mix(id,Ap,Kap,zet,Vx,etc,Vc,Sc,ReD,(id==1 ? du10:du20),&Lam);
      Em_R=Em_Reich(id,Kap,zet,etc,ReD*du10,ReD*du20);
      if (Em_m-Em_R >0.) zetx_hi=zet;
      else zetx_lo=zet;
   } while (zetx_hi-zetx_lo > 1.0e-6);
   return (zetx_hi+zetx_lo)/2.;
}
double intgrl(char what,int id,double Ap,double Kap,int N,double *zet,
              double *u,double *V,double *x,double etc,double B0,double Sc,
              double Sct,double ReD,double du10,double du20,double dx0) {
   int n;
   double del_zet,halfstep,K1V,K2V,K3V,K4V,K1x,K2x,K3x,K4x;
   double um,K1u,K2u,K3u,K4u,K1um,K2um,K3um,K4um;
   double zetX=find_zetX(id,Ap,Kap,N,zet,V,etc,V[N-1],Sc,ReD,du10,du20);
   if (what=='u') u[0]=um=0.;
   if (what=='V') V[0]=0.;
   if (what=='x') x[0]=(id==1 ? B0 : 0.);
   for (n=0;n<N-1;++n) {
      del_zet=zet[n+1]-zet[n];
      halfstep=(zet[n]+zet[n+1])/2.;
      if (what=='u') K1u=del_zet*du(id,Ap,Kap,zetX,zet[n],
                                    V[n],etc,V[N-1],Sc,ReD,du10,du20);
      K1V=del_zet*dV(id,Ap,Kap,zetX,zet[n],V[n],x[n],
                     etc,V[N-1],B0,Sc,Sct,ReD,du10,du20,dx0);
      K1x=del_zet*dx(id,Ap,Kap,zetX,zet[n],V[n],x[n],etc,V[N-1],
                     B0,Sc,Sct,ReD,du10,du20,dx0);
      if (what=='u') K2u=del_zet*du(id,Ap,Kap,zetX,halfstep,V[n]+0.5*K1V,
                                    etc,V[N-1],Sc,ReD,du10,du20);
```

```
      K2V=del_zet*dV(id,Ap,Kap,zetX,halfstep,V[n]+0.5*K1V,x[n]+0.5*K1x,
                  etc,V[N-1],B0,Sc,Sct,ReD,du10,du20,dx0);
      K2x=del_zet*dx(id,Ap,Kap,zetX,halfstep,V[n]+0.5*K1V,x[n]+0.5*K1x,
                  etc,V[N-1],B0,Sc,Sct,ReD,du10,du20,dx0);
      if (what=='u') K3u=del_zet*du(id,Ap,Kap,zetX,halfstep,V[n]+0.5*K2V,
                             etc,V[N-1],Sc,ReD,du10,du20);
      K3V=del_zet*dV(id,Ap,Kap,zetX,halfstep,V[n]+0.5*K2V,x[n]+0.5*K2x,
                  etc,V[N-1],B0,Sc,Sct,ReD,du10,du20,dx0);
      K3x=del_zet*dx(id,Ap,Kap,zetX,halfstep,V[n]+0.5*K2V,x[n]+0.5*K2x,
                  etc,V[N-1],B0,Sc,Sct,ReD,du10,du20,dx0);
      if (what=='u') K4u=del_zet*du(id,Ap,Kap,zetX,zet[n+1],V[n]+K3V,
                             etc,V[N-1],Sc,ReD,du10,du20);
      K4V=del_zet*dV(id,Ap,Kap,zetX,zet[n+1],V[n]+K3V,x[n]+K3x,etc,
                  V[N-1],B0,Sc,Sct,ReD,du10,du20,dx0);
      K4x=del_zet*dx(id,Ap,Kap,zetX,zet[n+1],V[n]+K3V,x[n]+K3x,etc,
                  V[N-1],B0,Sc,Sct,ReD,du10,du20,dx0);
      if (what=='u') {
         K1um=   del_zet*u[n];
         K2um=   del_zet*(u[n]+0.5*K1u);
         K3um=   del_zet*(u[n]+0.5*K2u);
         K4um=   del_zet*(u[n]+K3u);
         u[n+1]=u[n]+(K1u+2.*K2u+2.*K3u+K4u)/6.;
         um+=(K1um+2.*K2um+2.*K3um+K4um)/6.;
      }
      if (what=='V')
         V[n+1]=V[n]+(K1V+2.*K2V+2.*K3V+K4V)/6.;
      if (what=='x')
         x[n+1]=x[n]+(K1x+2.*K2x+2.*K3x+K4x)/6.;
   }
   if (what=='u') return um;
   return 0.;
}
void solve_x(double Ap,double Kap,int N,double *zet1,double *zet2,
             double *u1,double *u2,double *V1,double *V2,double *x1,
             double *x2,double B0,double Sc,double Sct,double ReD,
             double etc,double du10,double du20,double *dx0) {
   int n;
   double dx0_hi= 0.0,dx0_lo=-1000.0;
   do {
      *dx0=(dx0_hi+dx0_lo)/2.;
     intgrl('x',1,Ap,Kap,N,zet1,u1,V1,x1,etc,B0,Sc,Sct,ReD,du10,du20,*dx0);
     intgrl('x',2,Ap,Kap,N,zet2,u2,V2,x2,etc,B0,Sc,Sct,ReD,du10,du20,*dx0);
      if (x1[N-1]>x2[N-1]) dx0_hi= *dx0;
      else dx0_lo= *dx0;
   } while (fabs((dx0_hi-dx0_lo)/(*dx0)) > .00001);
   for (n=0;n<N;++n) x2[n]+=x1[N-1]-x2[N-1];
}
double solve_u(double Ap,double Kap,int N,double *zet1,double *zet2,
              double *u1,double *u2,double *V1,double *V2,double *x1,
              double *x2,double B0,double Sc,double Sct,double ReD,
              double *etc,double *du10,double *du20,double dx0) {
   int n;
   double log_min=-6.,um1,um2,etc_hi=1.,etc_lo=0.45;
   double log_del=(-log_min)/(N-2);
   zet1[0]=zet2[0]=0.;
   do {
      *etc=(etc_hi+etc_lo)/2.;
      zet1[N-1]= *etc;
      zet2[N-1]=1.- *etc;
      for (n=1;n<N-1;++n) {
         zet1[n]=(*etc)*pow(10.,log_min+(n-1)*log_del);
         zet2[n]=(1.- *etc)*pow(10.,log_min+(n-1)*log_del);
      }

      double du10_hi=10000.0,du10_lo=100.0;
      do {
         *du10=(du10_hi+du10_lo)/2.;
         *du20= (*du10)*(1.-(*etc))/(*etc)
               +(V2[N-1]-V1[N-1]*(1.-(*etc))/(*etc))/Sc;
         um1=intgrl('u',1,Ap,Kap,N,zet1,u1,V1,x1,*etc,B0,Sc,Sct,ReD,
                    *du10,*du20,dx0);
         um2=intgrl('u',2,Ap,Kap,N,zet2,u2,V2,x2,*etc,B0,Sc,Sct,ReD,
                    *du10,*du20,dx0);
         if (um1+um2 > 1.) du10_hi= *du10;
         else du10_lo= *du10;
      } while ((du10_hi-du10_lo)/(*du10) > .0001);
      if (u1[N-1]-u2[N-1] > 0.) etc_hi= *etc;
      else etc_lo= *etc;
   } while ((etc_hi-etc_lo)/(*etc) > .0001);
   return um1+um2;
}
int main() {
   int n,N=1000;
   double Kap=0.4,Ap=26.0,Sct=0.9,ReD=1.e6,Sc=1.,B0=0.75;
   double um,etc,dx0=0.,last_dx0,du10,du20,zet1[N],u1[N],V1[N];
   double x1[N],zet2[N],u2[N],V2[N],x2[N];
   for (n=0;n<N;++n) {V1[n]=V2[n]=0.; x1[0]=x2[0]= -1.;}
   solve_u(Ap,Kap,N,zet1,zet2,u1,u2,V1,V2,x1,x2,.5,Sc,Sct,ReD,
          &etc,&du10,&du20, dx0);
   if (B0!=.0) do {
      last_dx0=dx0;
      solve_x(Ap,Kap,N,zet1,zet2,u1,u2,V1,V2,x1,x2,B0,Sc,Sct,ReD,
              etc, du10, du20,&dx0);
      if (last_dx0!=0.) dx0=(last_dx0+dx0)/2.; // damp oscillations
      intgrl('V',1,Ap,Kap,N,zet1,u1,V1,x1,etc,B0,Sc,Sct,ReD,du10,du20,dx0);
      intgrl('V',2,Ap,Kap,N,zet2,u2,V2,x2,etc,B0,Sc,Sct,ReD,du10,du20,dx0);
      um=solve_u(Ap,Kap,N,zet1,zet2,u1,u2,V1,V2,x1,x2,B0,Sc,Sct,ReD,
                 &etc,&du10,&du20, dx0);
   } while (fabs((dx0-last_dx0)/dx0)>0.0001);
   if (fabs(u1[N-1]-u2[N-1]) > .01 || fabs(1.-um)>.01 || fabs(x2[0])>.01)
      printf("Solution failed: um=%f x2[N-1]=%e\n",um,x2[0]);
   printf("B0=%e etc=%f cf1=%e cf2=%e Sh=%e\n",
         B0,etc,4.*du10/ReD,4.*du20/ReD,-dx0/(1.-B0));
   return 1;
}
```

33.5 PROBLEMS

33-1 Some turbulent heat transfer correlations differ depending on whether the fluid is being heated or cooled. Do Nusselt number predictions from the turbulence models described in this chapter change depending on the direction of heat transfer? Justify your answer.

33-2 Beyond the thermal entry length of the pipe, is the convection coefficient h expected to be a function of Reynolds number if the flow is turbulent? Why or why not?

33-3 The heat equation for turbulent Poiseuille flow between smooth plates of constant heat flux was developed in Section 33.3.3. Numerically solve the heat equation to determine the Nusselt number when $\mathrm{Re}_D = 10^5$ and $\mathrm{Pr} = 5.9$. Plot T^+ versus $\log(y^+)$, and compare with Kander's correlation, Eq. (32-15).

33-4 Consider turbulent Couette flow between two smooth parallel plates as illustrated. The top plate moves with a velocity U and has a temperature T_2. The lower plate is stationary and has a temperature T_1. Use the mixing length model to determine the governing equations for heat and momentum transfer between the plates in terms of the mixing length model parameters and the variables

$$\eta = y/a, \quad u = \overline{v}_x/U, \quad \text{and} \quad \theta = (T - T_1)/(T_2 - T_1).$$

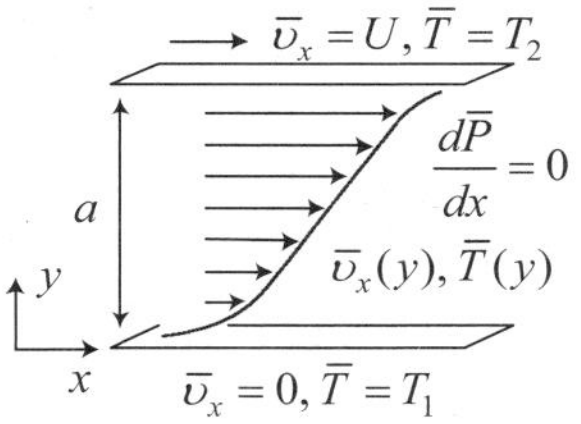

Using the boundary conditions $\theta(0) = 0$ and $\theta(1/2) = 1/2$, numerically solve for the temperature distribution between the plates. Defining a convection coefficient for the heat transfer between the plates $q = h(T_2 - T_1)$, numerically evaluate the Nusselt number $Nu_a = ha/k$. Determine the Nusselt number when $\mathrm{Re}_a = Ua/\nu = 10^6$ and $\mathrm{Pr} = 0.7$. Plot the wall-variable temperature distribution and contrast with Kander's correlation, Eq. (32-15).

33-5 Consider turbulent Poiseuille flow in a smooth pipe of radius R. The walls of the pipe are held at an isothermal temperature. For a flow that is fully developed thermally and hydrodynamically, derive the heat equation using the variables

$$\eta = 1 - \frac{r}{R}, \quad u = \frac{\overline{v}_z}{\overline{v}_m}, \quad \text{and} \quad \theta = \frac{\overline{T} - \overline{T}_m}{T_s - \overline{T}_m}.$$

Solve the turbulent heat and momentum transport equations for water with $\mathrm{Re}_D = 10^6$ and $\mathrm{Pr} = 5.9$ using the mixing length model. Evaluate A_α^+ from the Prandtl number using Figure 32-3. Find expressions for the wall variables

$$y^+ = \frac{\sqrt{\tau_w/\rho}}{\nu} y \quad \text{and} \quad T^+ = \frac{T_s - \overline{T}}{q_w/\rho C_p} \sqrt{\tau_w/\rho}$$

in terms of η, u, and θ. Plot T^+ versus $\log(y^+)$, and compare with Kander's correlation (32-15). What is $Nu_D(\mathrm{Re}_D = 10^6, \mathrm{Pr} = 5.9)$? Contrast this result with the result for the Nusselt number when making the assumption that $A_\alpha^+ = A_\nu^+$ in the mixing length model.

33-6 Consider turbulent Poiseuille flow in a smooth pipe of radius R. The walls of the pipe are subject to a constant heat flux q_w. For a flow that is fully developed both thermally and hydrodynamically, derive the heat equation. Solve the turbulent heat transport equations for air using Reichardt's formula for the turbulent diffusivity. What is $Nu_D(\mathrm{Re}_D = 10^6, \mathrm{Pr} = 0.7)$? Find expressions for the wall variables in terms of the dimensionless variables of the problem. Plot T^+ versus $\log(y^+)$, and compare with the result of Problem 32-2. What is the main source of the disagreement? How well does the result of Problem 32-2 work when the fluid is water ($\mathrm{Pr} = 5.9$)?

33-7 Consider the heat transfer $q = h(T_2 - T_1)$ between the parallel plates shown. There is a turbulent Poiseuille flow of air between the plates.

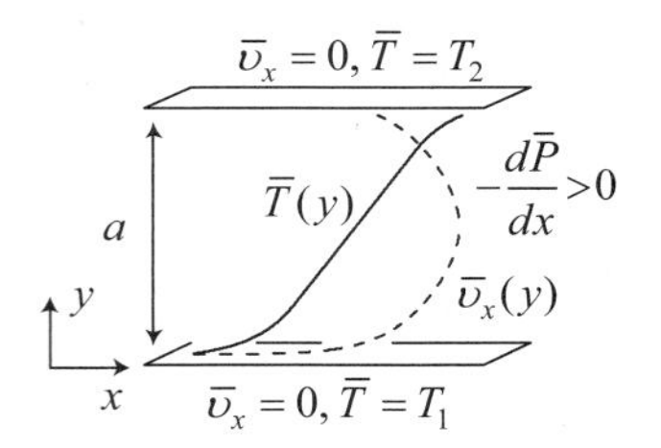

Defining

$$Nu_D = \frac{ha}{k} \quad \text{and} \quad \mathrm{Re}_D = \frac{\overline{v}_m 2a}{\nu},$$

determine the Nusselt number Nu_a for heat transfer between the plates when the Reynolds number is $\mathrm{Re}_D = 10^6$. Use Reichardt's formula to describe the turbulent diffusivity. Plot $u = \overline{v}_x/\overline{v}_m$ and $\theta = (\overline{T} - T_1)/(T_2 - T_1)$ from your solution.

REFERENCE

[1] V. Gnielinski, "New Equations for Heat and Mass Transfer in Turbulent Pipe and Channel Flow." *International Chemical Engineering*, **16**, 359 (1976).

Chapter 34

Turbulence over Rough Surfaces

34.1 Turbulence over a Fully Rough Surface

34.2 Turbulent Heat and Species Transfer from a Fully Rough Surface

34.3 Application of the Rough Surface Mixing Length Model

34.4 Application of Reichardt's Formula to Rough Surfaces

34.5 Problems

Laminar flows are largely unaffected by surface roughness, as long as the roughness scale ε is small in comparison with other geometric scales of the flow, as shown in Figure 34-1. In contrast, turbulent flows can be strongly affected if the surface roughness penetrates the viscous sublayer. In this case, the scale of surface roughness can influence the scale of turbulent mixing. As seen in Chapter 30, the transition between the viscous sublayer and the inner region of turbulence occurs over a range of distances from the wall $5 \leq y^+ \leq 70$ (see Figure 30-8), where

$$y^+ = y\sqrt{\tau_w/\rho}/\nu \tag{34-1}$$

is the wall variable introduced in Section 30.6.1. Consequently, surface roughness will begin to influence turbulence over this range of scales, $5 \leq y^+ \leq 70$. To make this assessment, it is useful to express the surface roughness as a hydrodynamic scale using the wall variable form:

$$\varepsilon^+ = \varepsilon\frac{\sqrt{\tau_w/\rho}}{\nu} = \frac{\varepsilon}{D}\mathrm{Re}_D\sqrt{c_f/2}. \tag{34-2}$$

The *hydrodynamic surface roughness* ε^+, also referred to as the *roughness Reynolds number*, can be categorized into three regimes [1]:

$$\varepsilon^+ < 5 \qquad \text{when surface is } \textit{smooth} \tag{34-3}$$

$$5 \leq \varepsilon^+ \leq 70 \qquad \text{when surface is } \textit{transitionally rough} \tag{34-4}$$

$$70 < \varepsilon^+ \qquad \text{when surface is } \textit{fully rough} \tag{34-5}$$

Turbulent BL

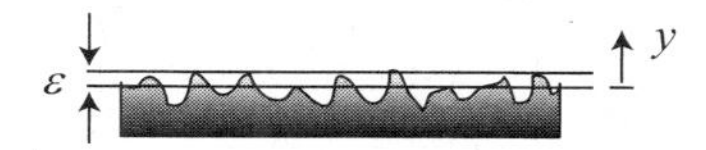

Figure 34-1 Rough surface underlying a turbulent boundary layer.

Whenever $\varepsilon^+ > 5$, the surface roughness will penetrate the viscous sublayer and become a hydrodynamic concern. A dimensional sense of this scale can be obtained by evaluating the coefficient of friction c_f in Eq. (34-2) with an empirical correlation. For turbulent Poiseuille flow between smooth parallel plates, Dean's correlation (31-17) for c_f applied to Eq. (34-2) suggests that when

$$\frac{\varepsilon}{D} > \frac{24}{\mathrm{Re}_D^{7/8}}, \tag{34-6}$$

the surface roughness will penetrate the viscous sublayer (i.e., the condition $\varepsilon^+ > 5$ will be satisfied). To illustrate further, according to Eq. (34-6), when $\mathrm{Re}_D = 10^5$ any surface roughness $\varepsilon/D > 0.001$ will be of hydrodynamic concern.

34.1 TURBULENCE OVER A FULLY ROUGH SURFACE

On a rough surface, Prandtl's mixing length model for the inner region of turbulence breaks down because the mixing length does not go to zero approaching the wall. Instead, as $y \to 0$, it is reasonable to expect that the mixing length approaches a scale that is proportional to the surface roughness ε. This suggests a model in which the mixing length is evaluated from

$$\ell = \kappa y + \lambda \varepsilon \tag{34-7}$$

throughout the inner region of turbulence, where both κ and λ are empirically fitted constants. With this model, as $y \to 0$ approaching the wall, the mixing length approaches a constant $\ell \to \lambda\varepsilon$. This model can only be valid when the scale of surface roughness ε is large enough to negate any influence of the viscous sublayer ($\varepsilon^+ \gtrsim 70$). This situation defines the *fully rough* surface condition. For such a surface, the velocity profile through the inner region of turbulence differs significantly from the law of the wall developed in Section 30.6.2 for a smooth surface. In the next section, the law of the wall for a fully rough surface is evaluated.

34.1.1 Inner Region: (Advection) $\ll$ (Diffusion) and $\varepsilon_M \gg \nu$

Since the viscous sublayer is absent from a flow over a fully rough surface, the inner region of turbulence is immediately bounded by the wall. Recalling that the inner region has a large turbulent diffusivity compared to molecular diffusivity ($\varepsilon_M \gg \nu$), the diffusion flux of momentum through this region is given by

$$\varepsilon_M \frac{\partial \overline{v}_x}{\partial y} = \ell^2 \left(\frac{\partial \overline{v}_x}{\partial y}\right)^2 \quad \left(= \frac{\tau_w}{\rho}\right). \tag{34-8}$$

Since advection transport remains negligible in close proximity to the wall, the diffusion flux of momentum described by Eq. (34-8) remains constant through the inner region of turbulence and equals the wall value of τ_w/ρ. Introducing the wall variable for velocity

$$u^+ = \overline{v}_x / \sqrt{\tau_w/\rho}, \tag{34-9}$$

Eq. (34-8) for the inner region becomes

$$\frac{\partial u^+}{\partial y} = \frac{1}{\ell} = \frac{1}{\kappa y + \lambda \varepsilon}. \tag{34-10}$$

Notice that the wall variable for position ($y^+ = y\sqrt{\tau_w/\rho}/\nu$) has not been utilized because the molecular diffusivity ν is no longer a relevant scale for turbulent flows over a fully rough surface. Integrating the momentum equation (34-10) for the inner region yields

$$u^+ = \frac{1}{\kappa}\ln\left(\frac{\kappa y}{\lambda\varepsilon} + 1\right), \tag{34-11}$$

where the integration constant is evaluated with the no-slip condition at the wall $u^+(0) = 0$. Equation (34-11) is the *law of the wall* for the inner region of turbulence over a fully rough surface. With detailed measurements of the turbulent velocity profile near a wall, this result can be used to evaluate the turbulent mixing length model constants for a fully rough surface.

Nikuradse [2] compared the law of the wall for flows over rough and hydrodynamically smooth surfaces. For this comparison, the wall variables

$$y^+ = \frac{y\sqrt{\tau_w/\rho}}{\nu} \quad \text{and} \quad \varepsilon^+ = \varepsilon\frac{\sqrt{\tau_w/\rho}}{\nu} \tag{34-12}$$

are introduced to Eq. (34-11). Assuming that $\kappa y \gg \lambda\varepsilon$, Eq. (34-11) becomes

$$u^+_{rough} \approx \frac{1}{\kappa}\ln\left(\frac{\kappa y}{\lambda\varepsilon}\right) = \frac{1}{\kappa}\ln(y^+) - \frac{1}{\kappa}\ln(\varepsilon^+) + \frac{1}{\kappa}\ln\left(\frac{\kappa}{\lambda}\right). \tag{34-13}$$

If this velocity profile for the rough surface is subtracted from the law of the wall for a smooth surface (see Section 30.6.2), the result is

$$u^+_{smooth} - u^+_{rough} = \frac{1}{\kappa}\ln(\varepsilon^+) - \underbrace{\left(\frac{1}{\kappa}\ln\left(\frac{\kappa}{\lambda}\right) + \frac{1}{\kappa}\ln(E^+_\nu) - E^+_\nu\right)}_{= \, const.}. \tag{34-14}$$

Grouped on the right-hand side of this result is a constant that is nominally independent of the surface roughness ε^+. For fully rough surfaces ($\varepsilon^+ > 70$), Nikuradse's experimental results suggest that this constant evaluates to approximately

$$\frac{1}{\kappa}\ln\left(\frac{\kappa}{\lambda}\right) + \frac{1}{\kappa}\ln(E^+_\nu) - E^+_\nu \approx 3.5. \tag{34-15}$$

Assuming that $E^+_\nu = 10.8$ and $\kappa = 0.41$, as found for turbulent flows over smooth surfaces, the result given by (34-15) indicates that the best choice for the empirical roughness coefficient is

$$\lambda = 0.0126 \quad \text{(for a fully rough surface)}. \tag{34-16}$$

The smallness of this roughness coefficient indicates that the smallest scale of turbulent mixing is considerably smaller than the length scale of the surface roughness.

34.2 TURBULENT HEAT AND SPECIES TRANSFER FROM A FULLY ROUGH SURFACE

By analogy to momentum transport, the mixing lengths for heat and species transfer in the presence of a fully rough surface should be evaluated with

$$\ell_H = \kappa_H y + \lambda_H \varepsilon \quad \text{(heat)} \tag{34-17}$$

$$\ell_C = \kappa_C y + \lambda_C \varepsilon \quad \text{(species)} \tag{34-18}$$

through the inner region of turbulence. The constants κ_H, λ_H, κ_C, λ_C are empirically determined. In Section 32.3.2, it was indicated that $\kappa/\kappa_H \approx 0.9$ through most of the inner region of a turbulent flow over a smooth surface. It is appropriate to apply the same result to rough surfaces ($\kappa/\kappa_H \approx \kappa/\kappa_C \approx 0.9$). It is reasonable as well to assume $\lambda/\lambda_H \approx \lambda/\lambda_C \approx 0.9$, if the relative scale of the mixing lengths for momentum and heat or species transport remain constant throughout the entire flow. In this case, the turbulent Prandtl number and the turbulent Schmidt number are both constants:

$$\mathrm{Pr}_t = \frac{\varepsilon_M}{\varepsilon_H} = \frac{\ell^2|\partial\overline{v}_x/\partial y|}{\ell_H\ell|\partial\overline{v}_x/\partial y|} = \ell/\ell_H \approx 0.9 \tag{34-19}$$

$$Sc_t = \frac{\varepsilon_M}{\varepsilon_C} = \frac{\ell^2|\partial\overline{v}_x/\partial y|}{\ell_C\ell|\partial\overline{v}_x/\partial y|} = \ell/\ell_C \approx 0.9. \tag{34-20}$$

The heat and species mixing lengths should approach a minimum value: $\ell_H \to \lambda_H\varepsilon$ and $\ell_C \to \lambda_C\varepsilon$ at the wall. The location of the wall ($y = 0$) can be defined as where the no-slip condition $\overline{v}_x = 0$ is valid. Unfortunately, there is some ambiguity about the average temperature or species concentration at $y = 0$, since this plane typically intersects both fluid and wall material due to the rough surface characteristics. Therefore, heat and species transfer from the bulk of the wall to the plane at $y = 0$ will likely be influenced by the molecular diffusivities of the fluid. This situation is unlike the case of momentum transport, where molecular diffusivity never enters the picture. The ambiguity of the fluid temperature at $y = 0$ is revisited when the thermal law of the wall for a rough surface is investigated in the next section. It is straightforward to extend the results of that investigation to species transport with a simple analogy between the two.

34.2.1 Heat Transfer through the Inner Region: (Advection) $\ll$ (Diffusion) and $\varepsilon_H \gg \alpha$

On a fully rough surface, the diffusion sublayer is absent from the description of heat transfer, making the inner region of turbulence the closest region to the wall. Within this region, advection transport is negligible and turbulent diffusivity ($\varepsilon_H \gg \alpha$) dominates transport. Without advection, turbulent heat diffusion remains constant across the inner region,

$$\varepsilon_H\frac{\partial\overline{T}}{\partial y} = \ell_H\ell\frac{\partial\overline{v}_x}{\partial y}\frac{\partial\overline{T}}{\partial y} \quad \left(= \frac{-q_s}{\rho C_p}\right), \tag{34-21}$$

and equals the flux from the wall. Introducing the wall variables for velocity and temperature,

$$u^+ = \frac{\overline{v}_x}{\sqrt{\tau_w/\rho}} \quad \text{and} \quad T^+ = \frac{T_s - \overline{T}}{q_s/\rho C_p}\sqrt{\tau_w/\rho}, \tag{34-22}$$

the heat flux across the inner region of turbulence (34-21) becomes

$$1 = \ell_H\ell\frac{\partial u^+}{\partial y}\frac{\partial T^+}{\partial y}. \tag{34-23}$$

However, the momentum equation (34-10) for the inner region requires $\partial u^+/\partial y = 1/\ell$. Therefore, the heat equation for the inner region simplifies to

$$\frac{\partial T^+}{\partial y} = \frac{1}{\ell_H} = \frac{1}{\kappa_H y + \lambda_H\varepsilon}. \tag{34-24}$$

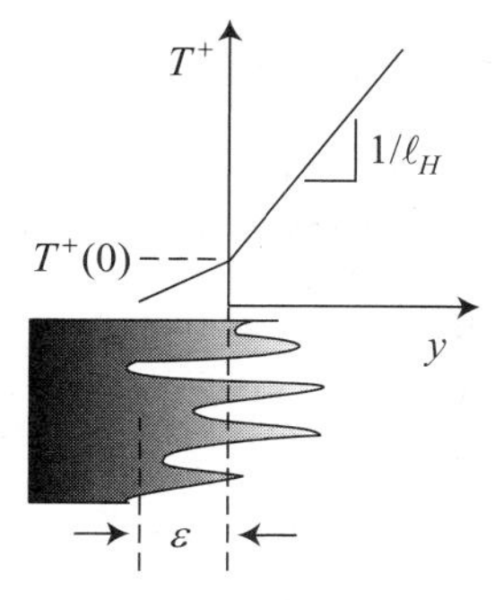

Figure 34-2 Effect of surface roughness on temperature profile.

Again, the wall variable for position ($y^+ = y\sqrt{\tau_w/\rho}/\nu$) is not utilized because the molecular diffusivity ν is not a relevant scale for flow over a fully rough surface. Integrating the heat equation over the inner region yields

$$T^+ - T_0^+ = \frac{1}{\kappa_H}\ln\left(\frac{\kappa_H y}{\lambda_H \varepsilon} + 1\right) \quad \left(= \frac{\mathrm{Pr}_t}{\kappa}\ln\left(\frac{\kappa y}{\lambda\varepsilon} + 1\right)\right), \tag{34-25}$$

where the integration constant $T_0^+ = T^+(0)$ is the wall variable temperature at $y = 0$. The second form of this result is expressed in terms of the turbulent Prandtl number (34-19) and makes use of the fact that $\kappa/\kappa_H = \lambda/\lambda_H$ when Pr_t is taken to be constant. This is the thermal *law of the wall* for the inner region of a turbulent boundary layer over a fully rough surface.

It would be convenient to always assume that $\overline{T}(y = 0) = T_s$ (the temperature of the wall), such that $T^+(0) = 0$. However, this is not substantiated by experimental data. For the purpose of integration, $y = 0$ is an effective position where $u^+(y = 0) = 0$. Since $y = 0$ can be physically removed from the bulk of the wall, as illustrated in Figure 34-2, it is understood that the fluid temperature at $y = 0$ may not be the same as the wall temperature. The magnitude of $T^+(0)$ depends on the surface hydrodynamic roughness and the thermal diffusivity of the fluid that carries heat from the wall to the plane at $y = 0$. The experimental data of Dipprey and Sabersky [3] suggests that the surface temperature at $y = 0$ may be estimated from

$$T^+(0) \approx k_f(\varepsilon^+)^{0.2}\,\mathrm{Pr}^{0.44} + A_f \quad \text{for } \varepsilon^+ \geq 70\,. \tag{34-26}$$

Dipprey and Sabersky determined that the constants $k_f = 5.19$ and $A_f = 8.48$ are suitable for granular close-packed surface roughness. For other types of surface roughness, these constants are likely to change.

34.3 APPLICATION OF THE ROUGH SURFACE MIXING LENGTH MODEL

Consider Couette flow between rough parallel plates as illustrated in Figure 34-3. The top plate moves with a velocity U and leaches small amounts of a contaminant "A". The lower plate is stationary and is a getter for the contaminant. Both surfaces are assumed to be fully rough. Turbulent Couette flow between smooth plates was solved previously in Section 31.2.

The Couette flow problem is solved assuming that the velocity profile between the plates is fully developed, such that $\overline{v}_y = 0$ and $\partial\overline{v}_x/\partial x = 0$ everywhere in the flow. Since advection does not contribute to cross-stream momentum transport in fully developed flows and no streamwise pressure gradient is present, these terms can be omitted from the momentum equation. Additionally, in the absence of a viscous sublayer,

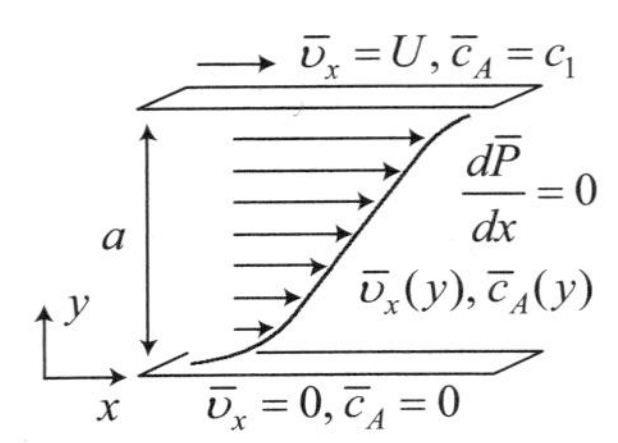

Figure 34-3 Turbulent Couette flow between rough parallel plates.

turbulence dominates diffusion transport ($\varepsilon_M \gg \nu$), and the steady momentum equation becomes

$$0 = \frac{\partial}{\partial y}\left[\varepsilon_M \frac{\partial \bar{v}_x}{\partial y}\right]. \tag{34-27}$$

The momentum equation is subject to the boundary conditions

$$\bar{v}_x(0) = 0 \quad \text{and} \quad \bar{v}_x(a/2) = U/2, \tag{34-28}$$

where specifying the velocity at the mid-position is made possible by recognizing the point symmetry in the solution. Integrating the momentum equation with respect to y yields

$$b_1 = \varepsilon_M \frac{\partial \bar{v}_x}{\partial y}. \tag{34-29}$$

The integration constant is determined by evaluating this result at the wall ($y = 0$), such that

$$b_1 = \left(\varepsilon_M \frac{d\bar{v}_x}{dy}\right)_{y=0} = \left((\lambda\varepsilon)^2 \frac{d\bar{v}_x}{dy}\bigg|_{y=0}\right)\frac{d\bar{v}_x}{dy}\bigg|_{y=0}. \tag{34-30}$$

Consequently, the momentum equation becomes

$$\varepsilon_M \frac{\partial \bar{v}_x}{\partial y} = \left(\lambda\varepsilon \frac{d\bar{v}_x}{dy}\bigg|_{y=0}\right)^2. \tag{34-31}$$

Introducing the dimensionless variables

$$\eta = y/a \quad \text{and} \quad u = \bar{v}_x/U, \tag{34-32}$$

the momentum equation can be expressed in the form

$$\varepsilon_M u' = Ua\left(\frac{\lambda\varepsilon}{a}u'(0)\right)^2, \tag{34-33}$$

where primes are used to denote derivatives with respect to η. Defining a dimensionless mixing length as $\Lambda = \ell/a$, the turbulent diffusivity becomes

$$\varepsilon_M = \ell^2\left|\frac{d\bar{v}_x}{dy}\right| = Ua\Lambda^2 u'. \tag{34-34}$$

Note that u' is everywhere positive. The dimensionless mixing length is determined from the rule

$$\Lambda = \ell/a = \begin{cases} \Lambda^* & \Lambda^* < \gamma \\ \gamma & otherwise \end{cases}, \quad \text{where } \Lambda^* = \kappa\eta + \frac{\lambda\varepsilon}{a}. \tag{34-35}$$

In this way, the dimensionless mixing length evaluates to either $\Lambda = \Lambda^*$ for the inner region, or $\Lambda = \gamma$ for the outer region of turbulence. Substituting the turbulent diffusivity into the momentum equation yields

$$u' = \lambda\frac{\varepsilon}{a}\frac{u'(0)}{\Lambda}, \tag{34-36}$$

which is subject the boundary conditions

$$u(0) = 0 \quad \text{and} \quad u(1/2) = 1/2. \tag{34-37}$$

The momentum equation (34-36) may be integrated between the limits of the lower wall and the centerline between the parallel plates:

$$\frac{1}{2} = \int_0^{1/2} \lambda\frac{\varepsilon}{a}\frac{u'(0)}{\Lambda}d\eta = \lambda\frac{\varepsilon}{a}u'(0)\int_0^{1/2}\frac{d\eta}{\Lambda} = \lambda\frac{\varepsilon}{a}u'(0)\left[\int_{\lambda\varepsilon/a}^{\gamma}\frac{d\Lambda}{\kappa\Lambda} + \int_{\frac{\gamma}{\kappa}-\frac{\lambda}{\kappa}\frac{\varepsilon}{a}}^{1/2}\frac{d\eta}{\gamma}\right] \tag{34-38}$$

or

$$\frac{1}{2} = \lambda\frac{\varepsilon}{a}u'(0)\left[\frac{1}{\kappa}\ln\left(\frac{\gamma}{\lambda\varepsilon/a}\right) + \frac{1}{2\gamma} - \frac{1}{\kappa} + \frac{\lambda}{\kappa\gamma}\frac{\varepsilon}{a}\right]. \tag{34-39}$$

This result can be used to determine the velocity gradient at the wall:

$$u'(0) = \frac{1}{2\lambda\frac{\varepsilon}{a}\left[\frac{1}{\kappa}\ln\left(\frac{\gamma}{\lambda\varepsilon/a}\right) + \frac{1}{2\gamma} - \frac{1}{\kappa} + \frac{\lambda}{\kappa\gamma}\frac{\varepsilon}{a}\right]}. \tag{34-40}$$

The coefficient of friction can be evaluated from

$$c_f = \frac{\tau_w/\rho}{U^2/2} = \frac{\left(\lambda\varepsilon\left.\frac{d\overline{v}_x}{dy}\right|_{y=0}\right)^2}{U^2/2} = 2\left(\lambda\frac{\varepsilon}{a}u'(0)\right)^2. \tag{34-41}$$

Substitution of the result for $u'(0)$ yields

$$c_f = \frac{1}{2\left[\frac{\lambda}{\kappa\gamma}\frac{\varepsilon}{a} - \frac{1}{\kappa}\ln\left(\frac{\varepsilon}{a}\right) + \frac{1}{\kappa}\left(\ln\left(\frac{\gamma}{\lambda}\right) - 1\right) + \frac{1}{2\gamma}\right]^2}. \tag{34-42}$$

When the mixing length parameters $\kappa = 0.41$, $\lambda = 0.0126$, and $\gamma = 0.085$ are used,

$$c_f = \frac{1}{\left[0.511\frac{\varepsilon}{a} - 3.45\ln\left(\frac{\varepsilon}{a}\right) + 11.5\right]^2}. \tag{34-43}$$

Neglecting the smallest term in the denominator simplifies the result to

$$c_f \approx \left[3.45 \ln\left(\frac{1}{\varepsilon/a}\right) + 11.5\right]^{-2} = \left[3.45 \ln\left(\frac{27.7}{\varepsilon/a}\right)\right]^{-2}. \tag{34-44}$$

The coefficient of friction is independent of the Reynolds number. This is necessary because the Reynolds number is by definition related to molecular diffusivity. However, molecular diffusivity only plays a physical role in the viscous sublayer of turbulence, which is absent from the hydrodynamic picture of a flow over a fully rough surface.

34.3.1 Species Transfer across Couette Flow

Dilute species transfer across the Couette flow bounded by rough surfaces, as shown in Figure 34-3, is solved assuming fully developed conditions ($\overline{v}_y = 0$ and $\partial \overline{c}_A/\partial x = 0$), such that advection does not contribute to species transport between the plates. In the absence of a diffusion sublayer, the turbulence diffusivity is always much greater than the molecular contribution ($\varepsilon_C \gg Đ_A$), and the steady-state species transport equation becomes

$$0 = \frac{\partial}{\partial y}\left(\varepsilon_C \frac{\partial \overline{c}_A}{\partial y}\right). \tag{34-45}$$

It is assumed that the lower getter plate removes all of the contaminant from the fluid it is in contact with, and that the top plate maintains a fluid concentration equal to c_1. Since the effective location of the wall surface can be physically removed from the bulk of the wall, as illustrated in Figure 34-2, the fluid concentration at the "surfaces" ($y = 0$ and $y = a$) may not be the same as specified by the wall values. Letting Δ_C denote the concentration offset from the wall value, the boundary conditions may be expressed as

$$\overline{c}_A(y = 0) = \Delta_C \quad \text{and} \quad \overline{c}_A(y = a) = c_1 - \Delta_C. \tag{34-46}$$

Integrating the species equation (34-45) with respect to y yields

$$b_1 = \varepsilon_C \frac{\partial \overline{c}_A}{\partial y}. \tag{34-47}$$

The integration constant is evaluated at the wall:

$$b_1 = \left(\varepsilon_C \frac{d\overline{c}_A}{dy}\right)_{y=0} = \frac{(\lambda\varepsilon)^2}{Sc_t} \frac{d\overline{v}_x}{dy}\bigg|_{y=0} \frac{d\overline{c}_A}{dy}\bigg|_{y=0}, \tag{34-48}$$

where the turbulent Schmidt number $Sc_t = \varepsilon_M/\varepsilon_C$ has been introduced. Substituting this result back into Eq. (34-47), the species transport equation becomes

$$\varepsilon_C \frac{\partial \overline{c}_A}{\partial y} = \frac{(\lambda\varepsilon)^2}{Sc_t} \frac{d\overline{v}_x}{dy}\bigg|_{y=0} \frac{d\overline{c}_A}{dy}\bigg|_{y=0} \quad \text{or} \quad \varepsilon_M \frac{\partial \overline{c}_A}{\partial y} = (\lambda\varepsilon)^2 \frac{d\overline{v}_x}{dy}\bigg|_{y=0} \frac{d\overline{c}_A}{dy}\bigg|_{y=0}. \tag{34-49}$$

Introducing the dimensionless variables

$$\eta = y/a \quad \text{and} \quad \phi = \overline{c}_A/c_1, \tag{34-50}$$

the species equation becomes

$$\frac{\varepsilon_M}{a}\phi' = \left(\lambda\frac{\varepsilon}{a}\right)^2 U\, u'(0)\,\phi'(0), \tag{34-51}$$

where primes are used to denote derivatives with respect to η. Evaluating the turbulent diffusivity with Eq. (34-34), the species equation becomes

$$\phi' = \frac{\left(\lambda\frac{\varepsilon}{a}\right)^2 u'(0)\,\phi'(0)}{\Lambda^2 u'}, \tag{34-52}$$

where the dimensionless mixing length $\Lambda = \ell/a$ is evaluated with the rule given by Eq. (34-35). Substituting the momentum equation (34-36) for u' into the species transport equation yields

$$\phi' = \lambda\frac{\varepsilon}{a}\frac{\phi'(0)}{\Lambda}. \tag{34-53}$$

The solution to the species transport equation (34-53) must satisfy the boundary conditions

$$\phi(0) = \Delta_C/c_1 \quad \text{and} \quad \phi(1/2) = 1/2. \tag{34-54}$$

The second boundary condition makes use of the point symmetry in the solution. This symmetry requires that the concentration at the mid-position between the plates is $c_1/2$, regardless of the value of Δ_C.

The species transport equation (34-53) can be solved in the same manner as the momentum equation (34-36) solved in the last section. However, for the special situation where $\Delta_C = 0$, the governing equation and boundary conditions for ϕ become identical to the problem for u. In this case, the solution to $\phi'(0) = u'(0)$ is given by Eq. (34-40).

The species flux J_A^* between the plates can be evaluated from the solution to the species transport equation (34-53). The result can be presented in terms of a Stanton number for species transfer:

$$St_A = \frac{h_A}{U} = \frac{J_A^*}{Uc_1} = \frac{1}{Uc_1}\left(\frac{\varepsilon_M}{Sc_t}\frac{\partial \bar{c}_A}{\partial y}\right)\bigg|_{y=0} = \frac{1}{Uc_1}\left(\frac{(\lambda\varepsilon)^2}{Sc_t}\frac{d\bar{v}_x}{dy}\bigg|_{y=0}\frac{d\bar{c}_A}{dy}\bigg|_{y=0}\right)$$
$$= \left(\lambda\frac{\varepsilon}{a}\right)^2\frac{u'(0)\phi'(0)}{Sc_t}. \tag{34-55}$$

For the special situation where $\Delta_C = 0$ (such that $\phi'(0) = u'(0)$), the Stanton number can be expressed simply in terms of the coefficient of friction using Eq. (34-41) for the result

$$St_A = \frac{c_f}{2Sc_t}. \tag{34-56}$$

The solution to the coefficient of friction c_f is given by Eq. (34-42). The close relation between the Stanton number and the coefficient of friction is noteworthy, because it addresses the same considerations that resulted in the Reynolds analogy between heat and momentum transport, discussed in Section 25.5.

34.4 APPLICATION OF REICHARDT'S FORMULA TO ROUGH SURFACES

As discussed in Section 31.6, the Reichardt's formula is useful for internal flows, when the value of the centerline turbulent diffusivity is important to quantifying transport. For fully rough surfaces, Reichardt's formula can be offset by the value of turbulent diffusivity at the wall. Turbulent diffusivity at the wall is evaluated from the mixing length model

$$\varepsilon_M(y=0) = (\lambda\varepsilon)^2(\partial\overline{v}_x/\partial y)\Big|_{y=0} = \nu(\lambda\varepsilon^+)^2(\partial u^+/\partial y^+)\Big|_{y^+=0}. \tag{34-57}$$

With this offset, Reichardt's formula can be expressed as

$$\left(\frac{\varepsilon_M}{\nu}\right)_{\substack{Reich,\\ rough}} = (\lambda\varepsilon^+)^2(du^+/dy^+)\Big|_{y^+=0} + \frac{\kappa y^+}{6}(1+r/R)\left(1+2(r/R)^2\right), \tag{34-58}$$

where, as in the case of pipe flow, r/R is the relative distance from the centerline of the flow.

34.4.1 Turbulent Poiseuille Flow between Rough Parallel Plates

To illustrate the application of Reichardt's formula to a turbulent flow bounded by rough surfaces, consider Poiseuille flow between parallel plates as illustrated in Figure 34-4. This problem was solved for smooth plates in Section 31.1. The velocity profile is assumed to be fully developed, such that $\overline{v}_y = 0$ and $\partial\overline{v}_x/\partial x = 0$ everywhere in the flow. Since advection does not contribute to cross-stream momentum transport in fully developed flows, and turbulence dominates diffusion transport ($\varepsilon_M \gg \nu$), the momentum equation simplifies to

$$0 = \frac{\partial}{\partial y}\left[\varepsilon_M \frac{\partial\overline{v}_x}{\partial y}\right] - \frac{1}{\rho}\frac{\partial\overline{P}}{\partial x}. \tag{34-59}$$

Upon integrating with respect to y, the momentum equation becomes

$$b_1 = \varepsilon_M\frac{\partial\overline{v}_x}{\partial y} - \frac{1}{\rho}\frac{\partial\overline{P}}{\partial x}y \quad \left(=\frac{\tau_w}{\rho}\right). \tag{34-60}$$

This result can be evaluated at the wall ($y=0$) for the integration constant $b_1 = \tau_w/\rho$. Furthermore, the streamwise pressure gradient must be balanced by the stresses transmitted to the walls. Consequently,

$$2\tau_w = a\left(-\frac{d\overline{P}}{dx}\right) \quad \text{or} \quad -\frac{1}{\rho}\frac{d\overline{P}}{dx} = \frac{2}{a}\frac{\tau_w}{\rho}. \tag{34-61}$$

Combining this last result with Eq. (34-60), the momentum equation becomes

$$\frac{\tau_w}{\rho} = \varepsilon_M\frac{d\overline{v}_x}{dy} + \frac{2}{a}\frac{\tau_w}{\rho}y. \tag{34-62}$$

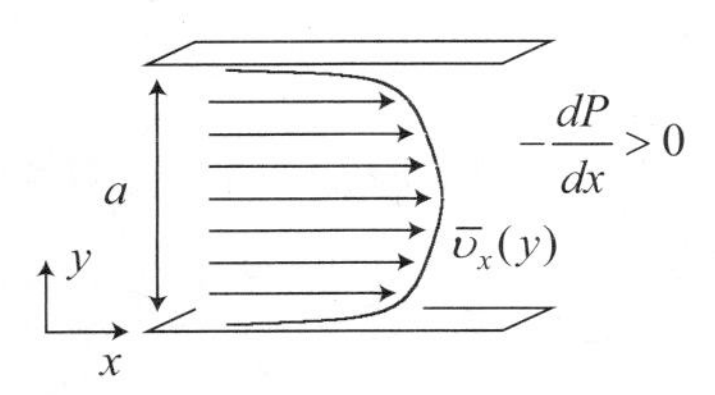

Figure 34-4 Turbulent Poiseuille flow between rough parallel plates.

At $y = 0$, Eq. (34-62) evaluates to

$$\frac{\tau_w}{\rho} = \left(\varepsilon_M \frac{d\overline{v}_x}{dy}\right)_{y=0} = \left(\lambda\varepsilon \frac{d\overline{v}_x}{dy}\bigg|_{y=0}\right)^2, \tag{34-63}$$

since the mixing length is $\ell = \lambda\varepsilon$ at the wall. Therefore, the momentum equation (34-62) can be written as

$$\varepsilon_M \frac{d\overline{v}_x}{dy} = \left(1 - \frac{2}{a}y\right)\frac{\tau_w}{\rho} = \left(1 - \frac{2}{a}y\right)\left(\lambda\varepsilon \frac{d\overline{v}_x}{dy}\bigg|_{y=0}\right)^2. \tag{34-64}$$

Using the dimensionless variables

$$\eta = y/a \quad \text{and} \quad u = \overline{v}_x/\overline{v}_m, \quad \text{where} \quad \overline{v}_m = \int_0^a \overline{v}_x dy/a, \tag{34-65}$$

the momentum equation (34-64) can be expressed as

$$u' = \frac{1 - 2\eta}{\varepsilon_M/\nu}\frac{\mathrm{Re}_D}{2}\left(\frac{\lambda\varepsilon}{a}u'(0)\right)^2, \tag{34-66}$$

where

$$\mathrm{Re}_D = \overline{v}_m(2a)/\nu \tag{34-67}$$

and primes are used to denote derivatives with respect to η. The solution to the momentum equation (34-66) is subject to the normalization requirement:

$$\int_0^{1/2} u d\eta = 1/2, \tag{34-68}$$

which results from the definition of $u = \overline{v}_x/\overline{v}_m$.

It is left as an exercise (see Problem 34-2) to solve Eq. (34-66) using the mixing length model described in Section 31.1. Here Reichardt's formula is used to evaluate turbulent diffusivity. To adapt Reichardt's formula to the problem at hand, it is observed that the turbulent diffusivity at the wall is

$$(\varepsilon_M/\nu)\big|_{y=0} = (\lambda\varepsilon^+)^2(du^+/dy^+)\big|_{y^+=0} = (\mathrm{Re}_D/2)(\lambda\varepsilon/a)^2 u'(0). \tag{34-69}$$

Furthermore, distances measured from the centerline of the flow may be expressed as $r/R = 1 - 2\eta$, and distances from the wall can be expressed by

$$y^+ = \frac{y\sqrt{\tau_w/\rho}}{\nu} = \eta\sqrt{\frac{\mathrm{Re}_D}{2}\frac{\varepsilon_M}{\nu}\bigg|_{y=0}}\,u'(0) = \eta\frac{\mathrm{Re}_D}{2}\frac{\lambda\varepsilon}{a}u'(0). \tag{34-70}$$

Therefore, Reichardt's formula (34-58) becomes

$$\left(\frac{\varepsilon_M}{\nu}\right)_{\substack{Reich,\\ rough}} = \frac{\mathrm{Re}_D}{2}\left[\frac{\lambda\varepsilon}{a} + \kappa\eta(1-\eta)\left(1 - \frac{8}{3}\eta(1-\eta)\right)\right]\frac{\lambda\varepsilon}{a}u'(0). \tag{34-71}$$

Substituting Eq. (34-71) into Eq. (34-66) yields the final form of the momentum equation to be integrated for Poiseuille flow between fully rough parallel plates:

$$u' = \frac{2\lambda(1-2\eta)(\varepsilon/D)u'(0)}{2\lambda(\varepsilon/D) + \kappa\eta(1-\eta)(1-(8/3)\eta(1-\eta))}, \quad \text{with } u(0) = 0. \tag{34-72}$$

Notice that the relative roughness has been expressed in terms of the hydraulic diameter $\varepsilon/D = \varepsilon/(2a)$. This will facilitate comparison of the parallel plate results with other flow geometries.

Code 34-1 is used to solve Eq. (34-72) for different values of the relative roughness ε/D. The mixing length model variables are evaluated with $\lambda = 0.0126$ and $\kappa = 0.41$. The spatial variable η is uniformly discretized with respect to the log of distance from the wall in the numerical solution. Runge-Kutta integration is used with shooting to determine $u'(0)$ in the governing equation (34-72) that satisfies the normalization requirement (34-68).

To verify adherence of the solution to the law of the wall, Eq. (34-11) is plotted in Figure 34-5 for contrast with the numerical solution of Eq. (34-72). It is seen that the numerical solution exhibits the correct law of the wall behavior through the inner region of turbulence.

From the solution, the friction coefficient for turbulent Poiseuille flow between rough plates can be calculated from

$$c_f = \frac{\tau_w/\rho}{\bar{v}_m^2/2} = 8(\lambda(\varepsilon/D)u'(0))^2. \tag{34-73}$$

A point of contrast for this result can be made by considering the Colebrook formula [4], which interpolates values of the friction coefficient between smooth-wall conditions and fully rough-wall conditions for turbulent pipe flow:

$$\frac{1}{\sqrt{c_f}} = -4\log\left(\frac{\varepsilon/D}{3.7} + \frac{1.26}{\mathrm{Re}_D\sqrt{c_f}}\right) \quad \text{or} \quad \frac{1}{\sqrt{c_f}} = -4\log\left[\frac{\varepsilon}{D}\left(0.27 + \frac{1.77}{\varepsilon^+}\right)\right]. \tag{34-74}$$

The Colebrook formula describes the turbulent flow behavior that was famously plotted by Lewis Ferry Moody [5], in what is now known as the Moody diagram. In the limit that

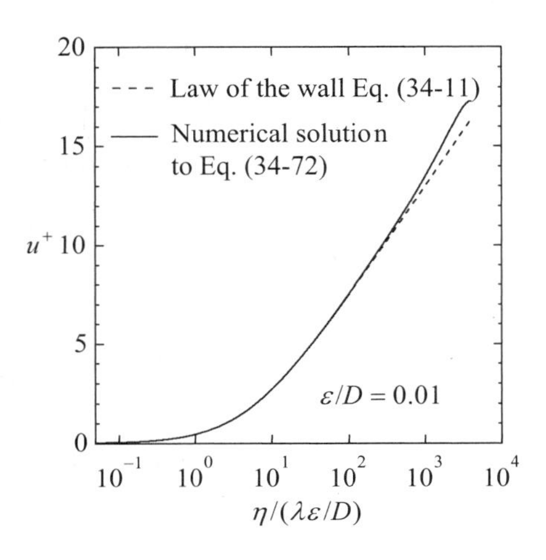

Figure 34-5 Numerical solution for u^+ contrasted with the law of the wall for a rough surface.

Code 34-1 Solves turbulent Poiseuille flow between parallel rough plates

```
#include <stdio.h>
#include <stdlib.h>
#include <math.h>

double du(double minLam,double Kappa,double du0,double eta) {
   return 2.*(1.-2.*eta)*minLam*du0/
          (2.*minLam+Kappa*eta*(1.-eta)*(1.-8.*eta*(1.-eta)/3.));
}

double solve_rough_u(int N,double *eta,double *u,double minLam,double Kappa)
{
   int n,iter=0;
   double del_eta,K1u,K2u,K3u,K4u;
   double du0,um,K1um,K2um,K3um,K4um;

   double du0_low=1.;                    // lower bound on du0
   double du0_high=1000000.0;            // upper bound on du0

   u[0]=0.0;                             // initial conditions
   do {                                  // RK integration
      um=0.;
      du0=(du0_low+du0_high)/2.0;        // use bisection method
      for (n=0;n<N-1;++n) {
         del_eta=eta[n+1]-eta[n];
         K1u=del_eta*du(minLam,Kappa,du0,eta[n]);
         K2u=K3u=del_eta*du(minLam,Kappa,du0,(eta[n]+eta[n+1])/2.);
         K4u=del_eta*du(minLam,Kappa,du0,eta[n+1]);

         K1um=   del_eta*u[n];
         K2um=   del_eta*(u[n]+0.5*K1um);
         K3um=   del_eta*(u[n]+0.5*K2um);
         K4um=   del_eta*(u[n]+K3um);

         u[n+1]=u[n]+(K1u+2.*K2u+2.*K3u+K4u)/6.;
         um+=(K1um+2.*K2um+2.*K3um+K4um)/6.;
         if (um>0.5) {
            du0_high=du0;                    // shoot lower
            break;
         }
      }
      if (n==N-1) du0_low=du0;           // shoot higher
   } while (++iter<200 && (du0_high-du0_low)>1.0e-7);
   if (fabs(um-0.5)>.00001) {
      printf("\nSoln. failed, du0=%e %e\n",du0,um);
      exit(1);
   }
   return du0;
}

int main()
{
   int n,N=20001;
   double du0,e_D,cf,ys,up;
   double lam=0.0126,Kappa=0.41; // model constants
   double eta[N],u[N];  // non-dimensional dependent variables

   eta[0]=0.;
   eta[N-1]=0.5;
   double log_min=-6.;   // use log steps
   double log_del=(log10(eta[N-1])-log_min)/(N-2);
   for (n=1;n<N-1;++n) eta[n]=pow(10.,log_min+(n-1)*log_del);

   for (e_D=0.00005;e_D<=0.05;e_D*=10.0) {
      du0=solve_rough_u(N,eta,u,lam*e_D,Kappa);
      cf=8.*lam*e_D*lam*e_D*du0*du0;
      printf("%e %e %e\n",e_D,cf,1./pow(4.*log10(3.7/e_D),2.));
   }

   e_D=0.01;
   du0=solve_rough_u(N,eta,u,lam*e_D,Kappa);
   FILE *fp=fopen("up.dat","w");
   for (n=0;n<N;n++) {       // output velocity
      ys=eta[n]/lam/e_D;
      up=u[n]/(2.*lam*e_D*du0);
      fprintf(fp,"%e %e %e\n",ys,up,log(Kappa*ys/2.+1.)/Kappa);
   }
   fclose(fp);

   return 0;
}
```

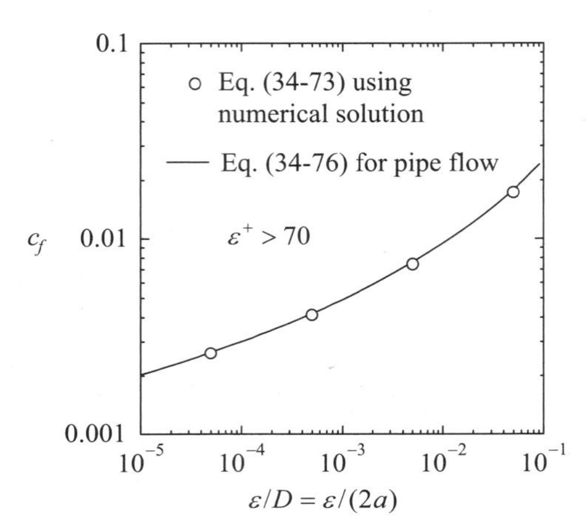

Figure 34-6 Friction coefficient for turbulent Poiseuille flow bounded by fully rough surfaces.

$\varepsilon/D \to 0$, the Colebrook formula reduces to Prandtl's correlation [6] for turbulent flow in a smooth-wall pipe:

$$\frac{1}{\sqrt{c_f}} = 4.0 \log\left(\frac{\mathrm{Re}_D\sqrt{c_f}}{1.26}\right) \quad \text{or} \quad \frac{1}{\sqrt{f}} = 2.0 \log\left(\mathrm{Re}_D\sqrt{f}\right) - 0.8, \tag{34-75}$$

where $f = 4c_f$. In the other limit, where the pipe wall is fully rough $\varepsilon^+ > 70$, the Colebrook formula (34-74) gives

$$c_f = \left(4.0 \log\left(\frac{3.7}{\varepsilon/D}\right)\right)^{-2}, \tag{34-76}$$

which is independent of the Reynolds number. It is worth emphasizing the fact that any degree of surface roughness can approach fully rough conditions as the Reynolds number Re_D becomes large. This is demonstrated by the definition of hydrodynamic surface roughness ε^+, as given by Eq. (34-2).

The solution to Eq. (34-72) for a few discrete roughness values is used to determine the friction coefficients indicated by open circles in Figure 34-6. The numerical results are contrasted with the correlation (34-76) for fully rough-wall pipe flow. Despite the geometric differences between the two flows, presenting the results in terms of the hydraulic diameter ($\varepsilon/D = \varepsilon/(2a)$) yields a quite similar comparison.

34.4.2 Turbulent Heat Convection in Flow between Fully Rough Parallel Isothermal Plates

Heat transfer between a turbulent Poiseuille flow and bounding parallel plates, as illustrated in Figure 34-7, is reconsidered for the case of fully rough isothermal walls. This problem was solved for smooth walls in Section 33.3.2. For fully rough surfaces, the presence of the diffusion sublayer is removed from the flow solution. In this case, turbulent diffusivity is always dominant over molecular diffusivity, and the heat equation becomes

$$\bar{v}_x \frac{\partial \bar{T}}{\partial x} = \frac{\partial}{\partial y}\left(\varepsilon_H \frac{\partial \bar{T}}{\partial y}\right). \tag{34-77}$$

As discussed in Section 33.3.2, when the temperature field is fully developed and the wall temperature T_s is constant, $\partial\bar{T}/\partial x$ can be related to changes in the mean flow temperature:

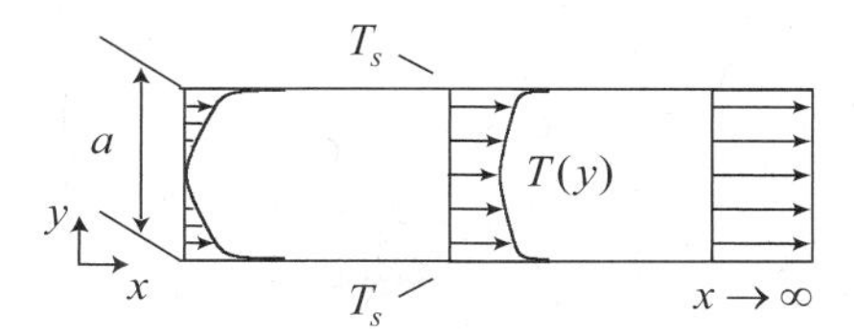

Figure 34-7 Fully developed temperature field between rough isothermal surfaces.

$$\frac{\partial \overline{T}}{\partial x} \approx \frac{T_s - \overline{T}}{T_s - \overline{T}_m} \frac{d\overline{T}_m}{dx}. \tag{34-78}$$

Furthermore, changes in the mean flow temperature may be evaluated from an energy balance over a differential distance dx downstream, yielding

$$\overline{v}_m \frac{d\overline{T}_m}{dx} = \frac{\Gamma q_s}{A\rho C_p} = \frac{\Gamma h}{A\rho C_p}(T_s - \overline{T}_m). \tag{34-79}$$

The mean flow quantities are defined by

$$\overline{T}_m = \int_A \overline{v}_x \overline{T} dA/(\overline{v}_m A) \quad \text{and} \quad \overline{v}_m = \int_A \overline{v}_x dA/A, \tag{34-80}$$

where A is the cross-sectional area of the flow and Γ is the peripheral length around A through which the heat transfer q_s occurs. Using Eq. (34-78) to relate $\partial\overline{T}/\partial x$ to $\partial\overline{T}_m/\partial x$, and Eq. (34-79) to eliminate $\partial\overline{T}_m/\partial x$, the heat equation (34-77) becomes

$$\frac{\overline{v}_x}{\overline{v}_m}(T_s - \overline{T})\frac{\Gamma h}{A\rho C_p} = \frac{\partial}{\partial y}\left(\varepsilon_H \frac{\partial \overline{T}}{\partial y}\right) \quad \text{(thermally and hydrodynamically fully developed turbulent flow with rough isothermal boundaries).} \tag{34-81}$$

The problem can be made dimensionless with the variables

$$\eta = \frac{y}{a}, \quad u = \frac{\overline{v}_x}{\overline{v}_m}, \quad \text{and} \quad \theta = \frac{\overline{T} - \overline{T}_m}{T_s - \overline{T}_m}, \tag{34-82}$$

where the definition

$$\overline{T}_m = \frac{\int_0^{a/2} \overline{v}_x \overline{T} dy}{\overline{v}_m a/2} \tag{34-83}$$

imposes the requirement

$$\int_0^{1/2} u\,\theta d\eta = 0. \tag{34-84}$$

With $\Gamma/A = 2/a$ for the parallel plate geometry, and the observation that

$$\frac{h(2a)}{\rho C_p} = \frac{2a}{T_s - \overline{T}_m}\left[-\varepsilon_H \frac{\partial \overline{T}}{\partial y}\right]\Bigg|_{y=0} = -2\left[\varepsilon_H \frac{d\theta}{d\eta}\right]\Bigg|_{\eta=0}, \tag{34-85}$$

the heat equation becomes

$$-2u(1-\theta)\left[\varepsilon_H \frac{d\theta}{d\eta}\right]\bigg|_{\eta=0} = \frac{\partial}{\partial\eta}\left(\varepsilon_H \frac{\partial\theta}{\partial\eta}\right). \tag{34-86}$$

The heat equation may be integrated once for

$$\theta' = \frac{1-g}{\varepsilon_H}\left[\varepsilon_H \frac{d\theta}{d\eta}\right]\bigg|_{\eta=0}, \tag{34-87}$$

where

$$g(\eta) = 2\int_0^{\eta} u(1-\theta)d\eta. \tag{34-88}$$

The mixing length model cannot be used to solve equation (34-87), since it predicts that $\varepsilon_H \to 0$ at the centerline of the flow. However, the turbulent heat diffusivity can be expressed in terms of Reichardt's formula. Using the turbulent Prandtl number, the heat diffusivity is related to the momentum diffusivity (34-71) for the result

$$\frac{\varepsilon_H}{\alpha} = \frac{\text{Pr}}{\text{Pr}_t}\left(\frac{\varepsilon_M}{\nu}\right)_{\substack{\text{Reich,}\\ \text{rough}}} = \frac{\text{Pr}}{\text{Pr}_t}\frac{\text{Re}_D}{2}\left[\frac{\lambda\varepsilon}{a} + \kappa\eta(1-\eta)\left(1-\frac{8}{3}\eta(1-\eta)\right)\right]\frac{\lambda\varepsilon}{a}u'(0). \tag{34-89}$$

Therefore, with

$$\frac{\varepsilon_H(\eta=0)}{\varepsilon_H(\eta)} = \frac{\lambda\varepsilon/a}{\lambda\varepsilon/a + \kappa\eta(1-\eta)(1-(8/3)\eta(1-\eta))}, \tag{34-90}$$

the heat equation becomes

$$\theta' = \frac{2\lambda(1-g)(\varepsilon/D)\,\theta'(0)}{2\lambda(\varepsilon/D) + \kappa\eta(1-\eta)(1-(8/3)\eta(1-\eta))}. \tag{34-91}$$

Notice that the relative roughness has been expressed in terms of the hydraulic diameter $\varepsilon/D = \varepsilon/(2a)$.

The constraints given by Eq. (34-68) and Eq. (34-84) require that Eq. (34-88) evaluates to $g(1/2) = 1$. Furthermore, the symmetry in the solution, stipulated by $\theta'(1/2) = 0$, is enforced in the governing equation (34-91) through the requirement that $g(1/2) = 1$. The fluid temperature at $y = 0$ can be related to the isothermal wall temperature using Eq. (34-26). To this end, the wall variable temperature T^+ is related to the fluid temperature θ by

$$T^+ = \frac{-\text{Pr}_t}{2\lambda(\varepsilon/D)}\frac{1-\theta}{\theta'(0)}. \tag{34-92}$$

Using Eq. (34-26) to estimate $T^+(y=0)$, the corresponding fluid temperature at $y = 0$ is determined with Eq. (34-92) to be

$$\theta(0) = 1 + 2\lambda\left(k_f(\varepsilon^+)^{0.2}\text{Pr}^{0.44} + A_f\right)\frac{\varepsilon}{D}\frac{\theta'(0)}{\text{Pr}_t}, \quad \text{for } \varepsilon^+ \geq 70, \tag{34-93}$$

where

$$\varepsilon^{+} = 2\lambda(\varepsilon/D)^{2}u'(0)\mathrm{Re}_{D}. \tag{34-94}$$

Although the governing equation is only explicitly dependent on the roughness ε/D, the empirical boundary condition for $\theta(0)$ introduces the added dependencies on Pr and Re_D (or ε^{+}).

The governing equations for momentum (34-72) and heat (34-91) transport, and constraint (34-88), may be expressed as three coupled first-order equations for u, θ, and g, as summarized here:

$$\frac{du}{d\eta} = \frac{2\lambda(1-2\eta)(\varepsilon/D)u'(0)}{2\lambda(\varepsilon/D) + \kappa\eta(1-\eta)(1-(8/3)\eta(1-\eta))}, \quad u(0) = 0 \tag{34-95}$$

$$\frac{\partial\theta}{\partial\eta} = \frac{2\lambda(1-g)(\varepsilon/D)\,\theta'(0)}{2\lambda(\varepsilon/D) + \kappa\eta(1-\eta)(1-(8/3)\eta(1-\eta))}, \quad \theta(0) = 1 + 2\lambda\left(k_f(\varepsilon^{+})^{0.2}\mathrm{Pr}^{0.44} + A_f\right)\frac{\varepsilon}{D}\frac{\theta'(0)}{\mathrm{Pr}_t} \tag{34-96}$$

$$\frac{\partial g}{\partial\eta} = 2u(1-\theta), \qquad g(0) = 0 \quad \text{and} \quad g(1/2) = 1. \tag{34-97}$$

Code 34-2 is used to determine the heat transfer between the isothermal rough parallel plates and the fully developed turbulent Poiseuille flow. Different values of the relative roughness ε/D are considered for the fully rough hydrodynamic condition $\varepsilon^{+} = 70$. The turbulence model is solved with the empirical constants $\lambda = 0.0126$, $k_f = 5.19$, and $A_f = 8.48$ for the rough surface model, and a turbulent Prandtl number of $\mathrm{Pr}_t = 0.9$. The value of $u'(0)$ is known from the previous solution of the momentum equation (Section 34.4.1) and the shooting method is used to determine $\theta'(0)$ that satisfies the constraint $g(1/2) = 1$.

The Stanton number for heat transfer can be calculated from the solution by

$$St_T = \frac{h}{\rho C_p \overline{v}_m} = \frac{1}{\overline{v}_m}\frac{1}{T_s - \overline{T}_m}\left[-\varepsilon_H\frac{\partial\overline{T}}{\partial y}\right]\Bigg|_{y=0} = -4\frac{u'(0)\theta'(0)}{\mathrm{Pr}_t}\left(\lambda\frac{\varepsilon}{D}\right)^{2}. \tag{34-98}$$

Notice that the Stanton number is independent of the molecular diffusivities of the fluid ν and α. Based on experimental measurements of pipe flow, Dipprey and Sabersky [3] developed a correlation for the Stanton number:

$$St = \frac{c_f/2}{1 + \sqrt{c_f/2}\left(5.19(\varepsilon^{+})^{0.2}\mathrm{Pr}^{0.44} - 8.48\right)}, \quad \varepsilon^{+} \geq 70. \tag{34-99}$$

Figure 34-8 contrasts the numerical results of the current model with the Dipprey and Sabersky correlation using Pr = 0.7 (air), Pr = 5.9 (water), and a hydrodynamic roughness of $\varepsilon^{+} = 70$ (fully rough). The numerically calculated Stanton number agrees well with the correlation over the range of surface roughness and Prandtl numbers investigated. In the current model, the only dependence on Prandtl number is through the empirically calculated surface temperature (34-93).

Code 34-2 Solves turbulent Poiseuille heat transfer between rough plates (uses subroutine solve_rough_u( ) and du( ) in Code 34-1)

```
#include <stdio.h>
#include <stdlib.h>
#include <math.h>

double solve_rough_u(int N,double *eta,double *u,double minLam,double Kappa);
double du(double minLam,double Kappa,double du0,double eta);

double dT(double minLam,double Kappa,double dT0,double eta,double g) {
   return 2.*(1.-g)*minLam*dT0/
         (2.*minLam+Kappa*eta*(1.-eta)*(1.-8.*eta*(1.-eta)/3.));
}

double solve_rough_T(int N,double *eta,double *T,double du0,double Pr,
                     double Prt,double ep,double minLam,double Kappa)
{
   int n,iter=0;
   double del_eta,dT0,u,g,K1u,K2u,K3u,K4u,K1T,K2T,K3T,K4T,K1g,K2g,K3g,K4g;
   double dT0_lo=-1000000.0;     // lower bound on dT0
   double dT0_hi=0.0;            // upper bound on dT0

   do {                          // RK integration
      u=g=0.;
      dT0=(dT0_lo+dT0_hi)/2.0;  // use bisection method
      T[0]=1.+2.*( 5.19*pow(Pr,.44)*pow(ep,.2)-8.48 )*minLam*dT0/Prt;
      for (n=0;n<N-1;++n) {
         del_eta=eta[n+1]-eta[n];

         K1u=del_eta*du(minLam,Kappa,du0,eta[n]);
         K1T=del_eta*dT(minLam,Kappa,dT0,eta[n],g);
         K1g=del_eta*2.*u*(1.-T[n]);
         K2u=del_eta*du(minLam,Kappa,du0,(eta[n]+eta[n+1])/2.);
         K2T=del_eta*dT(minLam,Kappa,dT0,(eta[n+1]+eta[n])/2.,g+0.5*K1g);
         K2g=del_eta*2.*(u+0.5*K1u)*(1.-T[n]-0.5*K1T);
         K3u=del_eta*du(minLam,Kappa,du0,(eta[n]+eta[n+1])/2.);
         K3T=del_eta*dT(minLam,Kappa,dT0,(eta[n+1]+eta[n])/2.,g+0.5*K2g);
         K3g=del_eta*2.*(u+0.5*K2u)*(1.-T[n]-0.5*K2T);
         K4u=del_eta*du(minLam,Kappa,du0,eta[n+1]);
         K4T=del_eta*dT(minLam,Kappa,dT0,eta[n+1],g+K3g);
         K4g=del_eta*2.*(u+K3u)*(1.-T[n]-K3T);
         u+=          (K1u+2.*K2u+2.*K3u+K4u)/6.;
         T[n+1]=  T[n]+(K1T+ 2.*K2T+ 2.*K3T+ K4T )/6.;
         g+=           (K1g+ 2.*K2g+ 2.*K3g+ K4g )/6.;
         if (g>1.) {
            dT0_lo=dT0;        // shoot higher
            break;
         }
      }
      if (n==N-1) dT0_hi=dT0; // shoot lower

   } while (++iter<200 && (dT0_hi-dT0_lo)>1.0e-7);

   if (fabs(g-1.0)>.00001) {
      printf("\nSoln. failed, dT0=%e g=%e\n",dT0,g);
      exit(1);
   }
   return dT0;
}

int main()
{
   int n,N=20001;
   double du0,dT0,e_D,StT;
   double lam=0.0126,Kappa=0.41,Prt=.9; // model constants
   double eta[N],u[N],T[N];      // non-dimensional dependent variables

   double Pr=0.7;
   double ep=70.;
   eta[0]=0.;
   eta[N-1]=0.5;
   double log_min=-6.;   // use log steps
   double log_del=(log10(eta[N-1])-log_min)/(N-2);
   for (n=1;n<N-1;++n) eta[n]=pow(10.,log_min+(n-1)*log_del);

   FILE *fp=fopen("StT.dat","w");
   for (e_D=0.00005;e_D<=0.05;e_D*=10.0) {
      du0=solve_rough_u(N,eta,u,lam*e_D,Kappa);
      dT0=solve_rough_T(N,eta,T,du0,Pr,Prt,ep,lam*e_D,Kappa);
      StT=-4.*du0*dT0*lam*e_D*lam*e_D/Prt;
      fprintf(fp,"%e %e\n",e_D,StT);
   }
   fclose(fp);

   return 0;
}
```

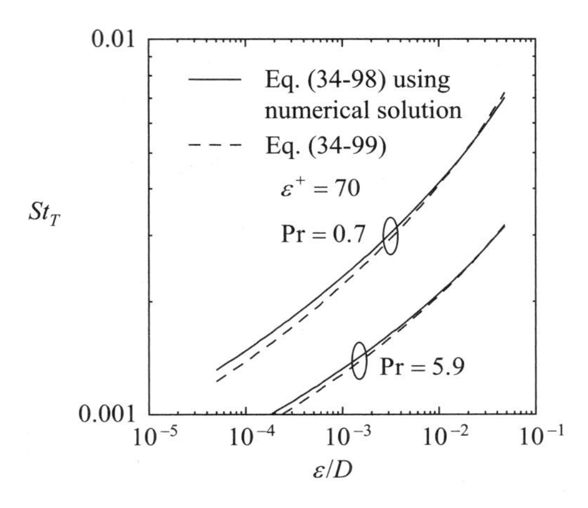

Figure 34-8 Stanton number for heat transfer between rough isothermal plates.

34.5 PROBLEMS

34-1 The Moody diagram shows the relation for friction factor: $f(\text{Re}, \varepsilon/D)$. The diagram also shows that $f(\varepsilon/D)$ as $\text{Re} \to \infty$; that is, the friction factor depends only on the pipe's relative roughness in the limit that Re becomes large. Suggest an explanation for this observation based on the mixing length model for rough surfaces.

34-2 Consider turbulent Poiseuille flow between fully rough parallel plates, as discussed in Section 34.4.1. Turbulent diffusivity in the mixing length model is evaluated with

$$\frac{\varepsilon_M}{\nu} = (\text{Re}_D/2)\Lambda^2 u',$$

where $\text{Re}_D = v_m(2a)/\nu$ is based on the hydraulic diameter. The dimensionless mixing length $\Lambda = \ell/a$ is evaluated with the rule

$$\Lambda = \begin{cases} \Lambda^* & \Lambda^* < \gamma/2 \\ \gamma/2 & \textit{otherwise} \end{cases}, \qquad \text{where } \Lambda^* = \kappa\eta + \frac{\lambda\varepsilon}{a}.$$

In this way, the dimensionless mixing length evaluates to either $\Lambda = \Lambda^*$ for the inner region, or $\Lambda = \gamma/2$ for the outer region of turbulence. Show that for the mixing length model the momentum equation for turbulent Poiseuille flow between rough parallel plates becomes

$$u' = \lambda\frac{\varepsilon}{a}\frac{\sqrt{1-2\eta}}{\Lambda}u'(0), \quad \text{with } u(0) = 0.$$

Solve the momentum equation for $\varepsilon/(2a) = 0.01$ and plot the velocity solution against the law of the wall. What is the friction coefficient for this wall roughness?

34-3 Consider turbulent Poiseuille flow in a rough pipe of radius R and wall roughness of $\varepsilon/D = 0.01$. Using a mixing length model for turbulence, calculate Darcy's friction factor $f = 4c_f$ for $\text{Re}_D = 10^6$. How does this result compared with the mixing length model result for a smooth-wall pipe? How do both results compare with numbers that you can read off the Moody diagram?

34-4 Consider turbulent Poiseuille flow in a rough pipe of radius R. The walls of the pipe are held at an isothermal temperature. For a flow that is fully developed, both thermally and hydrodynamically, develop the governing equations using Reichardt's formula for turbulent diffusivity. Demonstrate that the governing equations may be expressed as three coupled first-order equations for u, θ, and g:

$$\frac{du}{d\eta} = \frac{12\lambda(\varepsilon/D)(1-\eta)u'(0)}{12\lambda(\varepsilon/D) + \kappa\eta(2-\eta)(3-2\eta(2-\eta))}, \quad u(0) = 0 \quad \text{and} \quad \int_0^1 u(1-\eta)d\eta = 1/2$$

$$\frac{\partial\theta}{\partial\eta} = \frac{12\lambda(\varepsilon/D)(1-g)\,\theta'(0)/(1-\eta)}{12\lambda(\varepsilon/D) + \kappa\eta(2-\eta)(3-2\eta(2-\eta))}, \quad \theta(0) = 1 + 2\lambda\left(k_f(\varepsilon^+)^{0.2}\,\mathrm{Pr}^{0.44} + A_f\right)\frac{\varepsilon}{D}\frac{\theta'(0)}{\mathrm{Pr}_t}.$$

$$\frac{\partial g}{\partial\eta} = 2u\,(1-\eta)(1-\theta), \quad g(0) = 0 \quad \text{and} \quad g(1) = 1,$$

where

$$\eta = 1 - \frac{r}{R}, \quad u = \frac{\overline{v}_z}{\overline{v}_m}, \quad \text{and} \quad \theta = \frac{\overline{T} - \overline{T}_m}{T_s - \overline{T}_m}.$$

Solve the turbulent heat and momentum transport equations for $\varepsilon^+ = 70$ and $\mathrm{Pr} = 0.7$ to determine the Stanton number as a function of relative roughness for the range $0.00005 \le \varepsilon/D \le 0.05$.

REFERENCES

[1] H. Schlichting, *Boundary-Layer Theory.* New York, NY: McGraw–Hill, 1979.

[2] J. Nikuradse, *Strömungsgesetze in Rauhen Rohren*, VDI-Forschungsheft, No. 361, 1933. (English translation: National Advisory Committee for Aeronautics, Technical Memorandum 1292, Washington, DC, 1950.)

[3] D. F. Dipprey and R. H. Sabersky, "Heat and Momentum Transfer in Smooth and Rough Tubes at Various Prandtl Numbers." *International Journal of Heat and Mass Transfer*, **6**, 329 (1963).

[4] C. F. Colebrook, "Turbulent Flow in Pipes, with Particular Reference to the Transition Region Between Smooth and Rough Pipe Laws." *Journal of the Institution of Civil Engineers* (London), **11**, 133 (1939).

[5] L. F. Moody, "Friction Factors for Pipe Flow." *Transactions of the ASME*, **66**, 671 (1944).

[6] L. Prandtl, *Führer, durch die Strömungslehre.* Braunschweig, Germany: Vieweg, 1944.

Chapter 35

Turbulent Boundary Layer

35.1 Formulation of Transport in Turbulent Boundary Layer

35.2 Formulation of Heat Transport in the Turbulent Boundary Layer

35.3 Problems

In this chapter, the time-averaged turbulent transport equations are solved for an external boundary layer, as shown in Figure 35-1. The boundary layer equations are partial differential equations, and less simple to solve than the fully developed turbulent internal flow problems discussed in Chapters 31, 33, and 34. The mixing length model, introduced in the context of boundary layer flows in Chapter 30, is again used in this chapter. No consideration is given for a discrete transition from a laminar to a turbulent flow in the downstream description of the boundary layer. As the viscous sublayer envelops the thickness of the boundary layer approaching the leading edge of the plate, the mixing length model forces the magnitude of the turbulent flux to zero. Therefore, the flow description in this limit is given by Blasius' laminar flow solution. Although this is a useful engineering view of the problem, one should be mindful that, in reality, a more discreet transition to turbulence is expected somewhere in the vicinity of $\mathrm{Re}_{x,\,tran} \approx 5x10^5$, as measured from the leading edge of the plate. However, the transition to turbulence can be significantly lower by intentional tripping of the flow, or significantly higher if care is taken not to destabilize the flow.

35.1 FORMULATION OF TRANSPORT IN TURBULENT BOUNDARY LAYER

To illustrate analysis of a turbulent boundary layer, consider a situation in which the flow external to the boundary layer is at a constant velocity $\overline{U}$. A simplifying consequence is that $\partial\overline{P}/\partial x$ is zero in the boundary layer. The time-averaged turbulent equations governing the boundary layer are

$$\text{Continuity}: \qquad \frac{\partial \overline{v}_x}{\partial x} + \frac{\partial \overline{v}_y}{\partial y} = 0 \tag{35-1}$$

and

$$\text{Momentum}: \quad \overline{v}_x \frac{\partial \overline{v}_x}{\partial x} + \overline{v}_y \frac{\partial \overline{v}_x}{\partial y} = \frac{\partial}{\partial y}\left[(\nu + \varepsilon_M)\frac{\partial \overline{v}_x}{\partial y}\right] - \overbrace{\frac{1}{\rho}\frac{\partial \overline{P}}{\partial x}}^{=0}. \tag{35-2}$$

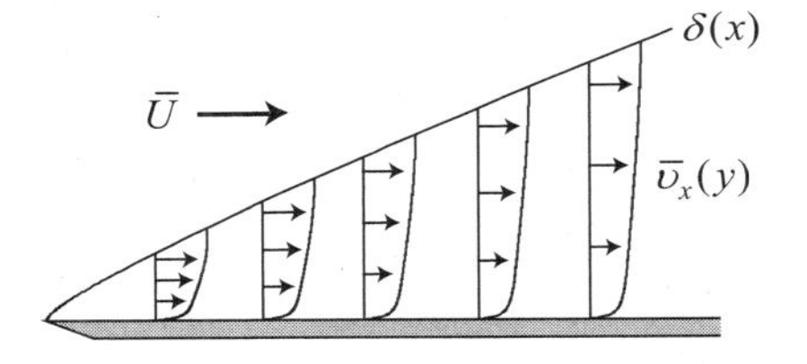

Figure 35-1 Development of a turbulent boundary layer on a smooth flat plate.

Notice that the streamwise diffusion term is also neglected in the boundary layer form of the momentum equation.

The boundary layer equations for a turbulent flow cannot be solved using the same similarity approach applied to laminar boundary layers because of the nonconstant turbulent diffusivity ε_M. However, the governing equations (35-1) and (35-2) can be expressed in finite differencing form, and numerically integrated. To this end, the physical domain of the boundary layer must be discretized, which is complicated by the downstream growth of the boundary layer thickness. A simple rectangular domain, which captures the downstream boundary layer thickness, will relegate a large number of upstream nodes to the uninteresting region outside the boundary layer. To reduce this waste, it is preferable to have a mesh that grows in thickness with the boundary layer. However, it is not yet known how the turbulent boundary layer will grow with distance. The investigation of laminar boundary layers in Chapter 25 demonstrated that for Blasius flat plate problems, the boundary layer thickness grows as $\delta(x) \sim \sqrt{\nu x/\overline{U}}$. In lieu of more specific information, this laminar relation is used for scaling of the turbulent domain to be discretized. Therefore, the spatial coordinates for the numerical solution are chosen to be

$$\eta = \frac{y}{\sqrt{\nu x/\overline{U}}} \quad \text{and} \quad X = \frac{x}{L}. \tag{35-3}$$

Equation 35-3 defines the relation between the physical and computational domains shown in Figure 35-2. The turbulent velocity components are nondimensionalized with the definitions

$$F' = \frac{\overline{v}_x}{\overline{U}} \quad \text{and} \quad V = \frac{\overline{v}_y}{\overline{U}}\sqrt{\frac{\overline{U}x}{\nu}}. \tag{35-4}$$

The variable F' is intended to be reminiscent of the derivative of dimensionless stream function f used in the laminar flow problem (Section 25.1). Recall that for the laminar flow:

$$\text{Blasius problem}: f''' + \frac{1}{2}f f'' = 0 \quad \text{with } v_x = \overline{U}f' \text{ (for laminar flow)} \tag{35-5}$$

$$f(0) = 0,\ f'(0) = 0, \quad \text{and} \quad f'(\infty) = 1. \tag{35-6}$$

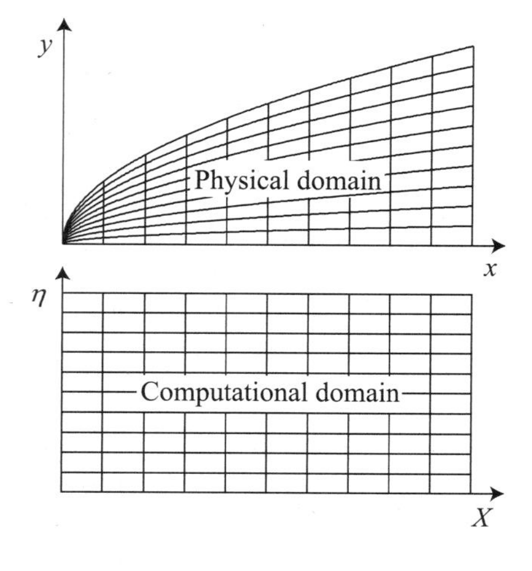

Figure 35-2 Mesh transformation.

However, unlike $f(\eta)$ for the laminar problem, $F(X,\eta)$ for the turbulent problem is dependent on two spatial variables.

The velocity gradients, appearing in both the continuity equation (35-1) and momentum equation (35-2), must be transformed into expressions of the new variables $F'(X,\eta)$ and $V(X,\eta)$. The required transformations are

$$\frac{\partial \overline{v}_x}{\partial x} = \overline{U}\frac{\partial F'}{\partial x} = \overline{U}\left[\left.\frac{\partial \eta}{\partial x}\right|_y \frac{\partial F'}{\partial \eta} + \left.\frac{\partial X}{\partial x}\right|_y \frac{\partial F'}{\partial X}\right] = \overline{U}\left[\frac{-\eta}{2x}\frac{\partial F'}{\partial \eta} + \frac{1}{L}\frac{\partial F'}{\partial X}\right], \tag{35-7}$$

$$\frac{\partial(\ )}{\partial y} = \left[\left.\frac{\partial \eta}{\partial y}\right|_x \frac{\partial(\ )}{\partial \eta} + \overbrace{\left.\frac{\partial X}{\partial y}\right|_x}^{=0} \frac{\partial(\ )}{\partial X}\right] = \sqrt{\frac{\overline{U}}{\nu x}}\frac{\partial(\ )}{\partial \eta}, \tag{35-8}$$

$$\frac{\partial \overline{v}_x}{\partial y} = \overline{U}\frac{\partial F'}{\partial y} = \overline{U}\sqrt{\frac{\overline{U}}{\nu x}}\frac{\partial F'}{\partial \eta}, \quad \text{and} \quad \frac{\partial \overline{v}_y}{\partial y} = \frac{\partial}{\partial y}\left(\sqrt{\frac{\nu \overline{U}}{x}}V\right) = \sqrt{\frac{\nu \overline{U}}{x}}\frac{\partial V}{\partial y} = \frac{\overline{U}}{x}\frac{\partial V}{\partial \eta}. \tag{35-9}$$

Expressing the equations for continuity (35-1) and momentum (35-2) in terms of these new variables yields

$$\text{Continuity:} \qquad X\frac{\partial F'}{\partial X} - \frac{\eta}{2}\frac{\partial F'}{\partial \eta} + \frac{\partial V}{\partial \eta} = 0 \tag{35-10}$$

and

$$\text{Momentum:} \quad XF'\frac{\partial F'}{\partial X} + \left(V - \frac{\eta F'}{2}\right)\frac{\partial F'}{\partial \eta} = \frac{\partial}{\partial \eta}\left[\left(1 + \frac{\varepsilon_M}{\nu}\right)\frac{\partial F'}{\partial \eta}\right]. \tag{35-11}$$

Notice that in the limit of $X \to 0$ (at the leading edge of the plate) the continuity equation becomes

$$\text{for } X = 0, \quad \text{Continuity}: \frac{\partial V}{\partial \eta} = \frac{\eta}{2}\frac{\partial F'}{\partial \eta}, \tag{35-12}$$

which integrates to

$$\text{for } X = 0, \quad V(0,\eta) = \frac{1}{2}(\eta F' - F). \tag{35-13}$$

Using this result for $V(0,\eta)$, in the limit that $X \to 0$ the momentum equation becomes

$$\text{for } X = 0, \quad \text{Momentum}: \frac{\partial}{\partial \eta}\left[\frac{\partial F'}{\partial \eta}\right] + \frac{F}{2}\frac{\partial F'}{\partial \eta} = 0, \tag{35-14}$$

subject to

$$F'(0,\eta = 0) = 0 \quad \text{and} \quad F'(0,\eta \to \infty) = 1. \tag{35-15}$$

It is assumed that $\varepsilon_M/\nu \to 0$ at the leading edge of the plate, since the diminishing scale of the boundary layer thickness forces the boundary layer velocity profile into the viscous sublayer as $X \to 0$.

Notice that the momentum equation (35-14) for $F(0,\eta)$ is the same as the Blasius equation for $f(\eta)$, which was solved in Chapter 25. Therefore, the boundary conditions imposed on the problem for $F'(X,\eta)$ and $V(X,\eta)$ are

$$\begin{array}{llll} \textit{at} \quad X=0 & F'(0,\eta)=f'(\eta) & \textit{and} & V(0,\eta)=(\eta F'-F)/2 \\ \textit{at} \quad \eta=0 & F'(X,0)=0 & \textit{and} & V(X,0)=0 \\ \textit{for} \quad \eta\to\infty & F'(X,\infty)=1. & & \end{array} \tag{35-16}$$

The mixing length model is introduced to the momentum equation, as discussed in Section 30.5. The turbulent diffusivity is evaluated from

$$\frac{\varepsilon_M}{\nu}=\frac{\ell^2}{\nu}\frac{\partial \overline{v}_x}{\partial y}=\eta_{99}^2 \mathrm{Re}_x^{1/2}\Lambda^2\frac{\partial F'}{\partial \eta} \tag{35-17}$$

where

$$\eta_{99}=\delta/\sqrt{\nu x/\overline{U}}, \quad \mathrm{Re}_x=\overline{U}x/\nu, \quad \text{and} \quad \Lambda=\ell/\delta. \tag{35-18}$$

The mixing length is selected by the rule

$$\Lambda=\begin{cases}\Lambda^* & \Lambda^*<\gamma \\ \gamma & \textit{otherwise}\end{cases} \tag{35-19}$$

where the Van Driest function is used for both the viscous sublayer and inner region of the boundary layer

$$\Lambda^*=\frac{\kappa\eta}{\eta_{99}}\left[1-\exp\left(\frac{-\eta}{A_\nu^+}\sqrt{\mathrm{Re}_x\, c_f/2}\right)\right], \tag{35-20}$$

where

$$c_f=\frac{2}{\mathrm{Re}_x^{1/2}}\left.\frac{\partial F'}{\partial \eta}\right|_{\eta=0}. \tag{35-21}$$

35.1.1 Finite Difference Representation of the Momentum Equation

It is useful to modify the differential form of the momentum equation one more time before casting it into a finite differencing equation. Using the relation that

$$\frac{\partial F'}{\partial \eta}\frac{\partial}{\partial \eta}\left(\frac{\varepsilon_M}{\nu}\right)=\frac{\varepsilon_M}{\nu}\left(\frac{2}{\Lambda}\frac{\partial \Lambda}{\partial \eta}\frac{\partial F'}{\partial \eta}+\frac{\partial^2 F'}{\partial \eta^2}\right), \tag{35-22}$$

the momentum equation (35-11) can be written as

$$XF'\frac{\partial F'}{\partial X}+\left(V-\frac{\eta F'}{2}\right)\frac{\partial F'}{\partial \eta}=\frac{\varepsilon_M}{\nu}\left(\frac{2}{\Lambda}\frac{\partial \Lambda}{\partial \eta}\frac{\partial F'}{\partial \eta}+\frac{\partial^2 F'}{\partial \eta^2}\right)+\left(1+\frac{\varepsilon_M}{\nu}\right)\frac{\partial^2 F'}{\partial \eta^2} \tag{35-23}$$

or, more simply, as

$$a\frac{\partial^2 F'}{\partial \eta^2} + b\frac{\partial F'}{\partial \eta} = XF'\frac{\partial F'}{\partial X} \tag{35-24}$$

with

$$a = \left(1 + 2\frac{\varepsilon_M}{\nu}\right) \quad \text{and} \quad b = \frac{2}{\Lambda}\frac{\partial \Lambda}{\partial \eta}\frac{\varepsilon_M}{\nu} - V + \frac{\eta F'}{2}. \tag{35-25}$$

To avoid creating a nonlinear system of equations, coefficients in the momentum equation (a, b, and XF') are treated as known information. In practice, this is accomplished by iteratively solving the momentum equation, using information from the last iteration to evaluate unknown information when necessary. Variables that reference the last iteration will be denoted with an asterisk (e.g., a^*, b^*, F'^*, . . .).

Figure 35-3 illustrates the point about which central differencing is performed on the momentum equation. It is centered between notes $n-1$ and $n+1$ in the vertical direction η, and between nodes m and $m+1$ in the downstream direction X. Across the boundary layer the node spacing $\Delta\eta$ is constant, but ΔX may change with downstream distance. The discretized momentum equation will be solved for unknown values of F' at the $m+1$ nodes; values of F' at the m nodes are known, as integration is marched forward in the X direction. The discretized evaluation of the momentum equation (35-24) is averaged between nodes m and $m+1$, such that

$$\begin{aligned} &\frac{1}{2}\left[a^*\frac{\partial^2 F'}{\partial \eta^2}\bigg|_n + b^*\frac{\partial F'}{\partial \eta}\bigg|_n\right]_{m+1} + \frac{1}{2}\left[a\frac{\partial^2 F'}{\partial \eta^2}\bigg|_n + b\frac{\partial F'}{\partial \eta}\bigg|_n\right]_m \\ &= \left[\frac{(XF')^*_{m+1} + (XF')_m}{2}\frac{\partial F'}{\partial X}\bigg|_{m+1/2}\right]_n. \end{aligned} \tag{35-26}$$

To attain a linear system of equations, some expressions at the $m+1$ nodes are evaluated using information from the previous iteration (as denoted with an asterisk). Using the center differencing forms of

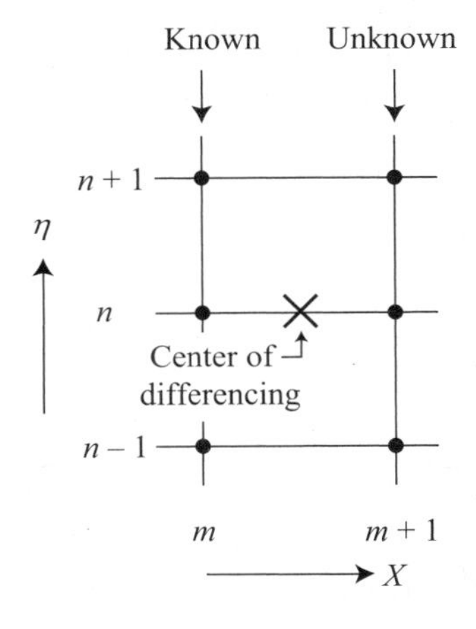

Figure 35-3 Center differencing scheme for momentum equation.

$$\left.\frac{\partial F'}{\partial \eta}\right|_n = \frac{F'_{n+1} - F'_{n-1}}{2\Delta\eta}, \quad \left.\frac{\partial^2 F'}{\partial \eta^2}\right|_n = \frac{F'_{n+1} - 2F'_n + F'_{n-1}}{\Delta\eta^2}, \quad \text{and} \quad \left.\frac{\partial F'}{\partial X}\right|_{m+1/2} = \frac{F'_{m+1} - F'_m}{\Delta X}, \tag{35-27}$$

the discretized momentum equation can be written as

$$\begin{aligned}(a^*_{m+1,n} - c^*_{m+1,n})F'_{m+1,n-1} - (2a^*_{m+1,n} + d_{m,n})F'_{m+1,n} + (a^*_{m+1,n} + c^*_{m+1,n})F'_{m+1,n+1} \\ = -(a_{m,n} - c_{m,n})F'_{m,n-1} + (2a_{m,n} - d_{m,n})F'_{m,n} - (a_{m,n} + c_{m,n})F'_{m,n+1}\end{aligned} \tag{35-28}$$

where

$$a_{m,n} = (1 + 2e_{m,n}), \; c_{m,n} = \frac{\Delta\eta}{2} b_{m,n} = \frac{\Lambda_{m,n+1} - \Lambda_{m,n-1}}{2\Lambda_{m,n}} e_{m,n} - \frac{\Delta\eta}{2}\left(V_{m,n} - \frac{\eta_n F'_{m,n}}{2}\right) \tag{35-29}$$

$$d_{m,n} = \frac{\mathrm{Re}_{m+1} F'^*_{m+1,n} + \mathrm{Re}_m F'_{m,n}}{\Delta \mathrm{Re}_{m+1/2}/\Delta\eta^2}, \quad \text{and} \quad e_{m,n} = (\eta_{99})^2_m \mathrm{Re}_m^{1/2} \Lambda^2_{m,n} \frac{F'_{m,n+1} - F'_{m,n-1}}{2\Delta\eta}. \tag{35-30}$$

Notice that the length of the plate can be discretized in terms of the Reynolds number $\mathrm{Re} = \overline{U}x/\nu$ rather than distance. To this end, substitutions of the form $X/\Delta X = \mathrm{Re}/\Delta\mathrm{Re}$ have been made in the discretized momentum equation, where

$$\Delta\mathrm{Re} = \overline{U}\Delta x/\nu. \tag{35-31}$$

Also, note that $e_{m,n}$ is the discretized form of Eq. (35-17) for the turbulent diffusivity.

The discretized momentum equation (35-28) can be written for every node n between the first and last, spanning the boundary layer at $m+1$. With the boundary conditions $F'_{m+1,0} = 0$ at the first node and $F'_{m+1,N-1} = 1$ at the last node, a system of equations can be prescribed for F'_{m+1}, in which the number of equations equals the number of unknowns.

35.1.2 Finite Difference Representation of the Continuity Equation

The discretized momentum equation is solved for the streamwise velocity component F' at $m+1$. It remains a task for the continuity equation (35-10) to solve for the transverse velocity component V. Since the continuity equation is linear, the procedure is straightforward.

Figure 35-4 illustrates the point about which central differencing is performed on the continuity equation. It is centered between notes $n-1$ and n in the vertical direction η from the surface, and between nodes m and $m+1$ in the downstream direction X. The

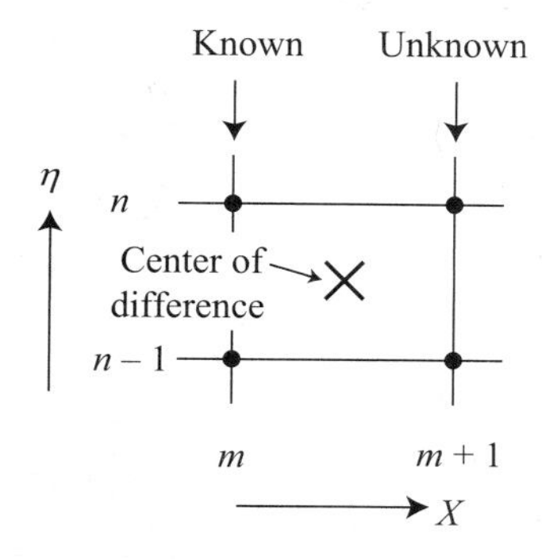

Figure 35-4 Center differencing scheme for continuity equation.

discretized continuity equation will be solved for the unknown value of V at the node $(m+1, n)$, marching away from the surface of the plate. Evaluation of the discretized continuity equation (35-10) is averaged between nodes m and $m+1$, and between nodes $n-1$ and n, such that

$$\frac{1}{2}\left[\frac{1}{2}\frac{\eta_{n-1}+\eta_n}{2}\frac{\partial F'}{\partial \eta}\bigg|_{n-1/2} - \frac{\partial V}{\partial \eta}\bigg|_{n-1/2}\right]_{m+1} + \frac{1}{2}\left[\frac{1}{2}\frac{\eta_{n-1}+\eta_n}{2}\frac{\partial F'}{\partial \eta}\bigg|_{n-1/2} - \frac{\partial V}{\partial \eta}\bigg|_{n-1/2}\right]_{m}$$
$$= \frac{1}{2}\left[\frac{X_{m+1}+X_m}{2}\frac{\partial F'}{\partial X}\bigg|_{m+1/2}\right]_{n} + \frac{1}{2}\left[\frac{X_{m+1}+X_m}{2}\frac{\partial F'}{\partial X}\bigg|_{m+1/2}\right]_{n-1}. \quad (35\text{-}32)$$

Using the center differencing forms of

$$\frac{\partial F'}{\partial \eta}\bigg|_{n-1/2} = \frac{F'_n - F'_{n-1}}{\Delta \eta}, \quad \frac{\partial V}{\partial \eta}\bigg|_{n-1/2} = \frac{V_n - V_{n-1}}{\Delta \eta}, \quad \text{and} \quad \frac{\partial F'}{\partial X}\bigg|_{m+1/2} = \frac{F'_{m+1} - F'_m}{\Delta X_{m+1/2}}, \quad (35\text{-}33)$$

the discretized continuity equation can be written in the form

$$V_{m+1,n} = V_{m+1,n-1} - V_{m,n} + V_{m,n-1} + (g_n - h_m)F'_{m+1,n} - (g_n + h_m)F'_{m+1,n-1} + (g_n + h_m)F'_{m,n} - (g_n - h_m)F'_{m,n-1} \quad (35\text{-}34)$$

where

$$g_n = \frac{\eta_{n-1}+\eta_n}{4} \quad \text{and} \quad h_m = \frac{\Delta\eta}{2}\frac{\mathrm{Re}_{m+1}+\mathrm{Re}_m}{\Delta \mathrm{Re}_{m+1/2}}. \quad (35\text{-}35)$$

Again, substitutions of $X/\Delta X = \mathrm{Re}/\Delta\mathrm{Re}$ have been made to eliminate the explicit reference to the spatial variable X.

35.1.3 Marching Scheme for Numerical Solution

The governing equations for momentum and continuity are solved simultaneously marching forward from the leading edge of the plate, where both F' and V are known from the laminar flow solution to the Blasius flat plate problem (see Chapter 28). A system of equations can be written with the discretized form of the momentum equation (35-28) to solve for F' in the first column of nodes ($m = 1$), as illustrated in Figure 35-5. Once F' is calculated for the column, V can be determined from the discretized form of the continuity equation (35-34) by marching up the column of nodes. However, each time the momentum equation is solved, it assumes knowledge of the asterisked terms

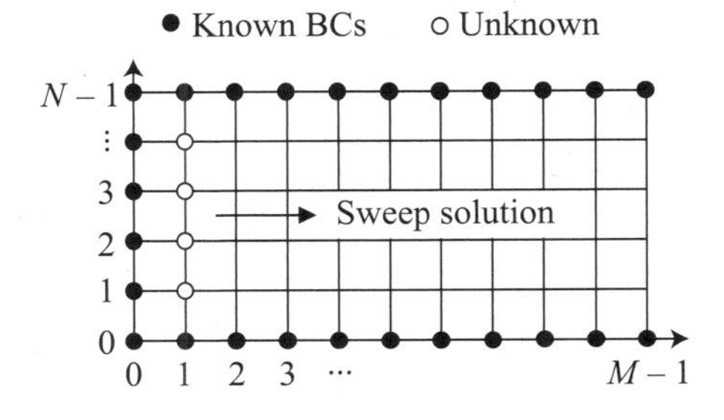

Figure 35-5 Propagation of numeric solution.

(F'^*, η^*_{99}, ...) that reference information not currently known. Therefore, the momentum equation must be solved repeatedly, each time using a better estimate of the asterisked terms from the previous iteration. Every time the momentum equation is solved, the continuity equation must be solved to update the solution for V across the column. Eventually, additional iterations will stop yielding any change in the calculated unknowns. Two characteristics of the solution that can be watched for convergence are the boundary layer thickness (η_{99}) and the local drag coefficient (c_f). After converging on the solution for the first column of nodes, the process is stepped forward to the next column and repeated.

To solve the momentum equation for each column of nodes requires solving a system of equations for F'. Since the discretized form of the momentum equation relates F' at n only to the neighboring nodes at $n-1$ and $n+1$, the resulting system of equations can be written as a tridiagonal matrix:

$$\begin{bmatrix} B_0 & C_0 & & & 0 \\ A_1 & B_1 & C_1 & & \\ & A_2 & B_2 & \ddots & \\ & & \ddots & \ddots & C_{N-2} \\ 0 & & & A_{N-1} & B_{N-1} \end{bmatrix} \begin{bmatrix} F'_{m+1,0} \\ F'_{m+1,1} \\ F'_{m+1,2} \\ \vdots \\ F'_{m+1,N-1} \end{bmatrix} = \begin{bmatrix} D_0 \\ D_1 \\ D_2 \\ \vdots \\ D_{N-1} \end{bmatrix}. \tag{35-36}$$

To satisfy the boundary condition $F'_{m+1,0} = 0$ requires that $C_0 = D_0 = 0$ and $B_0 = 1$. To satisfy the boundary condition $F'_{m+1,N-1} = 1$ requires that $B_{N-1} = D_{N-1} = 1$ and $A_{N-1} = 0$. For $n = 1, 2, \ldots N-2$, the terms in the matrix equation are evaluated using the coefficients defined in Eqs. (35-29) and (35-30) as

$$A_n = \left(a^*_{m+1,n} - c^*_{m+1,n}\right),\ B_n = -\left(2a^*_{m+1,n} + d_{m,n}\right),\ C_n = \left(a^*_{m+1,n} + c^*_{m+1,n}\right) \tag{35-37}$$

$$D_n = -(a_{m,n} - c_{m,n})F'_{m,n-1} + (2a_{m,n} - d_{m,n})F'_{m,n} - (a_{m,n} + c_{m,n})F'_{m,n+1}. \tag{35-38}$$

The tridiagonal matrix can be solved using the Thomas algorithm*. The first step of this algorithm consists of modifying the coefficients as follows:

$$C^*_0 = C_0/B_0 \quad \text{and} \quad D^*_0 = D_0/B_1 \tag{35-39}$$

and

$$C^*_n = C_n/\left(B_n - C^*_{n-1}A_n\right) \qquad \text{for} \quad n = 2, 3, \ldots, N-2 \tag{35-40}$$

$$D^*_n = \left(D_n - D^*_{n-1}A_n\right)/\left(B_n - C^*_{n-1}A_n\right) \quad \text{for} \quad n = 2, 3, \ldots, N-1. \tag{35-41}$$

The solution is then obtained by back substitution:

$$F'_{m+1,N-1} = D^*_{N-1} \tag{35-42}$$

---

*Named after the British physicist and applied mathematician Llewellyn Hilleth Thomas (1903–1992).

and

$$F'_{m+1,n} = D_n^* - C_n^* F'_{m+1,n+1} \qquad \text{for} \quad n = N-2, N-3, \cdots, 0. \tag{35-43}$$

35.1.4 Results of Momentum Transport

The discretized form of the governing equations (35-28) and (35-34) is solved with Code 35-1. The matrix form (35-36) of the momentum equation is solved using the Thomas algorithm. A large number of nodes is used to discretize distances from the surface ($N = 4001$) in order to adequately resolve the near wall region. This number could be reduced by using a log scale of distances from the wall (as was done for internal flows in Section 31.1); however, this somewhat complicates the discretization of $\partial^2 F'/\partial\eta^2|_n$ in the momentum equation.

The turbulent boundary layer solution is determined over a length of the flat plate prescribed by the range of Reynolds numbers $\text{Re}_x = 10^2 - 10^7$. The velocity profile in wall variables is plotted in Figure 35-6 at a few distances (Re_x) downstream of the leading edge of the plate. The wall variables are related to the variables of the solution by

$$y^+ = \eta\sqrt{\text{Re}_x c_f/2} \quad \text{and} \quad u^+ = F'/\sqrt{c_f/2}, \tag{35-44}$$

where the local friction coefficient is evaluated using Eq. (35-21). As expected, the flow initially exhibits the profile of the viscous sublayer region, followed by the law of the wall behavior. At distances further from the wall, the turbulent flow becomes dependent on the effects of advection. This is signified by a departure from the law of the wall, which is dependent on the local value of the Reynolds number.

The solution to the turbulent boundary layer equations can be used to evaluate the local friction coefficient from Eq. (35-21) as a function of distance (reported as Re_x) from the leading edge of the flat plate. Figure 35-7 presents the numerical results for the friction coefficient from the mixing length model and contrasts these results with the Schultz-Grunow [1] empirical correlation:

$$c_f = 0.37\left[\log\text{Re}_x\right]^{-2.584}. \tag{35-45}$$

In the limit of low Reynolds numbers ($\text{Re}_x < 10^3$), the friction coefficient acquires the laminar flow value found in Section 25.1. The Schultz-Grunow correlation adheres well to the results of the numerical calculation.

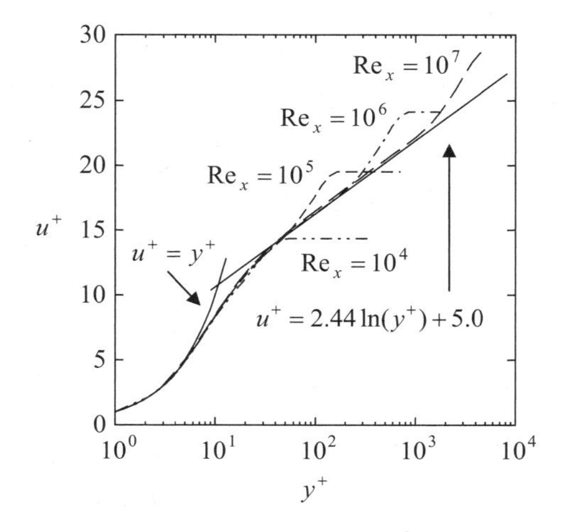

Figure 35-6 Law of the wall behavior in the boundary layer at different Reynolds numbers.

Code 35-1 Solves turbulent momentum boundary layer equations for flow over a smooth flat plate. (Blasius) subroutine from Code 25-1

```
#include <stdio.h>
#include <stdlib.h>
#include <math.h>
#define _N_ 4001
#define _M_ 301
double Lambda(double eta,double Y99,double Kappa,double Ap,double Re_Cf) {
   double Gamma=0.085;
   double Lam=Kappa*(eta/Y99)*(1.-exp(-eta*sqrt(Re_Cf/2.)/Ap));
   if (Lam>Gamma) return Gamma;
   return Lam;
}
void MomCoef(double Y99,double Cf,double Re,double *dF,double *V,
             int n,double dY,double Kappa,double Ap,double *an,double *cn) {
   double Lam=Lambda(n*dY,Y99,Kappa,Ap,Re*Cf);
   double delLam=Lambda((n+1)*dY,Y99,Kappa,Ap,Re*Cf)
                 -Lambda((n-1)*dY,Y99,Kappa,Ap,Re*Cf);
   double en=pow(Y99*Lam,2.)*sqrt(Re)*(dF[n+1]-dF[n-1])/2./dY;
   *an=1.+2.*en;
   *cn=delLam*en/2./Lam-dY*(V[n]-n*dY*dF[n]/2.)/2.;
}
double findY99(int N,double dY,double *dF) {
   int n=1;
   while (n<N && dF[n]<0.99) ++n;
   return dY*(n-1+(0.99-dF[n-1])/(dF[n]-dF[n-1]));
}
void SolveTurbMomBL(int M,int N,double Yinf,double Kappa,double Ap,
                    double *Y99,double *Re,double *Cf) {
   int m,n;
   extern double dF[_M_][_N_],V[_M_][_N_];
   double A[N],B[N],C[N],D[N],E[N],F[N],dY=Yinf/(N-1);
   Y99[0]=findY99(N,dY,dF[0]);
   Cf[0]=2.*dF[0][1]/dY/sqrt(Re[0] ? Re[0] : 1.);
   for (m=0;m<M-1;++m) { // loop over m, solve for dF[m+1][n] and V[m+1][n]
      for (n=1;n<N-1;++n) { // initial guess
        dF[m+1][n]=dF[m][n];
        V[m+1][n] =V[m][n];
      }
      Y99[m+1]=findY99(N,dY,dF[m+1]);
      Cf[m+1]=2.*dF[m+1][1]/dY/sqrt(Re[m+1]);
      int done,cnt=0;
      double an,cn,dn,gn,hm=dY*(Re[m+1]+Re[m])/(Re[m+1]-Re[m])/2.;
      do {
         for (n=1;n<N-1;++n) {
            dn=dY*dY*(Re[m+1]*dF[m+1][n]+Re[m]*dF[m][n])/(Re[m+1]-Re[m]);
            MomCoef(Y99[m],Cf[m],Re[m],dF[m],V[m],n,dY,Kappa,Ap,&an,&cn);
            D[n]= -(an-cn)*dF[m][n-1]
                  +(2.*an-dn)*dF[m][n]-(an+cn)*dF[m][n+1];
            MomCoef(Y99[m+1],Cf[m+1],Re[m+1],dF[m+1],V[m+1],
                    n,dY,Kappa,Ap,&an,&cn);
            A[n]=    an-cn;
            B[n]=-2.*an-dn;
            C[n]=    an+cn;
         }
         E[0]=F[0]=0.0; // solve system of equations for dF[m+1][n]
         for (n=1;n<N-1;++n) {
            E[n]  = C[n]/( B[n] - E[n-1]*A[n] );
            F[n] = ( D[n] - F[n-1]*A[n] )/( B[n] - E[n-1]*A[n] );
         }
         dF[m+1][N-1] = 1.0; // enforce boundary condition
         for (n=N-2;n>=0;--n)
            dF[m+1][n] = F[n] - E[n]*dF[m+1][n+1];
         V[m+1][0]=V[m][0]=0.0; /* enforce boundary condition */
         for (n=1;n<N;++n) { // solve continuity equation for V[m+1][n]
            gn=dY*(n-.5)/2.;
            V[m+1][n] = V[m+1][n-1] - V[m][n] + V[m][n-1]
                      + (gn-hm)*dF[m+1][n] - (gn+hm)*dF[m+1][n-1]
                      + (gn+hm)*dF[m][n]   - (gn-hm)*dF[m][n-1];
         }
         double Y99m_plus1=findY99(N,dY,dF[m+1]);
         double Cfm_plus1=2.*dF[m+1][1]/dY/sqrt(Re[m+1]);
         done=(fabs(Y99m_plus1-Y99[m+1])/Y99m_plus1 > 0.0001
               || fabs(Cfm_plus1-Cf[m+1])/Cfm_plus1 > 0.0001 ? 0 : 1);
         Y99[m+1]=Y99m_plus1;
         Cf[m+1]=Cfm_plus1;
      } while (!done && ++cnt<100);
      if (cnt==100 || dY*sqrt(Re[m+1]*Cf[m+1]/2.) > 5.0) {
         printf("\nError in SolveTurbMomBL()!\n");
         exit(1);
      }
   }
}

void blasius(double Yinf,int N,double *f,double *df,double *ddf);
double dF[_M_][_N_], V[_M_][_N_];
int main()
{
   int m,n,M=_M_,N=_N_;
   extern double dF[_M_][_N_],V[_M_][_N_];
   double Y99[M],Re[M],Cf[M],F[N],E[N];
   double Ap=26.0,Kappa=0.4;
   double Yinf=50.0,dY=Yinf/(N-1);
   double log_min=2.; // use log steps Re: 10^2-10^7
   double log_del=(7.-log_min)/(M-1);
   for (m=0;m<M;++m) Re[m]=pow(10.,log_min+m*log_del);
   blasius(Yinf,N,F,dF[0],E);
   for (n=0;n<N;++n) V[0][n]=(n*dY*dF[0][n]-F[n])/2.;
   SolveTurbMomBL(M,N,Yinf,Kappa,Ap,Y99,Re,Cf);
   FILE *fp=fopen("Cf.dat","w");
   for (m=1;m<M;++m) fprintf(fp,"%e %e %e %e\n",Re[m],Cf[m],
        0.37*pow(log10(Re[m]),-2.584),0.664/sqrt(Re[m]));
   fclose(fp);
   return 1;
}
```

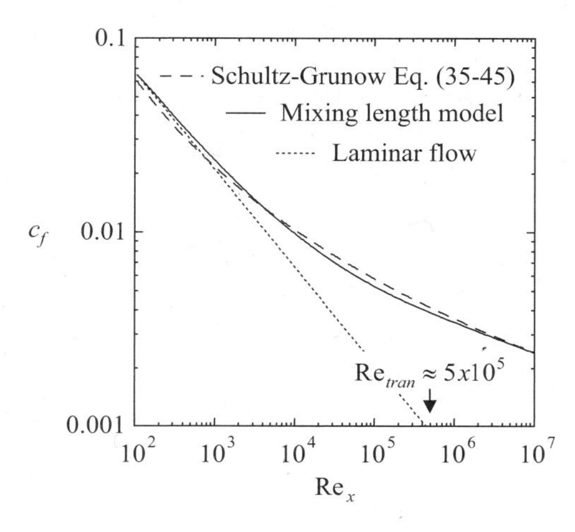

Figure 35-7 Friction coefficient for turbulent flow over a smooth flat plate.

35.2 FORMULATION OF HEAT TRANSPORT IN THE TURBULENT BOUNDARY LAYER

In the absence of viscous heating and compressibility effects, heat transport through a turbulent boundary layer is governed by

$$\text{Heat:} \qquad \overline{v}_x \frac{\partial \overline{T}}{\partial x} + \overline{v}_y \frac{\partial \overline{T}}{\partial y} = \frac{\partial}{\partial y}\left[(\alpha + \varepsilon_H)\frac{\partial \overline{T}}{\partial y}\right]. \tag{35-46}$$

Using the dimensionless spatial variables $\eta = y/\sqrt{\nu x/\overline{U}}$ and $X = x/L$, the velocity variables $F' = \overline{v}_x/\overline{U}$ and $V = \overline{v}_y\sqrt{x/(\nu\overline{U})}$, the turbulent Prandtl number $\mathrm{Pr}_t = \varepsilon_M/\varepsilon_H$, and the dimensionless temperature

$$\theta = \frac{T_w - \overline{T}}{T_w - \overline{T}_\infty}, \tag{35-47}$$

the heat equation is transformed into

$$XF'\frac{\partial \theta}{\partial X} + \left(V - \frac{\eta F'}{2}\right)\frac{\partial \theta}{\partial \eta} = \frac{\partial}{\partial \eta}\left[\left(\frac{1}{\mathrm{Pr}} + \frac{1}{\mathrm{Pr}_t}\frac{\varepsilon_M}{\nu}\right)\frac{\partial \theta}{\partial \eta}\right]. \tag{35-48}$$

In the limit of $X \to 0$ (at the leading edge of the plate), where $V(0,\eta) = (\eta F' - F)/2$ and $\varepsilon_M/\nu \to 0$, the heat equation becomes

$$\text{for } X = 0, \text{ Heat:} \qquad \frac{\partial^2 \theta}{\partial \eta^2} + \mathrm{Pr}\frac{F}{2}\frac{\partial \theta}{\partial \eta} = 0, \tag{35-49}$$

which is subject to $\theta(0, \eta = 0) = 0$ and $\theta(0, \eta \to \infty) = 1$. This is the same heat equation as solved for in the laminar Blasius flow problem. Therefore, the boundary conditions imposed on the problem for $\theta(X, \eta)$ are

$$\begin{array}{lll} \textit{at} & X = 0 & \theta(0,\eta) = \theta_B(\eta) \\ \textit{at} & \eta = 0 & \theta(X,0) = 0 \\ \textit{for} & \eta \to \infty & \theta(X,\infty) = 1 \end{array} \tag{35-50}$$

where $\theta_B(\eta)$ denotes the solution to the heat equation for the laminar boundary layer discussed in Section 25.3.

The solution to the heat equation (35-48) can be used to determine the local value of the Nusselt number. In terms of the variables of the heat equation, the turbulent boundary layer Nusselt number is

$$Nu_x = \frac{hx}{k} = \frac{\overline{q}_s}{(T_w - \overline{T}_\infty)}\frac{x}{k} = \sqrt{\frac{Ux}{\nu}}\frac{\partial\theta}{\partial\eta}\bigg|_{\eta=0} = \mathrm{Re}_x^{1/2}\frac{\partial\theta}{\partial\eta}\bigg|_{\eta=0}. \tag{35-51}$$

35.2.1 Finite Difference Representation of the Heat Equation

The turbulent boundary layer heat equation (35-48) is solved using the mixing length model expressions for the turbulent momentum diffusivity and its derivative:

$$\frac{\varepsilon_M}{\nu} = \eta_{99}^2 \mathrm{Re}_x^{1/2}\Lambda^2\frac{\partial F'}{\partial\eta} \quad \text{and} \quad \frac{\partial}{\partial\eta}\left(\frac{\varepsilon_M}{\nu}\right) = \left(\frac{2}{\Lambda}\frac{\partial\Lambda}{\partial\eta} + \frac{\partial^2 F'/\partial\eta^2}{\partial F'/\partial\eta}\right)\frac{\varepsilon_M}{\nu}. \tag{35-52}$$

If it is assumed that Pr_t is constant, the heat equation (35-48) can be expressed as

$$a\frac{\partial^2\theta}{\partial\eta^2} + b\frac{\partial\theta}{\partial\eta} = \mathrm{Pr}_t X F'\frac{\partial\theta}{\partial X}, \tag{35-53}$$

where

$$a = \left(\frac{\mathrm{Pr}_t}{\mathrm{Pr}} + \frac{\varepsilon_M}{\nu}\right) \quad \text{and} \quad b = \left(\frac{2}{\Lambda}\frac{\partial\Lambda}{\partial\eta} + \frac{\partial^2 F'/\partial\eta^2}{\partial F'/\partial\eta}\right)\frac{\varepsilon_M}{\nu} - \mathrm{Pr}_t\left(V - \frac{\eta F'}{2}\right). \tag{35-54}$$

Notice that the coefficients to the differential heat equation are all known functions of the spatial variables. These coefficients are evaluated with the solution to the momentum equation, which must precede evaluation of the heat equation. The heat equation is discretized using the same mesh and central differencing strategy employed for the momentum equation. The center point for differencing, shown in Figure 35-8, is between nodes $n-1$ and $n+1$ in the vertical direction η from the surface, and between nodes m and $m+1$ in the downstream direction X. The discretized heat equation will be solved for

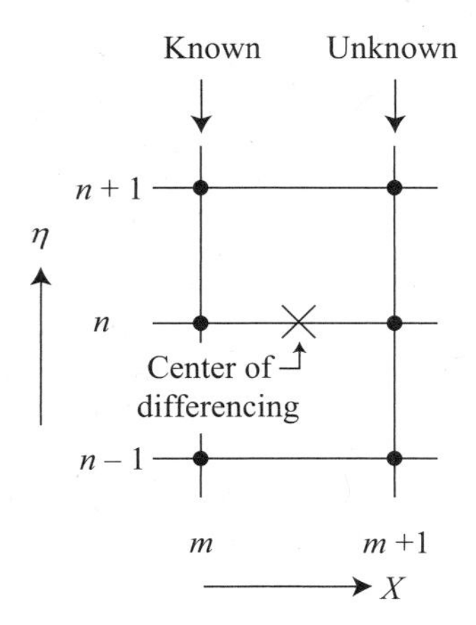

Figure 35-8 Center differencing scheme for the heat equation.

unknown values of θ at the $m+1$ nodes. The discretized evaluation of the heat equation (35-53) is averaged between nodes m and $m+1$, such that

$$\frac{1}{2}\left[a\frac{\partial^2\theta}{\partial\eta^2}\bigg|_n + b\frac{\partial\theta}{\partial\eta}\bigg|_n\right]_{m+1} + \frac{1}{2}\left[a\frac{\partial^2\theta}{\partial\eta^2}\bigg|_n + b\frac{\partial\theta}{\partial\eta}\bigg|_n\right]_m = \mathrm{Pr}_t\left[(XF')_{m+1/2}\frac{\partial\theta}{\partial X}\bigg|_{m+1/2}\right]_n. \quad (35\text{-}55)$$

Using the center differencing forms

$$\frac{\partial\theta}{\partial\eta}\bigg|_n = \frac{\theta_{n+1}-\theta_{n-1}}{2\Delta\eta}, \quad \frac{\partial^2\theta}{\partial\eta^2}\bigg|_n = \frac{\theta_{n+1}-2\theta_n+\theta_{n-1}}{\Delta\eta^2} \quad (35\text{-}56)$$

and

$$(XF')_{m+1/2} = \frac{(XF')_{m+1}+(XF')_m}{2}, \quad \frac{\partial\theta}{\partial X}\bigg|_{m+1/2} = \frac{\theta_{m+1}-\theta_m}{\Delta X_{m+1/2}}, \quad (35\text{-}57)$$

the discretized heat equation can be written as

$$\begin{aligned}(a_{m+1,n}-c_{m+1,n})\theta_{m+1,n-1} &- (2a_{m+1,n}+d_{m,n})\theta_{m+1,n} + (a_{m+1,n}+c_{m+1,n})\theta_{m+1,n+1} \\ &= -(a_{m,n}-c_{m,n})\theta_{m,n-1} + (2a_{m,n}-d_{m,n})\theta_{m,n} - (a_{m,n}+c_{m,n})\theta_{m,n+1}\end{aligned} \quad (35\text{-}58)$$

where

$$a_{m,n} = \left(\frac{\mathrm{Pr}_t}{\mathrm{Pr}} + e_{m,n}\right), \quad (35\text{-}59)$$

$$\begin{aligned}c_{m,n} = \frac{\Delta\eta}{2}b_{m,n} &= \left(\frac{\Lambda_{m,n+1}-\Lambda_{m,n-1}}{2\Lambda_{m,n}} + \frac{F'_{m,n+1}-2F'_{m,n}+F'_{m,n-1}}{F'_{m,n+1}-F'_{m,n-1}}\right)e_{m,n} \\ &\quad -\mathrm{Pr}_t\frac{\Delta\eta}{2}\left(V_{m,n}-\frac{\eta_n F'_{m,n}}{2}\right),\end{aligned} \quad (35\text{-}60)$$

$$d_{m,n} = \mathrm{Pr}_t\frac{\mathrm{Re}_{m+1}F'_{m+1,n}+\mathrm{Re}_m F'_{m,n}}{\Delta\mathrm{Re}_{m+1/2}/\Delta\eta^2} \quad \text{and} \quad e_{m,n} = (\eta_{99})_m^2\,\mathrm{Re}_m^{1/2}\,\Lambda_{m,n}^2\frac{F'_{m,n+1}-F'_{m,n-1}}{2\Delta\eta}. \quad (35\text{-}61)$$

Again, substitutions of the form $X/\Delta X = \mathrm{Re}/\Delta\mathrm{Re}$ have been made to eliminate the explicit reference to the spatial variable X.

The discretized heat equation (35-58) can be written for every node n spanning the boundary layer at $m+1$. Since the discretized heat equation relates θ at n only to the neighboring nodes at $n-1$ and $n+1$, the resulting system of equations can be written as a tridiagonal matrix:

$$\begin{bmatrix} B_0 & C_0 & & & 0 \\ A_1 & B_1 & C_1 & & \\ & A_2 & B_2 & \ddots & \\ & & \ddots & \ddots & C_{N-2} \\ 0 & & & A_{N-1} & B_{N-1}\end{bmatrix}\begin{bmatrix}\theta_{m+1,0} \\ \theta_{m+1,1} \\ \theta_{m+1,2} \\ \vdots \\ \theta_{m+1,N-1}\end{bmatrix} = \begin{bmatrix} D_0 \\ D_1 \\ D_2 \\ \vdots \\ D_{N-1}\end{bmatrix}. \quad (35\text{-}62)$$

To satisfy the boundary condition $\theta_{m+1,0} = 0$ requires that $C_0 = D_0 = 0$ and $B_0 = 1$. To satisfy the boundary condition $\theta_{m+1,N-1} = 1$ requires that $B_{N-1} = D_{N-1} = 1$ and $A_{N-1} = 0$. Using the coefficients defined in Eqs. (35-59) through (35-61), the terms in the matrix equation (35-62) are evaluated for $n = 1, 2, \ldots N-2$ as

$$A_n = (a_{m+1,n} - c_{m+1,n}), \quad B_n = -(2a_{m+1,n} + d_{m,n}), \quad C_n = (a_{m+1,n} + c_{m+1,n}) \tag{35-63}$$

$$D_n = -(a_{m,n} - c_{m,n})\theta_{m,n-1} + (2a_{m,n} - d_{m,n})\theta_{m,n} - (a_{m,n} + c_{m,n})\theta_{m,n+1}. \tag{35-64}$$

The tridiagonal matrix (35-62) can be solved using the Thomas algorithm discussed in Section 35.1.3.

35.2.2 Results of the Heat Equation

The numerical solution for θ is marched across the computational domain in a process similar to the momentum equation. However, because the governing equation for θ is linear, no iterative steps are required. The heat equation is solved with Code 35-2, over a length of a flat plate prescribed by the range of Reynolds numbers $\mathrm{Re}_x = 10^2 - 10^7$. The turbulent flow is air with a Prandtl number of $\mathrm{Pr} = 0.7$. The mixing length model uses the imperial constants $\kappa = 0.40$, $A_\nu^+ = 26.0$, $\gamma = 0.085$, $A_\alpha^+ = 32.0$, and $\mathrm{Pr}_t = 0.9$.

Figure 35-9 presents the numerical results for the Nusselt number evaluated from Eq. (35-51), and contrasts the results of the mixing length model with a correlation useful for gases ($0.5 < \mathrm{Pr} < 1.0$) [2]:

$$\frac{Nu}{\mathrm{Re}_x \mathrm{Pr}} = 0.0287 \mathrm{Re}_x^{-0.2} \mathrm{Pr}^{-0.4}. \tag{35-65}$$

In the limit of low Reynolds numbers ($\mathrm{Re}_x < 10^3$), the Nusselt number acquires the laminar flow value found in Section 25.3. The correlation (35-65) agrees well with the results of the numerical calculation for $\mathrm{Re}_x > 3 \times 10^4$. To obtain agreement between the mixing length model and experimental data over a wider range of fluid Prandtl numbers requires consideration of the nonconstant behavior of the turbulent Prandtl number (see Problem 35-1).

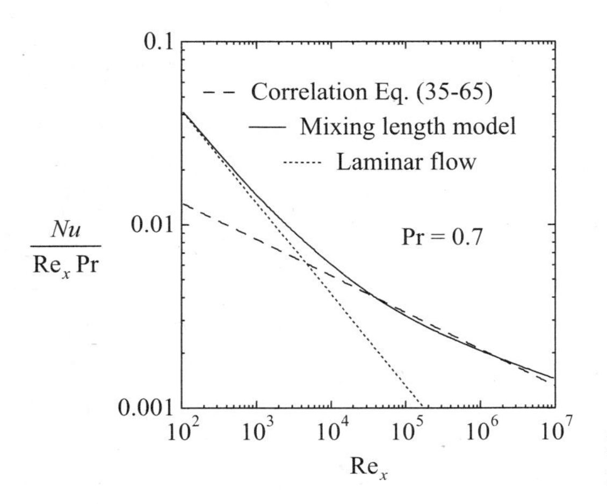

Figure 35-9 Nusselt number for turbulent flow over a smooth flat isothermal plate.

Code 35-2 Solves turbulent momentum boundary layer equations for flow over a smooth flat plate. (Uses SolveTurbMomBL( ) from Code 35-1 and blasiusT( ) from Code 25-2.)

```
#include <stdio.h>
#include <math.h>
#define _N_ 4001
#define _M_ 301

double Lambda(double eta,double Y99,double Kappa,double Ap,double Re_Cf);

void HeatCoef(double Pr,double Prt,double Kappa,double Ap,double ApH,
              double Y99,double Cf,double Re,double *dF,double *V,
              int n,double dY,double *an,double *cn) {
   double Lam=Lambda(n*dY,Y99,Kappa,Ap,Re*Cf);
   double delLam=Lambda((n+1)*dY,Y99,Kappa,Ap,Re*Cf)
                -Lambda((n-1)*dY,Y99,Kappa,Ap,Re*Cf);
   double en=pow(Y99*Lam,2.)*sqrt(Re)*(dF[n+1]-dF[n-1])/2./dY;
   *an=Prt/Pr+en;
   if (dF[n+1]==dF[n-1]) *cn=(delLam/2./Lam)*en;
   else *cn=(delLam/2./Lam
           +(dF[n+1]-2.*dF[n]+dF[n-1])/(dF[n+1]-dF[n-1]))*en;
   *cn -=Prt*dY*(V[n]-n*dY*dF[n]/2.)/2.;
}
void SolveTurbHeatBL(int M,int N,double Pr,double Yinf,double Kappa,
                     double Ap,double ApH,double KdivKH,double *Nu,
                     double *Y99,double *Re,double *Cf) {

   int m,n;
   double Prt=0.9;
   extern double dF[_M_][_N_],V[_M_][_N_],T[_M_][_N_];
   double A[N],B[N],C[N],D[N],E[N],F[N];
   double dY=Yinf/(N-1);
   Nu[0]=sqrt(Re[0])*T[0][1]/dY;
   for (m=0;m<M-1;++m) { // loop over m, solve for T[m+1][n]
      double an,cn,dn;
      for (n=1;n<N-1;++n) {
         dn=dY*dY*Prt*(Re[m+1]*dF[m+1][n]+Re[m]*dF[m][n])/(Re[m+1]-Re[m]);
         HeatCoef(Pr,Prt,Kappa,Ap,ApH,Y99[m],
                  Cf[m],Re[m],dF[m],V[m],n,dY,&an,&cn);
         D[n]= -(an-cn)*T[m][n-1]
               +(2.*an-dn)*T[m][n]-(an+cn)*T[m][n+1];
         HeatCoef(Pr,Prt,Kappa,Ap,ApH,Y99[m+1],
                  Cf[m+1],Re[m+1],dF[m+1],V[m+1],n,dY,&an,&cn);
         A[n]=    an-cn;
         B[n]=-2.*an-dn;
         C[n]=    an+cn;
      }
      E[0]=F[0]=0.0; // solve system of equations for T[m+1][n]
      for (n=1;n<N-1;++n) {
         E[n]  = C[n]/( B[n] - E[n-1]*A[n] );
         F[n] = ( D[n] - F[n-1]*A[n] )/( B[n] - E[n-1]*A[n] );
      }
      T[m+1][N-1] = 1.0; // enforce boundary condition
      for (n=N-2;n>=0;--n)
         T[m+1][n] = F[n] - E[n]*T[m+1][n+1];

      Nu[m+1]=sqrt(Re[m+1])*T[m+1][1]/dY;
   }
}

void blasius(double Yinf,int N,double *f,double *df,double *ddf);
void blasiusT(double Yinf,double Pr,int N,double *f,double *T,double *dT);
void SolveTurbMomBL(int M,int N,double Yinf,double Kappa,double Ap,
                    double *Y99,double *Re,double *Cf);

double dF[_M_][_N_], V[_M_][_N_], T[_M_][_N_];

int main()
{
   int m,n,M=_M_,N=_N_;
   extern double dF[_M_][_N_],V[_M_][_N_],T[_M_][_N_];
   double Y99[M],Re[M],Cf[M],Nu[M],F[N],E[N];
   double Yinf=50.0,dY=Yinf/(N-1);
   double log_min=2.; // use log steps Re: 10^2-10^7
   double log_del=(7.-log_min)/(M-1);
   for (m=0;m<M;++m) Re[m]=pow(10.,log_min+m*log_del);

   double Pr=0.7,Ap=26.0,ApH=32.0,KdivKH=0.9,Kappa=0.4;
   blasius(Yinf,N,F,dF[0],E);
   blasiusT(Yinf,Pr,N,F,T[0],E);
   for (n=0;n<N;++n) V[0][n]=(n*dY*dF[0][n]-F[n])/2.;

   SolveTurbMomBL(M,N,Yinf,Kappa,Ap,Y99,Re,Cf);
   SolveTurbHeatBL(M,N,Pr,Yinf,Kappa,Ap,ApH,KdivKH,Nu,Y99,Re,Cf);
   FILE *fp=fopen("St.dat","w");
   for (m=1;m<M;++m)
      fprintf(fp,"%e %e %e %e\n",Re[m],Nu[m]/Pr/Re[m],
         0.0287*pow(Re[m],-0.2)/pow(Pr,0.4),
         Nu[0]*sqrt(Re[m])/sqrt(Re[0])/Pr/Re[m]);
   fclose(fp);
   return 1;
}
```

35.3 PROBLEMS

35-1 Suppose that the turbulent Prandtl number is not constant. Show that the turbulent heat equation can be written in the form of Eq. (35-53) with the coefficients evaluated as

$$a = \left(\frac{\mathrm{Pr}_t}{\mathrm{Pr}} + \frac{\varepsilon_M}{\nu}\right)$$

and

$$b = \left(\frac{2}{\Lambda}\frac{\partial \Lambda}{\partial \eta} + \frac{\partial^2 F'/\partial \eta^2}{\partial F'/\partial \eta} - \frac{\partial}{\partial \eta}\ln(\mathrm{Pr}_t)\right)\frac{\varepsilon_M}{\nu} - \mathrm{Pr}_t\left(V - \frac{\eta F'}{2}\right).$$

Using the variable turbulent Prandtl number expressed by

$$\mathrm{Pr}_t = \frac{\kappa(1-\exp(-y^+/A_\nu^+))}{\kappa_H(1-\exp(-y^+/A_\alpha^+))},$$

show that

$$\frac{\partial}{\partial \eta}\ln(\mathrm{Pr}_t) = \frac{f_x^+/A_\alpha^+}{1-\exp(\eta f_x^+/A_\alpha^+)} - \frac{f_x^+/A_\nu^+}{1-\exp(\eta f_x^+/A_\nu^+)}, \quad \text{where } f_x^+ = \sqrt{\mathrm{Re}_x c_f/2}.$$

Using the empirical constants $\kappa = 0.40$, $A_\nu^+ = 26.0$, $\gamma = 0.085$, $A_\alpha^+ = 32.0$, and $\kappa/\kappa_H = 0.9$, solve the turbulent boundary layer problem for a flow of air $\mathrm{Pr} = 0.7$ using the mixing length model. Find expressions for the wall variables

$$y^+ = \frac{\sqrt{\tau_w/\rho}}{\nu}y \quad \text{and} \quad T^+ = \frac{T_s - \overline{T}}{q_w/\rho C_p}\sqrt{\tau_w/\rho}$$

in terms of η, Pr, Re_x, c_f, Nu_x, and θ. Plot T^+ versus $\log(y^+)$ at the downstream locations where $\mathrm{Re}_x = 10^4, 10^5, 10^6$, and 10^7. Compare these temperature profiles with Kander's correlation (32-15).

REFERENCES

[1] F. Schultz-Grunow, "Neues Widerstandsgesetz für glatten Platten," *Luftfahrtforschung* **17**, 239 (1940).

[2] W. Kays and M. Crawford, *Convective Heat and Mass Transfer*, Second Edition. New York, NY: McGraw-Hill, 1987.

Chapter 36

The K-Epsilon Model of Turbulence

36.1 Turbulent Kinetic Energy Equation

36.2 Dissipation Equation for Turbulent Kinetic Energy

36.3 The Standard K-Epsilon Model

36.4 Problems

The mixing length model developed in Chapters 30 through 35 pursues the physical idea that the turbulent diffusivity is related to the scale of eddy motion in the flow and the magnitude of velocity fluctuations through the scaling $\varepsilon_M \sim \ell\, v'$. For turbulence in close proximity to the wall—that is, the inner region—the mixing length scales well with distance from the wall $\ell \sim y$ and the magnitude of velocity fluctuations scales well with the velocity gradient of the mean flow $v' \sim \ell \partial \overline{v}_x/\partial y$. However, further from the wall, in the outer region of turbulence, the physical scales employed in the mixing length model have diminished significance. A clear failure in the mixing length model arises when $\partial \overline{v}_x/\partial y$ goes to zero, since the model then suggests that v' and ε_M also go to zero. This is physically not the case in such instances as along the centerline of internal flows. This failure and other shortcomings of the mixing length model have led to contemplation of alternate physical scales that are important to characterizing the turbulent diffusivity far from the wall. The k-epsilon model discussed in this chapter, pursues the notion that the level of turbulent kinetic energy $\overline{k} = \overline{v_i' v_i'}/2$ is an important scale. Through this light, the turbulent diffusivity scaling $\varepsilon_M \sim \ell\, v'$ is no longer directly connected to the velocity gradient of the mean flow, since velocity fluctuations can be related to turbulent kinetic energy by

$$v' \sim \sqrt{\overline{k}}. \tag{36-1}$$

However, it remains to be demonstrated that the rate of dissipation of turbulent kinetic energy ε is another important scale, and that the mixing length scales as $\ell \sim \overline{k}^{3/2}/\overline{\varepsilon}$. Once this is established, the scaling $\varepsilon_M \sim \ell\, v'$ leads to $\varepsilon_M \sim \overline{k}^2/\overline{\varepsilon}$ for turbulent diffusivity in the k-epsilon model.

The k-epsilon model is implemented in most general purpose computational fluid dynamics codes. Although imperfect, the model offers a good compromise in terms of accuracy, robustness, and simplicity. A broader look at the mathematical models of turbulence can be found in references [1] and [2].

36.1 TURBULENT KINETIC ENERGY EQUATION

The level of turbulent kinetic energy in a flow can be derived through Reynolds decomposition of the steady incompressible momentum equation:

$$v_j \partial_j v_i = \nu \partial_j \partial_j v_i - \frac{1}{\rho} \partial_i P \tag{36-2}$$

$$(\overline{v}_j + v_j')\partial_j(\overline{v}_i + v_i') = \nu \partial_j \partial_j (\overline{v}_i + v_i') - \frac{1}{\rho}\partial_i(\overline{P} + P') \tag{36-3}$$

$$\overline{v}_j \partial_j \overline{v}_i + \partial_j(\overline{v_j' v_i'}) = \nu \partial_j \partial_j \overline{v}_i - \frac{1}{\rho}\partial_i \overline{P}. \tag{$\overline{36\text{-}3}$}$$

The governing equation for velocity fluctuations in the flow is derived by subtracting the time-averaged momentum equation ($\overline{36\text{-}3}$) from the total momentum equation (36-3):

$$(36\text{-}3) - \overline{(36\text{-}3)} = \overline{v}_j \partial_j v_i' + v_j' \partial_j \overline{v}_i + v_j' \partial_j v_i' - \partial_j(\overline{v_j' v_i'}) = \nu \partial_j \partial_j v_i' - \frac{1}{\rho}\partial_i P'. \tag{36-4}$$

This resulting equation for v_i' is dotted with v_i' to yield a scalar equation for turbulent kinetic energy:

$$v_i' \cdot (36\text{-}4) = \underbrace{v_i' \overline{v}_j \partial_j v_i'}_{*} + v_i' v_j' \partial_j \overline{v}_i + \underbrace{v_i' v_j' \partial_j v_i'}_{***} - v_i' \partial_j(\overline{v_j' v_i'}) = \underbrace{v_i' \nu \partial_j \partial_j v_i'}_{**} - v_i' \frac{1}{\rho}\partial_i P'.$$

The turbulent kinetic energy equation is simplified with the relations:

a. $\overbrace{v_i' \partial_j v_i'}^{*} = \partial_j \left(\frac{v_i' v_i'}{2} \right)$

b. $\partial_j \partial_j \left(\frac{v_i' v_i'}{2} \right) = \partial_j (v_i' \partial_j v_i') = \overbrace{v_i' \partial_j \partial_j v_i'}^{**} + \partial_j v_i' \partial_j v_i'$

c. $\overbrace{v_i' v_j' \partial_j v_i'}^{**} = v_j' \partial_j \left(\frac{v_i' v_i'}{2} \right) = \partial_j \left(v_j' \frac{v_i' v_i'}{2} \right)$ (by continuity)

resulting in

$$\begin{aligned} &\overline{v}_j \partial_j \left(\frac{v_i' v_i'}{2} \right) + v_i' v_j' \partial_j \overline{v}_i + \partial_j \left(v_j' \frac{v_i' v_i'}{2} \right) - v_i' \partial_j (\overline{v_j' v_i'}) \\ &\quad = \nu \left[\partial_j \partial_j \left(\frac{v_i' v_i'}{2} \right) - \partial_j v_i' \partial_j v_i' \right] - v_i' \partial_i \left(\frac{P'}{\rho} \right). \end{aligned} \tag{36-5}$$

The turbulent kinetic energy equation (36-5) is then time averaged for the result

$$\begin{aligned} &\overline{v}_j \partial_j \left(\frac{\overline{v_i' v_i'}}{2} \right) + \left(\overline{v_i' v_j'} \right) \partial_j \overline{v}_i + \partial_j \left(\overline{v_j' \frac{v_i' v_i'}{2}} \right) - v_i' \partial_j (\overline{v_j' v_i'}) \\ &\quad = \nu \partial_j \partial_j \left(\frac{\overline{v_i' v_i'}}{2} \right) - \nu \overline{\partial_j v_i' \partial_j v_i'} - \partial_i \left(\frac{\overline{v_i' P'}}{\rho} \right) \end{aligned}$$

or

$$\overline{v}_j \partial_j \left(\frac{\overline{v_i' v_i'}}{2} \right) = \partial_j \left[\nu \partial_j \left(\frac{\overline{v_i' v_i'}}{2} \right) - \overline{v_j' \left(\frac{v_i' v_i'}{2} + \frac{P'}{\rho} \right)} \right] + \left(-\overline{v_i' v_j'} \right) \partial_j \overline{v}_i - \nu \overline{\partial_j v_i' \partial_j v_i'}. \tag{$\overline{36\text{-}5}$}$$

Defining k to be the kinetic energy of turbulent fluctuations, such that

$$\bar{k} = \frac{\overline{v_i'v_i'}}{2} \quad \text{and} \quad k' = \frac{v_i'v_i'}{2}, \tag{36-6}$$

and defining the dissipation rate of turbulent kinetic energy $\bar{\varepsilon}$, such that

$$\bar{\varepsilon} = \nu \overline{\partial_j v_i' \partial_j v_i'}, \tag{36-7}$$

the transport equation for turbulent kinetic energy becomes

$$\underbrace{\bar{v}_j \partial_j \bar{k}}_{\substack{advection \\ of\ \bar{k}}} = \partial_j \Big[\underbrace{\nu \partial_j \bar{k}}_{\substack{molecular \\ diffusion\ of\ \bar{k}}} + \underbrace{-\overline{v_j'(k' + P'/\rho)}}_{\substack{turbulent \\ diffusion\ of\ \bar{k}}} \Big] + \underbrace{(-\overline{v_i'v_j'})\partial_j \bar{v}_i}_{source\ of\ \bar{k}} - \underbrace{\bar{\varepsilon}}_{sink\ of\ \bar{k}}. \tag{36-8}$$

The turbulent kinetic energy equation has been organized to emphasize the familiar structure of a transport equation having advection and diffusion terms, with additional source and sink terms. The role of pressure fluctuations can be lumped into the turbulent diffusion flux since kinetic energy is a mechanical form of energy that is transferred through forces, such as pressure. At this point, it is unclear that the term identified in Eq. (36-8) as a source of kinetic energy is as such. This will be demonstrated later.

The turbulent kinetic energy equations can be investigated further in the context of a boundary layer problem. Expanding in two dimensions and making the usual boundary layer approximations (see Section 25.1), Eq. (36-8) becomes

$$\bar{v}_x \frac{\partial \bar{k}}{\partial x} + \bar{v}_y \frac{\partial \bar{k}}{\partial y} = \frac{\partial}{\partial y}\left[\nu \frac{\partial \bar{k}}{\partial y} - \overline{v_y'\left(k' + \frac{P'}{\rho}\right)} \right] + \left(-\overline{v_x'v_y'}\right)\frac{\partial \bar{v}_x}{\partial y} - \bar{\varepsilon}. \tag{36-9}$$

The flux of kinetic energy resulting from small-scale turbulent advection can be represented mathematically as a diffusion flux. To this end, an eddy diffusivity ε_K for turbulent kinetic energy is defined in close analogy to other turbulent diffusivities, such that

$$\varepsilon_k \frac{\partial \bar{k}}{\partial y} = -\overline{v_y'\left(k' + \frac{P'}{\rho}\right)}. \tag{36-10}$$

Recalling that the turbulent momentum flux is $\overline{v_x'v_y'} = -\varepsilon_M \partial \bar{v}_x / \partial y$, the boundary layer transport equation (36-9) for turbulent kinetic energy can be written in the form

$$\bar{k} - eqn: \quad \underbrace{\bar{v}_x \frac{\partial \bar{k}}{\partial x} + \bar{v}_y \frac{\partial \bar{k}}{\partial y}}_{advection} = \underbrace{\frac{\partial}{\partial y}\left[(\nu + \varepsilon_k) \frac{\partial \bar{k}}{\partial y} \right]}_{diffusion} \underbrace{+ \varepsilon_M \left(\frac{\partial \bar{v}_x}{\partial y} \right)^2}_{source} - \underbrace{\bar{\varepsilon}}_{sink}. \tag{36-11}$$

The turbulent kinetic energy equation (36-11) provides useful insight into the relation between ε_M, $\bar{k}$, and $\bar{\varepsilon}$. Notice that the source term in the kinetic energy equation is equal to the product of the turbulent diffusivity ε_M and the square of the time-averaged velocity field gradient $(\partial \bar{v}_x / \partial y)^2$. Since this can only be positive, the identity of this term as a source of turbulent kinetic energy is revealed. The velocity gradient is largest at the wall, and declines with distance from the wall. However, the turbulent diffusivity, which is zero at the wall, increases, at least initially, with distance from the wall. Consequently, the source term, which results from the product of these two trends, is expected to go through a maximum at some distance not too far from the wall, presumably in the inner region of the turbulent boundary layer. It is instructive to exploit the success of the mixing length model for the inner region to provide a better understanding of the characteristics of turbulent kinetic energy.

36.1.1 Inner Region Scaling of the Turbulent Kinetic Energy

In the inner region of the boundary layer, the momentum flux toward the surface is dominated by turbulent diffusion, with negligible contributions from advection and molecular diffusion. In the absence of advection, the momentum diffusion flux is constant and equal to the value at the wall:

$$\varepsilon_M \frac{\partial \overline{v}_x}{\partial y} = \nu \frac{\partial \overline{v}_x}{\partial y}\bigg|_{y=0} \quad \text{or} \quad \frac{\partial \overline{v}_x}{\partial y} = \frac{\nu}{\varepsilon_M} \frac{\partial \overline{v}_x}{\partial y}\bigg|_{y=0}.$$

The latter expression for the velocity gradient can be substituted into the mixing length expression for turbulent diffusivity for the result

$$\varepsilon_M = \ell^2 \left| \frac{\partial \overline{v}_x}{\partial y} \right| \quad \rightarrow \quad \varepsilon_M = \ell \sqrt{\nu \frac{\partial \overline{v}_x}{\partial y}\bigg|_{y=0}}. \tag{36-12}$$

This indicates that the turbulent diffusivity is directly proportional to the mixing length in the inner region. However, since $\varepsilon_M \sim \ell v'$ and $v' \sim \sqrt{\overline{k}}$, the turbulent diffusivity is expected to scale as $\varepsilon_M \sim \ell \sqrt{\overline{k}}$. Combining this expectation with Eq. (36-12) suggests that the turbulent kinetic energy should scale as the momentum flux,

$$\overline{k} \sim \nu \frac{\partial \overline{v}_x}{\partial y}\bigg|_{y=0} = \varepsilon_M \frac{\partial \overline{v}_x}{\partial y}, \tag{36-13}$$

which is constant through the inner region. With the insight that $\partial \overline{k}/\partial y \sim 0$ in the inner region, the plateau of turbulent kinetic energy requires, in the absence of advection, that the remaining source and sink terms in Eq. (36-11) be in balance. From this balance, the following relation occurs for the inner region:

$$\overline{\varepsilon} \sim \varepsilon_M \left(\frac{\partial \overline{v}_x}{\partial y} \right)^2. \tag{36-14}$$

However, from Eq. (36-13) it is expected that $\partial \overline{v}_x/\partial y \sim \overline{k}/\varepsilon_M$ in the inner region. Applying this condition to Eq. (36-14) yields

$$\varepsilon_M \sim \frac{\overline{k}^2}{\overline{\varepsilon}}. \tag{36-15}$$

This scaling expectation describes an explicit relation between the turbulent momentum diffusivity, the turbulent kinetic energy, and the turbulent kinetic energy dissipation rate. Although this scaling result was deduced for the inner region of the boundary layer, the k-epsilon model presumes that this relation holds in the outer region of the boundary layer as well, and defines

$$\varepsilon_M = C_\mu \frac{\overline{k}^2}{\overline{\varepsilon}}, \tag{36-16}$$

where C_μ is an experimentally determined constant. With the argument that $\varepsilon_M \sim \ell\, v' \sim \ell\, \sqrt{\overline{k}}$, Eq. (36-15) also reveals that for the k-epsilon model

$$\ell \sim \frac{\overline{k}^{3/2}}{\overline{\varepsilon}}. \tag{36-17}$$

To evaluate the turbulent diffusivity from Eq. (36-16) requires quantification of both $\overline{k}$ and $\overline{\varepsilon}$ as a part of the turbulent flow solution. Equation (36-11) governs the description of $\overline{k} = \overline{v_i' v_i'}/2$, but a transport equation for the dissipation rate of turbulent kinetic energy $\overline{\varepsilon} = \nu \overline{\partial_j v_i' \partial_j v_i'}$ remains to be established.

36.2 DISSIPATION EQUATION FOR TURBULENT KINETIC ENERGY

The transport equation for the dissipation rate of turbulent kinetic energy (the $\overline{\varepsilon} - eqn.$) can be obtained with the following steps:

$$\partial_\ell\,(36\text{-}4) = (36\text{-}18) \tag{36-18}$$

$$(\nu \partial_\ell v_i') \cdot (36\text{-}18) = (36\text{-}19) \tag{36-19}$$

$$\overline{(36\text{-}19)} = \overline{\varepsilon} - eqn. \tag{$\overline{36\text{-}19}$}$$

Equation (36-18) describes the turbulent flow quantity $\partial_\ell v_i'$, where Eq. (36-4) is carried over from the derivation of the $\overline{k} - eqn.$ in Section 36.1. Equation (36-19) describes the instantaneous dissipation rate of turbulent kinetic energy, $\nu \partial_\ell v_i' \partial_\ell v_i'$, and the desired time-averaged dissipation rate equation expressed by $(\overline{36\text{-}19})$ can be organized into the following form for boundary layer problems:

$$\overline{v}_x \frac{\partial \overline{\varepsilon}}{\partial x} + \overline{v}_y \frac{\partial \overline{\varepsilon}}{\partial y} = \frac{\partial}{\partial y}\left[\nu \frac{\partial \overline{\varepsilon}}{\partial y} - \overline{turbulent\ flux\ of\ \varepsilon}\right] + (source) - (sink). \tag{36-20}$$

The turbulent diffusion flux in the $\overline{\varepsilon} - eqn.$ is used to define an eddy diffusivity such that

$$\overline{turbulent\ flux\ of\ \varepsilon} = -\varepsilon_\varepsilon \frac{\partial \overline{\varepsilon}}{\partial y}. \tag{36-21}$$

Closure of the k-epsilon model requires that all expressions of correlated turbulent fluctuations are related back to the time-averaged variables of the flow: $\overline{v}_i$, $\overline{k}$, and $\overline{\varepsilon}$. However, the formal derivation of the $\overline{\varepsilon} - eqn.$, as outlined above, leads to new expressions of correlated turbulent fluctuating variables in the *"source"* and *"sink"* terms. Therefore, to avoid defining new variables and failing to achieve closure, simplistic source and sink terms are deduced by the scaling arguments:

$$\overline{\varepsilon}_{source} \sim \frac{d}{dt}(\overline{k}_{source}) \sim \underbrace{\frac{\overline{\varepsilon}}{\overline{k}}}_{1/\tau_\varepsilon} \underbrace{\left(\varepsilon_M \left(\frac{\partial \overline{v}_x}{\partial y}\right)^2\right)}_{\overline{k}\ source} \tag{36-22}$$

$$\overline{\varepsilon}_{sink} \sim \frac{d}{dt}(\overline{k}_{sink}) \sim \underbrace{\frac{\overline{\varepsilon}}{\overline{k}}}_{1/\tau_\varepsilon} \underbrace{(\overline{\varepsilon})}_{\overline{k}\ sink}. \tag{36-23}$$

The time scale associated with the creation and dissipation of turbulent kinetic energy, $\tau_\varepsilon \sim \ell/v' \sim \overline{k}/\overline{\varepsilon}$, is deduced using Eq. (36-1) and Eq. (36-17). Applying Eqs. (36-21) through (36-23) to the dissipation equation (36-20) for boundary layers yields

$$\overline{\varepsilon} - eqn.: \quad \underbrace{\overline{v}_x \frac{\partial \overline{\varepsilon}}{\partial x} + \overline{v}_y \frac{\partial \overline{\varepsilon}}{\partial y}}_{advection} = \underbrace{\frac{\partial}{\partial y}\left[(\nu + \varepsilon_\varepsilon)\frac{\partial \overline{\varepsilon}}{\partial y}\right]}_{diffusion} + \underbrace{C_{1\varepsilon}\frac{\overline{\varepsilon}}{\overline{k}}\left[\varepsilon_M\left(\frac{\partial \overline{v}_x}{\partial y}\right)^2\right]}_{source} - \underbrace{C_{2\varepsilon}\frac{\overline{\varepsilon}^2}{\overline{k}}}_{sink}. \tag{36-24}$$

Empirically determined constants $C_{1\varepsilon}$ and $C_{2\varepsilon}$ have been added as coefficients to the source and sink terms, respectively, in the dissipation equation.

36.3 THE STANDARD K-EPSILON MODEL

The solutions to the $\overline{k} - eqn.$ (36-11) and the $\overline{\varepsilon} - eqn.$ (36-24) are coupled to the boundary layer momentum equation and incompressible continuity equation:

$$\overline{v}_x - eqn.: \quad \overline{v}_x \frac{\partial \overline{v}_x}{\partial x} + \overline{v}_y \frac{\partial \overline{v}_x}{\partial y} = \frac{\partial}{\partial y}\left[(\nu + \varepsilon_M)\frac{\partial \overline{v}_x}{\partial y}\right] \tag{36-25}$$

$$continuity: \quad \frac{\partial \overline{v}_x}{\partial x} + \frac{\partial \overline{v}_y}{\partial y} = 0. \tag{36-26}$$

The simplistic expressions for "*source*" and "*sink*" terms in the dissipation equation (36-24) are adequate for the inner and outer regions of the boundary layer, but turn out to be physically incompatible with the approach to the viscous sublayer. Consequently, the "standard" k-epsilon model only solves the $\overline{k}$ and $\overline{\varepsilon}$ equations for the fully turbulent regions of the boundary layer (outside of the viscous sublayer), where $\varepsilon_M/\nu \gg 1$. This model is also referred to as the high Reynolds number form of the k-epsilon model [3].

The mixing length model can be used to provide the near wall boundary values of $\overline{k}$ and $\overline{\varepsilon}$ leading to a smooth transition between the mixing length model and the k-epsilon model, as illustrated in Figure 36-1. This transition is possible at some distance $y = y_0$ from the wall within the inner region of the turbulent boundary layer. The point of transition must satisfy two requirements: (1) a balance between the source and sink terms in the $\overline{k} - eqn.$, and

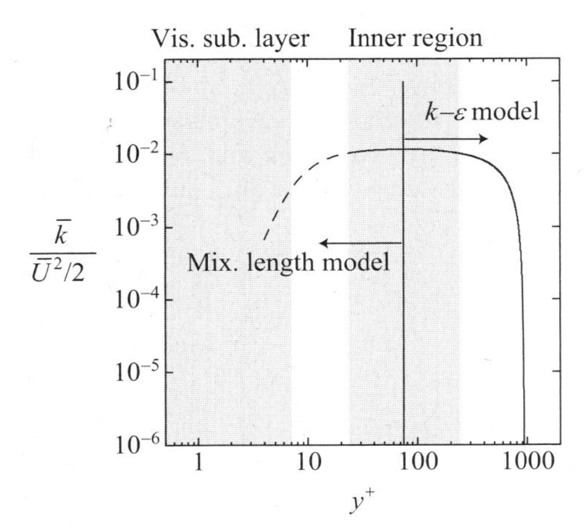

Figure 36-1 Turbulent kinetic energy near the wall.

(2) $\partial \bar{k}/\partial y = 0$. Both requirements are related to the expectation that minimal transport of $\bar{k}$ in the inner region causes the plateau in turbulent kinetic energy illustrated in Figure 36-1. The balance between source and sink terms in Eq. (36-11) dictates that

$$\left[\bar{\varepsilon} = \varepsilon_M \left(\frac{\partial \bar{v}_x}{\partial y}\right)^2\right]_{y=y_0} . \tag{36-27}$$

Using the k-epsilon model definition of turbulent diffusivity (36-16) to eliminate $\bar{\varepsilon}$ from Eq. (36-27), the turbulent kinetic energy can be expressed as

$$\left[\bar{k} = \frac{\varepsilon_M}{\sqrt{C_\mu}} \frac{\partial \bar{v}_x}{\partial y}\right]_{y=y_0} \quad \text{or} \quad \left[\bar{k} = \frac{1}{\sqrt{C_\mu}} \left(\kappa y \frac{\partial \bar{v}_x}{\partial y}\right)^2\right]_{y=y_0} . \tag{36-28}$$

The latter form makes use of the mixing length model definition of the turbulent diffusivity, in which κ is the von Kármán proportionality constant between the mixing length size and distance from the wall. Equation (36-28) is used to quantify the level of turbulent kinetic energy at $y = y_0$. Likewise, the turbulent kinetic energy dissipation rate at $y = y_0$ can be determined from Eq. (36-27) for the result

$$\left[\bar{\varepsilon} = \varepsilon_M \left(\frac{\partial \bar{v}_x}{\partial y}\right)^2 = (\kappa y)^2 \left(\frac{\partial \bar{v}_x}{\partial y}\right)^3\right]_{y=y_0} . \tag{36-29}$$

The location of $y = y_0$ from the wall is determined from the requirement that $\partial \bar{k}/\partial y = 0$. Using Eq. (36-28), this location is determined from

$$\left[\frac{\partial \bar{k}}{\partial y} = 0\right]_{y=y_0} \rightarrow \left[\frac{\partial}{\partial y}\left(\kappa y \frac{\partial \bar{v}_x}{\partial y}\right) = 0\right]_{y=y_0} . \tag{36-30}$$

Problem 36-2 illustrates that the condition (36-30) for y_0 can be evaluated for internal flows from the momentum equation in a straightforward manner. However, for boundary layer problems, the position y_0 changes as a function of downstream distance.

36.4 PROBLEMS

36-1 What is the weakest assumption in the mixing length model and how is it addressed in the k-epsilon model?

36-2 As was derived in Section 31.1, the momentum equation for turbulent Poiseuille flow between parallel plates separated by a distance a is

$$\left[1 + (\mathrm{Re}_D/2)\Lambda^2 u'\right]u' = (1 - 2\eta)u'(0)$$

where

$$\mathrm{Re}_D = \frac{\bar{v}_m(2a)}{\nu}, \quad \Lambda = \frac{\ell}{a}, \quad \eta = \frac{y}{a}, \quad \text{and} \quad u = \frac{\bar{v}_x}{\bar{v}_m}.$$

Demonstrate that a smooth transition between the mixing length model and k-epsilon model is possible at the distance from the wall where

$$\eta_0 \approx \frac{1}{\sqrt{\kappa\sqrt{2\,\mathrm{Re}_D\, u'(0)}}},$$

in which κ is the von Kármán proportionality constant between the mixing length size and distance from the wall.

36-3 Consider turbulent pipe flow. Using the definitions

$$u = \frac{\overline{v}_z}{v_m}, \quad \eta = 1 - \frac{r}{R}, \quad \text{and} \quad \mathrm{Re}_D = \frac{\overline{v}_m(2R)}{\nu},$$

demonstrate that a smooth transition between the mixing length model and k-epsilon model is possible at the distance from the wall, where

$$\eta_0 \approx \frac{1}{\sqrt{\kappa\sqrt{\mathrm{Re}_D\, u'(0)}}}.$$

Show that the corresponding boundary conditions for the k-epsilon model are

$$\overline{k}(\eta = \eta_0) = \frac{\overline{v}_m^2}{2}\frac{4(1-\eta_0)u'(0)}{\sqrt{C_\mu}\,\mathrm{Re}_D} \quad \text{and} \quad \overline{\varepsilon}(\eta = \eta_0) = \frac{\overline{v}_m{}^4/\nu}{\kappa\eta_0\,\mathrm{Re}_D/2}\left(\sqrt{\frac{(1-\eta_0)u'(0)}{\mathrm{Re}_D/2}}\right)^3.$$

REFERENCES

[1] B. E. Launder and D. B. Spalding, *Mathematical Models of Turbulence*. London, UK: Academic Press, 1972.
[2] C. J. Chen and S. Y. Jaw, *Fundamentals of Turbulence Modeling*. Washington, DC: Taylor & Francis, 1998.
[3] W. P. Jones and B. E. Lander, "The Prediction of Laminarization with a Two-Equation Model of Turbulence." *International Journal of Heat and Mass Transfer*, **15**, 301 (1972).

Chapter 37

The K-Epsilon Model Applied to Fully Developed Flows

37.1 K-Epsilon Model for Poiseuille Flow between Smooth Parallel Plates
37.2 Transition Point between Mixing Length and K-Epsilon Models
37.3 Solving the K and E Equations
37.4 Solution of the Momentum Equation with the K-Epsilon Model
37.5 Turbulent Diffusivity Approaching the Centerline of the Flow
37.6 Turbulent Heat Transfer with Constant Temperature Boundary
37.7 Problems

The principles behind the standard k-epsilon model were discussed in Chapter 36. In this chapter, application of the k-epsilon model is illustrated for fully developed turbulent internal flows, as was treated with the mixing length model in Chapter 31.

37.1 K-EPSILON MODEL FOR POISEUILLE FLOW BETWEEN SMOOTH PARALLEL PLATES

Consider Poiseuille flow between smooth parallel plates separated by a distance a, as illustrated in Figure 37-1. The flow is assumed to be fully developed, such that $\overline{v}_y = 0$ and $\partial\overline{v}_x/\partial x = 0$. As was derived in Section 31.1, the momentum equation for this flow takes the form of

$$\frac{\partial u}{\partial \eta} = \frac{(1-2\eta)u'(0)}{1+\varepsilon_M/\nu}, \tag{37-1}$$

where

$$\eta = \frac{y}{a} \quad \text{and} \quad u = \frac{\overline{v}_x}{\overline{v}_m} \tag{37-2}$$

are the dimensionless variables for position and velocity, respectively. The constant $u'(0) = \partial u/\partial\eta|_{\eta=0}$ in the momentum equation is the dimensionless velocity gradient at the

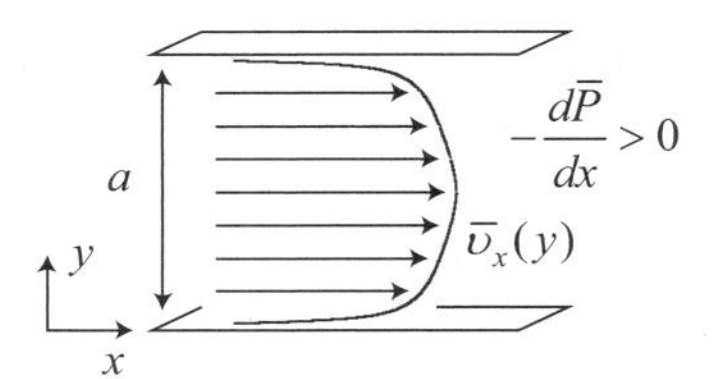

Figure 37-1 Turbulent Poiseuille flow between smooth parallel plates.

wall. For a given value of Re_D, the correct value of $u'(0)$ requires the solution to the momentum equation to satisfy the condition that

$$\int_0^1 u d\eta = 1. \tag{37-3}$$

In the k-epsilon model, turbulent diffusivity is evaluated from

$$\frac{\varepsilon_M}{\nu} = \frac{C_\mu}{\nu}\frac{\overline{k}^2}{\overline{\varepsilon}} = \frac{C_\mu}{4}\frac{K^2}{E}, \tag{37-4}$$

where K and E are introduced as dimensionless variables for the turbulent kinetic energy and the dissipation rate:

$$K = \frac{\overline{k}}{\overline{v}_m^2/2} \quad \text{and} \quad E = \frac{\overline{\varepsilon}}{\overline{v}_m^4/\nu}. \tag{37-5}$$

The transport equations for $\overline{k}$ and $\overline{\varepsilon}$ were developed in Chapter 36 for boundary layers. These equations are modified for internal flows by dropping the contributions of advection when fully developed conditions are considered, where $\overline{v}_y = 0$, $\partial\overline{k}/\partial x = 0$, and $\partial\overline{\varepsilon}/\partial x = 0$. The transport equations for $\overline{k}$ and $\overline{\varepsilon}$ can be made dimensionless with the definitions given in Eqs. (37-2) and (37-5). Subject to the standard k-epsilon model restriction that $\varepsilon_M/\nu \gg 1$ (see Section 36.3), the dimensionless governing equations become

$$u - eqn.: \quad \frac{\partial u}{\partial \eta} = \frac{(1-2\eta)u'(0)}{\varepsilon_M/\nu} \quad (\varepsilon_M/\nu \gg 1) \tag{37-6}$$

$$K - eqn.: \quad 0 = \frac{\partial}{\partial \eta}\left[\frac{\varepsilon_M}{\nu}\frac{\partial K}{\partial \eta}\right] + Sc_k\left(2\frac{\varepsilon_M}{\nu}\left(\frac{\partial u}{\partial \eta}\right)^2 - \frac{\mathrm{Re}_D^2}{2}E\right) \tag{37-7}$$

$$E - eqn.: \quad 0 = \frac{\partial}{\partial \eta}\left[\frac{\varepsilon_M}{\nu}\frac{\partial E}{\partial \eta}\right] + Sc_\varepsilon\left(2C_{1\varepsilon}\frac{E}{K}\frac{\varepsilon_M}{\nu}\left(\frac{\partial u}{\partial \eta}\right)^2 - \frac{1}{2}C_{2\varepsilon}\mathrm{Re}_D^2\frac{E^2}{K}\right) \tag{37-8}$$

where

$$\mathrm{Re}_D = \frac{\overline{v}_m(2a)}{\nu}, \quad Sc_k = \frac{\varepsilon_M}{\varepsilon_k}, \quad \text{and} \quad Sc_\varepsilon = \frac{\varepsilon_M}{\varepsilon_\varepsilon}. \tag{37-9}$$

Notice that the Reynolds number Re_D is based on the hydraulic diameter of the flow between the parallel plates. The turbulent Schmidt numbers Sc_k and Sc_ε are introduced to relate turbulent diffusivities back to the value for momentum transport ε_M. Using Eq. (37-4) for ε_M/ν and Eq. (37-6) for $\partial u/\partial\eta$, the K and E equations may be rewritten as

$$K - eqn.: \quad 0 = \frac{\partial}{\partial \eta}\left[\frac{C_\mu}{4}\frac{K^2}{E}\frac{\partial K}{\partial \eta}\right] + \frac{Sc_k}{2C_\mu}E\left[\left(\frac{4(1-2\eta)u'(0)}{K}\right)^2 - C_\mu \mathrm{Re}_D^2\right] \tag{37-10}$$

$$E-eqn.:\quad 0=\frac{\partial}{\partial\eta}\left[\frac{C_\mu}{4}\frac{K^2}{E}\frac{\partial E}{\partial\eta}\right]+\frac{Sc_\varepsilon}{2C_\mu}\frac{E^2}{K}\left[C_{1\varepsilon}\left(\frac{4(1-2\eta)u'(0)}{K}\right)^2-C_\mu C_{2\varepsilon}\mathrm{Re}_D^2\right]. \quad (37\text{-}11)$$

Notice that Re_D and $u'(0)$ are the only variables concerning momentum transport that the K and E equations depend on. Therefore, a solution to the K and E equations can be sought for given values of Re_D and $u'(0)$. However, the correct value of $u'(0)$ for a given value of Re_D cannot be determined without integrating the momentum equation, which relies on knowledge of the turbulent momentum diffusivity. In the k-epsilon model, the turbulent diffusivity is calculated from the results of the K and E transport equations.

The K and E variables can change by orders of magnitude in a turbulent flow. Consequently, there is some utility in changing the dependent variables to $\ln K$ and $\ln E$. This transformation allows K and E to approach zero in a way that excludes negative values from numerically occurring. Transforming the dependent variables to $\ln K$ and $\ln E$ yields the equations

$$\ln K-eqn.:\quad 0=\frac{\partial^2\ln K}{\partial\eta^2}+\left(3\frac{\partial\ln K}{\partial\eta}-\frac{\partial\ln E}{\partial\eta}\right)\frac{\partial\ln K}{\partial\eta}+S_K \quad (37\text{-}12)$$

$$\ln E-eqn.:\quad 0=\frac{\partial^2\ln E}{\partial\eta^2}+2\frac{\partial\ln K}{\partial\eta}\frac{\partial\ln E}{\partial\eta}+S_E \quad (37\text{-}13)$$

where the functions $S_K=S_K(Sc_k,C_{1k},C_{2k})$ and $S_E=S_E(Sc_\varepsilon,C_{1\varepsilon},C_{2\varepsilon})$ are defined for the index $j=K$ or E by

$$S_j(Sc_j,C_{1j},C_{2j})=\frac{Sc_jK}{8(\varepsilon_M/\nu)^2}\left[\overbrace{C_{1j}\left(\frac{4(1-2\eta)u'(0)}{K}\right)^2}^{source}-\overbrace{C_\mu C_{2j}\mathrm{Re}_D^2}^{dissipation}\right]. \quad (37\text{-}14)$$

Note that when $j=K$, $C_{1k}=C_{2k}=1$.

The function S_j represents the competing effects of the source and sink terms in the $\ln K$ and $\ln E$ equations. Notice that approaching the viscous sublayer, $K\to 0$, $\ln K\to -\infty$, and S_j becomes unbounded. Since the standard k-epsilon model is valid only when $\varepsilon_M/\nu \gg 1$, the law of the wall derived from the mixing length theory will be used to describe the near wall solution. However, some attention is needed to decide where the transition between the two models should occur.

37.2 TRANSITION POINT BETWEEN MIXING LENGTH AND K-EPSILON MODELS

As discussed in Section 36.3, the mixing length model and the k-epsilon model are both valid in the inner region of the turbulent boundary layer. A smooth transition between these models is possible within this region. The specific location of this transition, denoted by $\eta=\eta_0$, is suggested by two conditions. The first arises from the expectation that the level of turbulent kinetic energy reaches a maximum in the inner region:

$$\left[\frac{\partial K}{\partial\eta}=0\right]_{\eta=\eta_0}. \quad (37\text{-}15)$$

The second condition is the expectation that the plateau in turbulent kinetic energy through the inner region results in a balance between the source and dissipation terms of the K equation. From Eq. (37-7), this expectation yields

$$\left[2\frac{\varepsilon_M}{\nu}\left(\frac{\partial u}{\partial \eta}\right)^2 = \frac{\mathrm{Re}_D^2}{2}E\right]_{\eta=\eta_0} \quad \text{or} \quad \left[2\frac{\varepsilon_M}{\nu}\left(\frac{\partial u}{\partial \eta}\right)^2 = \frac{\mathrm{Re}_D^2}{2}\frac{C_\mu}{4}\frac{K^2}{\varepsilon_M/\nu}\right]_{\eta=\eta_0}, \tag{37-16}$$

where Eq. (37-4) is used to eliminate E in the second expression. The last result simplifies to the condition

$$\left[K = \frac{4}{\sqrt{C_\mu}}\frac{\varepsilon_M}{\nu}\frac{\partial u/\partial \eta}{\mathrm{Re}_D}\right]_{\eta=\eta_0}. \tag{37-17}$$

Using the mixing length evaluation of turbulent diffusivity (see Section 31.1),

$$\varepsilon_M/\nu = \ell^2\left|\frac{\partial \overline{v}_x}{\partial y}\right| = (\mathrm{Re}_D/2)\Lambda^2\left|\frac{\partial u}{\partial \eta}\right|, \tag{37-18}$$

where the dimensionless mixing length is defined by $\Lambda = \ell/a$, the level of turbulent kinetic energy at the transition point becomes

$$\left[K = \frac{2}{\sqrt{C_\mu}}(\Lambda u')^2\right]_{\eta=\eta_0}. \tag{37-19}$$

This result, combined with the requirement given by (37-15), dictates that the transition point must satisfy the relation

$$\left[\Lambda' u' + \Lambda u'' = 0\right]_{\eta=\eta_0}, \tag{37-20}$$

where the notation $()' = \partial()/\partial\eta$ and $()'' = \partial^2()/\partial\eta^2$ has been utilized. The momentum equation (37-1) written in the form

$$(1 - 2\eta)u'(0) = \left[1 + (\mathrm{Re}_D/2)\Lambda^2 u'\right]u' \tag{37-21}$$

can be differentiated:

$$-2u'(0) = \frac{\partial}{\partial \eta}\left[u' + (\mathrm{Re}_D/2)(\Lambda u')^2\right] = u'' + \mathrm{Re}_D(\Lambda u')[\Lambda' u' + \Lambda u''] \tag{37-22}$$

to show that, as a consequence of Eq. (37-20), the following relations hold at the transition point:

$$[u'' = -2u'(0)]_{\eta=\eta_0} \quad \text{or} \quad \left[\Lambda' u' = 2u'(0)\Lambda\right]_{\eta=\eta_0}. \tag{37-23}$$

Since $\left[\mathrm{Re}_D\Lambda^2 u'(0)\right]_{\eta=\eta_0} \gg 1$ and $\eta_0 \ll 1$, the momentum equation (37-21) can be evaluated for

$$\left[u' \approx \frac{1}{\Lambda}\sqrt{\frac{(1-2\eta)u'(0)}{(\mathrm{Re}_D/2)}} \approx \frac{1}{\Lambda}\sqrt{\frac{2u'(0)}{\mathrm{Re}_D}}\right]_{\eta=\eta_0}. \tag{37-24}$$

Therefore, Eq. (37-23) becomes

$$\left[\Lambda' = 2\Lambda^2\sqrt{u'(0)\mathrm{Re}_D/2}\right]_{\eta=\eta_0}. \tag{37-25}$$

With $\Lambda = \kappa\eta$, $\Lambda' = \kappa$, and $\eta = \eta_0$, this last expression can be solved for the transition point η_0:

$$\eta_0 = \frac{1}{\sqrt{\kappa\sqrt{2\mathrm{Re}_D u'(0)}}} \quad \text{corresponding to} \quad y_0^+ = \sqrt{\frac{\sqrt{u'(0)\mathrm{Re}_D/2}}{2\kappa}}. \tag{37-26}$$

At the transition point $\eta = \eta_0$, the turbulent kinetic energy is found by combining Eq. (37-19) and Eq. (37-24) for the result

$$K_0 = \frac{4(1-2\eta_0)u'(0)}{\sqrt{C_\mu}\mathrm{Re}_D} \quad \text{and} \quad (\ln K)_0 = \ln(K_0). \tag{37-27}$$

The turbulent diffusivity at the transition point is found from the mixing length model definition (37-18), which is evaluated with Eq. (37-24), yielding

$$\left.\frac{\varepsilon_M}{\nu}\right|_0 = \left[\frac{\mathrm{Re}_D}{2}\Lambda^2 u'\right]_{\eta=\eta_0} = \kappa\eta_0\sqrt{(1-2\eta_0)u'(0)\mathrm{Re}_D/2}. \tag{37-28}$$

The turbulent kinetic energy dissipation rate can be expressed in terms of the turbulent diffusivity using Eq. (37-4). Therefore, at the transition point,

$$E_0 = \frac{C_\mu K_0^2}{4(\varepsilon_M/\nu)\big|_0} \quad \text{and} \quad (\ln E)_0 = \ln(E_0), \tag{37-29}$$

where the turbulent diffusivity $(\varepsilon_M/\nu)\big|_0$ is given by Eq. (37-28).

37.3 SOLVING THE K AND E EQUATIONS

The near wall boundary conditions at $\eta = \eta_0$, required for integration of the $\ln K$ and $\ln E$ equations, are given by Eqs. (37-27) and (37-29). At the centerline of the flow, the boundary conditions are

$$\left.\frac{\partial K}{\partial \eta}\right|_{\eta=1/2} = \left.\frac{\partial \ln K}{\partial \eta}\right|_{\eta=1/2} = 0 \quad \text{and} \quad \left.\frac{\partial E}{\partial \eta}\right|_{\eta=1/2} = \left.\frac{\partial \ln E}{\partial \eta}\right|_{\eta=1/2} = 0. \tag{37-30}$$

To better resolve steep changes in the numerical solution near the wall, integration with respect to a logarithmic spatial variable is preferred. This is accomplished using $\chi = \ln\eta$ as the independent variable, such that the governing equations become

$$\ln K - eqn.: \quad 0 = \frac{\partial^2 \ln K}{\partial \chi^2} + \left(3\frac{\partial \ln K}{\partial \chi} - \frac{\partial \ln E}{\partial \chi} - 1\right)\frac{\partial \ln K}{\partial \chi} + e^{2\chi}S_K \tag{37-31}$$

$$\ln E - eqn.: \quad 0 = \frac{\partial^2 \ln E}{\partial \chi^2} + \left(2\frac{\partial \ln K}{\partial \chi} - 1\right)\frac{\partial \ln E}{\partial \chi} + e^{2\chi}S_E. \tag{37-32}$$

To simplify the presentation of the numerical solution, it is observed that the equations for $\ln K$ and $\ln E$ have a similar form. Again adopting the notation that $()' = \partial()/\partial\chi$ and $()'' = \partial^2()/\partial\chi^2$, the $\ln K$ and $\ln E$ equations can both be expressed as

$$0 = \phi'' + f_\phi \phi' + S_\phi, \tag{37-33a}$$

where

$$\begin{cases} \text{if } \phi = \ln K: & f_{\ln K} = 3(\ln K)' - (\ln E)' - 1, \quad S_{\ln K} = e^{2\chi} S_K \\ \text{if } \phi = \ln E: & f_{\ln E} = 2(\ln K)' - 1, \quad S_{\ln E} = e^{2\chi} S_E \end{cases}. \tag{37-33b}$$

When $\phi = \ln K$, Eq. (37-33) becomes identical to Eq. (37-31) for the $\ln K$, and when $\phi = \ln E$, Eq. (37-33) becomes identical to Eq. (37-32) for the $\ln E$. It should be noted that although the independent variable is now $\chi = \ln \eta$, the constant $u'(0)$ appearing in S_j retains the definition $u'(0) = \partial u/\partial \eta|_{\eta=0}$.

Numerical integration of the $\ln K$ and $\ln E$ equations is challenging because of the highly nonlinear algebraic relation describing the competition between the source and sink terms in the S_j function given by Eq. (37-14). This situation influences the treatment of boundary conditions when solving the governing equations. In principle, the derivatives $\ln K'$ and $\ln E'$ at $\eta = \eta_0$ could be guessed by shooting methods to satisfy the boundary conditions at the centerline of the flow. However, because of the highly nonlinear nature of the S_j function, $\ln K'$ and $\ln E'$ at $\eta = \eta_0$ have vanishing influence on the flow conditions at the centerline. This is unfavorable for Runge-Kutta type integration that requires that all initial conditions be specified at one boundary. Consequently, finite differencing methods are better suited for solving the $\ln K$ and $\ln E$ equations, since spatially separated boundary conditions pose no special difficulty in this approach. The only drawback is that a system of equations must be solved simultaneously. Because Eq. (37-33) for $\ln K$ and $\ln E$ is nonlinear, an iterative approach to solving this system of equations is pursued.

Let the residual associated with the function $\phi(\chi)$ be defined by

$$r_\phi = \phi'' + f_\phi \phi' + S_\phi. \tag{37-34}$$

The function $\phi(\chi)$ can be forced toward the solution of Eq. (37-33) by making the residual r_ϕ approach zero with each iteration. Suppose a pseudo time variable τ is introduced to integrate the residual r_ϕ toward zero. When $r_\phi(\chi) > 0$, the local value of $\phi(\chi)$ must increase with τ to reduce the residual:

$$\frac{\partial \phi}{\partial \tau} = r_\phi. \tag{37-35}$$

The amount by which ϕ changes with each integration step can be controlled by the step size in τ. The amount by which ϕ changes can also be weighted by any spatial function of χ, such that $\partial\phi/\partial\tau$ in Eq. (37-35) can be replaced with $h(\chi)(\partial\phi/\partial\tau)$. The specific nature of $h(\chi)$ is unimportant to the observation that $\phi(\chi)$ will stop changing with forward integration when the solution $r_\phi(\chi) = 0$ is attained. However, a judicious choice of $h(\chi)$ can improve the rate at which $\phi(\chi)$ approaches the solution. The result of Problem 37-1 is used to suggest a weighting function that leads to the time marching equation:

$$\frac{e^{2\chi}}{8(\varepsilon_M/\nu)} \frac{\partial \phi}{\partial \tau} = r_\phi. \tag{37-36}$$

Equation (37-36) can be discretized for numerical integration using the finite differencing scheme illustrated in Figure 37-2:

$$\frac{e^{2\chi}}{8(\varepsilon_M/\nu)^\tau} \frac{\phi^{\tau+\Delta\tau} - \phi^\tau}{\Delta\tau} = \frac{r_\phi^{\tau+\Delta\tau} + r_\phi^\tau}{2}. \tag{37-37}$$

The integration step $\Delta\tau$ should be suitably small to limit the magnitude of change in $\phi(\chi)$ with each time step. Equation (37-36) suggests that $\Delta\tau$ should be selected to ensure

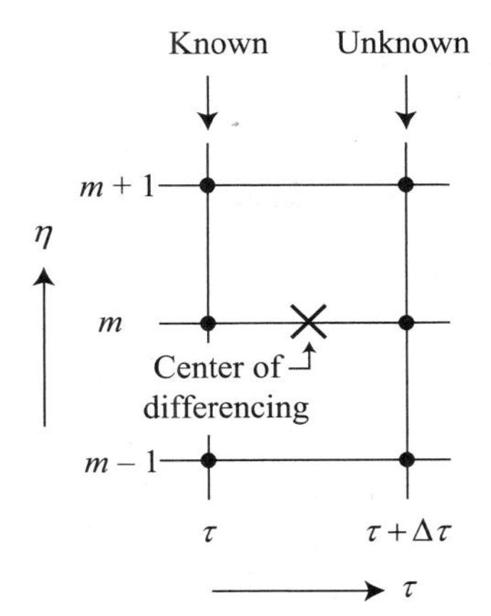

Figure 37-2 Center differencing scheme.

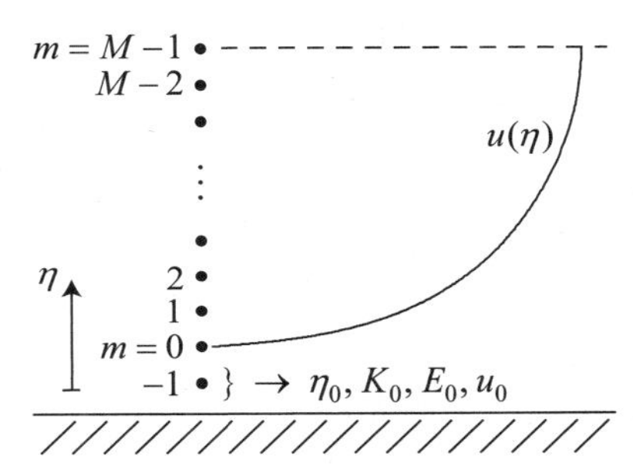

Figure 37-3 Spatial discretization of flow field.

$$\Delta\tau \ll \frac{e^{2\chi}}{8(\varepsilon_M/\nu)^\tau} \frac{\phi^\tau}{r_\phi^\tau} \tag{37-38}$$

for all values of χ.

Using Eq. (37-34) to evaluate the residual $r_\phi^{\tau+\Delta\tau}$, the time marching equation (37-37) can be reorganized into the form

$$\begin{aligned} &\phi^{\tau+\Delta\tau} - 4(\varepsilon_M/\nu)^\tau e^{-2\chi}\Delta\tau\left[(\phi'')^{\tau+\Delta\tau} + f_\phi^{\tau+\Delta\tau}(\phi')^{\tau+\Delta\tau}\right] \\ &\quad = \phi^\tau + 4(\varepsilon_M/\nu)^\tau e^{-2\chi}\Delta\tau\left(r_\phi^\tau + S_\phi^{\tau+\Delta\tau}\right). \end{aligned} \tag{37-39}$$

This is a nonlinear equation for $\phi^{\tau+\Delta\tau}$ by virtue of the ${f_\phi}^{\tau+\Delta\tau}$ and $S_\phi^{\tau+\Delta\tau}$ terms. Therefore, Eq. (37-39) is solved by an iterative approach, where with each iteration ${f_\phi}^{\tau+\Delta\tau}$ and $S_\phi^{\tau+\Delta\tau}$ are evaluated from the results of the previous iteration. This process is repeated until ${f_\phi}^{\tau+\Delta\tau}$ and $S_\phi^{\tau+\Delta\tau}$ are described correctly in the solution to Eq. (37-39).

Numerical integration of the governing equations is performed on the discretized space shown in Figure 37-3. The centerline of the flow occurs at the node $m = M - 1$. Notice that the node $m = -1$ corresponds to the solution at the transition node $\eta = \eta_0$, where K_0 and E_0 are given by Eqs. (37-27) and (37-29). Using the finite differencing expressions

$$\phi'' = \frac{\phi[m-1] - 2\phi[m] + \phi[m+1]}{\Delta\chi^2} \quad \text{and} \quad \phi' = \frac{\phi[m+1] - \phi[m-1]}{2\Delta\chi}, \tag{37-40}$$

the discretized time marching equation (37-39) can be expressed for all nodes, excluding the centerline, as

$$A_\phi \phi^{\tau+\Delta\tau}[m-1] + B_\phi \phi^{\tau+\Delta\tau}[m] + C_\phi \phi^{\tau+\Delta\tau}[m+1]$$
$$= \phi^\tau[m] + 4(\varepsilon_M/\nu)^\tau e^{-2\chi}\Delta\tau\left(r_\phi^\tau[m] + S_\phi^{\tau+\Delta\tau}[m]\right). \tag{37-41}$$

For nodes other than at the centerline, the coefficients are given by

$$A_\phi = +(\varepsilon_M/\nu)^\tau e^{-2\chi}\frac{\Delta\tau}{\Delta\chi^2}\left(g_\phi^{\tau+\Delta\tau}[m] - 2\Delta\chi - 4\right),$$
$$B_\phi = \left(1 + 8(\varepsilon_M/\nu)^\tau e^{-2\chi}\frac{\Delta\tau}{\Delta\chi^2}\right), \tag{37-42}$$
$$C_\phi = -(\varepsilon_M/\nu)^\tau e^{-2\chi}\frac{\Delta\tau}{\Delta\chi^2}\left(g_\phi^{\tau+\Delta\tau}[m] - 2\Delta\chi + 4\right)$$

where

$$g_\phi[m] = \begin{cases} 3(\ln K[m+1] - \ln K[m-1]) - (\ln E[m+1] - \ln E[m-1]) & (\text{for } \phi = \ln K) \\ 2(\ln K[m+1] - \ln K[m-1]) & (\text{for } \phi = \ln E). \end{cases} \tag{37-43}$$

The residual $r_\phi^\tau[m]$ in Eq. (37-41) is evaluated from the discretized form of Eq. (37-34). For nodes other than at the centerline, the residual is

$$r_\phi[m] = S_\phi[m] +$$
$$\frac{1}{\Delta\chi^2}\left(\phi[m-1] - 2\phi[m] + \phi[m+1] + \frac{1}{4}(g_\phi[m] - 2\Delta\chi)(\phi[m+1] - \phi[m-1])\right). \tag{37-44}$$

At the centerline of the flow, where $\phi' = 0$, the discretized time marching equation (37-41) is evaluated slightly differently. The finite difference representation of ϕ'' at the centerline node is given by

$$m = M-1: \quad \phi'' = 2\frac{\phi[m-1] - \phi[m]}{\Delta\chi^2}. \tag{37-45}$$

This changes the coefficients for the discretized time marching equation (37-41), which become

$$A_\phi = +\left(-8(\varepsilon_M/\nu)^\tau e^{-2\chi}\frac{\Delta\tau}{\Delta\eta^2}\right), \quad B_\phi = \left(1 + 8(\varepsilon_M/\nu)^\tau e^{-2\chi}\frac{\Delta\tau}{\Delta\eta^2}\right), \quad C_\phi = 0. \tag{37-46}$$

At the centerline, the residual $r_\phi^\tau[m]$ in Eq. (37-41) is evaluated as

$$m = M-1: \quad r_\phi^\tau[m = M-1] = \frac{2}{\Delta\eta^2}(\phi^\tau[m-1] - \phi^\tau[m]) + S_\phi^\tau[m]. \tag{37-47}$$

Equation (37-41) represents a tridiagonal system of equations that can be solved for $\phi^{\tau+\Delta\tau}$ using the Thomas algorithm discussed in Section 35.1.3. Once $\phi^{\tau+\Delta\tau}$ is known, the solution for $\phi(\chi)$ is integrated another step forward after indexing time. Forward integration continues until the condition $r_\phi^{\tau+\Delta\tau} = 0$ defined by the solution is satisfied. This condition can be inferred by monitoring changes in the centerline value of (ε_M/ν), as it tends to converge slowly. Integration of the $\ln K$ and $\ln E$ equations is performed for specified values of Re_D and $u'(0)$, which are related to each other through the solution to the momentum equation.

37.4 SOLUTION OF THE MOMENTUM EQUATION WITH THE K-EPSILON MODEL

The momentum equation (37-6) can be expressed in terms of the new independent variable $\chi = \ln \eta$ for the result

$$\frac{\partial u}{d\chi} = e^{\chi}\frac{(1 - 2e^{\chi})u'(0)}{\varepsilon_M/\nu}. \tag{37-48}$$

The momentum equation can be integrated for u using the usual Runge-Kutta method (see Chapter 23). However, integration of the momentum equation requires knowledge of $u'(0) = \partial u/\partial\eta|_{\eta=0}$ and the function (ε_M/ν). The turbulent diffusivity is found from Eq. (37-4) using the solutions to the $\ln K$ and $\ln E$ equations. However, a guess at $u'(0)$ is required to solve the K and E equations. The correctness of $u'(0)$ is evaluated by integrating the velocity solution obtained from the momentum equation to check the requirement that

$$\int_0^{1/2} u\,d\eta = \frac{1}{2}. \tag{37-49}$$

However, the standard k-epsilon model can only be solved over the range $\eta_0 \leq \eta \leq 1/2$. The initial part of the integration $(0 \leq \eta \leq \eta_0)$ could be carried out with the mixing length model. However, since the distance to η_0 is quite small, it is satisfactory to evaluate the initial part of the integration with

$$\int_0^{\eta_0} u d\eta \approx \frac{u_0\eta_0}{2}. \tag{37-50}$$

The value for u_0 is obtained from the law of the wall result (see Section 30.6.2):

$$u_0 = \sqrt{2u'(0)/\mathrm{Re}_D}\left(\frac{1}{\kappa}\ln(y_0^+/E_\nu^+) + E_\nu^+\right), \tag{37-51}$$

in which y_0^+ is evaluated with Eq. (37-26). Expressed in terms of the new independent variable $\chi = \ln \eta$, Eq. (37-49) becomes

$$\int_0^{1/2} u d\eta = \frac{u_0\eta_0}{2} + \int_{\ln(\eta_0)}^{\ln(1/2)} u e^{\chi} d\chi = 1/2. \tag{37-52}$$

If evaluation of Eq. (37-52) yields a result greater than 1/2, the value of $u'(0)$ is too high; if the result is less than 1/2, the value of $u'(0)$ is too low. This comparison allows the guess for $u'(0)$ to be refined, after which the process of solving the $\ln K$ and $\ln E$ equations to determine (ε_M/ν) is repeated. Once Eq. (37-52) is satisfied, the correct value of $u'(0)$ has been determined and the k-epsilon model of turbulence has been solved.

Code 37-1 implements the k-epsilon model solution for turbulent Poiseuille flow between smooth parallel plates. The model uses the coefficients $\kappa = 0.4$, $E_\nu = 11.6$, $C_\mu = 0.09$, $Sc_k = 1.0$, $C_{1\varepsilon} = 1.44$, $C_{2\varepsilon} = 1.92$, and $Sc_\varepsilon = 1.3$, which are commonly used values in the literature. The coefficients for the k-epsilon model transport equations are those suggested by Launder and Sharma [1].

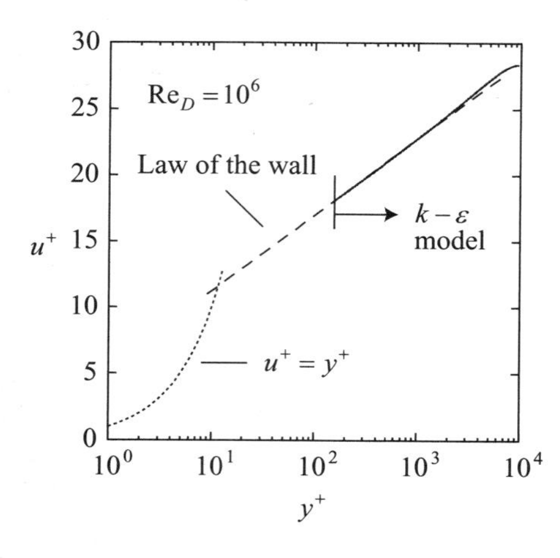

Figure 37-4 Law of wall behavior for k-epsilon model.

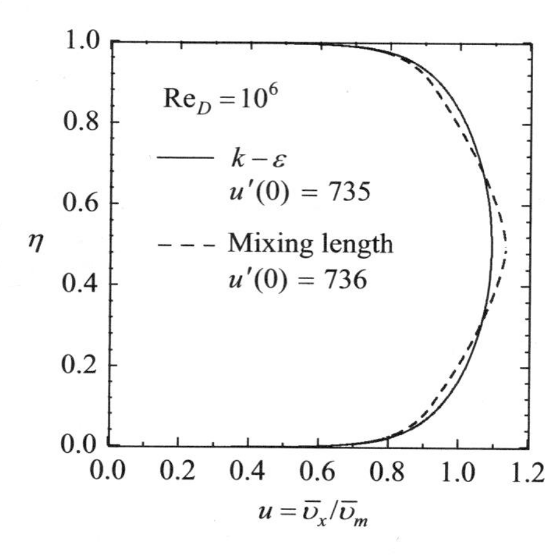

Figure 37-5 Comparison of velocity profiles between k-epsilon and mixing length models.

Figure 37-4 checks the law of the wall behavior of the solution, for $\text{Re}_D = 10^6$. Since the k-epsilon model commences in the inner region of turbulence, the solution should initially conform to the law of the wall, as it does. The expected departure from the law of the wall is seen in the outer region of turbulence.

Figure 37-5 contrasts the velocity solution of the k-epsilon model with that of the mixing length model. The normalized velocity $u = \bar{v}_x/\bar{v}_m$ is plotted as a function of distance $\eta = y/a$ across the channel. The k-epsilon model yields a smoother profile than the mixing length model, which exhibits a characteristically pointed profile near the centerline. Despite differences in the velocity profile, both models yield nearly identical values of $u'(0)$, which is related to the friction coefficient by

$$c_f = \frac{\tau_w}{\rho \bar{v}_m^2/2} = 2\frac{\nu}{\bar{v}_m R} u'(0) = \frac{4u'(0)}{\text{Re}_D}. \tag{37-53}$$

37.5 TURBULENT DIFFUSIVITY APPROACHING THE CENTERLINE OF THE FLOW

One failure of the mixing length model was the vanishing turbulent diffusivity approaching the centerline of the flow, where $u' = 0$. Figure 37-6 contrasts the performance of the k-epsilon with the mixing length model and the experimental measurements of Hussain and Reynolds [2] for channel flow. The normalized turbulent diffusivity

Code 37-1 Turbulent k-ε model for Poiseuille flow between smooth parallel flat plates

```
#include <stdio.h>
#include <stdlib.h>
#include <math.h>

inline double phiSource(double Sc,double Cu,double C1,double C2,double
                    ReD,double du0,double lnEta,double lnK,double lnE)
{
   double v1=4.*(1.-2.*exp(lnEta))*du0/exp(lnK);
   double Em=Cu*exp(2.*lnK)/(4.*exp(lnE));
   return exp(2.*lnEta)*(C1*v1*v1-Cu*C2*ReD*ReD)*Sc*exp(lnK)/8./Em/Em;
}

inline double du_ke(double ReD,double du0,double lnEta,double Em)
{  return exp(lnEta)*(1.-2.*exp(lnEta))*du0/Em;  }

inline double g_k(int n,double *lnK,double *lnE)
{  return 3.*(lnK[n+1]-lnK[n-1])-(lnE[n+1]-lnE[n-1]); }

inline double g_e(int n,double *lnK,double *lnE)
{  return 2.*(lnK[n+1]-lnK[n-1]);  }

void phiResid(double (*func)(int,double *,double *),double Sc,double Cu,
              double C1,double C2,double ReD,double du0,double *var,
              double *resid,int M,double *lnEta,double *lnK,double *lnE)
{
   int m;
   double del_lnEta=(lnEta[M-1]-lnEta[0])/(M-1);
   double del_lnEta2=del_lnEta*del_lnEta;
   for (m=0;m<M-1;++m) {
      resid[m] = ( 0.25*(func(m,lnK,lnE)-2.*del_lnEta)*(var[m+1]-var[m-1])
              +   var[m-1]-2.*var[m]+var[m+1] )/del_lnEta2;
      resid[m]+= phiSource(Sc,Cu,C1,C2,ReD,du0,lnEta[m],lnK[m],lnE[m]);
   }
   resid[m]=2.*(var[m-1]-var[m])/del_lnEta2
           +phiSource(Sc,Cu,C1,C2,ReD,du0,lnEta[m],lnK[m],lnE[m]);
}

void solve_phi(double (*func)(int,double *,double *),double Sc,double Cu,
               double C1,double C2,double ReD,double du0,double *Em,
               double *var,double *_D,int *not_converged,double dt,
               int M,double *lnEta,double *lnK,double *lnE)
{
   int m;
   double wght,A[M],B[M],C[M],D[M],GAM[M];
   double del_lnEta=(lnEta[M-1]-lnEta[0])/(M-1);
   double del_lnEta2=del_lnEta*del_lnEta;
   for (m=0;m<M;m++) {
      wght=Em[m]*exp(-2.*lnEta[m])*dt/del_lnEta2;
      if (m<M-1) A[m]=C[m]=func(m,lnK,lnE)-2.*del_lnEta;
      else A[m]=C[m]= -4.;
      A[m]= +wght*(A[m]-4.);
      C[m]= -wght*(C[m]+4.);
      B[m]=  1.+8.*wght;
      D[m]= _D[m];
   }
   D[0]-=A[0]*var[-1];
   A[0]= 0.;

   double BET=B[0];              // Solve tri-diagonal system of equations
   D[0]/=BET;
   for (m=1; m<M; m++)  {        // Decomposition and forward substitution
      GAM[m]=C[m-1]/BET;
      BET=B[m]-A[m]*GAM[m];
      D[m]=(D[m]-A[m]*D[m-1])/BET;
   }
   var[M-1]=D[M-1];
   for (m=M-2; m>=0; m--) {      // Back substitution
      D[m]-=GAM[m+1]*D[m+1];
      if (fabs((D[m]-var[m])/var[m]) > 1.e-5) *not_converged=1;
      var[m]=D[m];
   }
}

void solve_KE(double ReD,double du0,double Cu,double Sck,double C1,
double C2,double Sce,int M,double *lnEta,double *lnK,double *lnE,double *Em)
{
   int m,not_converged,iter=0;
   double DK[M],DE[M],rlnK[M],rlnE[M],last_EmCent=0.;
   Em[-1]=Cu*exp(2.*lnK[-1])/(4.*exp(lnE[-1]));

   do {
      phiResid(&g_k,Sck,Cu,1.,1.,ReD,du0,lnK,rlnK,M,lnEta,lnK,lnE);
      phiResid(&g_e,Sce,Cu,C1,C2,ReD,du0,lnE,rlnE,M,lnEta,lnK,lnE);

      double dt,dt_max=1.; // size step forward ...
      for (m=0;m<M;++m) {
         Em[m]=Cu*exp(2.*lnK[m])/(4.*exp(lnE[m]));
         if (isnan(Em[m])) { printf("\nSoln. blew!"); exit(1);}
         dt=fabs(exp(2.*lnEta[m])*lnE[m]/8./Em[m]/rlnE[m]);
         if (dt<dt_max) dt_max=dt;
         dt=fabs(exp(2.*lnEta[m])*lnK[m]/8./Em[m]/rlnK[m]);
         if (dt<dt_max) dt_max=dt;
      }
      dt=dt_max/10;
      if (fabs(Em[M-1]-last_EmCent)/Em[M-1] < .001) break;
      last_EmCent=Em[M-1];

      for (m=0;m<M;m++) {
         DK[m]=lnK[m]+4.*Em[m]*exp(-2.*lnEta[m])*dt*( rlnK[m] +
            phiSource(Sck,Cu,1.,1.,ReD,du0,lnEta[m],lnK[m],lnE[m]) );
         DE[m]=lnE[m]+4.*Em[m]*exp(-2.*lnEta[m])*dt*( rlnE[m] +
            phiSource(Sce,Cu,C1,C2,ReD,du0,lnEta[m],lnK[m],lnE[m]) );
      }
```

```
      int inside_iter=0;
      do {
         not_converged=0;
         solve_phi(&g_k,Sck,Cu,1.,1.,ReD,du0,
                   Em,lnK,DK,&not_converged,dt,M,lnEta,lnK,lnE);
         solve_phi(&g_e,Sce,Cu,C1,C2,ReD,du0,
                   Em,lnE,DE,&not_converged,dt,M,lnEta,lnK,lnE);
      } while (not_converged && ++inside_iter<100);
   } while (++iter<5000);
}

double solve_u(double ReD,int M,double *lnEta,double *u,double *Em)
{
   int m,iter=0;
   double Kappa=0.4,Ev=11.6,Cu=0.09,Sck=1.,C1=1.44,C2=1.92,Sce=1.3;
   double lnEta_half,Em_half,K1u,K2u,K3u,K4u,lnK[M],lnE[M];
   double du0,um,K1um,K2um,K3um,K4um;
   double du0_low=1.;                // lower bound on du0
   double du0_high=5.0e5;            // upper bound on du0

   do {
      du0=(du0_low+du0_high)/2.0;    // use bisection method

      double yp0=sqrt(sqrt(ReD*du0/2.)/2./Kappa);
      double eta0=yp0/sqrt(du0*ReD/2.);
      double k0=2.*(1.-2.*eta0)*du0/(ReD/2.)/sqrt(Cu);
      double lnK0=log(k0);
      double Em0=Kappa*eta0*sqrt(ReD*(1.-2.*eta0)*du0/2.);
      double lnE0=log(Cu*k0*k0/4./Em0);
      double lnEta0=log(eta0);
      double del_lnEta=(log(0.5)-lnEta0)/(M-1);

      for (m=0;m<M;++m) {
         lnEta[m]=lnEta0+m*del_lnEta; // uniform grid
         lnK[m]=lnK0;
         lnE[m]=lnE0;
      }
      solve_KE(ReD,du0,Cu,Sck,C1,C2,Sce,M-1,lnEta+1,lnK+1,lnE+1,Em+1);

      u[0]=sqrt(2.*du0/ReD)*(log(yp0/Ev)/Kappa+Ev);
      um=u[0]*eta0/2.;
      for (m=0;m<M-1;++m) {          // integrate u
         lnEta_half=(lnEta[m]+lnEta[m+1])/2.;
         Em_half=(Em[m]+Em[m+1])/2.;
         K1u=del_lnEta*du_ke(ReD,du0,lnEta[m],Em[m]);
         K2u=K3u=del_lnEta*du_ke(ReD,du0,lnEta_half,Em_half);
         K4u=del_lnEta*du_ke(ReD,du0,lnEta[m+1],Em[m+1]);
         K1um= del_lnEta*u[m]*exp(lnEta[m]);
         K2um= del_lnEta*(u[m]+0.5*K1u)*exp(lnEta_half);
         K3um= del_lnEta*(u[m]+0.5*K2u)*exp(lnEta_half);
         K4um= del_lnEta*(u[m]+K3u)*exp(lnEta[m+1]);
         u[m+1]=u[m]+(K1u+2.*K2u+2.*K3u+K4u)/6.;
         um+=(K1um+2.*K2um+2.*K3um+K4um)/6.;
      }
      printf("eta0=%e du0=%e um=%e\n",exp(lnEta0),du0,um);
      if (fabs(um-0.5) < .0001) break;
      if (um>0.5) du0_high=du0;  // shoot lower
      else du0_low=du0;              // shoot higher
   } while (++iter < 200 && (du0_high-du0_low)/du0 > 1.0e-4);

   if (fabs(um-0.5)>.001) {
      printf("\nSoln. failed, du0=%e\n",du0);
   }
   return du0;
}

int main()
{
   int m,M=500;
   double du0,ReD;
   double lnEta[M],u[M],Em[M];   // non-dimensional variables

   ReD=1.e6;
   du0=solve_u(ReD,M,lnEta,u,Em);

   FILE *fp;
   fp=fopen("out.dat","w");
   for (m=0;m<M;m++)
      fprintf(fp,"%e %e %e\n",exp(lnEta[m]),u[m],Em[m]);
   fclose(fp);

   return 0;
}
```

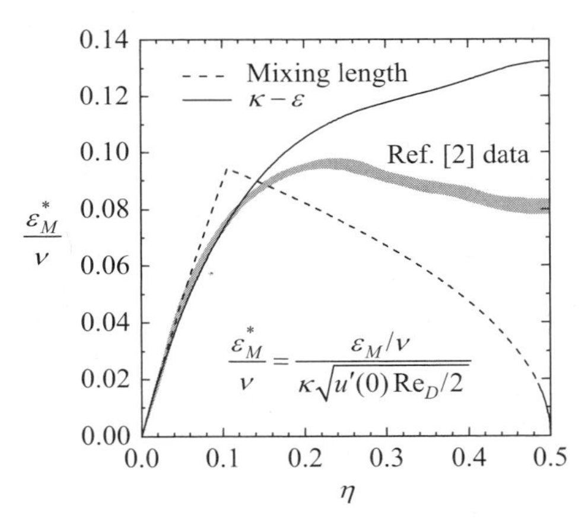

Figure 37-6 A comparison of calculated turbulent diffusivity with experimental data of Ref. [2].

$$\frac{\varepsilon_M^*}{\nu}=\frac{\varepsilon_M/\nu}{\kappa\sqrt{u'(0)\mathrm{Re}_D/2}}, \tag{37-54}$$

plotted in Figure 37-6, is insensitive to the Reynolds number of the flow. Compared with the experimental measurements, it is seen that the k-epsilon model over-predicts the centerline turbulent diffusivity by about 65% for the current set of constants used in the model. As was the case for the mixing length model, the impact of this failure is minimized by the small velocity gradients (and momentum transport) in the time-averaged flow near the centerline. Unlike the mixing length model, the k-epsilon model constants can be manipulated to yield better agreement with the centerline turbulent diffusivity data (see Problem 37-2). However, this degrades the performance of the model with respect to other measures of success, such as determining the friction coefficient. Therefore, although it offers some improvements over the mixing length model, it is clear that the k-epsilon model is still an imperfect model of turbulence.

37.6 TURBULENT HEAT TRANSFER WITH CONSTANT TEMPERATURE BOUNDARY

Consider the problem illustrated in Figure 37-7 of a flow between hot isothermal smooth plates. The turbulent flow is fully developed both hydrodynamically and thermally. Heat transfer between isothermal walls in this geometry of turbulent flow was analyzed in Section 33.3.2 using the mixing length model. Using the dimensionless variables

$$\eta=\frac{y}{a},\quad u=\frac{\overline{v}_x}{\overline{v}_m},\quad \text{and}\quad \theta=\frac{\overline{T}-\overline{T}_m}{T_s-\overline{T}_m}, \tag{37-55}$$

the heat equation (33-29) for fully developed transport is restated here for the $\varepsilon_H/\nu \gg 1$ condition assumed in the standard k-epsilon model:

$$\frac{\partial\theta}{\partial\eta}=\frac{1-g(\eta)}{\varepsilon_H/\alpha}\theta'(0)\quad (\eta_0\leq\eta\leq 1/2), \tag{37-56}$$

where

$$g(\eta)=2\int_0^\eta u(1-\theta)d\eta=u_0(1-\theta_0)\eta_0+2\int_{\eta_0}^\eta u(1-\theta)d\eta. \tag{37-57}$$

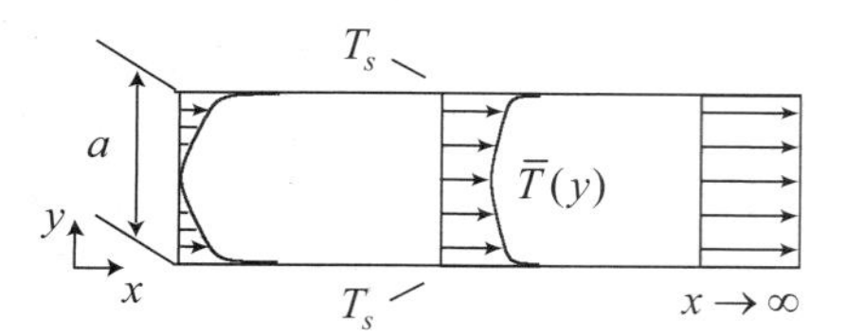

Figure 37-7 Fully developed temperature field between isothermal smooth wall plates.

Since the standard k-epsilon model cannot be used to integrate the solution to the wall, the thermal law of the wall is used to evaluate the temperature field at the transition point $\theta_0 = \theta(\eta = \eta_0)$. The location of this transition η_0 and the flow conditions at this point are determined as a part of solving the momentum equation in Section 37.4. As discussed in Section 32.3.2, the thermal law of the wall has the form

$$T^+ = \frac{\kappa/\kappa_H}{\kappa} \ln(y^+) + B(\mathrm{Pr}), \tag{37-58}$$

where the offset $B(\mathrm{Pr})$ for the logarithmic law is a function of the Prandtl number. For the current geometry of flow, the wall variables are given by

$$y^+ = \eta\sqrt{(\mathrm{Re}_D/2)u'(0)} \quad \text{and} \quad T^+ = \mathrm{Pr}\frac{\theta - 1}{\theta'(0)}\sqrt{(\mathrm{Re}_D/2)u'(0)}, \tag{37-59}$$

and the thermal law of the wall can be evaluated at $\eta = \eta_0$ for

$$\theta_0 = 1 + \frac{\theta'(0)/\mathrm{Pr}}{\sqrt{(\mathrm{Re}_D/2)u'(0)}}\left[\frac{\kappa/\kappa_H}{\kappa}\ln\left(\eta_0\sqrt{(\mathrm{Re}_D/2)u'(0)}\right) + B(\mathrm{Pr})\right], \tag{37-60}$$

where Kander's correlation [3] may be used to evaluate the offset:

$$B(\mathrm{Pr}) = (3.85\,\mathrm{Pr}^{1/3} - 1.3)^2 + 2.12\ln(\mathrm{Pr}). \tag{37-61}$$

Casting the heat equation (37-56) in terms of the new independent variable $\chi = \ln\eta$ yields

$$\frac{\partial\theta}{\partial\chi} = e^{\chi}\frac{\mathrm{Pr}_t}{\mathrm{Pr}}\frac{1 - g(\chi)}{\varepsilon_M/\nu}\theta'(0) \quad (\varepsilon_H/\nu \gg 1), \tag{37-62}$$

where

$$g(\chi) = u_0(1 - \theta_0)\eta_0 + 2\int_{\ln(\eta_0)}^{\chi} e^{\chi}u(1 - \theta)d\chi. \tag{37-63}$$

The heat equation makes use of the definition of the turbulent Prandtl number,

$$\frac{\varepsilon_H}{\alpha} = \frac{\mathrm{Pr}}{\mathrm{Pr}_t}\frac{\varepsilon_M}{\nu}, \tag{37-64}$$

which can be assumed constant over the region $\eta_0 \le \eta \le 1/2$ for which the standard k-epsilon model is valid.

In terms of the new independent variable, the two conditions to be imposed on the solution to the heat equation are

$$\theta(\chi = \ln \eta_0) = \theta_0 \quad \text{and} \quad g(\chi = \ln 1/2) = 1. \tag{37-65}$$

The first boundary condition is specified by the fluid temperature at the transition point $\eta = \eta_0$ approaching the isothermal wall. The second condition, discussed in Section 33.3.2, is a requirement of the normalized dependent variables. This second condition also enforces symmetry in the solution, $\theta'(\eta = 1/2) = 0$, through the governing equation (37-62).

The heat equation (37-62) can be integrated for θ using the usual Runge-Kutta method (see Problem 37-3). Since this task is preceded by determining the solution to the $u - eqn.$, $K - eqn.$, and $E - eqn.$, both the turbulent diffusivity (ε_M/ν) and the velocity profile u are known distributions in the heat equation. However, $\theta'(0) = \partial\theta/\partial\eta|_{\eta=0}$ is an unknown part of the solution. Therefore, the solution to the heat equation is determined by guessing $\theta'(0)$ to satisfy the requirement that $g(\chi = \ln 1/2) = 1$ using the shooting method.

Figure 37-8 checks the thermal law of the wall behavior of the solution for $\mathrm{Re}_D = 10^6$ and $\mathrm{Pr} = 0.7$. It is observed that the k-epsilon model initially conforms to the law of the wall, as given by Eq. (37-58), and departs from the law of the wall in the outer region of turbulence as expected.

Figure 37-9 contrasts the temperature solution of the k-epsilon model with that of the mixing length model. The k-epsilon model yields a smoother profile than the mixing

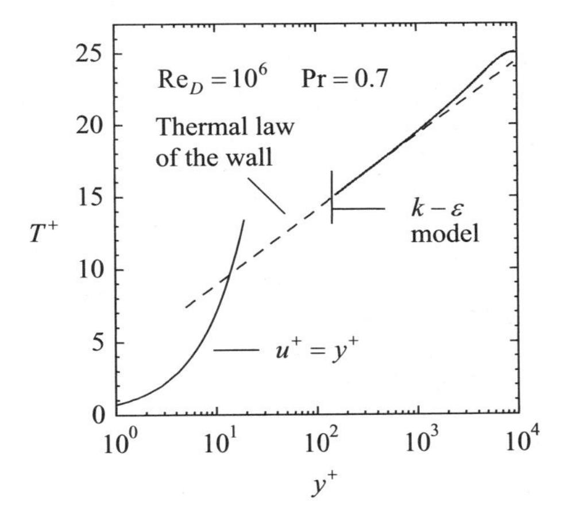

Figure 37-8 Thermal law of wall behavior for k-epsilon model.

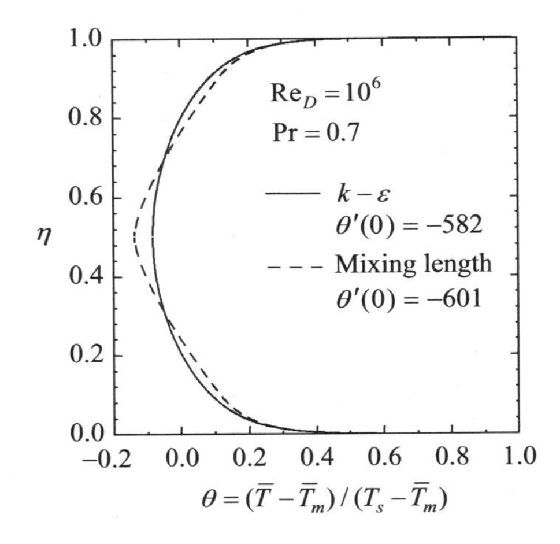

Figure 37-9 Comparison of temperature profiles between k-epsilon and mixing length models.

length model, which exhibits a characteristically pointed profile near the centerline. Despite differences in the temperature profile, both models yield similar values of $\theta'(0)$. The Nusselt number is related to the temperature gradient at the wall by

$$Nu_{D,T} = \frac{h(2a)}{k} = -\frac{2a}{T_s - \overline{T}_m} \frac{\partial \overline{T}}{\partial y}\bigg|_{y=0} = -2\theta'(0). \tag{37-66}$$

For $\text{Re}_D = 10^6$ and $\text{Pr} = 0.7$, the k-epsilon model solution and the mixing length model solution predict values of $\theta'(0)$ within about 3% of each other.

37.7 PROBLEMS

37-1 Derive the equations for $\ln K(\chi)$ and $\ln E(\chi)$ for a fully developed but transient flow, such that $\partial K/\partial t \neq 0$ and $\partial E/\partial t \neq 0$. With a suitable definition for τ, show that both equations can be expressed in the form

$$\frac{e^{2\chi}}{8(\varepsilon_M/\nu)} \frac{\partial \phi}{\partial \tau} = \frac{\partial^2 \phi}{\partial \chi^2} + f_\phi \frac{\partial \phi}{\partial \chi} + S_\phi$$

where

$$\begin{cases} \text{if } \phi = \ln K: & f_{\ln K} = 3\partial \ln K/\partial \chi - \partial \ln E/\partial \chi - 1, & S_{\ln K} = e^{2\chi} S_K \\ \text{if } \phi = \ln E: & f_{\ln E} = 2\partial \ln K/\partial \chi - 1, & S_{\ln E} = e^{2\chi} S_E \end{cases}.$$

37-2 Consider changing the k-epsilon constant $C_{2\varepsilon} = 1.92$ to a new value: $C_{2\varepsilon} = 1.7$. For Poiseuille flow between smooth parallel plates, assess the resulting prediction for the turbulent diffusivity near the centerline of the channel when $\text{Re}_D = 10^6$. What impact does this change have on the predicted friction coefficient?

37-3 Write a program to integrate the heat equation (37-62). Plot the thermal law of the wall behavior of the solution for $\text{Re}_D = 10^6$ and $\text{Pr} = 5.9$. Determine the Nusselt number for heat transfer under this condition.

37-4 Derive the governing equations for the k-epsilon model for a fully developed Poiseuille flow in a smooth-wall pipe. Using the definitions

$$\eta = 1 - \frac{r}{R}, \quad \chi = \ln \eta, \quad u = \frac{\overline{v}_z}{\overline{v}_m}, \quad \text{Re}_D = \frac{\overline{v}_m(2R)}{\nu}, \quad K = \frac{\overline{k}}{\overline{v}_m^2/2}, \quad \text{and} \quad E = \frac{\overline{\varepsilon}}{\overline{v}_m^4/\nu},$$

demonstrate that

$$u - eqn.: \quad \frac{\partial u}{\partial \chi} = e^\chi \frac{(1 - e^\chi)u'(0)}{\varepsilon_M/\nu} \quad (\eta_0 \leq \eta \leq 1/2),$$

where

$$\int_0^1 u(1 - e^\chi)e^\chi d\chi = \frac{u_0(1 - \eta_0)\eta_0}{2} + \int_{\ln(\eta_0)}^{\ln(1/2)} u(1 - e^\chi)e^\chi d\chi = 1/2$$

$$K - eqn.: \quad 0 = \frac{\partial^2 \ln K}{\partial \chi^2} + \left(3\frac{\partial \ln K}{\partial \chi} - \frac{\partial \ln E}{\partial \chi} - \frac{1}{1 - e^\chi}\right)\frac{\partial \ln K}{\partial \chi} + e^{2\chi} S_K$$

$$E - eqn.: \quad 0 = \frac{\partial^2 \ln E}{\partial \chi^2} + \left(2\frac{\partial \ln K}{\partial \chi} - \frac{1}{1 - e^\chi}\right)\frac{\partial \ln E}{\partial \chi} + e^{2\chi} S_E,$$

where the source function S_j ($j = K$ and E) is given by Eq. (37-14). Show that the transition point between the mixing length model and the k-epsilon model occurs at

$$\eta_0 \approx \frac{1}{\sqrt{\kappa\sqrt{\mathrm{Re}_D\, u'(0)}}}, \quad \text{where} \quad u_0 = \sqrt{\frac{2u'(0)}{\mathrm{Re}_D}}\left(\frac{1}{\kappa}\ln\left(\frac{(\mathrm{Re}_D\, u'(0))^{1/4}}{\sqrt{2}\kappa E_\nu^+}\right) + E_\nu^+\right),$$

$$\left.\frac{\varepsilon_M}{\nu}\right|_0 = \kappa\eta_0\sqrt{(1-\eta_0)u'(0)\mathrm{Re}_D/2}, \quad K_0 = \frac{4(1-\eta_0)u'(0)}{\sqrt{C_\mu}\mathrm{Re}_D}, \quad \text{and} \quad E_0 = \frac{C_\mu K_0^2}{4(\varepsilon_M/\nu)|_0}.$$

Solve for the coefficient of friction when $\mathrm{Re}_D = 10^6$. Plot u^+ versus y^+ using the k-epsilon model solution, and contrast with the law of the wall. Plot the turbulent diffusivity ε_M/ν over the region $\eta_0 \leq \eta \leq 1/2$ in the pipe.

37-5 Using the k-epsilon model, solve the heat equation for fully developed Poiseuille flow in a smooth isothermal pipe. Using the definitions

$$\eta = 1 - \frac{r}{R}, \quad \chi = \ln\eta, \quad u = \frac{\overline{v}_z}{\overline{v}_m}, \quad Nu_D = \frac{h(2R)}{k}, \quad \theta = \frac{\overline{T} - T_m}{T_s - T_m},$$

demonstrate that the heat transfer equation becomes

$$\frac{\partial\theta}{d\chi} = Nu_D e^{\chi}\frac{\mathrm{Pr}_t}{\mathrm{Pr}}\frac{g(\chi) - 1/2}{(1-e^{\chi})\varepsilon_M/\nu} \quad (\varepsilon_H/\nu \gg 1),$$

where

$$g(\chi) = \frac{u_0(1-\eta_0)(1-\theta_0)\eta_0}{2} + \int_{\ln(\eta_0)}^{\chi} e^{\chi}u(1-e^{\chi})(1-\theta)d\chi.$$

Using the solution developed in Problem 37-4, solve the k-epsilon model for fully developed heat transfer in the pipe. Plot the thermal law of the wall behavior of the solution for $\mathrm{Re}_D = 10^6$ and $\mathrm{Pr} = 5.9$. Determine the Nusselt number for heat transfer under this condition.

REFERENCES

[1] B. E. Launder and B. I. Sharma, "Application of the Energy Dissipation Model of Turbulence to the Calculation of Flow Near a Spinning Disc." *Letters in Heat and Mass Transfer*, **1**, 131 (1974).

[2] A. K. M. F. Hussain and W. C. Reynolds, "Measurements in Fully Developed Turbulent Channel Flow." *Journal of Fluids Engineering, Transactions of the ASME*, **97**, 568 (1975).

[3] B. A. Kander, "Temperature and Concentration Profiles in Fully Turbulent Boundary Layers." *International Journal of Heat and Mass Transfer*, **24**, 1541 (1981).

Appendix A

Table A-1 Fourier's diffusion law for heat transfer

(a) *Cartesian* (x, y, z)	(b) *Cylindrical* (r, θ, z)	(c) *Spherical* (r, θ, ϕ)
$q_x = -k\dfrac{\partial T}{\partial x}$	$q_r = -k\dfrac{\partial T}{\partial r}$	$q_r = -k\dfrac{\partial T}{\partial r}$
$q_y = -k\dfrac{\partial T}{\partial y}$	$q_\theta = -\dfrac{k}{r}\dfrac{\partial T}{\partial \theta}$	$q_\theta = \dfrac{-k}{r}\dfrac{\partial T}{\partial \theta}$
$q_z = -k\dfrac{\partial T}{\partial z}$	$q_z = -k\dfrac{\partial T}{\partial z}$	$q_\phi = \dfrac{-k}{r\sin\theta}\dfrac{\partial T}{\partial \phi}$

Table A-2 Fick's diffusion law for species transfer*

(a) *Cartesian* (x, y, z)	(b) *Cylindrical* (r, θ, z)	(c) *Spherical* (r, θ, ϕ)
$j_{A,x} = -\rho\, Đ_{AB}\dfrac{\partial \omega_A}{\partial x}$	$j_{A,r} = -\rho\, Đ_{AB}\dfrac{\partial \omega_A}{\partial r}$	$j_{A,r} = -\rho\, Đ_{AB}\dfrac{\partial \omega_A}{\partial r}$
$j_{A,y} = -\rho\, Đ_{AB}\dfrac{\partial \omega_A}{\partial y}$	$j_{A,\theta} = \dfrac{-\rho\, Đ_{AB}}{r}\dfrac{\partial \omega_A}{\partial \theta}$	$j_{A,\theta} = \dfrac{-\rho\, Đ_{AB}}{r}\dfrac{\partial \omega_A}{\partial \theta}$
$j_{A,z} = -\rho\, Đ_{AB}\dfrac{\partial \omega_A}{\partial z}$	$j_{A,z} = -\rho\, Đ_{AB}\dfrac{\partial \omega_A}{\partial z}$	$j_{A,\phi} = \dfrac{-\rho\, Đ_{AB}}{r\sin\theta}\dfrac{\partial \omega_A}{\partial \phi}$

*Expressions for the molar fluxes are obtained with the replacements $j_A \rightarrow J_A^*$, $\rho \rightarrow c$, and $\omega_A \rightarrow \chi_A$.

Table A-3 Newton's diffusion law for momentum transfer ($M_{ij} = -\tau_{ij}$)

(a) *Cartesian* (x, y, z)	(b) *Cylindrical* (r, θ, z)	(c) *Spherical* (r, θ, ϕ)
$M_{xx} = -2\mu\left[\frac{\partial v_x}{\partial x} - \frac{\partial_i v_i}{3}\right]$	$M_{rr} = -2\mu\left[\frac{\partial v_r}{\partial r} - \frac{\partial_i v_i}{3}\right]$	$M_{rr} = -2\mu\left[\frac{\partial v_r}{\partial r} - \frac{\partial_i v_i}{3}\right]$
$M_{yy} = -2\mu\left[\frac{\partial v_y}{\partial y} - \frac{\partial_i v_i}{3}\right]$	$M_{\theta\theta} = -2\mu\left[\frac{1}{r}\frac{\partial v_\theta}{\partial \theta} + \frac{v_r}{r} - \frac{\partial_i v_i}{3}\right]$	$M_{\theta\theta} = -2\mu\left[\frac{1}{r}\frac{\partial v_\theta}{\partial \theta} + \frac{v_r}{r} - \frac{\partial_i v_i}{3}\right]$
$M_{zz} = -2\mu\left[\frac{\partial v_z}{\partial z} - \frac{\partial_i v_i}{3}\right]$	$M_{zz} = -2\mu\left[\frac{\partial v_z}{\partial z} - \frac{\partial_i v_i}{3}\right]$	$M_{\phi\phi} = -2\mu\left[\frac{1}{r\sin\theta}\frac{\partial v_\phi}{\partial \phi} + \frac{v_r}{r} + \frac{v_\theta \cot\theta}{r} - \frac{\partial_i v_i}{3}\right]$
$M_{xy} = M_{yx} = -\mu\left[\frac{\partial v_x}{\partial y} + \frac{\partial v_y}{\partial x}\right]$	$M_{r\theta} = M_{\theta r} = -\mu\left[r\frac{\partial}{\partial r}\left(\frac{v_\theta}{r}\right) + \frac{1}{r}\frac{\partial v_r}{\partial \theta}\right]$	$M_{r\theta} = M_{\theta r} = -\mu\left[r\frac{\partial}{\partial r}\left(\frac{v_\theta}{r}\right) + \frac{1}{r}\frac{\partial v_r}{\partial \theta}\right]$
$M_{yz} = M_{zy} = -\mu\left[\frac{\partial v_y}{\partial z} + \frac{\partial v_z}{\partial y}\right]$	$M_{\theta z} = M_{z\theta} = -\mu\left[\frac{\partial v_\theta}{\partial z} + \frac{1}{r}\frac{\partial v_z}{\partial \theta}\right]$	$M_{\theta\phi} = M_{\phi\theta} = -\mu\left[\frac{\sin\theta}{r}\frac{\partial}{\partial \theta}\left(\frac{v_\phi}{\sin\theta}\right) + \frac{1}{r\sin\theta}\frac{\partial v_\theta}{\partial \phi}\right]$
$M_{zx} = M_{xz} = -\mu\left[\frac{\partial v_z}{\partial x} + \frac{\partial v_x}{\partial z}\right]$	$M_{zr} = M_{rz} = -\mu\left[\frac{\partial v_z}{\partial r} + \frac{\partial v_r}{\partial z}\right]$	$M_{zr} = M_{rz} = -\mu\left[\frac{1}{r\sin\theta}\frac{\partial v_r}{\partial \phi} + r\frac{\partial}{\partial r}\left(\frac{v_\phi}{r}\right)\right]$
$\partial_i v_i = \frac{\partial v_x}{\partial x} + \frac{\partial v_y}{\partial y} + \frac{\partial v_z}{\partial z}$	$\partial_i v_i = \frac{1}{r}\frac{\partial}{\partial r}(r v_r) + \frac{1}{r}\frac{\partial v_\theta}{\partial \theta} + \frac{\partial v_z}{\partial z}$	$\partial_i v_i = \frac{1}{r^2}\frac{\partial}{\partial r}(r^2 v_r) + \frac{1}{r\sin\theta}\frac{\partial}{\partial \theta}(v_\theta \sin\theta) + \frac{1}{r\sin\theta}\frac{\partial v_\phi}{\partial \phi}$

Table A-4 Continuity equation

(a) *Cartesian* (x, y, z)

$$\frac{\partial \rho}{\partial t} + \frac{\partial}{\partial x}(\rho v_x) + \frac{\partial}{\partial y}(\rho v_y) + \frac{\partial}{\partial z}(\rho v_z) = 0$$

(b) *Cylindrical* (r, θ, z)

$$\frac{\partial \rho}{\partial t} + \frac{1}{r}\frac{\partial}{\partial r}(\rho r v_r) + \frac{1}{r}\frac{\partial}{\partial \theta}(\rho v_\theta) + \frac{\partial}{\partial z}(\rho v_z) = 0$$

(c) *Spherical* (r, θ, ϕ)

$$\frac{\partial \rho}{\partial t} + \frac{1}{r^2}\frac{\partial}{\partial r}(\rho r^2 v_r) + \frac{1}{r\sin\theta}\frac{\partial}{\partial \theta}(\rho v_\theta \sin\theta) + \frac{1}{r\sin\theta}\frac{\partial}{\partial \phi}(\rho v_\phi) = 0$$

Table A-5 Momentum equation for a Newtonian fluid with constant density (ρ) and constant viscosity (μ)

(a) ***Cartesian (x, y, z)***

$$\frac{\partial v_x}{\partial t} + v_x \frac{\partial v_x}{\partial x} + v_y \frac{\partial v_x}{\partial y} + v_z \frac{\partial v_x}{\partial z} = \nu \left(\frac{\partial^2 v_x}{\partial x^2} + \frac{\partial^2 v_x}{\partial y^2} + \frac{\partial^2 v_x}{\partial z^2} \right) - \frac{1}{\rho} \frac{\partial P}{\partial x} + g_x$$

$$\frac{\partial v_y}{\partial t} + v_x \frac{\partial v_y}{\partial x} + v_y \frac{\partial v_y}{\partial y} + v_z \frac{\partial v_y}{\partial z} = \nu \left(\frac{\partial^2 v_y}{\partial x^2} + \frac{\partial^2 v_y}{\partial y^2} + \frac{\partial^2 v_y}{\partial z^2} \right) - \frac{1}{\rho} \frac{\partial P}{\partial y} + g_y$$

$$\frac{\partial v_z}{\partial t} + v_x \frac{\partial v_z}{\partial x} + v_y \frac{\partial v_z}{\partial y} + v_z \frac{\partial v_z}{\partial z} = \nu \left(\frac{\partial^2 v_z}{\partial x^2} + \frac{\partial^2 v_z}{\partial y^2} + \frac{\partial^2 v_z}{\partial z^2} \right) - \frac{1}{\rho} \frac{\partial P}{\partial z} + g_z$$

(b) ***Cylindrical (r, θ, z)***

$$\frac{\partial v_r}{\partial t} + v_r \frac{\partial v_r}{\partial r} + \frac{v_\theta}{r} \frac{\partial v_r}{\partial \theta} - \frac{v_\theta^2}{r} + v_z \frac{\partial v_r}{\partial z} = \nu \left(\frac{\partial}{\partial r} \left(\frac{1}{r} \frac{\partial}{\partial r} (r v_r) \right) + \frac{1}{r^2} \frac{\partial^2 v_r}{\partial \theta^2} + \frac{\partial^2 v_r}{\partial z^2} - \frac{2}{r^2} \frac{\partial v_\theta}{\partial \theta} \right) - \frac{1}{\rho} \frac{\partial P}{\partial r} + g_r$$

$$\frac{\partial v_\theta}{\partial t} + v_r \frac{\partial v_\theta}{\partial r} + \frac{v_\theta}{r} \frac{\partial v_\theta}{\partial \theta} + \frac{v_r v_\theta}{r} + v_z \frac{\partial v_\theta}{\partial z} = \nu \left(\frac{\partial}{\partial r} \left(\frac{1}{r} \frac{\partial}{\partial r} (r v_\theta) \right) + \frac{1}{r^2} \frac{\partial^2 v_\theta}{\partial \theta^2} + \frac{\partial^2 v_\theta}{\partial z^2} + \frac{2}{r^2} \frac{\partial v_r}{\partial \theta} \right) - \frac{1}{\rho} \frac{1}{r} \frac{\partial P}{\partial \theta} + g_\theta$$

$$\frac{\partial v_z}{\partial t} + v_r \frac{\partial v_z}{\partial r} + \frac{v_\theta}{r} \frac{\partial v_z}{\partial \theta} + v_z \frac{\partial v_z}{\partial z} = \nu \left(\frac{1}{r} \frac{\partial}{\partial r} \left(r \frac{\partial v_z}{\partial r} \right) + \frac{1}{r^2} \frac{\partial^2 v_z}{\partial \theta^2} + \frac{\partial^2 v_z}{\partial z^2} \right) - \frac{1}{\rho} \frac{\partial P}{\partial z} + g_z$$

(c) ***Spherical (r, θ, φ)***

$$\frac{\partial v_r}{\partial t} + v_r \frac{\partial v_r}{\partial r} + \frac{v_\theta}{r} \frac{\partial v_r}{\partial \theta} + \frac{v_\phi}{r \sin\theta} \frac{\partial v_r}{\partial \phi} - \frac{v_\theta^2 + v_\phi^2}{r}$$

$$= \nu \left(\frac{\partial}{\partial r} \left(\frac{1}{r^2} \frac{\partial}{\partial r} (r^2 v_r) \right) + \frac{1}{r^2 \sin\theta} \frac{\partial}{\partial \theta} \left(\sin\theta \frac{\partial v_r}{\partial \theta} \right) + \frac{1}{r^2 \sin^2\theta} \frac{\partial^2 v_r}{\partial \phi^2} - \frac{2}{r^2 \sin\theta} \frac{\partial}{\partial \theta} (v_\theta \sin\theta) - \frac{2}{r^2 \sin\theta} \frac{\partial v_\phi}{\partial \phi} \right) - \frac{1}{\rho} \frac{\partial P}{\partial r} + g_r$$

$$\frac{\partial v_\theta}{\partial t} + v_r \frac{\partial v_\theta}{\partial r} + \frac{v_\theta}{r} \frac{\partial v_\theta}{\partial \theta} + \frac{v_\phi}{r \sin\theta} \frac{\partial v_\theta}{\partial \phi} + \frac{v_r v_\theta}{r} - \frac{v_\phi^2 \cot\theta}{r}$$

$$= \nu \left(\frac{1}{r^2} \frac{\partial}{\partial r} \left(r^2 \frac{\partial v_\theta}{\partial r} \right) + \frac{1}{r^2} \frac{\partial}{\partial \theta} \left(\frac{1}{\sin\theta} \frac{\partial}{\partial \theta} (v_\theta \sin\theta) \right) + \frac{1}{r^2 \sin^2\theta} \frac{\partial^2 v_\theta}{\partial \phi^2} + \frac{2}{r^2} \frac{\partial v_r}{\partial \theta} - \frac{2 \cot\theta}{r^2 \sin\theta} \frac{\partial v_\phi}{\partial \phi} \right) - \frac{1}{\rho} \frac{1}{r} \frac{\partial P}{\partial \theta} + g_\theta$$

$$\frac{\partial v_\phi}{\partial t} + v_r \frac{\partial v_\phi}{\partial r} + \frac{v_\theta}{r} \frac{\partial v_\phi}{\partial \theta} + \frac{v_\phi}{r \sin\theta} \frac{\partial v_\phi}{\partial \phi} + \frac{v_\phi v_r}{r} + \frac{v_\theta v_\phi}{r} \cot\theta$$

$$= \nu \left(\frac{1}{r^2} \frac{\partial}{\partial r} \left(r^2 \frac{\partial v_\phi}{\partial r} \right) + \frac{1}{r^2} \frac{\partial}{\partial \theta} \left(\frac{1}{\sin\theta} \frac{\partial}{\partial \theta} (v_\phi \sin\theta) \right) + \frac{1}{r^2 \sin^2\theta} \frac{\partial^2 v_\phi}{\partial \phi^2} + \frac{2}{r^2 \sin\theta} \frac{\partial v_r}{\partial \phi} + \frac{2\cot\theta}{r^2 \sin\theta} \frac{\partial v_\theta}{\partial \phi} \right) - \frac{1}{\rho} \frac{1}{r \sin\theta} \frac{\partial P}{\partial \phi} + g_\phi$$

Table A-6 Heat equation for an isobaric flow ($DP/Dt=0$) or constant density (ρ) fluid, with constant thermal conductivity (k)

(a) ***Cartesian*** **(*x*, *y*, *z*)**

$$\frac{\partial T}{\partial t} + v_x \frac{\partial T}{\partial x} + v_y \frac{\partial T}{\partial y} + v_z \frac{\partial T}{\partial z} = \alpha\left(\frac{\partial^2 T}{\partial x^2} + \frac{\partial^2 T}{\partial y^2} + \frac{\partial^2 T}{\partial z^2}\right) + \frac{\mu}{\rho C_p}\Phi_v$$

$$\Phi_v = 2\left[\left(\frac{\partial v_x}{\partial x}\right)^2 + \left(\frac{\partial v_y}{\partial y}\right)^2 + \left(\frac{\partial v_z}{\partial z}\right)^2\right] + \left(\frac{\partial v_y}{\partial x} + \frac{\partial v_x}{\partial y}\right)^2 + \left(\frac{\partial v_z}{\partial y} + \frac{\partial v_y}{\partial z}\right)^2 + \left(\frac{\partial v_x}{\partial z} + \frac{\partial v_z}{\partial x}\right)^2 - \frac{2}{3}\left(\frac{\partial v_x}{\partial x} + \frac{\partial v_y}{\partial y} + \frac{\partial v_z}{\partial z}\right)^2$$

(b) ***Cylindrical*** **(*r*, θ, *z*)**

$$\frac{\partial T}{\partial t} + v_r \frac{\partial T}{\partial r} + \frac{v_\theta}{r}\frac{\partial T}{\partial \theta} + v_z \frac{\partial T}{\partial z} = \alpha\left[\frac{1}{r}\frac{\partial}{\partial r}\left(r\frac{\partial T}{\partial r}\right) + \frac{1}{r^2}\frac{\partial^2 T}{\partial \theta^2} + \frac{\partial^2 T}{\partial z^2}\right] + \frac{\mu}{\rho C_p}\Phi_v$$

$$\Phi_v = 2\left[\left(\frac{\partial v_r}{\partial r}\right)^2 + \left(\frac{1}{r}\frac{\partial v_\theta}{\partial \theta} + \frac{v_r}{r}\right)^2 + \left(\frac{\partial v_z}{\partial z}\right)^2\right] + \left(r\frac{\partial}{\partial r}\left(\frac{v_\theta}{r}\right) + \frac{1}{r}\frac{\partial v_r}{\partial \theta}\right)^2 + \left(\frac{1}{r}\frac{\partial v_z}{\partial \theta} + \frac{\partial v_\theta}{\partial z}\right)^2 + \left(\frac{\partial v_r}{\partial z} + \frac{\partial v_z}{\partial r}\right)^2 - \frac{2}{3}\left(\frac{1}{r}\frac{\partial}{\partial r}(rv_r) + \frac{1}{r}\frac{\partial v_\theta}{\partial \theta} + \frac{\partial v_z}{\partial z}\right)^2$$

(c) ***Spherical*** **(*r*, θ, ϕ)**

$$\frac{\partial T}{\partial t} + v_r \frac{\partial T}{\partial r} + \frac{v_\theta}{r}\frac{\partial T}{\partial \theta} + \frac{v_\phi}{r\sin\theta}\frac{\partial T}{\partial \phi} = \alpha\left[\frac{1}{r^2}\frac{\partial}{\partial r}\left(r^2\frac{\partial T}{\partial r}\right) + \frac{1}{r^2\sin\theta}\frac{\partial}{\partial \theta}\left(\sin\theta\frac{\partial T}{\partial \theta}\right) + \frac{1}{r^2\sin^2\theta}\frac{\partial^2 T}{\partial \phi^2}\right] + \frac{\mu}{\rho C_p}\Phi_v$$

$$\Phi_v = 2\left[\left(\frac{\partial v_r}{\partial r}\right)^2 + \left(\frac{1}{r}\frac{\partial v_\theta}{\partial \theta} + \frac{v_r}{r}\right)^2 + \left(\frac{1}{r\sin\theta}\frac{\partial v_\phi}{\partial \phi} + \frac{v_r + v_\theta\cot\theta}{r}\right)^2\right] + \left(r\frac{\partial}{\partial r}\left(\frac{v_\theta}{r}\right) + \frac{1}{r}\frac{\partial v_r}{\partial \theta}\right)^2$$

$$+ \left[\frac{\sin\theta}{r}\frac{\partial}{\partial \theta}\left(\frac{v_\phi}{\sin\theta}\right) + \frac{1}{r\sin\theta}\frac{\partial v_\theta}{\partial \phi}\right]^2 + \left[\frac{1}{r\sin\theta}\frac{\partial v_r}{\partial \phi} + r\frac{\partial}{\partial r}\left(\frac{v_\phi}{r}\right)\right]^2 - \frac{2}{3}\left[\frac{1}{r^2}\frac{\partial}{\partial r}(r^2 v_r) + \frac{1}{r\sin\theta}\frac{\partial}{\partial \theta}\left(v_\theta\sin\theta\right) + \frac{1}{r\sin\theta}\frac{\partial v_\phi}{\partial \phi}\right]^2$$

Table A-7 Species equation for a constant $\rho Ð_{AB}$ fluid†

(a) ***Cartesian*** **(*x*, *y*, *z*)**

$$\frac{\partial \omega_A}{\partial t} + v_x \frac{\partial \omega_A}{\partial x} + v_y \frac{\partial \omega_A}{\partial y} + v_z \frac{\partial \omega_A}{\partial z} = Ð_{AB} \left(\frac{\partial^2 \omega_A}{\partial x^2} + \frac{\partial^2 \omega_A}{\partial y^2} + \frac{\partial^2 \omega_A}{\partial z^2} \right) + \frac{r_A}{\rho}$$

(b) ***Cylindrical*** **(*r*, *θ*, *z*)**

$$\frac{\partial \omega_A}{\partial t} + v_r \frac{\partial \omega_A}{\partial r} + \frac{v_\theta}{r} \frac{\partial \omega_A}{\partial \theta} + v_z \frac{\partial \omega_A}{\partial z} = Ð_{AB} \left[\frac{1}{r} \frac{\partial}{\partial r} \left(r \frac{\partial \omega_A}{\partial r} \right) + \frac{1}{r^2} \frac{\partial^2 \omega_A}{\partial \theta^2} + \frac{\partial^2 \omega_A}{\partial z^2} \right] + \frac{r_A}{\rho}$$

(c) ***Spherical*** **(*r*, *θ*, *ϕ*)**

$$\frac{\partial \omega_A}{\partial t} + v_r \frac{\partial \omega_A}{\partial r} + \frac{v_\theta}{r} \frac{\partial \omega_A}{\partial \theta} + \frac{v_\phi}{r \sin\theta} \frac{\partial \omega_A}{\partial \phi} = Ð_{AB} \left[\frac{1}{r^2} \frac{\partial}{\partial r} \left(r^2 \frac{\partial \omega_A}{\partial r} \right) + \frac{1}{r^2 \sin\theta} \frac{\partial}{\partial \theta} \left(\sin\theta \frac{\partial \omega_A}{\partial \theta} \right) + \frac{1}{r^2 \sin^2\theta} \frac{\partial^2 \omega_A}{\partial \phi^2} \right] + \frac{r_A}{\rho}$$

†The species equations written in terms of the mole fraction of A are obtained with the replacements $\omega_A \to x_A$, $v \to v^*$, $\rho \to c$, and $r_A \to R_A$.

Index

A
Adiabatic, flow, 2, 10, 65, 262, 297, 333
  wall temperature in viscous flow, 395
Advection, 12, 14, 15
  bubble dynamics, 217, 219, 222, 349, 357
  compressible flow, 296, 315, 333
  fluxes (table of), 16
  ideal plane flow, 209, 224
  inviscid flow, 206, 297
  inviscid heat transport in a two dimensional box, 222
  inviscid species transport in a two dimensional box, 215
  open channel flow, 265, 284
Air bearing (on perforated surface), 204, 205
Airfoil, cambered (Joukowski), 240
  symmetric, 238
Alternating unit tensor, 30
Analogies, between transport, 390
Annulus flow, laminar heat transfer in, 445, 446
  turbulent, 490, 505

B
Bar, heat transfer in, 108, 122
  heat transfer in composite material, 117
  species transfer through, 115
  translation during heat treatment, 80, 359
Bearing lubrication (see also Step bearing, Journal bearing, Air bearing), 188
Bernoulli's equation, 209
  bubble dynamics, 218
  ideal plane flows, 215, 228, 231
  open channel flows, 265
Bessel functions, 128, 356
  table of differential properties, 129
Binary gas transport, 15
  turbulent flow, 532, 552
  water evaporation, 73, 80, 146
Biot number, 132, 448
Bisection method, 355
Blasius flow, 376
  heat transfer, 387, 389, 396, 457, 396, 575, 580
  species transfer, 383
  with viscous heating, 393, 463
Blowing boundary, 204, 205, 497, 505, 534
Boltzmann constant, 17
Boundary conditions, 67
  first, second, and third kind, 67
  floating, 260, 290, 325
  free surface, 68
Boundary layer, 206, 359
  approximation, 359
Boundary layer in developing laminar flow,
  Blasius flow (over a flat plate), 376, 457
  Falkner-Skan flow (over an inclined plate), 392, 396
  vertical plate (natural convection), 400, 411
Boundary layer in developing turbulent flow,
  Blasius flow (over a smooth surface), 565
Boundary layer heat convection in developing laminar flow,
  Blasius flow, 387, 389, 396, 457
  Blasius flow with viscous heating, 393, 396, 463
  Falkner-Skan flow (inclined plate), 393, 396
  vertical plate (natural convection), 400, 411
Boundary layer heat convection in developing turbulent flow,
  Blasius flow (over a smooth flat plate), 575, 580
Boundary layer heat convection in fully developed laminar flow,
  falling film on heated wall, 374
Boundary layer species convection in developing laminar flow,
  Blasius flow (over a flat plate), 383
Boundary layer species convection in fully developed laminar flow,
  fluid bounded by moving parallel plates, 374
  vertically conveyed liquid film, 369
Boundary layer thickness, 376, 383
  heat transfer, 388, 396,
  natural convection, 401
Boussinesq approximation, 400
Bubble dynamics (spherical), 217, 222
  effect of viscosity and surface tension, 219
  numerical integration of, 349, 357
Buoyancy, 399
Burgers' equation, inviscid, 250
  viscid, 255, 263

C
Calorically perfect gas, 10
Cartesian notation, 25
Cauchy-Riemann equations, 236
Chemical reaction, 51
  heating in a plane wall, 93
  heating in turbulent pipe flow, 517

Chemical reaction (*continued*)
  species transport in a plane wall, 358
  thermal oxidation, 172
Choked flow, 322
Circular tube (see Pipe flow)
Circulation, 229
Closed system, 3
Closure problem in turbulence, 470
Coating extrusion, 194
  porous surface, 198
  scaling in lubrication theory, 195
  solid surface, 195
  spin coating, 203
Coefficient of friction, 286
  entrance region laminar flow scaling, 413
  laminar boundary layer, 383, 460
  laminar flow through rectangular duct, 116
  laminar natural convection, 411
  laminar Poiseuille flow between parallel plates, 430, 456
  mixed turbulent Couette and Poiseuille flow between smooth walls, 504
  turbulent boundary layer over a smooth surface, 573
  turbulent Couette flow between smooth walls, 487
  turbulent Couette flow between rough walls, 551
  turbulent Poiseuille flow between rough parallel plates, 556
  turbulent Poiseuille flow between smooth parallel plates, 485
  turbulent Poiseuille flow between smooth parallel plates with blowing, 502, 540
  turbulent Poiseuille flow in rough wall pipe, 563
  turbulent Poiseuille flow in smooth wall annulus, 495, 505
  turbulent Poiseuille flow in smooth wall pipe, 489, 504, 604
Colburn analogy, 391
Colebrook turbulent pipe flow formula, 556
Complementary error function, 145
Complex combination, 245
Complex numbers, 234
Complex potentials, 235
  table of, 237
Complex temperature potential, 242, 247
Compressible, flow equations, 296, 334
  isentropic flow, 298, 319, 334
  quasi-one-dimensional flow, 315
  two-dimensional flow, 333
Compression wave, 300
Concentration, 9
Conservation equations (see Transport equations)
Conservation form (of transport equations), 249
  compressible one-dimensional flow, 307, 311
  compressible quasi-one-dimensional flow, 323, 330
  inviscid adiabatic flow, 262
  open channel flow, 276, 280, 281, 282, 283, 286, 288
Constitutive laws, 22
Continuity equation, 37, 39, 42, 48, 59
  across a shock wave, 302
  across hydraulic hump, 274
  decomposition in turbulent flows, 469
  in lubrication theory, 189, 196
  in open channel flow, 266, 286
  in quasi-one-dimensional compressible flow, 316
  in terms of the substantial derivative, 47
  of a simple surface wave, 268
  of a sound wave, 301
  of an incompressible fluid, 47
  satisfied by stream function definition, 210
  scaling in the momentum boundary layer, 377
  scaling in the thermal boundary layer, 389
Continuum theory, 13
Control volume analysis, 44, 59
  table of steps leading to differential equation, 45
Convection, 13
  of heat (see Heat transport by convection)
  of species (see Species transport by convection)
Convection coefficient (see also Newton's convection law),
  for heat transfer to external flow, 68, 388, 390, 409
  for heat transfer to internal flow, 413, 414, 426, 543
  for species transfer to external flow, 373, 386
  for species transfer to internal flow, 413, 424, 444, 517, 540
Converging/diverging nozzles, compressible flow, 320
  quasi-one-dimensional flow, 322, 327, 330, 331
  two-dimensional flow, 333, 342
  two-dimensional heat transport, 341, 342
Couette flow, 18
Couette flow (laminar), impulsively started, 101, 171
  with constant force on moving plate, 157
  with viscous heating, 79
Couette flow (turbulent),
  between rough walls, 549
  between smooth walls, 485, 504
  heat transfer between smooth walls, 543
  species transfer between smooth walls, 532
  species transfer between rough walls, 552
Courant, condition, 251
Critical flow, 267
Critical Reynolds number, 467
Curl, table of operators, 32
Cylinder (see also Pipe flow),
  inviscid flow around, 233
Cylinders (see also Annulus),
  heat transfer between nonconcentric cylinders, 242

Cylindrical, coordinates, 31
operators, 31
stream function defined in, 225

D

Dam-break in open channel flow, 273, 281
change in bed elevation, 282
chemical tracer, 280, 281
immiscible fluids of different density, 282, 283
Darcy, friction factor, 489
law, 198
Deal-Grove model, 173
Depression wave, 268
Differential equations, classifications, 66
Diffusion, 12, 17
fluxes (table of), 22
in binary gas, 20, 24, 40, 73
in turbulent flow, 472
irreversibility of, 22, 58
of heat, 18
of momentum (see also Momentum diffusion), 18
of species (see also Species diffusion), 20
steady two-dimensional, 103
transient one-dimensional, 82
with flux driven boundaries, 172
Diffusivity, 17
laminar heat transfer, 18
laminar momentum transfer, 18
laminar species transfer, 20
turbulent heat transfer, 509, 602
turbulent momentum transfer, 472, 495, 497, 581, 585
turbulent species transfer, 509
Dilatational viscosity, 20
Dilute species approximation, 21
Dissipation, equation (in turbulence), 585
viscous (see also viscous heating), 57
Divergence operator, 32
Duct flow (laminar), 111, 116, 125
Duhamel's theorem, 161
Dummy index, 26
Dynamic viscosity, 19

E

Eckert number, 394
Eddy diffusivity (see also turbulent diffusivity), 472
Eigencondition, 86
Eigenfunction, defined, 86, 119
expansion method, 119
Eigenvalue, 86, 119
Einstein notation, 25
Elevation wave, 268
Energy (see Internal energy, Kinetic energy, Transport equations)
Enthalpy, defined, 4
change of an ideal gas, 7
change of an incompressible liquid or solid, 8
of stagnation, 297
Entrance region, 412
of plug flow between plates of constant heat flux, 415
Entropy, 1
generation, 2, 22, 58
change of an ideal gas, 10
change of an incompressible liquid or solid, 10
second law of thermodynamics, 1
transport with heat, 2
Equation of state, 7
Equations of transport (see Transport equations)
Error function, 145
Euler equations, 297
Euler formula, 235
Eulerian frame, 46, 62
Evaporation into gas column, steady, 73, 80
transient, 146
Expansion wave, 300
Explicit numerical scheme, 250
Extensive properties, 3

F

Falkner-Skan flow, 392
Fanning friction factor (see Coefficient of friction)
Fick's law, 21
Fin, cooling, 131
Finite slab (see Plane wall)
First law of thermodynamics, 1
First-order reaction, 172
Flat plate, Blasius flow, 376
inclined surface, 392
laminar convection (see Laminar boundary layer)
natural convection near vertical wall, 400
turbulent convection, 565, 575, 580
Flat plates with inviscid flow between,
entrance region for heat transfer (constant wall heat flux), 415, 426
entrance region for species transfer (constant wall concentration), 427
fully developed heat transfer (constant wall heat flux), 421
fully developed heat transfer (constant wall temperature), 417
fully developed heat transfer (convective wall boundary), 427
fully developed species transfer (constant wall concentration), 417, 426
Flat plates with fully developed Poiseuille flow between,
heat transfer (constant wall heat flux), 435
heat transfer (constant wall temperature), 432
heat transfer (constant wall temperature, non-Newtonian fluid), 462

Flat plates with fully developed Poiseuille flow between, (*continued*)
heat transfer (constant wall temperature, temperature dependent diffusivity), 432
laminar flow, 430
laminar flow heat exchanger, 437
turbulent flow (smooth wall), 480, 496
turbulent flow with blowing, 497
Flat plates with turbulent Couette flow between smooth walls, 485, 504
Flux-conservative form of transport equations, 249
Force, lift, 229
buoyancy, 399
Forced convection, 376
boundary layers (see Boundary layer laminar/turbulent)
fully developed flows (see Fully developed laminar/turbulent)
Fourier's law, 18, 22
Free convection (see Natural convection)
Free index, 25
Friction factor,
Fanning (see also coefficient of friction), 286
Darcy, 489
Friction slope, 286
Friction velocity, 474
Froude number, 266
Fully developed condition, 68, 376, 412
profiles, 419
Fully developed laminar flow heat transfer,
between parallel plates, 432, 435, 444, 446, 451
in a parallel plate heat exchanger, 437, 445
in a pipe, 444
in an annulus, 445, 446
Fully developed laminar flow species transfer,
in a pipe, 441, 444
in surface flow, 444
Fully developed transport in inviscid flow,
heat transfer to flow between plates, 417, 421, 427
heat transfer to flow in a pipe, 427
species transfer to flow between plates, 424, 426, 427
Fully developed turbulent flow, 479
Fully developed turbulent flow between rough walls,
Couette flow between parallel plates, 549
Poiseuille flow between parallel plates, 554
Poiseuille flow in a pipe, 563
Fully developed turbulent flow between smooth walls,
Couette flow between parallel plates, 485, 504
Poiseuille flow between parallel plates, 480, 496, 589, 604
Poiseuille flow between parallel plates mixed with Couette flow, 504
Poiseuille flow between parallel plates with blowing, 497, 505
Poiseuille flow in a pipe, 488, 504, 604
Poiseuille flow in an annulus, 490, 505
Fully developed turbulent flow heat transfer between rough walls,
in Poiseuille flow between parallel plates, 558, 563
Poiseuille pipe flow, 563
Fully developed turbulent flow heat transfer between smooth walls,
in Couette flow between parallel plates, 543
in Poiseuille flow between parallel plates, 524, 529, 543, 544, 601
in Poiseuille pipe flow, 543, 604
Fully developed turbulent flow species transfer between rough walls,
in Couette flow between parallel plates, 552
Fully developed turbulent flow species transfer between smooth walls,
in Couette flow of binary mixture between parallel plates, 532
in Poiseuille pipe flow, 517
Fundamental equation, 2

G

Gas, relations from kinetic theory, 17, 459
Gas dynamics, 296
Gauss's theorem, 45
Gibbs equation, 3
Gradient operators, table of, 32
Grashof number, 404
Gravity potential, 208

H

Heat capacity, (see Specific heats)
Heat transport by advection, 14
in compressible flow, 341
in plane flow, 222
Heat transport by convection,
inviscid external flow, 364, 374
inviscid internal flow, 415, 417, 421, 424, 426, 427
laminar Blasius flow, 387, 389, 393, 396, 463, 457
laminar Falkner-Skan flow (over a wedge), 393, 396
laminar flow between plates, 432, 435, 437, 444, 445, 446, 451
laminar flow in a pipe, 444
laminar flow in an annulus, 445, 446
laminar natural convection from vertical plate, 400, 411
turbulent Blasius flow (over a smooth surface), 575, 580
turbulent Couette flow between smooth plates, 543
turbulent Poiseuille flow between smooth walls, 524, 529, 543, 544
turbulent Poiseuille flow between rough walls, 558, 563
Heat transport by steady diffusion,
between co-planar surfaces separated by a gap, 247
between nonconcentric cylinders, 242
in a plane wall, 78, 447
in laminar Couette flow with viscous heating, 79
in turbulent Couette flow, 543

in a composite rectangular bar, 117
in a cylinder of finite length, 165
in a pin fin, 131
in a rectangular bar, 108, 115, 122
Heat transport by unsteady diffusion,
in a brick, 171
in a plane wall, 93, 101, 167, 170
in a semi-infinite body, 142, 157, 159, 169, 356
in a sphere, 136, 139, 170
melting of a semi-infinite body, 183
solidification of a semi-infinite body, 174, 178, 187
Homogeneous, equation, 66
Hydraulic, diameter, 116, 413
jump, 267, 269, 273
radius of a channel, 287
Hydrodynamic, fully developed condition, 68, 412
surface roughness, 545
Hyperbolic functions, solution to ordinary differential equations, 128, 130

I
Ideal flow, 209
Ideal gas, law, 3, 7
thermodynamic relations for, 7
Ideal plane flow, 224
table of flows, 225, 231
Incompressible flow, 47
equation of continuity for, 47
thermodynamics relations, 8
Index, notation, 25
differential operators, 26, 31
Instability of turbulence, 466
Intensive properties, 3
Interface, contact plane, 274
shock wave, 302
Stefan condition, 174
Internal energy, 1, 5
of an ideal gas, 7
of an incompressible liquid or solid, 8
Internal flow, defined, 412
fully developed (see Fully developed)
Inviscid flow, 206
advection of heat, 222
advection of species, 215
compressible flow, 296, 315
convection of heat, 364, 374, 417, 421, 427
convection of species, 424, 426, 427
open channel flow, 265
through a two-dimensional box, 210, 215, 221, 222
Irreversibility, 2
of diffusion, 22, 58
of hydraulic jumps, 273
of shock waves, 301
second law of thermodynamics, 1
Irrotational flow, 208, 209
Isentropic flow, 10, 59, 298, 319, 334

J
Journal bearing, 204, 357
with cavitations, 357
Joukowski transform, 236, 247

K
K-epsilon model, standard, 581, 586
Kinematic equations for planar flows, 209
Kinematic viscosity, 19
Kinetic energy, 16, 23, 55
equation, 55, 64
in turbulence, 581
Kinetic theory, relations for gases, 17, 459
Kronecker delta, 28
Kutta condition, 238
Kutta-Joukowski theorem, 230

L
Lagrangian frame, 46, 62
Laminar boundary layer heat transfer, in developing viscid flow, 387, 389, 393
in entrance region, 413
in fully developed viscid flow, 374
in inviscid flow, 363, 364, 374
with temperature dependent gas properties, 457, 463
with viscous heating, 393, 463
Laminar boundary layer species transfer, in developing viscid flow, 383, 396
in fully developed viscid flow, 369, 374
Laminar boundary layer momentum transfer, 376, 400
coefficient of friction, 383, 410, 411, 573
Laplace equation, governing steady heat transfer, 242
governing ideal plane flow, 210,
satisfied by complex potentials, 236
Laplacian operator, 31
Latent heat, 174
Leibniz's theorem, 62, 188
Levi-Civita epsilon, 30
Lewis number, 181
Lift, 229
coefficient, 242
on an line vortex, 227
Line vortex, 224
in proximity to a wall, 233
in uniform flow, 227
Line source/sink, 224
in proximity to a wall, 233
in uniform flow, 224
Lubrication theory, 188
approximation, 191

M
MacCormack integration, 249
Mach number, 300
Manning's formula, 287

Mass,
  advection (see Advection)
  averaged property, 9
  averaged velocity, 15
  conservation (see Continuity equation)
  diffusion (see Diffusion)
  transport (see Species transport)
Material derivative, 46
Maxwell relations, 4
Mean temperature, transport average, 414, 431
Mechanical energy (kinetic energy), 16, 23, 55
  equation, 55, 64
Method of images, 167, 225
Mixing, entropy of, 24
Mixing length (turbulent), 471
  model, 471
  table of heat diffusivity model, 476
  table of momentum diffusivity model, 513
  table of species diffusivity model, 515
  Van Driest damping function, 477
Mixtures, advection transport of, 15
  binary diffusion of, 20
  binary transport, 73, 80, 146
  binary transport in turbulence, 532, 552
  dilute species approximation, 21
  properties of, 9
Modified Bessel functions, 128
Molar averaged property, 9
Molar averaged velocity, 16
Molar quantities, 9
Molecular speed, in a gas, 17
Momentum advection (see also Advection), 14
Momentum diffusion, 18
  flux tensor, 19, 20
  in boundary driven fluid layer, 167
  in Couette flow driven by constant force on moving plate, 157
  in Couette flow impulsively started, 101, 171
  in gravity and boundary driven flow in a rectangular channel, 116, 139
  in lubrication, 188
  in pipe flow with impulsively started wall rotation, 139
  in Poiseuille flow between parallel plates impulsively started, 101
  in Poiseuille flow in a rectangular duct, 111, 116, 125
  in Poiseuille pipe flow impulsively started, 139
  in Poiseuille pipe flow periodically driven, 248
  in semi-infinite fluid bounded by a wall in motion, 162
  in semi-infinite fluid bounded by a wall in periodic motion, 248
Momentum transport by convection,
  in boundary layers (see Boundary layer)
  in fully developed flows (see Fully developed)
Moody diagram, 556
Moving boundaries, 172
  contact plane (gas dynamics), 274
  diffusion species with concentration dependent diffusivity, 187
  drug release from concentrations exceeding the solubility limit, 186
  melting of a solid initially at the melting point (heat transfer), 183
  shock wave, 302
  solidification of a binary liquid, 178
  solidification of a liquid from above the melting temperature, 187
  solidification of a liquid from an undercooled state, 174, 178
  Stefan condition, 174
  thermal oxidation, 172

N

Natural convection, 399
  Boussinesq approximation, 400
  from a vertical wall, 400
  Grashof number, 404
Navier-Stokes equations, 26, 55
  Reynolds averaged, 470
Newton's convection law, for external flows, 68
  for internal flows, 414
Newton's viscosity law, 19, 20, 54
Newton-Raphson method, 355
Nikuradse equation, 479
No-slip boundary condition, 68
Nonconservative form of conservation equations, 249
Non-Newtonian flow between parallel plates, 462
Nozzle, two-dimensional inviscid flow, 333
  quasi-one-dimensional inviscid flow, 315
Nusselt number, 388
  in laminar boundary layer, 374, 388, 396, 409
  in laminar entrance region, 413, 416, 426
  in laminar fully developed flows, 436, 444, 462
  in turbulent boundary layer, 576, 578
  in turbulent fully developed flows, 525, 528, 532, 543, 544, 604

O

Open channel flow, 265
  dam break, 273
  dam break with fluids of different density, 283
  dam break with open end channel, 281
  dam break with sloped floor, 282
  dam break with tracer species, 280
Open channel flow with friction, 284
  flow past a sluice gate, 287, 293, 294
  variable width channel, 295
Open system, 1
Operators, table of, 32, 33

Orthogonality property, 86
Oscillating, pressure gradient in a pipe, 248
species advection into semi-infinite medium, 245
species convection into semi-infinite medium, 248
wall bounding fluid, 248

P

Partition coefficient (mass transfer), 178
Parallel plates – bounding laminar flow,
Couette flow, 79
heat transfer in Couette flow, 79
heat transfer in Poiseuille flow, 432, 435, 437, 444, 445, 446, 451
Poiseuille flow of non-Newtonian fluid, 462
Poiseuille flow with blowing, 80
species transfer in mixed Couette and Poiseuille flow, 374
species transfer in Poiseuille flow, 80
squeeze flow, 189
transient Couette flow, 101, 157, 171
transient Poiseuille flow, 101
Parallel plates – bounding inviscid flow,
heat transfer, 415, 417, 421, 427
species transfer, 424, 426, 427
Parallel plates – bounding turbulent flow,
Couette flow, 485, 504, 549
heat transfer in Couette flow, 543
heat transfer in Poiseuille flow, 524, 529, 543, 544, 558, 563
Poiseuille flow, 480, 496, 554
Poiseuille flow mixed with Couette flow, 504
Poiseuille flow with blowing, 497, 505
species transfer in Couette flow, 532, 552
Péclet number, for heat transfer, 360
for mass transfer, 217
Pipe flow,
inviscid heat transfer, 427
laminar heat transfer, 444
laminar species transfer, 441, 444
laminar transient, 139, 248
turbulent flow, 488, 504, 563
turbulent heat transfer, 543, 563
turbulent species transfer, 517
Pin fin, 131
Plane wall, diffusion of heat, 78, 89, 93, 101, 167, 170
diffusion of species, 358
Poiseuille flow (see also Annulus, Parallel plates, Pipe flow, Rectangular duct), 111, 429
Polytropic processes, 221
Porous medium, Darcy's law for flow in, 198
Potential flow, 208
line source/sink, 224, 233
line vortex, 224, 227, 233
over a wall, 246
over wedges, 231
rotating cylinder in a uniform flow, 233, 246
table of simple plane flows, 225
Power series solution, 148
Prandtl, mixing length, 471
number, 388
number (turbulent), 512
turbulent pipe flow correlation, 489, 558
Pressure, thermodynamic definition, 3, 4
Product superposition, 165
Pseudo-steady-state (see Quasi-steady-state)

Q

Quasi-steady-state, assumption, 173
periodic solutions, 245, 248
Quasi-one-dimensional, compressible flow, 315
open channel flow, 265, 284

R

Rankine body, 224
Rarefaction wave, 300
Rayleigh, equation, 219
number, 404
Rayleigh-Plesset equation, 220
Rectangular duct, laminar flow in, 111, 116, 125
Regular perturbation, 448
Reversible flow (see Isentropic flow)
Reynolds analogy, between heat and moment transport, 391,
between turbulent diffusivities, 508
Reynolds averaged Navier-Stokes equations, 470
Reynolds decomposition, 468
continuity equation, 468
heat boundary layer equation, 507
momentum boundary layer equation, 470
species boundary layer equation, 514
Reynolds lubrication equation, 202
Reynolds number, 190, 207, 378, 467
critical, 467
Reynolds stress, 470
Riemann variables, 302
Rotating cylinder, in inviscid flow around, 233, 246
unsteady viscous flow within, 139
Roughness Reynolds number, 545
Runge-Kutta integration, 344

S

Saint-Venant equations, 284
Scaling, estimations, 75
Scaling of the boundary layer equations,
for heat transfer, 360, 389
for momentum transfer, 377
for natural convection, 401, 411
for species transfer, 373, 375
Scaling of the lubrication equation, 189, 192, 195
Scanning laser heat treatment of solid, 359

Schmidt number, 385
Schultz-Grunow correlation, 573
Second law of thermodynamics, 1
Semi-infinite body, diffusion of heat, 142, 157, 159, 169, 356
  diffusion of momentum (laminar), 162, 248
  diffusion of species, 152
  diffusion with time dependent boundary condition, 152, 157, 159, 169
Separation of variables, 82
  eigenfunction expansion method, 119
  in cylindrical coordinates, 128, 131
  in non-Cartesian coordinates, 130
  in space and time, 83
  in spherical coordinates, 130, 136
  in two spatial dimensions, 103
Shear stress, 19
  relation to momentum diffusion, 19
  ordering of subscripts, 30
Sherwood number, 373
  fully turbulent flow developed, 518, 521, 540
  laminar boundary layer, 373, 375, 386
  laminar entrance region, 413, 427
  laminar fully developed flow, 425, 442, 445
Shock tube, 304, 307, 312
  with dissimilar gases, 311, 313
  with open end, 313
Shock wave, 302
Shooting method, 353
Similarity solutions, 140
  determining the similarity variable, 140
  in melting and solidification, 174, 178, 183, 187
  in transient diffusion of heat, 142, 157, 159, 169, 356
  in transient diffusion of momentum, 162
  in transient diffusion of species, 146, 152, 157, 178, 187
  of heat transfer boundary layer, 359, 374, 387, 389, 393, 396, 400, 411, 457, 463
  of momentum boundary layer, 376, 392, 396, 400, 411, 457
  of species transfer boundary layer, 369, 374, 393, 396, 384
Simple surface wave, 267
Skin-friction (see Coefficient of friction)
Sluice gate, in open channel flow, 281, 287, 293, 294, 295
Solidification, of a binary alloy, 178
  of an undercooled liquid, 174, 178
Solids, heat conduction in, 78, 89, 93, 101, 108, 115, 117, 122, 131, 136, 139, 142, 157, 159, 165, 167, 169, 170, 171, 242, 247, 356, 447
  melting/solidification of, 174, 178, 183, 187
  species transport in, 89, 101, 115, 139, 152, 157, 172, 186, 187, 358
Sonic Flow, 320
Sound wave, 299
  speed of, 300
Species advection, 15
  in compressible flow, 311, 313
  in open channel flow, 280, 281, 282, 283
  in plane flow, 215,
Species diffusion, 20
  during solidification of a binary liquid, 178
  in a plane wall, 89, 101
  in a plane wall with a chemical reaction, 358
  in a rectangular bar, 115, 139
  in a semi-infinite body, 152, 186
  in thermal oxidation, 172
  in two semi-infinite bodies placed in contact, 157
  in water evaporation, 73, 80, 146
  with concentration dependent diffusivity, 187
  with concentrations exceeding solubility limit, 186
Species transport in a binary mixture,
  turbulent flow, 532, 552
  water evaporation, 73, 80, 146
Species transport by convection,
  through boundary layer, 369, 384, 393, 396
  one-dimensional flow, 71, 80, 97, 256
  with periodic boundary condition, 248, 263
  internal laminar flows, 374, 424, 426, 427
  internal turbulent flows, 441, 444, 517, 532, 552
Specific heats, 6
Speed of sound, 300
Sphere, transient one-dimensional diffusion of heat, 136, 139, 170
Spherical, bubble dynamics, 217, 219, 222, 349, 357
  coordinates, 31
  operators, 32, 33, 34
  separation of variables, 130, 136, 139
Spin coating, 203
Square duct, flow in, 111, 116, 125
Squeeze flow, 48
  damping in an accelerometer design, 191
  scaling requirements of lubrication theory, 189
Stagnation point, 226, 233, 238
Stanton number, for heat transfer, 561, 564
  for species transfer, 553
State postulate, 4
State variables, 2
Steady one-dimensional convection of species, 71
  in laminar Poiseuille flow between parallel plates with blowing, 80
Steady one-dimensional diffusion of heat, plane wall, 78, 447
  in Couette flow with viscous heating, 79
Steady one-dimensional diffusion of species, plane wall with chemical reaction, 358
  water evaporation, 73, 80
Steady two-dimensional advection,
  compressible flow, 333
  heat in ideal plane flow, 222
  species in ideal plane flow, 215

Steady two-dimensional convection (see boundary layer, fully developed)
Steady two-dimensional diffusion of heat, between nonconcentric cylinders, 242
  between co-planar surfaces separated by a gap, 247
  composite rectangular bar, 117
  cylinder of finite length, 165
  pin fin, 131
  rectangular bar, 108, 115, 122
Steady two-dimensional diffusion of momentum (laminar),
  gravity and boundary driven flow in a rectangular channel, 116, 139
  Poiseuille flow in a rectangular duct, 111, 116, 125
Steady two-dimensional diffusion of species, rectangular bar, 115, 139
Steady quasi-one-dimensional advection,
  compressible flow, 315
  open channel flow, 265, 284
Stefan, condition, 174
  flow, 41, 74, 146, 533
Step bearing, 203
  on porous surface, 205
Stream function, applied to ideal plane flow, 209
  in Cartesian coordinates, 209, 225
  in cylindrical coordinates, 225
  in laminar boundary layers, 379, 392, 405
  imaginary part of the complex potential, 236
  of simple flows, 225
  of wedge flows, 231
Stress tensor, 29
Subcritical flow, 267
Subsonic flow, 319
Substantial derivative, 46
Supercritical flow, 267
Superposition, 107
  Duhamel's theorem, 162
  in space, 107, 164
  in time, 160
  of plane flows, 224
Supersonic flow, 300
Surface normal operator, 27
Surface roughness, hydrodynamic, 545
Surface tension, influence on bubble dynamics, 220
Surface wave speed, 268

T

Tensor, arithmetic (see Index notation)
  advection flux of momentum, 14
  diffusion flux of momentum, 20, 22, 54
  ordering of subscripts, 30
  stress, 29
Temperature dependent properties, in boundary layer heat transfer, 457
  in conduction through a plane wall, 447
  in fully developed heat transfer (convection), 451
Temporally periodic conditions, imposed on transport, 245
Thermal conductivity, 18
  with temperature dependence, 447, 452, 459
Thermal Oxidation, of a solid, 172
Thermodynamics, 1
  equation of state, 7
  first law of, 1
  fundamental equations of, 2
  Maxwell's equations of, 4
  of an ideal gas, 7
  of an incompressible fluid, 8
  second law of, 1
  table of relations, 10
Time averaged, quantities (in turbulence), 465, 468
Time derivative, 26
  substantial, 46
Transient one-dimensional advection,
  bubble dynamics, 217, 219, 222, 349, 357
  dam break, 273
  dam break with fluids of different density, 283
  dam break with open end channel, 281
  dam break with sloped floor, 282
  dam break with tracer species, 280
  shock tube, 304, 307, 312
  shock tube with dissimilar gases, 311, 313
  shock tube with open end, 313
Transient one-dimensional convection of species, 97, 256
  with periodic boundary condition, 248, 263
Transient one-dimensional diffusion of heat, plane wall, 101, 167, 170
  plane wall with heat generation, 93
  semi-infinite body, 142, 157, 159, 169, 356
  semi-infinite body melting, 183
  semi-infinite body solidification, 174, 178, 187,
  sphere, 136, 139, 170
Transient one-dimensional diffusion of species, plane wall, 89, 101
  release of concentration exceeding solubility limit, 186
  semi-infinite body, 152
  solidification of a binary liquid, 178
  thermal oxidation, 172
  two semi-infinite bodies placed in contact, 157,
  water evaporation, 146
  with concentration dependent species diffusivity, 187
Transient one-dimensional diffusion of momentum (laminar), boundary driven fluid layer, 167
  Couette flow driven by constant force on moving plate, 157
  Couette flow impulsively started, 101, 171
  pipe flow with impulsively started wall rotation, 139
  Poiseuille flow between parallel plates impulsively started, 101

Transient one-dimensional diffusion of momentum (laminar), boundary driven fluid layer (*continued*)
  Poiseuille pipe flow impulsively started, 139
  Poiseuille pipe flow periodically driven, 248
  semi-infinite fluid bounded by a wall in motion, 162
  semi-infinite fluid bounded by a wall in periodic motion, 248
Transient three-dimensional diffusion of heat, in a brick, 171
Transport analogies between heat, species, and momentum transfer, 390
Transport equations, continuity equation, 37, 51, 59
  energy, 61, 297
  entropy, 58
  flux-conservative form (conservation form), 249
  heat, 57, 64
  kinetic energy (mechanical energy), 55, 64
  momentum, 54, 60, 297
  open channel flow, 265, 276, 286
  quasi-one-dimensional compressible flow, 296, 297, 307
  species, 51
  table of, 44, 55
  table of transport terms, 60
Tube flow (see Pipe flow)
Turbulence, 465
  dissipation of, 583, 585
  flow over a rough surface, 545
  fully developed flow (see Fully developed flow)
  inner region, 473, 474
  k-epsilon model, 581
  kinetic energy of, 581
  law of the wall, 475, 547
  mixing length model, 471, 476
  outer region, 473, 475
  Prandtl's pipe flow correlation, 489, 558
  Reynolds analogy between turbulent diffusivities, 508
  thermal law of the wall, 510, 549
  Van Driest damping function, 477
  viscous sublayer, 473
  von Kármán constant, 473, 476
  wall variables, 474, 510
Turbulent, boundary layer (see Boundary layer),
  diffusivity, 472, 497, 509, 584, 598
  dissipation equation, 585
  flow, 465
  kinetic energy equation, 581
  Prandtl number, 512
  Schmidt number, 514
  transition, 466

V

Van Driest damping function, 477
Vector operators, table of, 32, 33
Velocity, fluctuations (in turbulence), 469
  mass averaged, 15
  molar averaged, 16
Velocity potential, 208, 236
Vertical plate, free convection, 400
Viscosity, dilatational, 20
  dynamic, 19
  kinematic, 19
  Newton's law of, 19, 20, 54
  with temperature dependence, 453, 459
Viscous heating, 57
Viscous sublayer, 473
von Kármán constant, 473, 476
von Neumann stability analysis, 256
Vortex, line in proximity to a wall, 233
  line in uniform flow, 227
  sheets, 466
Vorticity, 30, 208, 466
  equation, 466

W

Wall variables, 474, 510
Wall bounding fluid, set in motion, 162, 248
Wedge, flow over, 393, 396

Y

Young-Laplace pressure, 220